Direct Digital
Frequency Synthesizers

Books of Related Interest from IEEE Press . . .

NONLINEAR MICROWAVE CIRCUITS
Stephen Maas
1997   Hardcover   504 pp   IEEE Order No. PP5385   ISBN 0-7803-3403-5

RADIO FREQUENCY PRINCIPLES AND APPLICATIONS: The Generation, Propagation, and Reception of Signals
Albert A. Smith, Jr.
1998   Hardcover   240 pp   IEEE Order No. PC5704   ISBN 0-7803-3431-0

MONOLITHIC PHASE-LOCKED LOOPS AND CLOCK RECOVERY CIRCUITS: Theory and Design
Edited by Behzad Razavi
1996   Hardcover   508 pp   IEEE Order No. PC5620   ISBN 0-7803-1149-3

DELTA-SIGMA CONVERTERS: Theory, Design, and Simulation
Edited by Steven R. Norsworthy et al.
1997   Hardcover   512 pp   IEEE Order No. PC3954   ISBN 0-7803-1045-4

DIRECT DIGITAL FREQUENCY SYNTHESIZERS

Edited by

Vĕnceslav F. Kroupa

The Institute of Radio Engineering and Electronics,
Academy of Sciences of the Czech Republic

A Selected Reprint Volume

IEEE Ultrasonics, Ferroelectrics, and Frequency Control Society, *Sponsor*

The Institute of Electrical and Electronics Engineers, Inc., New York

ISBN 0-7803-3438-8

IEEE Order Number: PP4148

Library of Congress Cataloging-in-Publication Data
Direct digital frequency synthesizers / edited by Věnceslav F. Kroupa
    p.   cm.
  Includes bibliographical references and index.
  ISBN 0-7803-3438-8
  1. Frequency synthesizers. 2. Digital integrated circuits.
  3. Phase-locked loops. I. Kroupa, Věnceslav F., 1923– .
TK7872.F73D57   1998
621.3815'486—dc21   98-25067
                                     CIP

To Magda

Contents

Part VIII   Digital-to-Analog Converters   245

Part IX   State of the Art and Some Applications   289

Part X   New Ideas for the DDFS Design   319

Part XI   Mathematical Background   359

Preface

FOR centuries only mathematicians were interested in the problems of the approximation of real numbers, such as π, with rational fractions. However, with the advent of the modern age, practical applications were born. At the end of the seventeenth century, Huygens introduced continued fractions as a tool for the design of gear boxes, the real *mechanical synthesizers,* in his planetarium. During the Industrial Revolution, the problem became a concern for mechanical engineers who now confronted the practical restrictions posed by their hardware.

A new era began when electronic systems applied basic mathematical operations (addition, subtraction, multiplication, and division) to frequencies. After earlier simple frequency multipliers, mixers, and dividers, real *frequency synthesizers* were designed. In recent years, the perfection of these devices with the help of digital integrated circuits has resulted in fresh investigation of the problem of the approximations of real numbers.

For nearly two centuries, mechanical synthesizers—which change the angular velocity of the driving engine into the desired speed of the working shaft, that is gear boxes—were designed with the assistance of simple continued fractions. Because of practical limitations, the desired approximation was often too large. However, with experience engineers learned that many other suitable approximations were possible and they compiled tables. This approach was not possible with electronic frequency synthesizers. Reinvestigation of the problem with the assistance of computers resulted in finding an exact solution for both gear boxes and frequency synthesizers. Another incentive for modern direct digital frequency synthesizers* was the rapid technological progress made in the field of integrated circuits.

After arranging the material for this book, dedicated exclusively to DDFS, we finally decided to present a collection of reprinted papers with a few specially written articles. Our purposes were twofold: to provide a useful textbook for students seeking essential background theory; and to help practicing engineers obtain a better understanding of actual devices, without the prerequisite of a specialist's knowledge of the theory. In addition, we wanted to provide a collection of often cited papers.

Based on these considerations, we have divided the monograph into eleven parts.

Part I is an introduction into the problem.

Part II is dedicated to the theory and design of single-frequency DDFS and the new general approximation theorem useful for mechanical and electronic synthesizers.

Part III acquaints the reader with the principle and the state of the art of wide-range DDFS.

Parts IV and V collect papers dealing with the problems of spurious signals in DDFS, with Part V devoted to the possibilities of reducing these troublesome phenomena.

Since the range of the output frequencies of DDFS is limited (at present below 1 GHz), the extension of their advantages (such as the tuning speed, very small frequency steps, small weight, and low power consumption), to higher frequencies is possible. However, application of direct multiplications or mixing is practical in a limited number of cases only, but the assistance of *phase-lock loops* (PLLs) provides simple and elegant solutions. The basics of the PLL theory and combinations with DDFS are discussed in Part VI.

In some applications the knowledge of DDFS output phase noise may be needed. Therefore, after a general introduction into the problem, Part VII summarizes some practical, hard-to-find published results.

Digital-to-analog converters (DAC) on the one hand form one of the most important building blocks of DDFS and on the other hand limit speed resolution and spectral purity. For these reasons, Part VIII discusses their construction and properties at some length.

Parts IX and X briefly review the present state of the art, with several new ideas in designed DDFS.

The monograph concludes with Part XI which contains two specially written papers by the editor. Included is an explanation of a theory of quasiperiodic omission of pulses with a mathematical background. In addition, the editor provides several computer programs useful for theoretical investigations of DDFS properties.

We apologize to the many authors of valuable papers and articles which we had to leave out of the collection because of inevitable space limitations. To provide at least a partial remedy, we mentioned some of them in the introductions to the individual parts and in the references.

Finally, the editor would like to thank the IEEE Press for publishing this volume, as well as the IEEE Ultrasonics, Ferroelectrics, and Frequency Control Society for its support of this enterprise.

My thanks are also extended to the Institute of Radio Engineering and Electronics of the Academy of Sciences of the Czech Republic for its generous support of the present work.

The editor would like to express his gratitude to the unknown reviewers for their efforts, particularly to those who read all the specially written parts with such great attention and suggested many useful improvements.

Since this work also contains much new matter, we hope that it will be a useful textbook.

Věnceslav F. Kroupa
Prague, September 1998

---

*Please note that the acronym for *direct digital frequency synthesizers* is either DDFS or DDS; both are used in this book and in the literature.

Part I

Introduction and Tutorial

As a result of the steadily increasing congestion in communication bands of the electromagnetic spectrum, efforts are being made to utilize new ranges, particularly microwave and optical frequencies, and to better exploit existing frequency allocations. The latter tasks have led to improvements in communication techniques (SSB modulation, multiplexing, digital systems, spread spectrum application, etc.). Simultaneously, a perfection of frequency generators has been attempted in transmitter exciters and local oscillators in receivers, in most cases with the assistance of frequency synthesizers, that is, devices changing a given standard frequency, f_s, into the desired output frequency, f_x.

This approach is preferred to single-frequency oscillators (Xtals) because of larger tuning ranges, better short- and long-term stabilities (lower output noise), microprocessor-aided tuning or programming, and many other advantages.

The present status of frequency synthesis techniques in the whole range of the electromagnetic waves, from audio and RF to the visible frequencies, is shown in Fig. I-1. The lesson one learns from Fig. I-1 is that different approaches to frequency synthesis exist depending on the desired output frequency. Nevertheless, one basic principle is common to all systems.

As we have already mentioned, the task of frequency synthesis is to generate an arbitrary frequency, f_x, from a given reference or standard frequency, f_s, that is, to solve equation

$$f_x = \xi_x f_s \qquad [\text{I-1}]$$

where the factor ξ_x is the so-called normalized frequency.

The solution of eq. (I-1) is always based on algorithms for step-by-step approximations of real numbers, as will be explained in Part XI of this book. In instances where the factor ξ_x is a rational number, that is, a ratio of two relatively prime integers

$$\xi_x = \frac{X}{Y} \qquad [\text{I-2}]$$

the output frequency and the reference frequency will be in harmonic relation. However, in a more general case, ξ_x can belong to a much broader set of real numbers, and in this case we generally stop the approximation process as soon as the remainder falls below a prescribed or acceptable value.

Practical realizations of different types of frequency synthesizers are discussed in several books [I-1 through I-6]. How-

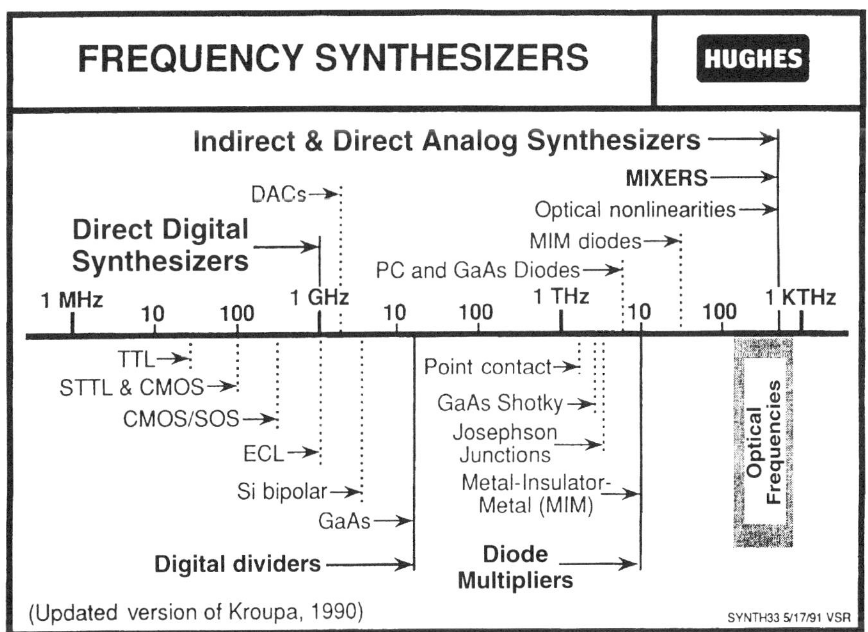

Fig. I-1. The state of the art of frequency synthesis from the lowest frequency ranges to the optical frequencies [from I-9].

ever, direct digital frequency synthesizers (DDFS) were mentioned to some degree only by Gorski-Popiel [I-2] about 20 years ago and later by Rohde [I-5]. Otherwise the reader was dependent on journal articles and corporate literature—Application Notes and Technical Data such as [I-7].

Some of the most important papers about DDFS, published mostly in the last ten years together with a few specially written by the editor, are reprinted in this volume.

In this connection, we would like to mention an excellent survey paper by Reinhardt et al. A major feature of that paper is the bibliography that lists all important books and papers published up to 1986 [I-8]. We would also like to mention three Tutorials read by Reinhardt, McNab, and Gould at the Frequency Control Symposia in the years 1991, 1992, and 1993 [I-9]. We also must mention two recent books, namely, that by Crawford (1994) where again only one short chapter is dedicated to DDFS [I-10]. A different situation obtains with the Goldberg (1996) monograph [I-11] in which a long Chapter 4 (DDS General Architecture) is dedicated to DDFS and is based on many of the papers reprinted in the present volume.

One of the latest books [I-12] is an expanded version of [I-5]; Chapter 3 of this work presents some introductory material dealing with DDS. Appendix C, the Philips Fractional-N Synthesizer # SA8025A, is reprinted from the company catalog.

To introduce the reader to the problem of frequency synthesis with the emphasis on DDFS, Part I begins with a specially written paper by the editor.

REFERENCES

[I-1] V. F. Kroupa. *Frequency Synthesis: Theory, Design and Applications.* London: Ch. Griffin, 1973; New York: John Wiley, 1973.

[I-2] J. Gorski-Popiel. *Frequency Synthesis: Techniques and Applications.* New York: IEEE Press, 1975.

[I-3] V. Manassewitsch. *Frequency Synthesizers, Theory and Design.* New York: John Wiley, 1976, 1980, 1983.

[I-4] W. F. Egan. *Frequency Synthesis by Phase Lock.* New York: John Wiley, 1981.

[I-5] U. L. Rohde. *Digital PLL Frequency Synthesizers, Theory and Design.* Englewood Cliffs, N.J.: Prentice-Hall, 1983.

[I-6] R. Stirling. *Microwave Frequency Synthesizers.* Englewood Cliffs, N.J.: Prentice-Hall, 1987.

[I-7] R. J. Zavrel and Gwyn Edwards, eds. *The DDS Handbook.* Santa Clara, Calif.: Stanford Telecommunications, 1990.

[I-8] V. S. Reinhardt, V. Gould, K. H. McNab, and M. Bustamante. "A short survey of frequency synthesizer techniques." *Proceedings of the 40th Annual Frequency Control Symposium,* pp. 355–365. 1986.

[I-9] V. S. Reinhardt, K. H. McNab, and V. Gould. "Frequency synthesizers—a tutorial." *45th Annual Symposium on Frequency Control,* Los Angeles, May 29–31, 1991 and *IEEE Frequency Control Symposia,* 1992 and 1993.

[I-10] J. A. Crawford. *Frequency Synthesizer Design Handbook.* Boston/London: Artech House, 1994.

[I-11] Bar-Giora Goldberg. *Digital Techniques in Frequency Synthesis.* New York: McGraw-Hill, 1996.

[I-12] U. L. Rohde. *Microwave and Wireless Synthesizers: Theory and Design.* New York: John Wiley, 1997.

Synthesis Techniques

VĚNCESLAV F. KROUPA

I. Historical Introduction

THE attempt to define particular events in the never-ceasing flow of time is as old as humankind itself. Daily alternations of light and dark, seasonal weather variations, and the like were among the first periodic occurrences met by men. However, the introduction of the term *frequency,* that is, the inverse value of the duration of one period or cycle,

$$f = \frac{1}{T_c} \qquad [1]$$

is rather recent.

On the contrary, the term *angular velocity,* connected with the rotation of the Earth, wheels, and so on is much older. Nevertheless, the physics of both phenomena are the same since the parametric equations of uniform circular motion (cf. Fig. 1) are given by

$$x = r * \cos(\omega t); \quad y = r * \sin(\omega t) \qquad [2]$$

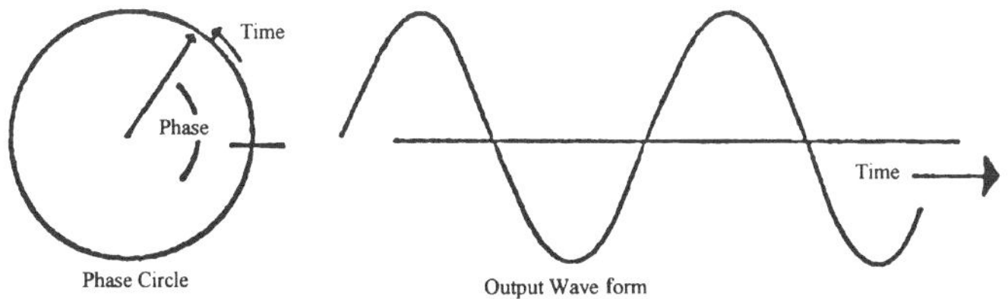

Fig. 1. Uniform circular motion and its projection on the horizontal time axis.

where r is the radius and the angular velocity ω is equal to

$$\omega = \frac{2\pi}{T_c} \qquad [3]$$

Evidently, devices for changing the angular velocity of the driving engine into another velocity of the working wheels can be considered as single-frequency mechanical synthesizers resting on the same mathematical design principles. The past and present situation of the mechanical and electronic synthesizers is shown schematically in Fig. 2.

II. Gear Boxes

A long time before the Industrial Revolution began, mechanical engineers had to solve the problem of changing the input angular velocity ω_s (i.e., some given *input revolutions/second, minute*) into a desired output angular velocity ω_x, or in other words, to solve the following basic equation

$$\omega_x = \xi_x * \omega_s \qquad [4]$$

by taking into account several restrictions that will be discussed later. An example provides the gear ratio of the compound train shown in Fig. 3.

The normalized output angular velocity, in accordance with eq. (4), is

$$\xi_x = \frac{\omega_x}{\omega_s} = \frac{t_1}{t_2} * \frac{t_3}{t_4} * \frac{t_5}{t_6} \qquad [5]$$

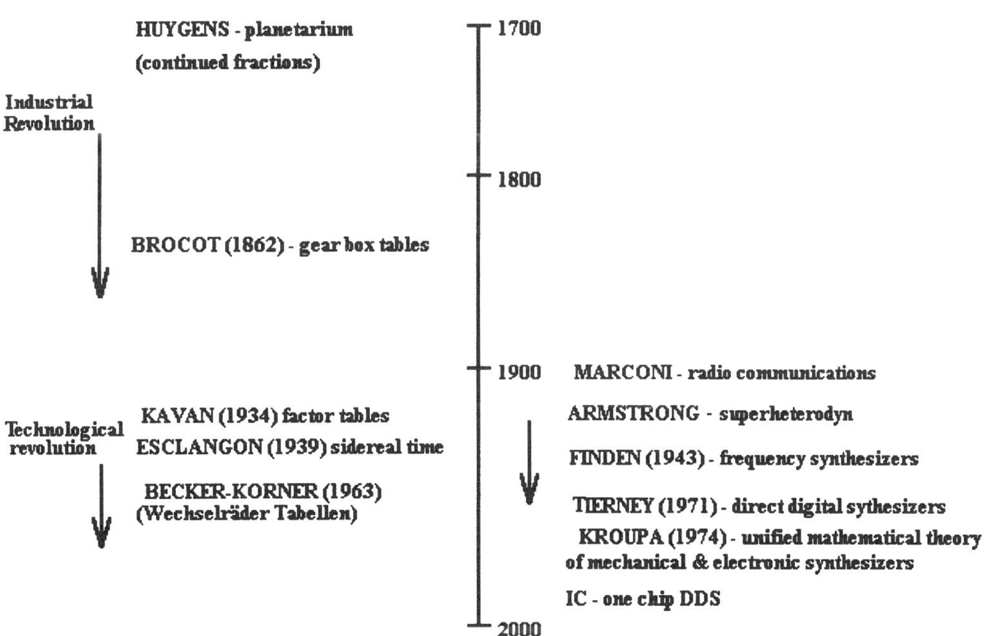

Fig. 2. The past and present situations of the mechanical and electronic synthesizers.

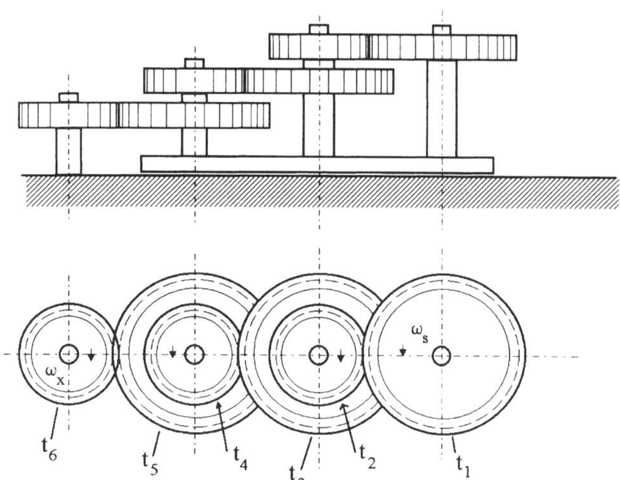

Fig. 3. Compound gear train.

where t_1 through t_6 are the number of teeth on individual wheels. Factors in the numerator represent multiplication, and those in the denominator represent dividers; cf. the equivalent circuit in Fig. 4.

The difficulty with the solution of the gear box equation is that only wheels with a limited number of teeth are used in practice. Furthermore, experience has taught mechanical engineers that a mere expansion of the ratio ξ_x into a continued fraction (first used by Huygens in the design of his planetarium [1]) generally does not work; either the approximation error or the number of teeth is too large.

Because of the lack of an appropriate theory, mechanical engineers compiled tables [e.g., 2,3] that made it possible to solve nearly all practical cases with sufficient accuracy.

In exceptional instances—for example, the gear-drive for changing the solar time into the sidereal time—suitable product approximations once laboriously found were published; for example, see [4].

III. Frequency Synthesis

The aim of frequency synthesis is to generate an arbitrary frequency, f_x, from a given standard frequency, f_s, that is, to solve equation

$$f_x = \xi_x * f_s \qquad [6]$$

in such a way that a practical device could be manufactured accordingly; note the similarity with eq. (5).

$$\xi_x = \frac{f_x}{f_s} = \frac{X_1}{Y_1}$$

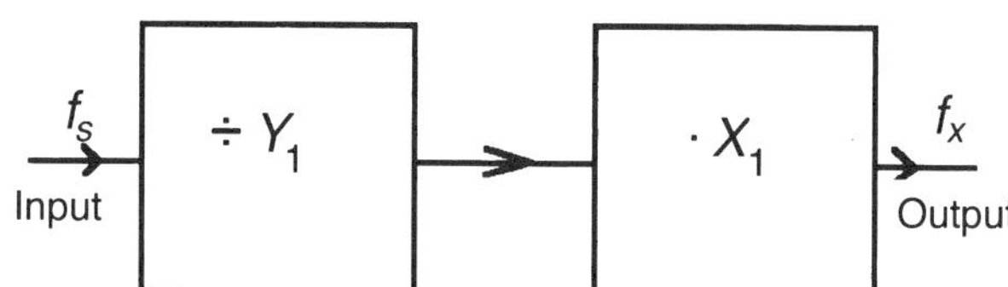

Fig. 5. Single-frequency synthesizer realized by a simple chain of one frequency divider and one frequency multiplier.

The term *frequency synthesis* was coined in the early 1940s when the first measuring generators were devised. The variable output frequency of these measuring generators was always harmonically related to a submultiple of the driving frequency f_s [5,6].

A. Single-Frequency Synthesizers

1. In the simplest case ξ_x is a fraction formed by small, relatively prime integers, that is,

$$\xi_x = X_1/Y_1 \qquad [7]$$

and the synthesizer is reduced merely to a chain of one frequency divider and one multiplier as illustrated in Fig. 5.

2. If X_1 and Y_1 in eq. (7) are products of small prime numbers, that is,

$$X_1 = \prod_1^n x_i \quad Y_1 = \prod_1^r y_j \qquad [8]$$

the synthesizer may be realized by a chain of frequency multipliers and dividers (see Fig. 6; note the similarity to Fig. 4). Factorization of integers X_1 and Y_1 is performed either with the assistance of tables [7] or by using a simple algorithm as noted in Section C.

3. In instances where X_1 but not the denominator Y_1 is the product of small and large prime numbers, we must start by approximating ξ_x by a ratio x_1/y_1 and then apply the approximation procedure to the relative residue X_2/Y_2, and so on. Consequently, step-by-step approximations are necessary. The process is simplified by the fact that basic algebraic operations (i.e., additions, subtractions, multiplications and divisions) are easily performed with frequencies, and we are in a position to propose a suitable mathematical model [8] (the software solution). However, there are difficulties with hardware solutions, namely, generation of spurious signals and often enhancement

$$\frac{\omega_x}{\omega_s} = \frac{t_1}{t_2} \cdot \frac{t_3}{t_4} \cdot \frac{t_5}{t_6}$$

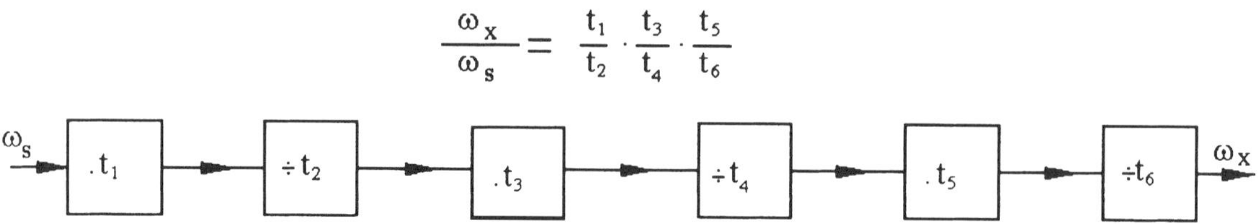

Fig. 4. Equivalent circuit of the compound gear train shown in Fig. 2.

$$\xi_x = \frac{\prod_1^n x_i}{\prod_1^n y_i}$$

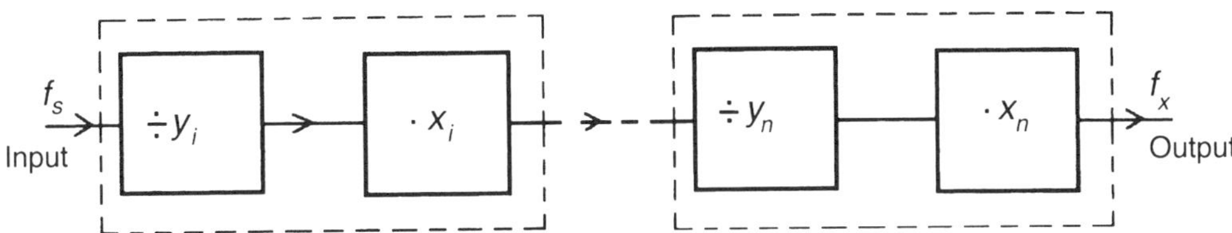

Fig. 6. Block diagram of frequency synthesizers formed by a chain of several frequency multipliers and dividers.

of the phase noise level. Practical realizations in different frequency ranges of the electromagnetic spectrum bring about additional limitations as discussed in recent books [9–14].

Often, for example, in atomic frequency standards, it happens that both the numerator and the denominator in eq. (7) are large prime numbers. In this case, we are in the situation of mechanical engineers, and we have to resort to approximations. However, in contrast to the limited number of teeth, the division and multiplication factors can be much larger due to the modern digital integrated circuits (ICs) [15].

B. Multiple-Output Frequency Synthesizers

1. Initially, "tunable" frequency synthesizers with many output frequencies in a desired frequency range were required in standard time and frequency laboratories for measurement purposes. This was soon followed by adaptations for communications where applications as transmitter exciters and as local oscillators in receivers were welcomed. For simplicity of operation in these first devices, the variable output frequency was set decimally. In some instances, the setting between the smallest frequency steps was provided with an incoherent interpolating oscillator. These first direct frequency synthesizers, manufactured in the 1960s and early 1970s (e.g., Hewlett-Packard type 5100) were bulky, but with rather good spectral purity and phase stability (e.g., Fluke type 645).

2. Reduction of the weight and size was provided by the application of phase-locked loops in frequency synthesizers. With the assistance of semiconductor technology, a "cubic inch" frequency synthesizer [16] was highlighted.

3. The most recent achievement in frequency synthesis design is the creation of direct digital frequency synthesizers (DDS or DDFS). The size and weight of these synthesizers were reduced to unprecedented values. The price paid for these reductions is limited output spectral purity and rounded (approximated) output frequencies. All these problems will be discussed in detail later in this book.

C. Computation of the Prime Factors

Finding prime factors or the power of any integer number N can be performed as follows: When N is an even number, we divide it by 2 as many times until we arrive at its odd factor

$$N = N/2^r \quad (r = 0, 1, \ldots);$$

Thereafter division by all odd numbers $(3, 5, 7, \ldots)$ smaller than $SQR(N)$ is tested, and the respective powers are noted. In this way, we perform the task of factoring any integer with the assistance of a computer or even a pocket calculator in a reasonable time.

IV. Mathematical Background

In the early years, the **trial-and-error method** was the workhorse for designers of frequency synthesizers. However, soon the editor, in his book [9], showed that the actual frequency synthesis process was nothing more than a step-by-step approximation of real numbers. Only the approximation process in the physical world of frequency synthesis is subjected to limitations caused by the required frequency range, the hardware technology, desired spectral purity, switching speed, cost, and so on.

We have investigated all important theorems for approximation of real numbers [17] and verified their advantages and disadvantages for applications in the frequency synthesis theory.

A. Approximation Theorems

1. Cantor series families
2. Lueroth series families
3. Cantor product expansions
4. Continued fraction approximations
5. Modulo-N approximations

1. Cantor Series Approximations. Before the widespread use of digital circuits, Cantor series expansions provided the general mathematical model, for both single and multiple-output frequency synthesizers. In the following we will explain their fundamentals.

For arbitrarily chosen integers $y_1, y_2, \ldots$ (equal or larger than 2), we can approximate any real number ξ_o with the assistance of the algorithm

$$x_i = \text{integer} \xi_i \quad (i = 0, 1, 2, \ldots)$$
$$\xi_{i+1} = (\xi_i - x_i) * y_{i+1} \qquad [9]$$

into the so-called Cantor series

$$\xi_o = x_o + \sum_{i=1}^{n} \frac{x_i}{y_1 y_2 \cdots y_i} \qquad [10]$$

with the approximation error

$$R_n \approx \frac{1}{y_1 y_2 \cdots y_n} \qquad [11]$$

In instances where all divisors y_i are the same, that is,

$$y_i = g \qquad [12]$$

the Cantor series is changed into the so-called g-dic or systematic fraction

$$x_o = \sum_{i=0} \frac{x_i}{g^i} \qquad [13]$$

Earlier "tunable" frequency synthesizers were based mainly on decimal fractions (i.e., on systematic fractions with the base $g = 10$). However, effective suppression of spurious signals generated by the mixing process required some modifications: namely, each x_i in the above equation is increased by the same large amount x_{io}. These additive parts form a geometric series the sum of which is subtracted in the output stage. The mathematical model is discussed in detail in [9].

Direct digital frequency synthesizers, however, are based on systematic fractions with the base $g = 2$. Since the useful normalized frequency is always $\xi_x < 0.5$, in this case we have the following mathematical model:

$$\xi_x = \frac{a_1}{2} + \frac{a_2}{2^2} + \frac{a_3}{2^3} + \cdots \qquad [14]$$

where a_1 is equal to 0 and the other a_i either to 1 or 0. The approximation error cannot exceed

$$R_n \le \frac{1}{2^n} \qquad [15]$$

Later we will see that in contrast to other approximation theorems, the convergence of the above series is rather slow.

B. Quasiperiodic Omission of Pulses

All direct digital frequency synthesizers are based on the process of changing the input clock rate frequency into the desired output frequency and waveform. The major difficulty with these systems is generation of spurious signals, often large ones. We can cope with this problem, from a mathematical point of view, by using approximation theorems that make possible quasiperiodic omission of pulses [18,19]. This condition is met with the modified Engel series (belonging to the Lueroth series family), with the continued fraction approximations for single-frequency DDFS, and with Modulo-N approximations for multiple-output frequency DDFS. Here we briefly summarize the respective theories.

1. Continued Fraction Approximations. As we have already mentioned, this was the first theorem used in "frequency synthesis." The advantage of this theorem is that it provides the best approximations of real number, i.e., with fractions with the smallest integers in numerators and denominators for a given approximation error. That is,

$$\xi_x = \frac{A_k}{B_k} + (-1)^{k-1} R_k \qquad [16]$$

where the approximation error or remainder is

$$|R_k| \le \frac{1}{B_k^2} \qquad [17]$$

Continued fraction expansion is based on Euclides' theorem, discussed in depth in Part II [15]. The uniqueness of the continued fraction expansions limits applications. See the block diagram in Fig. 7 in the paper by the editor [19] in Part XI.

In the most favorable cases, the frequency synthesizer can be realized in accordance with eq. (7) and Fig. 5. Another possibility is the gear box approach in instances where the numerator and the denominator can be factorized into suitable products—see Fig. 6 and the paper by Small in Part II [20].

The set of gear box solutions can be enlarged with the assistance of the General Approximation Theorem, which is based on combinations of different orders of the continued fraction expansion.

Shortened continued fraction expansion, important for appreciation of spurious signals in some types of DDFS, has been introduced by the editor and is dealt with in detail in the first reprinted paper in Part IV [21].

2. Modified Engel Series Approximations. This series, introduced by the editor, differs from original Lueroth series expansions with the alternating sign in the sums [15].

By assuming that the normalized output frequency $\xi_x = \gamma_o$ is smaller than one, the expansion of the modified Engel series is given by

$$\gamma_o = \sum_{i=1}^{n} \frac{(-1)^{i-1}}{p_1 p_2 \cdots p_i} \qquad [18]$$

where

$$\gamma_1 = \frac{1}{1 - \gamma_o} \qquad p_1 = \text{integer}(\gamma_1)$$

$$\gamma_2 = \frac{\gamma_1}{\gamma_1 - p_1} \qquad p_2 = \text{integer}(\gamma_2) \qquad [19]$$

until p_n.

Note that the partial divisors, p_i, are steadily increasing, which is the condition for convergence. At the same time, the remainder is very small, namely,

$$R_n \le \frac{1}{p_n^2} \qquad [20]$$

Practical frequency synthesizers, using this mathematical model, have been built—again see [21].

3. Modulo-N Approximations. The principle can be deduced from Fig. 7. At each cycle or pulse of the clock or reference frequency, a number, k, which represents a phase increment is added to that already stored in the accumulator. Every

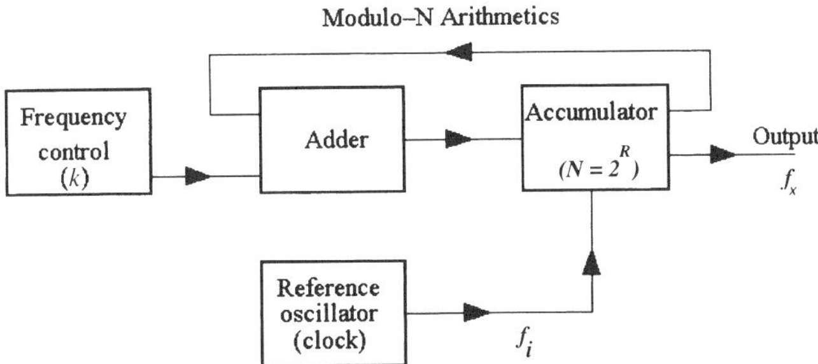

Fig. 7. Principle block diagram of overflowing Modulo-N device.

time the accumulator overflows, an output pulse is generated and a new period of accumulation is started, with a Modulo-N remainder (smaller than k) in the accumulator memory. Finally, after N clock periods, the remainder stored in the accumulator is zero; one fundamental period of the system ends and a new one starts.

Example 1. Let us synthesize a normalized frequency

$$\xi_x = 3/16$$

In this example, we use a 4-bits accumulator ($R = 4$) and put $k = 3$ into the frequency control circuit (see Fig. 7). The stored remainders in the accumulator will be successively

0,3,6,9,12,15,"18"
1st overflow   2,5,8,11,14,"17"
2nd overflow   1,4,7,10,13,"16"
3rd overflow   0,3,6, etc.,

Figure 8*a* illustrates the accumulation process in graphic form, and Fig. 8*b*, the respective output pulse train. We easily deduce that generally

$$NT_i = \sum_{r=1}^{k} T_{xr} = kT_{xo} \qquad [21]$$

where T_{xr} is duration of actual output periods in Fig. 8*b* and T_{xo} the idealized equidistant separation of output pulses; see Fig. 8*c*. As a consequence, the output pulse rate has the mean frequency

$$f_{xo} = \frac{k}{N} f_i \qquad [22]$$

which can easily be changed by reprogramming the "word" k in the frequency control (steering memory) block—see Fig. 7. Furthermore, in practical devices N is generally equal to a large power of 2. In this way, we arrive at very small frequency steps and simple hardware for the accumulators, that is, since

$$N = 2^R \quad (R = 24, 32, 48, \text{etc.}) \qquad [23]$$

Examination of the output pulse train in Fig. 8*b* reveals that individual periods T_{xr} are not equal. In the following examples, we will investigate this deficiency. By starting with zero in the accumulator, we arrive after m_1 clock pulses at its first overflow. As a consequence, we have the remainder

$$m_1 - N < k \qquad [24]$$

After the second overflow, we get

$$m_2 k - 2N < k \qquad [24a]$$

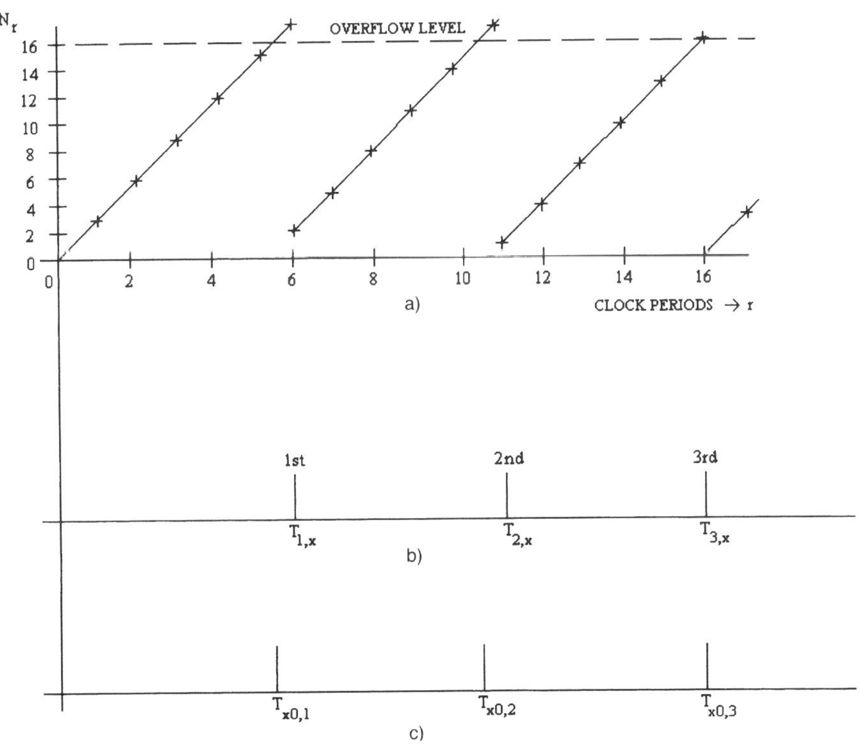

Fig. 8. Generation of a pulse train with a normalized frequency $\xi_x = 3/16$ in a Modulo-N system: (*a*) accumulator contents N_r; (*b*) modulated and (*c*) idealized output pulse train.

and so on (cf. Fig. 8a and Example 1) until

$$m_n k - nN = 0 \qquad [24b]$$

Generalization reveals

$$mk - rN < k \quad (r < k) \qquad [25]$$

from which

$$m - r\frac{N}{k} = \epsilon \qquad [26]$$

where

$$1 > \epsilon \geq 0 \qquad [27]$$

since m is an integer, $\epsilon = 0$ is met when

$$\text{integer}\left(m - r\frac{N}{k}\right) = 0 \qquad [28]$$

or

$$m = \text{integer}\left(r\frac{N}{k}\right) \equiv N \qquad [29]$$

After reverting to eq. (26) and multiplying it by the clock period T_i, we find with the assistance of (21) and (27) the timing error

$$mT_i - rT_{xo} \leq T_i \qquad [30]$$

and finally the modulation function [19,21]

$$s(t_r) = T_i\left[r\frac{N}{k} - \text{integer}\left(r\frac{N}{k}\right)\right] \qquad [31]$$

Evidently, the spurious (phase) time modulation does not exceed one clock period T_i. As a consequence, Modulo-N arithmetics also provides the mathematical model for DDFS, meeting the requirement for the quasiperiodic omission of pulses in the steering pulse train. Only the output pulses are on the places of the "swallowed" pulses.

V. DIRECT DIGITAL FREQUENCY SYNTHESIZERS

DDFS with single-output frequency (based on the continued fraction expansions or modified Engel series) are required only in rare instances. However, devices with variable-output frequencies are encountered much more often, and direct digital frequency synthesizers, based on IC technology, simplify the hardware considerably and at the same time provide very small tuning steps. These frequency synthesizers are based on Modulo-N arithmetics. At present, they are covering the range from "DC" to 500 MHz. However, recently, clock rates beyond 1 or 2 GHz have been reported.

The principle can be easily deduced from the basic block diagram recalled in Fig. 9. Note that it differs from Fig. 7 only by means for generation of the output sine wave, that is, by the ROM sine lookup table, digital-to-analog converter, and antialiasing filter.

In instances where the normalized frequency ξ_x is a ratio of small integers, the output sine wave is a staircase wave; see Fig. 10.

On the other hand, among the DDS advantages are small-output frequency steps since generally the accumulator capacity is very large (cf. 23) and consequently

$$\Delta f_x = \frac{1}{2^R} * f_i \qquad [32]$$

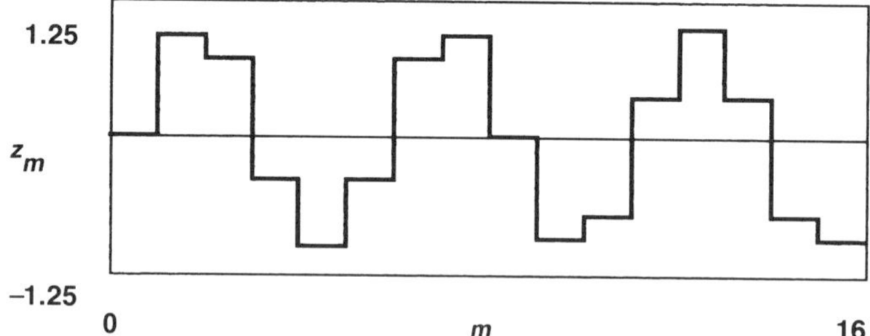

Fig. 10. The output sine wave for the normalized frequency $\xi_x = 3/16$.

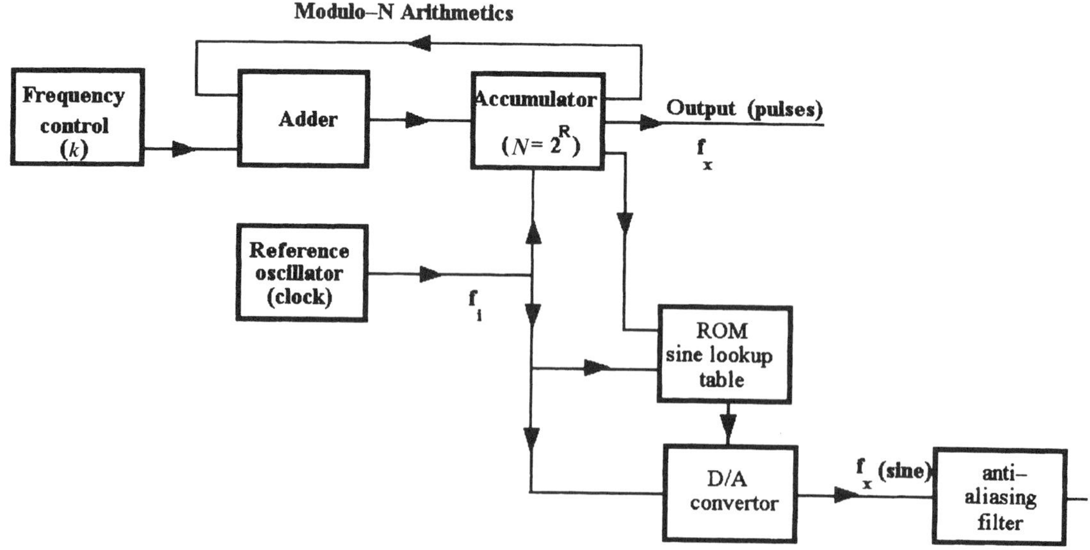

Fig. 9. Block diagram of the Modulo-N DDS.

A. Spurious Signals in DDS

The major difficulties associated with all direct digital frequency synthesizers are spurious signals accompanying the desired carrier. In some instances, these spurious signals are quite large. Therefore, knowledge of their sources and expected amplitudes and frequencies may be useful for both designers and users.

Since all DDSs are based on the sampling process, the output spectrum extends to multiples of the sampling frequency. This problem is solved by introducing antialiasing filters with the cutoff frequency smaller than the Nyquist frequency, that is, smaller than one-half the sampling frequency.

In some instances, however, application of bandpass filter makes it possible to extend the useful range to higher output frequencies.

B. Spurious Signals in DDS
with Rectangular Output Wave

The problem is discussed in detail by the editor in papers reprinted in this volume [21,22]. The largest spurious signals are

$$a_{sp,\max} \approx 20 \log \xi_x \qquad [33]$$

C. Spurious Signals in DDS
with Sine Output Wave

The most effective way to reduce the spurious signals in "tunable" DDS is to change the phase stored in the accumulator into a sine wave. However, even here we encounter several sources of spurious signals, which are illustrated schematically in Fig. 11.

1. Truncation of the Accumulator Bits. The high resolution of DDS requires a large Modulo-N accumulator capacity (e.g., 24, 32, 48, ... bits). As a consequence, it is impossible to use all the stored phase information for generating the output sine wave. Generally, we refer only some W most significant bits (MSB) from the R bit large accumulator and neglect all remaining B bits

$$W = R - B \qquad [34]$$

The consequence is a phase modulation of the output sine wave. A good estimation of the level of the largest spurious signals, $a_{sp,r}$, due to the truncation of the accumulator bits, in respect to the carrier, a_{car}, provides the following relation in dB measure [22]:

$$20 \log \left(\frac{a_{sp}}{a_{car}} \right) \approx -6W - 20 \log(r); \quad (r = 1, 2, \ldots) \qquad [35]$$

It has been proved that the number of these rather large spurious signals caused by the phase modulation in DDFS pass band is

$$2^B - 1 \qquad [36]$$

Consequently, the largest $r_{\max} = 2^{B-1}$.

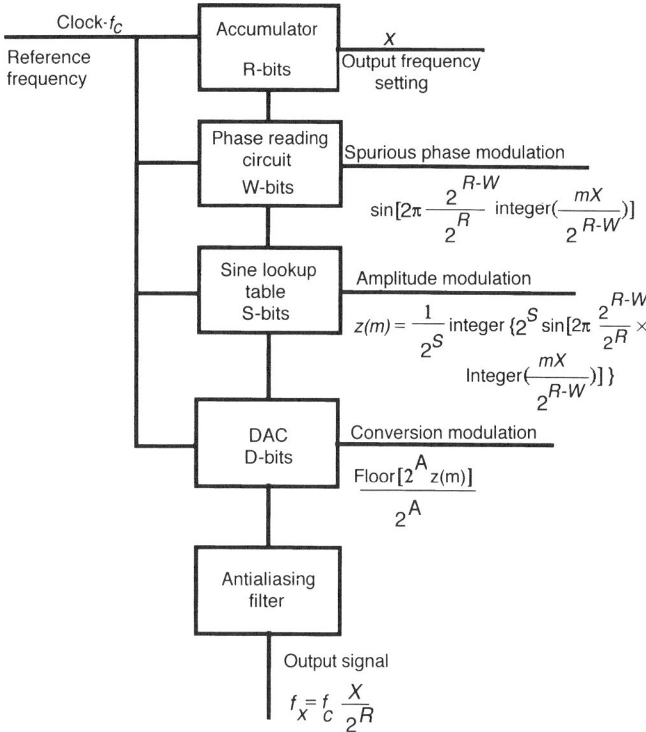

Fig. 11. Sources of spurious signals in DDS.

2. Truncation of Sine Values in Sine Lookup Tables. Generally, the sine values are only rational numbers with infinite decimal or binary places.

$$\sin x = \pm \left(\frac{a_1}{2} + \frac{a_2}{2^2} + \cdots \right.$$
$$\left. + \frac{a_{S-1}}{2^{S-1}} + \frac{a_S}{2^S} + \frac{a_{S+1}}{2^{S+1}} + \cdots \in \infty \right) \qquad [37]$$

However, to keep the memory size of sine lookup tables reasonable, we must limit the stored words to, say, S bits only. The consequence is a spurious amplitude modulation of the output signal.

$$\epsilon(m) = \sin(2\pi m \xi_x) - 2^{-S} \text{integer}[2^S \sin(2\pi m \xi_x)] \qquad [38]$$

By assuming that the error signal $e(m)$ has a uniform distribution (cf. Fig. 14 in [22]) in the interval from $-1/2^S$ to $+1/2^S$, we get for its variance

$$\sigma_e^2 \approx \frac{1}{3} 2^{-2S} \qquad [39]$$

However, computation of the respective mean power spectral density performed in the Appendix of [22] revealed

$$S(n) \approx \frac{1}{3Y} 2^{-2S} \qquad [40]$$

With the assistance of the computer simulations [22], we get for the mean power level of spurs even a bit lower value, namely, in dB measure [22]

$$\langle r_j \rangle = -7.8 - 6 * S - 10 \log(Y) \pm 10 [\text{dB}] \qquad [41]$$

In some instances, we encounter odd harmonics with amplitudes

$$a_h = 2 * \frac{2^{-S}}{\pi} * \frac{1}{h} \quad (h = 3, 5, \dots) \tag{42}$$

from which the level of the third harmonic of the output frequency in dB measure is

$$20 * \log\left(\frac{a_3}{a_1}\right) \approx -6 * S - 13.5[\text{dB}] \tag{43}$$

In a properly designed DDFS, the level of the expected largest spurious signals generated by the truncation of the accumulator and sine lookup tables should be approximately the same. Keeping this condition in mind, we have arrived in [22] to the relation between W and S

$$W \geq S + 2 \tag{44}$$

Addition of the constant

$$+\frac{1}{2^{S+1}} \tag{45}$$

to the absolute sine values before their truncation to the S-bits only results in rounded off values. As a consequence, the spurious rectangular modulation, discussed above, disappears, and in accordance with computer simulations the largest spurious signal is approximately

$$-6 * S - 20[\text{dB}] \tag{46}$$

However, the level of the mean power spectral density does not change much in respect to the relation (41).

D. Reduction of Sine Lookup Tables

Design of sine lookup tables depends on the required spurious spectral level at the DDFS output [cf. relation (41)] on the required speed (i.e., on the desired maximum output frequency), on the available memory area on the chip, and on the digital-to-analog converter (DAC) used. The first condition would be met with a large number S of bits in sine words. From the point of view of speed and chip area, however, small S bits words are preferred. Finally, as we will see in the next paragraph, it makes little sense to increase S above the DAC resolution.

1. Storing Only One Quarter of Sine Values. The first step in the reduction of sine lookup tables is to store only one quarter of the data, namely, those from 0 to $\pi/2$, and to use two MSBs from the accumulator for reconstruction of the whole sine wave from 0 to 2π. The first MSB determines its sign, and the next MSB decides whether the sine amplitude is increasing or decreasing. This procedure is used in DDFS intended for the highest working speed [e.g., 23 or 24].

The number of stored sine values is one quarter of the effective phase bits, namely,

$$2^{W-2} \tag{47}$$

and the total number of bits stored in the ROM lookup table

would be

$$2^{W-2} * S \approx 2^{W-2} * (W - 2) \tag{48}$$

Finally, to secure the symmetry for sine values for increasing and decreasing branch, as much as possible, in the neighborhood of $\pi/2$, we should store little-shifted sine values

$$\sin\left[2\pi\left(\frac{1}{2^W} * m + \frac{1}{2^{W+1}}\right)\right] \quad (m = 0, 1, \dots 2^{W-2} - 1) \tag{49}$$

2. Algorithmic Approximations. A further reduction of the sine lookup tables is possible with algorithmic approximations [e.g., 25] in which the effective phase is split into several components and trigonometric identities are used to assemble the desired result. However, such a straightforward approach would require the use of multiplications; as a result, it would be slow and complex. This difficulty can be avoided, as is discussed in [26–29].

VI. Digital-to-Analog Converters (DAC)

With the advent of digital technologies, the need for analog-to-digital and digital-to-analog converters started their use in practical applications.

The early devices were bulky and too slow for widespread use, particularly in DDFS. Today widespread use is connected with the introduction of integrated circuits (IC).

In selecting a good DAC for DDFS application, we have to consider several major criteria:

  a. Bit resolution

  b. Conversion accuracy

  c. Maximum clock frequency

  d. Power consumption

  e. And many others, in accordance with intended applications.

Note that speed and power consumption depend to a very high degree on the digital logic used and technology.

A. Digital Codes

In digital-to-analog converters, binary coding systems are prevalent. Any real number ξ_o can be expressed as a systematic fraction with the base 2; see eq. (14). However, in actual digital systems, the number of useful places or bits is limited, as discussed earlier in Section V. In addition, different applications require different coding systems.

1. Unipolar (straight) Binary Code. This code is shown in Table 1 for 4-bit systems, in the editor's introductory paper to Part VIII. It is very simple and instructive. Note that it contains 2^4 values from zero to 15/16. Generally, for n bits the maximum output value is

$$2^n - 1 \tag{50}$$

2. Bipolar Offset and Complementary Offset Binary Codes.
In DDFS applications, the desired output voltage of the DAC
should be bipolar. This requirement is met with offset binary
codes, which are discussed in more detail in the introductory
paper in Part VIII. Examination reveals that the MSB is responsible for the sign of the output voltage.

B. Basic DAC Circuits

Although digital-to-analog conversion can be realized in
many ways, only a few converters, most suited for monolithic
production, have survived. A basic introduction to the problem
prepared by the author has been published in a specially written paper in Part VIII. Here we will call the reader's attention
to some block diagrams published by several producers.

1. Voltage-Switching Converters. The converter reproduced
in Fig. 12 is of the voltage-switching type and uses an *R-2R*
resistor ladder network.

This 8-bit DAC (an early production by Plessey Semiconductors type ZN426-8 [30]) is simple but rather slow with the
settling time of 1 μs. (Note that currents can be switched more
rapidly, by at least one order of magnitude).

2. Current-Switching Converters. The block diagram of
this type of commercial DAC is reproduced in Fig. 13. The
DAC-08 Motorola series is a monolithic 8-bit high-speed multiplying DAC capable of settling to within 1/2 LSB (i.e., to
0.19%) in 85 ns. In addition, dual complementary current outputs with high-voltage compliance provide added versatility
and allow differential mode of operation to effectively double
the peak-to-peak output swing [31].

3. Segmentation of DAC. Segmentation of DAC makes it
possible to increase either speed or resolution and to reduce
spurious glitches. The device is composed of a set of nonweighted current cells and an ordinary part of weighted current
sources. Here, we illustrate the construction of such a DAC
with a block diagram of a commercial converter [32]
CX20051A in Fig. 14. This 10-bit DAC has the speed of 30
MSPS. It is intended for broadcasting applications as well as
high-definition TV.

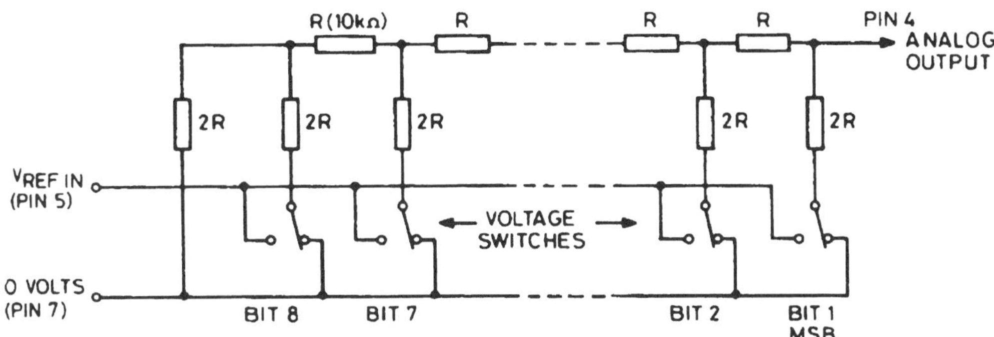

Fig. 12. Block diagram of the voltage-switching DAC with an R-2R resistor ladder network (reproduced from [30]). © Plessy Semiconductors. Used by permission.

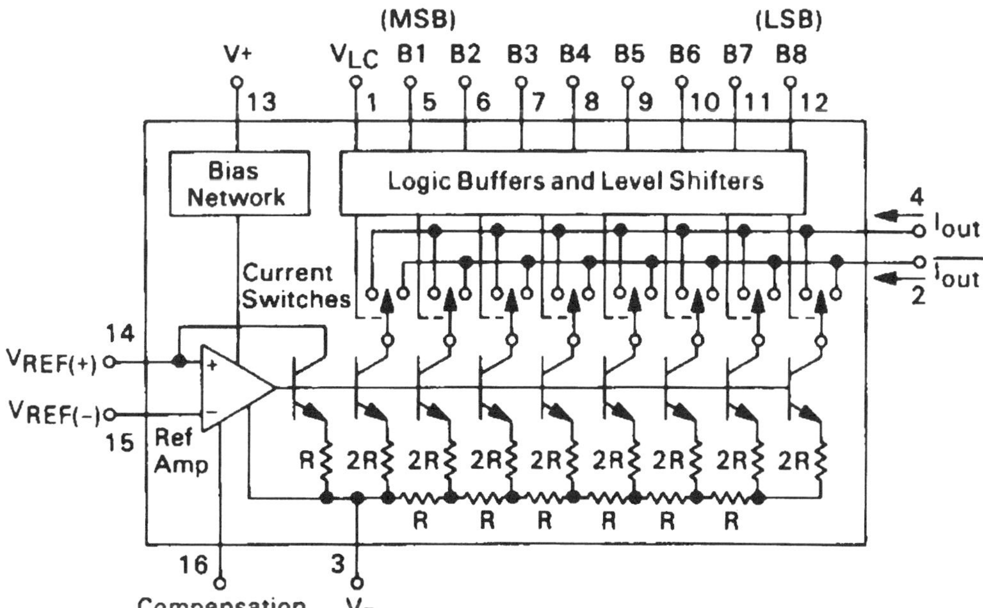

Fig. 13. Block diagram of the current-switching DAC with an R-2R ladder
network (reproduced from [31]). © Motorola. Used by permission.

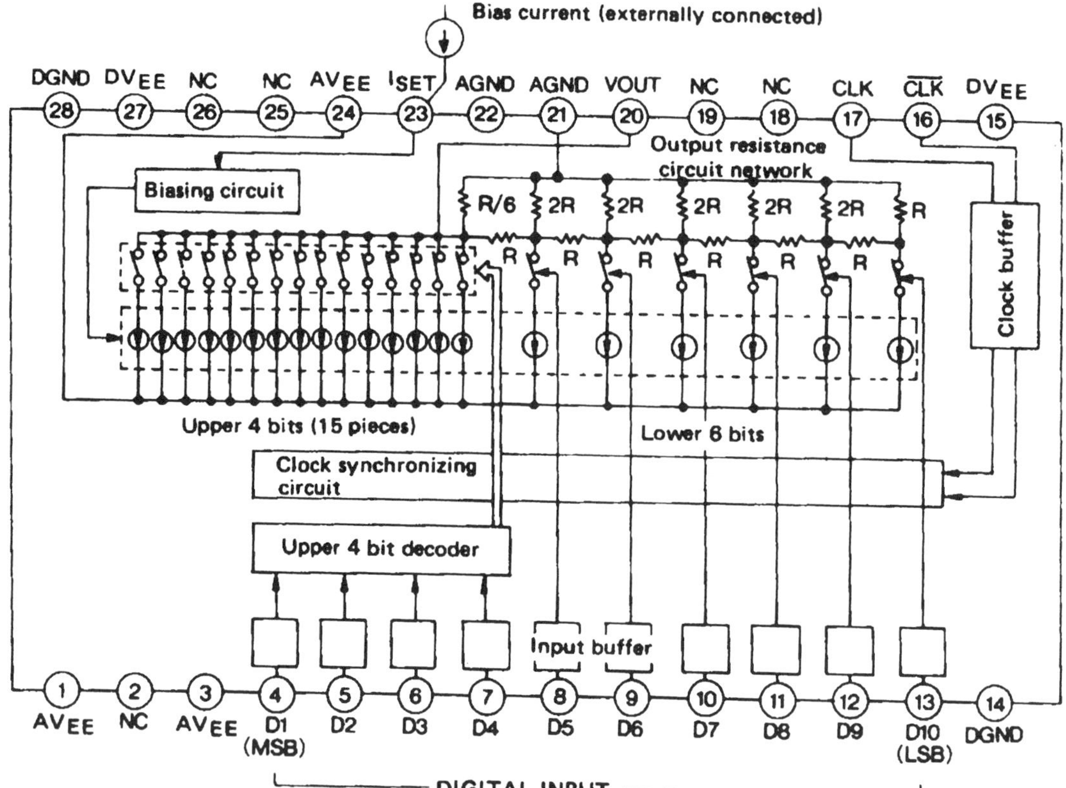

Fig. 14. Block diagram of two-stage segmented DAC (reproduced from
[32]). © Sony. Used by permission.

C. Signal-to-Noise Ratio (S/N)
Due Quantization

The quantization of the output sine wave due to the finite resolution of the lookup tables has been discussed earlier. However, the same problem is met in instances where DAC bit resolution is smaller than the length of sine words S.

Inspection of complementary binary codes in Part VIII reveals that a D-bit DAC can resolve in each quadrant only

$$2^{D-1} = 2^A \qquad [51]$$

values.

As a result, an ideal DAC introduces an error as large as

$$\frac{1}{2^A}$$

or as small as

$$0$$

In this case, the variance of the DAC truncation error has approximately a uniform distribution, and the ideal signal-to-noise ratio is

$$\frac{S}{N} = \frac{\frac{1}{2}\left(\frac{V_{pp}}{2}\right)^2}{\frac{1}{12}\left(\frac{V_{pp}}{2^D}\right)^2} = \frac{3}{2}2^{2D} \qquad [52]$$

or in dB measure

$$\frac{S}{N} = 1.76 + 6.02 * D[\text{dB}] \qquad [53]$$

This relation is generally given as the maximum level of spurious signal at the output of DDFS.

After reverting to relation (51), we conclude that matching between a sine lookup table storing S-bits long words and the DAC resolution requires that

$$S \geq D - 1 = A \qquad [54]$$

Computer simulations reveal that, theoretically, S need not exceed A. This is further emphasized by the presence of nonlinearities and spurious signals generated in DAC.

D. DAC Nonlinearities

The manufacturing processes are plagued with inevitable component inaccuracies, which are discussed in several papers reprinted in Part VIII. Together with the environmental influences, particularly temperature fluctuations of current weighting circuits, such as scaling resistors and base emitter voltages, beta drift of current switches, DACs exhibit additional nonlinearities that further decrease the output signal purity.

All manufacturers provide information about the stability and accuracy of their DAC. Unfortunately, the terminology used often differs from author to author and from producer to producer.

The editor has provided a comparison of different specifications and compiled a vocabulary of the most often used terms in a specially written paper in Part VIII. Here we list some of the most important terms.

12

1. The Offset Drift. The offset drift generates an actual DAC transfer function that is parallel to the ideal one and therefore has no effect on the output spectrum.

2. The Gain Error. The gain error rotates the V_{out} about the midpoint of the binary zero and is generally of little concern to frequency domain performance.

3. Differential Nonlinearity. Differential nonlinearity (DNL) is the worst-case deviation from one LSB output change from one adjacent state to the next; see Fig. 15. If DNL is less than or equal to $\pm$ LSB, the DAC will yield all voltage or current levels at the output—the respective DAC is monotonic.

4. Integral Nonlinearity. Integral nonlinearity (INL) is defined as the maximum deviation from a straight-line approximation of the DAC transfer function; see Fig. 15.

5. Superposition Errors. By considering bipolar binary codes, one encounters linearity errors and one's complements of these errors. In this instance, the DAC characteristic will appear symmetrical around midscale. Such behavior can easily be verified by simply measuring the output errors ε_i associated with n input codes instead of all of the 2^n possible combinations [34]

$$V_o = V_{FS}[2^{-1} + \epsilon_1 + 2^{-2} + \epsilon_2 + \cdots 2^{-n} + \epsilon_n] \qquad [55]$$

When the gain and offset errors have been removed, the summation of all ε_i should be zero. If the sum is greater than approximately 1/10 of the LSB, further testing of the DAC may become necessary.

Note that even when the midscale symmetry holds, higher harmonics of the output signal may be generated.

E. Glitches

Glitches appear in the DAC output during code transitions. A particularly large glitch is encountered at midrange change of binary code, for example, from

$$0111 \text{ to } 1000$$

Before the final transition is realized, some intermediate states

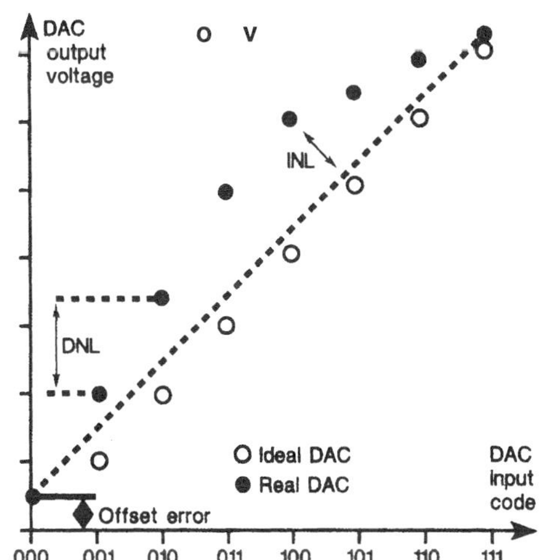

Fig. 15. Differential and integral nonlinearities of the DAC transfer characteristic (adapted from [33]).

may be formed, for example,

$$00000 \text{ or } 1111$$

for a while during the settling time. More detailed discussion may be found in the paper by the editor in Part VIII.

1. S/N Degradation Due to Glitches. Signal-to-noise degradation of the DDFS output is discussed in depth in the introductory paper to Part VIII. For cases where S/N should not be reduced more than by 3 dB due to glitches, we get a rule of thumb

$$V_g \tau \approx \frac{V_{pp}}{2^D} T_c \qquad [56]$$

2. Means for Glitch Reduction. More insight into the glitch problem is provided by Cremonesi et al. [35] who give guidelines for their reduction. One very effective means is segmentation of the DAC (cf. Fig. 14). Sometimes it helps to adjust the skew of driving circuitry and regulation of the logic threshold. Very effective deglitching circuit is a simple sampler.

VII. Phase Noise in DDFS

Every physical phenomenon is accompanied by smaller or larger fluctuations. In instances where these fluctuations are random, we speak about *noise*. The same is true for all frequency generators where we encounter both amplitude and phase or frequency variations that cause the so-called frequency instability of oscillators, frequency synthesizers, and so on.

The problem of the noise in DDFS is discussed in some depth in the introductory paper to Part VII [36]. From these results, as well as from our own measurements, we suggest for DDFS the following phase and noise characteristic

$$S_{\varphi,DDS}(f) \approx S_{\varphi,c}(f) * \xi_x^2 + \frac{10^{-10\pm2}}{f} + \frac{2^{-2(D-1)}}{3 * f_c} \qquad [57]$$

In instances where the numerical value of the f_c is larger than Y, the denominator of the normalized frequency ξ_x, then f_c in (57) must be replaced by Y. Evidently, both the flicker and white phase noise at the DDFS outputs may often be larger than one would expect from simple digital divider behavior.

VIII. Output Filters

All digital devices are based on clocking or sampling of the currents or voltages. As a consequence, we encounter the spectrum aliasing, that is, its repeating in the rate of the sampling frequency.

In accordance with the z transform theory (e.g., [37,38]) we can write the transfer function of the sampling system with the input signal $V_1(s)$

$$V_2(s) = H_h(s) * \hat{V}_1(s) \qquad [58]$$

where the first term on the right-hand side is

$$H_h(s) = T_c * e^{-sT_c/2} * \frac{\sinh(sT_c/2)}{sT_c/2} \qquad [59]$$

and the second one is

$$\hat{V}_1(j\omega) = \frac{1}{T_c} \sum_{n=0}^{\infty} V_1(j\omega - jn\omega_c) \qquad [60]$$

Investigation of the factor $H_h(s)$ reveals zeros for every

$$\omega = n\omega_c \qquad [61]$$

Similarly, from the relation (60), we conclude that the noise and spurious signals close to the fundamental carrier are repeated around every multiple of ω_c (see Fig. 16).

A. Antialiasing Filters

The major consequence of the above conclusions is that the closer the output frequency is to the

$$\omega = \frac{\omega_c}{2} \qquad [62]$$

the closer the spurious signal is to the desired one.

The difficulty is removed with the assistance of the low-pass filter having the corner frequency $f_c/2$. According to the steepness of the antialiasing filter, we can effectively use about 40 percent of the DDFS base band.

Example 2. In Fig. 16, where the normalized frequency is equal to $\xi_x = 13/32$, the desired signal is for $n = X = 13$ and is safely in the Nyquist range that extends from 0 to $Y/2 = 16$. The closest spurious signal is for $n = 19$ (since $32 - 13 = 19$). On the other hand, there would be the desired signal, $X = 15$; then the closest spurious would be for $n = 32 - 15 = 17$ and would have nearly the same amplitude difficulty to filter out.

Another drawback of the sampling process is the slope of the sampling transfer function $H_h(s)$ which at $f_c/2$ achieves

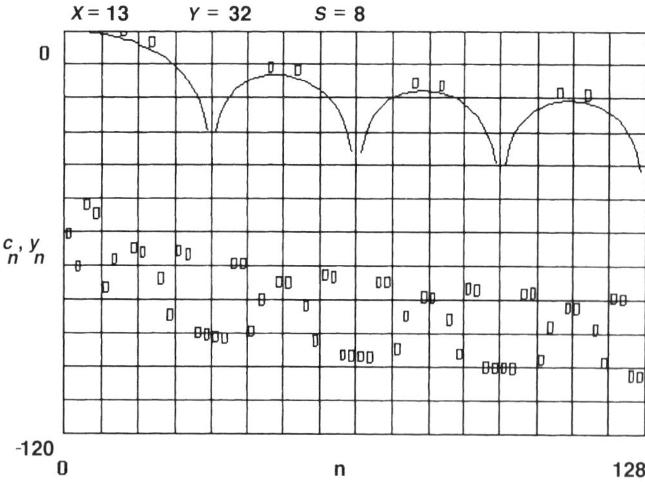

Fig. 16. Example of the output spectrum of the DDFS without antialiasing filter.

about -4 dB; see the continuous line in Fig. 16. This problem is not too important and can also be solved with proper design of the antialiasing filter.

B. Bandpass Filters

In some instances, we can extract from DDFS higher frequencies even above f_c with the assistance of the bandpass filters. However, the useful range is smaller than at base band because of the zeros of the $H_h(s)$ function.

IX. CONCLUSIONS

After a short historical survey, going back to the predecessors of DDFS and to the mechanical "angular velocity" synthesizers—the gear boxes—the basic mathematical theory common to both devices is discussed. The common problem is a step-by-step approximation of real numbers in the software approach and in keeping with physical limitations in hardware solutions, namely, with practicable numbers of teeth on cogwheels and frequency multiplication/division factors in frequency synthesizers.

The last condition is adroitly met with the Modulo-N procedure in modern DDFS, which generates large sets of output frequencies. Therefore, these systems are discussed in depth.

After a short introduction to the building blocks of DDFS, the most serious problem is discussed, namely, the spurious signals accompanying the desired signal. We pointed out their sources, briefly investigated their expected level below the carrier, and gave some guidelines for choice of the effective phase bits, for the length of words in the sine lookup tables, and for resolution of the DAC used.

Furthermore, we reviewed some commercial DACs and discussed their basic circuits together with digital codes.

Finally, we investigated the theoretical S/N ratio due to the principle of the operation of DAC and its decrease due to inevitable DAC nonlinearities. In this connection, we also focused on the aliasing problems encountered in DDFS as well as the expected background noise and phase noise of DDFS.

REFERENCES

[1] Ch. Huygens. *Opuscula posthuma. Descriptio automati planetarii.* Lugduni Batavorum (Leiden), 1703.

[2] A. Brocot. *Le Calcul des Rouages par Approximation.* Paris, 1862.

[3] F. Becher and A. Korner. *Pfauter Wechselraeder Tabellen.* Ludwigsburg: F. Becher, 1963.

[4] M. E. Esclangon. "Horloges et mecanismes a double indication de temps sideral et de temps solaire moyen." *Annales françaises de chronometrie,* vol. 9, pp. 323–30, 1939.

[5] H. J. Finden. "The frequency synthesizer." *Journal I.E.E.,* Part III, vol. 90, pp. 165–80, 1943.

[6] M. Boella. "Generatore di frequenze campione per misure di alta precisione." *Alta Frequenza,* vol. 14, pp. 183–94, September–December 1945.

[7] J. Kavan. *Tabula omnibus a 2 usque ad 256000 numeris integris omnes divisores primos preabens.* Stara Dala, 1934, Czechoslovakia. Reprinted

during World War II by Macmillan as *Factor Tables (Prime Factors of All Numbers up to 256000)*.

[8] V. F. Kroupa. "Theory of frequency synthesis." *IEEE Trans. on Instrum. Meas.,* vol. IM-17, pp. 56–68, March 1968.

[9] V. F. Kroupa. *Frequency Synthesis: Theory, Design et Applications.* London: Ch. Griffin, 1973; New York: John Wiley, 1973.

[10] J. Gorski-Popiel. *Frequency Synthesis: Techniques and Applications.* New York: IEEE Press, 1975.

[11] V. Manassewitsch. *Frequency Synthesizers, Theory and Design.* New York: John Wiley, 1976, 1980, 1983.

[12] W. F. Egan. *Frequency Synthesis by Phase Lock.* New York: John Wiley, 1981.

[13] U. L. Rohde. *Digital PLL Frequency Synthesizers, Theory and Design.* Englewood Cliffs, N.J.: Prentice-Hall, 1983.

[14] R. Stirling. *Microwave Frequency Synthesizers.* Englewood Cliffs, N.J.: Prentice-Hall, 1987.

[15] V. F. Kroupa. "Approximating frequency synthesizers." *IEEE Transactions on Instrumentation and Measurement.* December 1974. (Reprinted in Part II.)

[16] E. Ulicki. "Cubic Inch Frequency Synthesizers," *Proceedings of the 19th Annual Symposium on Frequency Control,* 1965, p. 580.

[17] O. Perron. *Irrationalzahlen,* 3rd ed. Berlin: de Gruyter, 1947.

[18] G. Becker. "Quasiperiodic frequency synthesis." *Proceedings of the 26th Annual Frequency Control Symposium,* June 1972, pp. 279–91.

[19] V. F. Kroupa. "Quasiperiodic omission of pulses." (A specially written paper for Part XI.)

[20] G. W. Small. "A frequency synthesizer for $10/2\pi$ kHz." *IEEE Transactions on Instrumentation and Measurement,* March 1973. (Reprinted in Part II.)

[21] V. F. Kroupa. "Spectra of pulse rate frequency synthesizers." *Proceedings of the IEEE,* vol. 67, pp. 1680–82, December 1979. (Reprinted in Part IV.)

[22] V. F. Kroupa. "Spectral properties of DDFS: Computer simulations and experimental verifications." Adapted from papers read at the *1993 and 1994 IEEE International Frequency Control Symposia.* (Specially written paper for Part V.)

[23] P. H. Saul and D. G. Taylor. "A high speed direct frequency synthesizer." *IEEE Journal of Solid-State Circuits,* February 1990. (Reprinted in Part III.)

[24] P. H. Saul and T. Coffey. "Current achievements in direct digital synthesis." *Microwave Engineering Europe,* December/January 1993, pp. 31–36.

[25] G. A. Sunderland, R. A. Strauch, S. S. Wharfield, H. T. Peterson, and Ch. R. Cole. "CMOS/SOS frequency synthesizer LSI circuit for spread spectrum communications." *IEEE Journal of Solid-State Circuits,* August 1984. (Reprinted in Part III.)

[26] H. T. Nicholas III and H. Samueli. "A 150-MHz direct digital frequency synthesizer in 1.25 μm CMOS with −90-dBc spurious performance." *IEEE Journal of Solid-State Circuits,* December 1991. (Reprinted in Part IV).

[27] H. T. Nicholas III, H. Samueli, and B. Kim. "The optimization of direct digital frequency synthesizer performance in the presence of finite word length effects." *Proceedings of the 42nd Annual Frequency Control Symposium* 1988. (Reprinted in Part IV).

[28] L. A. Weaver and R. J. Kerr. High resolution phase to sine amplitude conversion. U.S. Patent 4,905,177, February 27, 1990 (Assignee Qualcomm, Inc.).

[29] J. Vankka. "Methods of mapping from phase to sine amplitude in direct digital synthesis." *IEEE Transactions on Ultrasonics, Ferroelectrics and Frequency Control,* pp. 526–534, March 1997.

[30] *Data Converters and Voltage Reference IC Handbook.* Plessey Semiconductors, p. 129, 1989.

[31] Motorola. *Analog/Digital and Digital/Analog Conversion Manual,* pp. 2–3, 1985.

[32] *Sony Semiconductor IC Data Book,* 1993. A/D, D/A converters, p. 466, 1993.

[33] D. Buchanan. "Choosing DACs for direct digital synthesis." *Microwave Engineering Europe,* pp. 35–43, February 1993.

[34] Burr-Brown. *Applications Handbook.* Tucson: Burr-Brown Corporation, 1994.

[35] A. Cremonesi, F. Maloberti, and G. Polito. "A 100-MHz CMOS DAC for video-graphic systems." *IEEE Journal of Solid-State Circuits,* pp. 635–39, June 1989.

[36] V. F. Kroupa. "Introduction into the noise properties of frequency sources." (A specially written paper for Part VII.)

[37] J. R. Ragazzini and G. F. Franklin. *Sampled-Data Control Systems.* New York: McGraw-Hill, 1958.

[38] E. J. Angelo, Jr. "a tutorial introduction to digital filtering." *Bell Syst. Tech. J.,* September 1981.

Part II

Direct Digital
Single-Frequency Synthesizers

DIRECT digital frequency synthesizers (DDFSs) appeared nearly simultaneously with integrated circuits (ICs). However, at that time, DDFSs were designed for one output frequency only. At the beginning, the field for their application provided the standard time and frequency laboratories. As early as the mid-1960s, publications dealt with generation of sidereal time from the solar time — UT_2 [e.g., II-1]. The design was based on the trial-and-error method by adding pulses to a 100-kHz solar pulse train. The spurious signals, inherent to all DDFSs, were of no concern because they were filtered in the gears of the mechanical clock.

In the first of the reprint papers, the situation is different since Small (1973) needed a frequency $f_x = 10/2\pi$ kHz that is often used in precise measurements of resistance, capacitance, and inductance. He wanted to synthesize a signal of good spectral purity from a stable crystal oscillator with the assistance of simple ICs only. He says little about his search for the final mathematical model. Apparently, however, his approach was that of the early mechanical engineers; namely, he might provide a continued fraction expansion of the desired normalized frequency

$$\xi_x = \frac{5 * 10^3}{\pi * 5 * 10^6} = \frac{1}{\pi * 10^3} \qquad [1]$$

and looked for a possible factorization of the denominator. He also investigated spurious signals and put a crystal filter at the output of the DDFS to arrive at the desired spectral purity. His approach was that of the quasiperiodic omission of pulses.

This terminology was introduced by G. Becker in his paper "Quasiperiodic frequency synthesis" [II-2]. Becker discusses different approaches to the design of single-frequency DDFSs. His major contribution is his suggestion that the output pulse train or rectangular wave should be as periodic as possible, that is, quasiperiodic.

This idea has been pursued by Kroupa (1974) in the second reprint paper, "Approximating frequency synthesizers," the software of which is based on the number theory. Three different approximation algorithms are briefly discussed, namely, continued fractions, Cantor products, and Engel series, together with modifications suggested by the author.

In the paper specially prepared for this volume and reprinted in Part XI, Kroupa discusses in detail principles of the quasiperiodic omission of pulses and provides several practical examples.

For completeness, we only mention some other approaches to DDFS suggested in the past.

The paper [II-3] deals with early applications of DDFS in music by generating binary multiples of the audible tones in accordance with the temperate scale. This particular frequency synthesis was based on subtraction and addition of pulses in the master clock rate. Spectral purity was improved by digital dividers. From a more general point of view, we would like to emphasize that the same result can be arrived at by suppression of pulses only. [Note that the addition of pulses may be a vague operation; therefore, we do not recommend it even when it might simplify the hardware, for example, by applying a shortened continued fraction expansion. (Cf. the first paper by Kroupa in Part IV.)]

Systems based on the gear box model, such as that by Small, have been suggested in the past. As an example, we mention "A microprocessor controlled frequency synthesised signal source" by Jenkins [II-4] and a recently published paper by Ishihara [II-5] which discusses a fine-control digital frequency synthesizer for applications in rubidium atomic oscillators. The device contains three-stage phase-locked loop oscillators with variable dividers in the feedback path. The configuration resembles an electronic gear box.

Here, we feel obliged to correct the general belief that there exists a definite probability that the approximation error will be below a certain desired error. On the contrary, the editor [II-6] has proved that for the normalized gear ratio close to the small integer ratios such as 1/2 and 1/3, etc., the remainder is rather large — D being the greatest number of teeth on the wheels, or the greatest division and multiplication factors allowed. The remainder meets condition

$$R_{\text{max},k} \approx \frac{1}{2D^k} \qquad [2]$$

where k is the number of division/multiplication stages. The consequence is that a smaller remainder can be achieved only by increasing the number of stages in the mechanical or electronic gear box.

REFERENCES

[II-1] C. C. Bare and D. L. Thacker. "Sidereal time standard at the national radio astronomy observatory." *Proceedings of the 20th Annual Symposium on Frequency Control*, pp. 624–28, 1966.

[II-2] G. Becker. "Quasiperiodic frequency synthesis." *Proceedings of the 26th Annual Frequency Control Symposium*, pp. 279–91, June 1972.

[II-3] W. Adriaans and N. V. Franssen. "The sound of organ music inspires new bipolar efforts." *Electronics,* September 4, 1975, pp. 110–14.

[II-4] R. W. Jenkins and W. A. Evans. "A microprocessor controlled frequency synthesised signal source." *Proceedings of the Conference on Programmable Instruments* (IERE Conference Proceedings, No. 38), pp. 53–60, November 1977.

[II-5] N. Ishihara et al. "Development of a digitally controlled rubidium atomic oscillator." *Proceedings of the 44th Annual Symposium on Frequency Control,* pp. 59–65, 1990.

[II-6] V. Kroupa. "Design of gear boxes with the assistance of computers." *Strojirenstvi,* pp. 646–655, November 1974 (in Czech).

A Frequency Synthesizer
for 10/2 π kHz

GREIG W. SMALL

Abstract—A sinusoidal voltage source of frequency nominally $10/2\pi$ kHz ($\omega = 10^4$ rad/s) is derived from a 5-MHz standard. Digital counters are used to generate the required frequency, which differs from $10/2\pi$ kHz by less than 2 parts in 10^8, and is as stable as the standard.

Using a crystal filter and low-distortion amplifiers, an output of 10 V rms with total harmonic distortion less than 1 part in 10^3 is obtained.

I. Introduction

A SIGNAL frequency of $10/2\pi$ kHz ($\omega = 10^4$ rad/s) is often used in precise measurements of resistance, capacitance, and inductance. Some bridge configurations, such as the transformer ratio arm capacitance bridge and Maxwell inductance bridge, are in principle frequency independent. Others, such as the quad bridge,[1] in which capacitance and resistance are compared, require the frequency to be known to the accuracy of the measurement. The signal source for such a bridge should be of good spectral purity, for unwanted

Manuscript received October 21, 1971.

The author is with the Division of Applied Physics, Commonwealth Scientific and Industrial Research Organization, National Standards Laboratory, Sydney, Australia.

[1] A. M. Thompson, "An absolute determination of resistance based on a calculable standard of capacitance," *Metrologia*, vol. 4, pp. 1–7, 1968.

frequency components contribute to detector noise, and errors may arise through intermodulation of harmonics. The source to be described was designed to meet these requirements.

II. Synthesis of the Required Frequency

The system shown in Fig. 1, in which the 5-MHz source is divided in a preset digital counter, then multiplied in a frequency multiplier, was proposed as a means of generating the required frequency. A reasonable upper limit for the frequency ratio of a single stage multiplier is about 30, a search in this range for suitable values of divisor and multiplier revealed the combination 84 823.0016 $\cdots$ and 27. The prime factors of the integer 84 823 are 271 and 313. On this basis a synthesizer was built using two preset counters and three stages of multiplication, as shown in Fig. 2.

The overall division ratio is 3141.592 592 $\cdots$, which is less than the exact 1000π by 194 parts in 10^{10}. Hence the synthesized frequency is 194 parts in 10^{10} higher than nominal.

This synthesizer was used for some time, with a simple tuned amplifier as the output filter. Its performance was unsatisfactory due to insufficient spectral purity and unreliability in the multiplier stages.

These defects were eliminated by using a crystal filter to improve the spectral purity and a different synthesizer in

Reprinted from *IEEE Transactions on Instrumentation and Measurement*, Vol. IM-22, No. 3, pp. 34-37, March 1973.

Fig. 1. Proposed system of frequency synthesis.

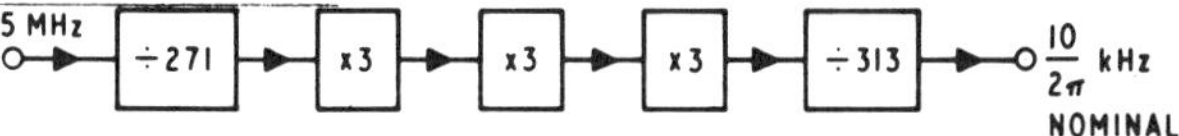

Fig. 2. First frequency synthesizer using multiplier stages.

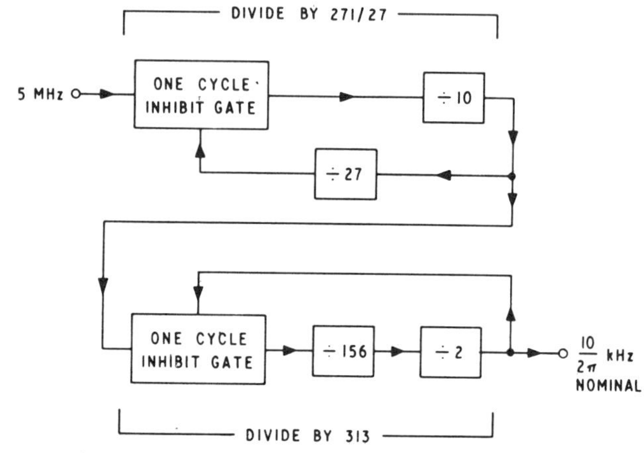

Fig. 3. Second frequency synthesizer using dividers only.

which the multiplier stages are avoided. This is achieved by observing that the process of division by 271 followed by multiplication by 27 can be implemented as a sequence of 26 cycles of division by 10 and one cycle of division by 11. Thus 271 input cycles are contained in 27 successive output cycles. Two synthesizers were constructed, using J–K flip-flops in two stages of division as shown in Fig. 3.

In the first stage, the output from the divide-by-10 counter drives the divide-by-27 counter, which drives the single-cycle inhibit gate. For each 27 cycles of output from the divide-by-10 stage, one input cycle is inhibited.

The output from the second stage is as nearly symmetrical as possible. The symmetrical divide-by-312 counter and the input inhibit gate result in a mark–space ratio of 156/157 in an output cycle of 313 counts.

III. DIVIDE-BY-271/27 STAGE

The complete circuit is shown in Fig. 4. Ripple-carry counting and gated clocking are combined to give a simple circuit in which delays are kept small.

The 5-MHz input signal and the output signal are buffered, using single NAND gates. Because the 5-MHz signal is sinusoidal, the input buffer is biased near to its point of transition, as shown in Fig. 5. The output from this buffer is therefore symmetrical and insensitive to variations of input level. The prototype synthesizers were found to perform correctly with input signals as small as 25 mV rms; in normal use the level is 1 V rms.

IV. DIVIDE-BY-313 STAGE

This stage uses mainly ripple-carry counting, with gated clocking of the final flip-flops. In this way the delay of an output transition from the corresponding clocking transition is only that of one flip-flop. By keeping the overall delay small, the effect of any variability in the delay is also small. Fig. 6 shows the complete stage. Both prototype synthesizers were tested and found to function correctly with input frequencies up to 40 MHz.

V. FREQUENCY SPECTRUM OF SYNTHESIZER OUTPUT

Because the synthesized frequency is derived from the input by counting processes only, there must be an integral number N of input cycles in each output cycle. The nonintegral frequency ratio results from N having other than a single value.

The number N_n of input cycles in the nth output cycle is given by

$$N_n = \text{quotient of } [(271 \times 313 + R_{n-1}) \div 27]$$

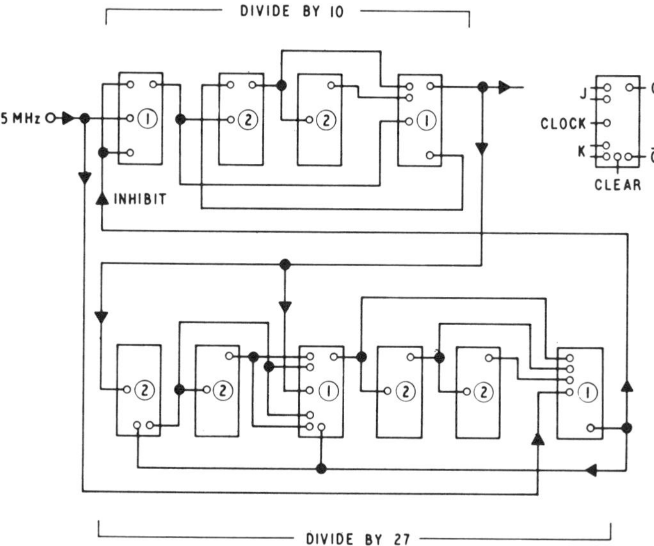

Fig. 4. Divide-by-271/27 stage.

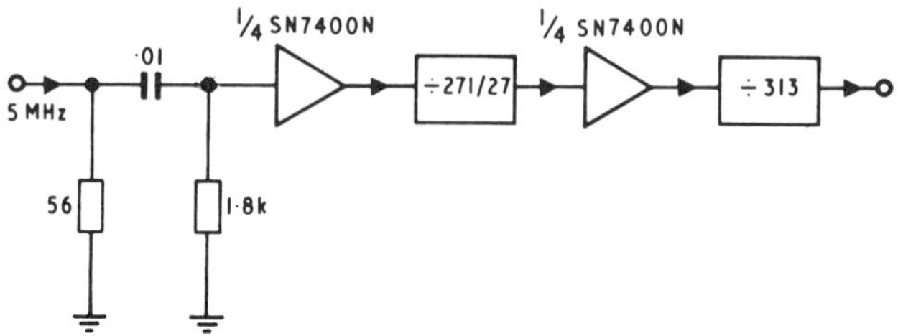

Fig. 5. Buffering of input and output signals.

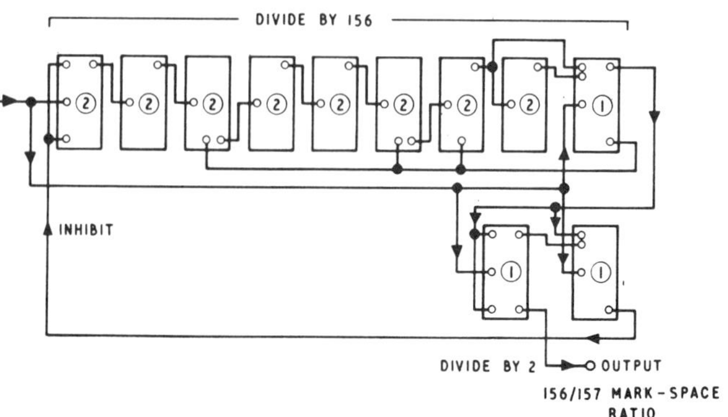

Fig. 6. Divide-by-313 stage.

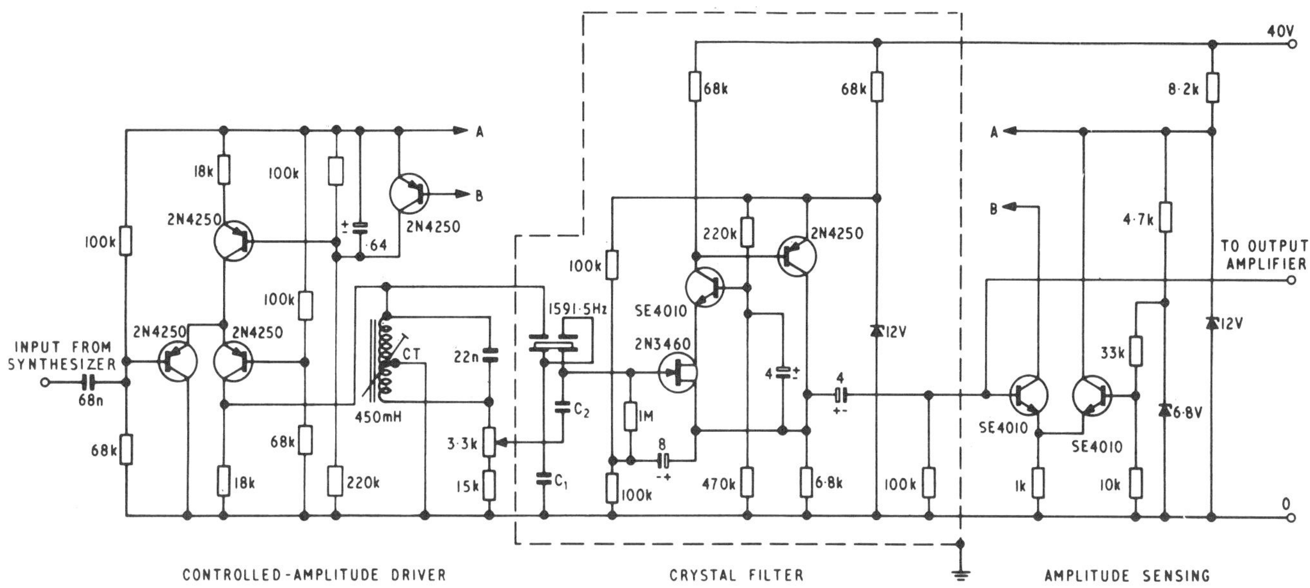

Fig. 7. Circuit diagram of crystal filter.

and

$$R_n = \text{remainder of } [(271 \times 313 + R_{n-1}) \div 27]. \qquad (1)$$

Two points may be observed from (1): R can have at most 27 different values; and since 271×313 and 27 are mutually prime, R will take all 27 values.

Thus the output is repetitive over 27 cycles, and is given by successive applications of (1) from an arbitrary initial value of R. Equation (1) may be written

$$N_n = 3141 + \text{quotient of } [(16 + R_{n-1}) \div 27] \qquad (2)$$

showing that N will be either 3141 or 3142.

Accounting for the 156/157 mark–space ratio of the divide-by-313 state, the numbers of input cycles in the alternate low and high states of the output are given by

$$N_n \text{ (low)} = \text{quotient of } [(271 \times 156 + R_{n-1} \text{ (high)}) \div 27]$$

$$N_n \text{ (high)} = \text{quotient of } [(271 \times 157 + R_n \text{ (low)}) \div 27] \qquad (3)$$

i.e.,

$$N_n \text{ (low)} = 1565 + \text{quotient of } [(21 + R_{n-1} \text{ (high)}) \div 27]$$

$$N_n \text{ (high)} = 1575 + \text{quotient of } [(22 + R_n \text{ (low)}) \div 27]. \qquad (4)$$

The frequency components of the synthesized signal can be obtained by Fourier series analysis of the sequence defined by (4). Because the fundamental period of this sequence contains 27 cycles of the main component, the frequency spectrum consists of the main component f_0 and sidebands spaced at $f_0/27$. The most significant components have been computed, and verified by measurement. These are listed in Table I.

The principle of modifying a counter so that its division ratio is varied in a specific pattern can be applied to other situations where a nonintegral frequency ratio is required. The most suitable application of this technique is in fixed frequency synthesis, where if necessary a narrow-band filter may be used to reduce the sideband components of frequency. The presence of sidebands is not peculiar to this method of frequency synthesis, but is also a property of synthesizers using frequency multipliers.

VI. Crystal Filter

To reduce the sideband and harmonic components of the synthesized signal to an acceptably low level, a quartz crystal filter is used. The low-distortion output amplifier of the filter is designed to present a high impedance to the crystal. Its impedance at 1592 Hz is greater than 100 MΩ and total harmonic distortion is 10^{-4}.

The amplitude of the filtered signal is locked to a dc reference voltage. A peak voltage detector compares the signal with the reference, producing a controlling signal that determines the amplitude of the current drive into the filter. The output of the filter is maintained at 1 V rms for a 10-to-1 range in crystal response. Fig. 7 shows the complete circuit of the filter, and its effect on unwanted frequency components is included in Table I.

TABLE I
FREQUENCY SPECTRUM OF SYNTHESIZED SIGNAL

Sideband Number [1]	Sideband Level Relative to Main Component				
	Synthesizer Output		Crystal Filter Response (c)	Filtered Signal (d) = (a) × (c)	At Detector of Quad Bridge [2] (e)
	Computed (a)	Measured (b)			
	$\times 10^{-6}$	$\times 10^{-6}$	$\times 10^{-6}$	$\times 10^{-9}$	$\times 10^{-12}$
1	64	65	420	27	11
2	34	35	180	6	2
11	318	330	22	7	3
27	5018	5300	21	≪ harmonic distortion	

[1] The sideband separation is 1/27 of the main component.
[2] The response of the particular quad bridge being quoted is quite flat over a wide range of sidebands and is equivalent to a bridge unbalance of 4×10^{-4} in the main component. Harmonic rejection filters are used to reduce the effects of harmonic components.

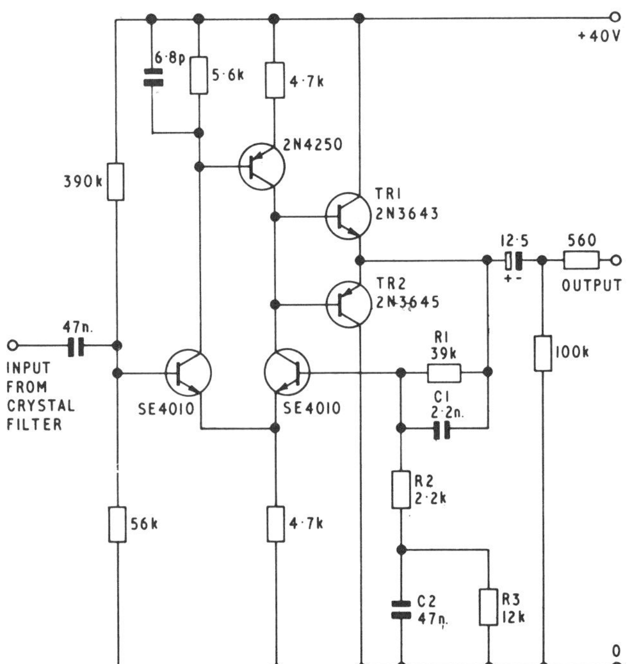

Fig. 8. Circuit diagram of output amplifier. C_1 and C_2 are selected to suit particular crystal; C_1 modifies resonant frequency; C_2 allows for cancellation of direct capacitance of crystal, and is approximately 1.5 pF.

TABLE II

HARMONIC CONTENT OF OUTPUT SIGNAL (TYPICAL)

Harmonic Number	Harmonic Amplitude Relative to Fundamental	
	on open circuit	with 60 ohm load
	$\times 10^{-6}$	$\times 10^{-6}$
2	170	760
3	130	200
4	60	60
5	40	50
Total harmonic distortion	230	790

VII. OUTPUT AMPLIFIERS

The complete signal source as constructed includes four output amplifiers, each providing 10 V rms on open circuit, with 600-Ω output impedance.

Fig. 8 shows the circuit of one amplifier. The input and driver stage has a high output impedance, and therefore appears as a current source to the output transistors $TR1$, $TR2$, which are connected as emitter followers in class B. This configuration possesses good linearity, and low crossover distortion because of the high gain of the driver stage during transition from one output transistor to the other.

The negative feedback network $R1$, $R2$, $R3$, $C1$, and $C2$ defines the signal gain and dc conditions of the amplifier, and subject to these constraints the network is optimized for minimum bandwidth. The capacitors in this network are responsible for a marked reduction in crossover distortion, not attributable to negative feedback only. The effect is due to the discharge of the capacitors through the load during crossover, maintaining continuity of waveform in both amplitude and gradient.

Some measured values of harmonic content of the output signal are given in Table II.

VIII. CONCLUSION

The synthesis of the required frequency is achieved using digital counters, so that the resulting frequency is as accurately defined as the 5-MHz source. The crystal filter and buffer amplifiers provide a low-distortion signal source for bridges, and particularly the quad bridge, used for measurements to 1 part in 10^8.

Approximating Frequency Synthesizers

VĚNCESLAV F. KROUPA

Abstract—The process of frequency synthesis is the step-by-step approximation to the normalized output frequency. In this paper, three different approximation algorithms are briefly discussed, namely continued fractions, Cantor products, and Engel series. Their common disadvantage is their uniqueness; it is shown that this difficulty may be overcome by application of the new general approximation theorem, but the solution can be provided only by computer. Approximation errors are generally small, sometimes smaller than 1×10^{-12}. The advantage of modified Cantor products or Engel series algorithms is the possibility of standardizing the hard ware of frequency synthesizers, particularly if the IC technology and the use of off-the-shelf circuits is emphasized.

I. INTRODUCTION

IT HAS BEEN shown earlier [1], [2, pp. 122–158] that the mathematical model of the frequency synthesis process is a step by step approximation to the normalized output frequency

$$\xi_x = f_x/f_s = X_1/Y_1. \tag{1}$$

In practice, ξ_x is generally a rational fraction and, in addition, the input standard frequency f_s is either 100 kHz, 1 MHz, or 5 MHz. In these cases, "the theory of frequency synthesis," based on the modified Cantor series or systematic fractions [1], is fully adequate for the solution of (1).

However, in instances where ξ_x is a rational number either "ex decisione" (for example generation of sidereal time from atomic time) or because its accuracy is limited by practical reasons (for example comparison of different atomic transition frequencies) the aforementioned "theory of frequency synthesis" may fail to provide a simple solution. This happens if both the numerator X_1 and the denominator Y_1 in (1) are products of large prime numbers or are prime numbers themselves. The difficulty can be overcome by *approximating the normalized output frequency* ξ_x by a ratio of smaller integers or products of smaller integers. But a new problem arises, namely, finding suitable approximating algorithms which impose a syste-

matic solution on any ξ_x, provide the remainder to the required tolerances, and ensure the simplest possible hardware.

II. GENERAL APPROXIMATION THEOREM

Approximating of the normalized frequency ξ_x by a convergent to the continued fraction expansion (see (15)) has the advantage of the highest accuracy for the smallest divisor B_n. Though this algorithm has been used in some frequency synthesis systems in the past [2, pp. 212–215] its application is by no means general. The main difficulty is the uniqueness of the expansion. We have therefore started to investigate the class of all rational approximations (of a real number ξ_x) that are within a given tolerance range $\pm R$. By changing both A_n and B_n in (15) by a small amount, we get

$$\xi_x + \Delta\xi_x = \frac{A_n + \Delta A_n}{B_n + \Delta B_n} + R_n \tag{2}$$

and after subtracting (15) from (2)

$$\Delta\xi_x = \frac{B_n\Delta A_n - A_n\Delta B_n}{B_n(B_n + \Delta B_n)}. \tag{3}$$

In cases where $\Delta\xi_x$ is small the numerator $B_n\Delta A_n - A_n\Delta B_n$ is nearly zero and any smaller order approximation of A_n/B_n substituted for $\Delta A_n/\Delta B_n$ may satisfy (3). Finally we arrive at

$$\xi_x = \frac{cA_n + dA_k}{cB_n + dB_k} + R_{n,k} \tag{4}$$

where $k < n$ and c and d are integers. We may conclude that any rational approximation of ξ_x is given by (4) and further that the class of all rational approximations satisfying the tolerance range $\pm R$ is given by [3]

$$\xi_x = \frac{cA_n + dA_{n-1}}{cB_n + dB_{n-1}} + R_{c,d} \tag{5}$$

where $|R_n| \geq R \geq |R_{n+1}|$ and $|R_{c,d}| \leq R$

$$R_{c,d} = (-1)^{n-1}d \frac{\xi_{n+1} - c/d}{(cB_n + dB_{n-1})(B_n\xi_{n+1} + B_{n-1})}. \tag{6}$$

Manuscript received July 3, 1974; revised September 5, 1974.

The author is with the Institute of Radio Engineering and Electronics, Czechoslovak Academy of Sciences, Prague 8, Czechoslovakia.

Reprinted from *IEEE Transactions on Instrumentation and Measurement*, Vol. IM-23, No. 4, pp. 521–524, December 1974.

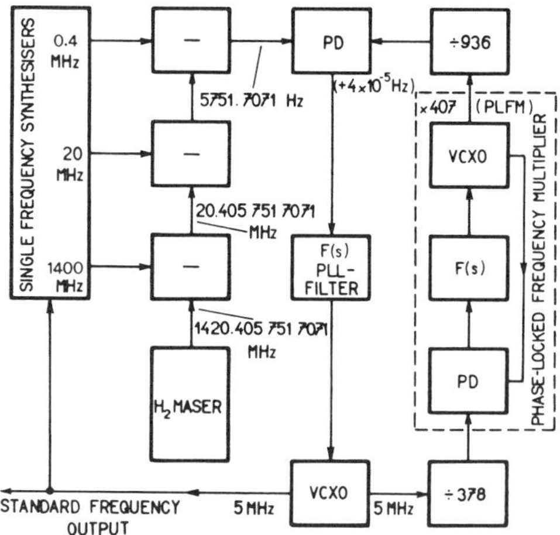

Fig. 1. Frequency synthesizer as used in hydrogen frequency standard.

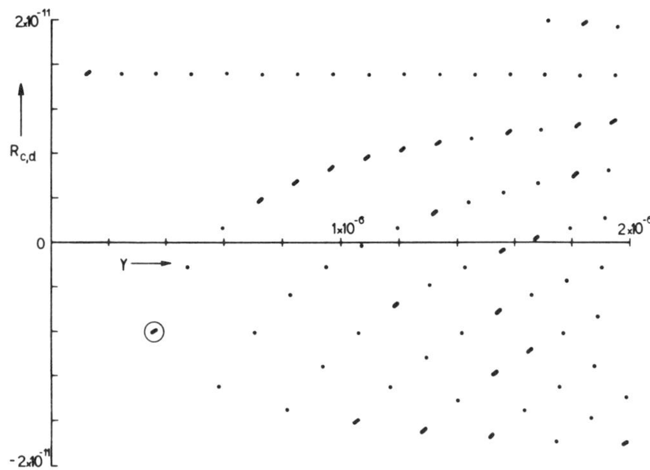

Fig. 2. All rational-number approximations of $f_x = 5751.7071/5 \times 10^6$ with the denominator $Y \leqq 2 \times 10^6$ and satisfying a tolerance range $\pm 2 \times 10^{-11}$.

For frequency synthesis applications both the numerator and the denominator in (5) must be factorized, i.e.,

$$\xi_x \approx \frac{x_0 \cdot x_1 \cdots x_n}{y_0 \cdot y_1 \cdots y_n} \tag{7}$$

where all x_k and y_k are smaller than a given upper bound D_u because neither multiplication nor division factors can be arbitrarily large in practical frequency synthesizers.

As an example, we have obtained the solution for the reference frequency generator in the hydrogen maser shown schematically in Fig. 1. In the first case we have chosen $f_x = 5\,751.7071$ Hz, the tolerance range $\Delta f_x = \pm 0.1$ mHz, and $f_s = 5$ MHz. A computer was used to find all rational-number approximations satisfying the given tolerance range and having the denominator smaller than 2×10^6. The results are plotted in Fig. 2. Short slant lines indicate successful factorizations while the small circle indicates the solution adopted in Fig. 1. Note that the frequency synthesizer is a mere chain of two frequency dividers and one phase-locked frequency multiplier. In the second case we have chosen $f_x = 20.405\,751\,7071$ MHz and left Δf_x and f_s unchanged. Possible mathematical models as found by computer are

$$\xi_x = \frac{3317}{2663} \times \frac{1268}{387} + 1.06 \times 10^{-11}$$

$$\xi_x = \frac{5141}{3107} \times \frac{846}{343} + 1.53 \times 10^{-12}$$

$$\xi_x = \frac{3833}{8291} \times \frac{3434}{389} - 7.93 \times 10^{-13}.$$

The hardware is a chain of two frequency dividers and two phase-locked frequency multipliers.

III. CANTOR PRODUCT APPROXIMATIONS

The disadvantage of the algorithm (7) is that no appreciable standardization of frequency synthesizer hardware is possible. This difficulty can be overcome by subjecting the relative remainder of a suitable additional approximation (5) to the Cantor product expansion (see Appendix

II). In this way, we get

$$\xi_x = \frac{cA_n + dA_{n-1}}{cB_n + dB_{n-1}}\bigg[1 + (-1)^n$$

$$\cdot \frac{d\xi_{n+1} - c}{(cA_n + dA_{n-1})(B_n\xi_{n-1} + B_{n-1})}\bigg]$$

$$= \frac{X}{Y}(1 \pm \rho). \tag{8}$$

Without any difficulty X, Y, and $1/\rho$ can be restricted to values smaller than some upper bound D_u. As a consequence also q_0 will be smaller than D_u. However, this will be no longer true with q_1 (see (21)) which will be of the order of D_u^2. The difficulty can be overcome by truncating the Cantor product expansion and approximating the second factor by the difference of two squares. Assuming the plus sign in (8), we get.

$$1 + \frac{1}{q_1} = \frac{1}{1 - 1/(q_1 + 1)} \approx \frac{1}{1 - 1/q_{11}} \cdot \frac{1}{1 + 1/q_{11}}. \tag{9}$$

In this way we shall achieve that all multiplication and division factors will be smaller than D_u. The approximation error R_c is

$$|R_c| \lesssim \frac{X}{Y} \cdot \frac{1}{q_{11}^3}. \tag{10}$$

But values even orders smaller can be found with the assistance of the computer.

The frequency synthesis system discussed in the preceding can be simplified in cases where X and Y are smaller than D_u, but $1/\rho$ is of the order of D_u^2. The approximation algorithm (9) can be imposed directly on the first term $(q_0 + 1)/q_0$ or $q_0/(q_0 + 1)$. The reference frequency generator for the hydrogen maser $f_x = 20.405\,751\,7071$ MHz, shown in Fig. 3, is of this type. Note that the hardware of the first and second "multiplier–divider module" is the same because the standard frequency f_s is here offset only by a small amount.

The advantage of these frequency synthesis systems is

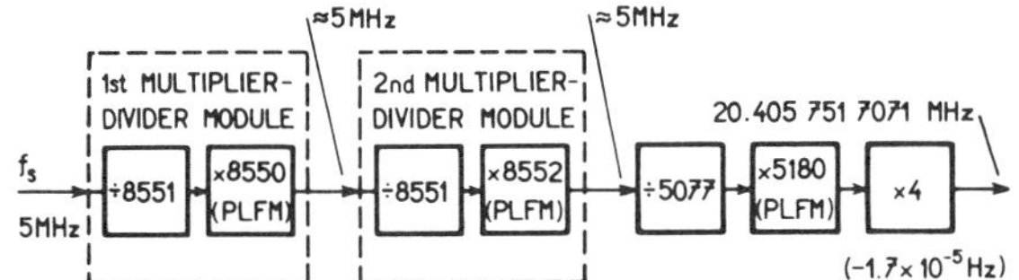

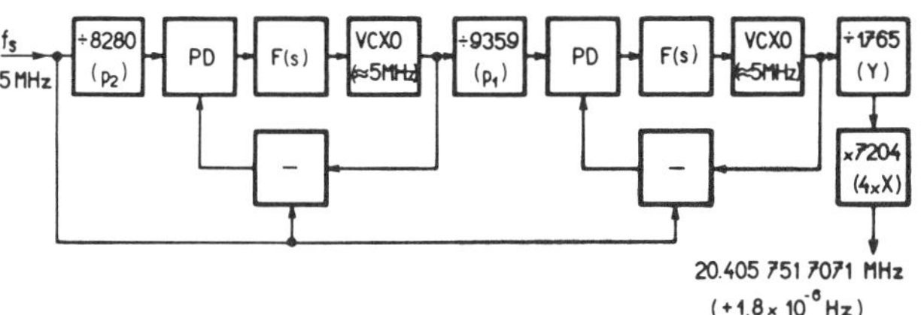

Fig. 3. Reference frequency generator for hydrogen maser based on Cantor product approximation.

Fig. 4. Reference frequency generator for hydrogen maser based on modified Engel series expansion.

that they do not contain mixers and consequently there are no direct intermodulation signals.

IV. ENGEL SERIES APPROXIMATIONS

In cases where poor spectral purity may be tolerated very simple frequency synthesizers can be assembled from off-the-shelf gates and digital dividers [4], [5]. The underlying algorithm is a series expansion. The most promising seems to be the modified Engel series in which the terms alternate in sign (see Appendix III). The difficulty with these expansions is again the uniqueness. In cases where coherence between input and output frequencies is required the algorithms must be exactly performed. On the other hand if there is a small error that can be tolerated a preliminary generalized continued fraction expansion followed by the modified Engel series expansion results in a very simple mathematical model and equally simple hardware. Evidently we must again start with (8) and restrict X, Y, and $1/\rho$ to values smaller than D_u. With the assistance of the computer division factors $p_1, p_2, \cdots$ may be kept in the interval, e.g., $\langle D_u; 10\, D_u \rangle$. In this instance the approximation error will be

$$|R_e| \leq \frac{X}{Y} \frac{1}{p_1 p_2 \cdots p_n p_{n+1}{}^2} \leq \frac{X}{Y} \frac{1}{D_u{}^{n+2}}. \qquad (11)$$

It is clear that even a two-stage approximation

$$\xi_x \approx \frac{X}{Y}\left[1 - \frac{1}{p_1}\left(1 - \frac{1}{p_2}\right)\right] \qquad (12)$$

may provide negligibly small errors. As a practical example we have chosen again the synthesis of the hydrogen maser reference frequency $-f_x = 20.405\,751\,7071$ MHz. Numerical results, got by a computer, are shown together with the block diagram in Fig. 4.

By considering (12), one easily arrives at the conclusion that the frequency synthesis system may be simplified by the mere addition or omission of pulses in instances where a good spectral purity is not required.

V. CONCLUSIONS

In this paper, we have discussed systematically the design of approximating frequency synthesizers. We have briefly recalled the new general approximation theorem and emphasized the advantages of its combination with Cantor products and Engel series expansions. But the software solution can be provided only by computer. On examples of practical approximating frequency synthesizers (reference frequency generators for hydrogen masers), we have shown the simplicity of the hardware and its compatibility with the IC technology.

APPENDIX I

CONTINUED FRACTIONS

Any real number ξ_0 expands into the continued fraction by application of Euclid's theorem

$$\xi_0 = b_0 + \frac{1}{\xi_1}, \qquad b_0 \leq \xi_0 < b_0 + 1 \qquad (13a)$$

$$\xi_1 = b_1 + \frac{1}{\xi_2}, \qquad b_1 \leq \xi_1 < b_1 + 1 \qquad (13b)$$

etc. By combining (13a), (13b), $\cdots$, we get

$$\xi_0 = b_0 + \cfrac{1}{b_1 + \cfrac{1}{b_2 + \cfrac{1}{\ddots}}}. \qquad (14)$$

The nth-order approximation is given by [6], [7]

$$\frac{A_n}{B_n} = [b_0, b_1, \cdots, b_n] \qquad (15)$$

where the numerator A_n and the denominator B_n are relatively prime and are found by the recurence formulas

$$A_{-2} = 0, \qquad A_{-1} = 1, \qquad A_n = A_{n-1}b_n + A_{n-2} \qquad (16a)$$

$$B_{-2} = 1, \qquad B_{-1} = 0, \qquad B_n = B_{n-1}b_n + B_{n-2}. \qquad (16b)$$

The remainder R_n is

$$|R_n| = \left|\frac{(-1)^n}{B_n(B_n\xi_{n+1} + B_{n-1})}\right| \leq \frac{1}{B_n B_{n+1}} < \frac{1}{B_n{}^2}. \qquad (17)$$

Note that both algorithms (13) and (16) are unique, consequently the expansion of any real number into the regular continued fraction is unique. Further there exists no other rational fraction P/Q with the denominator smaller than B_n which would approximate ξ_0 with a smaller error than $|R_n|$.

APPENDIX II

INFINITE PRODUCTS

Cantor [8] has shown that any real number α_0 larger than one can be developed into an infinite product

$$\alpha_0 = \prod_{k=0}^{\infty} (1 + 1/q_k) \qquad (18)$$

where all q_k are positive integers found by the following algorithms

$$1 \le q_0 \le \frac{\alpha_0}{\alpha_0 - 1} < q_0 + 1 \qquad (19)$$

$$\alpha_0 = (1 + 1/q_0)\alpha_1, \qquad \alpha_0 > \alpha_1 > 1. \qquad (20)$$

By imposing the algorithm (19) on α_1 one finds q_1, then q_2, etc., each next higher division factor satisfies the inequality

$$q_{k+1} \ge q_k{}^2. \qquad (21)$$

If from a certain $k + 1$ onwards the equivalence sign is permanent, then α_0 is a rational number. With the assistance of this conclusion, we can estimate the remainder if the expansion (18) is truncated after the nth factor

$$R_{n,\alpha} \le \alpha_0 \frac{1}{q_{n+1}} \le \alpha_0 \frac{1}{q_n{}^2}. \qquad (22)$$

If the approximated real number β_0 is smaller than one its expansion into an infinite product follows immediately from (18) by a mere inversion, i.e.,

$$\beta_0 = \prod_{k=0}^{\infty} \left(1 - \frac{1}{q_k + 1}\right) \qquad (23)$$

and further

$$q_0 \le \frac{1}{1 - \beta_0} < q_0 + 1 \qquad (24)$$

and eventually the remainder

$$|R_{n,\beta}| \le \left| -\beta_0 \frac{1}{q_{n+1} - 1} \right| \lesssim \beta_0 \frac{1}{q_n{}^2}. \qquad (25)$$

APPENDIX III

LÜROTH, ENGEL, AND SYLVESTER SERIES

If γ_0 is an arbitrary real number then the common form of these series is [7]

$$\gamma_0 = c_0 + \frac{1}{p_1}$$

$$+ \sum_{n=1}^{\infty} \frac{r_1 r_2 \cdots r_n}{p_1(p_1 - 1) p_2(p_2 - 1) \cdots p_n(p_n - 1)} \cdot \frac{1}{p_{n+1}} \qquad (26)$$

where c_0 and all p_k are integers

$$c_0 < \gamma_0 \le c_0 + 1 \qquad (27)$$

$$\gamma_0 = c_0 + 1/\gamma_1 \qquad (28)$$

$$p_k - 1 \le \gamma_k < p_k, \qquad k \ge 1 \qquad (29)$$

$$\frac{1}{\gamma_k} = \frac{1}{p_k} + \frac{r_k}{p_k(p_k - 1)} \cdot \frac{1}{\gamma_{k+1}}, \qquad k \ge 1 \qquad (30)$$

r_k is an arbitrary integer. The remainder is

$$R_{n,\gamma} \le \frac{r_1 r_2 \cdots r_n}{p_1(p_1 - 1) p_2(p_2 - 1) \cdots p_n(p_n - 1)}$$

$$\cdot \frac{1}{p_{n+1}(p_{n+1} - 1)}. \qquad (31)$$

Particularly simple frequency synthesizers result in cases where the sign in series (26) alternates, i.e., [9]

$$\gamma_0 = c_0 - \left[\frac{1}{p_1} + \sum_{n=1}^{\infty} \right.$$

$$\left. \cdot \frac{r_1 r_2 \cdots r_n (-1)^n}{p_1(p_1 + 1) p_2(p_2 + 1) \cdots p_n(p_n + 1)} \cdot \frac{1}{p_{n+1}} \right] \qquad (32)$$

This modification requires that

$$c_0 - 1 \le \gamma_0 < c_0 \qquad (33)$$

$$\gamma_0 = c_0 - 1/\gamma_1 \qquad (34)$$

$$p_k < \gamma_k \le p_k + p_k < \gamma_k + 1 \qquad (k \ge 1) \qquad (35)$$

$$\frac{1}{\gamma_k} = \frac{1}{p_k} - \frac{r_k}{p_k(p_k + 1)} \cdot \frac{1}{\gamma_{k+1}} \qquad (36)$$

$$|R_{n,\gamma}| \le \left| (-1)^n \frac{r_1 r_2 \cdots r_n}{p_1(p_1 + 1) \cdots p_n(p_n + 1)} \cdot \frac{1}{p_{n+1}{}^2} \right|. \qquad (37)$$

In accordance with the choice of $r_1, r_2, \cdots$, the series (26) or (32) becomes a Lüroth, Engel, or Sylvester series, respectively,

$$r_1 = r_2 = \cdots = 1 \qquad (38)$$

$$r_k = p_k \mp 1 \qquad (39)$$

$$r_\kappa = p_k(p_k \mp 1). \qquad (40)$$

REFERENCES

[1] V. F. Kroupa, "Theory of frequency synthesis," *IEEE Trans. Instrum. Meas.*, vol. IM-17, pp. 56–68, Mar. 1968.
[2] ——, *Frequency Synthesis, Theory, Design, and Applications.* London, England: Charles Griffin, 1973, pp. 122–158 and 212–215.
[3] ——, "Computer aided gear-drive design," Inst. Radio Eng. Electron., Prague, Czechoslovakia, Rep. Z-609, Nov. 1973.
[4] G. Becker, "Quasiperiodic frequency synthesis," in *Proc. 26th Annu. Frequency Control Symp.*, pp. 279–291, June 1972.
[5] ——, "Grundlagen der aperiodischen quasiperiodischen Frequenz-synthese," *Frequenz*, vol. 27, pp. 249–254, Sept. 1973 and pp. 279–283, Oct. 1973.
[6] O. Perron, *Irrationalzahlen*, 3rd ed. Berlin, Germany: de Gruyter, 1947, pp. 90–127.
[7] ——, *Die Lehre von der Ketten-Brüchen, Band I. Elementare Kettenbrüche.* Stuttgart, Germany: G. B. Teubner, 1954.
[8] G. Cantor, "Zwei Sätze über eine gewisse Zerlegung der Zahlen in unendliche Produkte," *Z. Math. Phys.*, vol. 14, pp. 152–158, 1869.
[9] V. F. Kroupa, Seminario sulla sintesi di frequenza, May 6–8, 1974, Turin, Italy.

Part III

Wide-Range Direct Digital Frequency Synthesizers

WE have seen in Part II the simplicity of frequency synthesizers built with the assistance of ICs only. No wonder that we soon encountered a project for a wide-range DDFS system. To the editor's knowledge, the fathers of this idea were Tierney, Rader, and Gold (1971); their paper is the first one reprinted in this section. (We should also mention a larger paper by Tierney [III-1] published later.) Most notably, their proposal was many years ahead of their time. The principle is the step-by-step accumulation of the phase information (Modulo-N), its changing into sine values in ROM lookup tables, and digital-to-analog conversion. This same system is still in use, and nearly all papers about DDFS cite Tierney in the first place.

A paper by Bramble (1981) [III-2] deserves attention for its clear explanation of the basic functions of DDFS, particularly with the assistance of Figs. 4 through 8. However, some conclusions must be corrected; for example, nowadays the useful output frequency range is nearing one-half of the clock frequency and not one-fourth as was required in applications in PLL systems at that time. Even the integration of circuitry ceased to be a solitary research effort but started to be part of concentrated research activities [III-3].

Since 1971, in only 20 years, the DDFS has grown from an engineering novelty to an important design tool. A real leap in use and applications provided the integration of all major parts into one single chip. As a consequence, the cost for application of DDFS decreased dramatically. However, many unresolved questions still remain.

The basic system requirements that have since motivated the designers of the chips were tuning bandwidth, tuning resolution, spectral purity, and, most often, the switching speed and power savings.

The useful frequency range of the DDFS is determined by (1) the desired output waveform, (2) the external filter used either to remove the alias from the Nyquist frequency range or to filter out higher frequency ranges in some cases, (3) the sampling speed of the IC, (4) the memory arrangement, and (5) the output circuit—generally by the speed and linearity of the available DAC converter.

Another design task is the frequency resolution that depends on the extent of the accumulator bits (R) and the clock frequency, f_c, that is,

$$\Delta f_{x,\min} = \frac{f_c}{2^R} \quad [\mathrm{Hz}] \qquad [1]$$

An efficient single-chip construction was discussed by Sunderland et al. in the mid-1980s (1984). By using 3.5-μm gate length complementary metal-oxide-semiconductor/silicon-on-sapphire (CMOS/SOS) technology, they succeeded in putting an accumulator with 20-bit capacity together with an efficiently reduced sine lookup table on a chip of about 50 mm^2. However, the clock frequency at that time was rather low—only 7.5 MHz. A great achievement was a low-power consumption of 300 mW, even though 14 phase bits were used to evaluate the output sine wave. However, savings in lookup tables were necessary. The MSB determines the sign, and the next MSB decides whether the sine amplitude is increasing or decreasing. Furthermore the remaining 12 bits are divided into 4-bit fractions—A, B, C—in such a way that the sine function is approximated as

$$\sin(A + B + C) \approx \sin(A + B) + \cos A * \sin C \qquad [2]$$

The same approach for optimization of DDFS performance in respect to the reduction of sine lookup ROMs is discussed by Nicholas et al. in [III-4 and III-5] and by Vankka [III-6]. All these papers are reprinted in Part IV.

The following paper by Saul and Taylor (1990) describes a UHF DDFS, which is intended primarily for radar and electronic warfare. The experimental device generates square-, sine-, and triangle-wave outputs in the range from 1 Hz to 500 MHz (maximum clock rate > 2.5 GHz). The device exhibits very fast update time of 20 ns (despite the accumulator length of 31 bits). To this end, an 8-bit DAC was built directly on the chip whose size was reduced to 4×5 mm. Because of the device speed, the sine ROMs are formed by 64 words 7 bits long. Phase and amplitude inversions are performed externally; evidently, only 8 or 9 MSB of the stored phase information are used. Finally, the high-output frequencies require technology that consumes a mode-dependent power, from 2 W to 5 W. Some additional information appears in the followup paper by Saul, Barber, Taylor, and Ward (1991).

Fischer and Hariharan (1993) [III-7] discuss a DDFS chip that realizes a 12-bit output sine wave (peak to peak) with a frequency resolution of 24 bits. The 2-μm CMOS implementation occupies a chip area of 2.5 mm^2. The maximum clock rate is 64 MHz, and the tuning latency is equal to 12 clock periods.

Thompson (1992) discusses a new configuration for his numerically controlled oscillator (NCO), which makes it possible to decrease latency to two clock cycles for a 130-MHz device. The system has 8-bit phase resolution into the DAC and 8-bit amplitude resolution.

Most DDFSs are based on binary phase accumulators, but in some instances a decimal base is required. Such a solution is discussed, for example, by Harris [III-8] or Goldberg [III-9].

The remaining references cite three papers that discuss DDFS which are not based on the Modulo-N arithmetics.

REFERENCES

[III-1] J. Gorski-Popiel. *Frequency Synthesis: Techniques and Applications.* New York: IEEE Press, 1975. Chapter V by J. Tierney, pp. 121–49.

[III-2] A. L. Bramble. "Direct digital frequency synthesis." *Proceedings of the 35th Annual Frequency Control Symposium,* pp. 406–14, May 1981.

[III-3] A. Bramble and J. Kesperies. "Army frequency agile synthesizer program." *Proceedings of the 40th Annual Frequency Control Symposium,* pp. 366–69, 1986.

[III-4] H. T. Nicholas III and H. Samueli. "An analysis of the output spectrum of direct digital frequency synthesizers in the presence of phase-accumulator truncation." *Proceedings of the 41st Annual Frequency Control Symposium* 1987. (Reprinted in Part IV.)

[III-5] H. T. Nicholas III, H. Samueli, and B. Kim. "The optimization of direct digital frequency synthesizer performance in the presence of finite word length effects." *Proceedings of the 42nd Annual Frequency Control Symposium* 1988. (Reprinted in Part IV.)

[III-6] J. Vankka. "Methods of mapping from phase to sine amplitude in direct digital synthesis." *IEEE Transactions on Ultrasonics, Ferroelectrics and Frequency Control,* pp. 526–534, March 1997.

[III-7] G. Fischer and A. Hariharan. "A wide-band digital frequency synthesizer with 24-bit resolution." *ECCTD 1993—Circuit Theory and Design, Proc.,* pp. 235–40.

[III-8] M. V. Harris. "A J-band spread spectrum synthesizer using a combination of DDS and phaselock techniques." Colloquium on Direct Digital Frequency Synthesis. *IEE Digest No: 1991/172,* November 1991. (Reprinted in Part VI.)

[III-9] Bar-Giora Goldberg. *Digital Techniques in Frequency Synthesis.* New York: McGraw-Hill, 1996, pp. 229–32.

[III-10] R. W. Jenkins and W. A. Evans. "A microprocessor controlled frequency synthesised signal source." Proceedings of the Conference on Programmable Instruments *(IERE Conf. Proceedings No. 38),* pp. 53–60, November 1977.

[III-11] E. Vasanatha and A. P. Shivaprasad. "A new approach to digital frequency synthesis." *IEEE Transactions on Instrumentation and Measurement,* IM-30, pp. 102–106, June 1981.

[III-12] A. Chrysafis. "Digital sine-wave synthesis using DSP56001." Motorola Inc., 1988.

A Digital Frequency Synthesizer

JOSEPH TIERNEY, Member, IEEE

CHARLES M. RADER, Member, IEEE

BERNARD GOLD, Senior Member, IEEE

M.I.T. Lincoln Laboratory
Lexington, Mass. 02173

Abstract

A digital frequency synthesizer has been designed and constructed based on generating digital samples of $\exp\left[j(2\pi nk/N)\right]$ at time nT. The real and imaginary parts of this exponential form samples of quadrature sinusoids where the frequency index k is allowed to vary $(-N/4)\leq K<(N/4)$. The digital samples drive digital to analog converters followed by low-pass interpolating filters to produce analog sinusoids. The method is superior to digital difference equations with poles on the unit circle since the noise or numerical inaccuracy remains bounded.

The digital technique used consists of factoring the exponential into two table look-ups from an efficiently organized small READ-ONLY memory table and performing a complex multiply to produce the real and imaginary components. A small array multiplier efficiently organized performs the multiplications.

The technique lends itself to the production of phase coherent or phase controlled sinusoids because of the indexing arrangement used. In addition finer frequency steps than the READ-ONLY memory allows are available by expanding the indexing register at no increase in inaccuracy.

Manuscript received July 22, 1970.

This work was sponsored by the Department of the Air Force.

Reprinted from *IEEE Transactions on Audio and Electroacoustics*, Vol. AU-19, No. 1, pp. 48-57, March 1971.

Introduction

The generation of many different frequencies from a single stable source frequency is commonly achieved with analog circuits [1], [2] or combinations of analog and digital circuits [3], [4]. All of these approaches [5] generate a large set of frequencies from a single source in the analog or continuous sense by division, phase lock, mixing, or a combination of such techniques. An attractive alternate approach is the generation of a set of sampled sinusoids by digital computation of some kind (including simple table look-up) at the sampling times. Fig. 1 indicates the sample trains produced for two different frequencies. The return to analog or continuous sinusoids is accomplished with simple realizable smoothing filters.

One can consider the conventional frequency synthesizer as processing a single source frequency to produce a large set of new frequencies. The digital frequency synthesizer uses a single frequency to establish a stable sampling time at which sample values are computed. The difference in approach is one of using the source as a frequency directly or as a time reference.

The Digital Approach

Given the problem of computing samples of sinusoids the most obvious choice is a digital recursion, a difference equation whose Z transform has poles on the unit circle. By starting such a recursion with the proper initial conditions one can produce sinusoidal samples. There are at least two problems with this approach (Fig. 2). The frequency of the sampled sinusoid produced by such a typical recursion is not linearly related to a settable coefficient but related by the function $\cos WT$, where T is the sampling interval, W is the produced frequency, and $2 \cos WT$ is the actual coefficient of the recursion. The noise produced by such a recursion is in general worse than can be obtained by other methods and may in some cases be a function of the number of iterations [6].

The chosen approach is simply a direct computation of the samples,

$$\cos (\omega n T + \phi), \qquad \sin (\omega n T + \phi) \qquad n = \text{time index.}$$

Consider values of

$$\cos 2\pi f n T = \text{Re} \ (e^{j2\pi f n T})$$

$$\sin 2\pi f n T = \text{Im} \ (e^{j2\pi f n T})$$

where $f = kfo$, $fo = $ lowest computed frequency. That is, we can compute multiples of some lowest frequency,

$$f_o = \frac{1}{NT}$$

where N is a design parameter. Then the exponential becomes

$$e^{j2\pi f n T} = \exp\left[j \frac{2\pi}{N} nk \right].$$

"

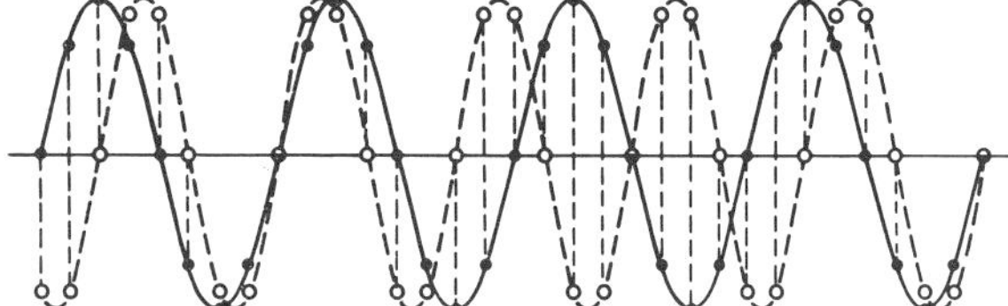

Fig. 1.   Digital frequency determination.

Fig. 2.   Pole position of the digital recursion for sinusoidal samples.

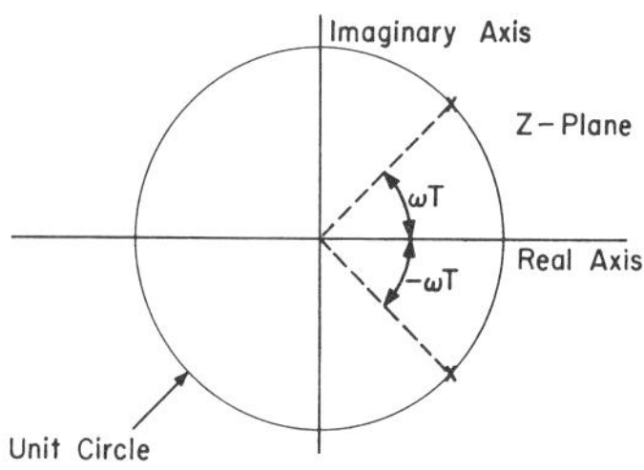

$$y_n = (2 \cos \omega T)\, y_{n-1} - y_{n-2}$$
$$y_0 = \cos \phi$$
$$y_{-1} = \cos [\phi - \omega T].$$

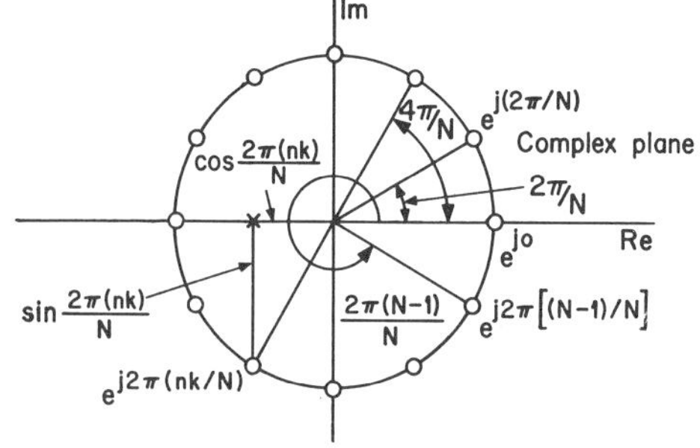

Fig. 3.   Equispaced samples on the unit circle.

Samples of this complex exponential are equivalent to values of sine and cosine of the argument $2\pi f n T$ with

$$f = \frac{k}{NT}.$$

Computing samples of this complex exponential indexed on a frequency index k, and a time index n, is equivalent to computing coordinates of one of N equispaced points around the unit circle (Fig. 3) in the complex plane described by

$$\exp [jo], \qquad \exp \left[j \frac{2\pi}{N} \right],$$

$$\exp \left[j \frac{2\pi}{N} 2 \right], \cdots, \exp \left[j \frac{2\pi}{N} (N-1) \right].$$

For a particular frequency index k, the argument of the exponential varies in increments of $(2\pi/N)k$ in successive time indices. The product nk is treated modulo N since $\exp[(j(2\pi/N)[X+N]) = \exp[j(2\pi/N)X]$ for any X. The generation of samples of a complex sinusoid consists then of accumulating multiples of k (i.e., nk at time n, $[n+1]k$ at time $n+1$), and using the accumulated value to calculate $\exp[j(2\pi/N)nk]$.[1]

---

[1] If in accumulating multiples of k, the accumulator is not initially zero but contains some constant C, the argument of the exponential becomes $(2\pi/N)(nk+C)$ which affects the phase of the result, but not the frequency. This can be useful in phase control.

The Calculation

To determine the value $\exp[j(2\pi/N)[nk+C] = \exp[j(2\pi/N)Y]$ consider the simplest case, namely, a table storing N values from $\exp[j(2\pi0/N)]$ to $\exp[j(2\pi(N-1)/N)]$. The computation then consists of using the value in the accumulator Y to index the table of N values producing a single complex sample [7]. As Y increases, the table is scanned producing sinusoidal samples. For a larger k value the table is covered faster (the interval between successive Ys, or nks is larger). Such an approach is suitable for small values of N, but more often we are interested in large N (a large set of frequencies) so that a single table look-up becomes impractical because of the size of the table.

For large N and $0 \leq Y \leq N-1$, Y may be broken into a sum of several words each of which represents a part of Y. If $Y = q + r + s$, then $\exp[j(2\pi Y/N)] = \exp[j(2\pi[q+r+s]/N)] = \exp[j(2\pi q/N)]\exp[j(2\pi r/N)]\exp[j(2\pi s/N)]$ where each factor takes on many fewer than N values and the overall storage has been reduced. For example, for N, a power of 2, say 2^b, then $0 \leq Y \leq 2^b - 1$ and can be represented as a binary number b digits long. $Y = \alpha_0 2^0 + \alpha_1 2^1 + \alpha_2 2^2 + \cdots + \alpha_{b-1} 2^{b-1}$. We can factor the exponential into b factors each of which is of the form $\exp[j(2\pi\alpha_i 2^i/N)]$ with $\alpha_i = 0$ or 1. Thus the table has been reduced from 2^b complex entries to b complex entries ($\log_2 2^b$) and we need $b-1$ complex multiplications to obtain the value $\exp[j(2\pi Y/N)] = \exp[j(2\pi/N)[nk+C]]$. Obviously the number of factors or the number of complex multiplies is one of the design parameters in this approach to frequency synthesis.

For our purposes factoring the complex exponential into two terms allows for a very efficient use of READ-ONLY memory. In addition we may take advantage of approximations to the sine and cosine of small angles as well as symmetries in the functions to effect further savings in READ-ONLY storage.

The Complete Synthesizer

The block diagram of a digital synthesizer which produces quadrature outputs is shown in Fig. 4. An input frequency control word k is stored in a register and used

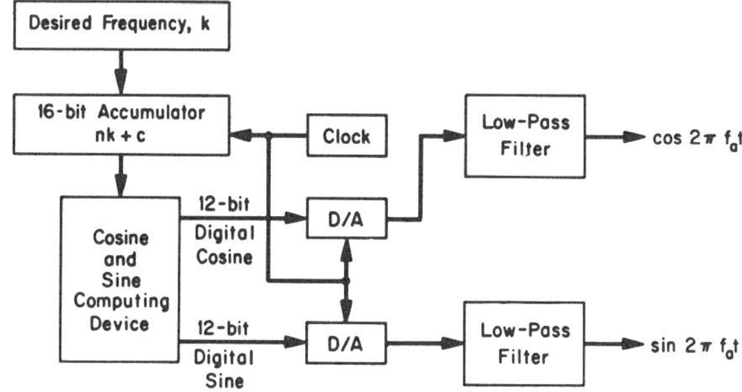

Fig. 4.   Synthesizer block diagram.

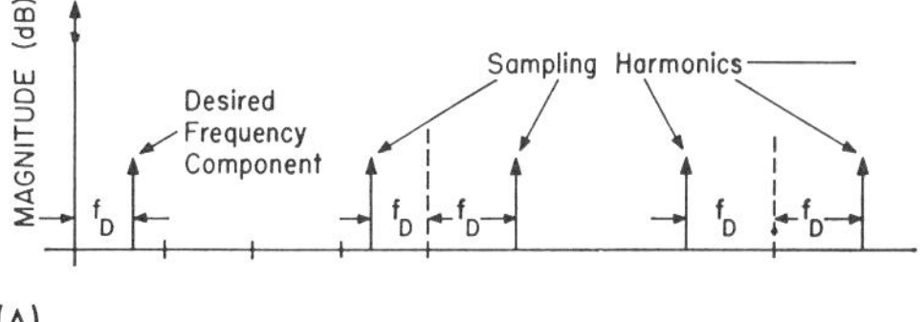

(A)

(B)

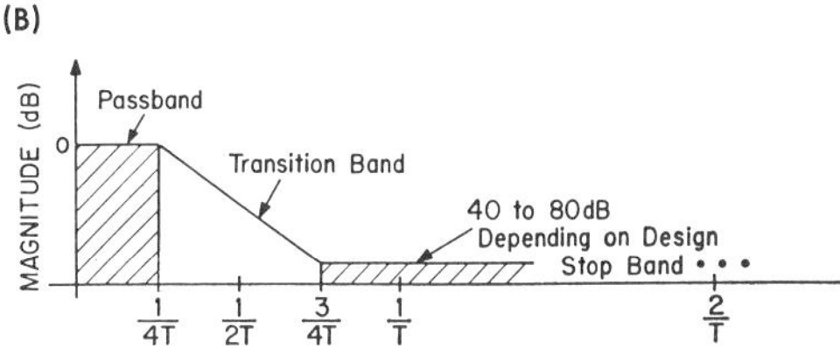

Fig. 5.   (A) Sampled sinusoid spectrum. (B) Smoothing filter response.

to update an accumulator every T seconds. Each time the accumulator is changed, the value $nk+C$ is used to compute the real and imaginary parts of $\exp\left[j(2\pi/N)Y\right]$ by one of the methods proposed in the previous section. The computed values of $\sin(2\pi/N)Y$, $\cos(2\pi/N)Y$ are used to drive a pair of D–A converters of the proper word length to produce analog samples. These samples are interpolated by the output smoothing filters. At a sampling interval of T seconds the output spectrum before smoothing of a sampled sine or cosine would look like that shown in Fig. 5. The Nyquist condition would allow us to produce frequencies of up to but less than $1/2T$ which we could recover with ideal LP filters with cutoff at $1/2T$. However, for ease of filtering consider using only $1/4$ of the sampling frequency as the band limit. Then we would like an output smoothing filter which passes all frequencies up to $1/4T$ with some design ripple, to have a transition band in the interval from $1/4T$ to $3/4T$, and to have some out of band attenuation depending on the allowed sampling harmonics. Such a filter is shown in Fig. 5(B). For an accumulator which overflows at some N and a sampling interval T, a digital frequency synthesizer as shown in Fig. 4 would produce a lowest frequency of $1/NT$, and a highest frequency of $1/4T$ for a total of $N/4$ different frequencies. If we take advantage of the quadrature outputs frum this realization we can double the bandwidth of $1/4T$ since we can modulate a carrier to $\pm 1/4T$ for a total of $N/2$ frequencies. Note that a single output synthesizer can be implemented using only one D–A converter if so desired. There is no constraint to produce quadrature outputs although they may in some cases be desirable, as shown above.

A Design Example

Up to this point the discussion of index accumulation and calculation of sine and cosine has been general. We could use any arithmetic and any size N. Consider now a design example using binary arithmetic and standard binary logic. The design specifications are: 1) 2^{15} frequencies; 2) 409.6 kHz bandwidth; 3) 12.5 Hz frequency

spacing (or lowest frequency); 4) 70 dB spectral purity (ratio of power in desired frequency to power in any other 100 Hz band). If we assume the quadrature outputs will be used to obtain a bandwidth of $1/2T$, then $T=1/2\times409.6$ kHz$\cong1.22$ μs. In addition $N/2=2^{15}$ or $N=2^{16}$ so that the accumulator will be a 16-bit binary register, and the frequency spacing of $1/NT=12.5$ Hz. If quadrature outputs are desired, the synthesizer must produce values of $\exp\left[j(2\pi/2^{16})Y\right]$, $0\leq Y\leq 2^{16}-1$, that is, one of 2^{16} points equispaced around the unit circle. Since the highest frequency allowed is $1/4T$, four points around the circle, the largest k value corresponding to this value of frequency is $2^{16}/4$ or 2^{14}. A two's complement negative frequency input will cause samples to occur in the opposite sense around the unit circle, which is meaningful as a negative frequency only if quadrature outputs are available.

The remaining problem is the computation of the samples. To meet the design requirement of 70 dB purity the computation must be carried to about 12 bits. That is, if a sample out of the computation consists of a sign and 11 binary digits and the accuracy is $\pm$ the last digit or $\pm 2^{-11}$, a worst case harmonic caused by such an error will be 66 dB down from the output. As shown in a later section this bound is quite pessimistic and 70 dB is more likely. Using this 12-bit arithmetic the following procedure, which takes considerable advantage of the symmetries of the sine and cosine, is used.

First note that neglecting the least significant bit in the 16-bit accumulator will cause an amplitude error of no more than $\sin(2\pi/2^{16})$ which is roughly 2^{-12}. In fact one could extend the accumulator register and the input frequency word on the low significance end, use these extra bits for accumulation, but ignore them for the computation of sine and cosine, and still be bounded by $\sin(2\pi/2^{16})$ as an amplitude error. This implies generating finer and finer frequency steps by adding only to the accumulator, a feature which is unique to this type of synthesis and efficient in terms of memory use. Computing sine and cosine of multiples of $2\pi/2^{15}$ (ignoring bit 16) is solved by table look-up and interpolation as follows. Consider Y represented in binary form as

$$
\begin{aligned}
Y &= 2^0 d_0 + 2^1 d_1 + 2^2 d_2 + 2^3 d_3 + 2^4 d_4 + 2^5 d_5 + 2^6 d_6 \\
&\quad + 2^7(d_7 + 2^1 d_8 + 2^2 d_9 + 2^3 d_{10} + 2^4 d_{11} \\
&\quad + 2^5 d_{12} + 2^6 d_{13} + 2^7 d_{14}) \\
&= 2^0 d_0 + 2^1 d_1 + 2^2 d_2 + 2^3 d_3 + 2^4 d_4 + 2^5 d_5 - 2^6 d_6 \\
&\quad + 2^7(d_6 + d_7 + 2^1 d_8 + 2^2 d_9 + 2^3 d_{10} + 2^4 d_{11} \\
&\quad\quad\quad + 2^5 d_{12} + 2^6 d_{13} + 2^7 d_{14})
\end{aligned}
$$

or

$$ Y = f + 2^7 e $$

where

$$
\begin{aligned}
f &= 2^0 d_0 + 2^1 d_1 + \cdots + 2^5 d_5 - 2^6 d_6 \\
e &= (d_6 + \cdots + 2^7 d_{14})
\end{aligned}
$$

so that the exponential can be factored as

$$
\left(\exp\left[j\frac{2\pi}{2^{15}} f \right] \right) \left(\exp\left[j\frac{2\pi}{2^{15}} 2^7 e \right] \right)
$$

$$
= \left(\exp\left[j\frac{2\pi}{2^{15}} f \right] \right) \left(\exp\left[j\frac{2\pi}{2^8} e \right] \right).
$$

The computation of the complex exponential is then reduced to two table look-ups $\exp[j(2\pi/2^{15})f]$, $\exp[j(2\pi/2^8)e]$ and a complex multiply. The index e consists of the eight high-order bits of the accumulator rounded by the bit d_6, while the index f consists of the six lower bits d_5 through d_0 if d_6 is zero, or these bits two's complemented if d_6 is one. From the point of view of computing the value of a point at a particular angle $(2\pi/N)Y$, around the unit circle, the value of e determines which of 2^8 equally coarsely spaced points is nearest the desired point, and the value of f determines which of 64 possible angular corrections should be added to or subtracted from the coarse point to get the desired value. The angular correction is, of course, a complex multiplication. For this two factor approach the complex multiply is implementing the trigonometric identities

$$ \sin\,(x + y) = \sin x \cos y + \cos x \sin y $$

$$ \cos\,(x + y) = \cos x \cos y - \sin x \sin y $$

with

$$ x = (2\pi/2^8)e, \qquad y = (2\pi/2^{15})f $$

where f may be positive or negative (according to bit d_6). Fig. 6 represents these operations on the unit circles.

To compute the value at X in Fig. 6, the eight high-order bits would be augmented by one (since $d_6 = 1$), giving e, and the value of f would be the two's complement of the distance above the center of the coarse interval. In other words we would go to a larger angle and decrement because $d_6 = 1$. If $d_6 = 0$ we would start at the lower angle and increment.

The bit d_6 is used to break the coarse intervals in half. If this bit is one, it means the computation needs the coarse value just bigger than the desired value. The fine correction is then clockwise. If d_6 is a zero, the coarse value just below the desired value will do, and the correction is counterclockwise.

Now we note that the cosine component of $\exp[j2\pi f/32\,768]$ is between 1.0 and 0.9999247, a difference in the 14th bit of its binary representation. Therefore we will approximate it by 1. The sine component is so small that its six most significant bits, not counting the sign bit, are equal to the sign bit, which is in turn equal to d_6. Thus the READ-ONLY memory indexed by f need only save the five least significant bits of the 11-bit plus sign representation of the sine corresponding to each of the 64 positive values of f; the sines corresponding to negative values of f can be found by taking the two's complement of f and changing the sign of the result. It is easier to take the one's complement, and this also results in an insignificant error.

The value of $\exp[j2\pi e/256]$ can also be found by look-up in a table with only 64 values corresponding to $\frac{1}{4}$ cycle of a sine wave. The sine and cosine components are addressed with the six least significant bits of e, and its two's complement, and the two most significant bits of e are used for exchanging the components and complementing either or both if necessary; mathematically, if the six least significant bits of e are g and the two most significant bits are h, we look up $\sin 2\pi(64-g)/256 + j \sin 2\pi g/256$ and multiply the result by j^h.

The final operation is the multiplication of the coarse estimate $\exp[j2\pi e/256]$ by the fine corrector $(1+j \sin 2\pi f/32\,768)$. This requires two 8 by 5 bit multiplications and two additions. Two answers are kept to an accuracy of 11 bits plus sign and fed to the D–A converters.

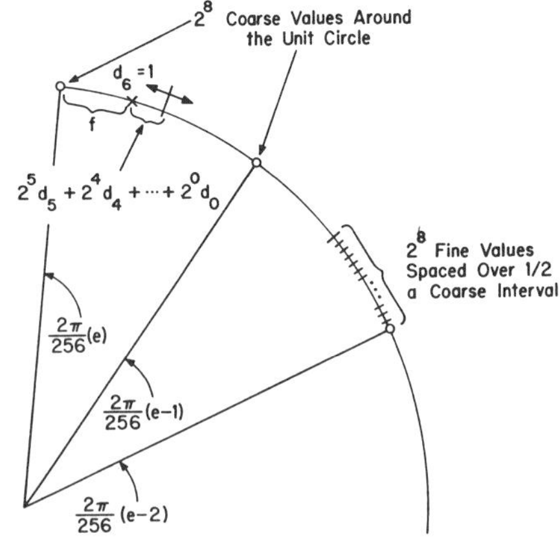

Fig. 6.

The READ-ONLY memory requirements are 64 words of 5 bits fine angle and 11 bits coarse angle which can be combined into a 64 word by 16-bit memory which is accessed three times in a computation. The detailed block diagram of this synthesizer is shown in Fig. 7.

The array multiplier indicated in the block diagram is used to perform the 5 by 8 multiplication (5×8 rather than 5×11 because of the accuracy required). It can be implemented in several ways using the 4-bit TTL adder packages currently available. Using only the bits needed for the final accuracy of $\pm 2^{-12}$ and using a tree-like interconnection between partial sums the multiplication time is about 160 ns.

The digital-to-analog converters driven by the final 12-bit sine and cosine outputs are updated every 1.22 μs and hold the data words between transfer times. This sample and hold operation provides smoothing and filtering in addition to that provided by the output low-pass filters.

A synthesizer designed and built according to the above discussion requires about 85 TTL logic packages, and dissipates about 12 W. If quadrature outputs are not needed, a still simpler device will suffice.

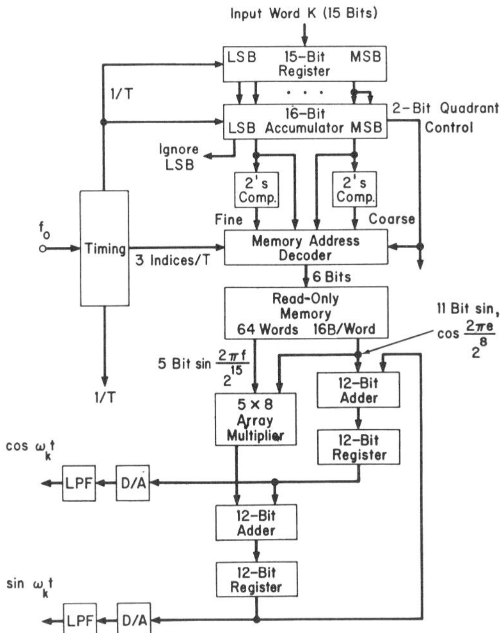

Fig. 7.

Output Noise and Time Response

The synthesizer noise output consists of three separate effects. The first contribution results from truncation or roundoff in computation of the sample value. The second effect is from the output digital to analog converter. Sampling harmonics passing through the smoothing filter contribute the third effect.

Truncation Noise

Since the calculation for a particular sample (that is, a particular value in the index accumulator) is always the same, the output samples are those of an exact sinusoid with some deterministic noise sample train added to it due to truncation. For an arbitrary generated frequency the truncation noise will have a period equal to NT, the largest period possible so that the truncation noise consists of a line spectrum with frequency spacing $f_0 = 1/NT$. If the generated frequency k/NT has one or more factors of 2 in k, or $k = 2^l k^*$, then the truncation noise harmonics are multiples of $2^l f_0$. This is a consequence of N being a power of 2. The limiting case occurs for $k = 2^a$, some power of 2. In that case the truncation harmonics are harmonics of the generated frequency. These cases are demonstrated for 16 equispaced samples around a unit circle in Fig. 8. Each sample has associated with it a truncation error ϵ_i and each line below represents the sequence of sample errors generated as the index k increases. The three cases are shown. Lines 1, 2, 4 are cases of a power of 2 times lowest frequency generated. Notice that the error period is the same as the generated period. Lines 3, 5, 7 represent noise periods of $16T$, the

Fig. 8.

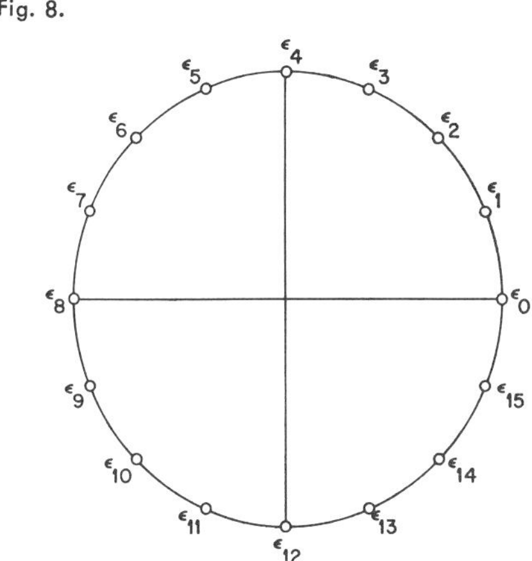

	Frequency		Noise Period
(1) Lowest	1/16T	Noise period same as generated period	= 16T
(2)	1/8T	Noise period same as generated period	= 8T
(3) •	3/16T	0 3 6 9 12 15 2 5 8 11 14 1 4 7 10 13	= 16T
(4) •	T/4	Noise period same as generated period	= 4T
(5) •	5/16T	0 5 10 15 4 9 14 3 8 13 2 7 12 1 6 11	= 16T
(6)	6/16T	0 6 12 2 8 14 1 10 0	= 8T
(7) Highest	7/16T	0 7 14 5 12 3 10 1 8 15 6 13 4 11 2 9	= 16T

longest possible because the frequency index K has no factors of 2. Line 6 shows one factor of 2. As a consequence of our particular arithmetic implementation the error waveform only contains odd harmonics, that is, it has the property $\epsilon(n) = -\epsilon[n + (P/2)]$ where P is the period.

To bound the noise contributed by the truncation error, consider a generated frequency which is an odd factor k

33

times $2\pi/NT$. In this case the error waveform is of period NT or N samples long. A Parseval's relation for a discrete Fourier transform over N samples can be written as

$$\sum_{i=0}^{N-1} |F_i|^2 = N \sum_{n=0}^{N-1} \epsilon_n{}^2$$

where F_i is a frequency amplitude defined by

$$F_i = \sum_{l=0}^{N-1} \epsilon_l \exp\left(-j\frac{2\pi}{N}il\right)$$

and ϵ_n is the error associated with a particular sample. The ϵ_n is hopefully a pseudorandom variable uniformly distributed over the interval -2^{-11} to $+2^{-11}$ because of the truncation. We may assume all of the error energy in one frequency to obtain a bound. For a particular

$$|F|^2 = N \sum_{n=0}^{N-1} \epsilon_n{}^2$$

and assuming worst case conditions on ϵ_n,

$$|F|^2 = N^2(2^{-11})^2 \quad \text{or} \quad |F| = N2^{-11}.$$

The desired generated frequency will have an amplitude of N in the discrete transform, so that the ratio of noise amplitude to signal amplitude is 2^{-11}, the case we referred to earlier. A more realistic "bound" for the cases when the noise period includes many samples should be

$$\sum_{i=0}^{N-1} |F_i|^2 = N^2\left(\frac{1}{N}\sum_{n=0}^{N-1} \epsilon_n{}^2\right)$$

where the quantity in parenthesis is the error variance. Since the variance is $[(2^{-11})^2/3]$ for uniformly distributed noise, we expect

$$\sum_{i=0}^{N-1} |F_i|^2 = N^2\frac{(2^{-11})^2}{3}.$$

So the bound obtained assuming all the energy in one harmonic is $2^{-11}/\sqrt{3}$ noise amplitude to signal amplitude ratio or about -71 dB. If one assumes that the noise waveform is white in one period and using the fact that it contains only odd harmonics we have

$$\frac{N}{2} |F|^2 = N^2\frac{(2^{-11})^2}{3}$$

so that noise to signal ratio $= \sqrt{2/3N}\, 2^{-11}$ which is very small for large N.

Digital to Analog Converter Noise

Even in the case of perfectly calculated samples driving the digital to analog converter, the output analog samples will produce noise from switching time disparities between bits, and differences in on and off switching. These so called "glitches" which occur at the transitions between sample outputs depend on the initial and final words upon which the converter is acting. A transition

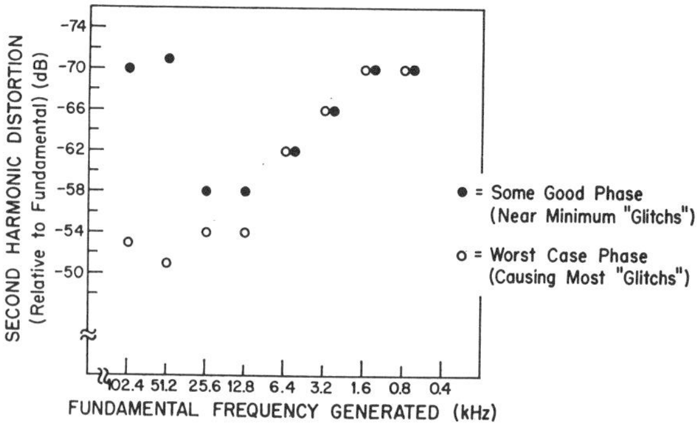

Fig. 9.

that changes many bits such as 10000 to 01111 will tend to produce more noise energy than a transition of only a few bits. This noise has the same periodicity properties that truncation noise has since it depends on transitions which are periodic if the sample train is periodic. However, all harmonics of the basic frequency are present since the "glitches" do not have odd symmetry. If one assumes a "glitch" amplitude can be as high as one-half of full scale out of the converter, then such a noise pulse of width Δ occurring even only once in each period will produce a noise-to-signal rate of approximately Δ/NT. For the higher generated frequencies such a ratio can be large. For example if $N=8$ (a frequency generated of $2^{13}\,(2\pi/NT)$) and we want to maintain a noise-to-signal ratio of 10^{-3} (60 dB), the "glitch" duration must be about 8 ns or less. Such a requirement is difficult to achieve even with current techniques. Basically then the time of nonlinearity must be small compared to the roughly 1-μs sample interval for low noise effects. In our design example a 12-bit converter with settling time of about 300 ns (presumably "glitches" of any consequence occupy only a small portion of this overall transient) in an interval of 1.22 μs produced a distortion curve as shown in Fig. 9. Since this curve represents only second-harmonic distortion it is not a result of truncation, but strictly converter distortion. Higher harmonics were smaller or equal to those shown on the curve. For generated frequencies which were not powers of 2 as given on the graph, the distortion products were less, falling between the two cases given. The large disparity between "good" and "bad" phase for the higher generated frequencies is a consequence of the small number of samples per period so that no averaging takes place. Either a good set of samples does or does not occur so that the noise is high or low.

Even for odd generated frequencies the predominant noise tends to be harmonics of the generated frequency rather than harmonics of the lowest frequency. This indicates that certain more significant bits of the D–A converter are delayed more or less than others, and differences between on and off switch times are significant. In order to reduce these transition effects, one is forced to gating or sampling devices on the output of the converter. However, for times of less than 1 μs and linearities extending into the greater than 60 dB range, such circuits require careful design. Using such a gate for shorting the converter· output to ground during transition time, the design presented earlier produced a worst case harmonic of 55 dB, an improvement upon the curve presented for the converter alone.

Smoothing Filter Time and Frequency Response

The third source of noise contributions to the synthesizer output is sampling harmonics of the generated frequency. As mentioned earlier a final smoothing filter is needed to interpolate computed sinusoidal samples or smooth the sampled spectrum. Since this smoothing filter is the energy storage device in the synthesizer the problem of time response arises. From Fig. 5(B) as previously seen, a filter was needed to pass the band up to 204.8 kHz and reject the band above 614.4 kHz. with a transition region between. The first tradeoff encountered is that between rejection or attenuation at 614.4 kHz and the time response of the filter. Consider a low-pass filter of fifth order (five poles) and examine its attenuation at 614.4 kHz as well as the time response. For several filter types the following table compares attenuation and step response. (Sample and hold response will add 10 dB to these figures.)

Type	Attenuation at 614.4 kHz	Step Response
Bessel (max flat phase)	$\sim$10 dB	$\sim$1 μs
Butterworth (max flat amp)	$\sim$48 dB	$\sim$4 μs
Chebyshev (1.0 dB ripple)	$\sim$65 dB	$\sim$7 μs
Cauer (elliptic function filter) 1.0 dB ripple	$\sim$85 dB	$\sim$9 μs

For this comparison the filters are 1.0 dB down at 204.8 and the step response is measured from 10 percent of final value to a $\pm$10 percent window around final value. In other words overshoots must settle down to within a $\pm$10 percent window. The obvious point made by this table is the tradeoff just mentioned. In addition, as the filter type moves from Bessel down, the phase or envelope delay characteristics become poorer. A similar trade can be effected between width of passband and step responses for a fixed attenuation at 614.4 kHz. If a certain attenuation is desired at that frequency a Bessel filter has a smaller passband than does a Butterworth, and so on down the list. For a fixed order of filter, and a fixed attenuation to be achieved at 614.4 kHz, the widest passband is obtained by using the sharpest filter and this in turn has the poorest time response.

The response times used in the above discussion have been step responses to dc rather than steps of sinusoidal input. It is clear that the time response of a step transition from one frequency to another consists of the response to two frequency steps; one to cancel the original frequency the other to start the new frequency. The amplitude response to each of these frequency steps reflects the natural frequencies of the driven filter. Therefore, the dc step response is still a measure of the time for amplitude response to settle down.

However, it is often the instantaneous frequency out of the synthesizer that is of importance, and the time this measure takes to settle down to some meaningful value. Obviously the instantaneous frequency must be influenced by the natural frequencies of the smoothing filter and must become some steady state value after the filter response time [8].

If an exact response for the instantaneous frequency is necessary, careful computation is necessary. If a rough response time is adequate, the table values will suffice. It is true that the smaller the frequency change, the smaller the frequency and amplitude perturbation.

It is also possible to reduce switching effects in both amplitude and phase response while still using a sharp cutoff filter such as an elliptic filter. This is accomplished by extending its cutoff and therefore its poor phase response into the transition region where no synthesizer outputs occur. In this way poorer properties of the filter are never manifested and some advantage is still taken of the sharp cutoff. Given any set of amplitude or phase criteria, there are a large set of filters to chose from satisfying smoothing requirements.

Producing RF Frequencies

Since the basic digital synthesizer cannot produce RF frequencies directly because of D–A speed limitations rather than logic problems, other methods must be used to produce an RF synthesizer output.

The method of single sideband modulation as mentioned earlier is useful if quadrature outputs are available. This is shown in Fig. 10 and requires a quadrature carrier as well. To move from upper to lower sideband the sign of one of the synthesizer outputs must change, and this is easily done. Since this technique requires a balanced subtraction at the output it is difficult to achieve high suppression of the opposite side frequency of more than 40–50 dB over a wide range.

A second approach to modulating the synthesizer output involves a single synthesizer output rather than quadrature. If the device need not produce quadrature outputs, it can work at a higher sampling or computation rate and produce a wider band of output frequencies ($\sim$1/4T). Then the lower generated frequency is limited to produce a band from 1/4T down to some f_{low} which requires a modest filtering after modulation as shown in Fig. 11.

A third approach which is implemented at the digital level may be useful. This approach consists of computing quadrature outputs but at different sampling times (same

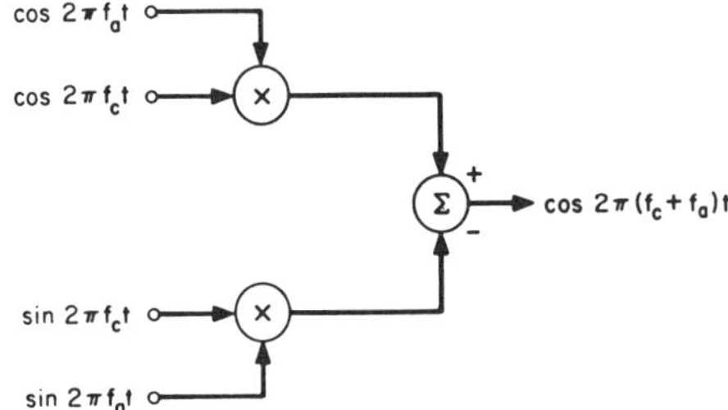

Fig. 10. Single sideband bandwidth doubling and frequency translation.

Fig. 11.

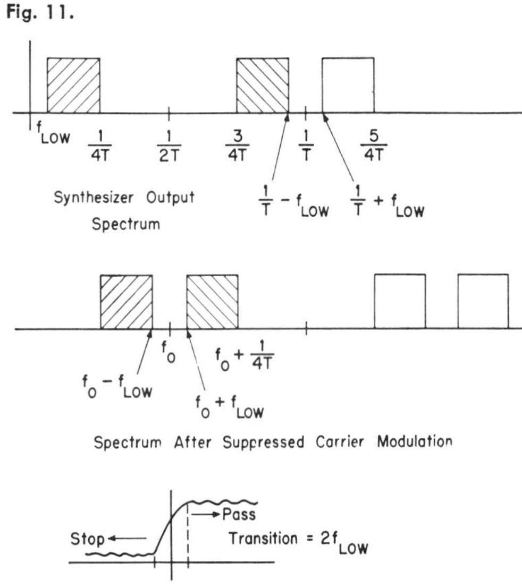

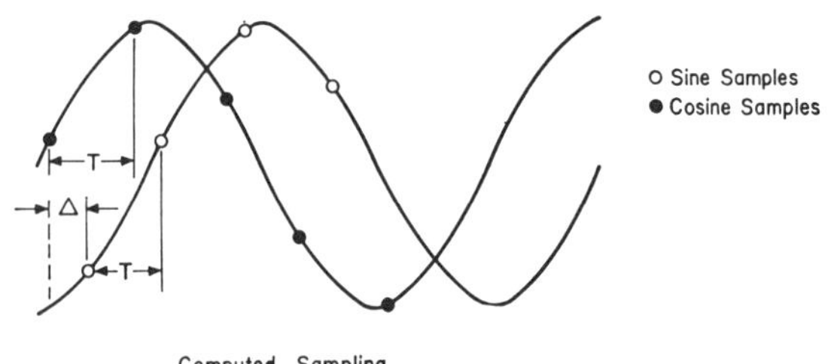

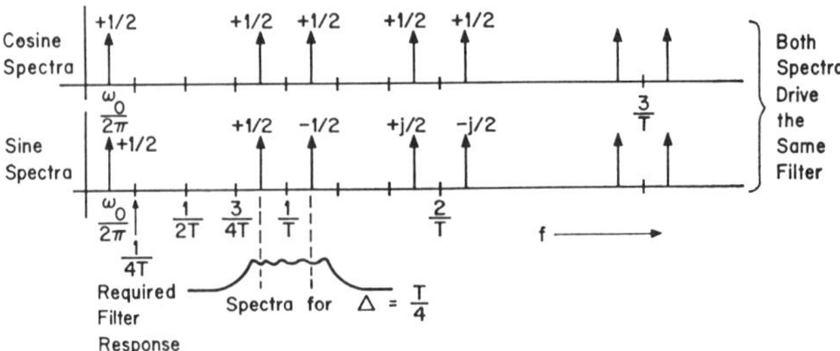

Fig. 12.

sampling interval) in order to cancel an upper or lower side frequency around some sampling harmonic (see Fig. 12). That is, if the cosine is computed at sampling times nT, then the sine must be computed at times $nT+\Delta$. To cancel the upper side frequency around the nth sampling harmonic $n2\pi/T$, Δ must be equal to $T/4n$, and the two computed samples are summed into the same smoothing filter (at their respective times). If the lower side frequency is to be eliminated, the difference between the two sample trains is used to drive the smoothing filter. This technique follows from the expressions for the spectra of sampled cosine computed at nT and sampled sine computed at $nT+\Delta$ as follows:

$$\text{cosine spectrum} = \frac{1}{2} \sum_{n=-\infty}^{\infty} \left[\delta\left(\omega - \omega_0 - \frac{n2\pi}{T} \right) + \delta\left(\omega - \omega_0 - \frac{n2\pi}{T} \right) \right]$$

$$\text{sine spectrum} = \frac{1}{2} \sum_{n=-\infty}^{\infty} \left[\delta\left(\omega - \omega_0 - \frac{n2\pi}{T} \right) - \delta\left(\omega + \omega_0 - \frac{n2\pi}{T} \right) \right] \exp\left[-j\frac{n2\pi\Delta}{T} \right].$$

This technique depends on generating narrow enough pulses out of the digital-to-analog converter to produce energy at some $n2\pi/T$ sampling harmonic.

Phase Control and Other Applications

The phase of the synthesizer output frequency at any computation time depends only on the stored value in the accumulator of Fig. 4 (ignoring any phase distortions introduced by the output smoothing filter). This allows for considerable phase control. In the normal mode of operation a change in frequency would be accomplished by changing only the input frequency word k and leaving unchanged the previous accumulator value. In this way the new frequency would be produced with no phase discontinuity since samples of the new frequency sinusoid would include the last sample of old frequency sinusoid. If on the other hand an arbitrary phase were desired at each change in frequency this means both a new frequency word k and a new accumulator value initial state. An example of such frequency switching would be resetting the accumulator (and phase) to zero whenever a new frequency is generated.

The ease with which phase can be controlled suggests many possible applications. The ability to frequency hop under phase control allows the implementation of new kinds of coherent communication systems if the channel phase is well behaved.

The synthesizer in the phase continuous mode can be used as a very linear frequency modulator. The modulating signal would drive an analog-to-digital converter (unless it is digital to begin with) whose output would be the digital frequency control word. The use of a large encoding range and a large encoding word would reduce distortion and noise. If the converter worked synchronous to the frequency synthesizer clock, no phase discontinuities could occur and a large deviation signal could be obtained with very high linearity.

Another class of applications is generation of sweep signals of various kinds. If the input frequency control word is incremented at certain fixed time intervals, a continuous stepped frequency waveform is generated. In the limit, this becomes a continuous frequency sweep so long as the sweep signal spectrum is not distorted by the sampling. The use of wired or stored frequency sequences could produce very complex sweep patterns.

Finally, the algorithm itself may be used in certain computational environments to produce digital samples of sine and cosine or complex coefficients for use in discrete Fourier transforms.

Some Design Considerations

A synthesizer of the type described here has several parameters which lead to design specifications. One is the time to produce a sample. This is composed of several elements, the accumulator time, the table look-up time, the multiplication time, and the D–A converter time. These times may be added, in some realizations, or they may be overlapped somewhat depending on the logic implementation. In our limited experience, the D–A time has been the largest element, and it is easily overlapped with other times, so that the time to produce a sample is the D–A time.

The reciprocal of the time to produce a sample is the sampling rate. This must be at least twice the highest frequency produced (or the bandwidth when a synthesizer design is intended to produce intermediate frequency sine waves), but it should be higher in order to simplify the requirements on the output smoothing filter. As discussed earlier there is, for any fixed sampling rate, a tradeoff between the complexity of the output filter required and the maximum frequency obtainable. This tradeoff is only interesting over a range of about 2:1 in maximum frequency, further simplification in the output filter as the sampling rate is increased becomes very small. The output filter also affects the switching time of the synthesizer output. Insofar as the sampling rate chosen affects the filter requirement, it also affects the obtainable switching time. This is an area where further work would be useful.

Another important set of parameters is the number of complex multiplications allowed in composing a sample. This is related to the number of subtables of constants required. With N words of storage, broken into $(n+1)$ subtables and combined by n multiplications (words, storage, multiplies all complex), the number of different complex exponentials (proportional to the number of frequencies which can be obtained) can be shown to be $(N/(n+1))^{n+1}$. By choosing n to maximize the number of exponentials, one can obtain a very large number indeed. For example, if $N=32$, it is quite straightforward to obtain 2^{16} exponentials with seven multiplies. In practice, we expect that it will not be desirable to use so many multiplications and probably one multiply and two subtables will be the most common choice.

The arithmetic precision needed, both in the stored constants and in the complex multiplications, will typically be determined by the capabilities of the output D–A converter. If the D–A converter is capable of resolving k bits, the computation should not attempt to produce numbers accurate to very much more than k bits. Note that the number of different frequencies obtainable is still unlimited, since the technique of using nearest samples is still available. If the sample used differs from the exact sample by less than 2^{-k}, the technique of using the nearest sample will not have any important error associated with it. This error is clearly bounded by $\sin 2\pi 2^{-k} \approx 2\pi 2^{-k}$. This implies using about $(k+3)$ bits of the accumulator in the computation, and using the lower bits only for accumulation to produce finer frequency increments. Many of these considerations become straightforward for particular design requirements.

Conclusion

A unique approach to the problem of frequency synthesis has been presented based on a sampled data realization. The simple factoring algorithm explained allows for very efficient use of storage and simple digital implementation. Finally, the phase properties of the synthesized signal and the ease of phase control allow the device to be used for generation of signals more complex than simple sinusoids.

References

[1] V. E. Van Duzer, "A 0–50 MHz frequency synthesizer with excellent stability, fast switching and fine resolution," *Hewlett-Packard J.*, vol. 15, May 1964, pp. 1–8.
[2] A. Noyes, Jr., "Coherent decade frequency synthesizers," *The Experimenter*, vol. 38, no. 9, Sept. 1964.
[3] E. Renschler and B. Welling, "An integrated circuit phase-locked loop digital frequency synthesizer," Motorola Semiconductor Products, Inc., Application Note 463.
[4] G. C. Gillete, "The digiphase synthesizer," *Frequency Technol.*, Aug. 1969, pp. 25–29.
[5] J. Noordanus, "Frequency synthesizers—A survey of techniques," *IEEE Trans. Commun. Technol.*, vol. COM-17, Apr. 1969, pp. 257–271.
[6] B. Gold and C. M. Rader, *Digital Processing of Signals.* New York: McGraw-Hill, 1969.
[7] A. W. Crooke, "A flexible digital waveform generator for use in matched filtering applications," presented at Arden House Workshop, Jan. 1970.
[8] H. Salinger, "Transients in frequency modulation," *Proc. IRE*, vol. 30, Aug. 1942, pp. 378–383.

CMOS/SOS Frequency Synthesizer LSI Circuit for Spread Spectrum Communications

DAVID A. SUNDERLAND, STUDENT MEMBER, IEEE, ROGER A. STRAUCH, STEVEN S. WHARFIELD,
HENRY T. PETERSON, AND CHRISTOPHER R. COLE, MEMBER, IEEE

Abstract —Using a 3.5 µm gate length complementary metal-oxide-semiconductor/silicon-on-sapphire (CMOS/SOS) technology, a single-chip, radiation-hardened, direct digital frequency synthesizer has been developed. The circuit is a critical component of a fast-tuning wideband frequency synthesizer for spread spectrum satellite communications. Each clock period, the chip generates a new digitized sample of a sine wave, whose frequency is variable in 2^{20} steps from dc to one-half the clock frequency. Operation at up to 7.5 MHz is possible in a worst case environment, including ionizing radiation levels up to 3×10^5 rads(Si).

A computationally efficient algorithm was chosen, resulting in 12-bit output precision with only 1084 logic gates and 3840 bits of on-chip read-only memory (ROM). The accuracy of the algorithm is sufficient to maintain in-band spurious frequency components below -65 dBc. At 300 mW, the chip replaces an MSI implementation which uses 25 integrated circuits and consumes 3.5 W.

Manuscript received November 21, 1983; revised March 30, 1984.
D. A. Sunderland and S. S. Wharfield are with the Hughes Aircraft Company, El Segundo, CA.
R. A. Strauch was with the Hughes Aircraft Company, El Segundo, CA. He is now with Teknekron Industries, Berkeley, CA.
H. T. Peterson is with the Hughes Aircraft Company, Carlsbad, CA.
C. R. Cole was with the Hughes Aircraft Company, El Segundo, CA. He is now with Lincoln Laboratory, Lexington, MA.

INTRODUCTION

A FREQUENCY synthesizer for a spread spectrum communications system must be capable of tuning quickly, in small frequency steps, over a wide band. Wide tuning bandwidth is best provided through indirect digital synthesis, in which an integer multiple of a reference frequency may be produced using a programmable divider in a phase-locked loop (see, for example, [1]). To achieve fine tuning resolution, however, requires a low reference frequency, which would yield a low tuning speed. For faster response, this method can be used for coarse tuning,

Reprinted from *IEEE Journal of Solid-State Circuits*, Vol. 19, No. 4, pp. 497-506, August 1984.

and the result mixed with that of a second synthesizer capable of finer resolution, which "fills in" each coarse tuning step. In this approach, a relatively wide bandwidth in the fine resolution stage permits the use of a simpler loop design.

The method best suited for use in the fine resolution stage is direct digital synthesis (DDS) [2]. In this method, a digital phase accumulator is incremented by a fixed amount at regular intervals. Because the fixed-length accumulator regularly overflows, the phase is periodic, with a frequency proportional to both the increment rate and the incremental amount. The digital phase word is used to produce an analog signal with the same frequency. Depending on the amount of hardware employed, a square wave, a triangle wave, or a sine wave may be synthesized.

A further requirement of the frequency synthesizer is spectral purity. This means that for square or triangle wave DDS, the bandwidth is limited to less than one octave by the filter which must remove second, third, or higher harmonics. For sine wave DDS, this means that the trigonometric calculation which changes phase into amplitude must be accurate. To increase the speed of the calculation, and hence the synthesizer tuning bandwidth, one or more read-only memory (ROM) look-up tables is typically employed. Unfortunately, ROM access time can itself pose a limit to operating speed. For spacecraft communications, a frequency synthesizer must also be extremely reliable, operate in a radiation environment, and have limited size, weight, and power requirements.

A radiation-hardened, CMOS/SOS implementation of a direct digital synthesizer has recently been developed. Through use of the sine wave method, spurious harmonic components are maintained below -65 dBc over a bandwidth extending to one-half the sampling rate. Reliability and operating speed are enhanced by implementing all digital functions, including ROM look-up tables, on a single large-scale integrated circuit. This is made possible through a trigonometric decomposition of the sine function, which is efficient, both in terms of ROM utilization and hardware computation.

Using a 7.5 MHz clock, the chip generates a 12 bit digitized sine wave sample every 133 ns, yet dissipates only 300 mW. The tuning bandwidth of up to 3.75 MHz may be covered in 2^{20} discrete steps. Developed for a wideband frequency synthesizer in a spread spectrum satellite communications system, it permits the replacement of an MSI implementation requiring 25 integrated circuits and dissipating 3.5 W. In this paper, we discuss the requirements and tradeoffs driving the design of the DDS LSI circuit, details of the design approach, and results of environmental and in system testing.

ARCHITECTURAL DESIGN

At the highest level, the chip consists of two parts: a 20 bit phase accumulator and a sine function ROM look-up table, as shown in Fig. 1. Since frequency is the rate of phase change, the command word which is added to the

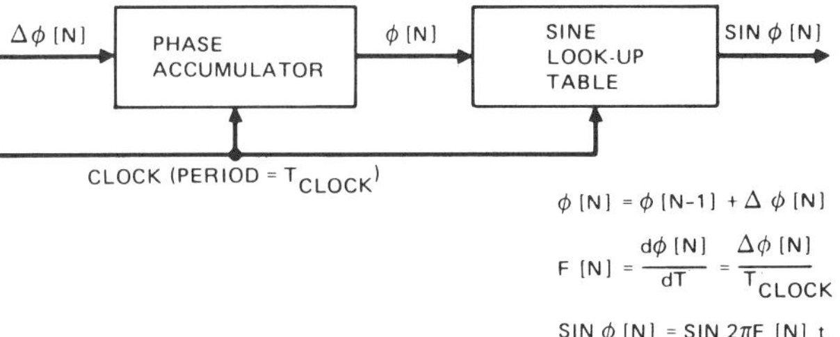

$$\phi [N] = \phi [N-1] + \Delta \phi [N]$$

$$F [N] = \frac{d\phi [N]}{dT} = \frac{\Delta \phi [N]}{T_{CLOCK}}$$

$$SIN \phi [N] = SIN \, 2\pi F [N] \, t$$

Fig. 1.   High-level block diagram of the direct digital synthesizer chip.

accumulator each clock period determines the output frequency. The largest value that can be stored in the accumulator represents 2π rads, so that overflow corresponds to a modulo 2π operation, and is not seen in the output. The accumulator addresses the ROM table to determine a 12 bit digitized sine amplitude value, which is fed to an external digital-to-analog converter (DAC). For output frequencies close to one-half the clock frequency, a low-pass filter at the DAC output may be used to eliminate the alias frequency component created by the sampling process.

Driving Requirements

The primary system requirements which drove the design of the chip were tuning bandwidth, tuning resolution, and noise power. The bandwidth of the DDS is determined by three factors: 1) the choice of synthesized waveform, 2) the sharpness of the external filter used to remove the alias frequency, and 3) the sampling speed. The choice of a sine-wave implementation permits the synthesizer to tune as close to one-half the clock frequency as the filter will allow. Thus, to maximize bandwidth, it remained for the chip design to maximize the clock speed.

For maximum throughput, the chip architecture has been divided into four pipeline stages: low order accumulator, high order accumulator, 1's complement/ROM, and sum/1's complement. A detailed block diagram of the chip is shown in Fig. 2. The purpose of the sum and complement blocks will be made clear in the next section. Although the processing of each sine-wave sample requires multiple clock periods, a new sample appears at the output each period. The time from the change of frequency select inputs to a change in output frequency (the data latency) is three clock periods.

For operation at 7.5 MHz, each stage of the pipeline must perform its function in one clock period, or 133 ns, under all combinations of voltage, temperature, processing parameters, and after exposure to at least 3×10^5 rads (Si) of ionizing radiation. This requirement made it particularly important that the ROM be implemented on the same chip as the logic. Three critical signal paths were identified and optimized for speed. To minimize the number of logic gate levels, the accumulator was broken into two pipeline stages. The low-order 4 bits are one pipeline stage behind the other bits, but are of low enough significance that this does not affect the output. They do, however, contribute to the frequency of the output over time.

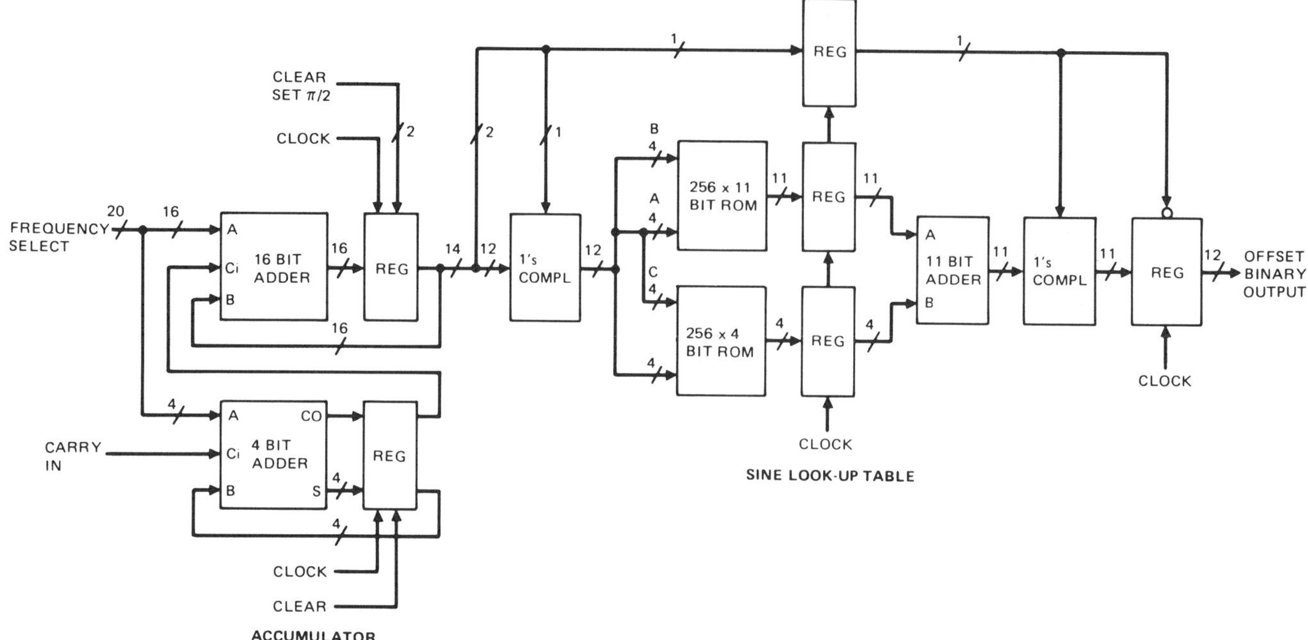

Fig. 2. Detailed block diagram of DDS chip.

TABLE I
NOISE POWER ANALYSIS

Algorithm error	±1.5 LSB
DAC error	±0.5 LSB
Total	±2 LSB
Total amplitude signal/noise	$\pm 2^{11}/\pm 2 = \pm 2^{10}$
Total power signal/noise (dB)	$20 \log (2^{10}) \approx 60$ dB

The size of the accumulator is a tradeoff between tuning resolution and resulting chip size. The chosen length of 20 bits is sufficient to obtain 7.2 Hz resolution from a 7.5 MHz clock ($7.5 \times 10^6/2^{20} \approx 7.2$). Finer resolution may be obtained by adding an external accumulator stage. The carry out of this stage is fed to a carry input of the chip, and has the same significance as the least significant bit (LSB) of the internal accumulator. This new circuit forms a fifth pipeline stage which is one clock behind the internal 4 bit least significant accumulator section. As with that section, the extra delay on the external stage would not affect the output accuracy.

Inaccuracy in the chip's sine wave algorithm represents a source of noise to the system. To maintain a signal-to-noise ratio at the DAC output of about 60 dB, the chip output needed to be accurate to ± 1.5 LSB at 12 bits, as summarized in Table I. In this analysis, note that of 12 output bits, one is a sign bit, and does not contribute to the signal amplitude. The level of algorithmic error that was acceptable determined the degree that approximations could be used in the sine calculation.

Algorithmic Approximations

In a brute force approach, a very large ROM must be used to encode the sine of each possible phase value. For the present case of 20 accumulator bits addressing an ROM with 12 output bits, 12 megabits of storage would be required. Instead, the phase is split into several components, and trigonometric identities are used to assemble the desired answer. The most straightforward of such identities call for the use of multiplication, and are therefore relatively slow and complex. By making suitable approximations, multiplication can be avoided, thereby significantly reducing both the amount of hardware and the time required for the calculation [3].

Although 20 bits of phase representation are required for frequency resolution, only 14 of these bits are needed at the ROM input to maintain the appropriate level of accuracy. This may be seen by noting that the amplitude error in truncating the remaining bits is no more than $\sin(2\pi/2^{14})$, which is about 2^{-11}, or 1 LSB.

The width of the ROM input word is further reduced by taking advantage of the quadrant symmetry of the sine function. The most significant two phase bits are used to decode the quadrant, while the remaining 12 bits are used to address a one-quadrant ROM. The most significant bit determines the required sign of the result, and the next most significant bit determines whether the sine amplitude is increasing or decreasing. If the amplitude is to decrease for increasing phase, a 1's complement is performed on the ROM address. For negative results, a 1's complement is performed on the output word. The error in using 1's complement in both of these places rather than 2's complement is at most 1 LSB.

Finally, the remaining 12 phase bits are divided into three 4 bit fractions: A, B, and C, such that $A < (\pi/2)$, $B < (\pi/2)(2^{-4})$ and $C < (\pi/2)(2^{-8})$. The desired sine function is given by (1).

$$\sin(A + B + C) = \sin(A + B)\cos C + \cos A \cos B \sin C$$
$$- \sin A \sin B \sin C. \quad (1)$$

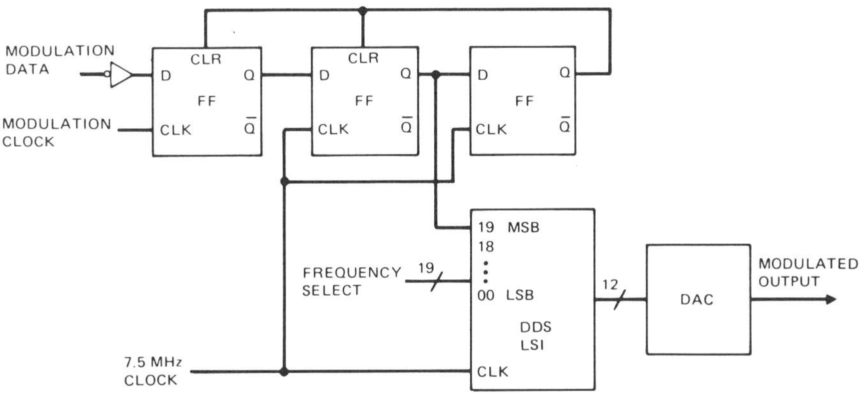

Fig. 3. DPSK biphase modulation circuit for encoding data on the DDS chip output.

Given the relative sizes of A, B, and C, (1) may be approximated by (2).

$$\sin(A + B + C) \approx \sin(A + B) + \cos A \sin C. \qquad (2)$$

The errors resulting from this approximation can be shown to be below 1 LSB. By using (2), what would have been the output of a $2^{12} \times 11 = 45\,056$ bit ROM can now be replaced by the sum of the outputs of two smaller ROM's. The address inputs of each ROM consist of two 4 bit words, and all the difficult mathematics (including the multiplication) is performed when the ROM data is calculated. The first term in (2) (a coarse approximation to the actual answer) requires $2^8 \times 11 = 2816$ bits, while $2^8 \times 4 = 1024$ bits are used for the second (a fine correction factor). Only 4 bits are stored for each word in the second ROM because the most significant 7 bits of that term are always zero. Note that the 4 bit fractions A, B, and C are labeled in Fig. 2.

Although none of the above approximations individually produces an error which is large compared to 1.5 LSB, the total algorithmic error required reduction through the use of three additional corrections [4]. The approximation in (2) is made closer by adding $(\pi/2)(2^{-5})$ (the average value of B) to A in the second term. Similarly, to adjust for the truncation of the 20 bit phase to 14 bits, the factor C includes a phase offset of $(\pi/2)(2^{-13})$. Because the final result must be less than one, both coarse and fine terms are multiplied by $(1 - 2^{-11})$, the largest number which can be represented in 11 bits. Note that each one of these factors is applied to data to be stored in ROM and has no effect on the calculation that must be performed in the chip. The coarse term was truncated to 11 bits and the fine term was rounded to the same precision (4 nonzero bits). The overall error budget of ± 1.5 LSB was verified for all possible values of A, B, and C by computer.

Advanced Features

Looking ahead to more sophisticated applications, a feature was added to the synthesizer design which permits operating two chips in parallel separated in phase by 90 degrees. These in-phase and quadrature signals may be used to develop single sideband modulation, or, in combination with other synthesizers synchronized at higher harmonics, to synthesize an arbitrary waveform. Synchronization is achieved by simultaneously resetting the accumulator of all chips. A special control input determines whether the accumulator is set to 0 phase or $\pi/2$ by the reset operation.

Although current systems do not require such frequency agility, we chose to implement the synthesizer LSI circuit in such a way that it could change frequency in a phase-continuous manner on every clock period. An outgrowth of this is the ability to introduce modulation on the output waveform by data [5]. The frequency control input is actually a 20 bit word which determines the amount by which the phase is to change on the next clock pulse. The most significant bit (MSB) of this word corresponds to a phase increment of π. Pulsing the MSB for one clock period will introduce an immediate 180 degree phase shift, which in differential phase shift keying (DPSK) modulation corresponds to a "zero" data bit. The circuit of Fig. 3 might be used to implement such a modulation scheme. The first and second flip-flops synchronize the data to the modulation clock, and the synthesizer clock, respectively. The third flip-flop cuts off the pulse.

An alternate use of the MSB is to specify the sense of the other 19 bits. When the MSB = 0, increasing the remaining bits increases the frequency. When the MSB = 1, increasing the remaining bits decreases the frequency, because each positive phase step of $(\pi + X)$ is equivalent to a negative phase step of $(\pi - X)$. Thus, an input word which produces a frequency of $Y \times f_{clk}/2$ when the MSB is low will produce a frequency of $(1 - Y) \times f_{clk}/2$ when the MSB is high.

Another advanced feature is related to testability. While verification of all circuit operations is required at functional test, exercising any arbitrary gate or ROM cell from the inputs is extremely difficult, considering the pipeline nature of the computation and the large number of possible states. To improve this situation, a built-in serial set and scan feature was added. Every flip-flop on the chip is preceded by a multiplexer which can select between the normal source of data for the flip-flop and the output of an

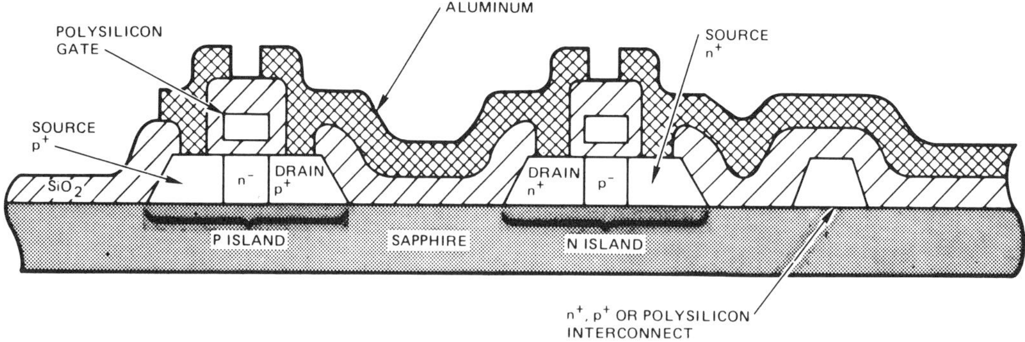

Fig. 4. Cross section of n-channel and p-channel CMOS/SOS transistors.

adjacent flip-flop. Thus, the entire chip can be configured as a long shift register, permitting the logical state of the chip to be loaded or verified. Each functional block may be tested separately, with 100 percent stuck-at fault detection and significantly reduced test time. During the design phase, the use of set/scan vectors simplified logic simulation, and had there been a design error, set/scan testing could have been used to track it down more easily. For our design, this testability feature added 14 percent additional random logic and one gate delay to the critical paths.

CMOS/SOS PROCESS

A 3.5 μm silicon-gate complementary metal-oxide-semiconductor/silicon-on-sapphire (CMOS/SOS) process was chosen to achieve a single-chip implementation of the direct digital synthesizer. The dielectric isolation of p-channel and n-channel transistors fabricated by this 8 mask process is illustrated in Fig. 4. As reported in the literature, this isolation in CMOS/SOS circuits results in higher speed and lower power through reduced parasitic capacitances, increased transient radiation hardness through small junction areas, and the prevention of latch-up by eliminating all vertical n–p–n and p–n–p structures [6].

Total dose radiation hardness is achieved through a combination of processing techniques, device parameter optimizations, and design techniques. The processes and parameters most affecting total dose hardness are listed in Table II. The n-channel threshold voltage is intentionally set higher than normal to compensate for the negative shift expected in the radiation environment, thereby preventing the device from going into depletion.

CIRCUIT AND LAYOUT DESIGN

Design of the synthesizer chip followed a standard cell layout approach, characterized by the use of a library of verified standard circuit elements combined using computer-aided design (CAD) techniques. With the exception of the ROM blocks, all layout was performed interactively on a graphics terminal at the symbolic level. Expansion of layer geometries was performed automatically. Because of the large number of cells within the ROM and the semi-random nature of the ROM data, manual placement of

these cells at the graphics terminal was deemed an undesirable, error prone process.

Instead, a computer program was written for the CAD host computer (PDP 11/45) which determined the ROM contents algorithmically and automatically placed the row decoders, column multiplexers, and memory cells. The program produced a list of the cells to be used in assembling the ROM blocks, along with the coordinates of each row of cells. This list was given to a standard cell placement module of the CAD layout system, which inserted the cells into the database containing cells placed interactively. An entire ROM block could then be moved and rotated interactively to conform to the rest of the layout.

The resulting ROM blocks are labeled in the photomicrograph of Fig. 5. The row decoders for both are located in a horizontal row in the middle of the chip, and the column multiplexers are back-to-back down the center. The same program that was used to assist in the layout of the ROM's was used to verify the algorithmic error budget and to produce a pair of ROM truth tables for use in logic simulation of the entire chip. Use of a common database for algorithm verification, chip layout, and logic simulation contributed substantially to the first-pass design success of the DDS chip.

An ROM tradeoff study selected a static memory with CMOS row decoders, active PMOS memory cells and column multiplexers, and passive NMOS load devices as the best choice for speed, power, density, and radiation hardness. A memory cell contains a PMOS transistor if the data stored in that location is a logical "1." The signal path through decoders, cell, and multiplexer is shown for both coarse and fine ROM's in Fig. 6. This circuit, unlike

TABLE II
KEY PROCESSES AND PARAMETERS FOR TOTAL DOSE RADIATION HARDNESS

Gate oxidiation	900 Å, wet, 925°C
Maximum temperature following gate oxidiation and anneal	875°C
Gate material	Phosphrous doped polysilicon
Source and drain formation	Ion implantation
Metalization	Sputtered Al/Si
P-channel V_t	−0.5 to −1.5 volts
N-channel V_t	1.2 to 2.4 volts

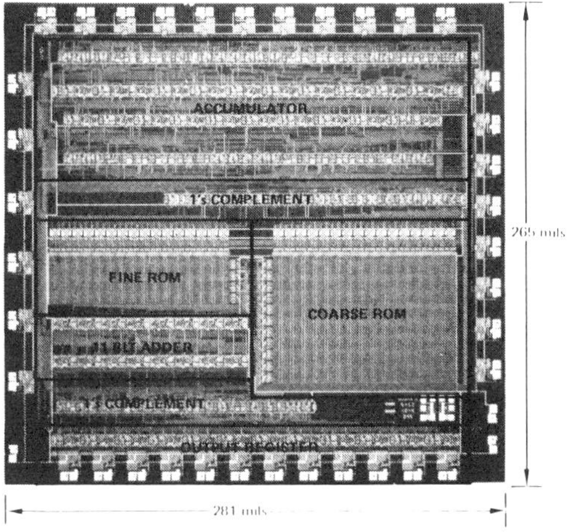

Fig. 5. Photomicrograph of CMOS/SOS Direct Digital Synthesizer LSI circuit.

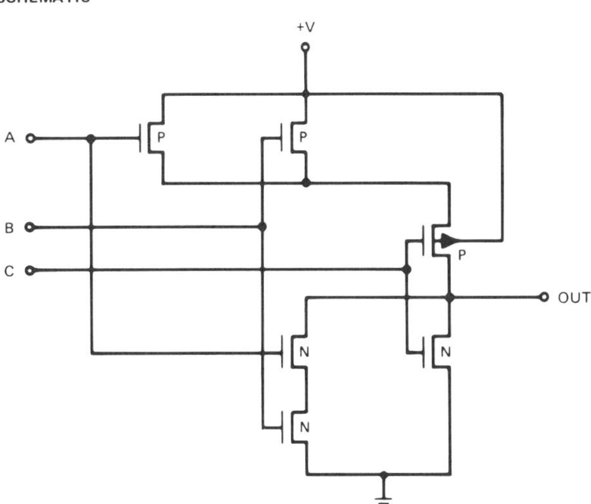

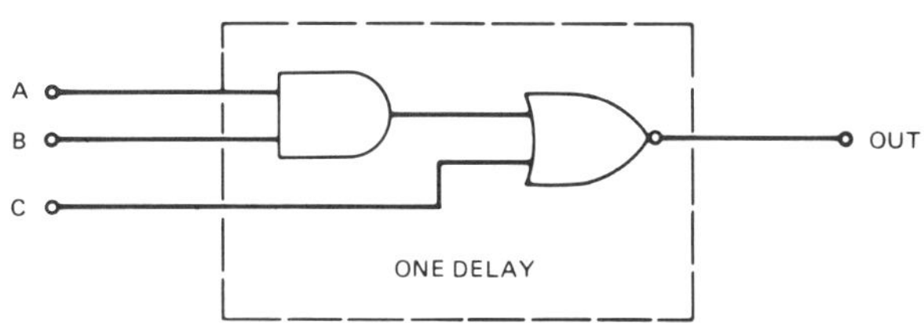

Fig. 7. Merged CMOS gate structure used to implement an AND/OR/INVERT function in one gate delay.

ordinary CMOS, does have a dc current path for output data in the "1" state. To construct a ROM without such a dc path that did not depend on dynamic circuitry, it would have been necessary to provide both NMOS and PMOS transistor memory elements as well as complementary row decode signals. This would have required considerably more area and, as such, was undesirable.

It was found that NMOS SOS transistors are too prone to radiation induced leakage currents to use them in the memory cells of a large (64 row) ROM. Similarly, diode memory cells, such as are used in bulk silicon commercial ROM's, are also unsuitable because of leakage currents in SOS gate-controlled diodes. Static circuitry was chosen over dynamic circuitry because of its relative insensitivity to leakage, simplicity of clocking, and absence of a minimum clock frequency. Final transistor sizes in the ROM circuits are the result of extensive optimization for speed under conflicting worst case conditions, using the SPICE circuit simulation program.

Considerable use was made of merged gate structures possible in CMOS, such as the one shown in Fig. 7, to

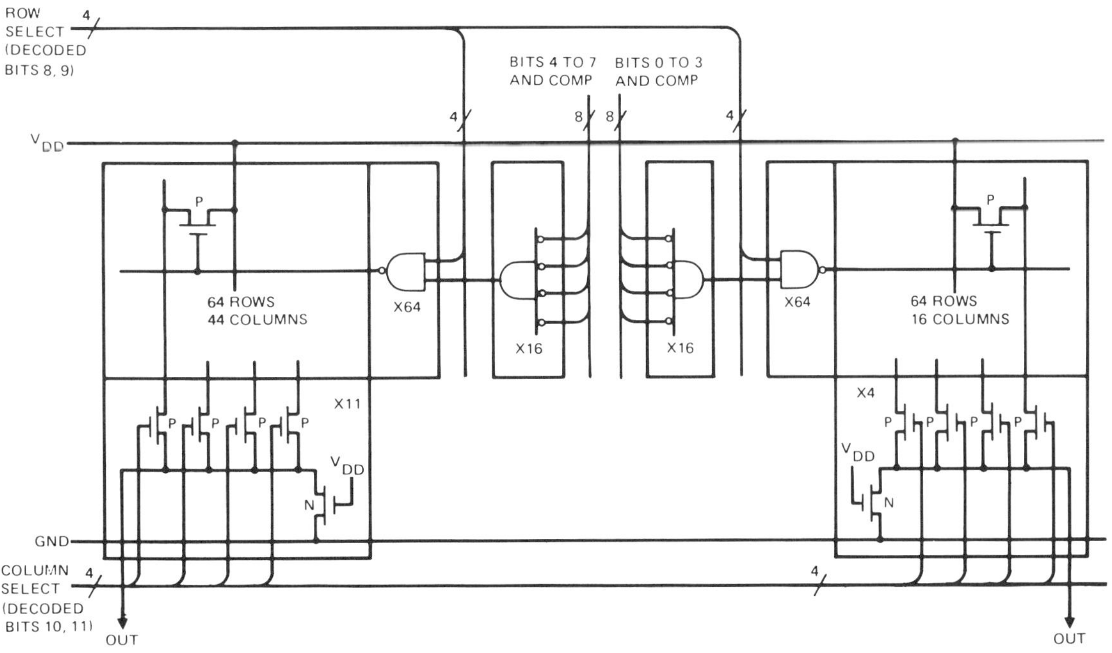

Fig. 6. Schematic diagram of the approach chosen for implementation of the two ROM's.

Chip size	265 x 280 mils
Complexity	
Number of transistors	6320
Number of equivalent two-input logic gates	1084
Number of ROM bits	3840
Performance at 7.5 MHz	
Frequency range	0 to 3.75 MHz
Frequency resolution	7.2 Hz
Functional throughput	3×10^{10} gate-Hz/cm^2
Radiation hardness level	3×10^5 rads(Si)
Operating temperature	-35°C to $+85^\circ$C
Power supply range	10.8 to 13.2 V
Typical power dissipation	300 mW

perform AND/OR/INVERT or OR/AND/INVERT functions with a single gate delay. Use of these cells permitted a considerable speed improvement in all three critical pipeline stages. Adder speed has been maximized through the use of carry look-ahead circuitry in 4 bit levels. This was the primary motivation for reducing the main accumulator stage to 16 bits.

Although the process is optimized for radiation hardness, the sensitivity of circuits is largely determined by design practice. Because p-channel transistor thresholds increase with radiation, while n-channel thresholds typically decrease, both speed and leakage considerations dictate the use of NAND rather than NOR configurations where large gate fan-in is required. In the DDS chip, the gate fan-in for NAND gates was limited to four, whereas the limit for NOR gates was two. An exception was the ROM address decoding, where a special 4 input NOR cell was utilized and SPICE simulation was performed to optimize overall ROM performance. High-speed operation in spite of increasing p-channel thresholds is ensured by use of a large (12 V) power supply voltage. Because the radiation-induced threshold shifts of p-channel SOS transistors are enhanced by positive gate-to-substrate bias [7], devices in which this bias condition is possible have their substrates, which are normally floating, tied to the positive rail.

A summary of chip statistics is given in Table III.

TEST RESULTS

Proof of design testing of the synthesizer LSI circuit involved tests on two wafer lots, with tests performed both at the wafer level and on packaged devices. These tests indicate that all design goals were met, and devices under nominal conditions operate at speeds of greater than 12 MHz. In these tests, maximum chip speed was determined by connecting the chip outputs to a DAC, and studying the DAC output with a spectrum analyzer.

Fig. 8 shows the effects of voltage and temperature extremes on the speed of the chip. CMOS devices generally exhibit a roughly linear relationship between maximum operating frequency and voltage, above some minimum operating voltage. For the DDS chip, this minimum voltage is about 2.5 V. This voltage increases with ionizing radiation exposure as shown in the Co^{60} tests of Fig. 9, and is a good measure of increasing PMOS transistor threshold.

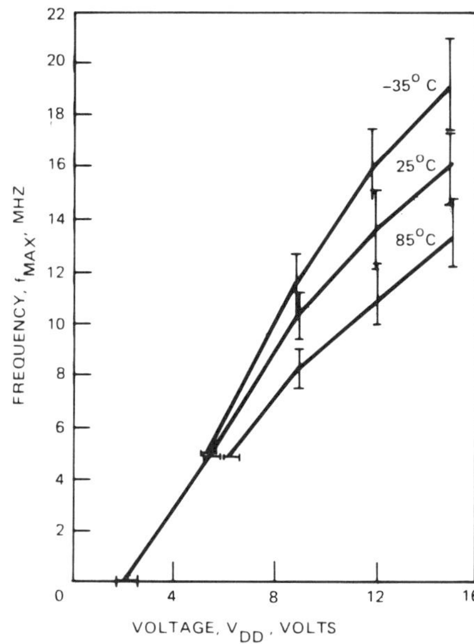

Fig. 8. Variation of maximum clock frequency with power supply voltage and temperature.

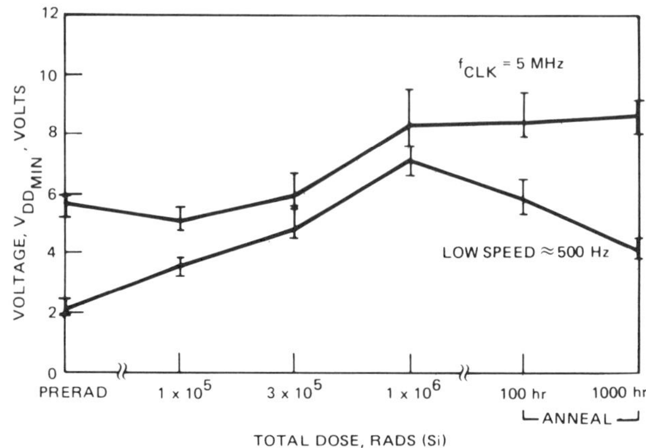

Fig. 9. Minimum power supply voltage for high- and low-speed operation as a function of radiation.

Normally, this increase, combined with a reduction in mobility, results in a decrease in speed with irradiation. In the case of this device, the speed-limiting mechanism is the time required for discharging the ROM output lines with the NMOS load transistors in the column multiplexer. Because NMOS transistor thresholds decrease with irradiation, the discharge current is increased, and the chip gets faster up to about 3×10^5 rads(Si), as shown in Fig. 10. From the data of Figs. 8–10, it has been determined that under the worst case conditions of ± 10 percent supply variation, ambient temperature between -35° and $+85^\circ$C, and ionizing radiation up to 3×10^5 rads(Si), the DDS LSI circuit exceeds the 7.5 MHz system requirement.

The decrease in NMOS transistor threshold with irradiation is accompanied by an increase in NMOS leakage current, particularly where the polysilicon gate crosses the silicon island edge. This causes a marked increase in the static power supply current of the chip, as shown in Fig. 11. Above about 3×10^5 rads(Si), the NMOS threshold begins to increase, causing the leakage to reach a maxi-

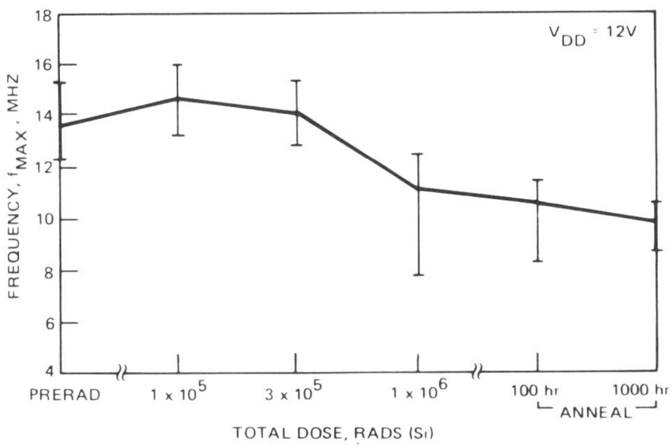

Fig. 10. Maximum clock frequency for $V_{DD} = 12$ V as a function of radiation.

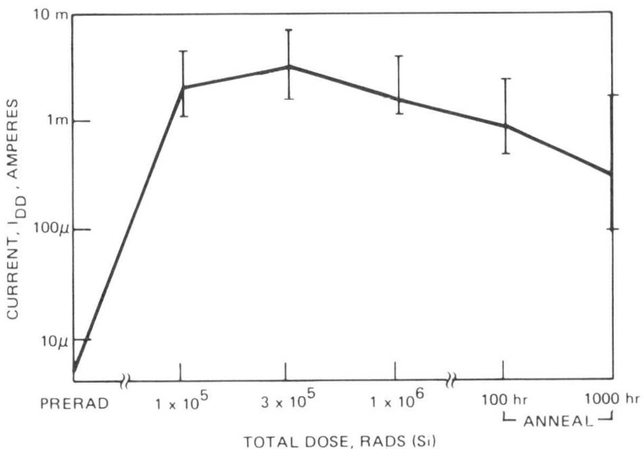

Fig. 11. Power supply current with all ROM bits nonconducting as a function of radiation.

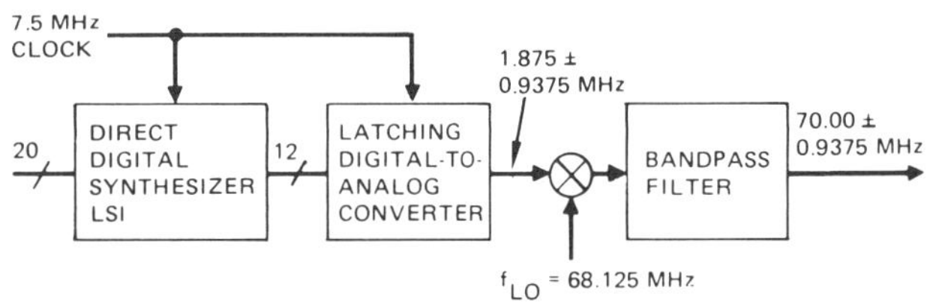

Fig. 12. Application of direct digital synthesizer chip in a wideband synthesizer fine resolution stage.

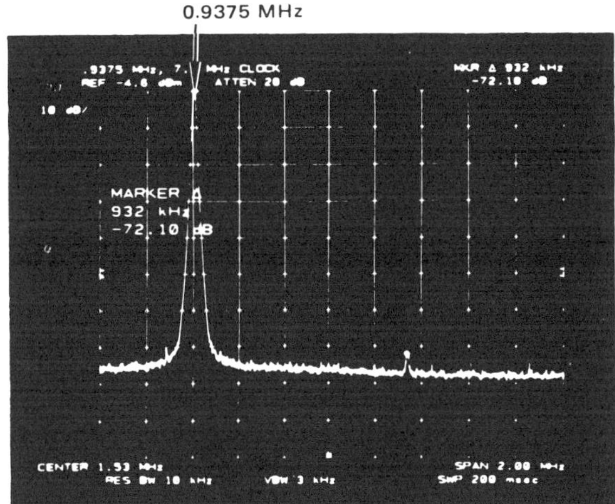

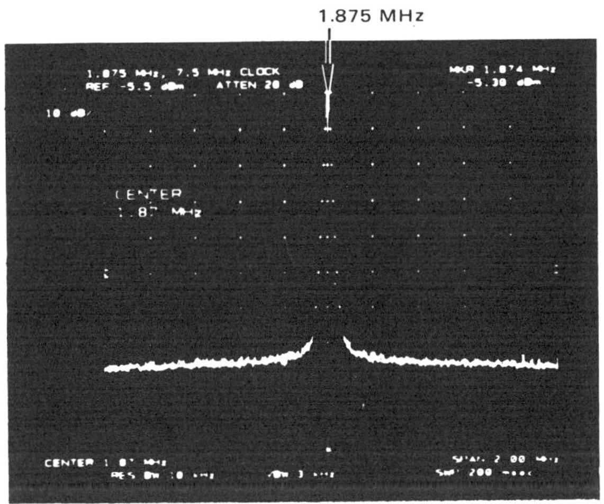

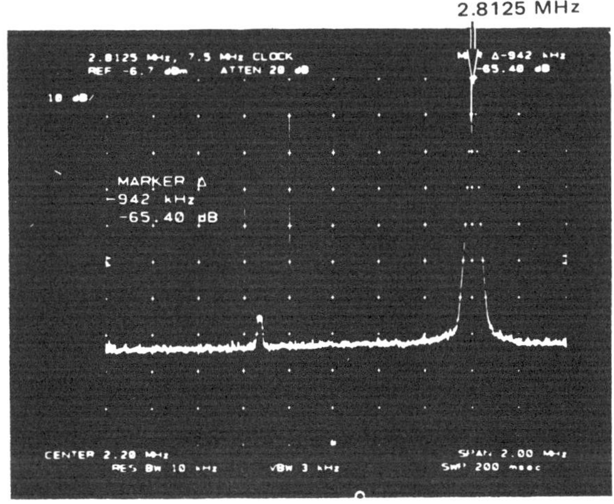

Fig. 13. Spectral analysis of the fine synthesizer output, prior to mixing.

mum. Additional power dissipation at this point is less than 100 mW, about one-third of the dynamic power consumption at 7.5 MHz.

Also indicated in Figs. 9–11 are the effects on radiation-induced parameter shifts of annealing over time. These data are included to allow for simulation of the low dose rate space environment. Unfortunately, no quantitative model yet exists for correlating annealing following a short irradiation period with irradiation over a long period. A short recovery period, however, is indicative of less severe degradation at low dose rates, as some annealing takes place concurrently with the irradiation at a low dose rate.

A set of 10 devices were placed on 125°C bias for 2000 h to verify the reliability of the design. No failures or significant parameter degradations were observed. Based on the worst case failure activation energy determined in earlier life test evaluations of the 3.5 μm CMOS/SOS process [8], a rate of no more than 250 failures per billion hours (FITS) at 55°C can be assigned to the device. A closer determination of the failure rate is planned as part of a further study involving approximately 80 parts from three different lots.

In the fine resolution stage of the spread spectrum communications frequency synthesizer, shown in Fig. 12, the DDS LSI circuit is mated with a commercially available digital-to-analog converter. This particular DAC incorporates an internal data register, which eliminates potential glitches due to skew between synthesizer chip outputs. The filtered output of this stage is mixed with that of the coarse resolution (indirect digital synthesis) stage, to produce a wideband output.

The chip was installed in a breadboard of the fine resolution stage to verify the predicted system performance. Baseband spectrum analysis results are displayed in Fig. 13, where the strongest spurious frequency component in the band of interest is seen at -65 dBc. In this particular application, the band extends to 3/8 of the clock frequency. Given the accuracy of the DDS chip, this could be extended to just under one-half of the clock frequency with a sufficiently precise filter.

Summary

A CMOS/SOS direct digital frequency synthesizer has been developed for use in the fine resolution stage of a wideband synthesizer for spread spectrum satellite communications. Through trigonometric decomposition and suitable approximations, 12 bit precision is achieved without the use of either a large ROM or a hardware multiplier. Processing speed, and hence tuning bandwidth, has been optimized by including ROM and logic on the same LSI chip. Several advanced features have been incorporated to make it a more general purpose communications building block. Interfacing a program which algorithmically determined the contents of the ROM with a standard cell CAD layout system permitted the circuit to be designed correctly without iteration. Substantial use of simulation resulted in an optimal ROM design for radiation hardened CMOS/SOS circuits.

Environmental testing has verified that the chip meets its worst case system design goals, including operation at 7.5 MHz following a radiation dose exceeding 3×10^5 rads(Si). A preliminary reliability study indicates that less than 250 failures per billion hours (FITS) can be expected for 55°C operation. Combined with a commercial DAC, the chip has demonstrated spectral purity of -65 dBc over a band extending to 3/8 of the clock frequency.

Acknowledgment

The authors wish to thank the following people for their contributions to this work: T. Sakuda, for chip layout and verification; P. Chang, for process development; R. Snow, for synthesizer processing; J. Storey, for system analysis; and K. McNab, for system integration and test.

References

[1] J. Gorski-Popiel, "Phase-locked loop frequency synthesizers," in *Frequency Synthesis: Techniques and Applications.* New York: IEEE Press, 1975.

[2] J. Tierney, "A digital frequency synthesizer," *IEEE Trans. Audio Electroacoust.*, vol. AU-19, p. 48, Mar. 1971.

[3] B. H. Hutchinson, Jr., "Contemporary frequency synthesis techniques," in *Frequency Synthesis: Techniques and Applications.* New York: IEEE Press, 1975.

[4] C. R. Cole, "Design of a direct digital frequency synthesizer," master's thesis, Mass. Inst. Technol., 1982.

[5] P. S. Angello and G. S. Des Brisay, Jr., "BPSK frequency modulator using a digital frequency synthesizer," Hughes Aircraft Co., patent disclosure 79227, March 13, 1981.

[6] T. B. Micholotti, "Forward to SOS special issue—SOS technology," *IEEE Trans. Electron Devices*, vol. ED-25, p. 857, Aug. 1978.

[7] D. P. Shumake, R. A. Kempke, and K. G. Aubuchon, "Hardened CMOS/SOS LSI circuits for satellite applications," *IEEE Trans. Nucl. Sci.*, vol. NS-24, p. 2177, Dec. 1977.

[8] S. S. Wharfield, "Evaluation of the Hughes Newport radiation hardened SOS-I silicon on sapphire technology," Hughes Aircraft Co., Tech. Intern. Corresp. 4141.50/03, Apr. 23, 1982.

A High-Speed Direct Frequency Synthesizer

PETER H. SAUL AND DAVID G. TAYLOR

Abstract—This paper describes a UHF direct frequency synthesizer (DFS), which is primarily intended for radar and electronic warfare applications. The device generates square-, sine-, and triangle-wave outputs, true and complement, in-phase and quadrature, over the range 1 Hz to 500 MHz. It is a fully integrated design, including two digital-to-analog converters (DAC's) which each have a faster operating specification than any DAC currently available. The circuit is composed of a number of structured circuit blocks, each of which can be tested independently on the chip. This approach has benefits both during device evaluation and as an aid to minimizing production test times.

I. INTRODUCTION

IN THE last two decades, frequency synthesis has progressed from the relatively crude "crystal bank" synthesizers to the modern multiloop phase-locked systems. This paper will describe new work on a class of device which will become the next generation in frequency synthesis.

There is an increasing trend towards "analog" circuits which are primarily digital in operation, but contain data conversion components at input or output. The advantages of this approach are that full digital control of the function is maintained as far as possible, and the limitations of analog design are minimized. The final product then needs little or no "alignment"; digital circuits are connected up and operate to strictly defined parameters, and, most importantly, this type of circuit is very reproducible.

Direct digital synthesis has been described in the literature [1] but has been considered as a low-speed technique only, since the achievable performance is limited by the digital-to-analog converters (DAC's) and high-speed logic available. This paper will show that direct frequency synthesis is now a practical proposition at VHF and UHF frequencies, and can be extended by mixing into the microwave region.

Block diagrams of a conventional phase-locked-loop (PLL) frequency synthesizer and a direct frequency synthesizer (DFS) are included in Fig. 1. The PLL type is more complex in concept, contains a significant amount of analog circuitry, and needs "setting up" after assembly. Unless the system is made appreciably more complex, the frequency range of operation is usually less than one octave. Channel spacing is also a limitation, since it directly affects the properties of the PLL, especially acquisition time, and leads to further design complexity. The main operational difference between a DFS and the PLL type is that the DFS does not contain feedback loops. This is a major advantage in settling to a new frequency; a good PLL has acquisition times of around 1 ms, whereas the DFS can acquire a new frequency in a time limited only by pipeline delays in the accumulator and the DAC settling time. The frequency shift in the DFS is phase coherent, which is very difficult to achieve in any other way. The primary source of stability is the clock oscillator, so that, in the limit, since the clock is always at a higher frequency than the output, the output phase noise is better than the clock itself. Finally, very narrow channel spacing is possible using DFS; this can be particularly problematic in PLL's.

Based on the results from an earlier DFS device [2], a multifunction DFS has been designed with applications in radar and electronic warfare. The new device was based on a 1-μm silicon bipolar process (Plessey Process HE). This process has transistor cutoff frequencies F_t of 22 GHz in the most advanced version, with a double-polysilicon structure and "trench" transistor isolated to produce transistors with very low parasitic capacitance. This new device is designed to have a low level of spurious outputs, offers in-phase and quadrature (I and Q) outputs with their complements, and has square-, triangle-, and sine-wave outputs available. To achieve this performance, a very high clock rate (over 2 GHz) is required, with very fast settling time 8-bit DAC's on the same chip. This technique avoids delays and line loading caused by interchip connections. The detailed circuit design of the new device included extensive "pipeline" latching and careful layout to equalize track delays. This is essential to avoid data skew at the extreme clock speeds required. Frequency setting inputs to the device were not deskewed. This limits the slew rate for true coherence to the accumulator through rate, i.e., 20 ns. In practice this was felt unlikely to be a problem. The block diagram of the DFS (Fig. 2) shows that, unlike many high-speed designs, the whole chip has to operate at the full clock rate. A particular advantage of the highly integrated structure used is that very few interfaces operate at high speeds; those which do are the clock (> 2 GHz) and the outputs (up to 500 MHz). Generation of the in-phase

Manuscript received July 17, 1989; revised October 9, 1989. This work was supported in part under the U.K. Alvey programme.

P. H. Saul is with Plessey Research and Technology, Caswell, Towcester, Northants NN128EQ, England.

D. G. Taylor was with Plessey Research and Technology, Caswell, Towcester, Northants, England. He is now with Plessey Semiconductors, Cheney Manor, Swindon, England.

IEEE Log Number 8932701.

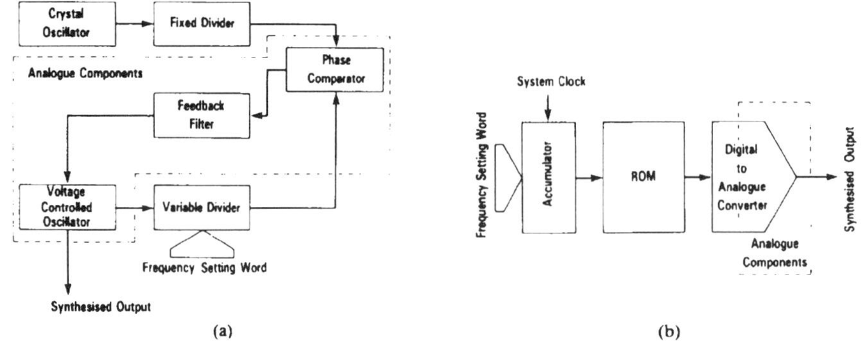

Fig. 1.   (a) Basic phase-locked loop system. (b) Direct frequency synthesizer.

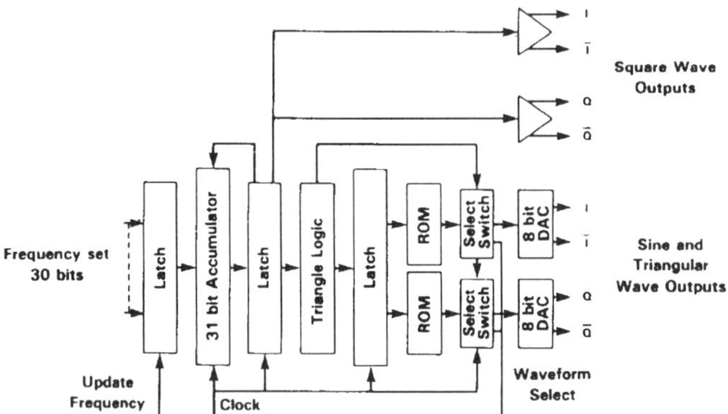

Fig. 2.   Block diagram of DFS.

and quadrature outputs is achieved by the design of a dual-port 400-ps access time ROM containing all data for both phases. On the DAC outputs, the open-collector current-source configuration has been very widely used in fast DAC's [3]. The new challenge was to operate at 500 MHz, but this was overcome by careful optimization of the DAC, clocked latches, and current switches.

II.   Circuit Design and Operations

The block diagram for the DFS is shown in Fig. 2. The device consists of a phase accumulator, a dual-port ROM, and two DAC's, with interstage latches to maintain data parallelism. A frequency setting word is applied to the accumulator, which is driven by the system clock. The frequency word is repeatedly added to the accumulator contents, through overflow and on through successive cycles. The length of the accumulator determines the output frequency resolution as

$$\text{accumulator length} = \text{clock frequency/frequency increment}.$$

The DFS has been designed with an accumulator length of 31 bits and 1-Hz frequency increment, so the nominal clock frequency is 2 147 483 648 Hz. Devices have been tested up to 2.5-GHz clock rate. Of course, lower clock

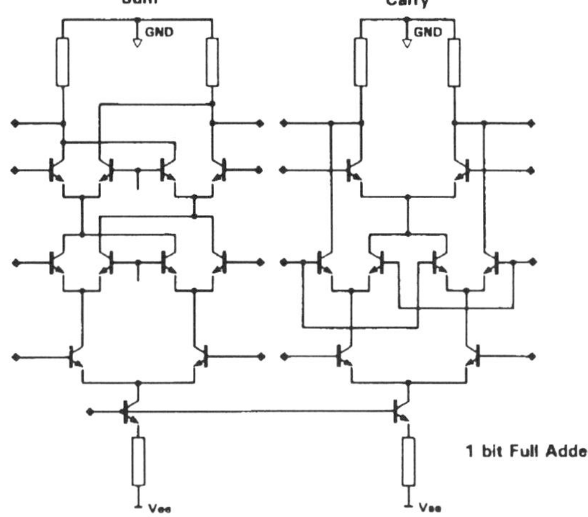

Fig. 3.   Example of stacked (multilevel) logic.

rates can be used, as can larger frequency increments, by ignoring lower bits on the accumulator input word.

Throughout the chip, differential current-mode logic (DCML) has been used with up to three levels of logic in the stack (Fig. 3). The power supply voltage applied to the chips was -4.5 V, i.e., positive rail grounded. Each current

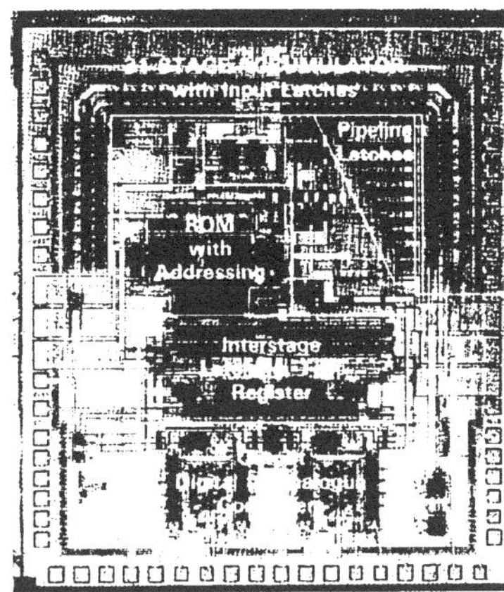

Fig. 4.   Chip photograph of DFS.

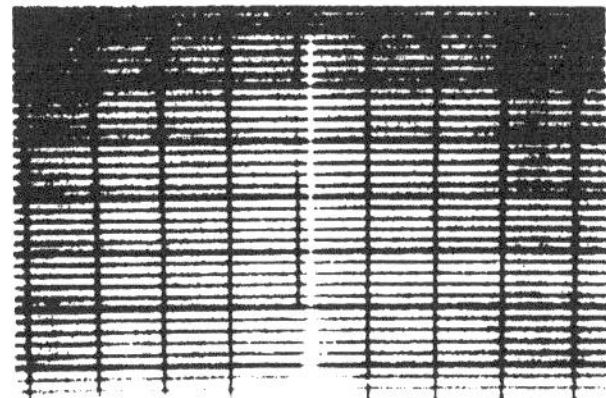

Fig. 5.   Synthesizer output at 70-MHz spur level < -55 dBc.

"tail" was set at 0.7 mA, with a further 0.5 mA in each level-shift stack. DCML has been found to offer the best compromise between speed, power consumption, and chip complexity; it can also be easily interfaced to conventional ECL. The added complexity in the routing needed by the differential tracks was offset by the avoidance of supplying three reference voltages at constant levels to all areas of the chip; DCML is tolerant of small variations from nominal levels caused by local ground and V_{EE} variations.

III.   The Chip Layout

The chip photograph is shown in Fig. 4. The chip size is 4×5 mm. The device is a literal translation of the block diagram (Fig. 2), with the inclusion of a serially readable and writable test register for simplification of the test procedure. Data on the frequency required are fed into the 31 parallel inputs of the input register. Although it would have been possible to load this register serially, a major requirement for this class of device is very fast frequency changing, so parallel load was preferred. The accumulator data are fed, via pipelining latches and some logic, to the read-only memory (ROM) which contains two part sine waves in a quadrature relationship. Both phase and amplitude inversion are used externally to the ROM, so that each sine wave contains 90°. The ROM has a single address port but dual outputs, thus simplifying the structure while guaranteeing accurate quadrature. ROM size is 896 bits, arranged as $2 \times 64 \times 7$ bits, which, with phase and amplitude inversions, is adequate for the 8-bit system used at the outputs to the DAC.

After the ROM, the amplitude and phase inversions are inserted into the data word, and the data are relatched. The test registers provide for serial readout of the data at this point, and for serial read-in of data to test the DAC's.

Longer word lengths were precluded by several practical factors. In the ROM, although the chip area consumed is minimal, a larger ROM pattern would have resulted in a slower access time. Although more words would have been available, the longer access time would have limited the maximum output frequency. Similarly, a DAC with more bits would have been of slower settling time and hence again limited the output frequency. This limitation is more serious than it would appear, because the critical factor in the DAC is settling time to a specified accuracy. In general, if we took say, 6 bits as the limiting case, an 8-bit DAC would settle faster to 6 bits than would a 10- or 12-bit DAC. Finally, on this as on most IC processes, untrimmed component matching accuracy is such that yields of 8-bit DAC's will be adequate, but 10 bits will not be so. This was the most decisive factor in this design. Of course, where either trimming facilities are available on chip, or where an external accurate DAC is a possibility, for example, at much lower maximum output frequencies, the bit economics of the design would be radically different.

IV.   Results

The DFS devices have recently been evaluated. These first-pass devices are fully functional with the exception of the sine-wave outputs. Some minor layout faults have been located which prevent operation of the sine wave, so no fully operational device is available from the samples tested so far.

Devices have been exercised at clock rates up to 2.5 GHz, although the results presented here are at 1.5-GHz clock. These initial measurements were carried out on a chip in a standard 68-pin LCC package mounted in a test jig.

Fig. 5 is a photograph of the spectrum analyzer displaying the fundamental of a triangular-wave output at 70 MHz. The very low level of spurious lines can be clearly seen, in this case primarily attributable to the clock source. The frequency chosen is one of the most favored ones in terms of spurious output, it is exactly one quarter of the clock frequency, and hence would not be expected to produce spurs. Less favored frequencies occur, and can be predicted by simulations. A general theory for spurs is given in [4]; a simplification is to assume the conventional (for ADC's) rule of $6N$ dB per bit of the DAC, where N is the number of effective DAC bits, i.e., the number of bits to which the DAC settles in one clock period. This figure is then modified by the effect of "oversampling," i.e., the number of successive accumulator outputs which lie within a single output frequency cycle. The effect of this is to add

Maximum clock rate	>2.5GHz	
Output waveform	Sine, triangular, square	
Output phases	I & Q, true and complement	
Output frequency	1Hz to 500MHz	
Channel spacing	1Hz	
Close to carrier noise	-135dBc/Hz at 10kHz	
Maximum spur level		
30 MHz output	-55dBc	
50MHz output	-55dBc	Sinewave output
500MHz output	-30dBc	
Chip size	4mm x 5mm	
Transistor count	5000	
Power consumption	2W to 5W, mode dependent	

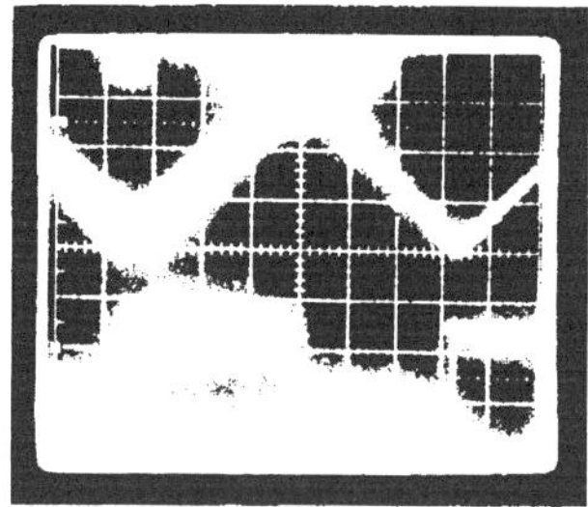

Fig. 6.   Square and triangle at 15-MHz output frequency.

a further 3 dB per octave of excess (over the minimum Nyquist rate) of the clock. In this particular example, this issue is slightly obscured by the requirement for a quadrature output presentation, so that 2 GHz represents the lowest clock frequency which can produce a 500-MHz output, i.e., a factor of 3 dB is lost compared to the simplest case. This oversampling has the effect of increasing the DAC resolution; it cannot improve the inherent accuracy of the DAC's. However, typical DAC's on this process are expected to show 9- to 10-bit linearity, so spur levels, at least at low frequencies, should show at -54 to -60 dBc. Close to carrier noise is a more difficult topic. Literally the noise terms, say at 10 kHz from a carrier of 1/4 clock frequency, are expected to be very good, around -135 dBc, based on the earlier device. More realistically, spurs of the levels described above are likely to form the major limitation to close to carrier noise floor in the general case.

Using the above factors, it is possible to estimate the worst-case output spectral purity at a number of output frequencies, always assuming that the frequencies are not subharmonically related to the clock and therefore having very low spurs. Full assessment of the devices in respect to spurs has not yet been carried out, but it seems from the preliminary results that the devices will more than meet the targets set out in Table I. Fig. 6 is an oscilloscope photograph of the square and triangular outputs at 15 MHz, while Fig. 7 shows the quadrature relationship of two triangles, also at 15 MHz. Fig. 8 shows the square and triangle outputs at a frequency of 375 MHz.

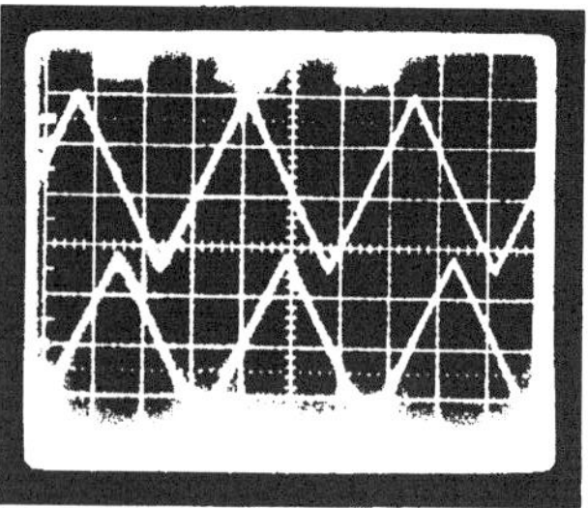

Fig. 7.   Quadrature triangles at 15 MHz.

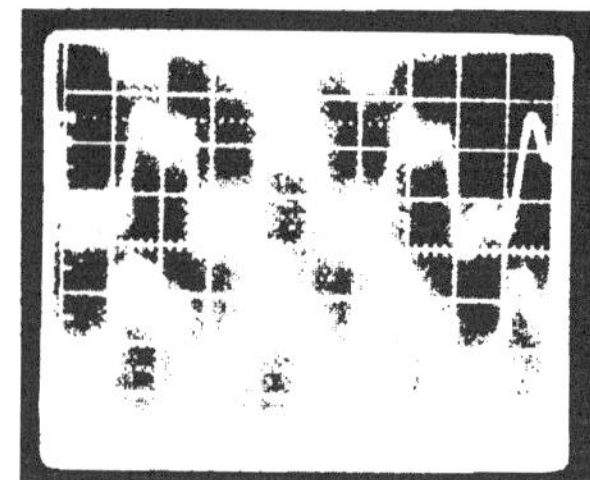

Fig. 8.   Square and triangle outputs at 375 MHz.

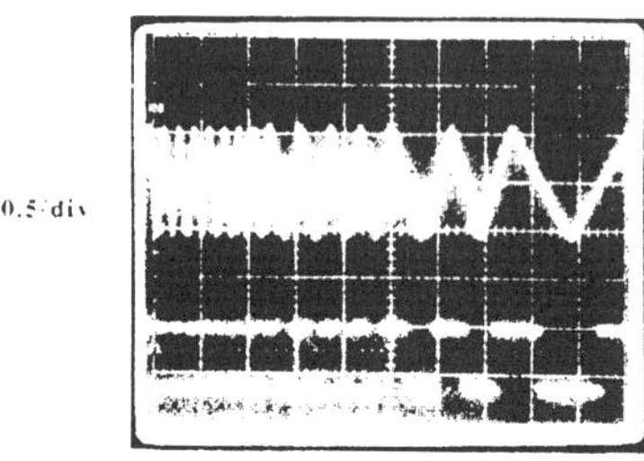

Fig. 9.   Phase-coherent frequency stepping.

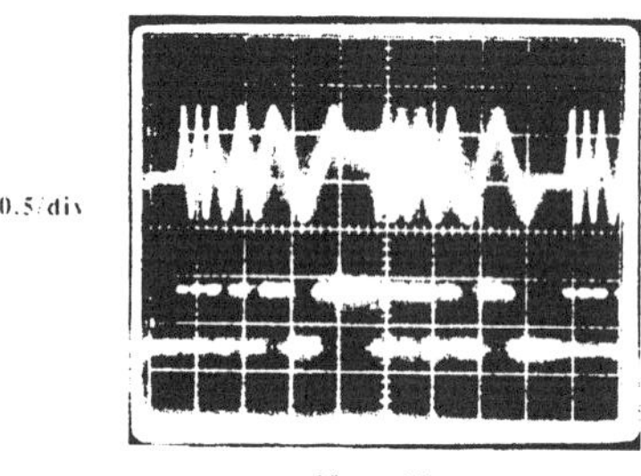

Fig. 10.   Chirp generation.

At lower frequencies, the various waveform generation possibilities are illustrated in Figs. 9, 10, and 11. These show, respectively, phase-coherent frequency stepping, chirp generation, and two-frequency burst generation. The limitation to the frequency shift rate is the rate at which the accumulator can be updated; all stages are pipelined, so that a total update time of 20 ns is required for the full accumulator. This represents a limit to the rate at which

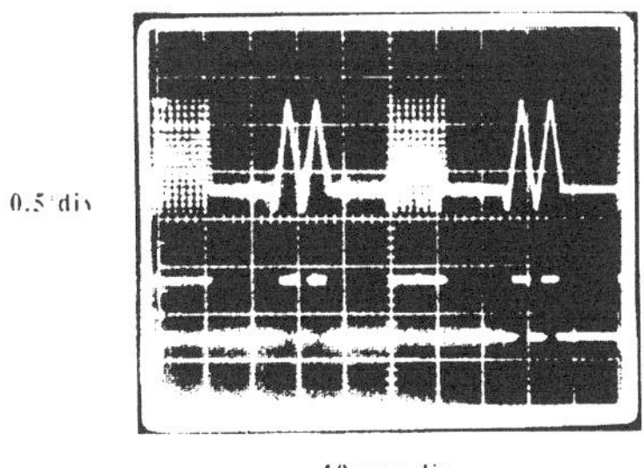

Fig. 11.   Two-frequency burst generation.

ACKNOWLEDGMENT

The authors acknowledge the efforts of M. Mudd for early discussions on the design, M. Golder and T. Ward for the bulk of the design work, and C. Powell and B. Barber for useful discussions on built-in test features and for test and evaluation.

REFERENCES

[1] V. Manassewitch, *Frequency Synthesis and Design*. New York: Wiley, 1980, pp. 37–49, 494–501.
[2] P. H. Saul and M. S. J. Mudd, "A direct digital synthesizer with 100-MHz output capability," *IEEE J. Solid-State Circuits*, vol. 23, no. 3, pp. 819–821, June 1988.
[3] P. H. Saul, P. J. Ward, and A. J. Fryers, "An 8-bit, 5-ns monolithic D/A converter subsystem," *IEEE J. Solid-State Circuits*, vol. SC-15, no. 6, pp. 1033–1039, Dec. 1980.
[4] H. T. Nicholas and H. Samueli, "An analysis of the output spectrum direct digital frequency synthesizers in the presence of phase accumulator truncation," in *Proc. 41st Annual Frequency Control Symp.*, May 1987, pp. 495–502.

discrete output frequencies can be selected. Beyond this, the output will sweep between successive frequency selections. Internal deskewing latches, visible on the chip photograph, ensure that data presented to the ROM maintain parallelism during such changes. This limitation also applies to digital modulation of the output frequency. No provision for any form of amplitude modulation has been made, but digital frequency modulation is straightforward and inherently very linear up to the limit imposed by the pipelining, i.e., 20 ns between successive frequencies, or 50-MHz maximum deviation.

V.   Conclusions

This paper has described a new class of UHF frequency synthesizer of unprecedented performance. The simplicity of operation, coupled with the complete absence of any "setting up" or alignment procedures, make this class of device very attractive. In some applications, the extreme speed of operation can make possible products which cannot be realized in any other way. Particular examples of this are in fast-hopping signal generation for EW and radar. Also, the device, when fully operational, will provide almost all of the functions for a multi-mode signal generator operating over the 1-Hz to 500-MHz frequency range. Predicted device parameters are included in Table I.

Single Chip 500 MHz Function Generator

P.H. SAUL, W. BARBER, D.G. TAYLOR, AND T. WARD

Indexing terms: Integrated circuits, Digital circuits, Waveform synthesis

Abstract: The paper describes a single-chip, digitally synthetised, analogue function generator. The device uses the 'direct frequency synthesis' (DFS) approach. It is expected to find applications in instrumentation, radar and electronic warfare. It has a capability for the generation of sine, triangle and square waves from 1 Hz to 500 MHz, with a minimum frequency increment of 1 Hz.

1 Introduction

This paper describes a single-chip, digitally synthetised, analogue function generator. Its specification is given in Table 1. Included on the chip are built-in test facilities for both its digital and analogue parts. The chip has an approximate gate-equivalent complexity of over 2000 gates, with 896 bits of dual-port ROM. All gates on the chip operate at the full clock speed of over 2 GHz, and both the ROM access time and the DAC settling time are better than 400 ps. All of these measures of performance

Paper 7679G (E3, E10), first received 24th January and in revised form 30th August 1990

Dr. Saul is with the GEC-Marconi Materials Technology Group, Plessey Research Caswell Limited, Caswell, Towcester, Northamptonshire NN12 8EQ, United Kingdom

Mr. Barber and Mr. Ward are with Plessey Semiconductors, Caswell, Towcester, Northamptonshire, United Kingdom

Mr. Taylor is with Plessey Semiconductors, Cheyney Manor, Swindon, Wiltshire, United Kingdom

Table 1: Specification

Maximum clock rate	>2 GHz
Output waveforms	Sine, triangle, square
Output frequency	1 Hz–500 MHz
Step size	1 Hz or any multiple
Spurious level	< -48 dB
Chip size	4 mm × 5 mm
Transistor count	5000
Power	2 W to 5 W, mode-dependent

exceed those previously reported, both in the literature (References 1 and 2) and in product data sheets.

2 Concept

The concept of direct frequency synthesis has been known for many years [3]. The advent of extremely high performance bipolar processes makes it possible to turn the technique from a low-frequency speciality into a realistic means for the generation of very wide frequency-range synthetised signals from an essentially very simple, but fast, chip. Direct frequency synthesis can be achieved in a conceptually elegant, all-digital chip. In the example given here, the only analogue components are those associated with the output digital-to-analogue convertors (DACs). The device described is a general-purpose laboratory function generator, and includes sine, triangle and square waves, with true, complement, in-phase and quadrature outputs.

3 Functional description

The block diagram is shown in Fig. 1. The frequency is set by a 30-bit binary word. Parallel loading is used to

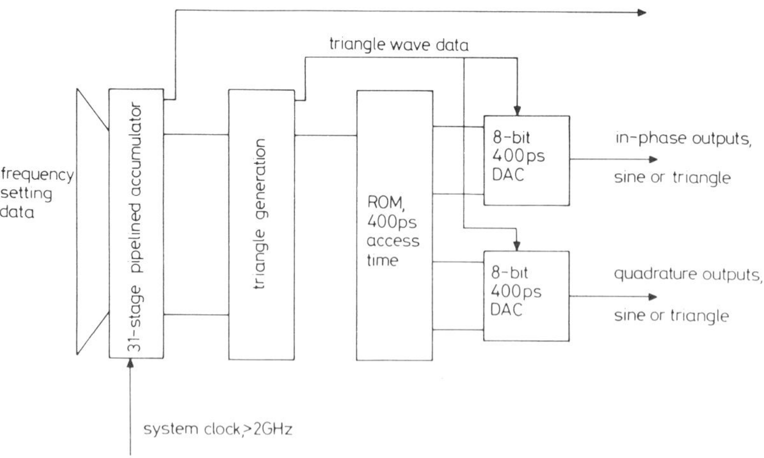

Fig. 1 *System block diagram*

maximise loading speed. One bit of the 32-bit accumulator is shown diagrammatically in Fig. 2. It consists of an input latch, level-shift stage, a full sum/carry adder and

30 fF and collector-base capacitance of 14 fF. The layout is designed around a standard height cell structure, but is essentially fully custom. A typical cell, that of the circuit

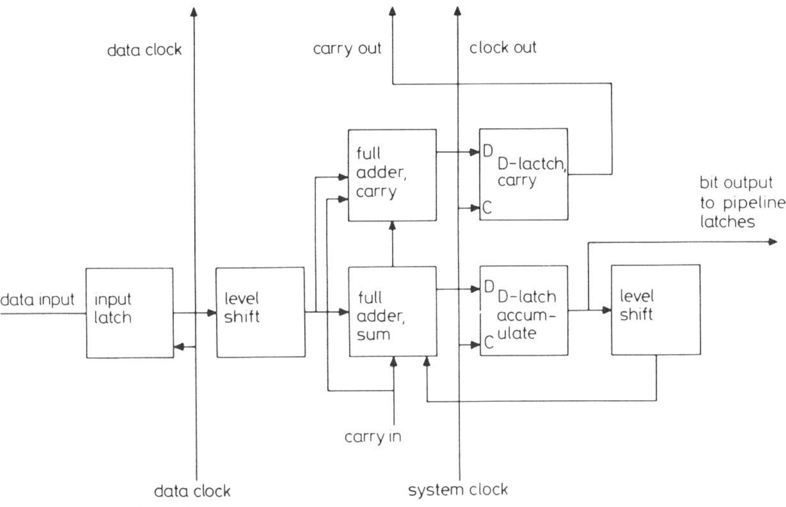

Fig. 2   *Accumulator bit*

two latches, one for accumulation and one for the pipelining of the data to the next stage of the accumulator. Fully differential, low voltage swing (250 mV) logic is used throughout the device [4]. An example is the half-adder circuit shown in Fig. 3. Transistors implemented in the one-micrometre self-aligned bipolar process show peak F_ts of 22 GHz. Under the bias conditions defined by the circuits used, F_t is 15 GHz. Base resistance is typically 220 Ω, with a collector-substrate capacitance of

of Fig. 3, is shown in Fig. 4. This cell, a half adder, is 117 micrometres by 56 micrometres in area, and uses 0.3 mA from the 4.5 V supply.

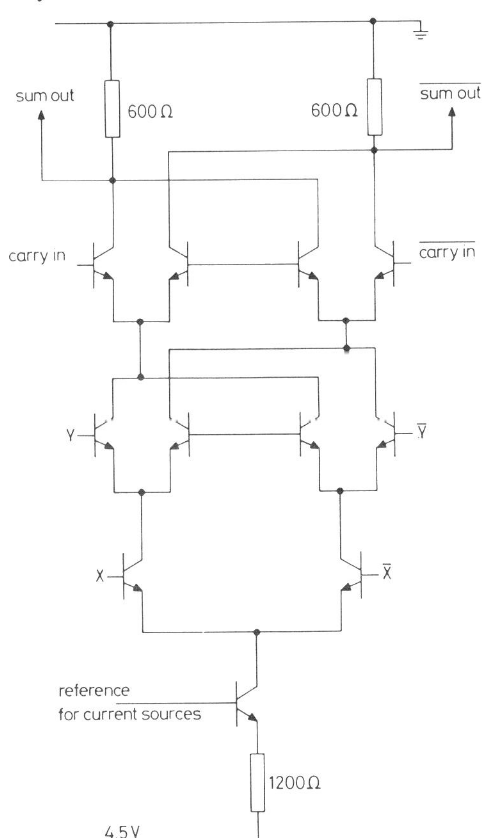

Fig. 3   *Half-adder circuit*

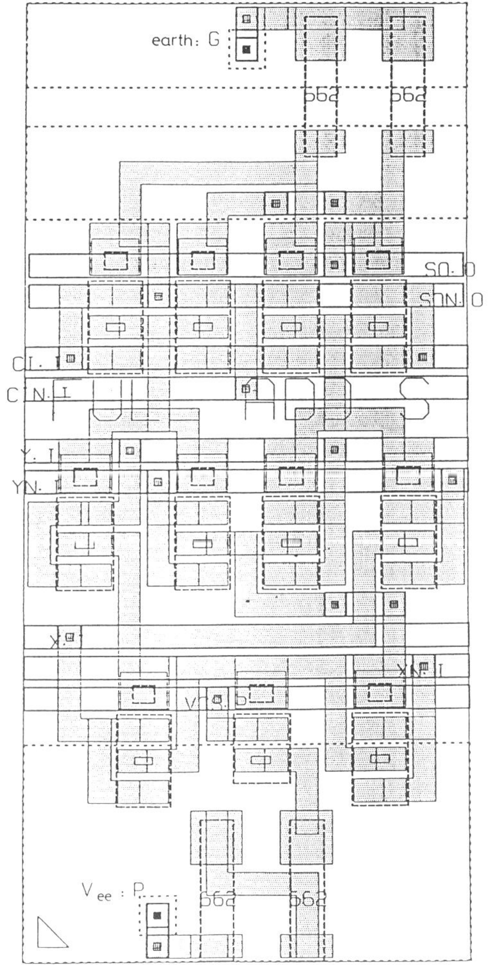

Fig. 4   *Half-adder layout*

Pipeline latching of the accumulator is used, which, with deskewing latches, provides a timed parallel digital representation of a sawtooth waveform. Pipelining of the data input to the input latches was omitted, since it is only of significance during the period of frequency transition. This does, however, introduce both a frequency-modulation limit (50 MHz) and a small hesitation in the takeup of the new frequency. This can be seen in Fig. 8 as a 3 ns delay between the output waveforms. The most significant bit (MSB) from the accumulator can be used as the square wave output. Alternatively, by the use of the bit complements of the LSBs in an array of exclusive OR gates, a triangle output can be derived, in digital form. This is illustrated in Fig. 5. The triangle is then fed

this leads to increased ROM access time, and so defeats the objective. The ROM size chosen was 896 bits, arranged as $2 \times 64 \times 7$ bits, which, with amplitude and phase inversion, is equivalent to 8 amplitude bits and 256 time increments, i.e. it is square. There is a strong case for a doubling, at least, of the number of time increments, but SPICE-predicted access times have precluded this option.

Although this device has a level of spurious signals that would make it unacceptable as a local oscillator in a high-performance single-frequency radio receiver, it has a frequency-hopping performancce that it about five orders of magnitude faster than that of a loop synthesiser, a valuable characteristic for radar and EW. Practical limits

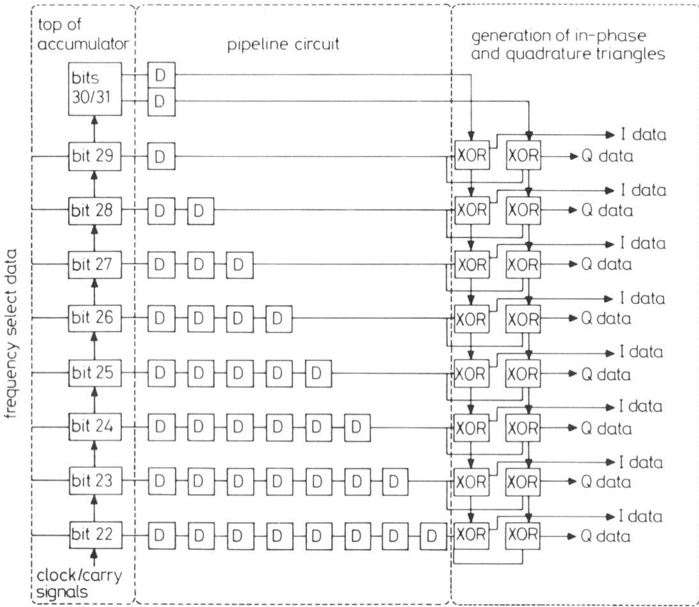

Fig. 5 *Accumulator, pipelining and triangle generation*

around the ROM, directly into the output DACs. In the third mode of operation, that for a sine wave, the triangle data is fed to the dual-port ROM, which contains two quadrature-related waveforms, sampled over a 90° space. Amplitude and phase inversions complete the 360° of the required output. This is not intended to save ROM chip area, but to enhance operating speed. Data entering the DACs is passed through an array of registers that can be switched to feed the data out serially, for functional checking of the digital parts of the circuit, or can be loaded with data to test the DACs.

4 Limitations and spurs

The limitation of a synthesiser of this type comes primarily from the truncation of the output frequency word, in both the ROM and the DAC. The provision of a relatively long (30-bit) phase word avoids problems from this source. Because very fast DACs are limited in achievable accuracy, they are the major limitation in this design. The effect of truncation of the digital word is the production of spurious output spectral lines. The magnitudes and spectral positions of these lines are described theoretically in References 5 and 6. The ROM design is also a compromise. A very large ROM would ideally be used, to give the very best representation of a sinewave, with a large number of time increments. However, in practice,

are imposed by the technology on the accurate word length that can be achieved. In addition, it should be realised that the critical parameter is the DAC accuracy achieved within the permitted settling time. At very high clock speeds, this is necessarily very short. There is a compromise between a very fast DAC of limited resolution, as here, at eight bits, and a much slower DAC of, say, 12 bits or more. As the latter device will almost certainly need some form of post-process trimming, it will inevitably be slow, and, hence, will severely limit the maximum output frequency available. In most IC processes, eight to ten bits represents the limit of inherent component matching accuracy on-chip.

The spurious peaks generated by the truncation depend on the exact frequency selected in relation to the clock frequency, but, in the worst case, are $6N$ dB below the output centre frequency, where N is the number of effective DAC bits. A very detailed analysis is available in Reference 5, with additional information in Reference 6, but the conclusion remains the same. In addition, there is a small advantage from multiple pulses per output cycle, to a limit of 3 dB per octave of 'oversampling' of the output by the clock. In a recent paper [7], it is shown that a more correct version relates N to the accuracy of the DAC, i.e. oversampling can be of some value, say 6 to 8 dB, but not, of course, at the extreme of output frequency.

This is closely analogous to the effect of true over-sampling in an ADC, indeed the mathematics is identical, with the same limitation that the added resolution gained from oversampling is limited by the absolute accuracy of the DAC. Typically, this would be expected to be about 9 to 10 bits, and so an output spurious level at low frequencies of -55 to $-60\,\mathrm{dB}$ is expected and obtained. The DAC design follows that of earlier work in this field [8]. The technique uses an array of current sources, arranged to minimise the effects of processing gradients across the device. The use of very high-speed transistors, as in this circuit, gives the advantage of speed. Since the circuit relies on current switching into external $50\,\Omega$ load resistors, there are no substantial voltage settling time problems to overcome. The output voltage swing is limited to $0.7\,\mathrm{V}$ below the earth rail. All outputs from the two DACs are used, to give true, complement, phase and quadrature output waves. The fully balanced arrangement reduces the number of power-rail transient induced 'glitches'.

5 Results

A photograph of the chip is shown in Fig. 6. The devices have been tested, primarily, at $2\,\mathrm{GHz}$, the design fre-

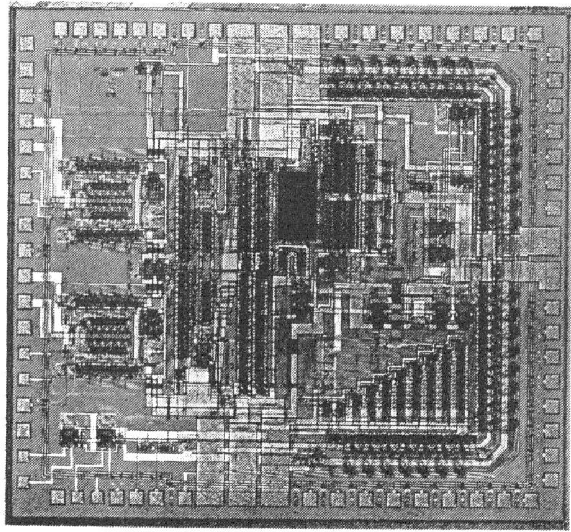

Fig. 6 *Chip photograph*

quency. Fig. 7a shows the output waveform and Fig. 7b shows the spectrum at $500\,\mathrm{MHz}$ output frequency, the maximum for the $2\,\mathrm{GHz}$ clock in the I/Q system. Fig. 8 shows the output frequency hopping from $25\,\mathrm{MHz}$ to $8\,\mathrm{MHz}$, with a delay of approximately $18\,\mathrm{ns}$. There is a further small delay of about $3\,\mathrm{ns}$ between the frequencies, as the new frequency data steps through the input register. DAC glitch levels are illustrated in Fig. 9. The glitch amplitude is $2.5\,\mathrm{mV}$ in a full scale (after attenuation) of $60\,\mathrm{mV}$. If we correct this to a full scale output of $1\,\mathrm{V}$, this is $41\,\mathrm{mV}$ for $1.6\,\mathrm{ns}$, i.e. $33\,\mathrm{psV}$, essentially equal on all edges. DAC settling time is targetted at $400\,\mathrm{ps}$. The oscilloscope measurements are primarily limited by parasitics of the measurement system. The devices will update at over $2\,\mathrm{GHz}$, although spurious performance deteriorates above $1.5\,\mathrm{GHz}$, suggesting that, at least on the sample tested, the settling time is better than $700\,\mathrm{ps}$ to 8 bits. A better test fixture would probably improve on this. At a nonideal frequency, about $225\,\mathrm{MHz}$, (Fig. 10) spurious peaks rise to the expected $-48\,\mathrm{dB}$ level (at $1\,\mathrm{GHz}$ clock frequency).

The low-frequency output waveforms are shown in Figs. 11a, b and c. Some amplitude jitter is evident from the DAC outputs, and rather more from the square wave,

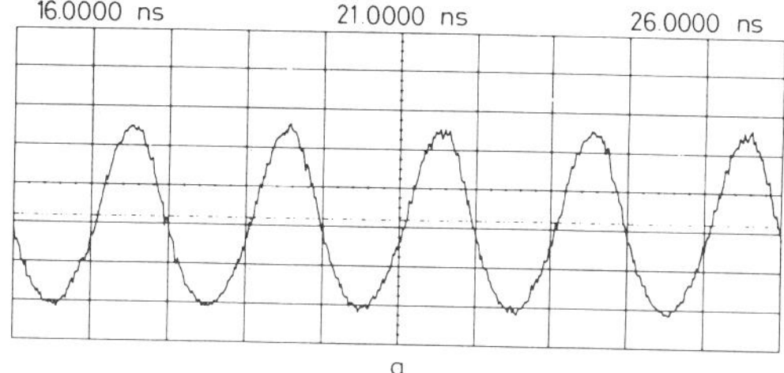

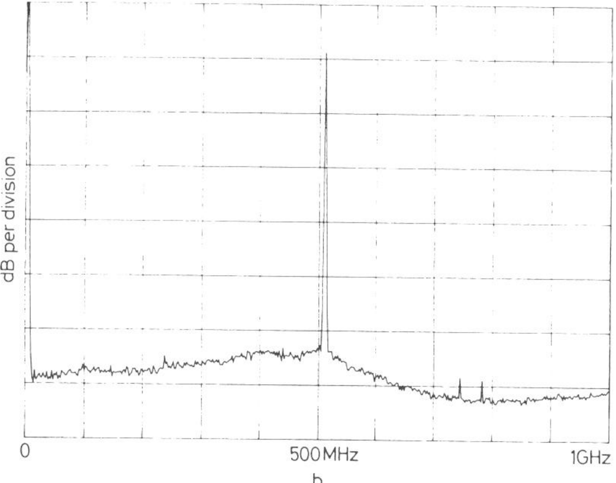

Fig. 7 *Output at 500 MHz with 2 GHz clock frequency*

a Waveform, amplitude = 10 mV/div, timebase = 1 ns/div, offset = -33.25 mV, delay = 16 ns, frequency = 501.102 MHz
b Spectrum

which does not come from the DACs. The source of this is probably decoupling deficiencies on the test fixture. These plots originate from a $20\,\mathrm{GHz}$ sampling oscilloscope, with no filtering, so all details are visible. The output frequency was $1/128$ of that of the clock, although the waveforms at these frequencies showed no visible deviation from ideal for nonintegral submultiples of the clock frequency. Performance is summarised in Table 1.

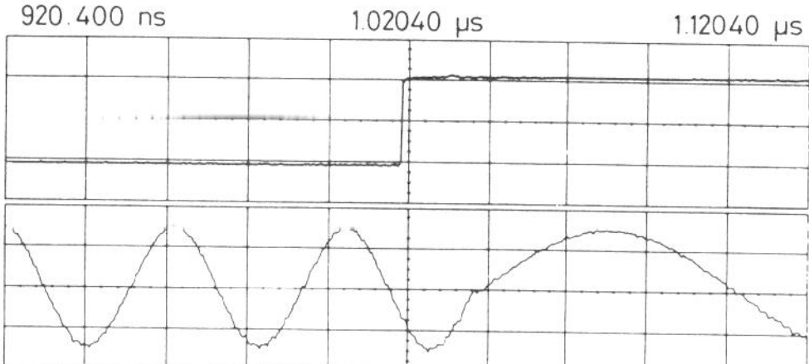

Fig. 8 *Frequency hopping from 25 to 8 MHz*
Channel 1 = 40 mV/div, channel 2 = 20 mV/div, timebase = 20 ns/div

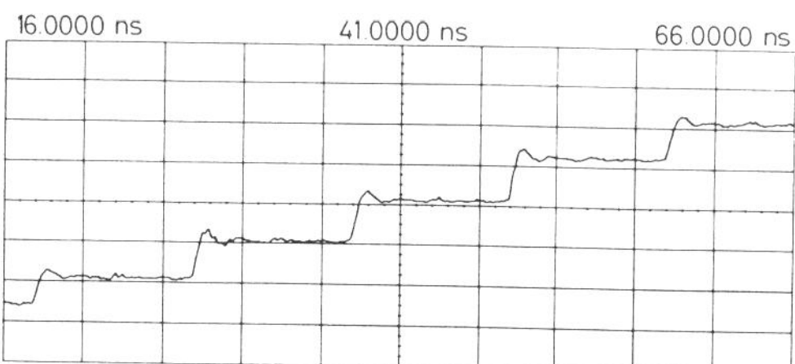

Fig. 9 *DAC glitch illustration*
10 mV/div, offset = -29.62 mV, timebase = 5 ns/div, delay = 16 ns

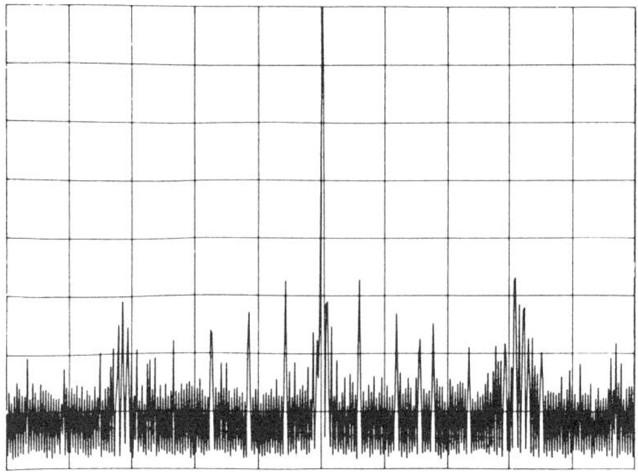

Fig. 10 *Output spectrum at 225 MHz with 1 GHz clock*

Centre frequency = 225 MHz, horizontal scale = 10 MHz/div, vertical scale = 10 dB/div

6 Conclusions

This paper has described a 500 MHz monolithic function generator. The devices described meet expected performance parameters. Applications for this class of device are in instrumentation, radar and EW. Further developments of the technique that make use of very high performance processes and innovative design may ultimately replace phase-locked loops as the general-purpose synthesiser.

7 Acknowledgement

This project was partially funded under the Alvey programme.

8 References

1 SUNDERLAND, D.: 'CMOS/SOS frequency synthesiser LSI circuit for spread spectrum applications', *IEEE J.*, 1984, **SC-19,** (4)
2 GIEBEL, B., LUTZ, J. and O'LEARY, P.L.: 'Digitally controlled oscillator', *ibid.*, 1989, **24,** (6), pp. 640–645
3 MANASSEWITCH, V.: 'Frequency synthesis and design' (Wiley, New York, USA, 1980) pp. 37–49 and pp. 494–501
4 SAUL, P.H., and TAYLOR, D.G.: 'A 2 GHz direct frequency synthesiser'. 1989 Symposium on VLSI Circuits, Digest of Technical Papers, Kyoto, Japan, pp. 89–90
5 NICOLAS, H.T., and SAMUELI, H.: 'An analysis of the output spectrum of direct digital frequency synthesisers in the presence of phase-accumulator truncation'. Proceedings of 41st IEEE Annual Frequency Control Symposium, pp. 495–502
6 MEHRGARDT, S.: 'Noise spectra of digital sine-generators using the table look-up method', *IEEE Trans.*, 1983, **ASSP-31,** (4), pp. 1037–1039
7 SAUL, P.H., and TAYLOR, D.G.: 'A high speed direct frequency synthesiser', IEEE J., 1990, **SC-25,** (1), pp. 215–219
8 SAUL, P.H., WARD, P.J., and FRYERS, A.J.: 'An 8 bit, 5 ns monolithic D/A converter subsystem', *IEEE J.*, 1980, **SC-15,** (6), pp. 1033–1039

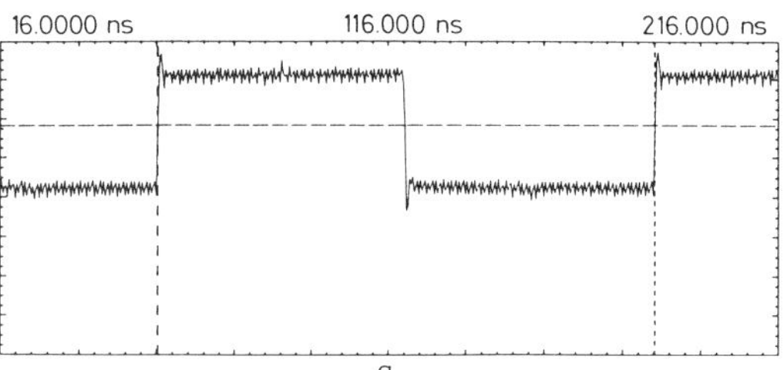

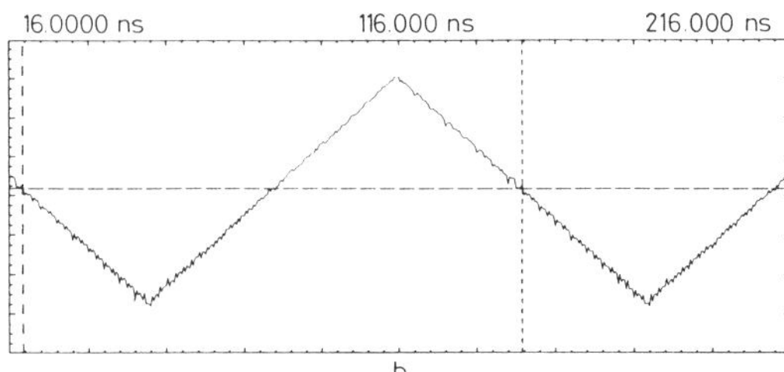

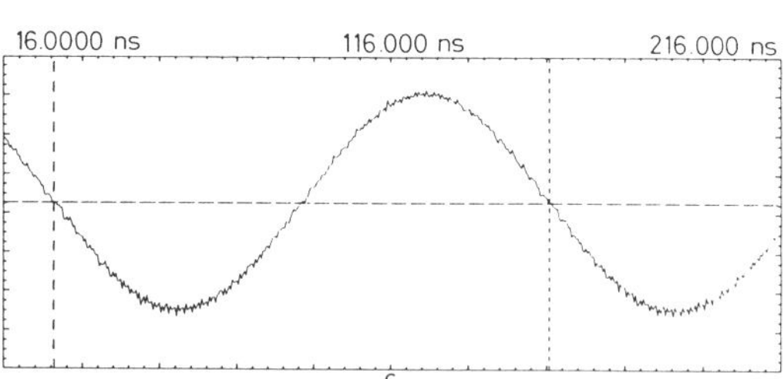

Fig. 11 *Output waveforms at 7.8 MHz*

a Square
b Triangle
c Sine
10 mV/div, offset = −30.25 mV, timebase = 20 ns/div, delay = 16 ns, frequency = 7.81172 MHz

Low-Latency, High-Speed Numerically Controlled Oscillator Using Progression-of-States Technique

MATTHEW THOMPSON

Abstract—A progression-of-states design technique for numerically controlled oscillators (NCO's) has measured a decrease in 16-b accumulator latency to two clock cycles for a 130-MHz device in a 1.0-μm CMOS gate array. An enhanced version derived and simulated directly from the fabricated device has been projected to accumulate at 267 MHz.

I. Introduction

IN THE LAST few years the role of digital techniques in frequency synthesis has been increasing. In the past, analog phase-locked loop (PLL) systems were primarily used due to their high output frequency capabilities and spectral purity [1]. Analog PLL techniques, however, exhibited a very small operating frequency range (typically less than one octave) [2], have had long frequency tuning times (1 ms) [3], and have not been easily produced (each unit requires special setup for proper operation). Digital synthesis techniques have solved these problems but, in the past, have been expensive to implement. With increasingly large die sizes used in ASIC technology, the cost of implementing digital frequency synthesizers has dramatically decreased. Digital frequency synthesis systems are now routinely incorporated into larger systems designs on ASIC chips [4].

The advantages of direct digital synthesis (DDS) systems are well known [5] and include fast switching times, smooth frequency transitions, and fine frequency steps. Direct digital synthesis systems convert a digital frequency setting word into an analog frequency. When the frequency setting word is very wide, a large number of frequencies can be generated (2^N) with very fine frequency resolution ($f_c/2^N$). By continuously accumulating the frequency setting word, the accumulator periodically overflows which generates a frequency at the output. The rate of overflow dictates the synthesized output frequency of the system.

In DDS systems, the numerically controlled oscillator (NCO) is directly limited by the accumulator speed [6]. Using the progression-of-states approach described in this paper, CMOS accumulators have been proven to operate at 130 MHz with two-clock-cycle latency. Simulated results indicate a 237-MHz five-clock-cycle latency NCO to be easily realizable.

Manuscript received February 27, 1991; revised July 22, 1991.

The author is with Interstate Electronics Corporation, Anaheim, CA 92803.

IEEE Log Number 9103861.

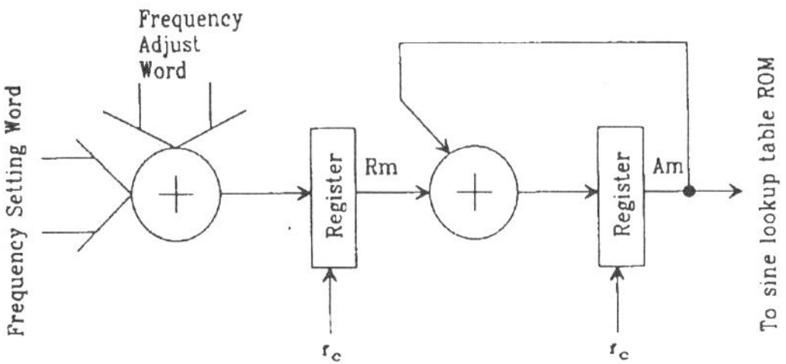

Fig. 1. Numerically controlled oscillator.

II. Background

In DDS systems, the accumulator is the main element of an NCO. A block diagram of an NCO is shown in Fig. 1. This oscillator, under numeric control, generates output frequencies based on the equation below:

$$f_o = R_n \ (f_c/2^N) \tag{1}$$

where

f_o output frequency of the NCO,
f_c accumulator clock frequency,
N number of bits in the accumulator,
R_n frequency setting word + frequency adjust word.

Equation (1) is derived from computing the coordinates of 2^N equispaced points around the unit circle in the z plane [7]. These coordinates are computed using the expression $\exp[j(2\pi/2^N)kR_n]$, where k is the time index analogous to f_c. The optimal implementation to generate f_o in (1) is an accumulator function whose output is given by

$$A_m = \left[\sum_{n=0}^{m} R_n \right]_{\mathrm{mod}\ 2^N} \tag{2}$$

where

A_m output state of the NCO,
R_n frequency setting word + frequency adjust word.

Typically, the frequency setting word is held constant and the frequency adjust word is a dynamic signal that is used as a modulator [8] or to close a feedback path for phase-locked systems.

In Fig. 1, the speed of the accumulator of an NCO is limited to the speed of the N-bit adder. Pipelining techniques are currently used to increase the speed of the accumulator [2], [9], [10]. These techniques, while increasing the speed of the accumulator, have also significantly increased accumulator latency.

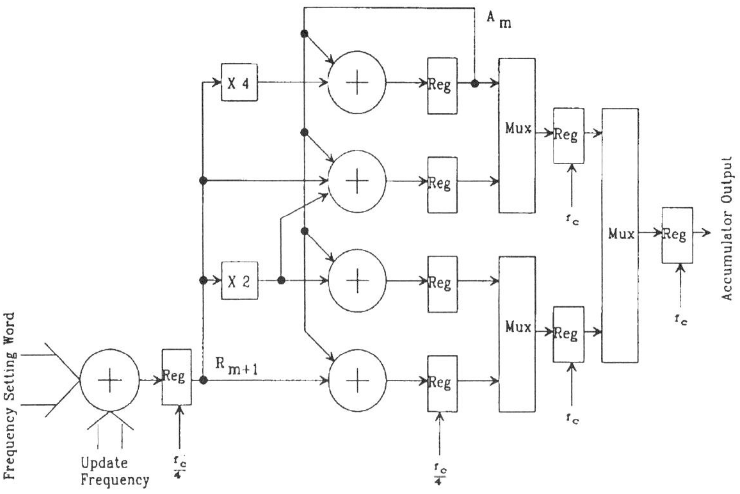

Fig. 2: Progression-of-states NCO implementation.

III. DESIGN AND IMPLEMENTATION

To increase the speed and reduce the latency of NCO accumulators, the output characteristics of an NCO were observed. Instead of considering more intensive pipelining schemes, a new approach was taken that concluded the NCO output states are additions of previous results, under some constraining conditions. These output states are shown in (3)–(6) and assume that the frequency adjust word is held constant for four accumulator clock cycles. Holding R_m constant for four accumulator clock cycles limits the rate at which the frequency adjust word may be updated but does not cause any imperfections in the accumulated values or the output frequency.

Each state in (3)–(6) is a modulo 2^N addition of R_{m+1} and A_m, where A_m is the previous state of the accumulator. If R_{m+1} is held constant for four accumulation clocks, the next four states may be reduced to the states shown in equations (3)–(6):

$$A_{m+1} = \left[\sum_{n=0}^{m+1} R_n \right]_{\text{mod } 2^N}$$
$$= A_m + R_{m+1} = A_m + R_{m+1} \qquad (3)$$

$$A_{m+2} = \left[\sum_{n=0}^{m+2} R_n \right]_{\text{mod } 2^N}$$
$$= A_{m+1} + R_{m+2} = A_m + 2R_{m+1} \qquad (4)$$

$$A_{m+3} = \left[\sum_{n=0}^{m+3} R_n \right]_{\text{mod } 2^N}$$
$$= A_{m+2} + R_{m+3} = A_m + 3R_{m+1} \qquad (5)$$

$$A_{m+4} = \left[\sum_{n=0}^{m+4} R_n \right]_{\text{mod } 2^N}$$
$$= A_{m+3} + R_{m+4} = A_m + 4R_{m+1}. \qquad (6)$$

To generate state A_{m+2}, consider (7) and (8), which were derived from (2):

$$A_{m+1} = A_m + R_{m+1} \qquad (7)$$

$$A_{m+2} = A_{m+1} + R_{m+2}. \qquad (8)$$

When the frequency word R_m is constant for four clocks, then $R_{m+1} = R_{m+2} = R_{m+3} = R_{m+4}$. Given that $R_{m+1} = R_{m+2}$, (7) may be substituted into (8) to generate (9) to obtain the final form of state A_{m+2}:

$$A_{m+2} = A_m + 2R_{m+1}. \qquad (9)$$

Once these substitutions are made, four consecutive output states of the NCO are then dependent on only two constants, A_m and R_{m+1}. Once A_m and R_{m+1} are stored, the NCO can take the next four clock cycles to produce the output states A_{m+1}, A_{m+2}, A_{m+3}, and A_{m+4}.

Fig. 2 is a block diagram of the implementation used to derive these four states. In Fig. 2, R_{m+1} is multiplied by 1, 2, 3, and 4 and then added to A_m to form the four consecutive states. To generate the A_{m+2} state, R_{m+1} is arithmetically shifted up 1 b to form $2R_{m+1}$ and then is added to A_m in a two-port adder. The arithmetic shift operation is a shift of bits into the adder and, therefore, requires no die area and has delay associated with it. To generate the A_{m+3} state, the $3R_{m+1}$ term must first be factored into easily generated components as shown in (10):

$$A_{m+3} = A_m + 3R_{m+1} = A_m + 2R_{m+1} + R_{m+1}. \qquad (10)$$

Factoring state A_{m+3} into (10) removes the need to multiply the R_{m+1} term by 3. Instead, a three-port adder is used to add the three factored terms in (10). Since the NCO adders are performing modulo 2^N addition, only the N LSB output bits of the three-port adder are of significance. Bit $N + 1$ is ignored. To generate the A_{m+4} state, R_{m+1} is arithmetically shifted up 2 b to form $4R_{m+1}$ and then is added to A_m in a two-port adder. Once state A_{m+4}

		Measured	Simulated			
Number of Adders		2	4	3	2	1
Maximum accumulator clock frequency	(MHz)	130	237	177	130	65
Maximum synthesized clock frequency	(MHz)	25	95	71	52	27
Channel spacing—frequency resolution	(Hz)	1983	3616	2700	1983	991
Accumulator latency	(clock cycles)	2	5	4	2	1
New frequency word adjust rate	(MHz)	25	59	59	65	65
Number of gates (accumulator only)		1684	2581	2214	1684	1317
Area of NCO	(mm^2)		1.98	1.71	1.29	1.02

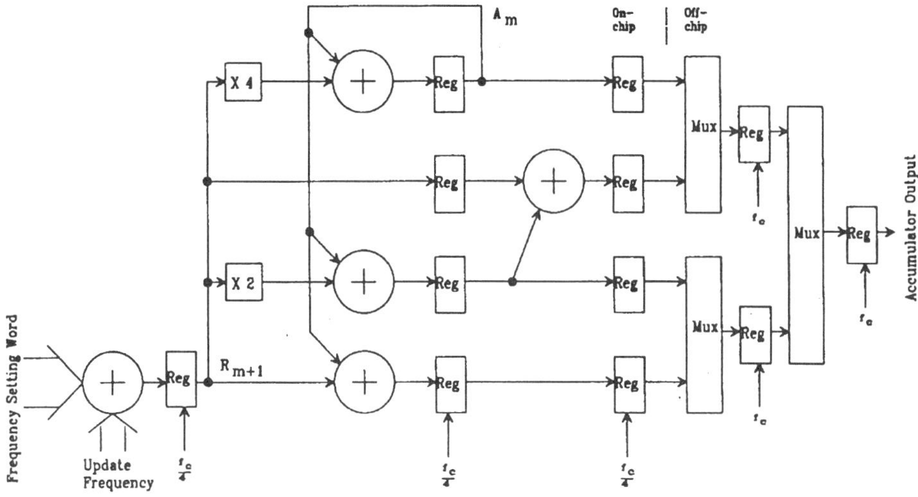

Fig. 3. Enhanced speed NCO implementation.

is generated and stored, it is the most current state and becomes state A_m for the next four clock cycles of the accumulator.

Four N-bit registers store the four generated states. These state registers receive new states once every four accumulation clock cycles ($f_c/4$). Scan registers with feedback are used to avoid gated clocks. Once these four states are stored, they are reconstructed into the valid output states by a recombining circuit that is implemented using 2:1 multiplexers running at the full accumulation rate f_c. On the same clock cycle that the state registers store the four states, a new value of R_{m+1} may be loaded into the NCO. This provides an update frequency rate of $f_c/4$ for the NCO.

These low latency accumulators are capable of operating at 237 MHz. To prove this operating speed, propagation delays for the adders and the 2:1 multiplexers must be observed. All delays are calculated under worst-case operating conditions ($V_{DD} = 4.75$ V, junction temperature = 70°C) for 1.0-μm technology and account for setup and hold times of the surrounding storage registers. The propagation delay for a 16-b three-port adder is 16.86 ns [11]. With four adders generating a state every 16.86 ns, a new state is generated every 4.21 ns (16.86/4). Since

the delay through the multiplexers in the recombine circuit is 3.74 ns [12], the adders limit the speed of the accumulators, yielding an accumulation rate of 237 MHz.

If the full 237-MHz speed of the four-adder NCO accumulator is not required, some of the adders in Fig. 2 may be removed to save area. A trade-off exists that allows a linear decrease in speed for a linear decrease in area used. For example, Table I shows that if an accumulation speed of 130 MHz is required, an NCO with only two adders can be used to meet this speed requirement. A 16-b NCO with two adders uses only 1684 gates whereas the faster four-adder NCO uses 2581 gates.

For implementations where latency is not critical, a faster enhanced accumulator design may be used. This design, shown in Fig. 3, increases the speed of the accumulator to 267 MHz, but also increases the latency to nine clock cycles. This implementation uses the same progression of states principles, but divides the slower three-port adder into 2 two-port add operations at different stages. Essentially, this adds one stage of pipelining to the accumulator. Generating the A_m term in the first stage of the pipeline ensures the accuracy of the accumulator.

In this enhanced implementation, the two-port adders generate the four states in 14.75 ns yielding a new state

TABLE II
NCO SIMULATED PERFORMANCE CHARACTERISTICS FOR ENHANCED
IMPLEMENTATION

NCO Width	16	32	bits
Number of adders	4	4	adders
Maximum accumulator clock frequency	267	234	MHz
Maximum synthesized clock frequency	106	93	MHz
Channel spacing—frequency resolution	4079	0.054	Hz
Accumulator latency	9	9	clock cycles
New frequency word adjust rate	66	58	MHz
Number of gates (accumulator only)	2741	5447	gates
Area of NCO	2.10	4.17	mm^2
Output buffer speed	66	58	MHz

every 3.69 ns. Using this implementation, the states are generated faster than the internal multiplexed registers can reconstruct them. Now the speed of the 2:1 multiplexer including setup and hold times of the registers limits the speed of the NCO. For a 1.0-μm LSI Logic gate array, this speed is 267 MHz. When the four states are reconstructed internally at 267 MHz, standard CMOS output buffers will experience problems transferring data off chip at this rate. To avoid these buffer problems, the four states may be sent directly off chip before the 2:1 multiplexer state recombine circuit. This allows the I/O buffers to operate at a 68-MHz rate. Once the states are off chip, external high-speed logic may then reconstruct the states at 267 MHz.

IV. RESULTS

Table I summarizes the measured and simulated performance parameters for the low-latency 16-b NCO's. Implementations are shown for accumulators using one, two, three, and four adders. Measured results are derived from the two-adder implementation. Table II summarizes the projected performance of the enhanced high-speed NCO's for accumulation widths of 16 and 32 b. The operating speeds of all NCO's in Tables I and II have been proven through timing references in Section III, and by a 10-ps resolution timing simulator. Simulation results indicated that the fabricated 16-b, 130-MHz, two-adder NCO would accumulate at 130 MHz under worst-case conditions. This NCO was proven to accumulate at 130 MHz, as shown in Table I, indicating the simulation results to be very accurate. Simulation and analysis showed a 267-MHz accumulator to be realizable. All parameters for these CMOS NCO's were observed under worst-case operating conditions.

Latency for the fabricated 130-MHz NCO was measured at two clock cycles. The maximum latency for any of the low-latency accumulators was simulated at five clock cycles. This is more than four times better than the typical 17-clock accumulator latency for pipelined designs [13]. Pipelining techniques break the accumulator into smaller adders to reduce the carry-propagate critical path. This technique increases latency by requiring a one clock additional delay for each additional stage of pipe-

lining. Pipelined accumulators typically trade off speed for latency. Using a custom 1.2-μm process pipelined down to the logic-gate level, a 24-b accumulator has demonstrated 700-MHz speed [14]. This device, however, requires an accumulator latency of 55 clock cycles to achieve this speed. Other 32-b full-custom pipelined devices have decreased latency to 13 clock cycles (including sine look-up table) using merged adder carry circuits but with a corresponding decrease in speed to 150 MHz [15].

The progression-of-states technique increases accumulator speed while decreasing latency. This technique reduces the critical path of an accumulator to the delay through a 2:1 multiplexer with minimal latency. A 130-MHz NCO has been proven and tested with two-clock-cycle accumulator latency using the progression-of-states technique.

Test conditions in the target system's laboratory were constrained to the output of the NCO for high-speed tests. A 130-MHz spectral plot of the MSB output of the NCO and sine look-up table is shown in Fig. 4. In this test, the NCO generated samples of a 65-MHz signal at exactly the Nyquist sampling rate, 130 MHz. This system has 8 b of phase resolution into the sine look-up table and 8 b of amplitude resolution into the DAC, giving a theoretical minimum spur level of this system of −48 dBc under ideal operating conditions. This design technique, though proven in CMOS technology, is directly applicable to ECL and GaAs technologies.

V. CONCLUSIONS

A new design technique has been developed and implemented that had decreased accumulator latency and increased accumulator speeds for CMOS NCO's, making a digital solution to frequency synthesis more attractive. The progression-of-states approach offers impressive latency advantages and speed advantages for gate array technology. Using the progression-of-states technique, an accumulator latency of two clock cycles was measured for a 130-MHz fabricated device. This technique allows NCO speeds to approach the speed of a 2:1 multiplexer with five-clock-cycle latencies independent of technology used. Through very accurate simulation and analysis, accumulator speeds have been shown to increase to 267 MHz.

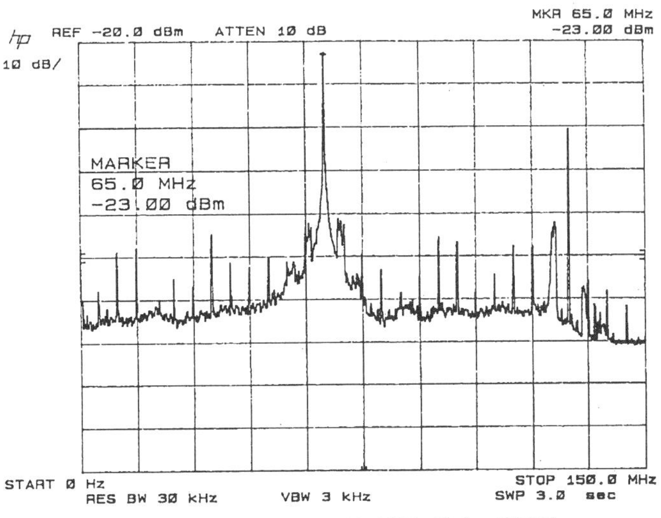

Fig. 4. MSB output of 16-b two-adder NCO with f_c = 130 MHz.

With the advantage of the progression-of-states NCO design technique, the performance of digital frequency synthesis systems can be increased.

REFERENCES

[1] A. Edwin, "Direct-digital synthesis applications," *Microwave J.*, vol. 33, no. 1, pp. 149–151, Jan 1990.

[2] P. Saul and M. Mudd, "A direct digital synthesizer with 100-MHz output capability," *IEEE J. Solid-State Circuits*, vol. 23, no. 3, pp. 819–821, June 1988.

[3] S. Goldman, "Phase locked VCOs for low noise UHF synthesizers," *Microwaves & RF*, vol. 28, no. 8, pp. 99–102, Aug 1989.

[4] M. Thompson and J. Luecke, "Configurable demodulator ASIC for the TDRSS communication system," in *Proc. 1990 IEEE ASIC Seminar and Exhibit*, Sept. 1990, pp. P2-1.1–P2-1.4.

[5] J. Browne, "DDS technology fulfills promise of speed and accuracy," *Microwaves & RF*, vol. 28, no. 9, pp. 153–158, Sept. 1989.

[6] V. Manassewitch, *Frequency Synthesis Theory and Design*. New York: Wiley, 1987, pp. 46–47.

[7] J. Tierney, C. Rader, and B. Gold, "A digital frequency synthesizer," *IEEE Trans. Audio Electroacoust.*, vol. AU-19, no. 1, pp. 48–56, Mar 1971.

[8] E. McCune, "Digital communications using direct digital synthesis," *RF Design*, pp. 39–46, Jan 1990.

[9] B. Giebel, J. Lutz, and P. O'Leary, "Digitally controlled oscillator," *IEEE J. Solid-State Circuits*, vol. 24, no. 3, pp. 640–645, June 1989.

[10] P. Saul and D. Taylor, "A high-speed direct frequency synthesizer," *IEEE J. Solid-State Circuits*, vol. 25, no. 1, pp. 215–219, Feb 1990.

[11] *Compacted Array Series Megafunction Databook*, LSI Logic, Milpitas, CA, 1990, pp. 122–125.

[12] *0.7-Micron Array-Based Products Databook*, LSI Logic, Milpitas, CA, 1990, pp. 2/86.

[13] *The DDS Handbook*, Stanford Telecom, Santa Clara, CA, 1990, pp. 79–90.

[14] F. Lu, H. Samueli, J. Yuan, and C. Svensson. "A 700-MHz 24-bit pipelined accumulator in 1.2 μm CMOS for application as a numerically controlled oscillator," in *Proc. 1991 IEEE Custom Integrated Circuits Conf.*, May 1991, pp. 25.3.1–25.3.4.

[15] H. T. Nicholas and H. Samueli, "A 150-MHz direct digital frequency synthesizer in 1.25 μm CMOS with −90 dBc spurious performance," in *1991 Int. Custom Integrated Circuits Conf. Dig. Tech. Papers*, Feb. 1991, pp. 42–43.

Part IV

Spurious Signals in Direct Digital Frequency Synthesizers

FROM the first applications of DDFS, both designers and users were faced with spurious signals—sometimes very strong and, for many, mysterious (cf. Becker [II-2]).

For the quasiperiodic rectangular-wave output signals, the first reprint paper by Kroupa (1979) provides an exact solution for the spurious phase modulation. Another paper by Kroupa [IV-1], reprinted in Part V, contains guidelines for estimating the amplitudes and frequencies of spurious signals.

Efficient reduction of spurious signals is provided by storing the instantaneous phase in an accumulator and changing it immediately into a sine wave with the assistance of sine lookup tables and digital-to-analog converters (DAC).

Ideally, one might expect that when all stored phase bits are used for generation of the sine wave and an "infinite" precision of lookup tables and DACc is assumed, we would generate a "pure" output sine wave. But even this is not the case because of phase quantization.

The first author who examined the "real-life" spectra of digital sine generators by using the table lookup method was S. Mehrgardt (1983), in the second reprint paper.

Later, in a more rigorous approach, Nicholas and Samueli (1987) tried to estimate the influence of the phase truncation, on the output spectrum of DDFS, by using only W MS-bits for the sine evaluation from the total R-bits. They have found the largest spurious signal to be

$$-6W \quad \text{[dB]} \qquad [1]$$

below the desired output signal

In a review paper by Kroupa [IV-1], published in Part V, the theory of the phase truncation error was simplified, deepened, and completed with the amplitude error due to finite sine word length or to the limitations caused by DAC resolution.

Another important problem in the design of DDFS is the reduction of the necessary memory of the lookup tables. The first steps are to use the MSB as the sign bit and the second MSB as the quadrant bit. This approach for reducing sine lookup ROMs, already discussed by Sunderland [IV-2], is examined in more detail by Nicholas, Samueli, and Kim (1988) in the fourth paper reprinted in this section. In addition, they suggest another, more effective compression scheme by storing the function

$$\sin(\phi) - 2\phi/\pi \qquad [2]$$

The reconstruction is performed with the assistance of an additional adder

$$\sin(\phi) = \left[\sin(\phi) - 2\phi/\pi\right] + 2\phi/\pi \qquad [3]$$

However, difficulties are also associated with the ROM reduction techniques, namely, longer access time due to the additional computations.

Next, we reprint a survey paper about sine-wave reconstruction from the stored phase information published by Vankka (1997).

Nicholas and Samueli (1991) refer to a monolithic CMOS DDFS with -90 dBc spurious performance and clock frequencies above 100 MHz, built on a chip 5×5 mm with power dissipation of 950 mW.

All three papers by Nicholas et al. suggest a simple accumulator extension to keep the background spurious signal level constant (see, e.g., Fig. 4 in the 1991 paper). With the assistance of an additional D flip-flop, they introduce a "one" every second clock pulse into the adder. In this way, the normalized output frequency

$$\xi_x = \frac{k}{2^R} = \frac{X}{Y} \qquad [4]$$

is changed to

$$\xi_x = \frac{2k + 1}{2^{R+1}} \qquad [5]$$

with the result that the numerator and denominator are always relatively prime. However, the effective accumulator capacity is increased by one and the actual output frequency by

$$\frac{1}{2^{R+1}} * f_c \qquad [6]$$

The frequency shift is small, and in the range of acceptable setting error; for a 32-bit accumulator it is nearly 10^{-4} pp million of the f_c. On the other hand, the background noise of DDFS is reduced to its minimum level, as we will see in the article [IV-1] by the editor in Part V.

Generalization of the aforementioned process (i.e., to increase the effective accumulator contents by more than 1 bit) is discussed by Ertl and Baier [IV-3].

In the last paper, Garvey and Babich (1990) introduce an exact spectral analysis of "number controlled oscillators" with the assistance of the Fourier transform and provide experimental measurements of the largest spurious signals as functions of the used phase bits and DAC (D) bits for the normalized output frequency ξ_x in the neighborhood of 1/3. Their results verify the validity of the relation (1) for the amplitudes of the largest spurious signals (note that their $D = W$).

REFERENCES

[IV-1] V. F. Kroupa. "Spectral properties of DDFS: Computer simulations and experimental verifications." Adapted from papers read at the *1993 and 1994 IEEE International Frequency Control Symposia* and the *8th European Frequency and Time Forum 1994,* March 1994. (Reprinted in Part V.)

[IV-2] G. A. Sunderland, R. A. Strauch, S. S. Wharfield, H. T. Peterson, and Ch. R. Cole. "CMOS/SOS frequency synthesizer LSI circuit for spread spectrum communications." *IEEE Journal of Solid-State Circuits,* pp. 497–506, August 1984. (Reprinted in Part III.)

[IV-3] R. Ertl and J. Baier. "Increasing the frequency resolution of NCO-systems using a circuit based on a digital adder." *IEEE Transactions on Circuit and Systems-II: Analog and DSP,* pp. 266–69, March 1996.

Spectra of Pulse Rate Frequency Synthesizers

VĚNCESLAV F. KROUPA

Abstract—First the formula for the output modulation function of pulse rate frequency synthesizers with quasiperiodic omission of pulses is derived, then it is used for the computation of the full output spectrum, and means are shown to find quickly those sideband ranges where the major spurious peaks are to be expected.

INTRODUCTION

Frequency synthesis systems based on subtracting (or adding) pulses from (or to) a pulse train generated by a reference source are not new [1], but their importance has been considered to be only marginal and, as a consequence, their properties have not been discussed in any of the recent books [2]–[4]. The situation has changed with the introduction of the integrated circuits, as these frequency synthesis systems can be assembled from the off-the-shelf circuits.

It was the author of this letter who first suggested that the mathematical model might be the modified Engel series expansion [5], [6]. Another algorithm successfullly tried was the continued fraction expansion [7] and, recently, even the systematic fraction expansions (generally binary additions [8]–[10]).

The difficulty with all the types of the pulse rate frequency synthesizers is a rather poor spectral quality of the output signal. Another problem which is of concern for designers is the position of the major spectral lines in respect to the desired signal, and it is the task of this letter to clarify this situation.

OUTPUT PHASE MODULATION FUNCTION

To keep the output phase fluctuations as small as possible, we have to remove pulses in nearly periodic intervals. As a consequence the phase-time error does not exceed one original period T_i, i.e.,

$$\frac{\varphi_x(t_k)}{\omega_x} = k \frac{\omega_i}{\omega_x} T_i - r T_i \leqslant T_i \tag{1}$$

where $kT_i\omega_i/\omega_x$ is the kth "zero crossing" of the idealized output wave, and rT_i is the nearest actual zero crossing (but always from the same side as is the case in simple systems working in real time). Evidently r must be equal to the largest integer in $(k\omega_i/\omega_x)$.

Manuscript received May 30, 1979.
The author is with the Institute of Radio Engineering and Electronics, the Czechoslovak Academy of Sciences, 182 51 Praha 8, Czechoslovakia.

In order to enhance the fundamental harmonic in the output signal, the rectangular output wave should have the spacemark ratio as close to $1:1$ as possible. This requirement can be met by different means, the simplest of which is to put a flip-flop or binary divider at the end of the synthesizing chain. In this instance both leading and trailing edges of the output wave are synchronized by the, say, leading edges of the original pulse train. Consequently the output phase modulation function (1) can be rewritten with the assistance of the normalized output frequency $f_x/f_i = X/Y$ (both X and Y are relatively prime)

$$\frac{\varphi_x(t_k)}{\omega_x} = T_i \left[k \frac{Y}{2X} - \text{int}\left(k \frac{Y}{2X} \right) \right]. \tag{2}$$

Evidently the fundamental period of the spurious modulation is either $T_i Y$ or $T_i Y/2$ in instances where Y is even (all odd harmonics in respect to $T_i Y$ are missing).

THE OUTPUT SPECTRUM

In the case of the rectangular output wave with an amplitude A and zero crossing in accordance with (1) or (2), the Fourier analysis supplies

$$a_n = \frac{2}{YT_i} A \sum_{k=0,2,4}^{2X-2} \int_{T_i \text{int}\,(k(Y/2X)}^{T_i \text{int}\,(k+1)\,Y/2X} \cos\left(n \frac{2\pi}{YT_i} t \right) dt$$

$$= \frac{1}{\pi n} A \sum_{k=0,1}^{2X-1} (-1)^{k+1} \sin\left[n \frac{2\pi}{Y} \text{int}\left(k \frac{Y}{2X} \right) \right] \tag{3}$$

and similarly

$$b_n = \frac{1}{\pi n} A \sum_{k=0,1}^{2X-1} (-1)^{k} \cos\left[n \frac{2\pi}{Y} \text{int}\left(k \frac{Y}{2X} \right) \right]. \tag{4}$$

Evidently, for evaluating the output spectrum, the use of a computer is inevitable. It is true that the respective software is very simple; however, in instances where X is large, the computation time can be considerable, and at the same time, one may lose orientation in the wealth of data.

APPROXIMATION OF THE MODULATION FUNCTION

We shall start with the shortened version of the continued fraction expansion of a real number ζ_0 arrived at by the modified Euclid's

Reprinted from *Proceedings of the IEEE*, Vol. 67, No. 12, pp. 1680-1682, December 1979.

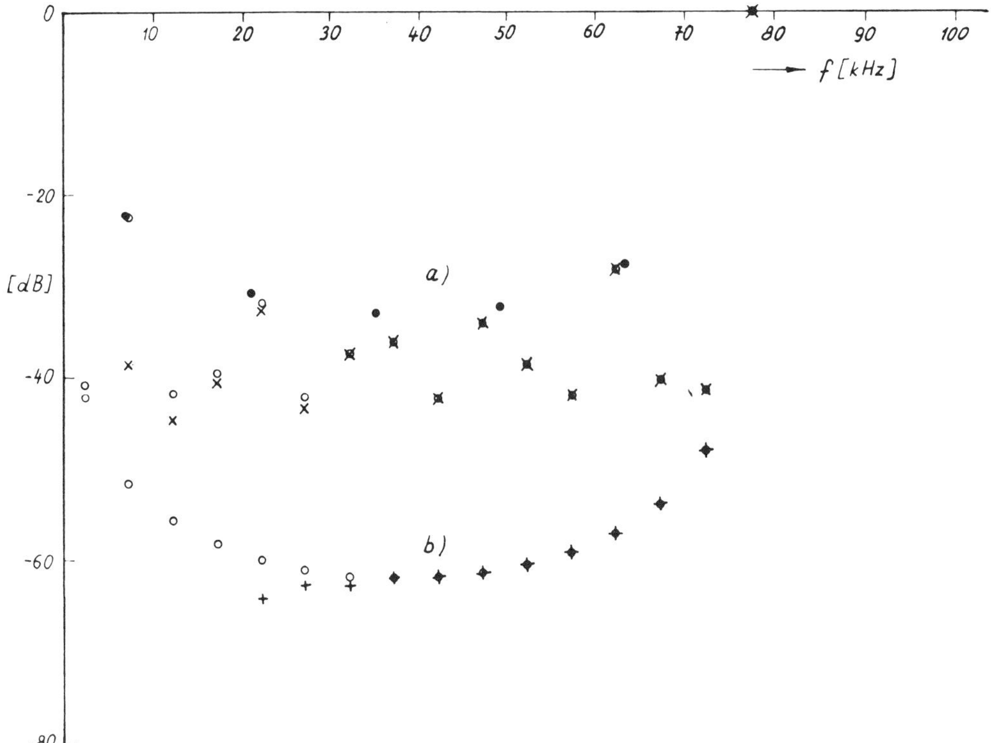

Fig. 1. Synthesis of f_x = 77.5 kHz from the standard frequency. a) 1 MHz (●) first-order (○) second-order approximations. b) from 10 MHz. Crosses indicate experimental results found by Becker [11].

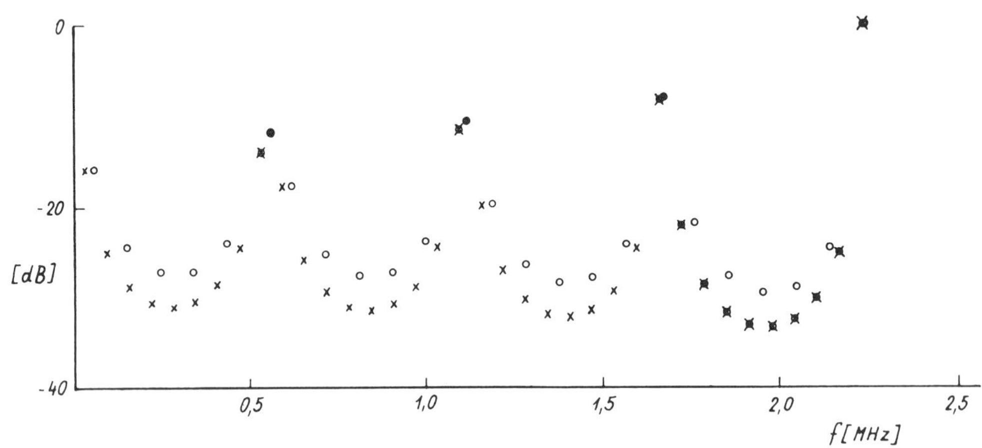

Fig. 2. Synthesis of the color subcarrier; spectral lines found by the first- (●) and second- (○) order approximations and theoretical (x) and experimental (●) results of a truncated modified Engel series expansion.

theorem

$$\zeta_0 = b_0 \pm \frac{1}{\zeta_1} \qquad \zeta_1 \geqslant 2 \qquad (5a)$$

$$\zeta_1 = b_1 \pm \frac{1}{\zeta_2} \qquad \zeta_2 \geqslant 2 \qquad (5b)$$

etc., all $b_0, b_1, \cdots$, being integers.

By combining (5a), (5b), $\cdots$, we get

$$\zeta_0 = b_0 + \cfrac{a_1}{b_1 + \cfrac{a_2}{b_2 + \cfrac{a_3}{\cdot \atop \cdot}}} \qquad (6)$$

$$= A_0 + \sum_{r=1}^{N} (-1)^{r-1} \left(\frac{a_1 a_2 \cdots a_r}{B_{r-1} B_r} \right) \qquad (7)$$

where $a_1, a_2, \cdots$, are equal either to +1 or −1. (See also (6).)

By introducing (7) for $Y/2X$ into (2) we get

$$s(t_k) \equiv \frac{\varphi(t_k)}{\omega_x} = T_i \left[k \left(A_0 + \frac{a_1}{B_1} - \frac{a_1 a_2}{B_1 B_2} + \cdots \right) - \text{int} \left(k \frac{Y}{2X} \right) \right]. \qquad (8)$$

At the first-order approximation of $Y/2X$ by A_1/B_1 the fundamental $s(t_k)$ is a simple sawtooth wave with the amplitude T_i. The second-order approximation by A_2/B_2 reveals another sawtooth wave with the amplitude T_i/B_1 and the new modulation period $A_2 T_i$ which is also imposed on the first sawtooth wave. The same happens with every other approximation.

The above discussion of (8) leads to the following conclusion.

1) There are always present components caused by the sawtooth waves with large amplitudes: T_i, T_i/B_1 etc.

2) To get the knowledge about the approximate distribution of major spectral lines it is sufficient to choose an appropriate approximation A_r/B_r of $Y/2X$ with B_r small enough to keep the computation time in reasonable limits.

66

3) The power level of the spurious components close to the carrier supplies the last sawtooth wave with the amplitude T_i/B_{n-1}; in some cases the last but one approximation should be also considered.

EXPERIMENTAL RESULTS

To demonstrate the disclosed theory we have chosen three examples.

1) The synthesis of the DCF carrier frequency from the standard frequency 1 MHz

$$\frac{f_x}{f_s} = \frac{77.5}{1000} = \frac{31}{400}$$

$$\frac{Y}{2X} = \frac{200}{31} = 6 + \frac{1}{2} - \frac{1}{2 \times 11} - \frac{1}{11 \times 31}.$$

Computed and experimental results are shown in Fig. 1.

2) The synthesis of the same frequency from the standard frequency 10 MHz

$$\frac{f_x}{f_s} = \frac{31}{4000}$$

$$\frac{Y}{2X} = \frac{2000}{31} = 65 - \frac{1}{2} + \frac{1}{2 \times 31}.$$

By comparing both expansions of $Y/2X$ one can see that they are different, and as a consequence, one cannot expect a mere reduction of the spurious signals by 20 dB. This is also demonstrated in Fig. 1.

3) The synthesis of the color subcarrier $f_x = 4433618.75$ Hz from the standard frequency $f_s = 5$ MHz. We have synthesized one half of f_x with the assistance of the modified Engel series expansion

$$\frac{f_x}{2f_s} = \frac{1}{2}\left(1 - \frac{1}{8}\left(1 - \frac{1}{10}\left(1 - \frac{1}{16}\left(1 - \frac{1}{156}\left(1 - \frac{1}{625}\right)\right)\right)\right)\cdots\right)$$

and compared the spectral lines computed by the first- and second-order approximations

$$\frac{Y}{2X} = \frac{800000}{709379} \approx 1 + \frac{1}{8} + \frac{1}{8 \times 47}$$

together with the computed and measured results of the truncated Engel series expansion

$$\frac{f_x}{2f_s} \approx \frac{1}{2}\left(1 - \frac{1}{8}\left(1 - \frac{1}{10}\right)\right).$$

The results are shown in Fig. 2.

REFERENCES

[1] M. W. Williard, and G. F. Anderson, "The use of a fractional Bi-stable multivibrator counter in the design of an automatic discriminator calibrator," *IRE Nat. Conv. Rec.*, Part 5, p. 176, 1959.
[2] V. F. Kroupa, *Frequency Synthesis, Theory, Design and Applications.* London, England, Griffin, 1973; New York: Wiley, 1973.
[3] J. Gorski-Popiel, *Frequency Synthesis: Techniques and Applications.* New York: IEEE Press, 1975.
[4] V. Manassewitsch, *Frequency Synthesizers, Theory and Design.* New York: Wiley, 1976.
[5] V. F. Kroupa, "Approximating frequency synthesizers," presented at Seminario sulla Sintesi di Frequenza (Torino, Italy, 1974).
[6] ——, "Approximating frequency synthesizers," *IEEE Trans. Instrum. Meas.*, vol. IM-23, p. 521, Dec. 1974.
[7] R. Schröder Zur quasiperiodischen Frequenzteilung, *Frequenz*, vol. 30, no. 3, p. 50, 1976.
[8] D. L. Duttweiler, *et al.*, "Analysis of digitally generated sinusoids with application to A/D and D/A converter testing," *IEEE Trans. Commun. Technol.*, vol. COM-26, p. 669, May 1978.
[9] A. D. Wilcox, "Noise prediction for rate multiplier and binary adder frequency synthesizers," *IEEE Trans. Aerosp. Electron Syst.*, vol. AES-14, p. 677, July 1978.
[10] D. G. Messerschmitt, "A new PLL frequency synthesis structure," *IEEE Trans. Commun. Technol.*, vol. COM-26, p. 1195, Aug. 1978.
[11] G. Becker, "Quasiperiodic frequency synthesis," in *Proc. 26th Annu. Frequency Control Symp.* (NJ), June 1972, p. 279.

Noise Spectra of Digital Sine-Generators Using the Table-Lookup Method

SOENKE MEHRGARDT

Abstract—The table-lookup method is a convenient and very flexible way to generate high-quality sinusoidal test signals for measurements in psychoacoustics, speech perception, etc. Exact spectra of such signals, which are of great interest in experiments, are derived, and the relation between spectral shape, table length, and sine frequency is shown.

In experiments on psychoacoustics or speech perception, sinusoidal waveforms are frequently used as test signals. Due to the extreme dynamic range of the human ear (more than 100 dB in certain experiments), these sinusoidal waveforms must be of high quality, i.e., as pure as possible. If adaptive testing procedures are used, not only are the testing strategy and collecting of results controlled by a computer, but even the generation of the test signals itself as well. One of the most flexible ways of generating sinusoidal signals with a minimum of computing power is the table-lookup method (see, e.g., [1], [2]). The exact spectral properties of signals generated by this method are of great interest; it could be troublesome in psychoacoustic experiments if the noise were concentrated in a few spectral lines instead of being distributed over the whole frequency range.

As Fig. 1 illustrates, by itself, computing of spectra does not satisfactorily depict the spectral features; all six examples exhibit spectra of sine functions with nearly equal frequencies ($\frac{1}{16}$ sampling frequency within less than 1 percent). The table length is even held constant in three of the spectra, varying only slightly at the remaining three examples. But nevertheless, the spectral shapes differ considerably! In order to elucidate these problems, we will derive the spectra of sinusoidal signals generated by the table-lookup method.

Consider a general-purpose computer with a digital-to-analog converter running at some constant sampling frequency f_s. If

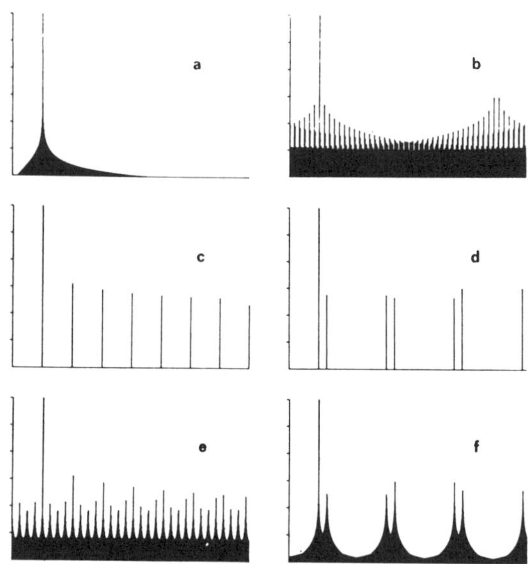

Fig. 1. Spectra of sinusoidal signals generated by the table-lookup method. The frequency scale (horizontal) has a range from 0 to half sampling frequency, and the vertical scale spans a 120 dB range. The table length varies between 1023 and 1027, and the sine frequencies are equal to $\frac{1}{16}$ sampling frequency within less than 1 percent.

one needs a sinusoidal signal of frequency f (with f less than $f_s/2$), the computer has to evaluate the numerical sequence

$$S_n = \sin (\Omega n), \qquad \Omega = 2\pi f/f_s. \tag{1}$$

The argument of the sine function is easily obtained by accumulating the value Ω. At low sampling frequencies the sine function can then be evaluated using a suitable polynomial approximation. But at higher frequencies (or for lowering the computational load) the function must be stored in a table;

Manuscript received June 21, 1982; revised December 6, 1982.

The author is with the Drittes Physikalisches Institut, Universität Göttingen, D-3400 Göttingen, West Germany.

a sine table of length M (usually a power of 2) contains the values

$$T_m = \sin\left(2\pi \frac{m}{M}\right), \qquad m = 0, \cdots, M - 1. \tag{2}$$

Instead of the exact S_n in (1), the table lookup yields the approximation

$$S_n' = T_{[nM\Omega/2\pi]\ \text{modulo}\ M} = \sin\left(\frac{2\pi}{M}\left[\frac{M}{2\pi}\Omega n\right]\right) \tag{3}$$

where the brackets denote rounding to the nearest integer. As a consequence of this approximation, the spectrum of S_n' contains not only energy at frequency Ω, but also noise components.

For simplicity we first consider continuous time functions. In the case of a continuous table-lookup (whatever that is), a sinusoidal time function

$$S(t) = \sin(2\pi f t) \tag{4}$$

is replaced by

$$S'(t) = T_{[Mft]\ \text{modulo}\ M} = \sin\left(\frac{2\pi}{M}[Mft]\right). \tag{5}$$

The bracket can be written as

$$[Mft] = Mft - x(t) \tag{6}$$

where $x(t)$ is a sawtooth of amplitude 0.5 and frequency Mf. For $M \gg 1$, this yields the approximation

$$S'(t) = \sin(2\pi f t) - \cos(2\pi f t)\frac{2\pi}{M}x(t). \tag{7}$$

Thus, $S'(t)$ consists of the pure sine wave and an amplitude-modulated sawtooth. The spectrum of $S'(t)$ is obtained using the Fourier series of the sawtooth $x(t)$ as follows:

$$x(t) = \sum_{k=1}^{\infty} (-1)^k \frac{\sin(2\pi k M f t)}{\pi k}. \tag{8}$$

From (7) and (8) we finally derive

$$S'(t) = \sin(2\pi f t) - \sum_{k=1}^{\infty} \frac{(-1)^k}{Mk}\{\sin 2\pi(kM + 1)ft$$
$$+ \sin 2\pi(kM - 1)ft\}. \tag{9}$$

The spectrum of $S'(t)$ contains two series of spectral lines, both with frequency spacing Mf. One series starts at frequency $(M - 1)f$ and the other starts at $(M + 1)f$, both having amplitudes falling with $1/k$ and phases alternating with k.

Coming back to the time-discrete case described in (1) and (3), one only has to realize that S_n and S_n' are the numerical sequences obtained by sampling $S(t)$ or $S'(t)$ with the sampling frequency f_s. (Replace t by n/f_s to get the result.) We first consider the samples x_n of the sawtooth $x(t)$. Replacing $2\pi f t$ by Ωn in (8) yields

$$x_n = \sum_{k=1}^{\infty} (-1)^k \frac{\sin(kM\Omega n)}{\pi k}. \tag{10}$$

Due to finite precision in real calculations, frequency Ω can certainly be written as

$$\Omega = 2\pi \frac{a}{b}, \qquad \text{with } a \text{ and } b \text{ incommensurable.} \tag{11}$$

Now the sine function in (10) is periodic in k with a period of $\hat{b}$, defined by

$$\hat{b} = \frac{b}{(M, b)}, \qquad (M, b) = \text{greatest common divisor of } M \text{ and } b. \tag{12}$$

Using the periodicity and antisymmetry of the sine function, we can rewrite (10) by collecting terms of equal frequency:

$$x_n = \sum_{k=1}^{\hat{b}/2} \alpha_k \sin\left(2\pi \frac{k\hat{M}a}{\hat{b}}n\right), \qquad \hat{M} = \frac{M}{(M, b)} \tag{13}$$

$$\alpha_k = \begin{cases} \dfrac{(-1)^k}{\hat{b}\,\sin(\pi k/\hat{b})} & \hat{b}\ \text{odd} \\[4mm] \dfrac{(-1)^k}{\hat{b}\,\tan(\pi k/\hat{b})} & \hat{b}\ \text{even.} \end{cases} \tag{14}$$

We note that the magnitude of the α_k decreases with k, independent of $\hat{b}$ even or odd. It is now convenient to arrange the sine functions in (13) in numerical order of frequency. We use a theorem of number theory stating that the equation

$$\langle k\hat{M}a\rangle_{\hat{b}} = 1, \qquad \text{with } \langle \cdot \rangle_{\hat{b}} \text{ denoting modulo } \hat{b} \text{ operation} \tag{15}$$

has a solution k' due to coprimality of the numbers $\hat{M}, a$, and $\hat{b}$. (For a proof see, e.g., [3].) We now use this solution to define

$$k_m = \langle mk'\rangle_{\hat{b}}. \tag{16}$$

It is easy to prove (but somewhat lengthy) that x_n finally can be written as

$$x_n = \sum_{m=1}^{(\hat{b}-1)/2} \alpha_{k_m} \sin\left(2\pi \frac{m}{\hat{b}}n\right) \tag{17}$$

where the coefficients α are taken from (14) using their property $\alpha_m = -\alpha_{\hat{b}-m}$. Combining (13) and (17) shows that x_n consists of the $(\hat{b} - 1)/2$ harmonics of the fundamental $2\pi/\hat{b}$ with amplitudes (except for sign) given by a permutation of the first $(\hat{b} - 1)/2$ coefficients α_k.

As final result, the Fourier series of S_n' is derived from (7) and (17) as follows:

$$S_n' = \sin\left(2\pi \frac{a}{b}n\right) - \frac{2\pi}{M}\cos\left(2\pi \frac{a}{b}n\right)x_n$$
$$= \sin\left(2\pi \frac{a}{b}n\right) - \frac{\pi}{M}\sum_{m=1}^{(\hat{b}-1)/2} \alpha_{k_m}$$
$$\cdot \left\{\sin 2\pi\left(\frac{m}{\hat{b}} + \frac{a}{b}\right)n + \sin 2\pi\left(\frac{m}{\hat{b}} - \frac{a}{b}\right)n\right\}. \tag{18}$$

In accordance with the Fourier series of the continuous signal $S'(t)$ in (9), the "noise" again consists of two series of spectral lines. But now some of the spectral lines in (18) may overlap, i.e., some of the sine functions may correspond to the same frequency. It can be shown that overlapping only occurs if (b, M) is less than 3. Otherwise, all spectral lines have distinct frequencies. The amplitude of the highest spectral line of the "noise" therefore is α_1/M if (b, M) is greater than 2. If overlapping occurs, it can further be shown that not more than two spectral lines overlap. Thus, the maximum amplitude value of a noise line can be $(\alpha_2 - \alpha_1)/M$ if (b, M) is less than 3.

Another effect not mentioned hitherto is the appearance of additional cosine terms to the Fourier series of S'_n. They are due to a difference between the true error of a table lookup and the error expressed by the Fourier series of a sawtooth. If a sine table is sampled exactly amidst two successive values, in practice, one of the (obviously not exact) table values has to be selected, whereas the Fourier series of the sawtooth yields no error. It can be shown that this type of error only occurs for even $\hat{b}$. It then adds a series e_n to x_n given by

$$e_n = \frac{1}{2\hat{b}} \sum_{k=0}^{\hat{b}-1} (-1)^k \cos\left(2\pi \frac{k}{\hat{b}} n\right). \tag{19}$$

In order to get the exact Fourier series of S'_n in the case of $\hat{b}$ even, we now have to replace x_n by $x_n + e_n$ in (18). The spectral lines resulting from e_n have the same features in respect to overlapping as already mentioned above for the spectral lines of x_n; all cosine terms have distinct frequencies if (b, M) is greater than 2, otherwise they overlap. The overlapping may either double the amplitude if $(b, M) = 1$, or (as can further be shown) it results in cancelling of the cosine terms if $(b, m) = 2$.

The spectral shape is mainly affected by the lines of the sawtooth x_n because cosine components (if they exist) usually have much lower amplitudes. The frequency spacing between successive lines of the sawtooth is seen from (13) to be

$$df = 2\pi \frac{\langle \hat{M}a \rangle_{\hat{b}}}{\hat{b}}. \tag{20}$$

If now df is near 0 or 2π, the first lines of the sawtooth (i.e., the highest!) lie close to the sine frequency $2\pi a/b$. Consequently, the next lines are still not far from the sine frequency — but are already lower in amplitude. Obviously, this kind of df yields a spectrum as shown in Fig. 1(a). The same considerations explain that a df close to π will yield spectra like that of Fig. 1(b). A more general df near $2\pi/n$ results in n maxima of the spectrum.

Only a few spectral lines exist if denominator b of the sine frequency and table length M have common divisors resulting in small numbers $\hat{b}$. Examples of such spectra can be seen in Fig. 1(c) and (d).

CONCLUSION

The spectra of sinusoidal signals generated by table-lookup in a sine table of length M are derived. They consist of spectral lines at the harmonics of the fundamental $2\pi/b$ if a sine of frequency $2\pi a/b$ is generated. The noise components of such signals are only sine components if $\hat{b} = b/(b, M)$ is odd. Maximum amplitude of these noise lines is $(\alpha_2 - \alpha_1)/M$ for (b, M) less than 3, and $-\alpha_1/M$ otherwise. Additional cosine components appear for even $\hat{b}$ and (b, M) not equal to 2. The maximum amplitude of these components is $2\pi/M\hat{b}$ if $(b, M) = 1$ and $\pi/M\hat{b}$ if (b, M) is greater than 2.

The spectral shape is determined by $df = 2\pi\langle \hat{M}a \rangle_{\hat{b}}/\hat{b}$. A df near 0 to 2π yields spectral concentration of noise energy around the sine frequency. More generally, if df lies close to $2\pi/n$, n maxima are found in the spectra. Finally, common divisors of M and b, resulting in a low $\hat{b}$, reduce the noise components to only a few lines.

ACKNOWLEDGMENT

The author wishes to thank H. Alrutz, H. W. Strube, and M. R. Schroeder for their fruitful discussions and help in preparing this article.

REFERENCES

[1] J. Tierney, C. M. Rader, and B. Gold, "A digital frequency synthesizer," *IEEE Trans. Audio Electroacoust.*, vol. AU-19, pp. 48–57, 1971.
[2] L. R. Rabiner and B. Gold, *Theory and Application of Digital Signal Processing.* Englewood Cliffs, NJ: Prentice-Hall, 1975.
[3] J. H. McClellan and C. M. Rader, *Number Theory in Digital Signal Processing.* Englewood Cliffs, NJ: Prentice-Hall, 1979.

An Analysis of the Output Spectrum of Direct Digital Frequency Synthesizers in the Presence of Phase-Accumulator Truncation

Henry T. Nicholas, III   and   Henry Samueli

Integrated Circuits and Systems Laboratory
Electrical Engineering Department
University of California, Los Angeles
Los Angeles, CA 90024

ABSTRACT

An algorithm is presented for the calculation of the spectrum of direct digital frequency synthesizers (DDFS's) as a result of phase accumulator truncation. This algorithm, which is derived using number theoretic methods, includes a closed form expression relating the magnitude, number, and position of the spurious noise lines in the output spectrum of a DDFS to the read-only memory (ROM) look-up table size, the amount of phase accumulator truncation and the input frequency control command. The combined finite word length effects of the ROM and the Digital-to-Analog converter (DAC) nonlinearities are also examined in the light of these new results and new design guidelines are developed. The spectrums predicted by these closed form expressions are compared against spectrums generated by a discrete Fourier transform (DFT) and are shown to have comparable accuracy.

As a result of obtaining an expression for the magnitude of the spurious noise frequencies, a relationship between the greatest common divisor of the input frequency command word and the ROM table size, the phase accumulator word lengths, and the magnitude of the worst case spur is obtained. This relationship is used as the basis for a novel modification to the conventional phase accumulator structure which results in a 3.922dB reduction in the magnitude of the worst case spurious response. This hardware modification is also shown to average out the error effects of DAC nonlinearities and roundoff in the stored sine ROM samples.

I. INTRODUCTION

Modern communication systems are placing increasing demands on the resolution and bandwidth of frequency synthesizer subsystems. Spread spectrum applications impose an extremely stringent set of specifications on a frequency synthesizer which include the achievement of extremely fast switching times over a wide bandwidth in a design that is compact, reliable and easily reproducible[1]. In order to meet these demanding requirements many designers are exploring the use of the direct digital approach to frequency synthesis as an effective alternative to conventional analog synthesizers. Direct digital frequency synthesizers (DDFS's) exhibit fast switching speed, excellent temperature and ageing stability, and the ability to switch frequencies while maintaining constant phase, which are all properties difficult to achieve with analog techniques.

The most popular technique for direct digital frequency synthesis is the sine look-up table method first introduced by Tierney, Rader, and Gold[2]. This method, which is well described in the literature, synthesizes a sinewave by successively scanning through a look-up table stored in a read-only-memory (ROM) and converting the recalled sine samples to an analog waveform via a digital-to-analog converter (DAC). As shown in Figure 1, one sample of the sinewave is recalled from the ROM every period of the stable frequency reference. Each of these recalled samples differs from the previous one by a constant phase increment; thus different frequencies may be synthesized by changing the phase difference between the recalled samples via a digital frequency control command.

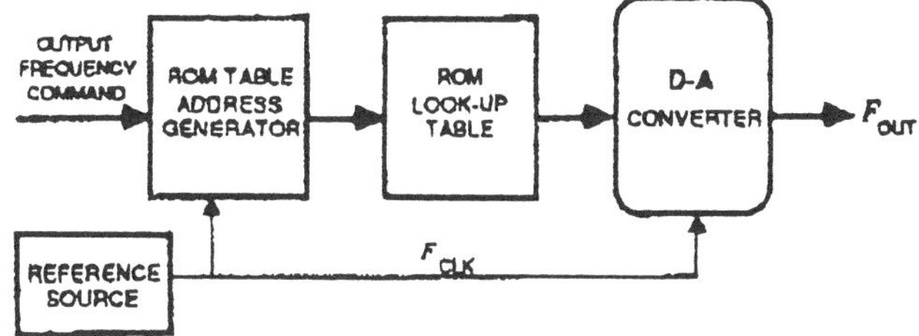

Figure 1.   Simplified block diagram of a DDFS.

The generation of the constantly increasing phase argument to the ROM is normally implemented by accumulating a binary frequency control word using a standard two's complement adder and register as shown in Figure 2.

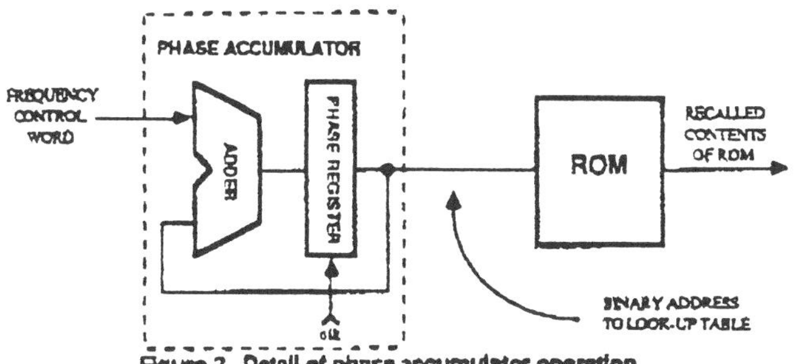

Figure 2.   Detail of phase accumulator operation.

Using this architecture, both the frequency and phase resolution of the synthesizer are determined by the wordlength of the phase accumulator. For most applications, a phase accumulator wordlength in excess of 22 bits is required to achieve acceptable resolution. The use of all 22 bits of the phase accumulator wordlength to address 2^{22} stored sine samples would require a prohibitively large ROM, even if the quarter wave symmetry and course-fine trigonometric reductions are applied to reduce the storage. Therefore, in actual practice, all sine DDFS's requiring fine phase resolution truncate part of the phase accumulator output when addressing the ROM.

The effective implementation look-up table based DDFS's has been seriously hampered by the lack of a precise theoretical characterization of the finite wordlength effects due to phase accumulator truncation. To illustrate this problem, three different output spectrums are depicted in Figures 3 through 5 showing the spurious response due to phase accumulator truncation for several different output frequencies. The three cases shown are the results of a hardware simulation of a DDFS with a phase accumulator word length of 12 bits and a ROM address length of 5 bits corresponding to the phase truncation of the 7 least significant bits of the phase register. These spectrums were generated by a discrete Fourier transform (DFT) of the numeric output of a hardware simulation of the digital logic, and thus show only the response due to algorithmic nonlinearities inherent in the system. They therefore represent the best case attainable given an "ideal" DAC.

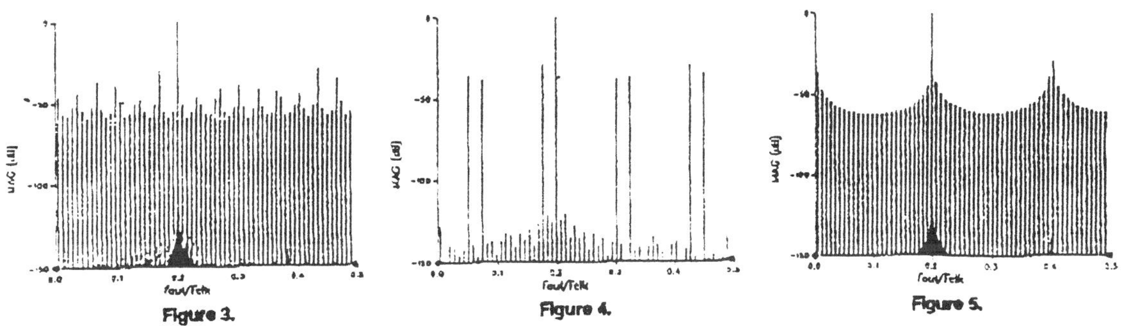

Figure 3.

Figure 4.

Figure 5.

Reprinted from *Proceedings of the 41st Annual Frequency Control Symposium,* pp. 495–502, 1987.

Figure 3 shows the output spectrum when the synthesizer is tuned to a frequency of .1989 times the clock frequency, F_{clk}. In contrast, Figure 4 shows the output spectrum for the next higher frequency resolvable by the synthesizer, .1992 F_{clk}. It is evident from comparing the spectrums in Figure 3 to Figure 4 that a change to the synthesizer's output frequency of only .12%, the smallest frequency change resolvable by this design, results in a completely different spurious response. The third plot in Figure 5 shows a change of only 1.1% (4 times the smallest resolvable frequency difference) and yet displays another response totally different from the first two. The dramatic change in the character of the response for all three cases demonstrates the importance of being able to explicitly describe the response of a DDFS for any combination of output frequency, ROM storage capacity, and phase accumulator length.

II. ANALYSIS

The operation of all direct digital frequency synthesizers that use some variation of the look-up table technique can be analyzed mathematically using the model shown in Figure 6. Although many techniques exist for performing the functional mapping from phase to sine, (i.e. Taylor series, ROM look-up, course/fine segmentation into multiple ROM's) the following model is valid for all methods not possessing feedback or memory elements other than pipeline registers.

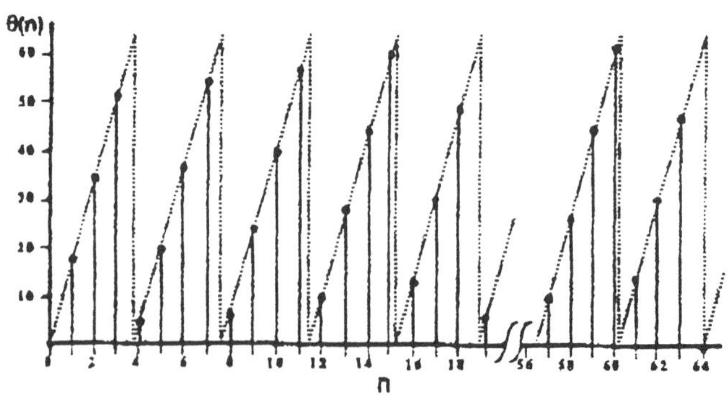

W – Word length of phase accumulator output used to address ROM.
L – Length in bits of phase accumulator register word.
D – Word length of sine values stored in look-up table.
F_r – Frequency control word.

Figure 6. Sources of noise in DDFS's

Contained within this model are three sources of distortion. The first source, ε_{DA}, is due to the DAC conversion noise at the output. The second, ε_T, is due to the finite wordlength of the sine samples stored in the ROM, and the third, ε_P, is due to the truncation of phase accumulator bits addressing the ROM. The most important of these finite wordlength effects is the truncation of phase accumulator bits. There are several reasons for this. First, in order to minimize ROM storage requirements, one would like to truncate as many of the phase bits as possible without incurring too much spurious distortion. In addition, the spurious response due to ε_T and ε_{DA} can be controlled by the designer with less of a hardware penalty, since the hardware complexity is a linear function of the wordlength of both the ROM output and DAC output, while it is an exponential function of the ROM address length. Finally, the characterization of the spurious response due to phase truncation is important because it involves a detailed analysis of the interaction of the phase accumulator output sequence and its properties under quantization, which once understood, yields insight into the spurious response due to all finite wordlength effects in the DDFS.

Efforts have been made to characterize the effects of phase accumulator truncation using statistical methods. These types of analyses have the drawback that results are only obtained for the average spur energy rather than for the magnitude of the worst case spur. Furthermore, a stochastic analysis gives no information about the relative distribution of large spurs in the output spectrum. This information is extremely important because some combinations of ROM table size, phase accumulator word length, and frequency control register contents result in worst case spurs occurring very close to the desired output frequency.

The development of the algorithm for the generation of the spurious spectrum of a DDFS will be divided into three sections. First, some observations are made on the behavior of the phase accumulator. Then, an algorithm is developed for the determination of the spectrum of the phase error sequence. Finally, the analysis techniques and results gained from generating the spectrum of the phase error are applied to develop an algorithm for the generation of the spectrum of the DDFS due to phase truncation.

PHASE ACCUMULATOR OPERATION

The behavior of the spurious spectrum a DDFS for different values of the frequency control word, F_r, is intrinsically linked to the numerical properties of the phase accumulator. For this reason, we will begin the analysis of the DDFS spectral response with a description of phase accumulator operation.

The phase accumulator operates on the principle of overflow arithmetic, using the modulo 2^L property of an L bit periodically overflowing accumulator register to simulate the modulo 2π property of the sine function by exploiting the relationship

$$\sin\left(2\pi \frac{\langle \theta(n) \rangle_{2^L}}{2^L}\right) = \sin\left(2\pi \frac{\theta(n)}{2^L}\right) \tag{1}$$

where $\langle \cdot \rangle_{2^L}$ represents taking the integer residue of a number modulo 2^L. The phase accumulator thus acts as a digital integrator followed by a modulo 2^L operator. The frequency of the output of the DDFS is related to the binary frequency control word by the relationship

$$F_o = \frac{\omega_0}{2\pi} = \frac{\frac{\Delta\theta(n)}{\Delta t}}{2^L} = \frac{F_r F_{clk}}{2^L} \tag{2}$$

Due to the discrete-time nature of the system, values of F_r larger than 2^{L-1} will result in an aliased output frequency equal to $F_{clk} \cdot F_o$. Thus there are $2^{L/2}$ discrete output frequencies that can be generated, each corresponding to a unique value of F_r less than 2^{L-1}.

The phase sequence $\theta(n)$ generated by the phase accumulator can be represented as the samples of an idealized sawtooth waveform, where, after normalizing with $F_{clk}=1$, the slope of the sawtooth is given by

$$\frac{\Delta\theta(n)}{\Delta t} = F_r \tag{3}$$

One concept that is key to the theoretical understanding of phase accumulator operation is the relationship between the period of the idealized continuous time waveform and the period of the discrete time phase accumulator output sequence. As shown in Figure 7, the phase accumulator outputs the samples of a hypothetical continuous time sawtooth waveform of amplitude 2^L and period $2^L/F_r$. Since the sine function is computed from the phase sequence using combinational logic without feedback, the period of the synthesized analog sinewave at the output of the DAC is also $2^L/F_r$. The numerical period of the phase accumulator output sequence is defined as the minimum value of N for which $\theta(n) = \theta(n+N)$ for all n. In general, the numeric period of the phase accumulator sequence is given by $2^L/(F_r,2^L)$, where $(F_r,2^L)$ represents the greatest common divisor of F_r and 2^L. Thus the period of the sampled sawtooth will equal the period of the idealized continuous time sawtooth only for the special case of $F_r = (F_r,2^L)$, i.e., when F_r is purely a power of two. This behavior is illustrated for the case of $L=6$ and $F_r=17$ in Figure 7. In this figure it is evident that the normalized output frequency of the continuous-time sawtooth wave is $2^L/F_r = 17/64 = .28563$, resulting in a period of $64/17 = 3.7647$. In contrast, however, the period of the numerical sequence of samples of this waveform is $64/(17,64) = 64$. The most important consequence of this phenomenon is that the numerical period of the sequence of samples recalled from the sine ROM will have the same value as the numerical period of the sequence generated by the phase accumulator. Therefore, the spectrum of the output waveform of the DDFS prior to digital-to-analog conversion is characterized by a discrete spectrum consisting of $2^L/(F_r,2^L)$ points.

Figure 7.

SPECTRUM OF THE PHASE ERROR SEQUENCE

In the ideal case of infinite precision in the look-up table word length and no phase quantization, the output sequence of a DDFS is given by

$$S(n) = \sin\left(2\pi\frac{F_r}{2^L}n\right) \tag{4}$$

which yields the sampled values of a sinusoid of frequency $F_r/2^L$. We now consider the operation of a DDFS in which the output of the phase accumulator is quantized to W bits by truncation. Under this condition, where we define B to be the number of bits truncated such that $L - W = B$, the output sequence of the DDFS is given by

$$S_t(n) = \sin\left(2\pi\frac{2^B}{2^L}\left[\frac{F_r}{2^B}n\right]\right) \tag{5}$$

where the operator $[\ ]$ represents truncation to integer values. Equation (5) can alternatively be expressed as

$$S_t(n) = \sin\frac{2\pi}{2^L}(F_r n - \mathcal{E}_P(n)) \tag{6}$$

where $\mathcal{E}_P(n)$ represents the phase error sequence. This phase error sequence can be modeled as the sampled values of a continuous-time sawtooth waveform which will be referred to as $\mathcal{E}_P(t)$. The amplitude of the sawtooth waveform, $\mathcal{E}_P(t)$, is 2^B, and the frequency is $F_r/2^B$ as exemplified in Figure 8.

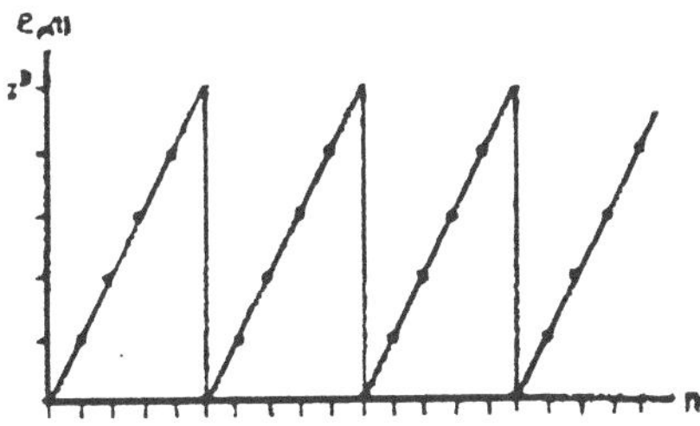

Figure 8. Phase accumulator error sequence

This sawtooth waveform is identical to the waveform that would be generated by a phase accumulator of word length B with a input frequency control command of $(F_r)_{2^B}$.

In order to characterize the spectral properties of the error sequence, $\mathcal{E}_P(n)$, the Fourier series must be obtained for the idealized continuous-time representation of the error waveform $\mathcal{E}_P(t)$. Unfortunately the sawtooth waveform defined above is specified to have a value of zero at the discontinuity which is in violation of the required Dirichlet conditions for a Fourier representation. Dirichlet conditions require that a function take on its average value at a point of discontinuity, which in this case is $2^B/2$.

Therefore, in order to represent the phase error as a Fourier series, the error waveform must be reexpressed as the superposition of an ideal "Dirichlet" sawtooth and a correcting waveform, such that the both waveforms satisfy Dirichlet conditions and such that the sampled values of the superposition of the two waveforms equals the actual error sequence, $\mathcal{E}_P(n)$.

SAWTOOTH: B=2 F_r=3 frequency=3/4

PULSE TRAIN: B=2 F_r=3 frequency=3/4 pulse width=1/3

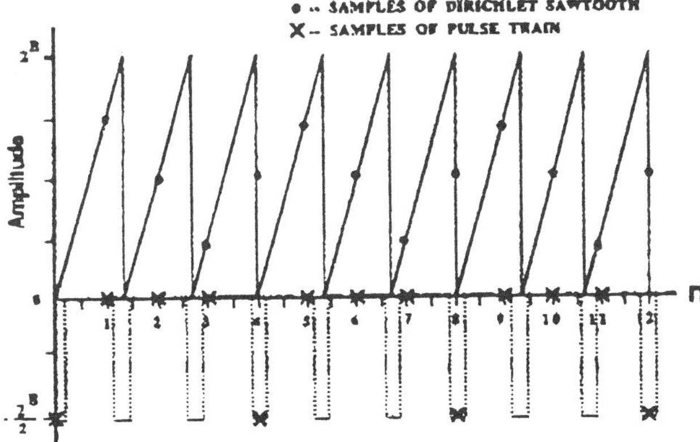

Figure 9.

The correcting waveform that is added to the Dirichlet sawtooth to make the sampled values agree with the actual phase error samples is a pulse train of frequency $F_r/2^B$ with each pulse having a width of $(2^B,F_r)/(2^B)$ as shown in Figures 9 and 10. This choice of frequency and pulse-width gives

the correcting waveform two important properties: first, that pulses are only sampled at the midpoint of the pulse, and second, that the only pulses sampled are those whose centers lie exactly on the discontinuity of the sawtooth.

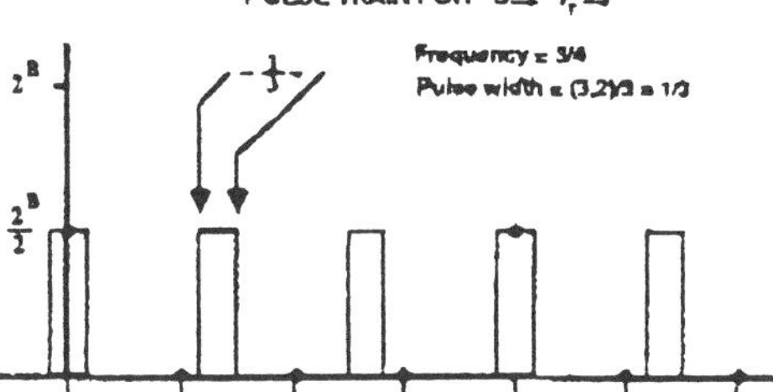

Figure 10.

The Fourier series for $\mathcal{E}_P(t)$ can now be expressed as the sum of the individual Fourier series for the "Dirichlet sawtooth" and the pulse train waveforms that were depicted in Figure 9. This representation is given in equation (7) where the sawtooth waveform is represented as a pure sine series and the pulse train as a pure cosine series (ignoring the D.C. term). In the term containing the cosine series the substitution $\Lambda = 2^B/2(2^B,F_r)$ is introduced, where Λ is equal to one half the reciprocal of the pulse train duty cycle.

$$\mathcal{E}_P(t) = \sum_{K=1}^{\infty}\frac{2^B}{\pi K}\sin(2\pi K\frac{F_r}{2^B}t) - \frac{2^B}{2\Lambda}\sum_{K=1}^{\infty}\frac{4\Lambda}{2\pi K}\sin(\frac{2\pi K}{4\Lambda})\cos(2\pi K\frac{F_r}{2^B}t) \tag{7}$$

This Fourier series may be sampled at integer time points to obtain

$$\mathcal{E}_P(n) = \sum_{K_o=1}^{\infty}\frac{2^B}{\pi K_o}\sin(2\pi K_o\frac{F_r n}{2^B}) - \frac{2^B}{2\Lambda}\sum_{K_o=1}^{\infty}\frac{4\Lambda}{2\pi K_o}\sin(\frac{2\pi K_o}{4\Lambda})\cos(2\pi K_o\frac{F_r n}{2^B}) \tag{8}$$

Because of the discrete nature of the time index, the error sequence, $\mathcal{E}_P(n)$, will be periodic with a period of at most 2^B. If F_r contains common divisors with 2^B, the sequence will repeat every $2^B/(F_n, 2^B)$ samples. This periodicity in the time domain will result in a sampled frequency domain representation whose magnitude contains only $2^B/2(F_r, 2^B)$ unique points because the magnitude spectrum of a real sequence is symmetric about the $F_{out} = 0$ axis. Therefore, $\Lambda = 2^B/2(F_r, 2^B)$ as defined in (7) can be interpreted as the number of unique spectral noise lines in the discrete magnitude spectrum of $\mathcal{E}_P(n)$. By exploiting the fact that the sine and cosine functions in (8) are periodic in K_o with period $2^B/2(F_r, 2^B)$, the sine and cosine series can be rewritten as double summations, where the inner summation enumerates the coefficients of the sine and cosine functions in (8) over one period of K_o.

If we define $\mathcal{E}_{sp}(n)$ as the sine series summation in the left term of (8) we obtain

$$\mathcal{E}_{sp}(n) = \frac{2^B}{\pi}\sum_{g=1}^{\infty}\left(\sum_{K=1}^{\Lambda-1}\frac{1}{(K+2\Lambda g)}\sin(2\pi(K+2\Lambda g)\frac{F_r}{2^B}n)\right.$$
$$\left. + \sum_{K=1}^{\Lambda-1}\frac{1}{(-K+2\Lambda g)}\sin(2\pi(-K+2\Lambda g)\frac{F_r}{2^B}n)\right)$$
$$+ \frac{2^B}{\pi}\sum_{K=1}^{\Lambda-1}\frac{1}{K}\sin(2\pi K\frac{F_r}{2^B}n) + \frac{2^B}{\pi}\sum_{g=1}^{\infty}\frac{1}{g\Lambda}\sin(2\pi g\Lambda\frac{F_r}{2^B}n) \tag{9}$$

Using trigonometric identities and noting that $F_r/(F_n, 2^B)$ is always an integer, and by interchanging the summations for g and K, equation (9) can be simplified to

$$\mathcal{E}_{sp}(n) = \frac{2^B}{2\Lambda}\sum_{K=1}^{\Lambda-1}\left(\left(\frac{2\Lambda}{K\pi} - 2\left(\frac{K\pi}{2\Lambda}\right)\sum_{g=1}^{\infty}\left((g\pi)^2 - \left(\frac{K\pi}{2\Lambda}\right)^2\right)^{-1}\right)\sin(2\pi K\frac{F_r}{2^B}n)\right) \tag{10}$$

Using Jolley [3] (Eqn.769) and noting that the argument of the summation is zero at $K = \Lambda$, equation (10) can be rewritten without loss of generality as

$$\varepsilon_{sp}(n) = \frac{2^B}{2\Lambda}\sum_{K=1}^{\Lambda}\left(\cot(\frac{K\pi}{2\Lambda})\sin(2\pi K\frac{F_r}{2^B}n)\right) \qquad (11)$$

Thus the sine portion of the phase error defined in (8) is expressed as a numerical summation of Λ different harmonics of the sampled sinewave $\sin(2\pi F_r n/2^B)$, with each amplitude determined by the coefficient $\cot(K\pi/2\Lambda)$.

Returning to the analysis of the Fourier series for $\varepsilon_p(n)$ in (8), let $\varepsilon_{cp}(n)$ be defined as the cosine part of the phase error sequence. This sequence is now rewritten in a manner similar to that in (9) as a double summation over K and g as follows:

$$\varepsilon_{cp}(n) = \frac{-2^B}{2\Lambda}\sum_{g=1}^{\infty}\left(\sum_{K=1}^{\Lambda-1}\text{sinc}(\frac{\pi K+2\pi\Lambda g}{2\Lambda})\cos(2\pi K\frac{F_r}{2^B}n)\right.$$
$$\left. + \sum_{K=1}^{\Lambda-1}\text{sinc}(\frac{-\pi K+2\pi\Lambda g}{2\Lambda})\cos(2\pi K\frac{F_r}{2^B}n)\right)$$
$$- \frac{2^B}{2\Lambda}\sum_{K=1}^{\Lambda-1}\text{sinc}(\frac{\pi K}{2\Lambda})\cos(2\pi K\frac{F_r}{2^B}n) - \frac{2^B}{\pi}\sum_{g=1}^{\infty}\text{sinc}(\frac{\pi g}{2})\cos(2\pi g\Lambda\frac{F_r}{2^B}n) \qquad (12)$$

After trigonometric and algebraic manipulation, and using Jolley (eqn. 770, pg. 146), we obtain

$$\varepsilon_{cp}(n) = \frac{-2^B}{2\Lambda}\sum_{K=1}^{\Lambda-1}\left(\cos(2\pi K\frac{F_r}{2^B}n) + \frac{1}{2}\cos(2\pi\Lambda\frac{F_r}{2^B}n)\right) \qquad (13)$$

In order to obtain a more concise representation of $\varepsilon_{cp}(n)$ that is in the same form as $\varepsilon_{sp}(n)$ in (11), the $\cos(2\pi\Lambda F_r n/2^B)$ term of the summation in (13) is multiplied by a "correction factor" of two. This operation is justified later by showing that the $K = \Lambda$ coefficient is always doubled due to aliasing. Using this correction factor, $\varepsilon_{cp}(n)$ is rewritten as

$$\varepsilon_{cp}(n) = \frac{-2^B}{2\Lambda}\sum_{K=1}^{\Lambda}\cos(2\pi K\frac{F_r}{2^B}n) \qquad (14)$$

Upon substituting back into (8) and using (11) we obtain

$$\varepsilon_p(n) = \frac{-2^B}{2\Lambda}\sum_{K=1}^{\Lambda}\left(\cot(\frac{K\pi}{2\Lambda})\sin(2\pi K\frac{F_r}{2^B}n) - \cos(2\pi K\frac{F_r}{2^B}n)\right) \qquad (15)$$

The error sequence is now expressed as a sum of Λ distinct frequencies with unique amplitudes corresponding to the discrete spectrum of the phase error. By combining the terms from the sine series and cosine series into a complex Fourier series representation we obtain

$$\varepsilon_p(n) = \sum_{K=1}^{\Lambda}\zeta_k e^{j(2\pi K\frac{F_r}{2^B}n)}e^{j\Psi(K,\Lambda)} \qquad (16)$$

where the magnitude of the complex phasor, ζ_K, and the phase angle, $\Psi(K,\Lambda)$, are defined as

$$\zeta_K \triangleq \frac{2^B}{2\Lambda}\text{cosec}(\frac{K\pi}{2\Lambda}) \qquad \Psi(K,\Lambda) \triangleq -\cot(\frac{K\pi}{2\Lambda}) \qquad (17)$$

The variable $\Gamma = F_r/(F_n 2^B)$ is now defined so that the frequency $F_r/2^B$ can be expressed as a rational fraction of two mutually prime integers. Using this definition, each of the spur magnitudes, ζ_K, corresponds to a unique frequency $K\Gamma(F_n 2^B)/2^B$. This, however, is only a partial solution, since an explicit result is required that will define the exact location of the spur frequencies F_{sp} in the range $0 < F_{sp} < F_{clk}/2$. In general $K\Gamma(F_n 2^B)/2^B$ will

range over several decades of the frequency range of interest. Because aliasing, the position of each spur in the spectral range $0 < (F_{sp}/F_{clk}) < 1$ is equal to the residue of the frequency $K\Gamma$ modulo $2^B/(F_n 2^B)$.

We now introduce the normalized integer frequency, F_n, which is defined such that all frequencies in the discrete spectrum (as would be generated by a DFT) are represented by integers. From the definition of the DFT, we know that the discrete spectrum of a sequence contains the same number of points that are guaranteed to be non-zero as the numerical period of that sequence. Using this and realizing that the period of the sequence $\varepsilon_p(n)$, is 2Λ, we define the relationship between the un-normalized output frequency and F_n as $F_n = \frac{f_{sp}2\Lambda}{f_{clk}}$. Where this definition guarantees that all possible spur locations in the spectrum are represented by integer values.

Using this definition for frequency axis of the discrete spectrum we can determine the frequency of the K-th spur from

$$\langle K\Gamma\rangle_{2\Lambda} = F_n \qquad (18)$$

Although (18) calculates the frequency that each spur, ζ_K, maps to in the range $0 < (F_{sp}/F_{clk}) < 1$, a much more powerful description of the spectrum would yield the amplitude (derived from K) as a function of the frequency, F_n. Furthermore, when using (18), it must be ensured that more than one value of K does not yield the same F_n. Thus an assurance must be gained that each of the frequencies represented by the summation in (16) corresponds to a unique frequency in the range $0 < (F_{sp}/F_{clk}) < 1$, and, given that assurance, a solution to the congruence in (18) for K must be obtained given any F_{sp} in that range. The congruence in (18) is actually a representation of the integral Diophantine equation,

$$K\Gamma + bY = F_n \qquad (19)$$

where Y is an integer, $b = 2\Lambda$ is an integer, and all solutions are defined to be an equivalence class modulo b. Because it is a Diophantine equation, integral values of K that solve (19) can be found if and only if $\langle F_n\rangle_{\langle\Gamma,b\rangle}=0$. From number theory, by the coprimality of Γ and $2^B/(F_n 2^B)$, it can be shown that each set of K's less than $2^B/(F_r 2^B)$ will map into a unique set of F_n's less than $2^B/(F_n 2^B)$. Thus, each frequency will map uniquely into the range $0 < (F_{sp}/F_{clk}) < 1$. Therefore, exploiting the fact that $(\Gamma, 2^B/(F_n 2^B)) = 1$, the solution to (18) can be expressed without loss of generality by either Euclid's algorithm or by Euler's Theorem. From Euler's Theorem, $\langle F_r^{\Phi(x)}\rangle_x=1$ for all F_r and X coprime, where $\Phi(x)$ is defined as the Euler Totient function. Therefore by multiplying both sides of (18) by $F_r^{\Phi(x)}$ with $X=2^B/(F_n 2^B)$ and then factoring out F_r one obtains:

$$\langle F_n\Gamma^{\Phi(2\Lambda)-1}\rangle_{2\Lambda} = K \qquad (20)$$

where $\Phi(n)$ is the Euler totient function which is defined as the number of integers smaller than n which are relatively prime to n. There is no explicit function that can be used to calculate $\Phi(n)$, therefore a table must normally be used to obtain its values. Fortunately, the Totient function can be easily obtained for the special case of n strictly a power of 2. Thus for $n = 2^x$ where x is an integer; all even numbers less than n will not be mutually coprime to n and all odd numbers less than n will be coprime to n. Therefore the Totient function for $n=2^x$ is simply the number of odd integers less than n, which is $n/2$ or 2^{x-1}. In the above analysis, $n=2^B/(F_r 2^B)$, which is by definition always a pure power of 2. Therefore we obtain

$$\Phi(\frac{2^B}{(F_n 2^B)}) = \frac{2^B}{2(F_n 2^B)} = \Lambda \qquad (21)$$

and thus,

$$K = \langle F_n\Gamma^{\Lambda-1}\rangle_{2\Lambda} \qquad (22)$$

Therefore, by using (22) to get the value of K to substitute into (18), the amplitude of the spurious response can be calculated at any point in the digital spectrum. A problem, however, arises when one considers that (22) was derived from (18), which maps each spur from a spectrum in which each ζ_K lies on a frequency $F_{sp}= K\Gamma(F_r 2^B)/2^B$ to one in which all of the spurs lie in the frequency range $0 < F_{sp}/F_{clk} < 1$. This "post mapped" frequency domain can be thought of as consisting of 2Λ possible spur locations. In this "post mapped" spectrum the spur location denoted by $F_n = 2\Lambda$ corresponds to the frequency F_{clk}, and the location $F_n = \Lambda$ corresponds to $F_{clk}/2$. This mapping is shown graphically in Figure 11 for the case of 3 bits of truncation and $F_r = 11$.

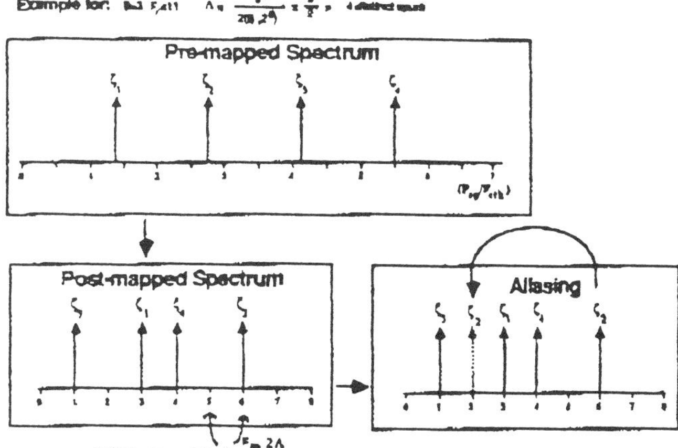

Figure 11.

In some cases one or more of the Λ different spurs will map onto the frequency numbers between $F_n = \Lambda$ and $F_n = 2\Lambda$. This is illustrated in Figure 11 by the spur which was mapped to $F_n = 6$, as predicted by (18). One might question then, how a value for K could be extracted using (22) for all values of F_n in the range $0 < F_n < \Lambda$ (corresponding to $0 < F_{sp} < F_{clk}/2$) when the mapping defined by (18), from which (22) was derived, leaves some of the spur locations in that range empty. This is exemplified by the empty $F_n = 2$ location for the example in Figure 11. The answer to this question comes from sampling theory. Because the spur location $F_n = \Lambda$ actually corresponds to the frequency $F_{clk}/2$, the spur lying on $F_n = 6$ in the "post mapped" spectrum in Figure 11 is aliased back to the frequency $F_n = 2$, as shown in the "Aliasing" box in Figure 11.

Ideally, it would be desirable to associate a spur number directly with each spur location in the range $0 < F_{sp} < F_{clk}/2$, rather than calculating the entire spectrum and then aliasing back the spurs in the range $F_{clk}/2 < F_{sp} < F_{clk}$. To calculate the value of K for the spurs lying on these "empty" spur locations, an aliasing phenomenon is used similar to that encountered in sampling theory. Since the summation expressing the number of spurs is defined to have Λ terms, and since the modulus operator in (22) is defined modulo 2Λ, calculations using (22) can yield a value of K for an empty F_n that is larger than Λ (which is impossible since there are only Λ different ζ_K's). This is similar to the aliasing of sampling theory because the frequency associated with the spur, $K\Gamma(F_n 2^B)/2^B$ can be associated with $(2\Lambda - K)\Gamma(F_n 2^B)/2^B$ for $\Lambda > K > 2\Lambda$. Therefore, if (22) yields a value for K that is in the range $\Lambda > K > 2\Lambda$, the true K value can be computed by subtracting the old K from 2Λ. This eliminates the need to determine whether a frequency location was empty or not when applying (22). In order to find the spur magnitude from the frequency number, simply use (22) and subtract K from Λ for K's greater than Λ. Fortunately, the expression for the spur amplitudes given in (17) is symmetric about the $K = \Lambda$ axis so the aliased values of K generated by (22) produce the correct amplitudes when substituted into (17) without calculating $K - 2\Lambda$ for values of K in the range $\Lambda > K > 2\Lambda$.

The procedure for obtaining the discrete spectrum for the phase error function, $\mathcal{E}_P(n)$, can now be summarized as follows:

1. Find the number of spurs from $\Lambda = 2^{B-1}/(F_r, 2^B)$ where:

   B = number of bits truncated
   F_r = contents of the frequency control register

2. Divide the frequency range of 0 to $F_{clk}/2(F_n 2^B)$ into Λ equally spaced spur locations.

3. Find the spur number K for ζ_K from the sequential spur number F_n using the equation:

   $$K = \left\langle F_n \Gamma^{-1} \right\rangle_{2\Lambda}$$

4. Calculate the amplitude of the spur using the value of K derived in step 3 from:

   $$\zeta_K \triangleq \frac{2^B}{2\Lambda} \operatorname{cosec}(\frac{K\pi}{2\Lambda})$$

DDFS Output Spectrum Due to Phase Truncation

Now that an algorithm has been developed for computing the discrete spectrum of the phase error, $\mathcal{E}_P(n)$, we can determine the effect of this phase error sequence on the output spectrum of the DDFS. Applying simple trigonometric identities, the ROM output sequence, $S_t(n)$, of (8) can be written as

$$S_t(n) = \sin(2\pi\frac{F_r n}{2^L})\cos(2\pi 2^{B-L}\frac{\mathcal{E}_P(n)}{2^B}) - \cos(2\pi\frac{F_r n}{2^L})\sin(2\pi 2^{B-L}\frac{\mathcal{E}_P(n)}{2^B}) \quad (23)$$

Because $\left|\frac{\mathcal{E}_P(n)}{2^B}\right| \le 1$ for all $\mathcal{E}_P(n)$, the supremum of the arguments to the sine and cosine terms containing $\mathcal{E}_P(n)$ in (23) is $2\pi 2^{B-L}$. Since it is reasonable to assume that $2^{B-L} \ll 1$ for all realistic DDFS implementations, (23) can be approximated with very little error by

$$S_t(n) = \sin(2\pi\frac{F_r n}{2^L}) - 2\pi\frac{\mathcal{E}_P(n)}{2^L}\cos(2\pi\frac{F_r n}{2^L}) \quad (24)$$

Therefore, to a reasonable approximation, the output spectrum of the DDFS will be composed of a sinewave at the desired output frequency corrupted by the cosine modulated harmonics of the sawtooth waveform $\mathcal{E}_P(n)$. Substituting the summation for the phase error spectrum obtained in (16) for $\mathcal{E}_P(n)$ yields the result

$$S_t(n) = \sin(2\pi\frac{F_r n}{2^L}) - \frac{2\pi}{2^L}\left(\sum_{K=1}^{\Lambda}\zeta_k\, e^{j(2\pi K\frac{F_r}{2^B}n)}\, e^{j\Psi(K,\Lambda)}\right)\cos(2\pi\frac{F_r n}{2^L}) \quad (25)$$

which can be rewritten as

$$S_t(n) = \sin(2\pi\frac{F_r n}{2^L}) - \frac{2\pi}{2^L}\sum_{K=1}^{\Lambda}\zeta_k\left(e^{j2\pi(\frac{F_r}{2^L} + K\frac{F_r}{2^B})n} + e^{-j2\pi(\frac{F_r}{2^L} - K\frac{F_r}{2^B})n}\right)e^{j\Psi(K,\Lambda)} \quad (26)$$

The spurious response now consists of two series of spectral lines whose amplitudes are proportional to the same set of ζ_K's that were derived for the the spectrum of the sawtooth. However, the discrete spectrum will consist of $2^L/(F_n 2^L)$ points onto which the two sets of Λ different ζ_K's will be aliased. Because 2Λ is in general less than $2^L/(F_n 2^L)$ for $L > 8$, there will always be points in the discrete spectrum containing no aliased spurs. Additionally it is possible for a single frequency to contain a noise line that results from the superposition of two different aliased ζ_K's. The variable $\zeta_{K\pm}$ is now defined to be the magnitude of the individual spurs in the output spectrum of a DDFS. This is expressed in (27) as a function of ζ_K, the magnitude of the spurs in the spectrum of the phase error sequence $\mathcal{E}_P(n)$.

$$\zeta_{K\pm} = \frac{\pi}{2^L}\zeta_K = \frac{\pi 2^{B-L}}{2\Lambda}\operatorname{cosec}(\frac{K\pi}{2\Lambda}) \quad (27)$$

As part of the definition in (27) we define notation ζ_{K+} to represent the spurs mapped from the first term, $e^{j2\pi(F_r/2^L + KF_r/2^B)}$, in (26), and ζ_{K-} to be the spurs mapped from the $e^{-j2\pi(F_r/2^L - KF_r/2^B)}$ term. The location of these spurs can be calculated using the same techniques derived earlier for the sawtooth spectrum. However the procedure for calculating the spectrum is now complicated by the fact that a noise spur does not lie on each frequency location. Therefore, in addition to the calculation of spur amplitudes for a given frequency, a criteria for determining if a spur exists at all at a given F_n must be included in the algorithm. To accomplish this, the frequency range $0 < F_{sp} < F_{clk}$ must be divided into $2^L/(F_n 2^L)$ frequency locations. As a result, F_n is now redefined to be the integer frequency number of the frequency location $0 < F_n < 2^L/(F_r, 2^L)$ where $F_n = 2^L F_{sp}/F_{clk}(F_r, 2^L)$. If equation (26) is rewritten to express the frequency $F_r/2^L$ as the ratio of two mutually prime integers using $\Gamma = F_r/(F_r, 2^B)$ we obtain

$$S_t(n) = \sin(2\pi\frac{F_r}{2^L}n) - \sum_{K=1}^{\Lambda}\zeta_{K\pm}\left(e^{j2\pi\frac{(F_n 2^B)}{2^L}(\Gamma + K\Gamma 2^{L-B})n} + e^{-j2\pi\frac{F_n 2^B}{2^L}(\Gamma - K\Gamma 2^{L-B})n}\right)e^{j\Psi(K,\Lambda)} \quad (28)$$

From (28) it is evident that each $\zeta_{K\pm}$ is mapped onto two different frequencies with frequency numbers given by

$$F_n = \left\langle \Gamma + 2^{L-B}K\Gamma \right\rangle_{\frac{2^L}{(F_r, 2^{L-B})}} \qquad \text{or} \qquad F_n = \left\langle \Gamma - 2^{L-B}K\Gamma \right\rangle_{\frac{2^L}{(F_r, 2^{L-B})}} \quad (29)$$

Thus using (29) it can be determined, for a given K, onto what frequency locations ζ_{K+} and ζ_{K-} are mapped. This is helpful in determining where in the frequency spectrum the largest spurs lie. However, normally when evaluating the spectral response, one is more interested in determining the magnitude of the response for a given frequency. Thus, to completely mathematically characterize the spectrum of the DDFS one would like to determine for a given frequency whether any spur exists there, and if so, whether that spur is due to the superposition of one or more $\zeta_{K\pm}$'s and the value of K for each one. From (29) it is evident that spurs exist only at frequency numbers for which the fraction $(F_n - \Gamma)/2^{L-B}$ or $(F_n + \Gamma)/2^{L-B}$ is an integer, or equivalently, for which the following congruences are satisfied:

$$\langle F_n \rangle_{2^{L-B}} = \langle \Gamma \rangle_{2^{L-B}} \quad \text{or} \quad \langle F_n \rangle_{2^{L-B}} = -\langle \Gamma \rangle_{2^{L-B}} \tag{30}$$

If a spur does exist at a given frequency it is possible that this spur is due to the superposition of two $\zeta_{K\pm}$'s. There are in general two ways in which $\zeta_{K\pm}$'s can overlap. The first is when the spurs from one family are aliased on top of spurs from another family (i.e. when a ζ_{K+} spur lies on top of a ζ_{K-} spur). The second way is when spurs from the same family are aliased on top of each other (i.e. ζ_{K-} on top of ζ_{K-}, or ζ_{K+} on top of ζ_{K+}). In general ζ_{K+} and ζ_{K-} will overlap only when one of the following equivalence relations are satisfied:

$$-\langle F_r - 2^{L-B}K_1 F_r \rangle_{2^L} = \langle F_r + 2^{L-B}K_2 F_r \rangle_{2^L} \tag{31}$$

or

$$\langle F_r - 2^{L-B}K_1 F_r \rangle_{2^L} = \langle F_r + 2^{L-B}K_2 F_r \rangle_{2^L} \tag{32}$$

Equation (31) can be rearranged and simplified to obtain

$$\left\langle \langle 2^{L-B-1} \rangle_{\frac{2^{L-1}}{(F_r,2^L)}} \langle K_1 - K_2 \rangle_{\frac{2^{L-1}}{(F_n,2^L)}} \right\rangle_{\frac{2^{L-1}}{(F_n,2^L)}} = 1 \tag{33}$$

It can be seen that (33) can only be satisfied when 2^{L-B-1} is an odd number. Since 2^{L-B-1} is odd only for $L - B - 1 = 0$, the overlap will occur when L and B satisfy $L - B = 1$. A second condition for equivalence in (33) is that $K_1 - K_2 = 1$. Therefore for the case of $L - B = 1$ (corresponding physically to a DDFS with all but the most significant bit truncated) the ζ_{K+} spurs are always aliased on top of the ζ_{K-} spurs. Thus the only spurs that do not fall on top of other spurs for this case are ζ_1 and ζ_Λ. Fortunately the special case of $L - B = 1$ is a condition that would never be realized in a practical DDFS.

The second condition for ζ_{K+} and ζ_{K-} overlap as expressed in (32) can be rearranged and simplified to obtain

$$\langle K_1 + K_2 \rangle_{2\Lambda} = 0 \tag{34}$$

Since $MAX(K_1) = MAX(K_2) = \Lambda$, (34) is satisfied only for

$$K_1 + K_2 = 2\Lambda \tag{35}$$

In other words overlap always occurs for the case of $K_1 = \Lambda$ and $K_2 = \Lambda$. Therefore independent of the conditions on L and B, $\zeta_{\Lambda+}$ is always aliased on top of $\zeta_{\Lambda-}$. This means that when calculating the magnitude of the spectral response at a frequency where it is determined that $\zeta_{\Lambda+}$ resides, the calculated magnitude of $\zeta_{\Lambda+}$ must be doubled since $\zeta_{\Lambda-}$ will always be aliased on top of it. This, it turns out, is the justification for the factor of 2 "correction factor" that multiplied the $(1/2)\cos(2\pi\Lambda F_r n/2^B)$ term in (15) to get the summation in (16). Therefore no provision needs to be made for the doubling of the $\zeta_{\Lambda\pm}$ terms when (27) is used to calculate the spur amplitudes.

As mentioned earlier, spurs can also overlap when spurs from the same family are aliased on top of one another (ζ_{K+} on top of ζ_{K+} or ζ_{K-} on top of ζ_{K-}). The conditions for this type of overlap are expressed as

$$\langle F_r + 2^{L-B}KF_r \rangle_{2^L} = \langle F_r + 2^{L-B}(K+n)F_r \rangle_{2^L} \tag{36}$$
$$n = \pm 1,2,3,...,\Lambda$$

or

$$\langle F_r + 2^{L-B}KF_r \rangle_{2^L} = 2^L - \langle F_r + 2^{L-B}(K+n)F_r \rangle_{2^L} \tag{37}$$
$$n = \pm 1,2,3,...,\Lambda$$

where spurs of the same family overlap if and only if either (35) or (37) is satisfied. Equation (31) can be rewritten as

$$\langle K \rangle_{\frac{2^B}{(F_n,2^B)}} = \langle K + n \rangle_{\frac{2^B}{(F_r,2^B)}} \tag{38}$$

Therefore the equivalence in (37) is satisfied only for n a multiple of $2^B/(F_r,2^B)$. Since n is bounded by Λ because there are only Λ spurs, (31) yields no self overlap conditions.

The other condition for self overlap, given by (37), can be rearranged and simplified to obtain

$$\langle 2^{L-B-1}(2K+n) \rangle_{\frac{2^L}{(F_n,2^L)}} = \langle 1 \rangle_{\frac{2^L}{(F_n,2^L)}} \tag{39}$$

Since -1 is an odd number, equivalence (39) can only be satisfied if 2^{L-B-1} is an odd number. 2^{L-B-1} is odd only for $L-B-1=0$ or if $L=B+1$. This yields the result that self overlap only exists for the trivial case of truncation of all bits addressing the ROM except the phase accumulator MSB.

Therefore, all of the conditions for spur overlap have now been exhaustively determined. Summarizing the results, it has been determined that there is always overlap among spurs of the same family for the condition $L-B=1$, a condition that is not realized for any practical DDFS design. Furthermore, other than for the case $L-B=1$, there is no overlap between spurs of the same family. For the case of overlap among spurs of different families it has been shown that the spurs $\zeta_{\Lambda+}$ and $\zeta_{\Lambda-}$ overlap under all conditions and that for the special case of $L-B=1$ all of the spurs are aliased on top of one another. In conclusion, for any practical DDFS design, it has also been proven that the only spur overlap in the spectrum will be between the $\zeta_{\Lambda+}$ and the $\zeta_{\Lambda-}$ spurs. It has also been shown that when calculating the spectrum of a DDFS no adjustment has to be made to add these two spurs together since their values are "pre-doubled" when their amplitude is calculated using (27).

Now that the conditions for spur overlap and the criteria for spur existence at a frequency number have been determined, the congruences given in (29) can be solved for K. This will enable the amplitude of the spurious response to be calculated from the frequency number in the digital spectrum. Equation (29), which describes the frequency location in terms of K, can be rewritten as

$$F_n = \langle \Gamma \pm 2^{L-B}K\Gamma \rangle_{\frac{2^L}{(F_r,2^L)}} \tag{40}$$

By subtracting and applying the modulus operator to both sides and simplifying we obtain

$$\left\langle \frac{F_n-\Gamma}{2^{L-B}} \right\rangle_{\frac{2^L}{(F_r,2^L)}} = \langle \pm K\Gamma \rangle_{\frac{2^L}{(F_n,2^L)}} \tag{41}$$

By applying Euler's Theorem using the same justification as in (18) and exploiting the fact that $2^B/(F_r,2^L) = 2^B/(F_r,2^B)$ in all cases where spurs due to phase truncation exist

$$\left\langle \frac{F_n-\Gamma}{2^{L-B}} \Gamma^{\Phi(2\Lambda)} \right\rangle_{2\Lambda} = \pm K \tag{42}$$

As shown earlier $\Phi(2\Lambda) = \Lambda$ for Λ a power of 2 therefore

$$\left\langle \frac{F_n-\Gamma}{2^{L-B}} \Gamma^{\Lambda-1} \right\rangle_{2\Lambda} = \pm K \tag{43}$$

This yields an expression for the K value of a spur (from which its amplitude can be calculated) from its frequency location in the digital spectrum. This result however presumes that one is interested in the spectrum modulo $2^L/(F_r,2^L)$. In reality the spectrum from 0 to $2^{L-1}/(F_r,2^L)$ is desired, with all of the spurs whose frequency numbers lie between $2^{L-1}/(F_r,2^L)$ and $2^L/(F_r,2^L)$ aliased back into the range 0 to $2^{L-1}/(F_r,2^L)$. If the location of a spur were calculated using (40), and if the resulting spur number, F_n, was larger than $2^L/(F_r,2^L)$, then the true aliased position of the spur could be calculated by $F_n = -F_n + 2^L/(F_r,2^L)$. Modifying (40) to account for the aliasing of the spurs in the range $2^{L-1}/(F_r,2^L)$ to $2^L/(F_r,2^L)$ we obtain

$$\frac{2^L}{(F_r,2^L)} - F_n = \langle \Gamma = 2^{L-B}K\Gamma \rangle_{\frac{2^L}{(F_r,2^L)}} \tag{44}$$

which simplifies to

$$\left\langle \frac{-F_n - \Gamma}{2^{L-B}} \Gamma^{\Lambda-1} \right\rangle_{2\Lambda} = \pm K \qquad (45)$$

Combining the result from (45) with that obtained in (43) gives an explicit expression for the spur amplitude for the two conditions given in (30) that determine if a spur exists at a frequency number. Therefore, if the fraction $F_n - \Lambda$ is an integer, then (43) can be used to calculate the spur amplitude, and if $F_n - \Lambda$ is an integer, then (45) can be used.

Therefore, the complete algorithm for determining the digital spectrum of a DDFS in the presence of phase truncation can be summarized as follows

> 1) Divide the frequency range from 0 to $F_{clk}/2$ into a sequence of $2^{L-1}/(F_r 2^L)$ evenly spaced potential spur locations, each sequentially numbered. The sequential frequency number of a spur location is related to the actual analog spur frequency, F_{sp}, by
>
> $$F_n = \frac{F_{sp} 2^L}{(F_r 2^L) F_{clk}}$$
>
> 2) Find the spur number, K, from the frequency number, F_n, using the following rules:
>
>   a) If both $F_n - \Lambda$ and $-F_n - \Lambda$ are not divisible by 2 then the magnitude of the spur at F_n is 0.
>
>   b) If 2 divides $F_n - \Lambda$ then
>
> $$\left\langle \frac{F_n - \Gamma}{2^{L-B}} \Gamma^{\Lambda-1} \right\rangle_{2\Lambda} = K$$
>
>   If 2 divides $-F_n - \Lambda$ then
>
> $$\left\langle \frac{-F_n - \Gamma}{2^{L-B}} \Gamma^{\Lambda-1} \right\rangle_{2\Lambda} = K$$
>
>   where we define $\Lambda = \frac{2^{B-1}}{(F_r 2^B)}$
>
> 3) The magnitude of the spurious noise line at F_n is
>
> $$\zeta_{K2} = \frac{\pi 2^{B-L}}{2\Lambda} \operatorname{cosec}\left(\frac{K\pi}{2\Lambda}\right)$$

III. IMPLICATION OF RESULTS

The spectrums generated using this algorithm were compared to those generated using a DFT and were found to exactly predict the spectrum of the DDFS to within the roundoff error involved in the FFT algorithm. All spur magnitudes larger than -120dB agreed with the DFT results to within 1%.

Now that an explicit relationship has been derived for the output spectrum due to phase truncation, several observations can be made about the interaction of the finite word length effects and the frequency control command, F_r. From (27) it is evident that the spectrums generated from values of F_r that have the same value of $(F_r, 2^B)$ have a one to one correspondence between the magnitudes of the spurs. In other words, in the two different spectrums the magnitudes of all the spurs will be the same, and only their position in the spectrum will change. This behavior provides a clue as to the number theoretic operation of the phase accumulator. For simplicity we will consider only odd values of F_r such that $(F_r, 2^B) = 1$. If one considers the output of one numerical period of the phase accumulator to be a vector, $\underline{V}$, of length 2^L where the basic vector, $\underline{V}_r$, is provided by $F_r = 1$ and is composed of the consecutive integers from 0 to $2^L - 1$. Under this model the phase accumulator can be considered as a permutation generator, where each value of F_r provides a different permutation of the values from 0 to $2^L - 1$ as given by

$$\left\langle F_r n \right\rangle_{2^L} = \underline{V}_r[n]$$

Where $\underline{V}_r[n]$ is defined as the nth element of the vector $\underline{V}_r$.

This property provides considerable insight into the nature of the spectrum of the DDFS. It can be shown that permutation operators of this class are commutative with the DFT operator. Therefore, a permutation of this type to the input vector to a DFT will result in an identical permutation of the transformed output vector. Therefore, the spectrum of a DDFS due to any distortions in the system which are the result of a memoryless phenomenon can be generated by using one DFT for each class of $(F_r, 2^B)$, and permuting the results for the desired F_r.

Returning to the analysis of the spurs due to phase truncation only, a relationship for the magnitude of the largest spur in the spectrum may be

obtained from (27). Because ζ_K is a monotonically decreasing function of K, the magnitude of the largest spur may be obtained by substituting $K = 1$. Using this fact the magnitude of the largest spur in the spectrum is given by

$$\zeta_{worst} = \zeta_{1\pm} = 2^{B-L} \frac{\frac{\pi(F_r, 2^B)}{2^B}}{\sin\left(\frac{\pi(F_r, 2^B)}{2^B}\right)} \qquad (48)$$

From (47) we see an important relationship between the magnitude of the worst case spur and $(F_r, 2^B)$. For values of F_r such that $(F_r, 2^B) = 2^B$, spurs due to phase truncation are nonexistent. However for values of F_r such that $(F_r, 2^B) < 2^B$, the magnitude of the largest spurs in the spectrum is a decreasing function of $(F_r, 2^B)$ with the maximum value provided by $(F_n, 2^B) = 2^{B-1}$. This relationship is shown graphically in Figure 12.

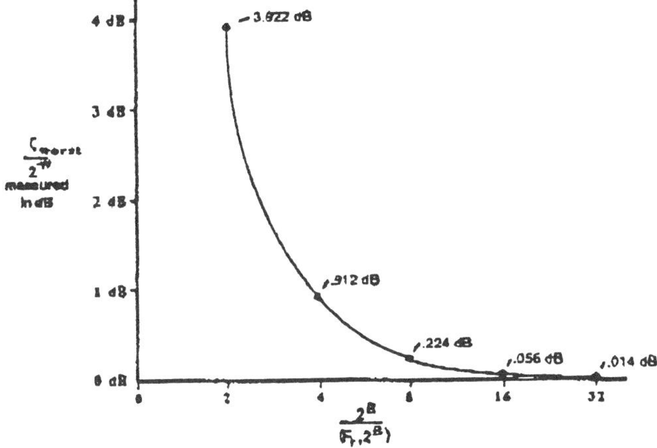

Figure 12. Worst Case Spur Amplitude vs. $\frac{2^B}{(F_r, 2^B)}$

Note from Figure 12 that if $(F_r, 2^B) = 1$, then as B gets larger, the magnitude of the worst case spur asymptotically approaches a value of $2^{B-L} = 2^{-W}$. This is 3.922dB less than the largest worst case spur which would result from the same B, but unrestricted values of F_r. The magnitude approaches this limit very quickly, so that for B larger than 4 bits and with $(F_n, 2^B) = 1$, the magnitude of the worst case spur can be considered to be 2^{-W}.

PHASE ACCUMULATOR MODIFICATION

This phenomenon suggests a simple method for improving the worst case spur performance. If a scheme could be developed to prevent values of F_r that would result in $(F_r, 2^B) = 2^{B-1}$ from being input to the phase accumulator, then the performance of the DDFS could be significantly improved. One brute force approach would be to only allow odd values of F_r to be used. This would provide a performance improvement of 3.922 dB predicted above for $B > 4$ at the expense of halving the frequency resolution of the synthesizer. However, a much more elegant solution is possible involving only a small amount of additional hardware that provides the 3.922 dB improvement, without reducing the frequency resolution. This solution is obtained by observing the behavior of the least significant bit of the phase accumulator when the input frequency control command, F_r, is odd. This condition, as mentioned previously, corresponds to the minimum worst case spur condition. Under these circumstances, since the carry-in to the adder in the least significant bit of the phase accumulator is zero, the least significant bit of the phase accumulator register will oscillate between one and zero every clock cycle. Therefore, the goal of the hardware addition is to modify the existing L bit phase accumulator structure to emulate the operation of a phase accumulator with a word length of L+1 bits under the assumption that the least significant bit of the frequency control word is always a one. Such a hardware addition is shown in Figure 13. As seen in the figure, the additional flip-flop and inverter synthesize the same carry-in sequence to the L bit adder that the least significant bit of an L+1 bit adder with an odd F_r input would propagate to its L most significant bits. Thus, this circuit simulates the doubling of the frequency resolution of the phase accumulator by adding an extra bit to its word length, while simultaneously halving the resolution by restricting the input F_r to be an odd number. The net result is a phase accumulator with the same frequency resolution but with a 3.922 dB performance improvement and no restriction on the input value of F_r. Fortunately this spur performance gain is implemented without adding any hardware that would increase the critical path delay through the adder therefore the throughput of the DDFS is not degraded by the modification.

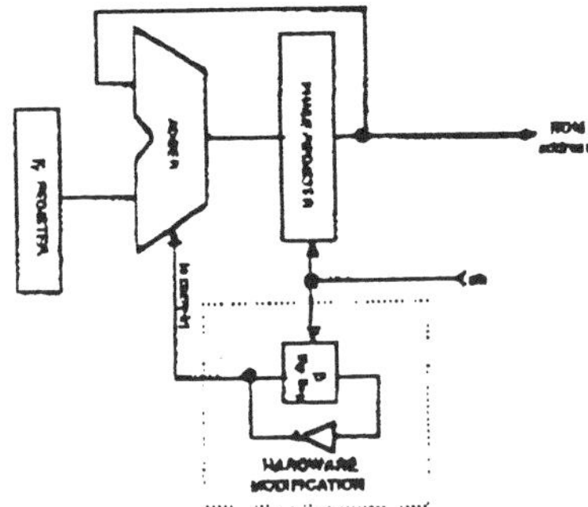

Figure 13. Hardware Modification to Force $(F_r, 2^B) = 1$

INTERACTION OF ALL FINITE WORD LENGTH EFFECTS

This modification also has the benefit of simplifying the analysis of the DDFS since the magnitude of the worst case spur is now independent of F_r. Using previously derived relations for the spurious response due to phase truncation we may now graph the worst case response due to all finite word length effects. In this graph it is assumed that all error effects are additive and that the errors due to finite word length effects in the ROM and DAC both contribute a potential spur of magnitude 1/2 LSB.

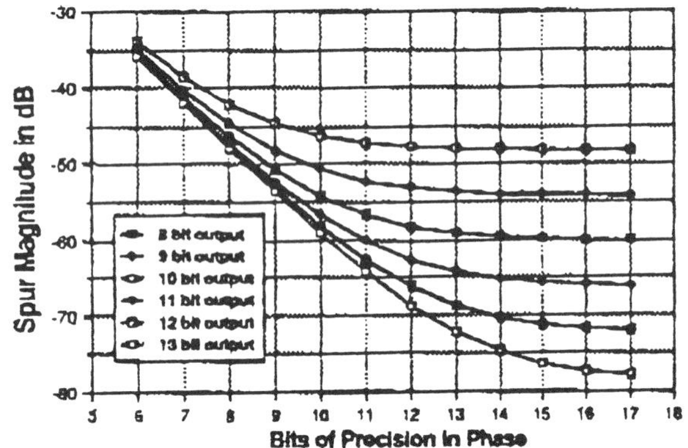

Figure 14. Worst Case Spurious Response of DDFS for all Finite Word Length Effects.

IV. CONCLUSION

In this paper, a simple algorithm is presented for generating the output spectrum of DDFS's resulting from phase accumulator truncation. This algorithm is fast and requires very little memory since the spectrum is computed sequentially. The computational burden of this algorithm is small enough that all the significant spurs in spectrums containing 2^{12} points have been generated on a hand held calculator (HP-41C). These results have been compared to those obtained by a DFT and have been shown to have comparable accuracy.

In the process of deriving the algorithm, several important insights were gained into the relationship between the amount of phase truncation, the choice of frequency control word, and the resulting output spectrum. One of these was the realization that input values of F_r with the same $(F_r, 2^B)$ will generate DDFS spectrums that differ only in the positions of the spurs. The relative amplitudes of these spurs and the total number of spurs in these spectrums will be the same. Another realization was that all phase accumulator output sequences with the same value of $(F_r, 2^B)$ are related to each other by a number theoretic permutation which is commutative with the DFT operator. This property thereby allows the spectrum of all DDFS output sequences with the same $(F_r, 2^B)$ to be generated from a single basic spectrum.

The derivation of Equation (27) for the DDFS spur amplitudes due to phase truncation yielded an important relationship between the value of $(F_r, 2^B)$ and the amplitude of the worst case spur. This relationship was used to propose a modification to the conventional phase accumulator structure to reduce the magnitude of the worst case spur. This hardware modification also reduced the detrimental effects of roundoff error and DAC nonlinearities on the spectral purity of the output sinewave. The modification also allows the output spectrum for any value of F_r to generated from a permutation of a singe spectrum without worrying about each different class of $(F_r, 2^B)$.

The use of the new phase accumulator structure also allowed the generation a design table which generalizes the worst case spurious behavior due to the additive effects of all finite wordlength phenomena. This table gives the worst case spurious response as a function of phase accumulator precision and the wordlength of the stored ROM samples.

In conclusion, the relationships derived in this paper showing the interaction of phase accumulator truncation, ROM table length, and output spurious noise level should greatly facilitate the design of DDFS's with the optimum hardware configuration for any requirements. Also, the implementation of the novel phase accumulator structure should greatly improve DDFS performance by reducing the worst case spur levels and by increasing the tolerance to roundoff errors and DAC nonlinearities.

V. ACKNOWLEDGMENTS

This research was supported in part by the University of California MICRO Program in cooperation with TRW Inc. under Grant 86-110.

The authors would like to thank Dr. Douglas N. Green of TRW for originally introducing them to the problems associated with the analysis of direct digital frequency synthesizers and also Mr. Gerald R. Fischer, Mr. Peter J. Hadinger, and Mr. William F. McDonnell of TRW for their valuable discussions and comments on this work.

VI. REFERENCES

[1] Albert L. Bramble, "Direct Digital Frequency Synthesis", Proc. 35th annual Frequency Control Symposium, USERACOM, Ft. Monmouth, N.J., May 1981, pp.406-414

[2] J. Tierney, C.M. Rader, and B. Gold, "A digital frequency synthesizer," IEEE Trans. Audio Electroacoust., vol, AU-19, pp. 48-57, 1971

[3] L. Jolley, Summation of Series, Chapman and Hall, London, 1925.

[4] H.T. Nicholas, "The Determination of the Output Spectrum of Direct Digital Frequency Synthesizers in the Presence of Phase Accumulator Truncation", Thesis, Masters of Science in Electrical Engineering, UCLA, 1985.

[5] C.E. Wheatley III and D.E. Phillips, "Spurious suppression in direct digital synthesizers," Proc. 35th annual Frequency Control Symposium, USERACOM, Ft. Monmouth, N.J., May 1981, pp.428-435

[6] D.A. Sutherland, R.A. Strauch, S.S. Wharfield, H.T. Peterson, and C.R. Cole, "CMOS/SOS Frequency Synthesizer LSI Circuit for Spread Spectrum Communications" IEEE Journal of Solid State Circuits, vol. SC-19, pp.497-505, Aug. 1984

[7] J. Gorski-Popiel, Frequency Synthesis: Techniques and Applications, New York: IEEE press, 1975

[8] V. Manassewitsch, Frequency Synthesizers, Theory and Design, New York: John Wiley, 2nd Ed., 1979

The Optimization of Direct Digital Frequency Synthesizer Performance in the Presence of Finite Word Length Effects

Henry T. Nicholas, III
TRW Digital Applications Laboratory
One Space Park, M5/1017
Redondo Beach, CA 90278

Henry Samueli
Electrical Engineering Department
University of California, Los Angeles
Los Angeles, CA 90024

Bruce Kim
Intel Corp.

Abstract

Techniques for optimizing the performance of Direct Digital Frequency Synthesizers (DDFS) that use the Tierney, Rader, Gold architecture [1] are presented. These techniques permit the optimal partitioning of DDFS hardware to minimize the effects of finite phase resolution and finite output word-length. Using these results, given constraints on the phase resolution and output word-length, DDFS performance can be optimized under either a worst case spurious response criterion or under a total signal to noise ratio criterion. These optimization techniques are also used to analyze the errors introduced by the compression of the required look-up table storage size. Two look-up table compression techniques are examined: the Sunderland technique [3][4], and a proposed modification of the Sunderland technique. It is shown that by using a different optimization technique to choose the look-up table samples, the algorithmic error incurred by the Sunderland architecture can be improved by approximately 12 dB. Furthermore, a new compression technique is proposed, based upon an extension of the Sunderland technique, which provides additional improvements in the look-up table storage compression. Finally an actual DDFS design is presented that utilizes this new compression technique and which provides a simulated digital spurious rejection of 90.2 dB, a 14 bit digital output, and a frequency resolution of .023 Hz at the maximum simulated clock frequency of 100 Mhz. This design, which will be fabricated in a radiation hardened 1.25µM CMOS process is presently in the layout phase of design and will enter fabrication in November '88.

Introduction

Direct Digital Frequency Synthesis is an indispensable technique for generating reference frequencies whenever extremely precise frequency resolution and fast switching speed are required. The most common DDFS architecture is the Tierney, Rader, Gold, architecture [1]. This architecture, which has been well documented in the literature [5][6][7], synthesizes a sine wave by using a periodically overflowing 2's complement phase accumulator to generate and store phase information, and uses a Read-Only Memory (ROM) based look-up table to compute the sine function. There are three sources of noise which are inherent to all DDFS implementations, in addition to the noise generated in the D/A conversion process. The first source of noise is $\Gamma(n)$, the distortion due to phase truncation at the input to the sine function computing hardware, which is usually a ROM. The second is $g(\cdot)$, which is a nonlinear distortion that is usually present when methods of compressing the storage requirements of the look-up table are employed, and the third is $A(n)$, the noise introduced by the finite precision of the sine samples stored in the look-up table. These noise sources are depicted symbolically in Figure 1. This paper will be concerned with the characterization of these sources and their relation to the hardware requirements.

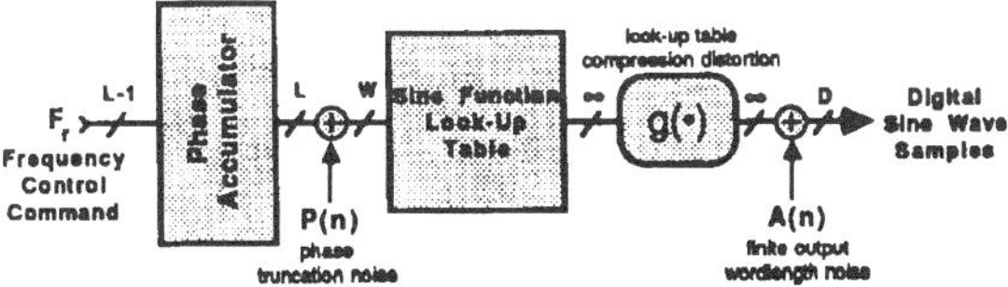

Figure 1. DDFS Finite Word-Length Effects.

Effects of Finite Phase Word-Length

There are two finite word-length effects in the phase accumulator that effect DDFS performance: the word-length of the phase accumulator register, denoted as L in Figure 1, and the truncation of the phase accumulator output, which is used to index the sine function look-up table, which is denoted by B. An understanding of these two finite word-length effects is extremely important because they have a profound impact upon the spectral characteristics of all other system nonlinearities. An exact spectral analysis of DDFS spurious performance was presented in [2], which provided an algorithm for determining the magnitude and position of all spurs generated by phase accumulator truncation. This algorithm is restated in Figure 2 to emphasize some of the subtleties of phase accumulator operation that impact the overall DDFS spurious spectrum.

Number of spurs $= 2\Lambda - 1$

Magnitude of all spurs $\varsigma_K = \dfrac{\pi \, 2^{-L} \, (F_r, 2^B)}{\sin\left(\dfrac{\pi \, K \, (F_r, 2^B)}{2^B}\right)}$

Position of the spurs in the spectrum (between 0 and $\dfrac{2^L}{(F_n, 2^L)}$)

$K = \left\langle \dfrac{F_n - \Gamma}{2^{L-B}} \Gamma \Lambda^{-1} \right\rangle_{2\Lambda}$ for $F_n - \Lambda$ divisible by 2

$K = \left\langle \dfrac{-F_n - \Gamma}{2^{L-B}} \Gamma \Lambda^{-1} \right\rangle_{2\Lambda}$ for $F_n - \Lambda$ divisible by 2

$\varsigma_K = 0$ otherwise

B = number of phase bits truncated

L = phase accumulator word length $\qquad \Gamma = \dfrac{F_r}{(F_r, 2^B)} \qquad \Lambda = \dfrac{2^{B-1}}{(F_r, 2^B)}$

(a,b) = greatest common divisor of a and b

Figure 2. Exact Spectrum of Phase Truncation Spurs

This algorithm provides an expression for the number of spurs generated, their amplitude, and their position in the spectrum as a function of B, the number of bits truncated from the phase accumulator, L, the phase accumulator word-length, and the frequency control word, F_r. These results illuminate some important properties of phase accumulator operation that effect the overall DDFS error spectrum. The most important property to note is that the number of spurs and the magnitude of the spurs depend on F_r only through the greatest common divisor function of F_r and 2^B, which is denoted by $(F_r, 2^B)$. As a result of this, for input values of F_r that have the same value of $(F_r, 2^B)$, the number of spurs and their respective amplitudes remain the same. Only the position of each spur in the spectrum changes. This phenomenon is a result of certain number theoretic properties of the phase accumulator. These properties are most easily demonstrated if we view the sequence of numbers formed by one numerical period of the phase accumulator as a single time series vector, $\mathbf{X}$. In this representation the length of one numerical period of the phase accumulator is defined as the length of the sequence of numbers that may be output by the phase accumulator before the first number is repeated. As was shown in [2], the length of this

Reprinted from *Proceedings of the 42nd Annual Frequency Control Symposium*, pp. 357-363, 1988.

sequence is $2^L/(F_r 2^L)$. We then define $_1X$ as the vector of length $2^L/(F_r 2^L)$, which is formed by the consecutive output samples of the phase accumulator when the input is $F_r = 1$. The elements of this vector are then defined using the notation:

$$_1X = [\ _1X_0\ ,\ _1X_1\ ,\ _1X_2\ ,\ _1X_3\ ,\ \bullet\bullet\bullet\ ,\ _1X_n\ ]^T \qquad (1)$$

This vector is printed as the transpose of a row vector for typographical convenience. Using this notation, the vector $_{F_r}X$ can be formed by using:

$$_1X_n = \left\langle F_r n \right\rangle_{2^L} \qquad (2)$$

to denote the value of each element of the vector from $n = 0$ to $2^L/(F_r 2^L) - 1$. To illustrate this concept, the example of a 3 bit accumulator is shown in figure 3. For this example, if F_r is odd then $(F_r 2^L)$ will be 1, thereby generating a time vector of length 8. The column vector on the left represents the time vector that results when $F_r = 1$. For this simple example, the 3 bit accumulator starting with an initial phase accumulator register value of zero, simply adds "1" to the register contents every clock cycle until the register overflows, and returns to zero. This result agrees with equation (2) with the elements $_1X_0 = 0$, $_1X_1 = 1$, $_1X_2 = 2$, ect. The column vector to the right in figure 3 represents the time vector formed when $F_r = 3$. Note that although the phase accumulator accumulates and overflows three times faster for $F_r = 3$ than for $F_r = 1$, the numeric period of the two output sequences are the same.

An extremely important "side effect" of having the same numerical periodicity is that the column vector for $F_r = 3$ can be formed from a permutation of the values of the $F_r = 1$ vector. This permutation is represented by the arrows in Figure 3. It can be shown from the definition of the time vector in equation (2) that any phase accumulator output vector, $_aX$, can be formed from the permutation of another output vector, $_bX$, if $(a, 2^L) = (b, 2^L)$. The set of permutations that relate these output vectors belongs to a very special class of number theoretic relationships called affine permutations. Equation (2) thus defines the set of affine permutations that relate all time output vectors to the basic output vector, $_1X$, when $(F_r 2^L) = 1$.

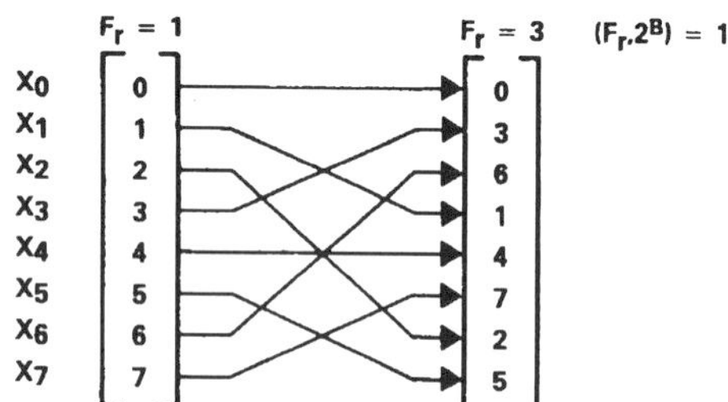

Figure 3. Time Series Vectors for a 3-Bit Accumulator for $F_r = 1$ and $F_r = 3$.

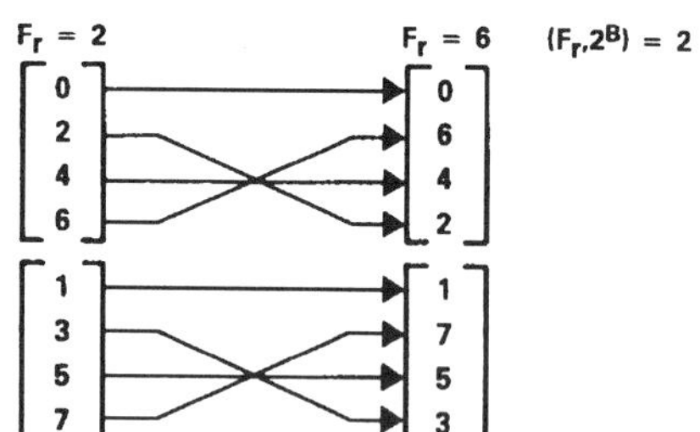

Figure 4. Time Series Vectors for a 3-Bit Accumulator for $F_r = 2$ and $F_r = 6$.

In Figure 4 we see the time vectors formed from values of $F_r = 2$ and $F_r = 6$. From the figure it is evident that the phase accumulator is now characterized by having two different sets of possible output vectors depending upon the initial contents of the phase accumulator. The top set of vectors in the figure represent the sequence of numbers output from the accumulator with an initial accumulator contents of 0. The bottom set of vectors represent the case for an initial accumulator contents of 1. In general, the phase accumulator will be characterized as having $(F_r 2^L)$ different "families" of output sequences to characterize its response, where output sequences belonging to the same "family" are defined as those having the same initial phase accumulator contents. As shown in Figure 4, time vectors belonging to the same family are related by a set of affine permutations of the same form as defined in equation (2).

There are several fundamental implications of these number theoretic properties of the phase accumulator. First, the worst case spurious response of the DDFS as a result of all memoryless system error effects will have exactly $(F_r 2^L)$ different values, one for each initial phase accumulator content from 0 to $(F_r 2^L) - 1$. This means that simulations of the spectrum of DDFS nonlinearities are only valid for initial phase accumulator contents belonging to the same time series vector.

Another important implication is associated with the affine permutation relationship that links time output vectors of the same family. In equation (3) we see that the time output vector for any value of F_r can be formed from a permutation of the individual elements of the vector for $F_r = 1$ by permuting the vector indices using $i' = \left\langle F_r i \right\rangle_{2^L}$.

$$_1X = [\ _1X_0\ ,\ _1X_1\ ,\ _1X_2\ ,\ _1X_3\ ,\ \bullet\bullet\bullet\ ,\ _1X_n\ ]^T$$
$$\text{vector } F_r = 1 \qquad (3)$$

$$_{F_r}X = [\ _1X_{\langle F_r 0 \rangle_{2^L}}\ ,\ _1X_{\langle F_r 1 \rangle_{2^L}}\ ,\ _1X_{\langle F_r 2 \rangle_{2^L}}\ ,\ \bullet\bullet\bullet\ ,\ _1X_{\langle F_r n \rangle_{2^L}}\ ]^T$$
$$\text{vector for arbitrary } F_r$$

It can be shown that the discrete Fourier transform operator is invariant under permutations of this class applied in both the time and frequency domain. If we assume that in general the DDFS operates by applying some memoryless nonlinear function $S\{\bullet\}$ to the phase accumulator output to produce the sine function, then the DDFS output at every point in time for a given F_r may be represented as function of one of the indices of $_{F_r}X$, $S\{_{F_r}X_n\}$. Therefore the spectrum of the DDFS output may be represented as:

$$N = \frac{2^L}{(F_r 2^L)} \qquad S\{K\} = \sum_{n=0}^{N-1} S\{_{F_r}X_n\}\ e^{-j2\pi Kn/N} \qquad (4)$$

As was shown in equation (3), all input time vectors may be formed from a permutation of another time vector by permuting the indices. If this is applied to the definition of the DDFS output spectrum in (), it can be shown that the following identity holds for affine permutations of the class $\langle F_r i \rangle_{2^L}$:

$$S\{\langle F_r K \rangle_{2^L}\} = \sum_{n=0}^{N-1} S\{_{F_r}X_{\langle F_r n \rangle_{2^L}}\}\ e^{-j2\pi Kn/N} \qquad (5)$$

Therefore a permutation of the samples in the time domain results in an identical permutation of the DFT samples in the frequency domain. In practical terms this means that the spurious spectrum due to all system nonlinearities can be generated from a permutation of another spectrum of the same number theoretic class. Thus, if we restrict the allowable values of F_r such that $(F_r 2^L)$ is always the same (such as by only allowing F_r to be odd) then all spurious spectrums may be generated from permutations of a single basic spectrum. If it is desired to optimize the worst case spurious response, then only one spectrum need be calculated for all values of F_r with common values of $(F_r, 2^L)$, since each spectrum will differ only in the position of the spurs and not in the magnitudes.

A simple modification to the basic phase accumulator structure was introduced in [2] that causes all phase accumulator output sequence to belong to the number theoretic class $(F_r, 2^{L+1}) = 1$, regardless of the value of F_r. By using this phase accumulator modification, which is depicted in Figure 5, only one simulation need be performed to determine the value of the worst case spurious response or the total signal to noise ratio due to memoryless system nonlinearities. In addition, this modification has the added advantages of eliminating any dependence of the output spectrum on the initial phase accumulator contents, and of reducing the magnitude of the worst case spurious response by 3.92 dB.

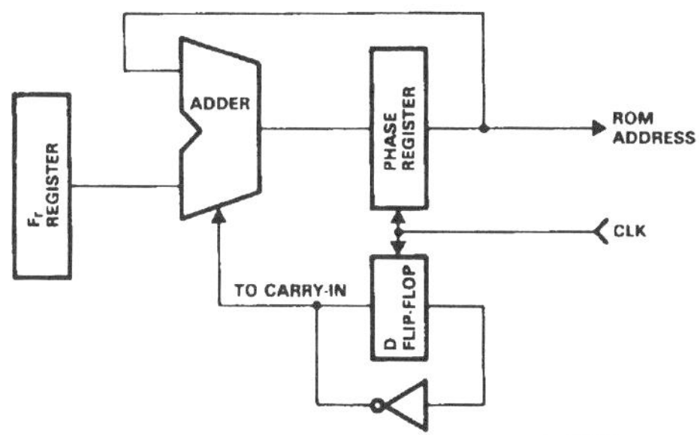

Figure 5. Modified Phase Accumulator For $(F_r, 2^{L+1}) = 1$.

Finite Amplitude Word-Length Effects

The results relating the number theoretic behavior of the phase accumulator finite word-length effects to the DDFS output spectrum may now be applied to help bring about the optimization of the sine sample computing hardware. Two error effects are introduced in computing the sine samples from the phase argument. These are the error from the storage compression nonlinearity, and the error from the amplitude quantization of the sine samples, which are depicted in Figure 1. In the past these error effects have been treated as random phenomena. However, using the results from the previous section, it is now computationally feasible to treat them as a deterministic process.

Look-up Table Storage Compression

Exploitation of Sine Function Symmetry

The most elementary method of sine storage compression is to exploit the symmetry of the sine function about π and $\pi/2$. By properly inverting the phase and the amplitude of the sine function, look-up table samples need only be stored for phase values between 0 and $\pi/2$. The details of this method are shown in Figure 6. As shown in the figure, the sine function between 0 and π may be synthesized from the samples between 0 and $\pi/2$ by taking the phase modulo $\pi/2$ and then taking the absolute value of the phase. This is easily implemented in hardware by truncating the phase MSB and then using the second MSB to full wave rectify the magnitude of the phase. As shown in Figure 6, the sampled waveform at the output of the look-up table is a full wave rectified version of the desired sine wave. The final output sine wave is then generated by multiplying the full wave rectified version by -1 when the phase is between π and 2π. This is accomplished simply by multiplying by the negative of the phase accumulator MSB.

In most practical DDFS digital implementations, numbers are represented in 2's complement format. Therefore the 2's complementor must be used to take the absolute value of the phase and multiply the output of the look-up table by -1. However, it can be shown that if a 1/2 LSB offset is introduced into the number that is to be complemented, then a 1's complementor may be used in place of the 2's complementor without introducing an error. This provides a considerable savings in hardware since a 1's complementor may be implemented as a simple exclusive-or gate. This 1/2 LSB offset is provided by choosing the look-up table samples such that there is a 1/2 LSB offset in both the phase and the amplitude of the samples.

Compression of $\mathrm{Sin}(\theta)$ for $0 < \theta < \pi/2$

Prior to the publication of this paper, the most effective method of compressing the look-up table storage requirements for $0 < \theta < \pi/2$ was introduced by Sunderland et. al. [3][4]. This architecture, which is shown in Figure 7, provides a reduction in the look-up table storage requirements by replacing the storage requirements of one large ROM of size 2^{A+B+C} words with two smaller ROM's of sizes 2^{A+B} words and 2^{A+C} words, whose outputs are added together to reconstruct the sine function. This provides a storage compression ratio of 45,056 bits to 3,840 bits, or 11.7:1. The assignment of look-up table samples in this architecture is based upon several trigonometric approximations. First, the bits representing the phase argument, θ, of one quarter period of the sine function are decomposed into the sum of three functions: $\alpha < (\pi/2)$, $\beta < (\pi/2)(2^{-A})$, and $\chi < (\pi/2)(2^{-(A+B)})$ such that

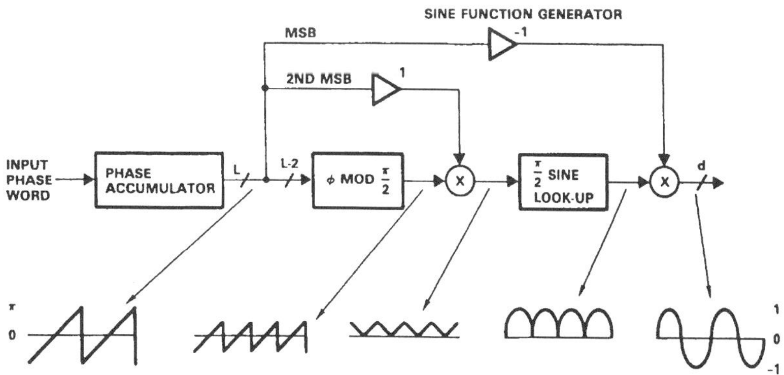

Figure 6. Logic to Exploit Quarter Wave Symmetry.

$\theta = \alpha + \beta + \chi$ [3]. Using trigonometric identities, the sine function is given by

$$\sin(\alpha+\beta+\chi) =$$
$$\sin(\alpha+\beta)\cos(\chi) + \cos(\alpha)\cos(\beta)\sin(\chi) - \sin(\alpha)\sin(\beta)\sin(\chi) \quad (6)$$

using the relative magnitudes of α, β, and χ, (6) may be approximated by

$$\sin(\alpha+\beta+\chi) \approx \sin(\alpha+\beta) + \cos(\alpha)\sin(\chi) \quad (7)$$

The contents of the first ROM is then chosen to be $\sin(\alpha+\beta)$ and the second ROM stores the result of $\cos(\alpha)\sin(\chi)$. The error resulting from this approximation for a 12 bit output with a phase word segmentation of A = 4, B = 4, and C = 4, which will be denoted as a (4,4,4) segmentation, can be shown to be approximately -72.2 dB.

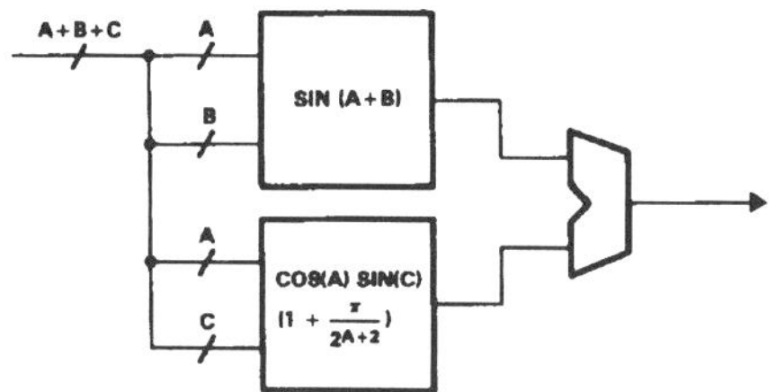

Figure 7. Sunderland Architecture.

However, significant improvements can be achieved, over the former results by using an alternate methodology for choosing the samples to be stored in the ROM's. Improvements in the spurious performance can be gained by modeling the segmentation of the phase arguments into the two ROM's as a low phase resolution course ROM, with additional phase resolution added by providing interpolation between the course phase samples using the fine ROM. We define as before, $\theta = \alpha + \beta + \chi$, with the word-length of the variable α to be A, the word-length of β to be B, and χ to be C. The sine function between 0 and $\pi/2$ is divided into 2^B different regions. In each region the same 2^B fine ROM interpolation samples are used for interpolation between course ROM samples given the same value for the variable χ. An example of this type of coarse-fine interpolation is shown in Figure 8 for the case of a 6 bit phase argument using a segmentation of A = 2, B = 2, and C = 2.

In this example, the course ROM samples are represented by the "heavy dots" along the dashed line, and the fine ROM samples are chosen to be the difference between the value of the sine function along the dashed line and the value value of the "error bars" directly below or above that point on the dashed line. In the example, the sine function is divided into 2^α quadrants, correspond to $\alpha = 00, 01, 10,$ and 11. Within each quadrant, only one correction value may be used between the error bars and the dashed line for all χ offsets from a coarse sample. The value of the offset used for each value of χ is chosen to minimize either the mean square or the maximum error of all offsets for each value of β within the quadrant. The actual algorithm for choosing the fine ROM samples to minimize the mean square error is shown in (8).

$$F_c(\alpha,\beta) = \sin\left(\frac{\pi}{2}\left(\frac{\alpha 2^B+\beta}{2^{A+B}} + \frac{1}{2^{A+B+C}}\right)\right) \quad (8)$$

$$F_f(\alpha,\chi) = \sum_{n=0}^{N-1}\frac{1}{2^B}\left[\sin\left(\frac{\pi}{2}\left(\frac{\alpha 2^{B+C}+\beta 2^C+\chi}{2^{A+B+C}} + \frac{1}{2^{A+B+C+1}}\right)\right) - F_c(\alpha,\beta)\right]$$

Where in (8) $F_c(\alpha,\beta)$ is defined to be the coarse ROM samples and $F_f(\alpha,\chi)$ is defined to be the fine ROM samples. The optimization criterion for choosing the samples to minimize the maximum absolute error is shown in (9).

$$F_f(\alpha,\chi) = \quad (9)$$
$$\frac{1}{2}MAX\left\{\sin\left(\frac{\pi}{2}\left(\frac{\alpha 2^{B+C}+\beta 2^C+\chi}{2^{A+B+C}} + \frac{1}{2^{A+B+C+1}}\right)\right)-F_c(\alpha,\beta)\right\}$$
$$-\ \frac{1}{2}MIN\left\{\sin\left(\frac{\pi}{2}\left(\frac{\alpha 2^{B+C}+\beta 2^C+\chi}{2^{A+B+C}} + \frac{1}{2^{A+B+C+1}}\right)\right)-F_c(\alpha,\beta)\right\}$$

Based upon simulation results, it was determined that the minimum-maximum error criterion tended to result in lower values of the maximum spur, although this result has not been explicitly proven. The simulations also showed that the mean square error criterion, provides the lowest total spur energy, in agreement with theory.

This methodology of choosing the coarse and fine ROM samples provides two advantages over previous methods. First, it provides superior spurious performance because the fine ROM samples may be chosen by computer optimization rather than by linear interpolation. Secondly, because this method of storage compression is not derived from trigonometric identities, it may be used to compress the storage of any arbitrary function, not just the sine function.

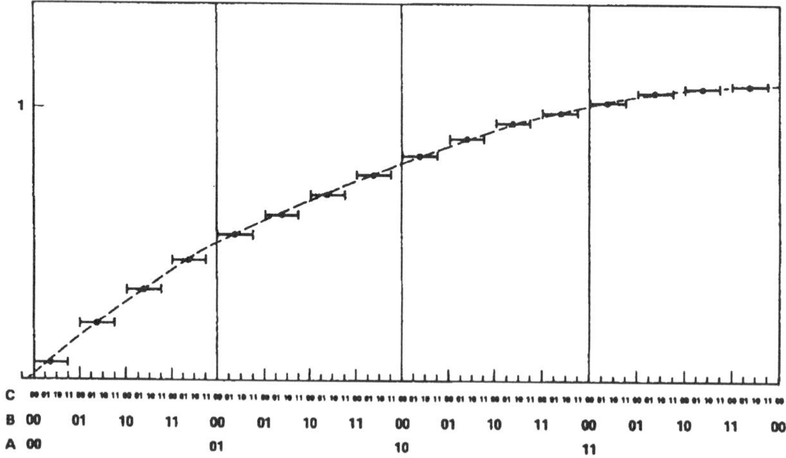

Figure 8. Coarse-Fine Interpolation for (2,2,2) Segmentation of the Phase.

A comparison was performed between the two different methodologies of choosing the ROM samples for the case of a hypothetical design with a phase precision of 12 bits. The segmentation was (4,4,4) for both cases. In the trigonometric derivation, the coarse ROM size was 256 x 11 bits and the fine ROM size was 256 x 4 bits. For the coarse/fine derivation, the coarse ROM size was 256 x 9 bits and the fine ROM size was 256 x 4 bits. The result of the simulations showed a worst case spurious performance of -72.2 dB for the trigonometric generation of the ROM samples and -84.2 dB for the coarse-fine generation of ROM samples. This demonstrates an improvement of 12 dB for the coarse-fine technique while simultaneously requiring less storage.

Sine-Phase Difference Algorithm

Since the coarse-fine algorithm can be used to store any arbitrary function, another technique may be employed to achieve even greater reduction in ROM size. This may be accomplished by storing the the function

$$f(x) = \sin(\pi x/2) - \pi x/2 \qquad (10)$$

instead of the sine function in the look-up table. The sine function is then generated by adding the phase argument from the phase accumulator to the output of the ROM's, as shown in Figure 9. The advantage of this technique is that it reduces the maximum amplitude of the function to be stored in the coarse ROM. Therefore the coarse ROM output word-length may be reduced and the output adder may be simplified. This modification reduces the look-up table storage requirements by at least of 2^{A+B+1} bits. The application of this technique to the hypothetical (4,4,4) segmented 12 bit output DDFS design considered earlier results in a savings of 512 bits in the coarse ROM and an increase in the compression ratio from 11.7:1 to 13.5:1. The penalty for this reduction in ROM storage is the addition of another adder. However, this is an advantageous tradeoff in the high speed VLSI implementation of a DDFS. The ROM propagation delay, which cannot be pipelined, is reduced, increasing the maximum clock frequency of the DDFS. The expense of another adder, which is readily pipelineable, is inconsequential in the full custom VLSI implementation.

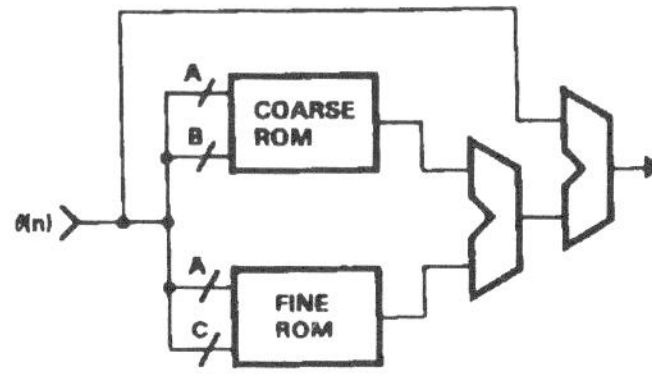

Figure 9. Sine-Phase Difference Algorithm.

Exploitation of Fine ROM Symmetry

The fine ROM size may also be reduced by exploiting the symmetry in the samples. As shown in Figure 10, if the coarse ROM samples are chosen in the middle of the interpolation region, then the fine ROM samples will be approximately symmetric about the $\chi = (2^C - 1)/2$ axis. Thus by using an adder/subtracter instead of and adder to sum the coarse and fine ROM values, the size of the fine ROM may be halved. Some additional complexity must be added to the adder/subtracter control logic if this technique is to be used with the sine-phase difference algorithm, since the slope of the function in equation (10) changes sign at a non symmetric point between 0 and $\pi/2$ on the x-axis. Fortunately, the digital logic required to perform this can be accomplished with less than four logic gates for the 12-bit phase case considered previously.

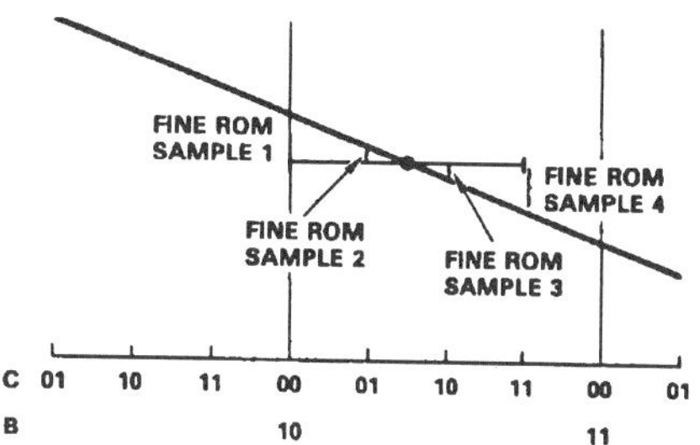

Figure 10. Exploitation of Fine ROM Symmetry.

Since the fine ROM is not generally in the critical speed path, the effective resolution of the fine ROM may be doubled at no extra hardware penalty, rather than halving the ROM. This is an extremely powerful technique because it allows the segmentation of the compression algorithm to be changed, effectively adding an extra bit of phase resolution to the look-up table, which thereby reduces the magnitude of the worst case spur due to phase accumulator truncation.

100 MHz DDFS Architecture Demonstration Vehicle

These architectural techniques were used to design a DDFS to meet a performance goal of better than -90 dB of spurious rejection with the potential of being pipelined to achieve a 100 Mhz clock rate in a standard 1.25 µM CMOS process. A 15-bit phase precision in the look-up table address was chosen, which from Figure 2 is shown to provide -90.3 dB of rejection in the magnitude worst case spur due to phase accumulator truncation. Next, the 15-bit phase must be segmented to address the coarse and fine ROM's. In this design it is assumed that the sine phase difference algorithm will be used, exploiting the symmetry of the fine ROM samples to double the effective fine ROM resolution. It is also assumed that the phase accumulator modification shown in Figure 5 is used, and that the quarter wave symmetry of the sinewave will be exploited, reducing the effective look-up table address requirements to 13 bits of phase precision for $0 < \theta < \pi/2$. To obtain the optimal segmentation of the 13-bit phase address to the coarse-fine ROM's, infinite output word-lengths were assumed in the simulations. Furthermore, it was heuristically determined that optimal segmentations of the phase argument have values of B and C that differ by at most 1, and that segmentations with values of $B > C$ perform better than $B < C$. Therefore, the optimization space of the word-lengths A, B, and C, were constrained such that $B = C$ or $B = C + 1$. Subject to these constraints, and that $A + B + C = 13$, the spurious response of the possible segmentations of the 13-bit phase were simulated. The results show the magnitude of the worst case spur in the spectrum plotted against the word-length variable A, and are pictured in Figure 11. A key concept to note in the graph is that the amount of storage required increases exponentially with increasing A, since the total ROM storage is given by $2^A(2^B + 2^C)$. Therefore the minimum value of A that meets the spurious rejection requirement should be chosen. The value of $A = 4$ corresponding to -95.33 dB meets the -90 dB requirement, which results in a phase segmentation of $A = 4$, $B = 4$, and $C = 5$. One anomaly in the graph in Figure 11 is the decreasing spurious performance as A is increased past 6 bits. This is a result of the fine ROM symmetry assumptions breaking down for small values of B. A similar set of simulations for a design without the exploitation of fine ROM symmetry show a monotonically increasing spurious performance with increasing values of A, as expected from theory.

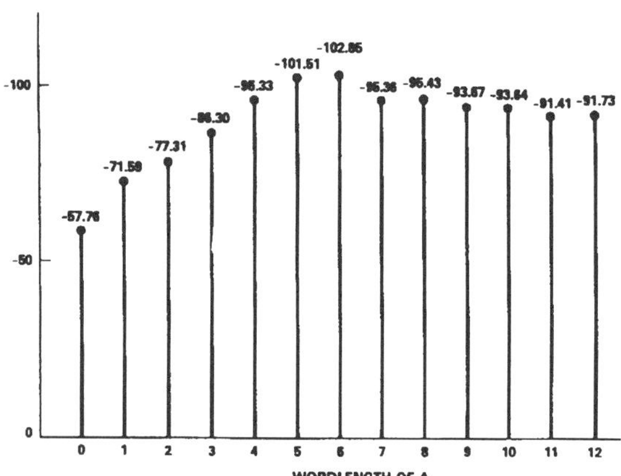

Figure 11. Worst Case Spurious Response for Different Phase Segmentations.

For the (4,4,5) segmentation of the phase, the effect of various output quantizations of the coarse ROM were simulated to determine the minimum word-length required to achieve the -90dB performance goal. The results of these simulations are shown in Figure 12. From these simulations, an output word-length of 9 bits was chosen for the coarse ROM, providing -89.9dB of spurious rejection including the effects of fine ROM quantization. The fine ROM word-length is provided as part of the simulation algorithm, where it is derived from the maximum amplitude of the samples calculated using equation (8). This resulted in a fine ROM word-length of 4 bits for a coarse ROM word-length of 9 bits.

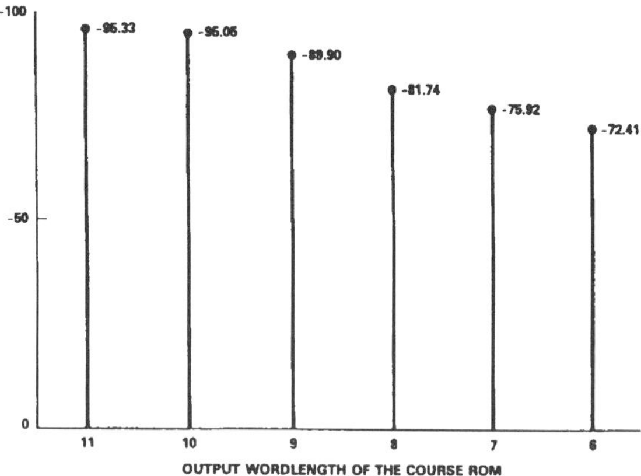

Figure 12. Spurious Response for Different Coarse ROM Output Word-Length Quantizations.

As a final optimization step, the amplitude values of the function to be stored in the look-up table can be scaled to provide improved performance in the presence of amplitude quantization. To accomplish this, the coarse and fine ROM samples were regenerated after multiplying the pre-quantized sinewave by a scaling constant that was varied in very small increments in the neighborhood of unity. The results of these simulations, which are graphed in Figure 13, show that a 1.75 dB improvement in the worst case spurious response can be achieved by using a scaling constant of .999994 instead of 1.0 to generate the coarse and fine ROM samples. This results in a negligible reduction in the amplitude of the synthesizer output, while providing a worst case spurious that is now limited by the spurs due to phase accumulator truncation rather than amplitude quantization.

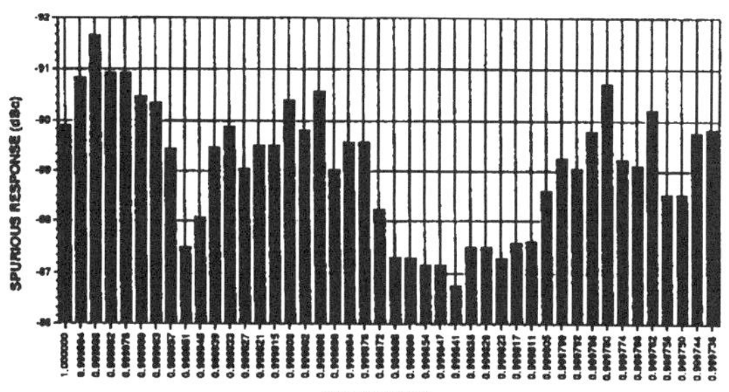

Figure 13. Spurious Improvement From Scaling Sinewave Amplitude.

DDFS Architecture Demonstration Vehicle

Figure 14 shows the detailed logic block diagram of the DDFS design exploiting all architectural innovations discussed in this paper. The design has a simulated worst case spurious response of -90.3 dB using only 3,072 bits of ROM storage, providing a total storage compression ratio of 37:1. The simulated output spectrum of this design is depicted in Figure 15, showing the high degree of digital spurious rejection attained using this architecture. This design provides a 14-bit sinewave output, where the most significant 12 bits represent the ideal 12-bit rounded samples of a sinewave. This design is presently being implemented as a single VLSI chip in TRW's VHSIC 1.25μM CMOS process. Detailed circuit simulations predict the chip will be capable of operating at a clock rate of 100 Mhz, providing a .023 Hz frequency resolution over a 33 MHz bandwidth. The key circuit structures have already been fabricated in TRW's CMOS process as a test chip, shown in Figure 16, and have validated all logic and circuit simulations. The final design is scheduled to enter fabrication in November '88.

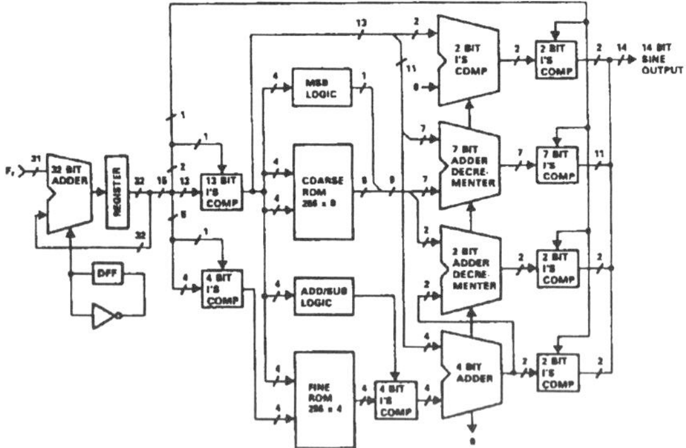

Figure 14. -90.3 dB Spurious DDFS Architecture.

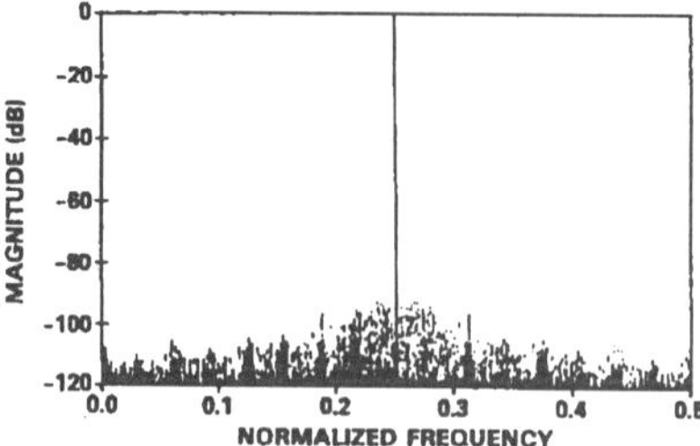

Figure 15. Simulated DDFS Output Spectrum.

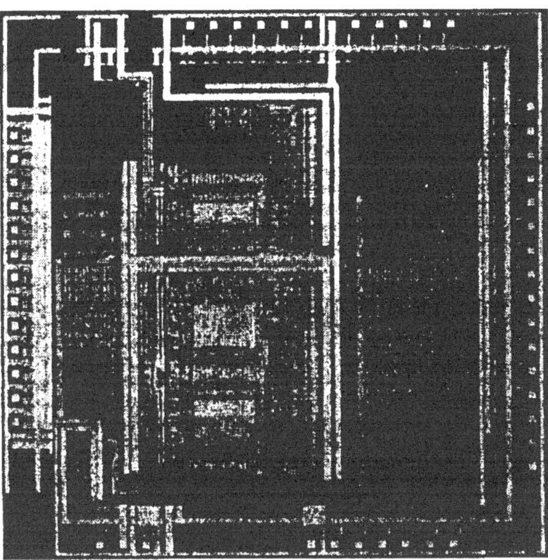

Figure 16. 1.25 µM CMOS DDFS Test Chip.

Summary

Techniques for the design of VLSI architectures for direct digital frequency synthesis have been introduced, which allow for the optimization of the spurious response in the presence of finite word length effects. These optimization techniques exploit certain number theoretic properties of the phase accumulator to make the exhaustive simulation of DDFS performance in the presence of different system nonlinearities computationally feasible. These design techniques have been applied to design a 14-bit output DDFS with a simulated spurious performance of -90.3 dB and a level of pipelining capable of allowing a 100 MHz clock rate in a 1.25 µM CMOS process.

Acknowledgements

The authors wish to thank M. Miscione of TRW Microelectronics Center for providing VLSI fabrication services and partial funding for this research. P. Hadinger and R. Smith of TRW Defense Communications Division also provided partial financial sponsorship and many valuable technical discussions on the systems applications of DDFS's.

References

[1] J. Tierney, C.M. Rader, and B. Gold, "A Digital Frequency Synthesizer," IEEE Trans. Audio Electroacoust., vol. AU-19, pp. 48-57, 1971

[2] H.T. Nicholas III and H. Samueli, "An Analysis of the Output Spectrum of Direct Digital Frequency Synthesizers in the Presence of Phase-Accumulator Truncation," Proc. 41st annual Frequency Control Symposium USERACOM, Ft. Monmouth, N.J., May 1987, pp.495-502

[3] D.A. Sunderland, R.A. Strauch, S.S. Wharfield, H.T. Peterson, and C.R. Cole, "CMOS/SOS Frequency Synthesizer LSI Circuit for Spread Spectrum Communications" IEEE Journal of Solid State Circuits, vol. SC-19, pp.497-505, Aug 1984

[4] R.E. Lundgren, V. S. Reinhardt, and K. W. Martin, "Designs and Architectures for EW/Communication Direct Digital Synthesizers", Research and Development Technical Report SLCET-TR-85-0424-1, U.S. Army Laboratory Command, Fort Monmouth, New Jersey, Aug 1986

[5] Albert L. Bramble, "Direct Digital Frequency Synthesis", Proc. 35th annual Frequency Control Symposium, USERACOM, Ft. Monmouth, N.J., May 1981, pp.406-414

[6] J. Gorski-Popiel, _Frequency Synthesis: Techniques and Applications,_ New York: IEEE press, 1975

[7] V. Manassewitsch, _Frequency Synthesizers. Theory and Design_, New York: John Wiley, 2nd Ed., 1979

[8] H.T. Nicholas, "The Determination of the Output Spectrum of Direct Digital Frequency Synthesizers in the Presence of Phase Accumulator Truncation", Thesis, Masters of Science in Electrical Engineering, UCLA, 1985.

Methods of Mapping from Phase to Sine Amplitude in Direct Digital Synthesis

Jouko Vankka

Abstract—There are many methods for performing functional mapping from phase to sine amplitude (e.g., ROM look-up, coarse/fine segmentation into multiple ROM's, Taylor series, CORDIC algorithm). The spectral purity of the conventional direct digital synthesizer (DDS) is also determined by the resolution of the values stored in the sine table ROM. Therefore, it is desirable to increase the resolution of the ROM. Unfortunately, larger ROM storage means higher power consumption, lower reliability, lower speed, and greatly increased costs. Different memory compression and algorithmic techniques and their effect on distortion and trade-offs are investigated in detail. A computer program has been created to simulate the effects of the memory compression and algorithmic techniques on the output spectrum of the DDS. For each memory compression and algorithmic technique, the worst case spurious response is calculated using the computer program.

I. Introduction

THIS PAPER first gives a description of the conventional direct digital synthesizer. The most elementary technique of compression is to store only $\pi/2$ rad of sine information, and to generate the ROM samples for the full range of 2π by exploiting the quarter-wave symmetry of the sine function. Beyond that, the methods of compressing the quarter wave memory include: a trigonometric identity, the Nicholas method, the use of Taylor series and the CORDIC algorithm. A different approach of phase-to-sine-amplitude mapping is the CORDIC algorithm, which uses an iterative computation method [1]. The penalties paid for different methods are the increased circuit complexity and the distortion due to the sine memory compression. Because the possible number of generated frequencies is large, it is impossible to simulate all of them to find the worst case situation. If the Nicholas modified phase accumulator is used, then only one simulation is needed to determine the worst case signal-to-spur level [2]. In this paper a case of 14-bit phase to 12-bit amplitude mapping is analyzed. A computer program has been created to simulate the effects of the memory compression and algorithmic techniques on the output spectrum of the DDS. For each memory compression and algorithmic technique, the worst case spurious response is calculated using these premises. Some examples of commercial circuits using the above methods are also presented.

Manuscript received April 23, 1996; accepted August 26, 1996.

J. Vankka is with the Helsinki University of Technology, Laboratory of Signal Processing and Computer Technology FIN-02150 Espoo, Finland (e-mail: jvan@mariso.hut.fi).

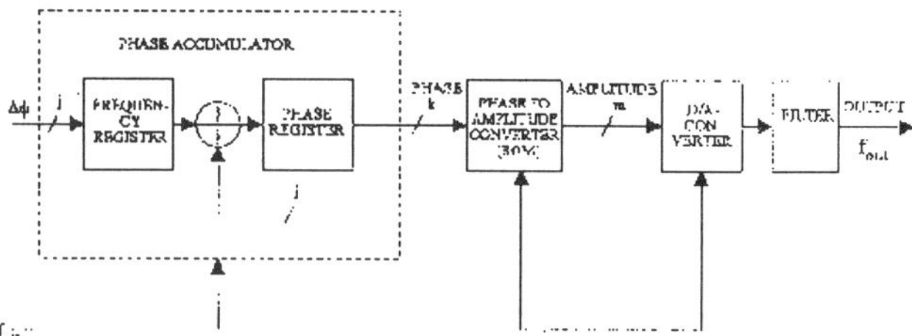

Fig. 1. Simplified block diagram of the direct digital synthesizer.

II. Conventional Direct Digital Synthesizer

The direct digital synthesizer is shown in simplified form in Fig. 1. The DDS has the following basic blocks: phase accumulator, phase-to-amplitude converter (generally a sine ROM), digital-to-analog converter, and filter [3]. The phase accumulator consists of a j-bit frequency register which stores a digital phase increment followed by a j-bit full adder and a phase register. The digital input phase increment is entered in the frequency register. At each clock pulse this value is added to the value previously held in the phase register. The phase increment represents a phase angle step that is added to the value of the previous step at every $1/f_{\text{clk}}$ seconds to produce a linearly increasing digital value. The phase value is generated using the modulo 2^j overflowing property of the j-bit phase accumulator. The rate of the phase accumulator overflows is the output frequency

$$f_{\text{out}} = \frac{\Delta\phi f_{\text{clk}}}{2^j}, \quad \Delta\phi < 2^{j-1}, \tag{1}$$

where $\Delta\phi$ is the phase increment word, f_{clk} is the clock frequency and f_{out} is the output frequency. The constraint in (1) comes from the sampling theorem. The phase increment word in (1) is an integer; therefore, the frequency resolution is found by setting $\Delta\phi = 1$

$$\Delta f = \frac{f_{\text{clk}}}{2^j}. \tag{2}$$

The read only memory (ROM) is a sine lookup table which converts the phase information into the values of a sine wave. In the ideal case of infinite precision of the phase and with no amplitude quantization, the output sequence of the table is given by:

$$\sin\left(2\pi\frac{\phi(n)}{2^j}\right). \tag{3}$$

Reprinted from *IEEE Transactions on Ultrasonics, Ferroelectrics, and Frequency Control*, Vol. 44, No. 2, pp. 526–534, March 1997.

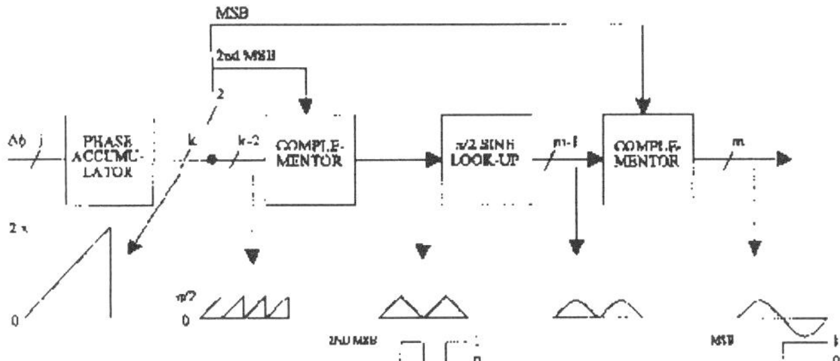 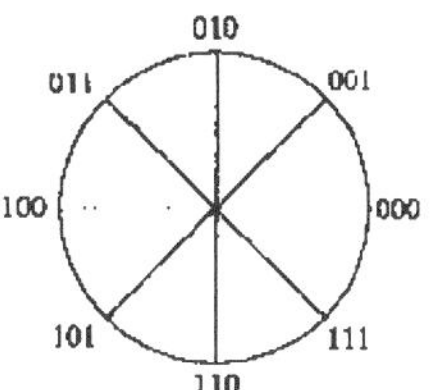 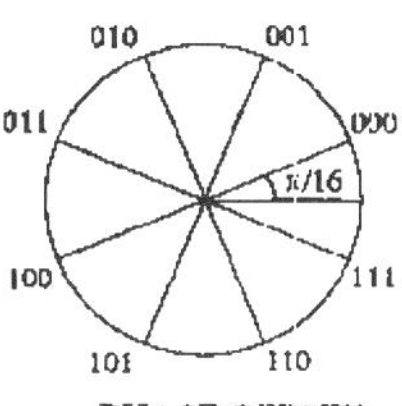

Fig. 2. Logic to exploit quarter wave symmetry.

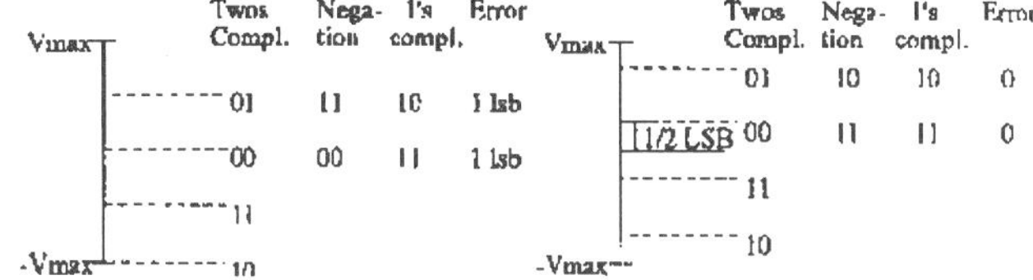

Fig. 3. 1/2 LSB phase offset is introduced in all phase addresses. In this case 1/2 LSB correspond $\pi/16$. The 1/2 LSB phase offset is added to all the sine lookup table samples. In this figure, 1's complementor maps the phase values to the first quadrant without error.

Fig. 4. 1/2 LSB offset is introduced into the amplitude that is to be complemented; then the negation can be carried out with 1's complementor without error in Fig. 2. There must also be a 1/2 LSB offset in the ROM output to the D/A-converter.

where $\phi(n)$ is (the 2^j-bit) phase register value (at the nth clock period). The numerical period of the phase accumulator output sequence is defined as the minimum value of P for which $\phi(n) = \phi(n - P)$ for all n. The numerical period of the phase accumulator output sequence (in clock cycles) is:

$$P = \frac{2^j}{\mathrm{GCD}(\Delta\phi, 2^j)}, \tag{4}$$

where $\mathrm{GCD}(\Delta\phi, 2^j)$ represents the greatest common divisor of $\Delta\phi$ and 2^j. The numerical period of the sequence of the samples fetched from the sine ROM will have the same value as the numerical period of the sequence generated by the phase accumulator [2], [4]. Therefore, the spectrum of the output waveform of the DDS prior to the digital-to-analog conversion is characterized by a discrete spectrum consisting of P points. The ROM output is presented to the D/A-converter, which develops a quantized analog sine wave. The filter removes the high frequency sampling components and provides a pure sine wave output.

The size of the ROM is $(2^k \times m)$, where k is the wordlength of the truncated phase address, and m is the wordlength of the ROM. The number of words in the ROM will determine the phase quantization error and the number of bits in each word will determine the amplitude quantization error. It is desirable to increase the resolution of the ROM, but larger ROM storage means higher power consumption, lower reliability, lower speed, and greatly increased costs.

III. EXPLOITATION OF SINE FUNCTION SYMMETRY

A well-known technique is to store only $\pi/2$ rad of sine information, and to generate the sine lookup table samples for the full range of 2π by exploiting the quarter-wave symmetry of the sine function. The decrease in lookup table capacity is paid by the additional logic necessary to generate the complements of the accumulator and lookup table output.

The details of this method are shown in Fig. 2. The two most significant phase bits are used to decode the quadrant, while the remaining $k - 2$ bits are used to address a one-quadrant sine lookup table. The most significant bit determines the required sign of the result, and the second most significant bit determines whether the amplitude is increasing or decreasing. The accumulator output is used "as is"

for the first and third quadrants. For the second and fourth quadrant, the phase bits must be complemented so that the slope of the saw tooth is inverted. As shown in Fig. 2, the sampled waveform at the output of the lookup table is a full rectified version of the desired sine wave. The final output sine wave is then generated by multiplying the full wave rectified version by -1 when the phase is between π and 2π.

In most practical DDS digital implementations, numbers are represented in 2's complement format. Therefore, the 2's complementing must be used to take the absolute value of the quarter phase and multiply the output of the look-up table by -1. However, it can be shown that, if a 1/2 LSB offset is introduced into a number that is to be complemented, then a 1's complementor may be used in place of 2's complementor without introducing an error [5]. This provides savings in hardware since a 1's complementor may be implemented as a set of simple exclusive-or gates. This 1/2 LSB offset is provided by choosing the lookup table samples such that there is a 1/2 LSB offset in both the phase and amplitude of the samples [5], Figs. 3 and 4. If there is no phase offset in Fig. 3, then 0 and $\pi/2$ have the same phase address to the quarter wave memory, and one more address bit is needed to distinguish these two values.

IV. COMPRESSION OF THE QUARTER-WAVE SINE FUNCTION

In this section the quarter wave memory compression will be investigated. The width of the sine lookup table is reduced before taking advantage of the quadrant symme-

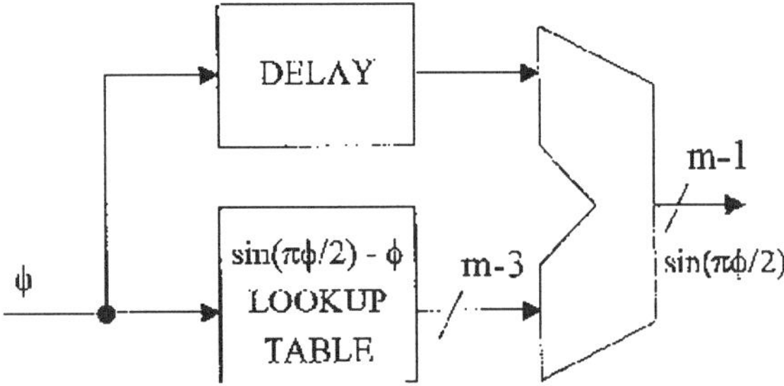

Fig. 5. The sine-phase difference algorithm.

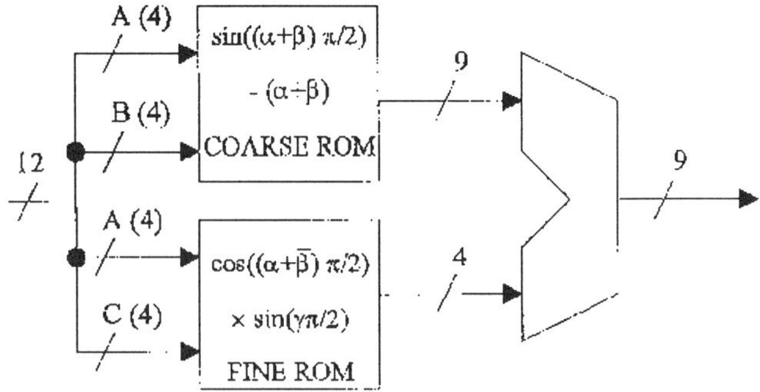

Fig. 6. The block diagram of modified Sunderland architecture for quarter wave sine function compression.

try of the sine function. The two most significant phase bits are used to decode the quadrant, while the remaining $k-2$ bits are used to address a one-quadrant lookup table. First, a sine-phase difference algorithm will be presented [5]. This algorithm is used in all the subsequent quarter wave compression techniques except for the CORDIC algorithm. The compression techniques are: a trigonometric approximation, the so-called Nicholas' architecture, the Taylor series method, and the CORDIC algorithm. In each method the total compression ratio, the size of the memory, the worst case spur level, and additional circuits are presented in Table 1. The amplitude values of the compressed quarter wave memory could be scaled to provide improved performance in the presence of amplitude quantization [5]. However, the optimization of the value scaling constant provides only negligible improvement in the amplitude quantization spur level, so it is beyond the scope of this paper.

A. Sine-Phase Difference Algorithm

Compression of the storage required for the quarter-wave sine function is obtained by storing the function

$$f(\phi) = \sin\left(\frac{\pi\phi}{2}\right) - \phi \qquad (5)$$

instead of $\sin(\pi\phi/2)$ in the lookup table (Fig. 5). Because

$$\max\left[\sin\left(\frac{\pi\phi}{2}\right) - \phi\right] \approx 0.21 \max\left[\sin\left(\frac{\pi\phi}{2}\right)\right], \qquad (6)$$

this saves 2 bits of amplitude in the storage of the sine function [5]. The penalty for this storage reduction is the introduction of an extra adder at the output of the look-up table to perform the operation

$$\left[\sin\left(\frac{\pi\phi}{2}\right) - \phi\right] + \phi. \qquad (7)$$

However, this is an advantageous trade-off in the high speed VLSI implementation of the DDS. The sine lookup table propagation delay, which cannot be easily pipelined, is reduced, increasing the maximum clock frequency of the DDS. The expense of another adder, which is readily pipelineable, is negligible in the full custom VLSI implementation.

B. Modified Sunderland Architecture

The original Sunderland technique is based on simple trigonometric identities [6]. There are two modifications of the original Sunderland paper. After the paper, a method for performing the two's complement negation function with only an exclusive-or, which does not introduce errors, has been published [5]. This method works by introducing the 1/2-LSB offsets into the phase and amplitude of the sine ROM samples as described in Section III. The sine-phase difference algorithm has been published after Sunderland paper [6], too.

The phase address of the quarter of the sine wave is decomposed to $\phi = \alpha + \beta + \gamma$, with the word-lengths of the variables: $\alpha \to A$, $\beta \to B$, and $\gamma \to C$ (Fig. 6). In this way the 12 phase bits are divided into three 4 bit fractions such that $\alpha < 1$, $\beta < (2^{-4})$, $\gamma < (2^{-8})$. The desired sine function is given by:

$$\sin\left(\frac{\pi}{2}(\alpha + \beta + \gamma)\right) - \sin\left(\frac{\pi}{2}(\alpha + \beta)\right)\cos\left(\frac{\pi}{2}\gamma\right) + \cos\left(\frac{\pi}{2}(\alpha + \beta)\right)\sin\left(\frac{\pi}{2}\gamma\right). \qquad (8)$$

Given the relative sizes of α, β, and γ, this expression may be approximated by:

$$\sin\left(\frac{\pi}{2}(\alpha + \beta + \gamma)\right) \approx \sin\left(\frac{\pi}{2}(\alpha + \beta)\right) - \cos\left(\frac{\pi}{2}\alpha\right)\sin\left(\frac{\pi}{2}\gamma\right). \qquad (9)$$

The approximation is made closer by adding (2^{-5}) (the average value of β) to α in the second term. In Fig. 6 the size of the upper memory is reduced by the sine difference algorithm. The access time of the upper memory is the most critical due to its larger size. If γ is subtracted from the lower memory, the result will be negative, and more logic is needed in the adder. The coarse ROM provides low resolution samples, and the fine ROM gives additional resolution by interpolating between the low resolution samples in Fig. 6.

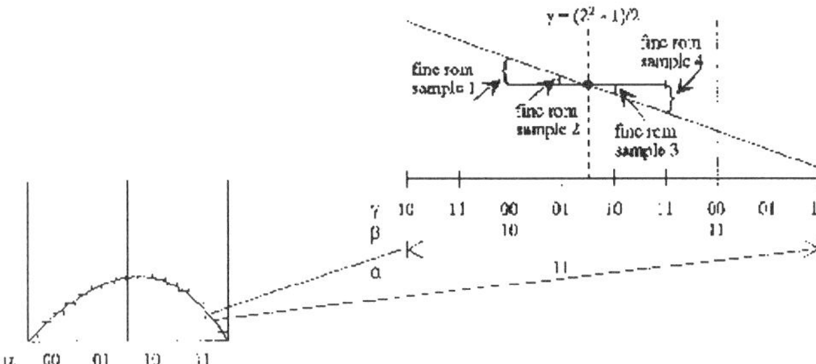

Fig. 7. The fine ROM samples are used to interpolate between the coarse samples, and the symmetry in the fine ROM samples around the $\gamma = (2^C - 1)/2$ axis. Here $C = 2$.

C. Nicholas Architecture

An alternative methodology for choosing the samples to be stored in the ROMs is based on numerical optimization [5]. The phase address of the quarter of the sine wave is defined as, $\phi = \alpha + \beta + \gamma$, with the word-length of the variable α to be A, the word-length of β to be B, and γ to be C. The variables α, β form the coarse ROM address, and the variables α, γ form the fine ROM address. In Fig. 7 the coarse ROM samples are represented by the dot along the solid line, and the fine ROM samples are chosen to be the difference between the value of the "error bars" directly below and above that point on the solid line. In Fig. 7, the function is divided into four regions, corresponding to $\alpha = 00, 01, 10,$ and 11. Within each region, only one interpolation value may be used between the error bars and the solid line for all the same γ values. The interpolation value used for each value of γ is chosen to minimize either the mean square or the maximum absolute error of the interpolation within the region [5].

Further storage compression is provided by exploiting the symmetry in the fine ROM correction factors, Fig. 7. If the coarse ROM samples are chosen in the middle of the interpolation region, then the fine ROM samples will be approximately symmetric around the $\gamma = (2^C - 1)/2$ axis. Thus, by using an adder/subtractor instead of an adder to sum the coarse and fine ROM values, the size of the fine ROM may be halved. Some additional complexity must be added to the adder/subtractor control logic if this technique is used with the sine-phase difference algorithm, since the slope of the function in (5) changes sign at a nonsymmetry point between 0 and $\pi/2$ on the x-axis. For example, the digital logic required to perform this can be accomplished with less than four logic gates for the 13-bit phase case [5]. Since the fine ROM is generally not in the critical speed path, the effective resolution of the fine ROM may be doubled, rather than halving the fine ROM. It allows the segmentation of the compression algorithm to be changed, effectively adding an extra bit of phase resolution to the look-up table, which thereby reduces the magnitude of the worst case spur due to phase accumulator truncation.

Computer simulations showed that the optimum partitioning of the ROM address word lengths to provide a

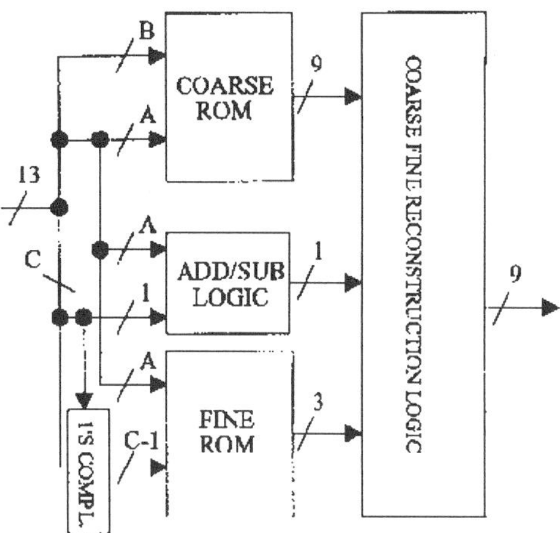

Fig. 8. The sine function generation logic of Nicholas' architecture.

13-b phase resolution was $A = 4$, $B = 4$, and $C = 5$, using the notation in Fig. 8 [5]. The simulations showed that the mean square and the minimum-maximum error criterion give the same maximum spur level in this segmentation. The architecture for this ROM compression technique is shown in Fig. 8. The amplitude values of the coarse and fine ROMs could be scaled to provide improved performance in the presence of amplitude quantization [5]. The optimization of the value scaling constant provides only a negligible improvement in the amplitude quantization spur level, so it is beyond the scope of this paper.

In a modified version of the above architecture the symmetry in the fine ROM samples is not utilized [7], so the extra bit of the phase resolution to the ROM address is not achieved. Therefore, the modified Nicholas architecture uses a 14-to-12-bit instead 15-to-12-bit phase to amplitude mapping in our case. Some hardware is saved, because an adder instead of an adder/subtractor is used to sum the coarse and fine ROM values, and the adder/subtractor control logic is not needed. The difference between the modified Nicholas architecture and the modified Sunderland architecture is that the samples stored in the sine ROM are chosen using the numerical optimization in the modified Nicholas architecture.

The IC realization of the Nicholas architecture is presented in [8], where a CMOS chip has the maximum clock frequency of 150 MHz. Analog Devices has also used this sine memory compression method in their CMOS device, which has the output word length of 12 bits and 100 MHz clock frequency [9]. The IC realization of the modified Nicholas architecture is presented in [7], where the CMOS quadrature digital synthesizer operates at 200 MHz clock frequency. The modified Nicholas architecture has also been used in [10], where a CMOS chip has four parallel ROM tables to achieve four times the throughput of a single DDS. The chip that uses only one ROM table has the clock frequency of 200 MHz [7]. Using the parallel architecture with four ROM tables, the chip attains the speed of 800 MHz [10].

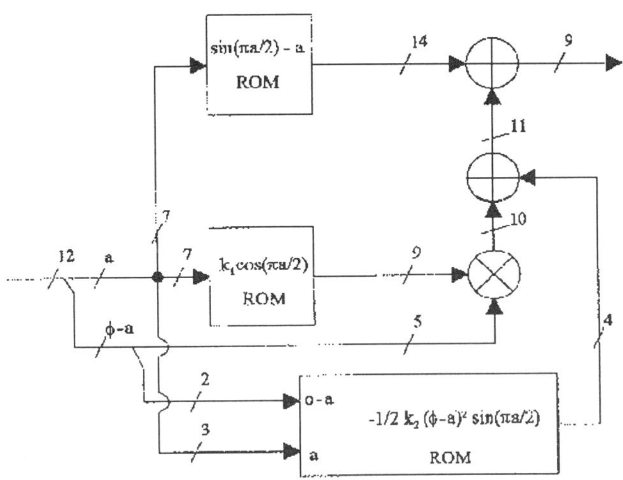

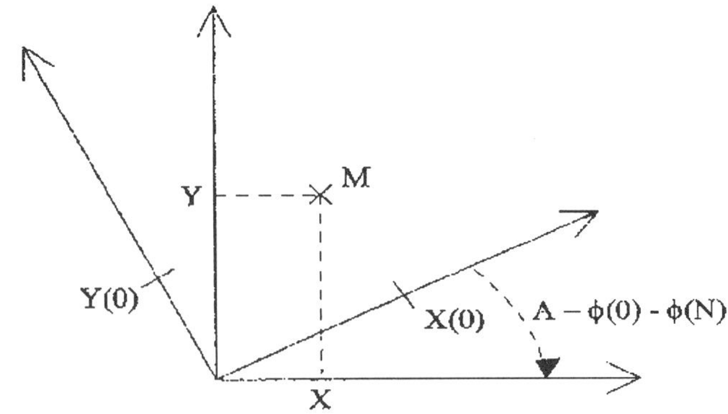

Fig. 11. "Rotation," given $X(0)$, $Y(0)$ and $\phi(0)$ find X and Y.

Fig. 9. Taylor series approximation for the quarter sine converter.

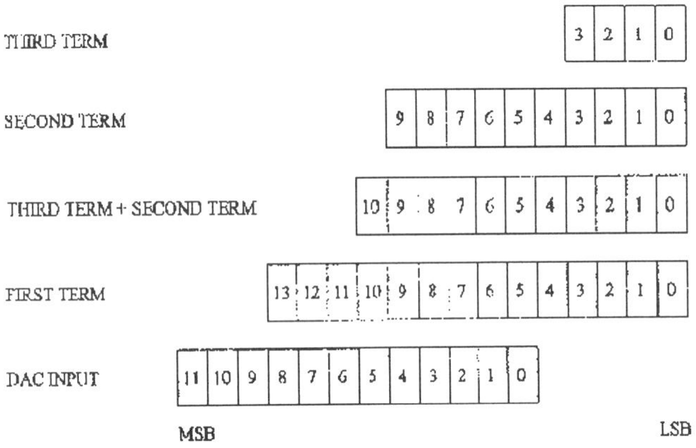

Fig. 10. Relative bit positions of multi-bit data words used in implementing the circuit of Fig. 9.

D. Taylor Series Approximation

The phase address "ϕ" is divided into the upper phase address "a" and the lower phase address "$\phi - a$" [11]. The Taylor series is performed around the upper phase address ($\phi = a$)

$$\sin\left(\frac{\pi}{2}\phi\right) \quad \sin\left(\frac{\pi}{2}a\right) + k_1(\phi - \alpha)\cos\left(\frac{\pi}{2}a\right)$$
$$- \frac{k_2(\phi - \alpha)^2 \sin\left(\frac{\pi}{2}a\right)}{2} + \cdots, \qquad (10)$$

where k_n represents a constant used to adjust the units of each series term. The adjustment in units is required because the phase values are given in angular units. Therefore, it is necessary to have a conversion factor k_n, which includes a multiple of $\pi/2$ to compensate for the phase units. The Taylor series (10) is approximated by taking three terms in Fig. 9. While additional terms can be employed, their contribution to accuracy is very small as shown in Fig. 10 and, therefore, of little weight in this application. Other inaccuracies present in the operation of current DDS designs override the finer accuracy provided by successive series terms.

The seven most significant bits of the input phase are selected as the upper phase address "a", which is transferred simultaneously to a sine ROM and a cosine ROM as address signals as shown in Fig. 9. The output of the sine ROM is the first term of the Taylor series and it is taken to the first adder, where it will be summed with the remaining terms. The size of the sine ROM is reduced by the sine difference algorithm. The output of the cosine ROM is configured to incorporate the predetermined unit conversion value k_1. The cosine ROM output is the first derivative of the sine. The least significant bits ($\phi - a$) are multiplied by the output of the cosine ROM to produce the second term. The third term is computed in a ROM by combining the second derivative of $\sin((\pi a)/2)$ and the square of the lower phase address "$\phi - a$". This is done by selecting the upper bits of "$\phi - a$" and "a" values as a portion of the address for the ROM. This is possible since the last term only roughly contributes 1/4 LSB to the D/A-converter input, as shown in Fig. 10. As with the cosine ROM, the unit conversion factor is included in the values stored in the ROM. The third term ROM output is combined with the multiplier output in a second adder, and subsequently combined with the first term ROM output in the first adder.

QUALCOMM has used the Taylor series approximations in their device, which has the output word length of 12 bits and 50 MHz clock frequency [12]. This DDS is CMOS device.

E. CORDIC Algorithm

The CORDIC algorithm performs vector coordinate rotations by using simple iterative shifts and add/subtract operations, which are easy to implement in hardware [1]. In Fig. 11, a pair of rectangular axes is rotated clockwise by the angle A by the CORDIC algorithm, then the coordinates of a point M transform from $(X(0), Y(0))$ to (X, Y)

$$X = X(0)\cos(A) - Y(0)\sin(A)$$
$$Y = Y(0)\cos(A) + X(0)\sin(A), \qquad (11)$$

where A is the total rotation angle after N CORDIC iterations

$$A = \phi(0) - \phi(N) - \sum_{n=0}^{N-1} u(n)a(n), \qquad (12)$$

where $\phi(0)$ is the desired rotation angle, $a(n) = \arctan(2^{-n})$ and $u(n) = \text{sign}(\phi(n))$. The absolute value of the angle approximation error can be bounded by [13]

$$|\phi(N)| = \left| \phi(0) - \sum_{n=0}^{N-1} u(n)a(n) \right| \leq \arctan(2^{-N+1}). \qquad (13)$$

The total number of iterations, N, is determined by the accuracy desired. CORDIC is a bit-recursive algorithm, which means that each iteration increases the accuracy of the results by approximately one bit. The truncation due to the finite precision in fixed point arithmetic causes errors, too. Both the scaling operation (see below) and the truncation errors must be taken into account when selecting the optimal number of iterations [13]. The CORDIC algorithm can be easily derived from (11). For angles 0 to $\pi/2$ it can be described as follows

Initiation: Given $X(0), Y(0), \phi(0)$
for $n = 0 : N - 1$ /* CORDIC equations */
$u(n) = \text{sign}(\phi(n))$ /* sign detection */
$X(n+1) = X(n) - u(n) \times Y(n) \times 2^{-n}$ /* iteration */
$Y(n+1) = Y(n) + u(n) \times X(n) \times 2^{-n}$ /* iteration */
$\phi(n+1) = \phi(n) - u(n) \times \arctan(2^{-n})$ /* angle updating */
End n-loop
$X = X(N)/K(N)$ /* scaling operation */
$Y = Y(N)/K(N)$ /* scaling operation */
$$\qquad (14)$$

The CORDIC processor either adds or subtracts a series of known angle values so that the value of $\phi(n)$ is driven to 0. The known angles (arctan) could be precomputed and stored in the CORDIC processor. The decision of whether to add or subtract the next angle in the series is based on the sign of $\phi(n)$, i.e., if $\phi(n)$ is negative, then the next angle is added to $\phi(n)$ to drive it closer to 0. If $\phi(n)$ is positive, then the next angle is subtracted from $\phi(n)$. The $(X(n), Y(n))$ coordinate pair is put through the series of angular rotations as $\phi(n)$ but with opposite sign (i.e., if $45°$ is subtracted from $\phi(n)$, then $(X(n), Y(n))$ is rotated $+45°$). The rotation is not a pure rotation but a rotation-extension. The scaling operations compensate for this magnitude change. The scaling factor for a particular value of N is constant, so it could be precomputed and stored in the CORDIC processor. The scaling factor is

$$K(N) = \prod_{n=0}^{N-1} \sqrt{1 + u_n^2 2^{-2n}}. \qquad (15)$$

If the initial values are chosen to be $X(0) = 1$, $Y(0) = 0$ and $\phi(0)$ is formed using the remaining $k - 2$ bits of the

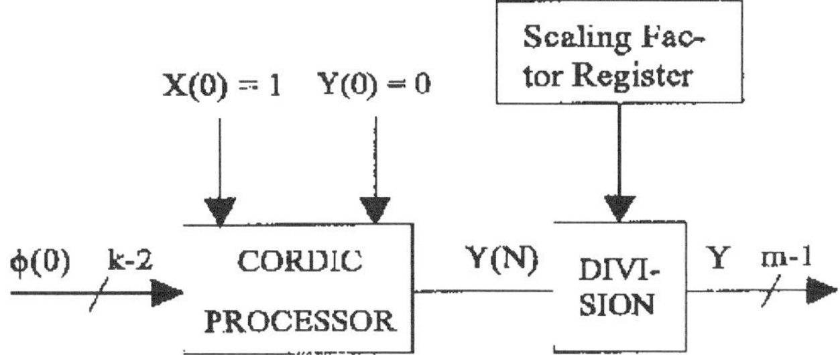

Fig. 12. The CORDIC processor for quarter sine wave converter.

phase register value from the DDS. From (11) the result will be $X = \cos(\phi(0) - \phi(N))$, $Y = \sin(\phi(0) - \phi(N))$, where $\phi(N)$ is the angle approximation error. If the initial values are chosen to be $X(0) = 1/K(N)$, $Y(0) = 0$, and $\phi(0)$ contains the remaining $k - 2$ bits of the phase register value from the DDS, then there is no need for the scaling operation after the CORDIC iterations. However, in the simulations the signal-to-spur level was worse than the case of the first initial values. Because $1/K(N)$ is approximately $0.607253(N > 13)$ and not power of two, the worse spur level was due to accumulated truncation errors in fixed arithmetic. The wordlength of the CORDIC iterations should be larger than in the case of the first initial values. The amplitude of the output waveform could be modulated by changing the scaling factor, but it is better to do the scaling after the iterations in fixed arithmetic. The value of Y may then be transferred to an appropriate D/A-converter. The architecture for quarter sine wave generation employing this technique is shown in Fig. 12.

The implementation of the CORDIC cell requires adders, barrel shifters, multiplexers and registers. The most speed-critical component will be the adder/subtractor due to the carry propagation. The throughput of the CORDIC datapath can be improved by introducing redundant number representation in the internal computation and eliminating the carry propagation from each addition/subtraction [14]–[16], but then the required chip area will increase [16]. The sign detection in the CORDIC algorithm would also be more complicated than in the nonredundant case, and the conversion and the reconversion to/from the redundant representation reduces the gain in the computation speed. To increase throughput, one could use pipelined processors with the penalty that the pipelined structure would require a large amount of chip area [17], [18].

The CORDIC algorithm is suitable for the DDS with more than 14 output bits, where the sine look-up table even with compression is still too large for high speed operation. It is also effective in solutions where there is the need of in-phase and quadrature components simultaneously, because the algorithm calculates both. For example GEC-Plessey I/Q splitter has 20 MHz clock frequency with 16-bit phase and amplitude accuracy [19], and Raytheon Semiconductor's DDS has 25 MHz clock frequency with 16-bit phase and amplitude accuracy [20].

TABLE I

SUMMARY OF MEMORY COMPRESSION AND ALGORITHMIC TECHNIQUES.

Method	ROM Needed	Total compression ratio	Additional circuits (not including quarter and sine difference logic)	Worst case spur (below carrier)	Comments
Uncompressed memory	$2^{14} \times 12$ bits	1 : 1	-	-97.23 dBc	reference
Mod. Sunderland architecture	$2^8 \times 9$ bits $2^8 \times 4$ bits	59 : 1	adder	86.91 dBc	good spur level and simple
Nicholas architecture	$2^8 \times 9$ bits $2^8 \times 3$ bits	128 : 1 Note $(k = 15, m = 12)$	adder/subtractor adder/subtractor control logic	-88.94 dBc	best compression ratio
Taylor series approximation with three terms	$2^7 \times 14$ bits $2^7 \times 9$ bits $2^5 \times 4$ bits	64 : 1	2 adders multiplier	-97.04 dBc	need multiplier
CORDIC algorithm	-	-	14 pipelined stages, 18-bit inner wordlength	84.25 dBc	much computation

V. SIMULATION

A computer program (in Matlab) has been created to simulate the direct digital synthesizer in Fig. 1. The memory compression and algorithmic techniques have been analyzed with no phase truncation (the phase accumulator length = the phase address), and the spectrum is calculated prior to the D/A-conversion. The number of points in the DDS output spectrum depends on $\Delta\phi$ (phase increment word) via the greatest common divisor of $\Delta\phi$ and 2^j ($\mathrm{GCD}(\Delta\phi, 2^j)$) (4). Any phase accumulator output vector can be formed from a permutation of another output vector regardless of the initial phase accumulator contents, when ($\mathrm{GCD}(\Delta\phi, 2^j)$) 1 for all values of $\Delta\phi$ (see Appendix). A permutation of the samples in the time domain results in an identical permutation of the discrete Fourier transform (DFT) samples in the frequency domain [2]. This means that the spurious spectrum due to all system nonlinearities can be generated from a permutation of another spectrum, when ($\mathrm{GCD}(\Delta\phi, 2^j)$) = 1 for all $\Delta\phi$, because each spectrum will differ only in the position of the spurs and not in the magnitudes [2]. When the least significant bit of the phase accumulator is forced to one (Nicholas modified phase accumulator [2]), it causes all of the phase accumulator output sequences to belong to the number theoretic class ($\mathrm{GCD}(\Delta\phi, 2^j)$) = 1, regardless of the value of $\Delta\phi$. Only one simulation is needed to be performed to determine the value of the worst case spurious response due to the system nonlinearities. In addition, the modified phase accumulator averages errors introduced by the D/A-converter, which is, however, beyond the scope of this text. The number of samples has been chosen (an integer number of cycles in the time record) so that problems of leakage in the fast Fourier transform (FFT) analysis can be avoided and unwindowed data can be used. The FFT was performed over the output period (4). The size of the FFT was 16384 points (except in Nicholas architecture 32768 points).

VI. CONCLUSION

Table I comprises the summary of memory compression and algorithmic techniques. The best spur level (almost the same level as with the ideal 12 bit rounded samples of a sine wave) is achieved by the Taylor series approximation with three terms. The Nicholas architecture has the best compression ratio in Table I, but it uses a 15-to-12-bit instead 14-to-12-bit phase to amplitude mapping in our case. The original Sunderland architecture produced worst case spur level of about -72.2 dBc [5], but the modified Sunderland architecture gives about 14 dB performance improvement. In the CORDIC algorithm the spur level is high, because the quantization step sizes are not equal (nonlinearity) due to the rotations by $\arctan(2^{-n})$. The number of the iterations (14) in the CORDIC processor for the quarter sine wave converter are more than needed for the 11-bit accuracy because of the high spur level.

Table II shows how much memory and additional circuits are needed in each memory compression and algorithmic technique to meet the spectral requirement for the worst case spur level, which is about -85 dBc due to the sine memory compression. In the DDS, most spurs are normally not generated by digital errors but rather by the analog errors in the D/A-converter. The spur level (-85 dBc) from the sine memory compression is not significant in DDS applications because it will stay below the spur level of a high speed 12-bit D/A-converter [21]. Differing from Table I, two terms are used for Taylor series approximation in Table II. Then, all memory compression and algorithmic techniques in Table II are comparable with almost the same worst case spur. In Table II the modified Nicholas architecture [7] is used and therefore the compression ratio and the worst case spur level are different from that in the Nicholas architecture [5] in Table I. The difference between the modified Nicholas architecture and the modified Sunderland architecture is that the samples stored in the sine ROM are chosen according to the numerical optimization

TABLE II

MEMORY COMPRESSION AND ALGORITHMIC TECHNIQUES WITH WORST CASE SPUR LEVEL DUE TO THE SINE MEMORY COMPRESSION SPECIFIED TO BE ABOUT -85 dBc.

Method	Needed ROM	Total compression ratio	Additional circuits (not including quarter and sine difference logic)	Worst case spur (below carrier)	Comments
Uncompressed memory	$2^{14} \times 12$ bits	1 : 1	-	-97.23 dBc	reference
Mod. Sunderland architecture	$2^8 \times 9$ bits $2^8 \times 4$ bits	59 : 1	adder	-86.91 dBc	simple
Mod. Nicholas architecture	$2^8 \times 9$ bits $2^8 \times 4$ bits	59 : 1	adder	-86.81 dBc	simple
Taylor series approximation with two terms	$2^7 \times 9$ bits[†] $2^7 \times 5$ bits[††]	110 : 1	adder multiplier	-85.88 dBc	need multiplier
CORDIC algorithm	-	-	14 pipelined stages, 18-bit inner wordlength	-84.25 dBc	much computation

[†] The first term ROM size, which is reduced by the sinedifference algorithm.

[††] The cosine ROM Size.

in the modified Nicholas architecture. In the 14-to-12-bit phase-to-amplitude mapping the numerical optimization gives no benefit, because the modified Sunderland architecture and modified Nicholas architecture give the same spur levels.

APPENDIX

A. Phase Accumulator as a Permutation Generator

The phase accumulator can be considered as a permutation generator, where each value of $\Delta\phi$ provides a different permutation of the values from 0 to $2^j - 1$ given by

$$\Delta_\phi \phi(n) = (n\Delta\phi) \mod 2^j. \tag{16}$$

Fig. 13 shows that any phase accumulator output vector can be formed from the permutation of another output vector regardless of the initial phase accumulator contents, when $(\text{GCD}(\Delta\phi, 2^j)) = 1$ for all values of $\Delta\phi$. Fig. 14 shows that the time vectors are formed from values, which have the property $(\text{GCD}(\Delta\phi, 2^j)) = 2$. From Fig. 14 it is evident that the phase accumulator is now characterized by having two different sets of possible output vectors depending upon the initial contents of the phase accumulator.

The time output vector for $\Delta\phi$ can be formed from a permutation of the individual elements of the vector for $\Delta\phi = 1$

$$\Delta_\phi \phi(n) = {}_1\phi((n\Delta\phi) \mod 2^j), \tag{17}$$

when $\Delta\phi$ and 2^j are relatively prime (Fig. 13). As was shown in (17), each input time vector may be formed from a permutation of another time vector by permuting the indices using $(n\Delta\phi) \mod 2^j$. The converse follows from

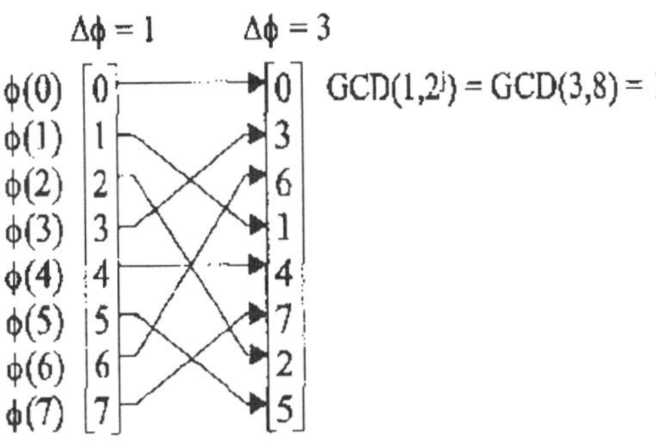

Fig. 13. Time series vectors for a 3-bit phase accumulator for $\Delta\phi = 1$ and $\Delta\phi = 3$. The column vector for $\Delta\phi = 3$ can be formed from a permutation of the values of the $\Delta\phi = 1$ vector, regardless of the initial phase accumulator contents.

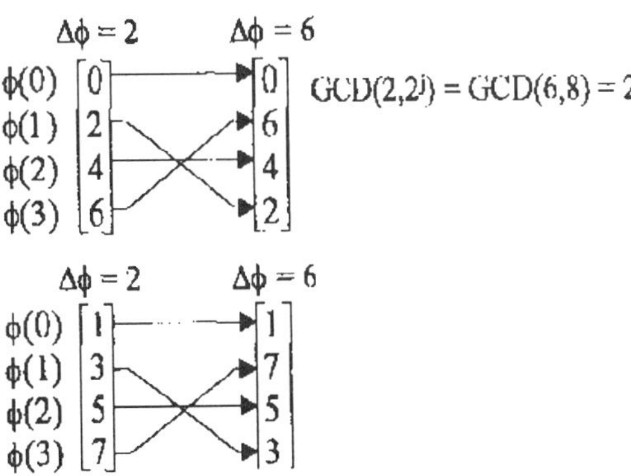

Fig. 14. Time series vectors for a 3-bit accumulator for $\Delta\phi = 2$ and $\Delta\phi = 6$.

the existence of a unique integer $0 \leq J < 2^j$ satisfying the relation

$$\Delta\phi J \mod 2^j = 1. \tag{18}$$

This is a fundamental result of number theory which requires that $\Delta\phi$ and 2^j are relatively prime [22]. In a sense J is the multiplicative inverse of $\Delta\phi$. From the above equation it follows that $\Delta\phi$ and J must be odd because 2^j is even. Therefore, J and 2^j are too relatively prime.

The sine output DDS operates by applying some memoryless nonlinear function $s\{\}$ to the phase accumulator output to produce the sine function. The DFT of the phase-to-amplitude converter output using (17) is:

$$S\{_{\Delta\phi}\phi(m)\} = \sum_{n=0}^{2^j-1} s\{_1\phi((n\Delta\phi) \mod 2^j)\}$$
$$\times W_{2^j}^{mn} \quad m = 0, 1, \ldots, 2^j - 1, \tag{19}$$

where the period of the phase accumulator is 2^j when $\Delta\phi$ and 2^j are relatively prime (4), and $W_{2^j} = e^{-j2\pi/2^j}$. Equation (18) can be used to show that the permutation samples in the time domain result in the same type permutation in the frequency domain by defining the new index

$$q = (n\Delta\phi) \mod 2^j, \tag{20}$$

and noting that

$$qJ \mod 2^j = J((n\Delta\phi) \mod 2^j) \mod 2^j$$
$$= n\Delta\phi J \mod 2^j. \tag{21}$$

Substituting from (18), (21) becomes

$$n = qJ \mod 2^j. \tag{22}$$

Reindexing (19) using (20) and (22), then

$$S\{_{\Delta\phi}\phi(m)\} = \sum_{q=0}^{2^j-1} s\{_1\phi(q)\} W_{2^j}^{m(qJ \mod 2^j)}$$
$$= \sum_{q=0}^{2^j-1} s\{_1\phi(q)\} W_{2^j}^{q(mJ \mod 2^j)} \tag{23}$$
$$= S\{_1\phi((mJ) \mod 2^j)\} m = 0, 1, \ldots, 2^j - 1.$$

The above equation establishes that the permutation of the samples in the time domain results in the same type permutation of the DFT samples in the frequency domain, because J and 2^j are relatively prime. This means that the spurious spectrum due to all system nonlinearities can be generated from a permutation of another spectrum, when $GCD(\Delta\phi, 2^j) = 1$ for all $\Delta\phi$, because each spectrum will differ only in the position of the spurs and not in the magnitudes.

REFERENCES

[1] J. E. Volder, "The CORDIC trigonometric computing technique," *IRE Trans. Electron. Comput.*, vol. EC-8, pp. 330–334, Sept. 1959.

[2] H. T. Nicholas and H. Samueli, "An analysis of the output spectrum of direct digital frequency synthesizers in the presence of phase-accumulator truncation," in *Proc. 41st Annu. Freq. Contr. Symp.*, 1987, pp. 495–502.

[3] A. L. Bramble, "Direct digital frequency synthesis," in *Proc. 35th Annu. Freq. Contr. Symp.*, USERACOM (Ft. Monmouth, NJ), May 1981, pp. 406–414.

[4] D. L. Duttweiler and D. G. Messerschmitt "Analysis of digitally generated sinusoids with application to A/D and D/A converter testing," *IEEE Trans. Commun.*, vol. COM-26, pp. 669–675, May 1978.

[5] H. T. Nicholas, H. Samueli, and B. Kim, "The optimization of direct digital frequency synthesizer performance in the presence of finite word length effects," in *Proc. 42nd Annu. Freq. Contr. Symp.*, 1988, pp. 357–363.

[6] D. A. Sunderland, R. A. Strauch, S.S. Wharfield, H. T. Peterson, and C. R. Cole, "CMOS/SOS frequency synthesizer LSI circuit for spread spectrum communications," *IEEE J. Solid-State Circuits*, vol. SC-19, pp. 497–505, Aug. 1984.

[7] L. K. Tan, and H. Samueli, "A 200-MHz quadrature digital synthesizer/mixer in 0.8 μm CMOS," *IEEE J. Solid-State Circuits*, vol. 30, pp. 193–200, Mar. 1995.

[8] H. T. Nicholas, and H. Samueli, "A 150-MHz direct digital frequency synthesizer in 1.25-μm CMOS with −90-dBc spurious performance," *IEEE J. Solid-State Circuits*, vol. 26, pp. 1959–1969, Dec. 1991.

[9] Analog Devices AD 9955 Data Sheet, Rev. 0, 1994.

[10] L. K. Tan, E. W. Roth, G. E. Yee, and H. Samueli, "A 800-MHz quadrature digital synthesizer with ECL-compatible output drivers in 0.8 μm CMOS," *IEEE J. Solid-State Circuits*, vol. 30, pp. 1463–1473, Dec. 1995.

[11] L. A. Weaver, and R. J. Kerr, "High resolution phase to sine amplitude conversion," U. S. Patent 4,905,177, Feb. 27, 1990.

[12] Qualcomm Q2334, Technical Data Sheet, June 1991.

[13] Y. H. Hu, "The quantization effects of the CORDIC algorithm," *IEEE Trans. Signal Processing*, vol. 40, pp. 834–844, Apr. 1992.

[14] M. D. Ercegovac, and T. Lang, "Redundant and on-line CORDIC: application to matrix triangularization and SVD," *IEEE Trans. Comput.*, vol. 39, pp. 725–740, June 1990.

[15] N. Takagi, T. Asada, and S. Yajima, "Redundant CORDIC methods with a constant scale factor for sine and cosine computation," *IEEE Trans. Comput.*, vol. 40, pp. 989–995, Sept. 1991.

[16] J. Lee, and T. Lang, "Constant-factor redundant CORDIC for angle calculation and rotation," *IEEE Trans. Comput.*, vol. 41, pp. 1016–1025, Aug. 1992.

[17] G. Gielis, R. van de Plassche, and J. van Valburg, "A 540-MHz 10b polar-to-cartesian converter," *ISSCC Digest Technical Papers*, pp. 160–161, Feb. 1991.

[18] A. Madisetti, A. Kwentus, and A. N. Wilson, Jr., "A sine/cosine direct digital frequency synthesizer using an angle rotation algorithm," *ISSCC Digest Technical Papers*, pp. 262–263, Feb. 1995.

[19] GEC-Plessey Semiconductors, Data Sheet PDSP16350 I/Q Splitter/NCO, Dec. 1993.

[20] Raytheon Semiconductor Data Book, Data Sheet TMC2340, 1994.

[21] Burr-Brown IC Data Book-Data Conversion Products, Data Sheet DAC650, 1994.

[22] J. H. McClellan and C. M. Rader, *Number Theory in Digital Signal Processing*. Englewood Cliffs, NJ: Prentice-Hall, 1979.

A 150-MHz Direct Digital Frequency Synthesizer in 1.25-μm CMOS with -90-dBc Spurious Performance

Henry T. Nicholas, III, and Henry Samueli, *Member, IEEE*

Abstract —A monolithic CMOS direct digital frequency synthesizer (DDFS) is presented which synthesizes a 12-b output sine wave at 150 Msamples/s. This sine wave is spectrally pure to -90.3 dBc over the entire tuning bandwidth. Phase noise of the output sine wave is equivalent to or better than that of the 150-MHz reference clock. The synthesizer covers a bandwidth from dc to 75 MHz in steps of 0.035 Hz with a switching speed of 6.7 ns and a tuning latency of 13 clock cycles. An efficient look-up table method for calculating the sine function is used, which reduces ROM storage requirements by a factor of 128:1. All circuit designs are fully static and are tolerant to transistor threshold shifts caused by radiation or process variations. The DDFS is fabricated in a 1.25-μm radiation-hardened double-level metal bulk P-well CMOS process which is tolerant to over 10^6 rd(Si) of total dose radiation. The die size is 195 mil $\times$ 195 mil with a device count of 35 000 transistors. Power dissipation is 950 mW at a clock rate of 100 MHz.

I. Introduction

MODERN communication systems are placing ever-increasing demands on the resolution, bandwidth, and switching speed of frequency synthesizers. In the past, these requirements have been satisfied by the conventional phase-locked loop (PLL) synthesizer. The fundamental advantage of PLL's has been their ability to synthesize an output sine wave of high spectral purity which may be tuned over a wide bandwidth. However, the switching speed and resolution of synthesizers are becoming critically important as frequency agile communications systems such as CDMA digital cellular telephones, spread spectrum wireless LAN's, and military frequency-hopped communications systems become widespread. Conventional PLL's are ill-suited to these applications because they suffer from an inherent inability to simultaneously provide fast frequency switching and high resolution without substantial design complexity. This is evident in the simplified block diagram of the PLL in Fig. 1. The resolution of the PLL is determined primarily by the frequency at the input to the phase detector. Lower reference

Manuscript received May 10, 1991; revised August 15, 1991. This work was supported in part by grants from the University of California MICRO Program and TRW Inc.

H. T. Nicholas, III is with PairGain Technologies, Torrance, CA 90501.

H. Samueli is with the Integrated Circuits and Systems Laboratory, Electrical Engineering Department, University of California, Los Angeles, CA 90024.

IEEE Log Number 9103595.

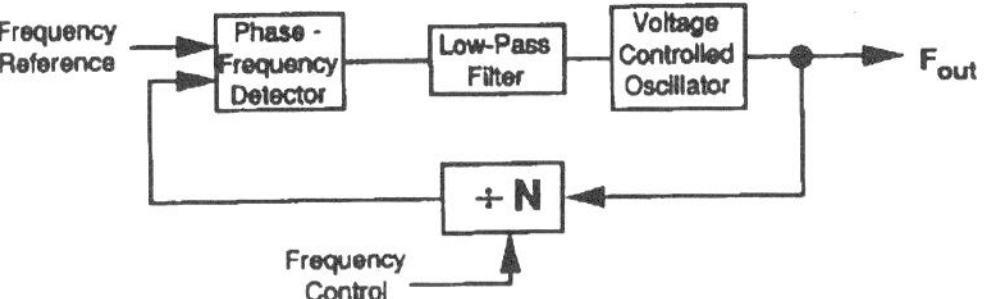

Fig. 1.   Simple PLL block diagram.

frequencies result in closer spacing between the output frequencies of the synthesizer, but also require a reduction in the loop filter bandwidth. This reduced loop filter bandwidth increases the time constant of the loop transient response and therefore degrades the switching speed of the synthesizer. Multiple-loop PLL techniques such as fractional-N have been employed in an attempt to improve frequency resolution without prohibitive increases in switching speed. However, these techniques substantially increase hardware complexity and usually require many precision analog components [1].

In the past few years, improvements in digital-to-analog converter (DAC) technology have made direct digital frequency synthesis feasible at RF frequencies. In direct digital frequency synthesizers (DDFS's), all of the signal processing operations that synthesize and tune the sine wave are performed digitally. The conversion from digital to analog takes place at the output of the synthesizer. The DAC and the preceding digital logic are clocked by a stable frequency reference, providing an output sine wave with extremely low phase noise. This phase noise is generally less than or equal to the phase noise of the reference itself. The output sine wave from a DDFS is fundamentally limited by the Nyquist criterion to a bandwidth of $f_{\text{clk}}/2$. Therefore it is imperative that f_{clk} be made as high as possible to provide a wide-output tuning bandwidth.

Almost without exception, DDFS's are designed using an architecture developed by Tierney *et al.* [2]. This powerful yet simple architecture, which is shown in Fig. 2, exploits the modulo 2^L overflowing property of an L-bit accumulator to generate the phase argument to the sine function generation logic. The 2^L words of the accumulator are mapped into phase values such that

$$\theta(n) = 2\pi\big(\phi(n)/2^L\big)$$

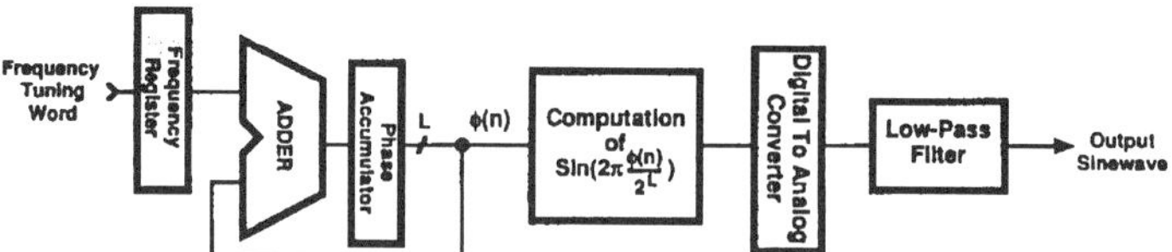

Fig. 2. Tierney *et al.* architecture for DDFS.

where $\phi(n)$ represents the contents of the phase accumulator at time $t = n/f_{clk}$. On each clock cycle, a digital frequency control word F is added to the previous contents of the accumulator such that $\phi(n+1) = F + \phi(n)$. The output frequency of the DDFS is therefore given by

$$f_{out} = \frac{\theta(n) - \theta(n-1)}{2\pi \Delta t_{clk}} = \frac{f_{clk}F}{2^L}. \tag{1}$$

There are several significant advantages that DDFS's have over other frequency synthesis techniques. As seen in (1), extremely fine frequency resolution can be obtained using the Tierney *et al.* architecture since the frequency spacing is halved with each additional phase accumulator bit. Several commercially available DDFS's exploit this property to provide frequency resolutions of less than 1 Hz [3], [4]. Another significant advantage of a DDFS is fast switching speed. A DDFS may tune between any two frequencies in one reference clock period, which is typically less than 100 ns. In contrast, 1 ms is considered an aggressive switching speed for a conventional PLL. A DDFS also has the ability to switch frequencies with absolute phase coherence, a property that is almost impossible to achieve using any other technique.

While high-frequency resolution and fast switching speed are natural byproducts of the DDFS technique, wide bandwidth and high spectral purity have in the past been mutually exclusive. For this reason most of the innovations in DDFS architectures have focused on improving the spectral purity/bandwidth product [5]. Efforts have been made to extend the bandwidth of DDFS designs by using GaAs [11] and silicon bipolar [9] processing technologies. These 8-b output DDFS designs have achieved clock rates of 1.0 and 1.5 GHz, respectively. Although CMOS DDFS's provide substantial cost and power advantages over those in silicon bipolar and GaAs technologies, their use has been restricted by their limited bandwidth. The fastest previous CMOS DDFS design has a maximum commercially derated clock rate of 80 MHz [3].

In CMOS, the problem in attaining wide bandwidth without compromising spectral purity is twofold. First, the design of a high-speed phase accumulator in a low-cost commercially available CMOS process can introduce excessive pipeline latency and clock load. Second, the computation of the sine function from the phase argument $\phi(n)$ is very difficult to pipeline and therefore becomes a bottleneck in the maximum achievable system clock rate. In recent DDFS designs, the computation of the sine function has been a fundamental limitation in system throughput, causing the usable output bandwidth of DDFS

designs to lag behind the capabilities of state-of-the-art DAC's. DAC's for DDFS applications have been reported with resolution of 12 b at 1 GHz in GaAs [12] and 12 b at 100 MHz in silicon bipolar [13], demonstrating performance that significantly exceeds the capabilities of previously published monolithic CMOS DDFS designs.

Most previous attempts at improving DDFS clock rates have sacrificed the digital spurious performance of the synthesizer by approximating the sine function in an effort to reduce the size and access time of the look-up table [5], [6]. The reason for this is the difficulty in calculating the sine function. Because it is a transcendental function, the sine function is computationally intensive to calculate with high accuracy. Therefore, sine computation is typically performed by ROM-based table look-up, as opposed to other published methods such as Taylor series and CORDIC algorithms [4], [6]. A straightforward implementation of the sine function look-up table requires more than 4 Gsamples of storage for a phase accumulator resolution of 32 b. Therefore to provide a practical ROM size, the phase argument $\phi(n)$ must be truncated and storage compression algorithms applied to minimize the look-up table size.

The spurious and bandwidth requirements of DDFS's are not limited solely by the performance capabilities of existing DAC's. There are applications for DDFS's that do not convert the digitized sine samples directly to analog. These applications, which include tunable digital bandpass filters, mixers for digital receivers, and real-time digital spectrum analysis, often require a digital spurious performance which is as close as possible to an ideal sine wave for a given output word length.

A monolithic DDFS is presented herein which simultaneously achieves high spectral purity and wide bandwidth through the use of several innovative architectural and circuit design techniques. The maximum clock rate, bandwidth, and switching speed of this device are all superior to any previously reported CMOS DDFS, and the spectral purity/bandwidth product is higher than any DDFS previously reported in any technology. This DDFS is also the first 12-b output DDFS of any speed or technology to achieve better than -90-dBc digital spurious performance over the entire tuning bandwidth.

II. DESIGN REQUIREMENTS

The primary considerations in the design of this DDFS were the bandwidth, spectral purity, and power dissipation. The output bandwidth required of the DDFS was 50 MHz at end of life. This corresponds to a clock rate well

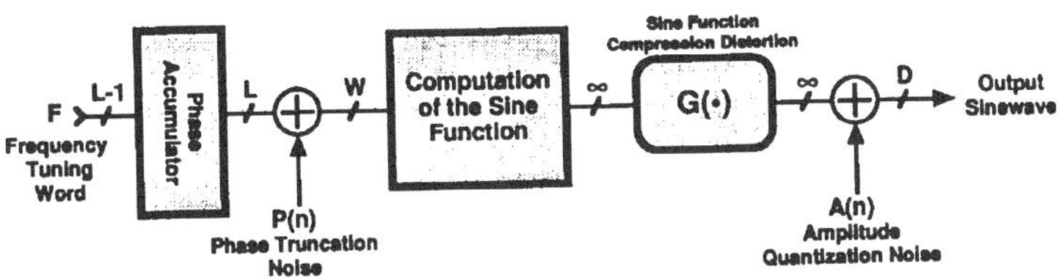

Fig. 3.   Sources of noise from DDFS finite word-length effects.

in excess of 100 MHz. The spectral purity of the digital sine wave was required to be -90 dBc in order to make the design as general purpose as possible. This level of performance allows the DDFS to be used with lower speed, higher spectral purity DAC's as well as without a DAC in digital mixer and tunable digital filter applications.

The first application of the DDFS was as part of a spaceborne frequency agile synthesizer. This program dictated that the DDFS be radiation hardened and space-flight qualified. Therefore, all circuits within the DDFS were required to be tolerant to transistor threshold shifts caused by radiation or process variations. All circuits were required to function at a 100-MHz maximum clock rate under the worst-case conditions of a 4.5-V power supply voltage, a 125° C operating temperature, and 1×10^6 rd(Si) of total dose radiation. The target reliability for the spaceflight parts was less than 100 FITS.

III. ARCHITECTURE

A. DDFS Finite Word-Length Effects

The optimization of DDFS performance involves trading off the finite word lengths and sine computation methods against the sine-wave spectral purity and maximum clock rate. Fig. 3 shows a basic block diagram of a DDFS that identifies the three basic sources of noise inherent to all DDFS designs. These noise sources are phase accumulator truncation $P(n)$, output amplitude quantization $A(n)$, and sine function compression distortion $G(\cdot)$. In Fig. 3, L is the phase accumulator word length, W is the phase resolution of the sine function, and F is the frequency command word to the phase accumulator. The amount of phase accumulator truncation B may be calculated as $B = L - W$. An analytical understanding of these finite word-length effects is crucial to the proper partitioning of internal word lengths to achieve high performance. The most important source of distortion affecting the spectral purity/bandwidth product is phase accumulator truncation. Phase accumulator truncation refers to the truncation of the least significant bits of the phase accumulator output when computing the sine function. The distortion introduced by this truncation must be traded off against the benefits gained by a reduction in the look-up table size. This trade-off is summarized by the relationship between the phase accumulator word length, phase accumulator truncation, and the magnitude

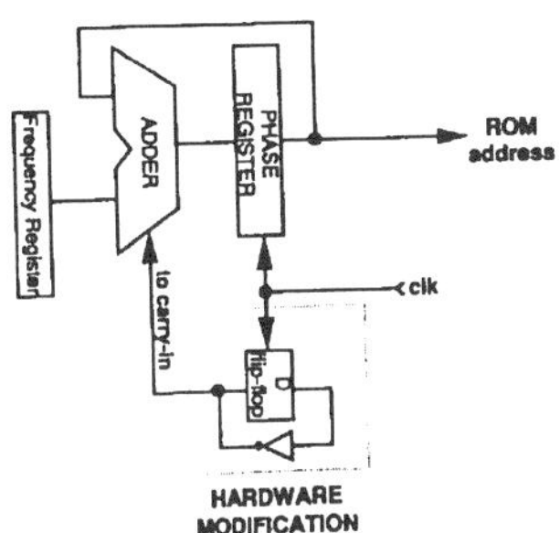

Fig. 4.   Simple modification to the phase accumulator to provide 3.9-dBc spurious improvement.

of the worst-case spurious response introduced by phase accumulator truncation, which was analytically determined in [7] to be

$$\zeta = 2^{B-L} \frac{\dfrac{\pi\left(F_r, 2^B\right)}{2^B}}{\sin\left(\dfrac{\pi\left(F_r, 2^B\right)}{2^B}\right)} \tag{2}$$

where ζ is the worst-case spur amplitude normalized to a signal amplitude of unity, and $(F, 2^B)$ denotes the greatest common divisor of F and 2^B. From (2), it is evident that the worst-case spur amplitude can be reduced in magnitude by 3.922 dB by forcing $(F_r, 2^B)$ to be equal to unity, i.e., by forcing the frequency command word to be relatively prime to 2^B. A very simple modification to the phase accumulator structure that accomplishes this is shown in Fig. 4. This modification, which was first introduced in [7], has been incorporated as a selectable option into this design. This modified phase accumulator has the additional advantage of eliminating the dependence of the output spectrum on the value contained in the phase accumulator at the time of a frequency change. Furthermore, the modification causes the output of the phase accumulator to behave as if the phase accumulator word length were $L + 1$ with a relatively coprime F and 2^B such that $(2F + 1, 2^{B+1}) = 1$. This causes the phase accumulator output sequence to have a maximal numerical period for all values of F, i.e., all 2^{L+1} possible values of the phase accumulator output sequence are generated before any of the values are repeated. This is a desirable property since it has the effect of randomizing the errors introduced by the quantized ROM samples and the D/A nonlinearities. Therefore a very consistent and predictable DDFS performance results for all values of the frequency command word F. This solves a significant problem in past DDFS designs, because now the combined effects of phase accumulator truncation, amplitude quantization, and sine compression nonlinearities can be minimized through computer optimization. In the past, the interactions between the phase accumulator truncation noise and the other finite word-length effects were so unpredictable that some designs have dithered the phase accumulator [10] and sine-wave amplitude [4] with pseudorandom numbers in order to randomize the error effects over many different output frequencies.

The only disadvantage of the modification to the phase accumulator used in this design is that it introduces an offset of $f_{clk}/2^{L+1}$ into the output frequency of the DDFS. However, for a phase accumulator word length of 32 b and a clock rate of 100 MHz this offset is only 0.0116 Hz, which is negligible for most applications. In any event, slightly shifting the clock frequency will compensate for this offset.

To meet the resolution requirements at a 100-MHz clock rate, a 32-b phase accumulator was used in this design, providing 0.0233 Hz of frequency resolution. Using (2) and assuming a 32-b phase accumulator incorporating the modification shown in Fig. 4, a phase precision of 15 b was selected to provide -90.3 dBc of rejection of spurious noise due to phase accumulator truncation. The computation of the sine function from the 15-b phase information was accomplished using a modified look-up table approach.

B. Sine Function Computation

Several new compression techniques were applied to reduce the size of the ROM look-up tables and to optimize the ROM contents. A well-known technique is to store only $\pi/2$ rad of sine information, and to generate the ROM samples for the full range of 2π by exploiting the quarter-wave symmetry of the sine function. Since the most significant two bits of the phase accumulator represent the quadrant of the sine function, the most significant bit may be used as the sign bit of the result, and the next most significant bit may be used to control whether the phase between 0 and $\pi/2$ should be increasing or decreasing. Ideally, these bits should control a two's complement negation function. Previous DDFS designs have simplified the hardware required for two's complement negation at the expense of a 1-LSB error in the result by using only an EXCLUSIVE-OR, rather than an EXCLUSIVE-OR followed by an adder, as required under two's complement arithmetic [5], [6]. A method for performing the two's complement negation function with only an EXCLUSIVE-OR has been developed which does not introduce a 1-LSB error when reconstructing a sine wave.

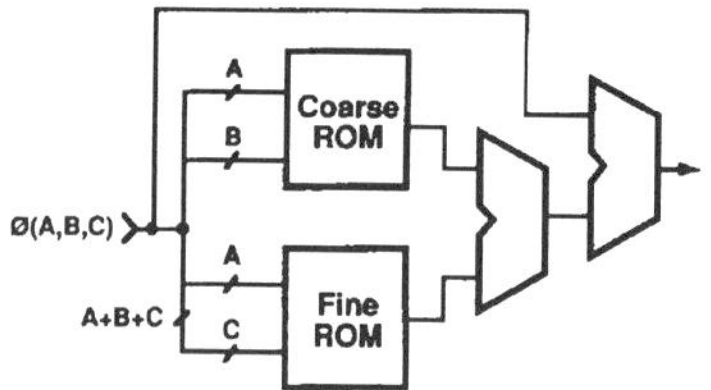

Fig. 5.   Coarse–fine segmentation of the function $\sin(\Theta)-\Theta$.

This method works by introducing the 1/2-LSB offsets into the phase and amplitude of the sine ROM samples as described in [8]. This allows the elimination of the 15-b and 12-b two's complement adders with no degradation in the sine-wave spectral purity.

Compression of the storage required for the quarter-wave sine function is obtained by storing the function $\sin(\phi)-2\phi/\pi$ in the look-up table instead of $\sin(\phi)$. Because $\max[\sin(\phi)-2\phi/\pi] = 0.21 \max[\sin(\phi)]$, this saves 2 b of amplitude in the storage of the sine function. The penalty for this storage reduction is the introduction of an extra adder at the output of the look-up table to perform the operation $[\sin(\phi)-2\phi/\pi]+2\phi/\pi$. Further storage compression is achieved by replacing the 2^{12} sample look-up table by a much smaller coarse ROM and fine interpolation ROM. Techniques have been published in which trigonometric approximations have been used to segment a large sine ROM into two smaller ROM's [5], [6]. However, these techniques have resulted in a peak distortion of 1.5 LSB's for a 12-b output sine wave, and are limited in application to the compression of the sine function. A generalized algorithm [8] for the compression of arbitrary functions into coarse and fine ROM samples has been employed in this design. This algorithm allows for the compression of the samples of the function $\sin(\phi)-2\phi/\pi$ into the coarse and fine ROM's, which are configured as shown in Fig. 5. Computer simulation determined that the optimum partitioning of the ROM address word lengths to provide a 13-b phase resolution was $A = 4$, $B = 4$, and $C = 5$, using the notation in Fig. 5. The ROM samples generated using this compression algorithm are optimized such that the mean square error due to algorithm errors and amplitude quantization is minimized. Further storage compression is provided by exploiting the symmetry in the fine ROM correction factors. Fig. 6 depicts how the fine ROM samples are used to interpolate a higher phase resolution function from the coarse samples and demonstrates the symmetry in the fine ROM. This symmetry allows the size of the fine ROM to be halved, at the expense of a bank of EXCLUSIVE-OR gates at the output of the fine ROM.

The final architecture for sine wave generation employing these look-up table compression techniques is shown in Fig. 7. This architecture compresses $2^{15} \times 12$ sine samples into a $2^8 \times 9$-b coarse ROM and a $2^8 \times 3$-b fine ROM, for a total compression ratio of 128:1. The worst-case distortion resulting from the combined effects of all

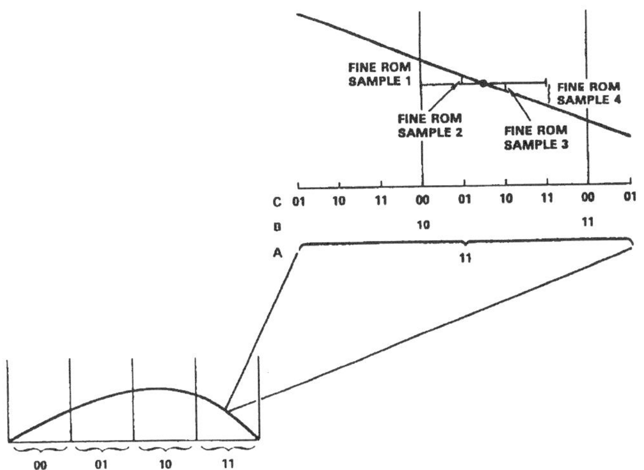

Fig. 6.   Symmetry in fine ROM interpolation samples.

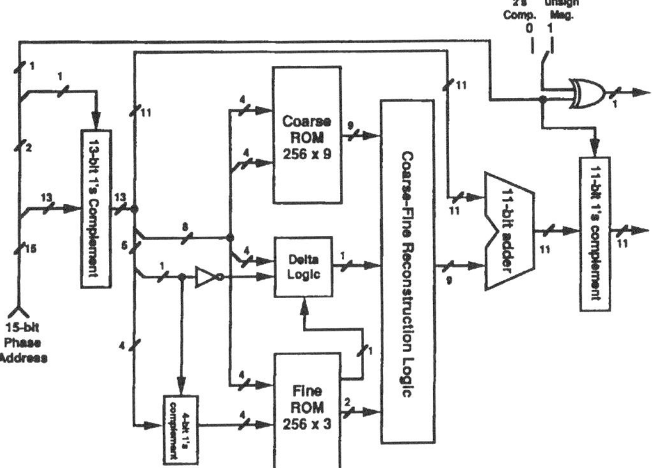

Fig. 7.   Low-spurious sine function generation logic.

compression techniques is -92 dBc. This results in a worst-case spurious response for the DDFS that is limited by phase accumulator truncation noise to -90.3 dBc, which is nearly an ideal result for a 12-b output sine wave. This performance is analytically guaranteed for all output frequencies.

C. 32-b Phase Accumulator and Input Registration

In order to provide operation at greater than 100 MHz, the 32-b phase accumulator was pipelined in 4-b stages. The input deskewing registers on the least significant 16 bits of the phase accumulator were eliminated to save chip area. This simplification only affects the phase co-herent switching property of the phase accumulator and has no effect on the output frequency of the device. The non-deskewed 16 LSB's of the frequency control word introduce a maximum phase discontinuity of 0.02°, which is insignificant for any foreseeable application. Further hardware simplification is provided by eliminating the deskewing registers for the least significant 17 bits of the phase accumulator output. This is possible because only the 15 most significant phase bits are used to calculate the sine function. The logic diagram of the 32-b phase accumulator is shown in Fig. 8.

The frequency control inputs to the phase accumulator are normally generated by circuitry that runs from a clock that is much lower in frequency than, and often asyn-

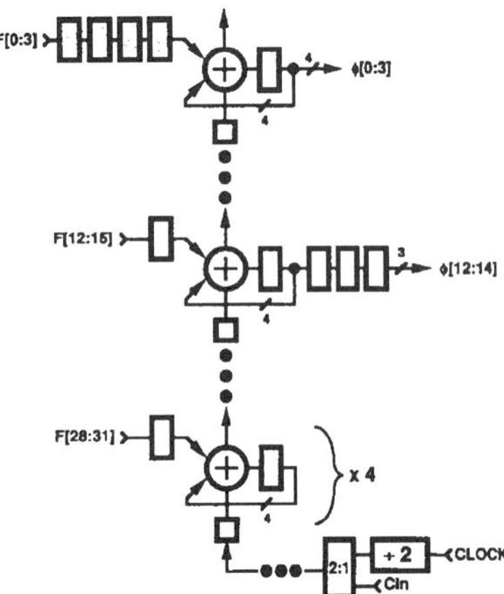

Fig. 8. Pipelined 32-b phase accumulator logic.

chronous to, the DDFS reference clock. To allow this asynchronous loading of the frequency control word, double buffering is used at the input to the phase accumulator.

IV. Circuit Design Issues

A. Clocking and Global Design Considerations

The circuit designs and layout of the DDFS were developed to meet the conflicting requirements for radiation hardness, low power, and speed. Spaceflight qualification and testability requirements dictated that no dynamic logic be used in this design, so many conventional high-speed design techniques could not be considered. A significant problem in CMOS circuit designs operating at over 100 MHz is clock distribution and clock skew. This problem is magnified in the presence of radiation effects, which can cause significant changes in the ratio of NMOS-to-PMOS drive current. Because of these radiation-induced device threshold shifts, the use of a standard two-phase clocking scheme in this design would have resulted in a large variation in the relative delays of true and false clock signals, making high-speed operation almost impossible. Therefore, to provide reliable operation at high clock rates, a true single-phase clocking scheme was chosen for the DDFS. The single-phase master clock is driven from a single-global-inverter clock buffer.

Several significant advantages are provided by this clocking scheme. First, it provides total immunity to changes in clock rise and fall times due to device threshold shifts. Because of the fast propagation delay of the single-inverter clock buffer design, negative hold times are maintained on all inputs, and clock-to-data output delays are reduced. This allows for chip-to-chip propagation delays of less than 6 ns with a 30-pF interchip capacitance. The elimination of distributed clock buffers also prevents clock skew problems due to propagation delay mismatches, providing more reliable high-speed operation. The localization of the clock driver allows for

control over power-supply clock switching noise, which is of critical importance in a design with over 400 simultaneously clocking register elements. Clock noise is also minimized by surrounding the global clock buffer with separate power and ground bond pads. Wide second-layer metal routing on clock, power, and signal lines is used extensively to reduce susceptibility to electromigration and to minimize voltage spikes due to large current transients.

B. Register and 4-b Pipelined Adder Design

The basic single-phase static D-type register element used in this design is shown in Fig. 9. Unlike other common register elements which use pass transistors or pull-up devices, this circuit contains no ratioed logic; i.e., its correct operation is never dependent upon a high-drive device overwhelming a low-drive device. Therefore extremely robust operation is obtained in the presence of device threshold shifts.

A disadvantage of this register design is its comparatively large setup delay. This problem was eliminated in the phase accumulator by integrating the last stage of the critical path of a 4-b adder into the master of the register, thus partially sharing the carry propagation delay with the register setup time. The adder was designed using a canonic delay ripple-carry approach. Furthermore, all logic functions requiring series NMOS or PMOS devices were implemented as separate logic gates, allowing each stage of the ripple-carry critical path to have the minimum number of drain diffusion contacts, thus achieving a propagation delay very close to that of an inverter. This simple and compact design also allowed the layout to be aggressively optimized for speed by minimizing diffusion area and interconnect capacitances. The last two stages of the ripple-carry path of the 4-b adder employing the carry-integrated register stage are shown in Fig. 10.

C. ROM Design

The ROM look-up table access time is the pipeline delay that limits the maximum achievable clock rate in this design. Careful optimization of the ROM architecture and circuits resulted in a 10-ns ROM access time over the worst-case extremes of temperature, process, and radiation. The $2^8 \times 9$ b of coarse ROM storage and $2^8 \times 3$ b of fine ROM storage were partitioned into four 16×48 ROM arrays to achieve maximum throughput and the optimum data-path floor plan. Each array stores the data for 256 three-bit output words. Using this organization, the word-line address decoders are identical for the coarse and fine ROM's, and thus may be shared. Therefore, it is economical to place a pipeline stage between the shared 4:16 word-line decoder and the word-line amplifiers for each ROM. Fig. 11 shows the simplified circuit diagram of the ROM word-line buffers, ROM array, and sense amps.

A major consideration in the design of the ROM was the minimization of the sensitivity to changes in device

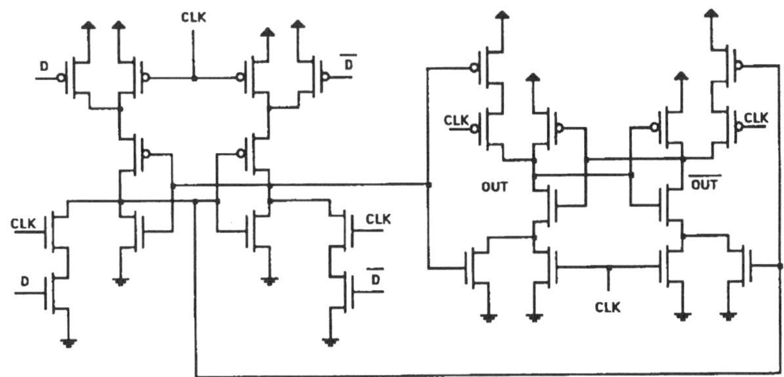

Fig. 9. Static D-type register element.

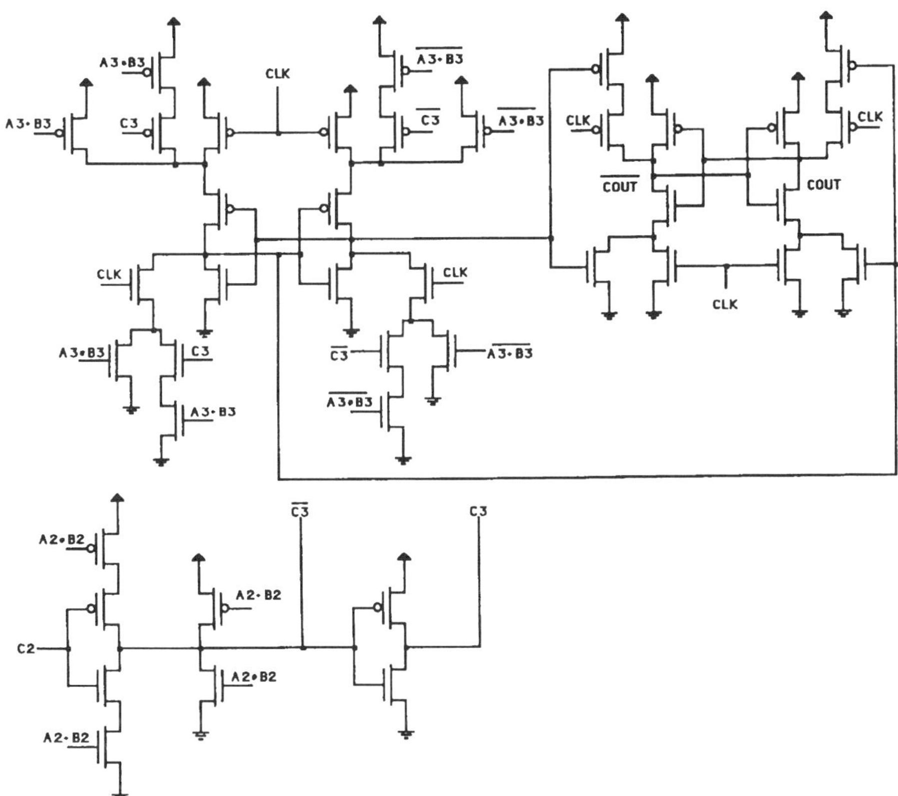

Fig. 10. Ripple-carry adder and carry-integrated register circuit.

thresholds and current drive. Therefore, early in the design it was decided that timed pulses would not be used for bit-line precharging, and that a simple inverter would be used for the read sense amp. The circuit shown in Fig. 11 satisfies these requirements without compromising the speed of the design. The operation of this circuit is described as follows: when the clock goes low, there is 5 ns for the data to propagate out of the pipelined 4:16 address decoders to the input of the word-line amp pass transistor, shown as $M1$ in Fig. 11. The entire 5-ns period is required in this case since this line shares the capacitive loading of all four ROM arrays and the associated inter-

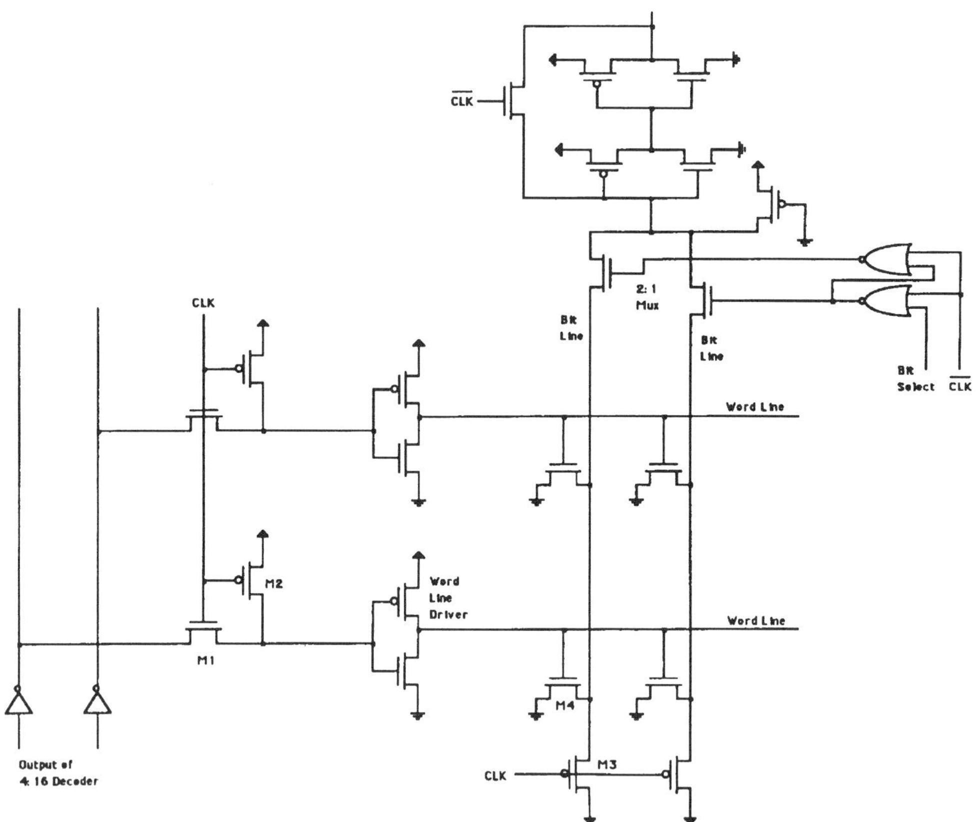

Fig. 11. Circuit digram of the radiation-tolerant ROM design.

connect bus. During the clock low period, the precharge transistor $M2$ forces the word-line driver to discharge the word line to ground. During this time the bit line is precharged high by $M3$. Because the circuit is precharged, there is no delay to read a high value at the input to the word-line driver. Therefore, the only speed-critical path is driving a word line high. When the clock goes low, the selected address line at the input to $M1$ has been discharged to ground. The capacitive address line then quickly discharges the input to the word-line amplifier through $M1$, which brings the word line high. If a zero is programmed into a ROM location, then a transistor such as $M4$ is connected to the word line, which discharges the bit line to ground. This signal then propagates through a 2:1 MUX, which also serves as a pass transistor for the sense-amp latch. The 2:1 MUX is the first stage of the 16:1 bit-line decoder. By placing the 2:1 MUX in the ROM array, the remaining 8:1 MUX operation to decode the ROM contents may easily be accomplished in a single pipeline stage. When the clock goes low, both pass transistors of the 2:1 MUX are forced low by the two NOR gates, thereby latching the read data into the sense-amp latch. The simulated maximum cycle time of this ROM is 9.9 ns over worst-case temperature, radiation, and process variations.

V. DEVICE LAYOUT AND FABRICATION

A photomicrograph of a packaged DDFS is shown in Fig. 12. The chip has a die size of 195 mil × 195 mil and a complexity of 35 000 transistors. The DDFS was fabricated in TRW's 1.25-μm radiation-hardened bulk P-well LOCOS process with double-layer AlCu metalization. The process employs a P^+ guard ring and a retrograde P-well on N/N^+ epi to achieve a total dose radiation hardness of over 1×10^6 rd(Si).

The chip operates from a nominal 5-V power supply, and all outputs are standard CMOS levels (5 V rail to rail). Careful attention was paid to the layout to minimize voltage spikes from driving highly capacitive output loads. One V_{DD} or GND pad is located next to every four output buffers to minimize power supply noise. Sine data outputs are located in the center of one side of the chip to assure minimum bond-wire lengths.

The chip layout was performed in process-portable design rules using MAGIC, a symbolic layout system developed by the University of California, Berkeley. This

Fig. 12. Photomicrograph of the packaged DDFS.

TABLE I
DDFS Chip Specifications

Technology	1.25-μm rad-hard DLM CMOS
Max. Clock Frequency	150 MHz
Switching Time	6.7 ns (at 150-MHz f_{clk})
Tuning Latency	13 clock cycles
Tuning Bandwidth	75 MHz (at 150-MHz f_{clk})
Freq. Resolution	0.035 Hz (at 150-MHz f_{clk})
Worst-Case Spurious	-90.3 dBc
SNR	74 dB
Output Word Length	12 b
Power Dissipation	950 mW at 100 MHz
Transistor Count	35 000 (including I/O's)
Die Size	195 mil $\times$ 195 mil

layout system uses a technology file to provide a mapping from the symbolic layout to the process specific artwork. Because all logic in the DDFS is fully synchronous, and because the circuit designs are tolerant to large changes in transistor characteristics, the DDFS could be fabricated in any standard double-level-metal CMOS process with minimal redesign. Compatibility with a new set of design rules may be accomplished through a simple translation of the layout artwork using a new MAGIC technology file.

VI. Test Results

The measured performance of the DDFS is summarized in Table I. Several devices were packaged for speed testing, and the digital sine-wave output sequences of

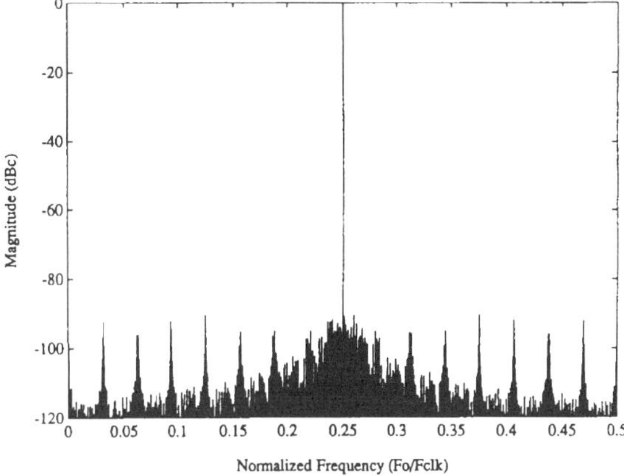

Fig. 13. Analytically guaranteed worst-case digital output spectrum.

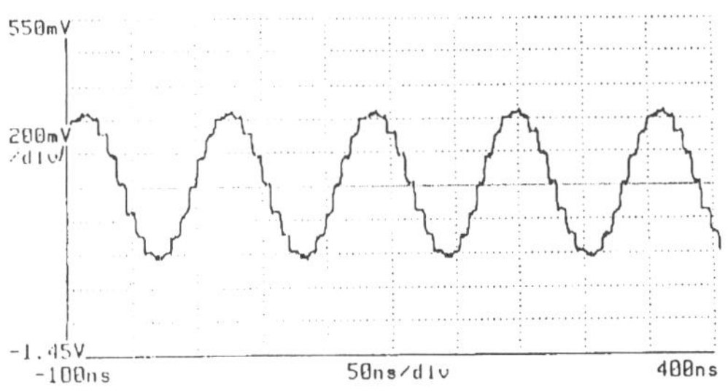

Fig. 14. DDFS analog output using an 8-b DAC for $f_{\text{out}} = 9.37$ MHz and $f_{\text{clk}} = 150$ MHz.

these devices were exhaustively verified at 100 MHz on a high-speed digital tester. Power dissipation of 950 mW was measured at 100 MHz. All functional devices are analytically guaranteed to have a -90.3-dBc worst-case spurious rejection over all tunable frequencies, which was verified by FFT tests conducted over a representative sample of the 2^{32} different output frequencies. All output frequencies have a spectral purity equivalent to the spectrum measured in Fig. 13, which is the test result for $f_{\text{out}} = 0.2512 f_{\text{clk}}$. Because the available high-speed digital tester was limited to 100 MHz, speed testing was performed on two devices in *in-situ* with an Analog Devices 8-b DAC. Measurements were taken with a Tektronix 11402 digitizing oscilloscope and a HP 3588A spectrum analyzer. A maximum clock frequency of 150 MHz was measured for $f_{\text{out}} < 3/8 f_{\text{clk}}$, and a maximum clock frequency of 145 MHz was measured for output frequencies close to $f_{\text{clk}}/2$. An 8-b output sine wave is shown in Fig. 14 for $f_{\text{out}} = 9.37$ MHz and $f_{\text{clk}} = 150$ MHz. The output spectrum for $f_{\text{out}} = 11.1$ MHz and $f_{\text{clk}} = 150$ MHz is shown in Fig. 15. The -51.48-dBc spurious components in this spectrum are due to the 8-b resolution of the DAC. An output spectrum for the DDFS is shown in Fig. 16 using an Analog Devices 9713A 12-b DAC for $f_{\text{out}} = 1.2$ MHz and $f_{\text{clk}} = 100$ MHz, which demonstrates the spectral pu-

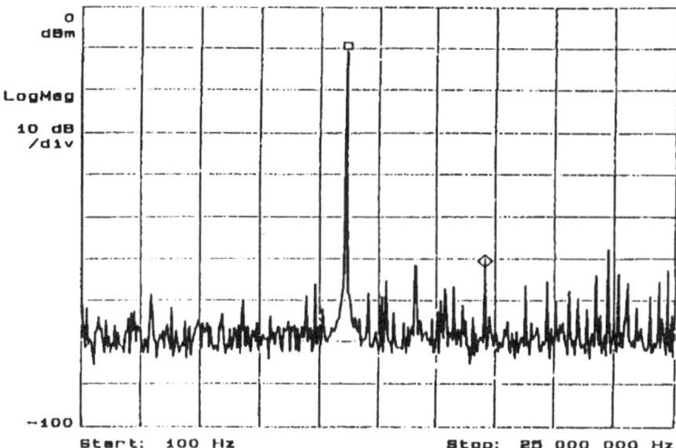

Fig. 15. DDFS analog output spectrum using an 8-b DAC for f_{out} = 11.1 MHz and f_{clk} = 150 MHz.

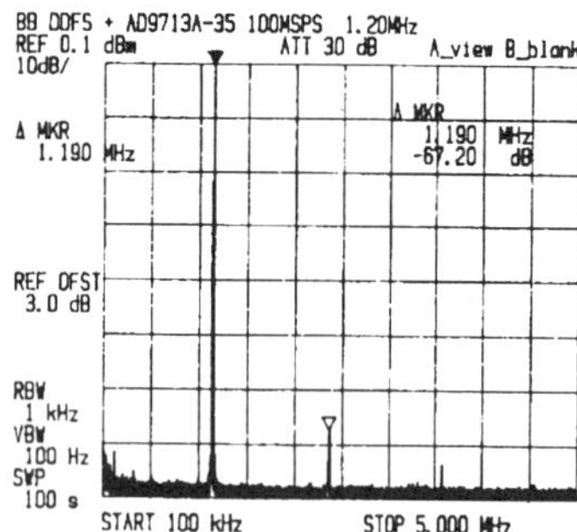

Fig. 16. DDFS output spectrum using a 12-b DAC for f_{out} = 1.2 MHz and f_{clk} = 100 MHz.

rity that may be obtained using a higher resolution DAC. A −67.2-dBc spurious response is achieved in this case. Measurements of the C_{out} signal of the phase accumulator have also been made at 170 MHz, demonstrating that the 32-b phase accumulator does not limit the speed of the DDFS chip.

Devices were irradiated with an Aracor flash X-ray source to a total dose of 2×10^6 rd(Si) of ionizing radiation. The devices were biased with a 5-V supply and static inputs and clock during irradiation. At 1×10^5 rd(Si) dose intervals the devices were tested with an exhaustive test vector set at 75 MHz, the limit of the test fixture. No device failures occurred in all measurements taken up to a total dose of 2×10^6 rd(Si), and the maximum clock frequency was not degraded below the 75-MHz limitation of the test fixture.

The 150-MHz maximum clock rate and −90.3 dBc of spectral purity achieved by this design represent a significant advance over the fastest previously published CMOS DDFS, which demonstrated −75 dBc of spectral purity at an 80-MHz maximum clock rate [3] when fabricated in a commercial CMOS process. The fastest previously reported radiation-hardened (1×10^6 rd(Si)) CMOS DDFS

had a maximum post-radiated clock rate of 45 MHz and −75 dBc of spectral purity [3], [14].

VII. CONCLUSIONS

A radiation-hardened high-speed CMOS monolithic direct digital frequency synthesizer chip has been presented which generates a 90-dBc spectrally pure digitized sine wave at 150 Msamples/s. Through new architectural techniques and optimized circuit designs, this DDFS is able to achieve substantially improved bandwidth and spectral purity over any previously reported CMOS DDFS. Extremely fast switching speed and sub-1-Hz frequency resolution have also been obtained in this design. These aggressive design goals have been met while satisfying extremely rigorous radiation hardness and spaceflight qualification standards.

The switching speed, phase noise, and frequency resolution of DDFS's are the key to making many frequency agile communications systems possible. The wide-bandwidth, low-spurious DDFS presented herein allows these benefits to be provided to either commercial or military systems in a low-cost manufacturable CMOS technology.

ACKNOWLEDGMENT

This work would not have been possible without the technical contributions of J. Cable, A. Coulson, and M. Miscione of TRW's Microelectronics Center. Special thanks to B. Kim and K. Diep for assistance in the DDFS design and test. High-speed testing was accomplished through a generous equipment donation by Hewlett Packard and Tektronix. D. Buchanan and J. Studders of Analog Devices are also gratefully acknowledged for providing D/A converter evaluation boards and for providing DDFS test results using the 12-b DAC. This research was made possible by TRW's Microelectronics Center through the donation of VLSI wafer fabrication in their radiation hardened CMOS process.

REFERENCES

[1] V. Manassewitsch, *Frequency Synthesizers, Theory and Design*, 2nd ed. New York: Wiley, 1989.
[2] J. Tierney, C. M. Rader, and B. Gold, "A digital frequency synthesizer," *IEEE Trans. Audio Electroacoust.*, vol. AU-19, pp. 48–57, 1971.
[3] *Direct Digital Synthesizer Handbook*, Stanford Telecom, Santa Clara, CA, Mar. 1990.
[4] "Q2334 dual direct digital synthesizer technical data sheet," Qualcom Inc., San Diego, CA, June 1990.
[5] D. A. Sutherland, R. A. Strauch, S. S. Wharfield, H. T. Peterson, and C. R. Cole, "CMOS/SOS frequency synthesizer LSI circuit for spread spectrum communications," *IEEE J. Solid-State Circuits*, vol. SC-19, pp. 497–505, Aug. 1984.
[6] R. E. Lundgren, V. S. Reinhardt, and K. W. Martin, "Design and architectures for EW/communication direct digital synthesizers," U.S. Army Lab. Command, Fort Monmouth, NJ, Res. Develop. Tech. Rep. SLCET-TR-85-0424-1, Aug. 1986.
[7] H. T. Nicholas, III, and H. Samueli, "An analysis of the output spectrum of direct digital frequency synthesizers in the presence of phase-accumulator truncation," in *Proc. 41st Annual Frequency Control Symp. USERACOM* (Ft. Monmouth, NJ), May 1987, pp. 495–502.

[8] H. T. Nicholas, III, H. Samueli, and B. Kim, "The optimization of direct digital frequency synthesizer performance in the presence of finite word length effects," in *Proc. 42nd Annual Frequency Control Symp. USERACOM* (Ft. Monmouth, NJ), May 1988, pp. 357–363.

[9] P. H. Saul and D. G. Taylor, "A high-speed digital frequency synthesizer," *IEEE J. Solid-State Circuits*, vol. 25, no. 1, pp. 215–220, Feb. 1990.

[10] C. E. Wheatley, III, and D. E. Phillips, "Spurious suppression in direct digital synthesizers," in *Proc. 35th Annual Frequency Control Symp. USERACOM* (Ft. Monmouth, NJ), May 1981, pp. 428–435.

[11] "GaAs direct digital frequency synthesizer chip set applications notes," Gigabit Logic Inc., Newbury Park, CA, June 1990.

[12] M. Gold, "Four call on Triquint to give D/A the GaAs," *Elec. Eng. Times*, p. 1, Jan. 15, 1990.

[13] "AD9713A D/A converter technical data sheet," Analog Devices, Inc., Greensboro, NC, May 1991.

[14] "STEL-1173RH technical data sheet," Stanford Telecom, Santa Clara, CA, Oct. 1989.

An Exact Spectral Analysis of a Number Controlled Oscillator Based Synthesizer

JOSEPH F. GARVEY AND DANIEL BABITCH

CALIFORNIA MICROWAVE INC., COMMUNICATIONS TECHNOLOGY DIVISION

990 ALMANOR AVE., SUNNYVALE, CALIFORNIA, 94086

e-mail: cmic!garvey
cmic!babitch

ABSTRACT

A Direct Digital Synthesizer (DDS) based on a Number Controlled Oscillator (NCO) has advantages in many applications due to its wide frequency range, quick frequency changes, and fine resolution. However, because of the digital nature of the NCO, conventional wisdom has deemed the DDS spectrally unclean. In addition, because of the digital nature of the NCO, it has been difficult to determine what the ultimate performance levels might be. This paper presents an exact analysis of an NCO-based synthesizer. We believe this is the first exact analysis, and will be extremely important to those who design DDS', NCOs, and DACs.

This paper describes the architecture of a DDS, presents an exact spectral analysis and discusses some actual performance observations.

1.0   INTRODUCTION

This paper examines the ultimate performance of an NCO-based frequency synthesizer by going through a number of steps:

- A description of an NCO-based synthesizer is provided for those unfamiliar with the architecture of these units;

- A continuous Fourier Transform is developed for an arbitrary stepped waveform like the DDS output;

- A method for generating the mapping PROM used by a DDS is described, so that the output waveform distortion is minimized;

- Computer-based calculations are presented so that the reader may gain insight to the workings of a DDS;

- A final discussion is presented to summarize what the authors have learned and observed from their efforts.

2.0   DESCRIPTION OF AN NCO-BASED SYNTHESIZER

2.1   Description Of A DDS

A DDS based on an NCO takes a fixed frequency reference clock and generates an analog waveform of variable frequency, based on digital frequency commands.

An NCO-based frequency synthesizer, as shown in Figure 1, consists of the following pieces:

<u>NCO</u>: Really just a big accumulator, which we describe as holding the "phase" of the output signal. As the NCO accumulator changes, the values represent a linear phase ramp. The NCO accumulator has P bits, and is incremented by S every time the NCO is clocked.

<u>Mapping PROM</u>: This Programmable Read Only Memory maps the most significant bits (MSB) of the NCO into a sinusoid. In other words, the linear phase from the NCO is converted to a series of values representing a sinusoidal waveform. In our case a cosine is used. It has A address bits, and A <= P, always. All data lines from the mapping PROM go to the DAC.

<u>DAC</u>: The Digital to Analog Converter is used to generate the analog output voltage that is the output of the synthesizer. It has D bits.

2.2   Further Description Of The NCO

As previously mentioned, the NCO is simply a large accumulator (typically 32 bits, or more). The value of the accumulator at any given time corresponds to a phase. It is useful to picture the accumulator value as an angle, which, like the second hand of some watches, steps from second to second with each "tick" of the fixed frequency reference clock. Unlike the second hand of a watch, however, the size of each step can be varied by changing the phase step size. For the second hand of a watch the <u>phase step size</u> is 1 sec, with a corresponding output frequency of 1/60th of a Hz. If we could change the step size to 2, then the output frequency would be 1/30 of a Hz. Changing the step size to 10 would result in an output frequency of 1/6 Hz.

Reprinted from *Proceedings of the 44th Annual Frequency Control Symposium*, pp. 511-521, 1990.

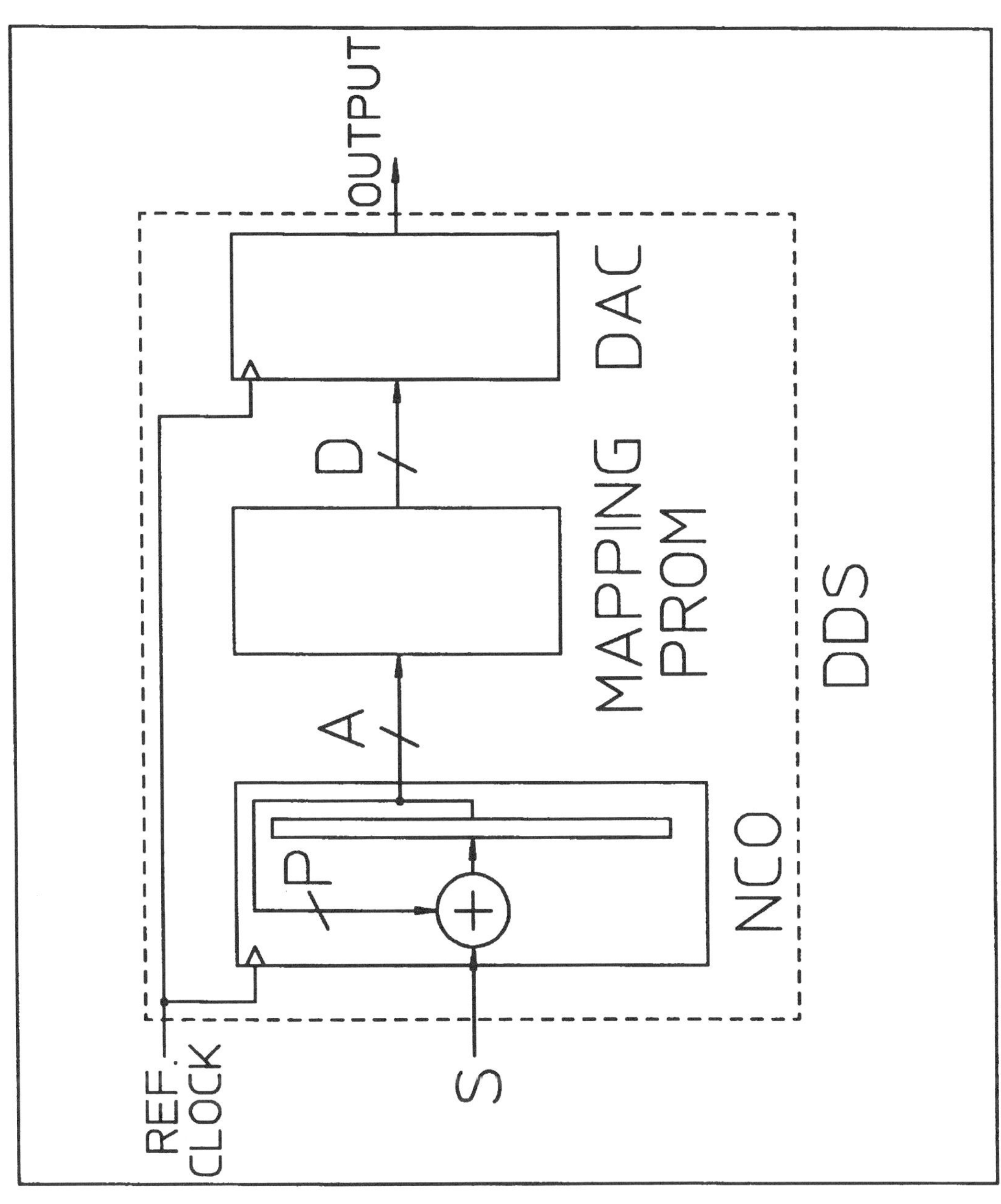

Figure 1.
DDS Block Diagram

Using this analogy, we can readily deduce the simple ratios that relate the phase step size (S), size of the accumulator (P), reference clock frequency (F_c), and output frequency (F_o).

$$\frac{\text{Step Size}}{\text{Accumulator Size}} = \frac{S}{2^P} = \frac{F_o}{F_c} = \frac{\text{Output Frequency}}{\text{Clock Frequency}} \quad (1)$$

From this we can see that frequency resolution is of the NCO synthesizer is simply:

$$\text{Resolution} = \frac{\text{Clock Frequency}}{\text{Accumulator Size}} = \frac{F_c}{2^P} \quad (2)$$

3.0 FOURIER TRANSFORM OF A REPEATING PULSE TRAIN

3.1 Grand Repetition Period

Though it may not be immediately obvious, the output of the synthesizer is a stepped waveform that at some point in time repeats. This occurs when the NCO contains exactly the same phase value as it started with. During the time it takes for this to occur, many sinusoidal cycles may occur, each slightly different from the preceding cycle. The minimum time for the NCO accumulator value to repeat is known as the Grand Repetition Period.

First, let's verify a repeating pattern occurs, and put an upper bound on how long this takes. To start, assume an initial NCO accumulator value of zero. Then take advantage of the fact that addition of a value to itself is equivalent to shifting the value "left" by one bit, and filling the least significant bit (LSB) with a zero.

To load the step size into the accumulator would take one tick. The next tick would add this value to itself. The value in the accumulator is the step size value shifted one bit to the left with a 0 filling in the LSB. To add this value to itself would take two more ticks for a total of four ticks. If we repeat the process, of adding the current accumulator value to itself, we keep doubling the number of ticks required for each iteration, and shifting left, and filling in with zeros. After P iterations of this procedure, we will have filled the accumulator with zeros, which is where we started. Counting up the number of ticks used to do this gives us an upper limit to the grand repetition period. So, the number of ticks in the worst-case grand repetition period is 2^P.

In the worst case grand repetition period, we may have crossed zero one or more times during these 2^P ticks. As we find out later, this has a direct effect on computational complexity, so let's pursue this issue a little further and find the grand repetition period exactly.

To find the grand repetition period, continue to think about what happens in the accumulator for different step sizes. Consider the case of P=4, for each possible step size.

Table 1

	Tick	P	P-1	...	4	3	2	1	0
Start, 0 phase	0	0	0	...	0	0	0	0	0
Add 3	1	0	0		0	0	0	1	1
Add 3	2	0	0		0	0	1	1	0 [1]
Add 3	3	0	0		0	1	0	0	1
Add 3	4	0	0		0	1	1	0 [2]	0 [2]
Add 3	5	0	0		0	1	1	1	1
Add 3	6	0	0		1	0	1	0	1
Add 3	7	0	0		1	0	1	0	1
Add 3	8	0	0		1	1	0 [3]	0 [3]	0 [3]

Note 1:	After 2^1 ticks, the first LSB is 0.
Note 2:	After 2^2 ticks, the first and second LSBs are 0.
Note 3:	After 2^3 ticks, the first, second and third LSBs are 0.

Table 2

Ticks	Step Size (P=4)							
	1	2	3	4	5	6	7	8
	0	0	0	0	0	0	0	0
1	1	2	3	4	5	6	7	8
2	2	4	6	8	10	12	14	0
3	3	6	9	12	15	2	5	
4	4	8	12	0	4	8	12	
5	5	10	15		9	14	3	
6	6	12	2		14	4	10	
7	7	14	5		3	10	1	
8	8	0	8		8	0	8	
9	9		11		13		15	
10	10		14		2		6	
11	11		1		7		13	
12	12		4		12		4	
13	13		7		1		11	
14	14		10		6		2	
15	15		13		11		9	
16	0		0		0		0	
Ticks to Repeat	16	8	16	4	16	8	16	2

If the step size is a multiple of two, then the waveform repeats multiple times in 2^P ticks. In fact for each factor of two in the step size the wave form repeats.

Thus, if we eliminate all the factors of two from the step size, we can determine the grand repetition period exactly. One simple way to find the number of factors of two in the step size is to find the <u>Greatest Common Divisor</u> between the number needed to roll the accumulator over, and the step size. Then the grand repetition period can be directly calculated.

$$\begin{array}{c}\text{Min. Number}\\\text{of Ticks}\\\text{in Grand}\\\text{Repetition Period}\end{array} = \frac{2^P}{\text{GCD}\left(2^P,S\right)} \equiv \mu \qquad (3)$$

μ = min. number of "ticks" in Grand Repetiton Period

GCD = greatest common divisor

3.2 Fourier Transform Of A Pulse

Now that we know the interval on which the NCO is periodic, we can perform a classical Fourier Transform.

To do this, let's consider each step of the output waveform as a separate pulse. Adding all these pulses up with proper phase and amplitude will create the original stepped waveform. Furthermore, let's just consider one of those pulses at this time:

As noted in Figure 2, the amplitude of the pulse is a function of its position (n). Notice that this will make the

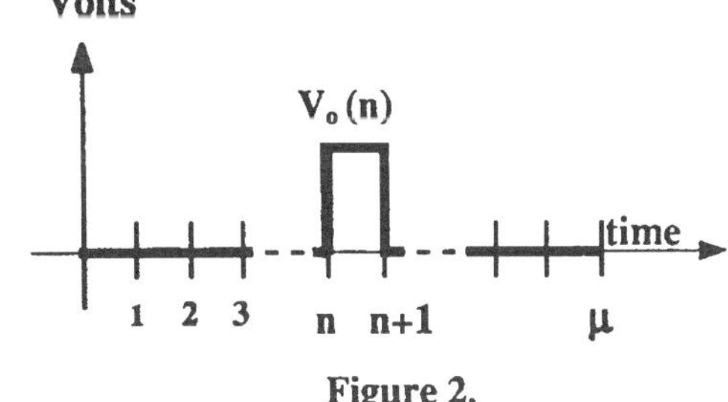

Figure 2.

analysis general – that is, it will work for any set of repeating pulses. Later, we'll substitute in the sinusoidal terms of the mapping PROM for the output voltage $V_o(n)$.

Now for the Fourier Transform itself:

Assume the pulse can be made of complex sinusoids:

$$V_o(t) = \sum_{i=-\infty}^{\infty} C_i\, e^{j\omega_i t} \quad \text{where } \omega_i = \frac{2\pi i}{\mu T_c} \tag{4}$$

$$T_c \equiv \frac{1}{F_c}$$

C_i are complex coefficients

Multiplying both sides by $e^{-j\omega_b t}$ and integrating over a period of $V_o(t)$:

$$\int_0^{\mu T_c} V_o(t)\, e^{-j\omega_b t}\, dt = \int_0^{\mu T_c} C_i \sum_{i=-\infty}^{\infty} e^{j(\omega_i - \omega_b)t}\, dt \tag{5}$$

Since this is an integration of a bounded well-behaved function, over a finite interval, we can fearlessly interchange the order of integration and summation:

$$\int_0^{\mu T_c} V_o(t)\, e^{-j\omega_b t}\, dt = \sum_{i=-\infty}^{\infty} C_i \int_0^{\mu T_c} e^{j(\omega_i - \omega_b)t}\, dt \tag{6}$$

The integral over a full period of two sinusoidal functions $e^{j\omega_i t}$ and $e^{-j\omega_b t}$ is always zero unless $i = b$. That's the definition of orthogonal functions. This elimi-

nates all terms of the summation, except when $i = b$. When $i = b$, the complex exponential cancels to 1, $e^{j(\omega_i - \omega_i)t}$

$$\int_0^{\mu T_c} V_o(t)\, e^{-j\omega_i t}\, dt = C_i\, \mu\, T_c \tag{7}$$

Rearranging terms we get:

$$C_i = \frac{1}{\mu T_c} \int_0^{\mu T_c} V_o(t)\, e^{-j\omega_i t}\, dt \tag{8}$$

For the pulse this becomes:

$$C_i = \frac{V_o(n)}{\mu T_c} \int_{n T_c}^{(n+1)T_c} e^{-j\omega_i t}\, dt \tag{9}$$

After integration, factoring out the exponential in the result, and a little algebra, we get:

$$C_i = \frac{V_o(n)}{\mu}\, \text{sinc}\, \frac{\pi i}{\mu}\, e^{-j\omega_i\left(n+\frac{1}{2}\right)T_c} \tag{10}$$

<u>Fourier Transform Of A Pulse Train</u>

Knowing the Fourier Transform of a single pulse with a period of μ allows us to use the simple concept of superposition to find out what the entire spectrum would look like. For a series of μ pulses, phased to create the stepped output waveform:

$$C_i = \sum_{n=0}^{\mu-1} \frac{V_o(n)}{\mu}\, \text{sinc}\, \frac{\pi i}{\mu}\, e^{-j\omega_i\left(n+\frac{1}{2}\right)T_c} \tag{11}$$

Rearranging terms, and factoring out part of the exponential, we get:

$$\boxed{\begin{aligned} C_i &= \frac{e^{-j\frac{\pi i}{\mu}}}{\mu}\, \text{sinc}\, \frac{\pi i}{\mu} \sum_{n=0}^{\mu-1} V_o(n) W_\mu^{ni} \\[2mm] W_\mu &\equiv e^{-j\frac{2\pi}{\mu}} \end{aligned}} \tag{12}$$

Notice when stated in this form that the summation is just the Discrete Fourier Transform of the synthesizer output over the grand repetition period, and since μ is composed of factors of two only, the Radix Two Fast Fourier Transform can be used. This makes computation of the coefficients quick and easy. After performing a Fast Fourier Transform, the coefficients are scaled by the term in front of the summation to get the Continuous Time Fourier Transform. An exact spectrum!

3.4 Verification

3.4.1 By Summation: A computer program was created to generate the output spectrum of the NCO based on equation 12. The program generated two primary lines, the upper and lower sidebands, which are centered on half the clock frequency, as expected. In order to verify the spectrum generated by the Fourier Transform of the NCO output was correct, a program that did a Fourier Summation was used. This allowed the spectrum to be converted back into a time domain waveform, and thereby verifying that the proper coefficients had been calculated.

One of the important features observed is that harmonics above the NCO clock frequency are required for accurate reproduction of the time waveform. In order to get a reasonably clean waveform harmonics up to 64 times the NCO clock rate were required.

3.4.2 Known Spectrum: A second test was made by checking what happens to the equation for C_i when half the output frequency was requested. In this case the formula must and does simplify to that of a square wave.

In addition, proper behavior is observed when a very low or very high frequency is requested. The lower sideband amplitude approaches unity (0 dB) at low frequency, and the upper sideband goes to 0 ($-\infty$ dB) as it approaches the clock frequency. Near half the clock frequency, each sideband is 3 dB down, as it should be.

3.4.3 By Measurement: DAC clock leakage in our experimental DDS prevented us from measuring the individual spurious frequencies. However, the DAC clock leakage is too small to significantly affect the amplitudes of the upper and lower sidebands. These were checked and were in excellent agreement with the predictions of the model (hundredths of a dB), over a wide range of frequencies. In addition, the measured spurious level of -65 dBc was consistent with theoretical predictions of -67.8 dBc cleanliness.

4.0 GENERATING THE MAPPING PROM

4.1 Need For Symmetry

To generate the cleanest possible output signal, the digitized waveform should be quantized as symmetrically as possible. This includes symmetry about the time (phase) and amplitude axes. Figure 3 shows a sinusoid, and it's corresponding quantization.

Notice, by selecting the quantization levels in the middle of each time period (pulse), the amplitude error for half the pulse is in one direction, and the error in the second half of the pulse is in the opposite direction.

Also notice that by drawing a line through the 90, and 270 degree marks, the quantized waveform is symmetric about these points as well. This again minimizes distortion, and preserves the symmetry one expects of a sinusoid.

4.2 Quantized Sinusoid - Mathematical Description

If we quantize a cosine wave form, then we want to put one cycle of the cosine into the Mapping PROM. The phase resolution of this PROM is determined by the number of address bits it has:

$$\text{PROM Phase Resolution} = \frac{\text{One Cycle}}{\#\ \text{of Points}} = \frac{2\pi}{2^A} \qquad (13)$$

In order to maintain symmetry, we need to sample the cosine wave at the mid-point between two pulses. This is obtained simply by averaging the two times.

$$\begin{aligned}\text{Midpoint between two pulses} &= \frac{\text{phase of point 1 + phase of point 2}}{2} \\ &= \frac{m + (m + 1)}{2}\end{aligned} \qquad (14)$$

$m \equiv$ address applied to PROM at time n

Finally, examining the amplitude of the sinusoid, we notice one bit can be considered a sign bit, and the remaindering bits are the magnitude of the amplitude. Also notice, the 0 value out of the DAC is never used.

$$\text{Amplitude of PROM} = 2^{(D-1)} \qquad (15)$$

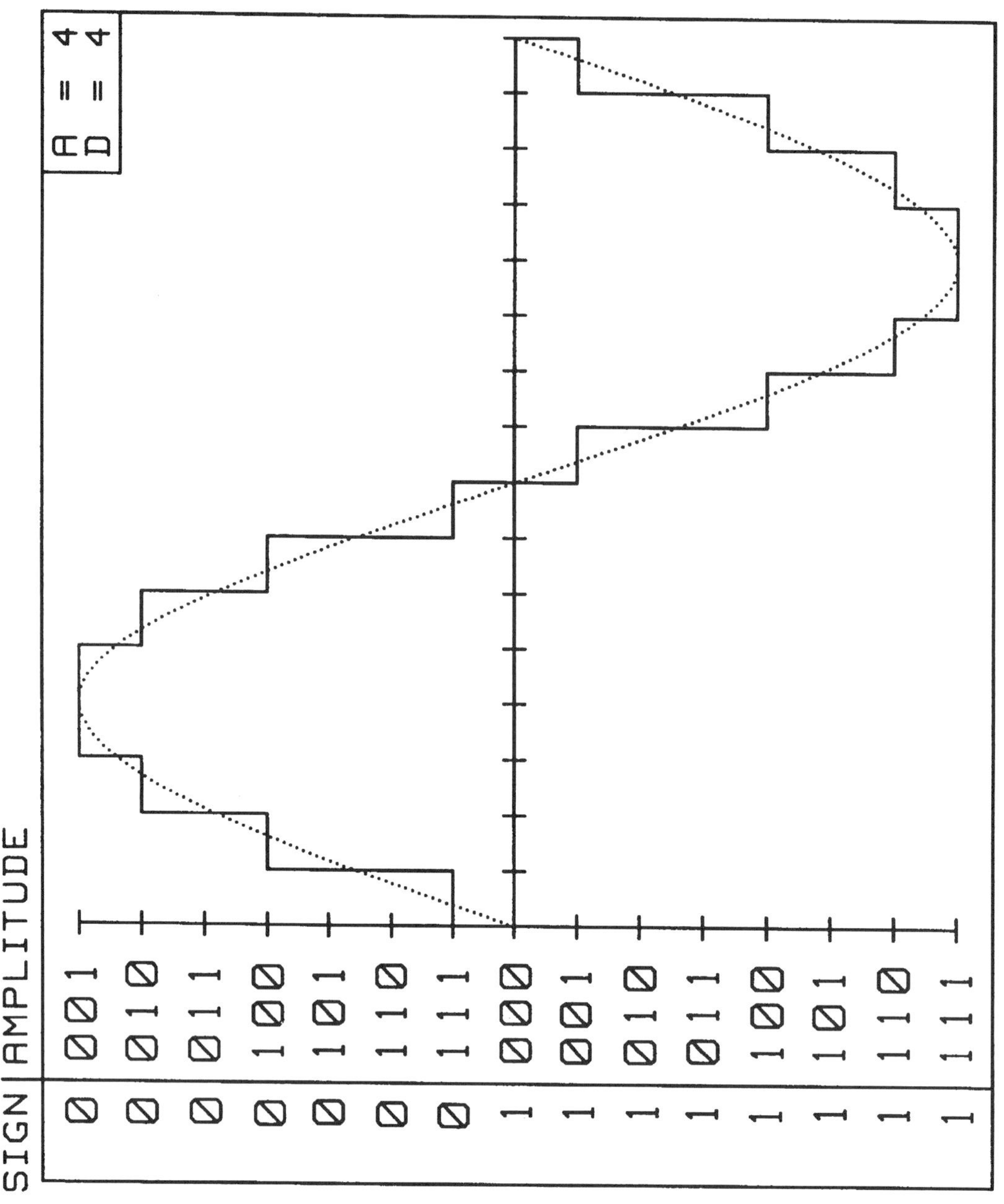

Figure 3.
Symmetrically Quantized Sinusoid

Putting this all together we have the mathematical description of the contents of the Mapping PROM.

$$V_o(n) = 2^{(D-1)} \cos\left(\frac{n + (n+1)}{2} \frac{2\pi}{2^A} \right) \qquad (16)$$

A word of caution: the value actually used in the PROM is a rounded integer value of $V_o(n)$. To avoid inducing any extra distortions, these values should be rounded symmetrically. This can be achieved by rounding the absolute value of $V_o(n)$, and then attaching the proper sign.

5.0 PRESENTATION OF RESULTS

5.1 Spectral Cleanliness

Using simulation software based on the equations in this paper, we modeled various DDS configurations. Figures 4 and 5 graphically present the results of these analyses.

All analyses presented here were performed with the output frequency set to 1/3 of the reference clock frequency. This is where the worst spurious occurs based on empirical evidence.

Figure 4 shows that the output cleanliness is not a function of the number of phase bits in the NCO. Rather, the phase bits in the NCO simply represent the resolution with which frequencies may be selected. There is a section of Figure 4 where the cleanliness actually goes up as the number of phase bits goes down. However, this range is extremely impractical to use, since it is characterized by the condition P=A=D. As a result of this condition, there is limited frequency resolution for real world values, and output slew rates tend to be too great to implement cleanly.

The big surprise is Figure 5. It shows that the cleanliness is a 6 dB/bit function of the number of mapping PROM address bits A, and an 8.5 dB/bit function of the number of DAC bits D.

Figure 5 clearly indicates there is a range of values where having extra DAC bits results in NO performance gain. This means that designers who push to incorporate large mapping PROMs into NCOs may besurprised that performance doesn't improve, if they haven't selected the right Address-Data bit combinations (A=8,D=6), (A=9,D=7), (A=10,D=7), (A=11,D=8), (A=12,D=9), (A=13,D=10), etc.

6.0 ADDITIONAL THOUGHTS

6.1 DAC Clock Leakage

One of the reasons an actual spectrum is not compared against a predicted spectrum is DAC clock leakage causes a severe distortion of the actual spectrum at the DAC output. When observed through filtering and limiting circuits the clock leakage sidebands were -65 dBc.

We used a 25 MHz NCO with a high-speed 8-bit DAC in our experimental DDS. This DAC had clock leakage 45 dB below the output signal at the clock frequency. This clock leakage signal mixed with the switched current sources in the DAC, so that mixing signals appeared at multiple points across the output spectrum. This is consistent with the concept of the DAC acting as a switching mixer, where the DAC MSB is acting as the switch control and the clock leakage (main spectral line) acting as the input.

The original analysis program was modified to verify this behavior. It correctly indicated where the DAC spurious would occur, but was not good at estimating spectral magnitude. To accurately model the magnitude of DAC spurious, the internal layout, parasitics, and process parameters would have to be included. Nevertheless, we believe the basic mechanism of clock leakage mixing sidebands is correctly modeled because of the match between predicted frequencies and observed frequencies.

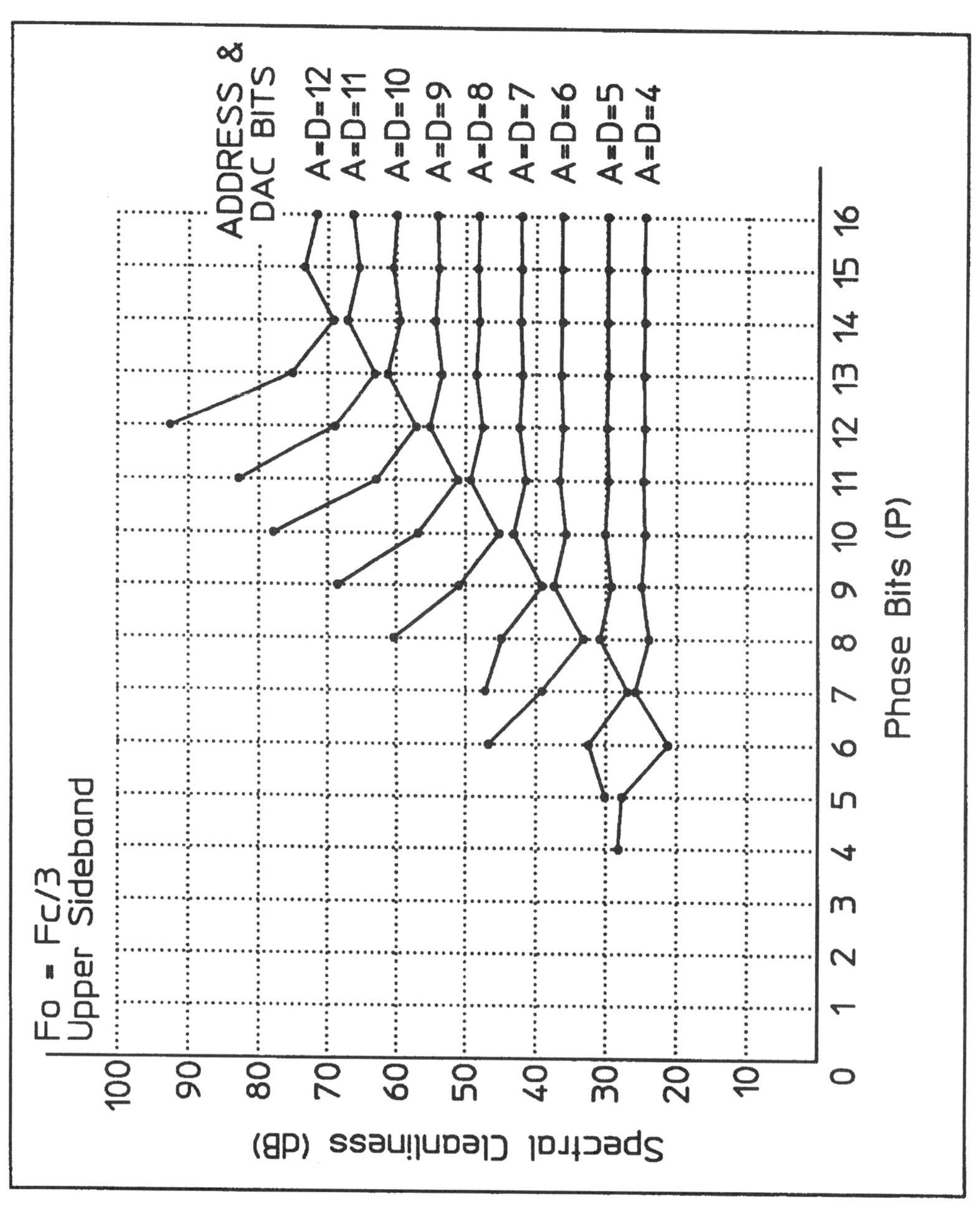

Figure 4.
Number of Phase Bits vs. Spectral Cleanliness

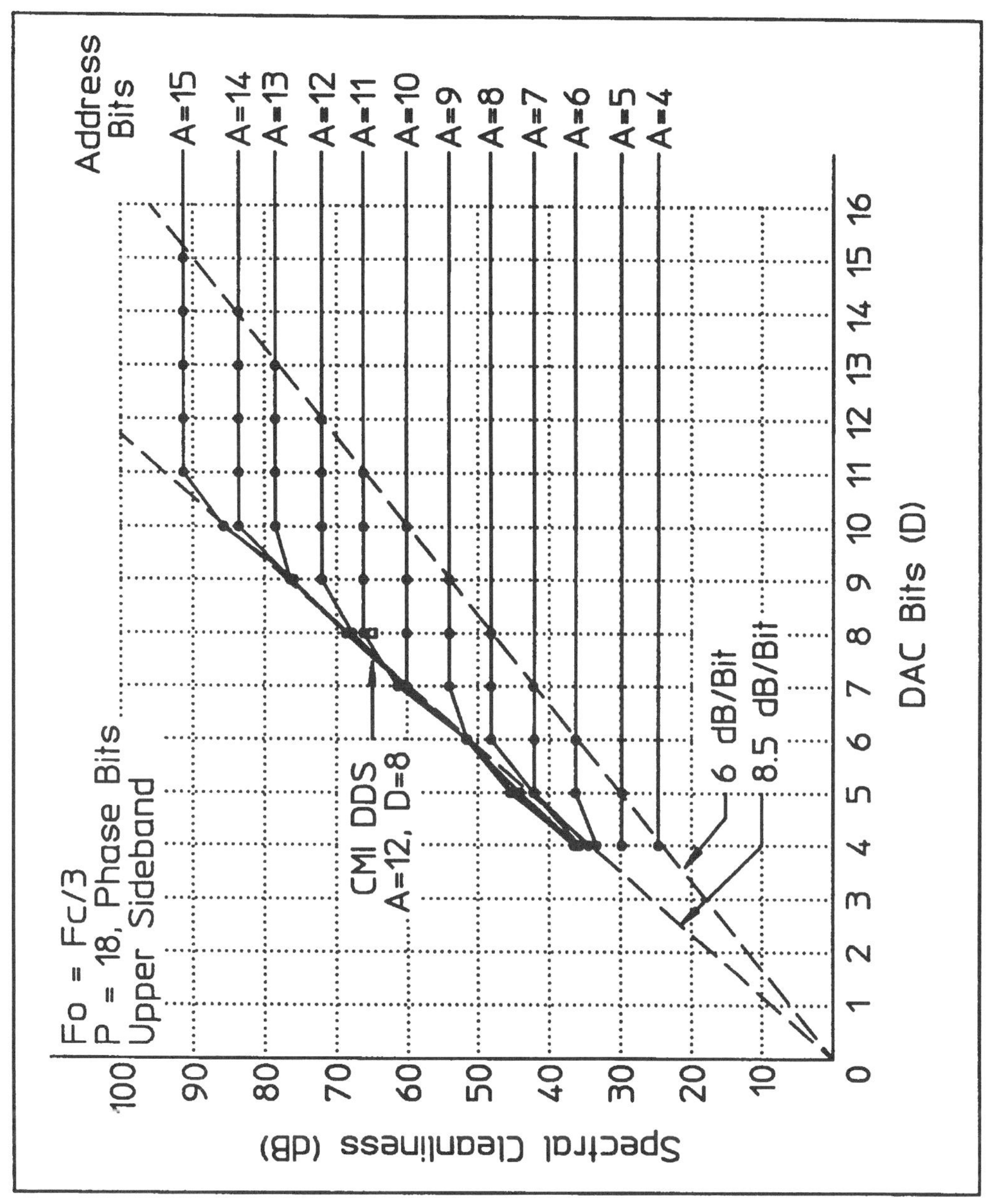

Figure 5.
Number of DAC Bits vs. Spectral Cleanliness

7.0   CONCLUSION

A DDS based on an NCO <u>can</u> be spectrally clean, which is contrary to conventional wisdom. Furthermore, its spectrum can be calculated exactly using a conventional radix-2 DFT, and applying a frequency dependent correction factor. Examining the results of these calculations, one discovers there are very specific configurations of the DDS architecture that will achieve maximum spectral purity with the least circuitry.

Currently, much work remains to be done in the actual implementation of an NCO-based DDS. Specifically, modulation of the DAC output by the clock is a significant problem. This is especially so with available high-speed DDS' having clock rates of hundreds of megahertz, where performance is still somewhat limited.

ACKNOWLEDGEMENTS

The authors would like to thank Kathy Stanley for her diligent efforts in typesetting this paper and her discerning editorial comments. A great deal of the paper's organization is due to the efforts of Rod Kronschnabel. He prodded us to distill our unnecessarily complex ideas down to the simple, straightforward form presented here.

ANNOTATED BIBLIOGRAPHY

<u>An Analysis of the Output Spectrum of Direct Digital Frequency Synthesizers in the Presence of Phase-Accumulator Truncation</u> by Nicholas and Samueli, Proceedings of the 41st Annual Frequency Control Symposium, 1987, presents what appears to be a less computationally intensive method for estimating spurious due solely to phase accumulator truncation. Because of the approximations used, the reader is also left without detailed knowledge of the trade-offs between DAC resolution and size of the mapping PROM. Of course with modern PC's and the widespread availability of DFT software an exact analysis is relatively straightforward to accomplish, now that one knows how to "correct" the DFT values.

<u>The Optimization of Direct Digital Frequency Synthesizer Performance in the Presence of Finite Word Length Effects</u> by Nicholas, Samueli, and Kim, Proceedings of the 42nd Annual Frequency Control Symposium, 1988. This is a follow up paper by Nicholas and Samueli. It describes and analyzes a hardware implementation of the sinusoid lookup table for those interested in alternatives to the conventional mapping PROM.

<u>Performance Analysis of the Numerically Controlled Oscillator</u> by Kisenwether, Proceedings of the 40th Annual Frequency Control Symposium, 1986, presents an analysis of a DDS using the NCO's most significant bit as the output. For many applications this is impractical since the required high slew rates cannot be achieved.

<u>Phase Noise in Direct Digital Synthesizers,</u> by Mattison and Coyle, Proceedings of the 42nd Annual Frequency Control Symposium, 1988, presents an interesting set of experimental results. Here the author compares the results of a DFT with actual measurements. He also observes that Kisenwether's approach of using the MSB has slew rate problems.

<u>Number Controlled Modulated Oscillator Theory and Applications,</u> by McCune of Digital RF Solutions in Santa Clara, Ca presents an interesting discussion on modulating the NCO output digitally.

Part V

Reduction of Spurious Signals
in Direct Digital Frequency Synthesizers

SINCE for many practical applications spurious signals, generated in DDFS, may be one of the limiting factors, we should not be surprised to encounter efforts to suppress them (or at least reduce the largest ones). Here, we repeat some major sources of spurious signals: (1) omission of a number of low significant phase bits, (2) finite length of words in lookup tables, (3) reduction of effective ROMs, (4) limitations of sine values in DAC, (5) intermodulation processes, and (6) DAC nonlinearities and glitches.

One of the first papers dedicated to this problem is that by Wheatley et al. (1981). Authors considered a rectangular output wave and suggested reducing large spurious discrete signals by jittering zero crossings of the rectangular output wave through the addition and subtraction of a random number (from 0 to $k - 1$) to the accumulator content. The price to be paid was increased background spurious noise and increased system complexity [V-1].

Based on the last paper by the editor in this section, the largest spurious signal in the instance of the rectangular output wave is given by

$$-20 * \log \frac{f_x}{f_c} \qquad [1]$$

Evidently, the smaller the normalized frequency is, the smaller the largest spurious signal is.

Bozic and Gardiner [V-2] have published some experimental measurements of spurious signals in the sine-wave DDFS as a function of different normalized frequencies $\xi_x = k/2^{10}$ and a different number of DAC bits. Comparing their results with the computer simulations, provided by the editor, revealed that the influences of DAC and of the measurement system nonlinearities are not negligible. The last phenomenon can be particularly important in instances where spectral properties of DDFS are evaluated with the assistance of spectrum analyzers.

The same result was achieved by Giffard and Cutler (1992) who discuss a high-resolution DDFS designed to provide modulation and fine frequency interpolation in a Cesium frequency standard. They emphasize the importance of suppressing glitches due to DAC either with the assistance of sample-and-hold circuits [V-3] or by blanking its output while it settles after each data transition. (Compare their Figs. 5 and 6 and the paper by the editor in Part VIII [V-4].) The reader may be puzzled by the evaluation of the last extremely high intermodulation order of the spurious signal in the last line of Table 1 by Giffard and Cutler. Kroupa provides the explanation in the last paper in this section.

Next we mention a surveying paper by Reinhardt [V-5] in which he investigates the theory of DDFS spurs, presents the means for spur reduction techniques, and introduces his own randomized DAC DDS (patented [V-5]).

The Qualcomm Company [V-6] reduces spurious signals in DDFS with the assistance of dithering by the proprietary noise reduction circuit connected before DAC [V-7].

New possibilities for reducing the effect of DAC nonlinearities and glitches in DDFS are discussed by Kushner and Ainsworth (1996), who propose a balanced configuration of the DDFS output.

The paper by O'Leary and Maloberti (1991) introduces the means for noise shaping to reduce the effects of accumulator truncation on the output spectrum close to the carrier in instances where ξ_x is very small. However, our own and Crawford's [V-8] experience suggest a small rearrangement of Fig. 4 in O'Leary et al.

For the reader's comfort, we include an often cited paper by Flanagan and Zimmerman, with some objections to their Sections III and IV. Their estimate of the phase error due to retaining only b most significant bits from the accumulator is about 13 dB higher than those of all other authors. In addition, their DDFS in Fig. 13 is a very complicated system.

In the closing paper, the editor summarizes all his previous results [V-9 through V-14]. He provides information about inevitable spurious signals in instances of square-wave, triangular wave, and sine-wave outputs of DDFS. He describes a simple closed-form solution for spurious phase noise function $s(m)$ due to the accumulator truncation, computes the expected level and number of the respective spurious signals, points to the importance of a large denominator in the normalized frequency for reducing the background spurious signals, investigates the dependence of the third harmonic of the output frequency in respect to DAC bits, and addresses the problem of generating actual intermodulation signals.

REFERENCES

[V-1] C. E. Wheatley. Digital frequency synthesizer with random jittering for reducing discrete spectral spurs. U.S. Patent No. 4,410,954, October 18, 1983.

[V-2] M. Bozic and J. G. Gardiner. "Direct digital synthesis spurious noise properties." *Colloquium on Direct Digital Frequency Synthesis. IEE Digest No. 1991/172,* November 1991.

[V-3] Thomas M. Higgins. "Analog output system design for a multifunction synthesizer." *Hewlett-Packard J.,* pp. 66–69, February 1989.

[V-4] V. F. Kroupa. "Digital to analog converters." (A specially written paper for Part VIII.)

[V-5] Victor S. Reinhardt. "Spur reduction techniques in direct digital synthesizers." *1993 IEEE International Frequency Control Symposium, Proc.* pp. 230–42; see also U.S. Patent Number 4,815,018, March 21, 1989.

[V-6] Qualcomm, Inc. *Direct Digital Synthesis, 21 Questions & Answers for RF Engineers.* San Diego.

[V-7] Richard J. Kerr and Lindsay A. Weaver. Pseudorandom dither for frequency synthesis noise. U.S. Patent No. 4,901,265, February 13, 1990.

[V-8] J.A. Crawford. *Frequency Synthesizer Design Handbook.* Boston/London: Artech House, 1994 (Figure 7.42).

[V-9] V. F. Kroupa. "Spectral purity of direct digital frequency synthesizers." *Proceedings of 44th Annual Symposium on Frequency Control,* 1990, pp. 498–510.

[V-10] V. F. Kroupa. "Principles of direct digital frequency synthesizers." *Kleinheubacher Berichte* 36 (1992), pp. 663–78.

[V-11] V. F. Kroupa. "Discrete spurious signals and background noise in direct digital frequency synthesizers." *1993 IEEE Frequency Control Symposium,* Proc., pp. 242–50.

[V-12] V. F. Kroupa. "Intermodulation signals in direct digital frequency synthesizers." *Proceedings of the 8th European Frequency and Time Forum,* held March 9–11, 1994, pp. 627–45.

[V-13] V. F. Kroupa. "Spectral properties of DDFS: Computer simulations and experimental verifications." *1994 IEEE International Frequency Control Symposium,* Proc., pp. 613–23.

[V-14] V. F. Kroupa. "Synthesis techniques (DDS)." *1995 IEEE International Frequency Control Symposium, Tutorial Sessions.*

Spurious Suppression in Direct Digital Synthesizers

C.E. WHEATLEY, III
ROCKWELL INTERNATIONAL
ANAHEIM, CA 92803

D.E. PHILLIPS
ROCKWELL INTERNATIONAL
CEDAR RAPIDS, IOWA 52406

Summary

This paper describes a unique method for the removal of discrete line spurs normally present in the output of direct digital synthesizers (DDS). The type of DDS considered consists primarily of an N-bit accumulator, whose contents are incremented by a value, k, at a clocking rate, F_c. Although many variations exist on the use of this technique to synthesize a periodic waveform, the simplest possible version uses only the time of occurrence of accumulator overflow to define the edges of a "square" wave. For input numbers, k, that are powers of 2, the output frequency is a direct submultiple of the clock frequency. However, for all other numbers, the circuit behaves like a fractional divider with a sequence of non-uniform output periods, which repeat with a definable pattern to give an average period equal to the desired signal's period. Unfortunately, line spurs or sidebands are created by this periodic phase modulation of the output. Historically, these lines have been reduced by using the numerical contents of the accumulator to synthesize an output waveform which has more than two levels -- e.g., a sinusoidal shape. This requires the use of digital-to-analog (D/A) converters and read-only memories (ROM) to obtain the desired wave shape, and the results are well described in the literature.

The technique described here provides for complete elimination of these discrete spurs (while retaining a binary output) by randomly varying or dithering the periods of the alternating output, while keeping the average of these periods unchanged. The method consists of adding a sequence of pseudo-random numbers to the content of the accumulator in a prescribed manner, and this process is shown to convert the discrete line spurs into a continuous noise floor, which is white out to frequencies approaching the clock frequency.

As the speed of digital accumulators continue to increase with new devices and new techniques, the noise floor realizable by this mechanization will continue to decrease at a rate of 6 dB for every doubling of the clock frequency.

Details involved in the dither process and expressions for deriving the expected value of carrier power to noise density of this synthesis technique are described. Also, expressions useful in determining the noise density associated with square waves that exhibit edge time jitters are provided. This information should prove useful to designers of logic systems where the effects of edge jitter are often not well accounted for. An important example is the generation of noise in an indirect digital synthesizer by edge instabilities of the phase detectors.

Two versions of this synthesizer using the dither technique have been breadboarded, and results confirm the predictions of the analysis. Data obtained from the breadboarded synthesizers is provided to show some of the more interesting results of this technique, as well as to define its limitations.

The Basic Concept of a Dithered Accumulator

The technique of using a digital accumulator to synthesize a selected frequency is quite well known[1], and Figure 1 shows this process in simple block diagram form. The basic idea is to increment the content (phase) of the digital accumulator at a regular clock rate, generating a frequency equal to the rate of the phase change with time. It is also well known that the resultant fundamental frequency output suffers from the presence of nonharmonic spurious spectral lines. These lines can be reduced by using D/A converters on the output of the accumulator (Figure 1) to generate sine waves. However, the spurious level will increase as the number of bits in the D/A converters decreases, reaching a maximum when only one bit is retained and the output waveform is simply a square wave.

An explanation for the spurious signals that result from the use of only one bit can be visualized by referring to Figure 2, the time process of the accumulator for a very simple example (i.e., with N = 4 and k = 3). For this example, the "ideal" output square wave would have three edges every 16 input clocks.

Reprinted from *Proceedings of the 35th Annual Frequency Control Symposium*, pp. 428-435, May 1981.

From Figure 2, one can observe that the synthesized output is exactly the same as that which one would obtain by sampling the "ideal" (no jitter) square wave at a clock frequency, $1/T_c$, and then passing these samples through a sample-and-hold device. Due to aliasing, this sampling process takes the spectrum of the ideal square wave and "folds" it around the clock frequency. Since the power of the harmonics of the square wave decreases rather slowly, a great many lines reappear near the desired fundamental output. Knowing the clock frequency and the output frequency, one can easily compute the magnitude and location of all spurs.

A slightly different but more common method of viewing the cause of the spurs is to observe that the output of the direct digital synthesizer can occur only at a clock edge. If the output frequency is not a direct submultiple of the clock, a phase error between the ideal output and the actual output slowly increases (or decreases) until it reaches one clock period, at which time the error returns to zero and starts to increase (or decrease) again. This ramp-like phase modulation generates the same spurious spectrum as determined from the sampling viewpoint.

By periodically adding a random number to the accumulator of the direct digital synthesizer, one can destroy the coherence of the phase error from edge to edge, thereby destroying the discrete inharmonic line spectrum that results. Figure 3 illustrates the method, and the following describes why the edge-to-edge jitter becomes uncorrelated.

At each overflow of the accumulator, a random value, X, where X ranges uniformly from 0 to K-1, is added to the content of the accumulator. As seen in Figure 4, this has the effect that the accumulator will next overflow at one of two clock times, depending upon the accumulator phase error, (τ_e of Figure 2) and the random number X selected.

From Figure 4, the normal phase error, τ_e, can be seen to be related to an error value, e, contained in the accumulator just prior to the normal overflow event. Clearly, e must be less than k. By adding a random number, X, the accumulator will still overflow at its normal, late edge if X is less than e, but will overflow one clock time early if X is greater than e. Now, if X is selected as specified from uniform random distribution from 0 to k-1, the probability density function of the two events is:

$$p(\tau_e) = \frac{e}{k} \; ; \; p(\tau_e - T_c) = \frac{1-e}{k}$$ (1)

The expected value of the overflow point is:

$$\langle \tau \rangle = \tau_e \, p(\tau_e) + (\tau_e - T_c) \, p(\tau_e - T_c)$$ (2)

$$= \tau_e \left\{ \frac{e}{k} \right\} + (\tau_e - T_c) \left\{ 1 - \frac{e}{k} \right\}$$

But, since $\frac{k-e}{k} = \frac{\tau_e}{T_c}$, (from Figure 4)

$$\langle \tau \rangle = 0$$ (3)

Thus, the <u>average</u> overflow event occurs at the time where the ideal phase ramp intersects the maximum value of the accumulator. This means that the average period synthesized by the process is the correct period, and the fact that any one period is not exact introduces phase noise.

More importantly, however, the expected overflow time is not dependent upon the original phase error, τ_e. This means that, even if adjacent errors are originally correlated, adding the random value X causes adjacent accumulator resets to be independent, and all coherent spurious lines originally generated by this correlation are eliminated. The penalty will be a continuous noise spectrum due to the random choice between the two possible reset times. The following paragraphs compute the density of this noise spectrum, given the probability distribution of the overflow event.

<u>Computation of the Spectrum for Gaussian Edge Jitter</u>

In Appendix 1, it is shown that the spectrum of a square wave that exhibits a small amount of edge time jitter may be computed if one knows the probability density function (pdf) of the time jitter, and the mathematics are relatively simple if the jitter is independent from edge to edge. Specifically, the spectrum depends upon the characteristic function of the time jitter process which is defined as:

$$F(jw) \triangleq E \left\{ e^{jw\tau} \right\}$$
$$= \int_{-\infty}^{\infty} p(\tau) e^{jw\tau} \, d\tau$$ (4)

and where $p(\tau)$ is the pdf of the time of occurrence of the edge, with respect to the edge of the "true" square wave being synthesized. In the case of the jittered accumulator synthesis technique, the actual values of τ can take on only two values, one ahead of the average edge, and one behind.

The complete spectrum of the square wave as derived in Appendix 1 (Eq A-12 and A-13), including both discrete and continuous components, is computed from $F(j\omega)$ as:

$$S(\omega) = \frac{A^2}{\omega^2 T_o} \left\{ 1 - |F(j\omega)|^2 + \frac{2\pi |F(j\omega)|^2}{T_o} \right.$$
$$\left. \cdot \sum_{m=-\infty}^{\infty} \delta \left(\omega - \frac{m\pi}{T_o} \right) \right\} \quad -\infty < \omega < \infty$$ (5)

where A is the peak-to-peak value and $2T_0$ is the complete time period of the fundamental output square wave.

As an example, consider the case of a logic device having jitter which can be modeled as a zero mean Gaussian distribution with variance σ_τ^2. In this case,

$$p(\tau) = \frac{1}{\sqrt{2\pi\sigma_\tau^2}} e^{-\tau^2/2\sigma_\tau^2}$$

By using Equation (4) to determine $F(j\omega)$, and then substituting this into Equation (5), one can show that the output spectrum of the square wave will be:

$$S(\omega) = \frac{A^2}{T_0} \left\{ \frac{1-e^{-\omega^2\sigma_\tau^2}}{\omega^2} \right\} + 2\pi \frac{A^2}{T_0^2} \frac{e^{-\omega^2\sigma_\tau^2}}{\omega^2}$$

$$\sum_{\substack{m=-\infty \\ m\ odd}}^{\infty} \delta(\omega - \frac{m\pi}{T_0}) \tag{6}$$

At the fundamental output frequency,

$(f_0 = \frac{\omega_0}{2\pi} = \frac{1}{2T_0})$ the noise power density to

carrier power will be (Note 1):

$$\frac{N_0}{C} = 2\pi^2 f_0 \sigma_t^2 \quad \text{for} \quad \sigma_t << T_0 \tag{7}$$

From this relationship it can be shown that for a digital device to provide an $\frac{N_0}{C}$ of -150 dBc/Hz,

at a typical frequency of 1 MHz, it must exhibit an edge instability of less than 7 picoseconds.

The Output Spectrum of a Synthesizer Using Accumulator Dither

Returning to the synthesized waveform where the distribution is not continuous, $F(j\omega)$ may still be computed from Equation (4) using Equation (1) for the discrete pdf of the early and late edges. After some computational labor, the result is:

$$|F(j\omega)|^2 = 1 - 4 \sin^2(\frac{\omega T_c}{2}) \left\{ \frac{\tau_e}{T_c} - (\frac{\tau_e}{T_c})^2 \right\} \tag{8}$$

Note that this depends upon the error value τ_e, which can vary from 0 to T_c. This means that although the average frequency is correct and independent of τ_e, the continuous noise density term of Equation (5) will depend upon τ_e, being maximum at $\tau_e = T_c/2$. If τ_e were always identically 0 (or T_c), no noise would be introduced by

---

NOTE 1: This is also known as $\mathscr{L}(f)$, the frequency domain measure of phase fluctuation sidebands.

the dither. In the general case, τ_e can be expected to vary uniformly from 0 to T_c, resulting in an average value for $|F(j\omega)|^2$ of:

$$<|F(j\omega)|>^2 = 1 - 2/3 \sin^2 \frac{\omega T_c}{2} \tag{9}$$

Finally, under the condition that $\omega_0 = \frac{\pi}{T_0} << \frac{1}{T_c}$, the spectrum of the synthesized square wave near this fundamental frequency as determined by substitution of $|F(j\omega)|^2$ into Equation (5) is:

$$S(\omega) = \frac{A^2 T_c^2}{6T_0} + \frac{2A^2}{\pi} \left\{ \delta(\omega - \frac{\pi}{T_0}) + \delta(\omega + \frac{\pi}{T_0}) \right\} \tag{10}$$

From the above, the noise power density to carrier power at the fundamental frequency will be:

$$\frac{N_0}{C} = \frac{\pi^2}{3} \frac{f_0}{f_c^2}, \tag{11}$$

Again, it must be emphasized that non-harmonic lines will not be present, only the noise floor defined by the first term of (10).

The power of this entirely digital technique ultimately lies in the ability of technology to provide high-speed accumulators, and provides a 6 dB N_0/C improvement for every doubling of the clock frequency. As an example, a present-day ECL synthesizer using a 100-MHz clock could deliver a 1-MHz output with an N_0/C of -95 dBc/Hz, which, although not nearly as large as can be obtained from fundamental sources, is quite sufficient to meet many practical needs. Ultimately, GaAs devices that are expected to be able to be clocked at GHz rates could provide N_0/C measures well below -100 dBc/Hz.

Experimental Results

A breadboard direct digital synthesizer (DDS) was constructed of low-power Shottky logic to test the dither effect (Figure 5). In this design a pipelined 24-bit accumulator was clocked at a 16-MHz rate and a pseudo-random number generator and control logic allowed a pseudo-random number to be periodically added to the accumulator. A dither size-select circuit was used to adjust the dither modulation to an amount just great enough to eliminate the spurious sidebands, for each value of input frequency.

The breadboard was built primarily to test circuits that will be incorporated into an LSI design using CMOS/SOS technology, rather than to optimize a TTL design. The results are not optimum for an LSI design either, but enable the principle to be checked out.

Figure 6 shows the spectrum of a typical undithered output with spurious sidebands present,

and Figure 7 shows the same output with dither.
As predicted, the spurious sidebands vanish, being
replaced by a uniform phase noise spectrum, flat
to within less than one Hertz of the carrier.

Figure 8 plots the noise and spurious levels
vs output frequency. The expected undithered
spurious sideband level is commonly expressed as

$$20 \log \left(\frac{f_o}{f_c}\right) \text{dBc} \qquad (12)$$

which is plotted as the theoretical value. Running
above it in a zig-zag course is the measured spurious level of the most significant bit "square
wave" output. A computer analysis also shows this
same pattern. This is due to the fact that the
spurious modulation pattern is the same on the
leading and trailing edges, except for a varying
relative phase, so there is a varying amount of
reinforcement and cancellation between edges.
Also, the spurious (modulation) frequency increases
and decreases, going through a zero beat each time
the f_c / f_o ratio is exact power of 2.

An important application of DDS is to serve
as a reference for a phase-lock loop. With
digital phase discriminators, only one edge of the
reference wave is used, the other being supplied
from the VCO, divider, etc. This results in 6 dB
less noise modulating the loop (because of the
correlation between edges). The spurious level is
now a uniform function of a carrier frequency,
since the edge interaction effects have been
removed.

When dither was applied, the amount of dither
modulation was determined by the most significant
bit in the frequency control word, resulting in a
3-dB step per octave. There is also a 3-dB drop in
noise when only one edge is used, because the two
edges are uncorrelated.

The developed formula for carrier-to-noise
density

$$\frac{C}{N_o} = \frac{3}{\pi^2} \frac{f_c^2}{f_o} \qquad (13)$$

is based upon a normalized 1-Hz noise bandwidth.
As shown in Figure 8, this formula is rearranged
and includes the system noise bandwidth. It can be
seen that the noise-bandwidth-to-clock ratio is the
dithered "improvement factor," as the original
spurious sideband energy is uniformly spread over
the band 0 to f_c, but only a portion is recognized
in the system bandwidth. The graph is plotted for
the commonly used voice communications bandwidth of
3 kHz.

Applications of the Dithered Accumulator

The dithered accumulator is particularly suited
to LSI integrated circuits, being all digital and
requiring no large ROM memory or D/A converters.
Its practical value will increase as the maximum
clock speeds increase with smaller-geometry
CMOS/SOS and gallium arsenide circuits, reducing
the noise levels. For the same output frequency,
1-MHz for example, the dithered single-edge phase
noise of -49 dBc (3-kHz bandwidth) with a 16-MHz
clock would drop to -62 dBc with a 64-MHz clock!
Even higher clock frequencies will be possible in
the future, which will allow either lower noise or
higher output frequencies.

If the concept of adding noise to a system
still seems strange (aren't we always trying to
reduce it?), one might consider the advantage that
reducing the bandwidth (for example, by following
the DDS with a narrowband phase-lock loop) lowers
the random noise, but close-in discrete spurious
levels remain unchanged. This is an important
factor in some narrowband systems such as HF tone-
coded transmissions and CW telegraphy.

As direct digital synthesizers find increasing use in frequency-hopping systems, a dithered
accumulator in a small, low-power, high-speed
integrated circuit could be of great value in HF,
VHF, and UHF communications.

References

1. Gorski-Popiek, J., Editor, "Frequency
   Synthesis: Techniques and Applications,
   IEEE Press, 1975.

Appendix 1

Duration of the spectrum of a square wave with edge time jitter

Acknowledgement

The authors wish to thank Dr. A. K. Nandi for
his contributions to the analysis contained in
this appendix.

Introduction

The case considered here is applicable to
situations wherein the dither time process is
uncorrelated (white phase noise) from edge to
edge, which is appropriate for many cases of
interest. In order to further simplify the mathematics, we start by analyzing an alternating
sequence of time-dithered impulses. Then, be noting
that a square wave can be viewed as the time
integral of a train of alternating impulses, we
easily obtain the desired prediction for the square-
wave power spectrum. A most important consequence
of the edge jitter phenomenon is that a continuous
spectrum (noise) around dc is created in the case of
the square wave.

Spectrum of Randomly Jittered Impulses (Uncorrelated Samples)

Consider the situation in Figure A-1 which
shows a periodic alternating sequence of impulses
of strength $\pm$ A. Successive impulses are separated
by t_o. In Figure A-2, the same impulse sequence is
shown with edge jitter added. The spacing between
successive impulses is now a random variable.

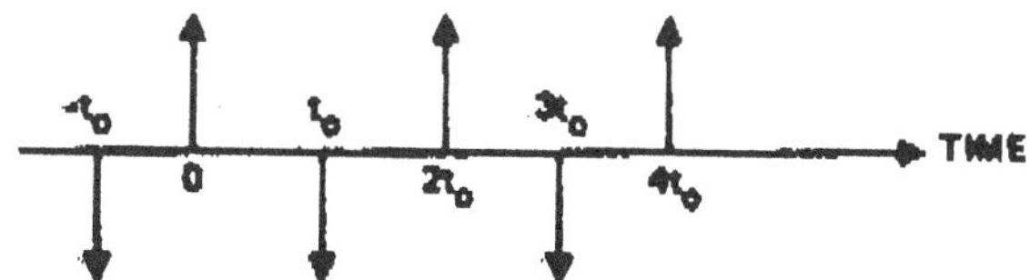

Figure A-1. Ideal Sequence
of Impulses

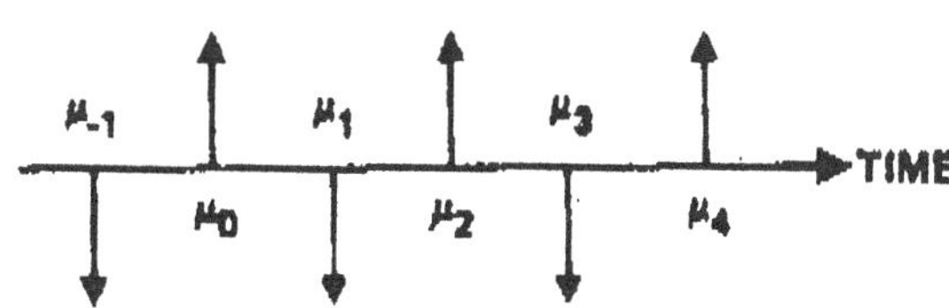

Figure A-2. Jittered Sequence
of Impulses

For the jittered sequence, one can write

$$u(t) = \sum_{n=-\infty}^{\infty} (-1)^n A\delta(t-nt_o - \mu_n) \qquad (A-1)$$

As a specific example, we will consider the case where the μ_n's are uncorrelated, identically distributed continuous Gaussian random variables such that

$$E\mu_n = 0 \qquad (A-2)$$

$$E\mu_n^2 = \sigma^2; \quad \sigma \ll t_o \qquad (A-2)$$

$$E\mu_k\mu_\ell = 0 \quad \text{if } k \neq \ell \qquad (A-2)$$

However, other distributions are acceptable and can be handled using the procedures described below. For the ideal situation in Figure A-1, all μ_n's are zero.

An often used expression (Reference A-1) for computing power spectral density of a wide-sense stationary process is

$$S_u(\omega) = \lim_{T \to \infty} \frac{1}{2T} E\left\{U_{2T}(j\omega)\, U_{2T}(-j\omega)\right\} \quad -\infty < \omega < \infty \quad (A-3)$$

where units are in power per Hertz and:

$U_{2T}(j\omega)$ = Fourier transform a finite section of length 2T of a sample function

$U_{2T}(-j\omega)$ = complex conjugate of $U_{2T}(j\omega)$

$E \equiv$ statistical expectation

Now, let

$$T = Nt_o \qquad (A-4)$$

On the average, one is considering a truncated stream of jittered impulses ($\approx(2N+1)$impulses) during the computation of the Fourier transform mentioned above. Thus,

$$U_{2T}(j\omega) = \int_{-\infty}^{\infty} U_{2T}(t)\, e^{-j\omega t}dt$$

$$= \int_{-\infty}^{\infty} \left(\sum_{n=-N}^{N}(-1)^n A\,(t-nt_o-\mu_n)\right) e^{-j\omega t}dt$$

$$= \sum_{n=-N}^{N}(-1)^n A\, e^{-j\omega(nt_o-\mu_n)} \qquad (A-5)$$

$$\therefore\ U_{2T}(-j\omega) = \sum_{n=-N}^{N}(-1)^n A\, e^{+j\omega(nt_o-\mu_n)} \qquad (A-6)$$

and

$$E\left\{U_{2T}(j\omega)\, U_{2T}(-j\omega)\right\}$$

$$= E\left\{A^2 \sum_{k=-N}^{N}\ \sum_{\ell=-N}^{N}(-1)^{\ell+k}\, e^{-j\omega(k-\ell)t_o} e^{-j\omega(\mu_k-\mu_\ell)}\right\}$$

$$= A^2 \sum_{k=-N}^{N}\ \sum_{\ell=-N}^{N}(-1)^{\ell+k}\, e^{-j\omega(k-\ell)\, t_o}$$

$$\cdot\ E\left\{e^{-j\omega(\mu_k-\mu_\ell)}\right\}$$

In order to evaluate the above expectation involving summations over k and ℓ, it is convenient to examine Figure A-3.

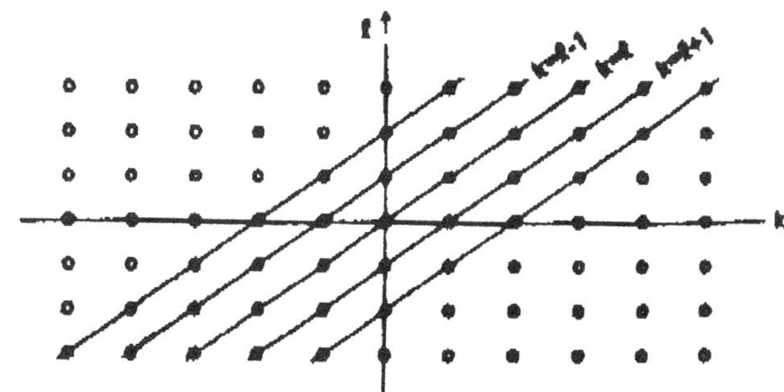

Figure A-3

Reference A-1: J. A. Aseltine, "_Transform Methods in Linear Systems Analysis_," McGraw Hill, New York, 1958.

One can write equation (A-7) as follows:

$$E\left\{U_{2T}(j\omega)\ U_{2T}(-j\omega)\right\} = A^2\left[(2N+1)\right.$$
$$+ \sum_{k=-N}^{N}\ \sum_{\ell=-N}^{N}(-1)^{\ell+k}\ e^{-j\omega(k-\ell)t_0}$$
$$\left.\cdot\ E\left\{e^{-j\omega(\mu_k-\mu_\ell)}\right\}_{\ell\neq k}\right] \qquad (A-8)$$

With $k \neq \ell$, one obtains

$$E\left\{e^{-j\omega(\mu_k-\mu_\ell)}\right\} = E\left\{e^{-j\omega\mu_k}\right\}\ E\left\{e^{-j\omega\mu_\ell}\right\}$$
$$= F(-j\omega)\ F(+j\omega) \qquad (A-9)$$

where $F(j\omega) \triangleq E\left\{e^{j\omega\mu_\ell}\right\}$
= characteristic function of μ_ℓ

Using Equation (A-9) in Equation (A-8) and summing along lines of constant $(k-\ell)^*$, one easily finds

$$E\left\{U_{2T}(j\omega)\ U_{2T}(-j\omega)\right\} = A^2\left[(2N+1) + |F(j\omega)|^2\right.$$
$$\cdot\left\{-2N\left(e^{-j\omega t_0}+e^{+j\omega t_0}\right)\right. \qquad (A-10)$$
$$\left.\left.+ (2N-1)\left(e^{-2j\omega t_0}+2j\omega t_0\right)..\right\}\right]$$

Adding and subtracting a term corresponding to $n=0$, one obtains

$$E\left\{U_{2T}(j\omega)\ U_{2T}(-j\omega)\right\} = A^2\left[(2N+1)\ (1-|F(j\omega)|^2)\right.$$
$$+ |F(j\omega)|^2\left\{(2N+1)\right.$$
$$\qquad (A-11)$$
$$- 2N\left(e^{-j\omega t_0}+e^{+j\omega t_0}\right)$$
$$\left.\left.+ (2N-1)\left(e^{-2\omega t_0}+e^{+2\omega t_0}\right)..\right\}\right]$$

Recalling now the basic formula for spectral density given in Equation (A-3):

$$S_u(\omega) = \lim_{T\to\infty}\frac{1}{2T}E\left\{U_{2T}(j\omega)\ U_{2T}(-j\omega)\right\}$$
$$= \lim_{T\to\infty}\frac{1}{2Nt_0}E\left\{U_{2T}(j\omega)\ U_{2T}(-j\omega)\right\}$$
$$= \frac{1}{t_0}A^2\left[(1-|F(j\omega)|^2) + |F(j\omega)|^2\right.$$
$$\left.\cdot\ \sum_{m=-\infty}^{\infty}(-1)^m\ e^{-j\omega m t_0}\right]$$
$$S_u(\omega) = \frac{A^2}{t_0}\left[(1-|F(j\omega)|^2) + |F(j\omega)|^2\right.$$
$$\left.\cdot\ \frac{2\pi}{t_0}\ \sum_{n=-\infty}^{\infty}\delta\left(\omega-\frac{n\pi}{t_0}\right)\right]\ \text{n odd} \qquad (A-12)$$

This is a most general result, from which the power spectrum can be evaluated. Note that both discrete and continuous (noise) terms exist in the spectrum.

Spectrum of Square Wave with Independently Jittered Edges

Consider a square wave of magnitude $\pm A/2$ indefinitely in time in both directions whose edges are jittering, and let the edge jitter statistics be described by Equation (A-2). Since the impulse train of Equation (A-1) can be obtained by differentiating this square wave, the power spectrum of the square wave is easily obtained by multiplying the impulse spectrum (A-12) by $1/\omega^2$. Thus,

$$S_{sq\ wave}(\omega) = \frac{1}{\omega^2}\ S_u(\omega) \qquad (A-13)$$

$$S(\omega) = \frac{A^2}{t_0}\left(\frac{1-e^{-\omega^2\sigma^2}}{\omega^2}\right)$$
$$+ \frac{2\pi A^2}{t_0^2\omega^2}\ e^{-\omega^2\sigma^2}\sum_{n=-\infty}^{\infty}\delta\left(\omega-\frac{n\pi}{t_0}\right) \qquad (A-14)$$
$$(n\ \text{odd})$$

Note that the period, T, of the fundamental square wave = $2T_0$. As above, this spectrum has both a noise term and discrete terms, but the noise term is maximum at $\omega=0$, reaching a value of

$$\frac{2A^2\sigma^2}{T}$$

---

*From Figure 3, it is clear that for
$(k-\ell) = \pm 1, \pm 3, \ldots \Rightarrow (-1)^{k+\ell} = -1$ and for
$(k-\ell) = \pm 2, \pm 4, \ldots \Rightarrow (-1)^{k+\ell} = +1$

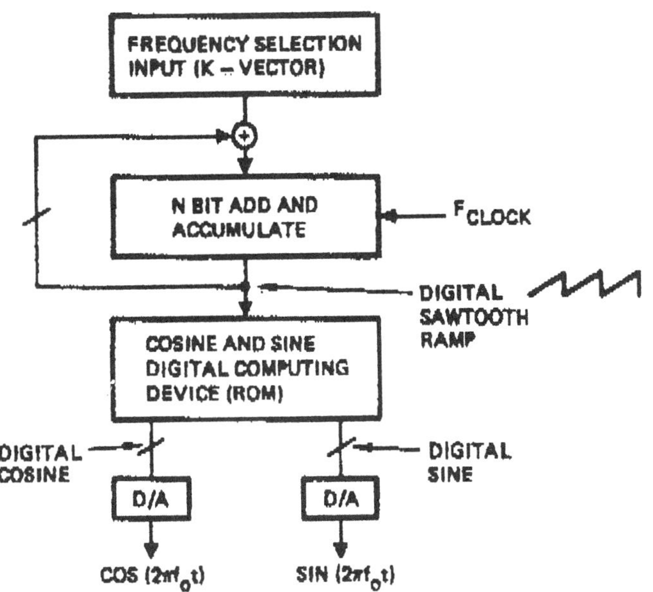

Figure 1. Direct Digital Synthesizer — Simplified Standard Block Diagram

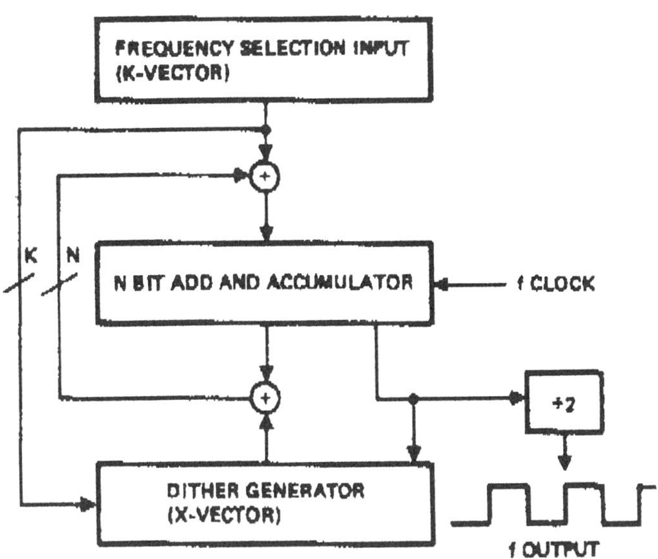

Figure 3. Direct Digital Synthesizer — Dithered Accumulator

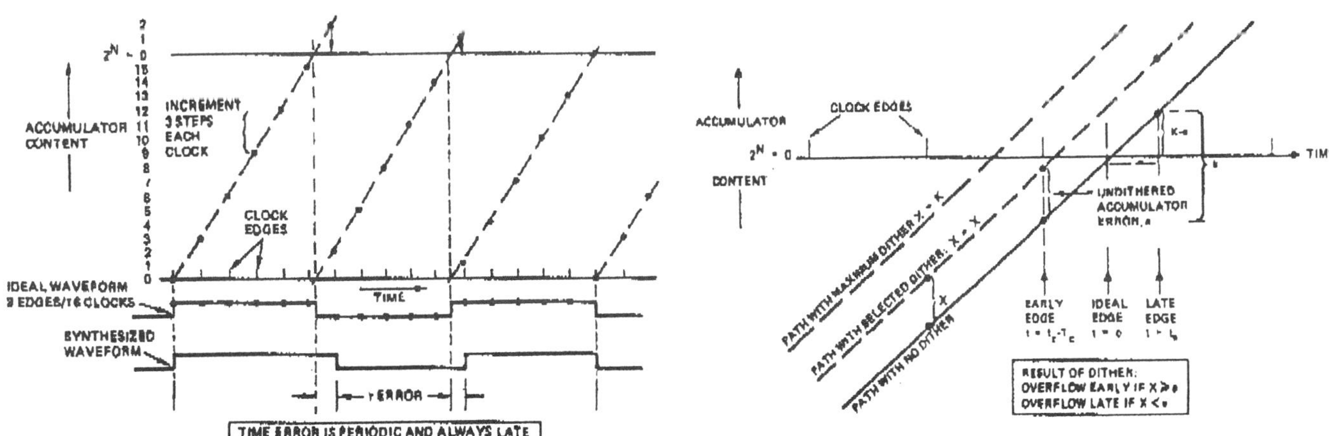

Figure 2. Accumulator Contents vs Time — No Dither (Simple Case N = 4, K = 3)

Figure 4. Accumulator Contents vs Time (Result of Random Dither)

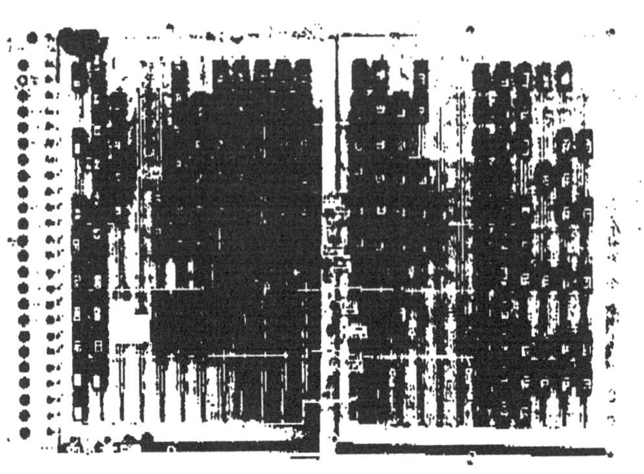

Figure 5. Dithered Accumulator Breadboard

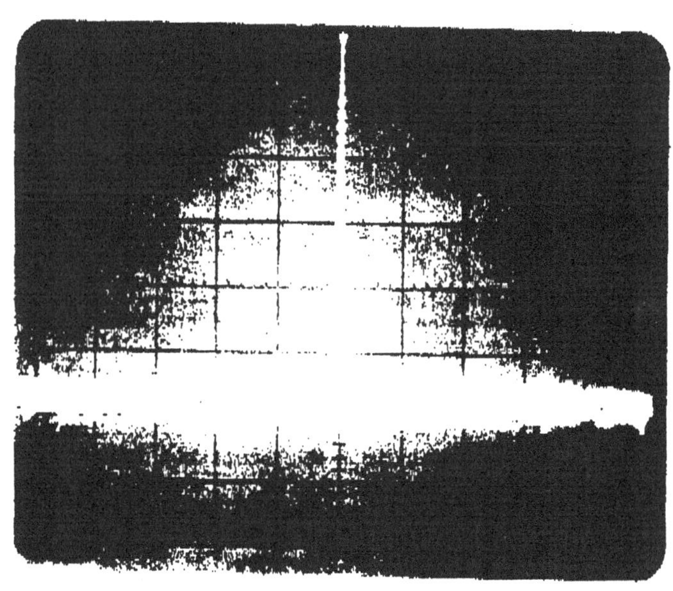

Figure 7. Digital Synthesizer
Output with Random Dither

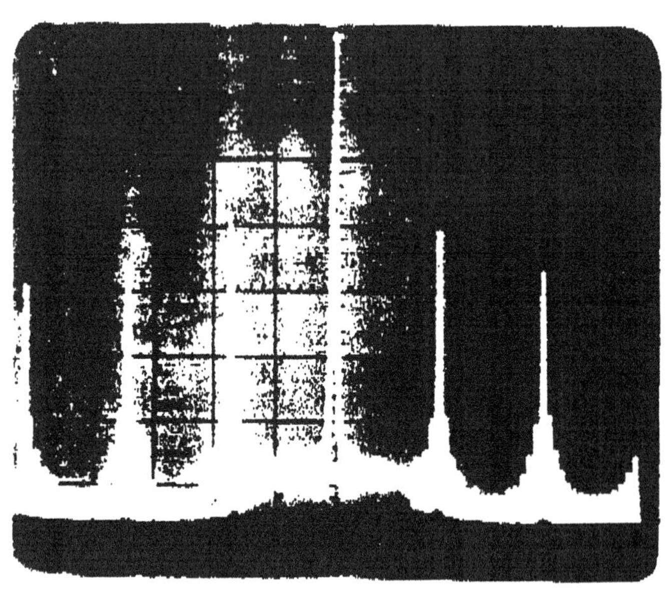

Figure 6. Digital Synthesizer
Output without Dither

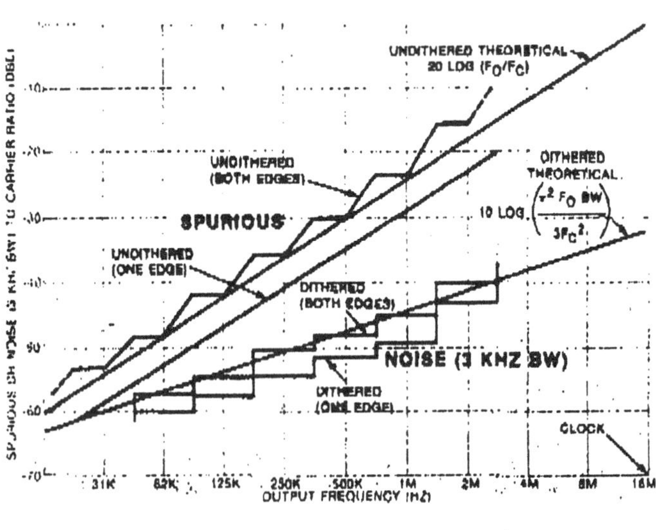

Figure 8. Carrier-to-Noise Density

A Low-Frequency, High Resolution Digital Synthesizer

ROBIN P. GIFFARD AND LEONARD S. CUTLER
HEWLETT PACKARD LABORATORIES
P.O. BOX 10350, PALO ALTO, CA 94303-0867

Abstract-A high resolution digital frequency synthesizer was designed to provide modulation and fine frequency interpolation in a new Cesium frequency standard. Requirements to be satisfied include resolution, spectral purity, phase-continuous frequency modulation, simplicity and low power. In the following paper we describe how these requirements were satisfied with a Direct Digital Frequency Synthesizer (DDFS). This type of frequency synthesizer was originally described by Tierney, Rader, and Gold [1], and has since been widely used and discussed [2,6]. The application to a hydrogen maser has been described by Matisson and Coyle [7].

Design Requirements

In the new Cesium standard, the microwave probe frequency has to be tuned to both the F=3, Mf=0 to F=4, Mf=0 "main" transition near 9192.632 MHz, and the F=3, Mf=1 to F=4, Mf=1 "Zeeman" line. At the chosen magnetic field, this requires the synthesizer to have a range of about 50 kHz. The necessary resolution is fixed at better than 92 microhertz by the requirement that the standard be steerable with a settability of 1 part in 10^{14} or less. To carry out effective square-wave frequency modulation, the synthesis chain must change frequency and settle to an accuracy of 1 part in 10^{14} in a time of about 2 ms, the duration of the transient generated by the tube when the frequency changes. For frequency changes of less than 1 kHz, the frequency should change without any phase discontinuity. Finally, to avoid offensive frequency pulling, the synthesizer should be free of sidebands close to the resonance line to a level of about -70 dBc. The synthesizer output has to be free of sidebands coherent with the modulation process and its harmonics to about -120 dBc. Because of the need for fast phase-continuous frequency switching, the requirements for this synthesizer are somewhat different from those relevant to hydrogen maser design [7].

A Direct Digital Frequency synthesizer was chosen to obtain the necessary resolution. In order not to degrade phase noise, the output frequency of the synthesizer is translated without multiplication to the probe frequency. To minimize power consumption and cost, the DDFS operates at the lowest satisfactory clock frequency. The overall synthesis in the cesium standard is defined by the following equation:

$$Fp = 29 \times 320 \text{ MHz} - ((35/32) \times 80 \text{ MHz} - Fs), \tag{1}$$

where Fp is the probe frequency, and Fs is the frequency of the digital synthesizer. The required range of the synthesizer is thus 125 kHz to 175 kHz. The architecture and overall synthesis scheme of the new Cesium standard are discussed in other papers [8,9].

Direct Digital Synthesis

The principal of the DDFS is well-known [1-3]. Its performance depends on the operation of the digital phase accumulator shown in Fig 1. The adder A causes the contents of the phase register R to increase by the value Na of the addend at each clock cycle. If the adder is binary and the phase register contains P bits, after N clock cycles the phase register will contain the number $N \cdot Na$ MODULO 2^P. The output of the phase accumulator is a sampled saw-tooth wave with a repetition period of $2^P/(Fc \cdot Na)$. It is well known that if the exact regularly spaced phase samples are used to calculate the inputs Vi to an ideal digital-to-analog converter (DAC) using the equation:

$$Vi = Const \times \sin\left[\frac{2\pi Ai}{2^P}\right], \tag{2}$$

the DAC output signal after an ideal anti-alias filter will be a pure sinusoid at a frequency $Fs = Fc \cdot Na/2^P$, completely free of non-harmonic distortion. The Nyquist condition: $Fs < Fc/2$, must be satisfied.

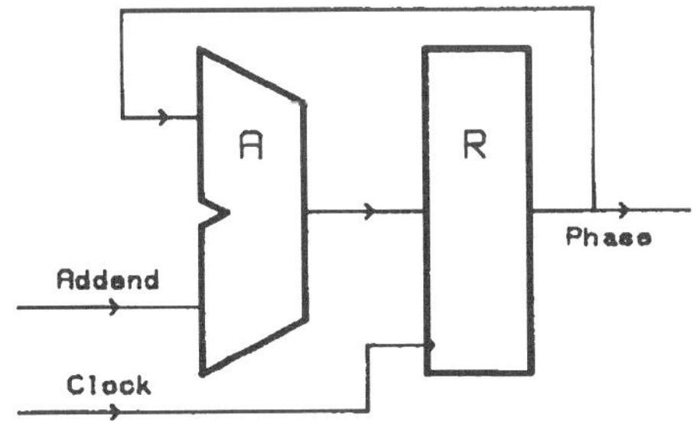

Fig 1. Digital phase accumulator consisting of a binary full-adder A and a clocked register R.

Reprinted from *Proceedings of the IEEE International Frequency Control Symposium*, pp. 188-192, 1992.

127

In practice, the conversion of linear phase to sinusoidal DAC input values is usually carried out by a "look-up table" or "waveform-map" consisting of a static read-only memory (ROM), avoiding the need for real-time calculation. Since the amount of memory required to encode the entire width of the phase accumulator would usually be prohibitive, only the Q most significant bits of the accumulator output are generally used to calculate the sine-wave samples. The truncation of $P-Q$ bits of the phase leads to errors in the computation, and the output waveform cannot be a pure sine-wave, although its average frequency is not changed. The errors cause spurious frequencies known as "phase truncation sidebands" to appear [4,5].

It is well known in the field of digital signal processing that the finite resolution of the DAC used to construct the waveform introduces sidebands due to quantization error. Non-ideal static and dynamic DAC effects lead to further degradation.

We will now consider how a synthesizer was designed to meet the requirements of the frequency standard.

<u>Design</u>

It is shown above that the DDFS resolution df is related to the clock frequency and the phase register length by $df=Fc/2^P$, so that register length and adder width are minimized by choosing the lowest clock frequency that will allow satisfactory anti-alias filtering. The requirements of the instrument are satisfied by a 34-bit register and a 1 MHz clock frequency. The clock frequency is more than 4 times the highest output frequency required, allowing the two highest bits of the adder to be implemented as a synchronous counter, and

making the anti-alias filter less critical.

The choice of the DAC resolution is dictated by the -70 dB spectral purity requirement. The quantization noise power in an ideal uniform DAC is easily shown to be of the order of $S^2/12$, where S is the step size. For a DAC having D bits, the maximum sine-wave output power is $2^{(2D-3)}S^2$, so that the total noise power is theoretically $-1.76-6.02 D$ dBc [10]. In practice this power is distributed over a number of frequencies which are m/n intermodulation products of the clock and synthesizer frequencies. The power in any given sideband can be minimized to some extent by choice of the ROM program [5]. A 12-bit DAC was chosen, and a computer program was used to calculate the level of the dominant spurious sideband and choose the ROM coding. Allowing for static errors, an ideal 11-bit DAC was simulated, and it was found that the level of the unwanted sidebands should be below -90 dBc.

In order to obtain sideband levels approaching these values in real DACs, it is normally necessary to suppress glitches either by following the DAC with a sample-and-hold circuit, or by blanking its output while it settles after each data transition. Large-scale non-linearity in the DAC transfer function is expected to cause harmonic distortion, which is not important in this application.

The magnitude and spectral distribution of truncation sidebands has been explored theoretically by Nicholas and Samueli [4] and by Nicolas, Samueli and Kim [5]. A very large number of sidebands is produced in general, with a complex distribution of frequencies. The amplitude of the sidebands is found to depend on the width of the phase accumulator P, and the number of bits Q input to the waveform map. A small modification of

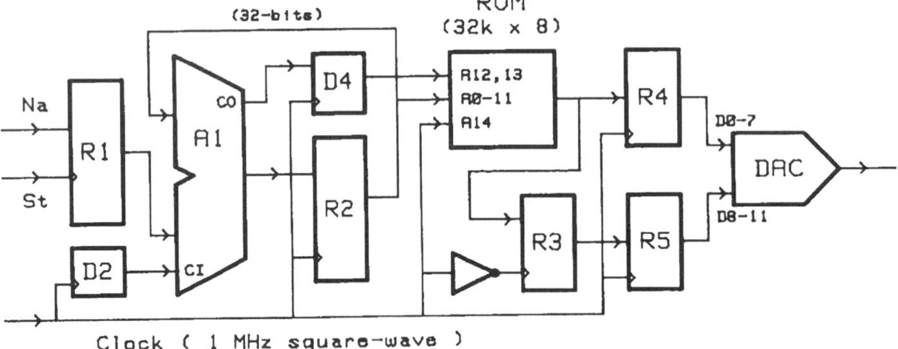

Fig 2. Block diagram of the Direct Digital Frequency synthesizer. The clock input is a 1 MHz square wave. The operation of the circuit is explained in the text. The DAC de-glitcher, and the anti-alias filter are not shown.

the accumulator reduces the worst possible sideband amplitude by about 4 dB. It was decided to use an address width of 14 bits in the synthesizer since this was a convenient match to a 256-k ROM with two bytes for each phase value. With an effective P value of 35, and $Q=14$, worst-case truncation sidebands are calculated to be below -84 dBc. We expect that as a result of the frequency switching sequence used in our application, the chance of a spurious sideband coherent with the modulation process is negligible.

Hardware Design

The hardware design of the synthesizer is shown in fig. 2. The 32 bit addend word Na is transferred to the register R1 by a hardware-derived timing strobe. The adder A1 is a 32-bit full binary adder whose output is latched into the 32-bit phase register R2 at each clock cycle. The carry output of the adder forms the enable of a 2-bit synchronous binary counter D4 whose output forms the two most significant bits of phase. This configuration is equivalent to a 34-bit phase accumulator with a 32 bit input word, limiting the output frequency to $Fc/4$. As suggested by Nicholas and Samueli, the carry input of the adder is at half the clock frequency. The upper 14 bits of the phase word are fed to address bits 0 through 13 of the ROM, and bit 14 alternates at the clock frequency. Registers R3-R5 align the upper and lower byte outputs of the ROM in time, and ensure that the DAC inputs all change simultaneously. To obtain good phase noise it is essential that the clock input to R4 and R5 has very low jitter. The DAC is a 12-bit, bipolar device with symmetric current outputs. De-glitching is implemented by gating off the DAC output for 200 nS after each data change with a balanced current switch. A

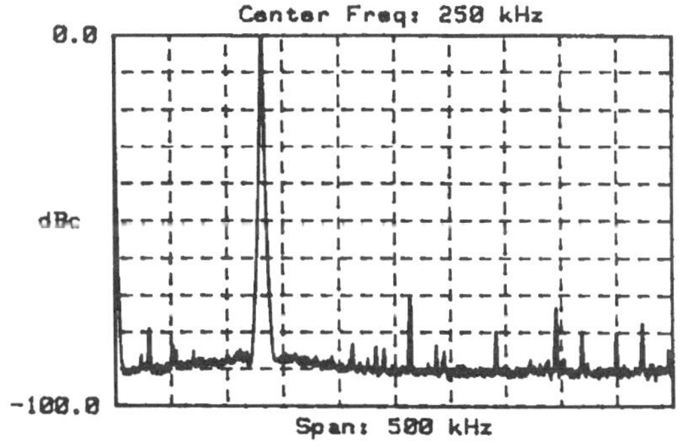

Fig 3. Output spectrum of the DDFS up to the Nyquist limit for a programmed output frequency of 131770 Hz. The lines in the spectrum are listed in Table I.

Table I

Frequency (Hz)	Level (dBc)	Origin
263540	-72.5	$2Fs$
395310	-74.1	$3Fs$
472920	-78.2	Fc-$4Fs$
30710	-78.4	$23Fs$-$3Fc$
341150	-79.2	Fc-$5Fs$
449470	-79.7	$11Fs$-Fc
418760	-80.1	$2Fc$-$12Fs$
51450	-83.0	$(2^{14}+1)Fs$-$2159Fc$

three-pole anti-alias filter with a cut-off of about 300 kHz is used. The power consumption of the complete synthesizer, implemented in the HCMOS logic family, is 760 mW. With no large-scale integration, the circuit is made on a single-sided surface-mount circuit board 10 x 19 cm in size.

Performance

Figure 3 shows the output spectrum of the synthesizer, programmed for a frequency of 131770 Hz. The plot is from zero frequency to 500kHz, the Nyquist limit. The output is taken before the anti-alias filter. Table I shows the amplitudes and numerical structure of the major spurious lines in the spectrum. The most prominent lines apart from the desired output are the second and third harmonics. A few m/n intermodulation products are visible at a level of about -80 dBc and below. An unexpected line which is an extremely high order mixing product with $m=1+2^{14}$ is seen at -83 dBc.

In our application, the critical frequency range for spurious signals is the region of the main resonance line, corresponding to frequencies between 120 and 145 kHz. The required synthesizer output frequencies are between 131.0 and 132.5 kHz. A computer program was used to find which orders of intermodulation would generate significant spurious sidebands. The output spectrum was then examined in detail to find the amplitudes of these products. Figure 4 shows how the sideband amplitude varies with order n up to 100. The lowest significant order is 37, at which the amplitude has fallen to about -90 dBc.

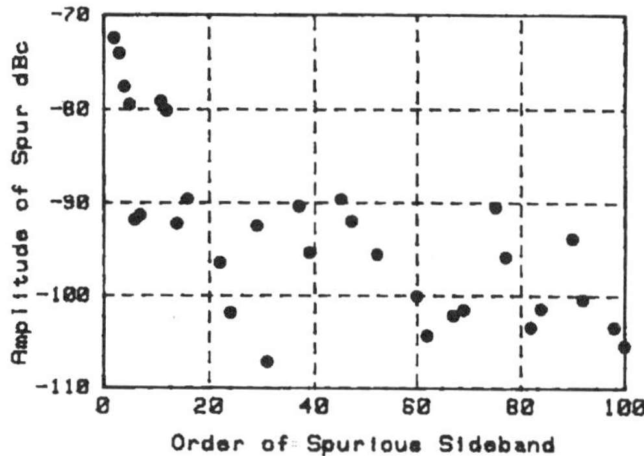

Fig 4. The observed variation of the amplitude of spurious products with their intermodulation order n. The frequency of these products is given by ABS($m \cdot Fc - n \cdot Fs$) where m and n are integers. m runs form 0 to 13.

Figure 5 shows the frequency region around a 131770 Hz output signal with greater resolution. A pair of spurious sidebands at $Fs \pm (38Fs-5Fc)$ is present at a level of about -90 dBc. At the level of about -93 dBc no truncation sidebands are seen for this addend value.

Figure 6 shows, for comparison, the same frequency range as Figure 5 but with the DAC de-glitching disabled. A large number of sidebands have appeared with amplitudes up to -66 dBc. Careful examination shows that these are m/n products. Clearly the de-glitching is useful.

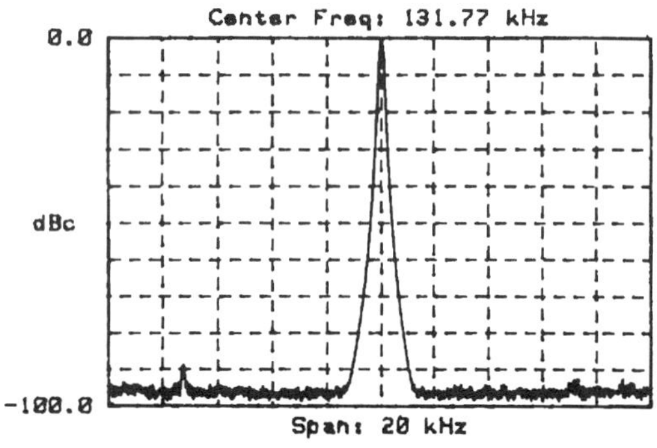

Fig 5. Output spectrum within ± 10 Khz of a programmed output of 131770 Hz. The effective noise bandwidth of the spectrum analyzer is 100 Hz. The frequencies of the spurious sidebands are given by $5Fc-37Fs$ and $38Fs-5Fc$.

Summary

A DDFS consisting of a 34-bit phase accumulator feeding a 14-bit wide address map and a 12-bit de-glitched DAC has been evaluated. For output frequencies in the range 131.0 to 132.5 kHz, which are required to excite the main Cesium resonance line, the level of spurious sidebands is less than -85 dBc. This is compatible with the expected performance of the DAC. No truncation sidebands have been seen down to -90 dBc. A set of sidebands with frequencies given by $2154Fc-(2^{14}-1)Fs$ and $(2^{14}+1)Fs-2154Fc$ are detectable with an amplitude of -83 dBc. The origin of these sidebands is not understood, and they do not occur at significant output frequencies.

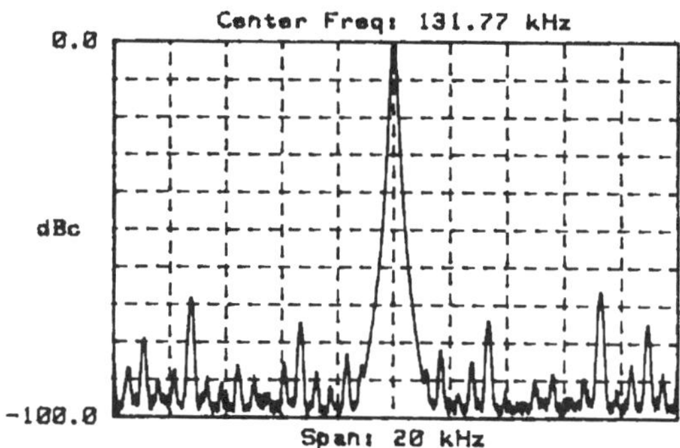

Fig 6. Same as Fig 5, with DAC de-glitcher disabled. The spurious sidebands visible are m/n intermodulation products with values of m and n between 37 and 260.

References

[1] J. Tierney, C.M.Rader, and B. Gold, "A digital frequency synthesizer" IEEE Trans. Audio Electroacoust., Vol AU-19, pp. 48-56, Mar. 1971.

[2] A.L. Bramble, "Direct digital frequency synthesis", in Proceedings of the 35th Annual Symposium on Frequency control, pp 406-414, IEEE, 1981.

[3] V. Reinhardt, K. Gould, K. McNab, and M. Bustamante, " A short survey of frequency synthesizer techniques", in Proceedings of the 40th Annual Frequency Control Symposium, pp 355-365, IEEE 1986.

[4] H.T. Nicholas, III and H. Samueli, "An analysis of the output spectrum of direct digital frequency synthesizers in the presence of phase-accumulator

truncation", in Proceedings of the 41st Annual Frequency Control Symposium, pp 495-502, IEEE 1987.

[5] H.T. Nicholas, III, H. Samueli, and B. Kim, "The optimization of direct digital frequency synthesizer performance in the presence of finite word length effects", in Proceedings of the 42nd Annual Frequency Control Symposium, pp 357-363, IEEE, 1988.

[6] J.F. Garvey, and Daniel Babitch, "An exact spectral analysis of a number controlled oscillator based synthesizer", in Proceedings of the 44th Annual Frequency Control Symposium, pp 511-521, IEEE, 1990.

[7] E.M.Matisson and L.M.Coyle, "Phase noise in direct digital synthesizers", in Proceedings of the 42nd Annual Frequency Control Symposium, pp 352-356, IEEE, 1988.

[8] L.S. Cutler and R.P. Giffard, "Architecture and algorithms for a new cesium beam frequency standard electronics", to be published in Proceedings of the 1992 IEEE Frequency Control Symposium, IEEE 1992.

[9] R. Karlquist, "A new RF architecture for Cesium frequency standards", to be published in Proceedings of the 1992 IEEE Frequency Control Symposium, IEEE 1992.

[10] "Digital Signal Processing", by A.V. Oppenheim and R.W. Shafer, Prentice-Hall, Englewood Cliffs, 1974.

A Spurious Reduction Technique for High-Speed Direct Digital Synthesizers

LAWRENCE J. KUSHNER AND MARCUS T. AINSWORTH

LINCOLN LABORATORY, MASSACHUSETTS INSTITUTE OF TECHNOLOGY

244 WOOD STREET

LEXINGTON, MASSACHUSETTS 02173-9108

Abstract – **Digital-to-Analog Converters (DACs) are the primary source of spurious in high-speed Direct Digital Synthesizers (DDS).   DAC glitches, DC-linearity errors, and digital noise feedthrough can corrupt the DDS output.   This paper proposes a balanced-DAC configuration at the DDS output that can dramatically reduce these effects.   A pair of single-ended DACs are employed, driven from a set of inverters that generate the out-of-phase drive signals. An on- or off-chip subtractor combines the two out-of-phase DAC outputs, canceling distortion, noise, and glitches.   Simulations along with experimental results from an 800 MHz balanced-DAC DDS breadboard are included.   Greater than 10 dB reduction in harmonics, aliased-harmonics, and noise floor have been achieved.   A monolithic version of this balanced-DAC DDS has been fabricated and is currently in test.**

I. BACKGROUND

The sources of spurious spectral components (spurs) at the output of a high-speed Direct Digital Synthesizer (DDS) are of great interest to the DDS designer, as spectral purity is one of the most critical and challenging requirements of any DDS (Figure 1).

Much attention has been paid to the issues of phase truncation (between the accumulator and ROM) and amplitude quantization (in the DAC) over the past 25 years [1] - [10].   By employing a sufficient number of

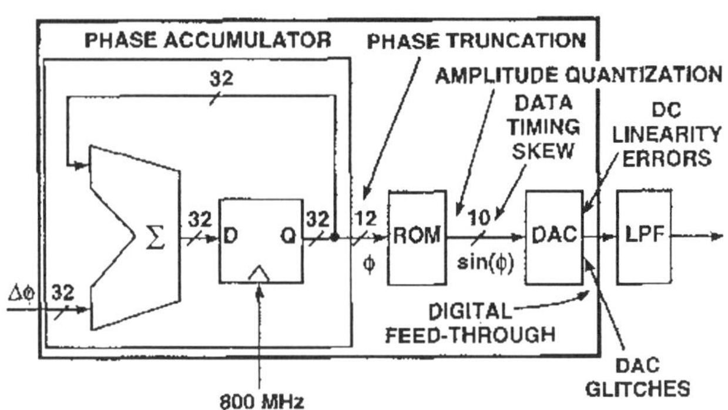

Fig. 1.   DDS spurious sources.

bits in the datapath, these numeric spurious sources have been made negligible in most well designed DDSs.   VLSI technological advances combined with sine-ROM compression techniques have afforded DDS designers this "luxury", no longer requiring them to skimp on the digital circuitry at the expense of spectral purity.   Twelve or more bits of effective ROM address and ten or more bits of DAC amplitude are common-place these days, even in high-speed DDSs.   The DAC is now the primary determinant of spectral purity.

This is especially true for DDSs clocking near 1 GHz, which typically have a few spurs in the -35 dBc to -45 dBc range (Figure 2).   As seen in the simulated spectrum of Figure 2a, with 12 bits of ROM address and

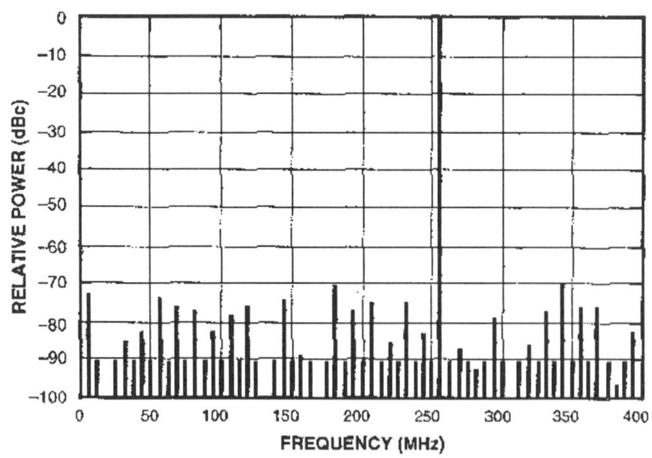

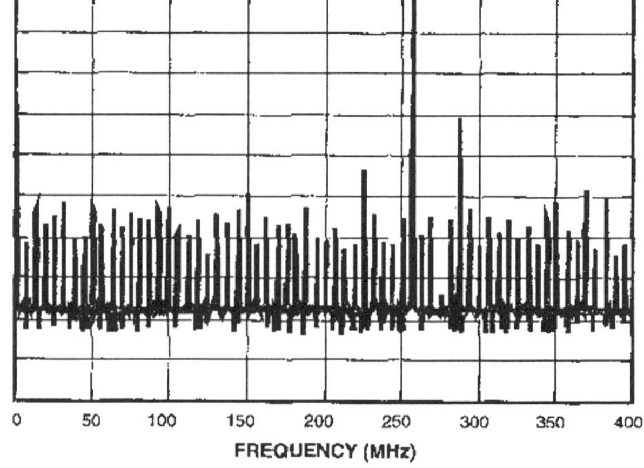

Fig. 2   DDS output   (a) Ideal DAC   (b) real-world DAC

Reprinted from *Proceedings of the IEEE International Frequency Control Symposium*, pp. 920-927, 1996.

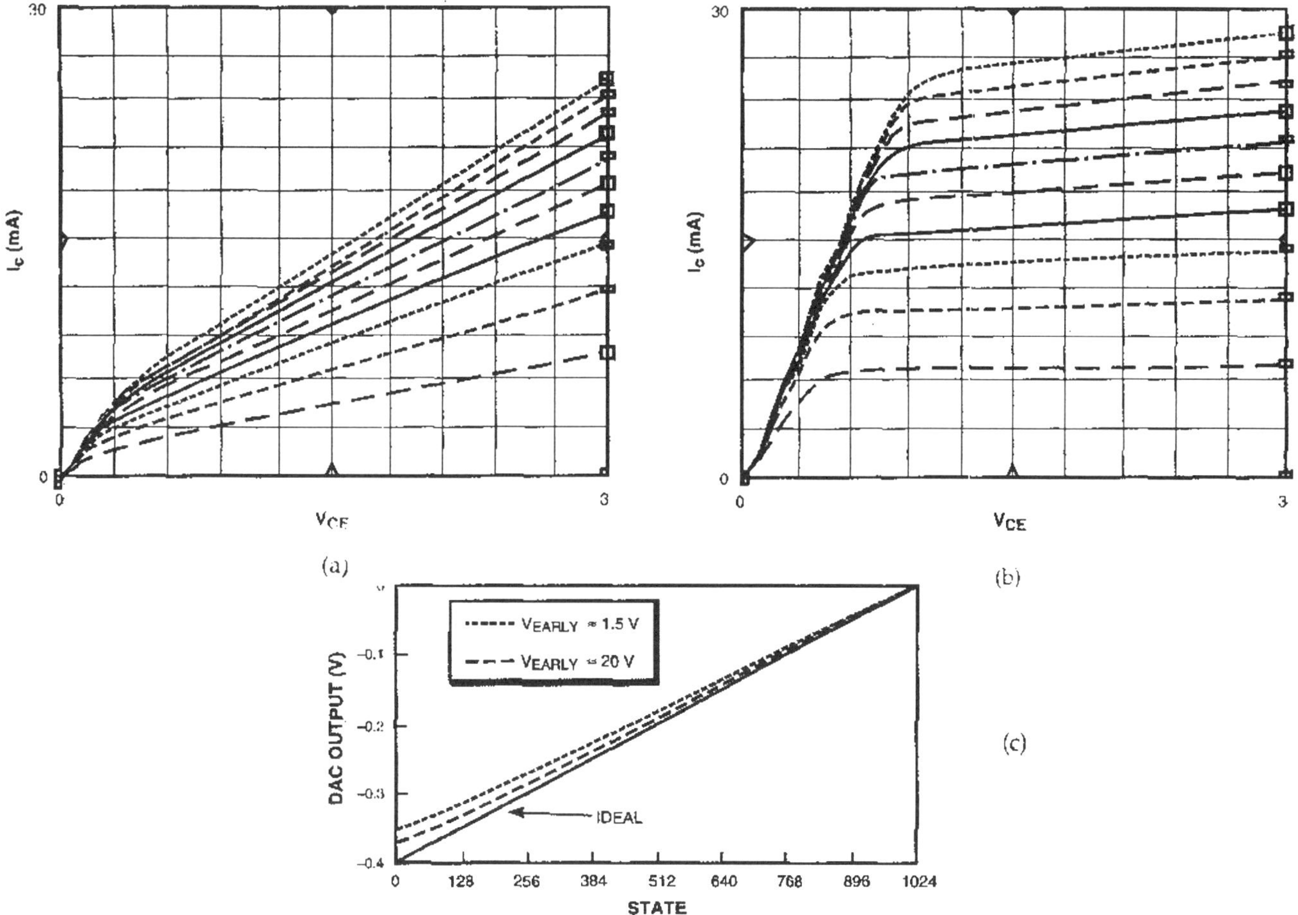

Fig. 3. (a) IV curves of transistor with Early voltage of 1.5V. (b) IV curves of transistor with Early voltage of 15V. (c) Measured and ideal DAC output voltage vs. input code.

a 10-bit DAC, all phase-truncation and amplitude quantization spurs are down below - 70 dBc for this particular DDS phase increment (52000000)16. This is in sharp contrast to the measured spectrum of Figure 2b, which exhibits a large number of spurs significantly stronger than -70 dBc, with a few (aliased-harmonics) in the -30 to -40 dBc range. These real-life spurs come from a combination of sources, including DAC glitches, DAC nonlinearities, and digital-noise feedthrough.

DAC glitches arise from data timing skew into the DAC (unequal bit arrival times), logic asymmetries (unequal rise and fall times), and bit-switch asymmetries. For a fixed glitch duration, these effects play a larger role as the clock speed is increased. A number of techniques have been employed over the years to reduce these effects, including careful layout, the use of differential (e.g., ECL, CML) logic, and a novel glitch-control feedback loop [11]. While these techniques are effective, it is not always possible to employ them, as in the case of designs based on TTL

and HI2L logic [12], [13] which are inherently single-ended. While the glitch-control feedback loop technique is helpful in reducing glitch-energy in these logic families, it does not eliminate glitches entirely. The feedback loop minimizes net glitch energy, allowing equal area positive and negative glitches, creating a glitch doublet.

Digital noise feedthrough is not surprising in DDSs, since they are inherently mixed-mode devices employing high-speed digital circuits (accumulator, ROM) in close-proximity to precision analog circuits (DACs). The digital noise can be impressed on the analog output through a common-mode path, such as a power supply or ground, can be capacitively coupled through the bit-switches, or electromagnetically coupled through neighboring IC traces, bond-wires, or pc board traces.

Finally, the DAC is never perfectly linear, even at DC. Most DAC designs employ a number of current sources whose currents are selectively steered into the

DAC output as determined by the DAC input bit-pattern. While an ideal current source would have infinite output impedance, in practice, due to finite transistor output impedance, the current source output currents are somewhat dependent on their output voltage, causing distortion (Figure 3). For bipolar transistors, Early voltage is a measure of the transistors' output impedance [14]. Larger transistor Early voltage results in higher current source output impedance and thus a more linear DAC.

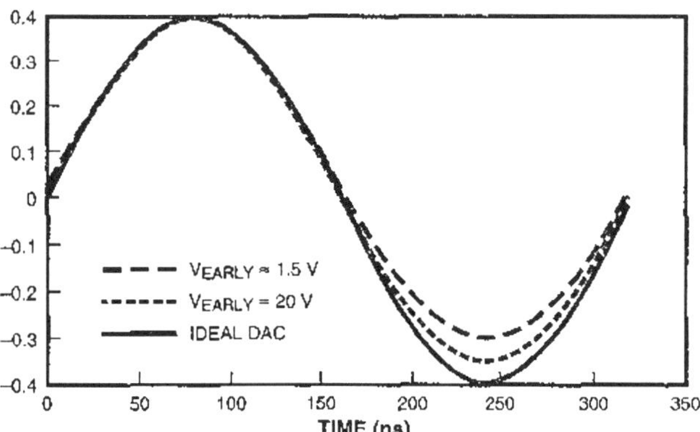

Fig. 4. DDS output – time domain.

The particular DAC linearity-error of Figure 3 manifests itself in poor half-wave symmetry (Figure 4), as the DACs all behave close to ideal on the positive half-cycle but deviate significantly from ideal on the negative half cycle. Half-wave asymmetry in the time-domain leads to even-order distortion in the frequency domain (Figure 5). While the 20-volt Early voltage transistor produces a more linear DAC and therefore better spurious than the 1.5 V transistor, neither real-life DAC is linear enough for many applications. One approach to improving spurious would be to increase the transistor Early voltage further by modifying the

underlying IC fabrication process, but this is expensive and time-consuming. Additionally, increased Early voltage often comes at the expense of reduced gain, which is undesirable. What is really needed is a circuit or architectural change that would reduce the effect these non-linear, real-world transistors have on DAC and DDS performance.

II. BALANCED-DAC DDS

The proposed Balanced-DAC DDS architecture (formerly called the Lincoln DDS architecture, or LDDS) of Figure 6 addresses all of the DAC spurious sources described in the previous section. A pair of imperfect (differential or single-ended) DACs are combined in a balanced topology. The input balun/splitter (unbalanced-to-balanced converter) is implemented digitally at the DAC inputs while the output balun/combiner is implemented in the analog realm at the DAC outputs. Even-order distortion caused by the finite Early voltage is essentially canceled out, as is a good portion of the DAC glitch energy and common-mode digital noise. This architecture is simple to implement as all that is required is a second *identical* DAC, digital inverters, and an on- or off-chip subtractor. Additionally, this technique works fine in single-ended logic families such as TTL and HI2L. In contrast to many proposed "dithering" techniques [2], [5], [10], this balanced technique reduces the overall spur-power level rather than "smearing" it out into a higher pseudo-noise floor.

Simulations show a dramatic improvement in the half-wave symmetry of the balanced-DAC DDS time-domain output (Figure 7) and thus improved even-order distortion (Figure 8). In particular, the aliased-second harmonic, often the dominant spur, can be reduced to below the noise floor (with well matched DACs and a high quality subtractor). Note that odd-order spurs are unaffected, as expected.

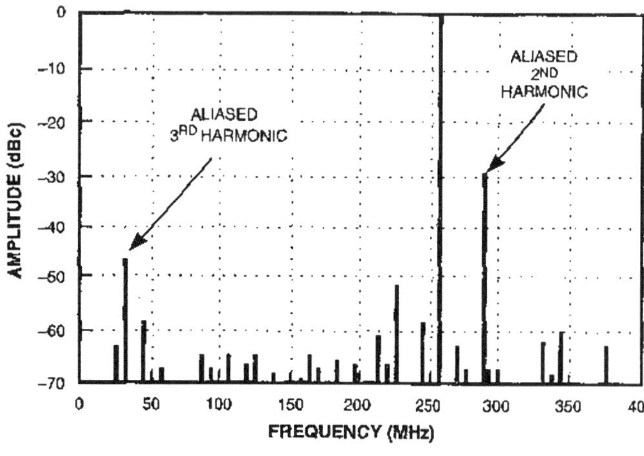

(a) $V_{EARLY} \approx 1.5V$.

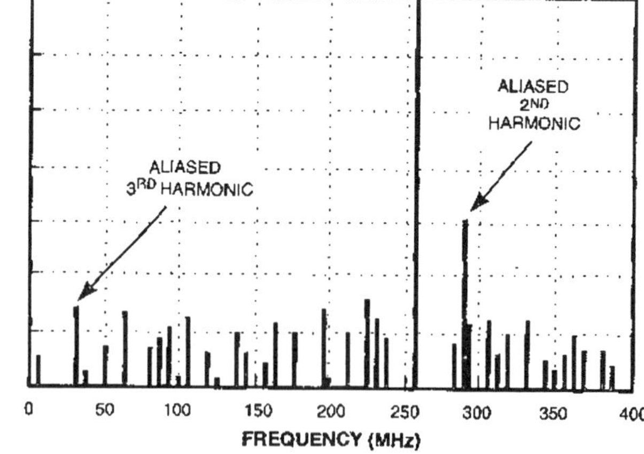

(b) $V_{EARLY} \approx 20V$.

Fig. 5. DDS output – frequency domain.

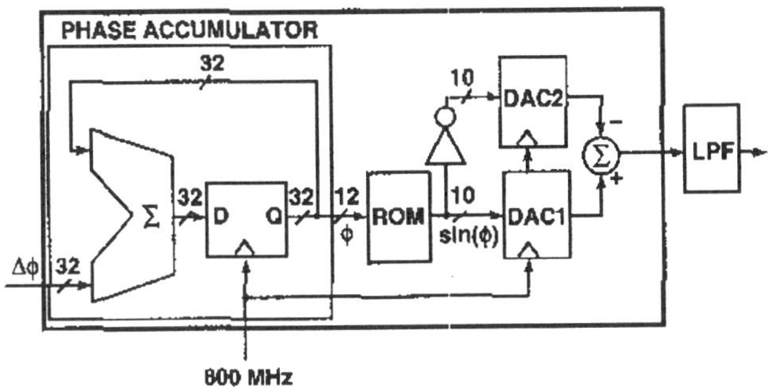

Fig. 6. Balanced-DAC DDS architecture.

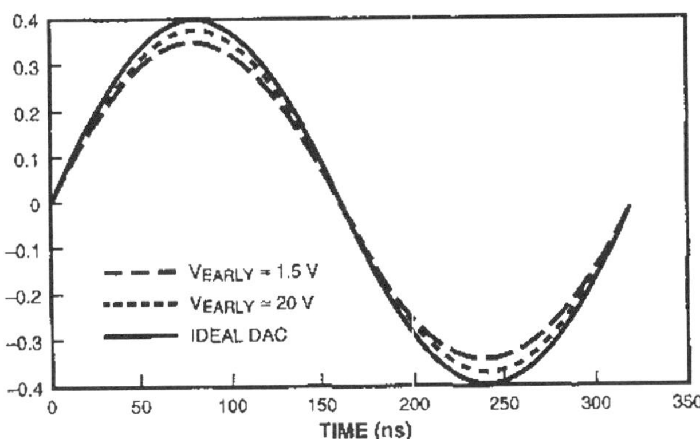

Fig. 7. Balanced-DAC DDS output – time domain.

An accurate time-domain simulation of a particular DAC could be employed to investigate DAC glitches. Instead, the following more general qualitative argument is made. Glitches often arise from logic asymmetry, such as the 0-to-1 logic transition being faster than the 1-to-0 transition. Now consider the transition through the mid-range state, where the most-significant bit (MSB) and all of the LSBs toggle (Figure 9). Momentarily all of the bits are effectively at logic 1, causing an upward glitch. This is true on both half-cycles of the sinewave. Therefore in the balanced-DAC DDS, even though the two DAC outputs are 180° out-of-phase, their glitches are in-phase and thus canceled by the output subtractor. A similar argument can be made in the case of segmented-MSB DACs.

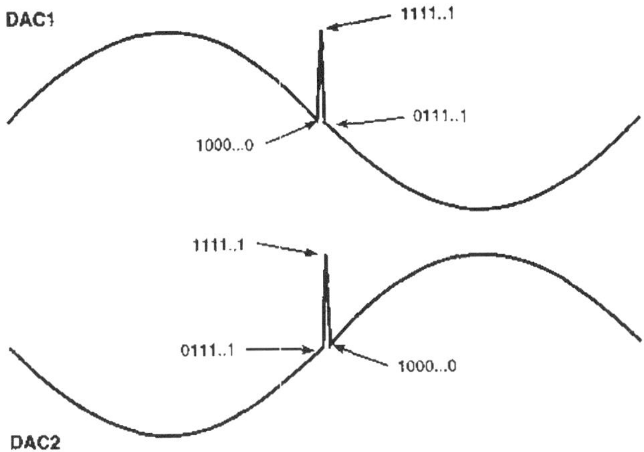

Fig. 9. Balanced-DAC DDS mid-range glitch.

The balanced-DAC architecture can offer improvements even in the case where a glitch-control feedback loop is employed. As mentioned in the previous section, this feedback loop forces the net glitch area to be zero but allows equal area positive and negative glitches, forming a glitch doublet (Figure 10). Typically this glitch doublet is in the same direction on every DAC output transition, allowing the output subtractor to reduce the spur energy caused by this doublet.

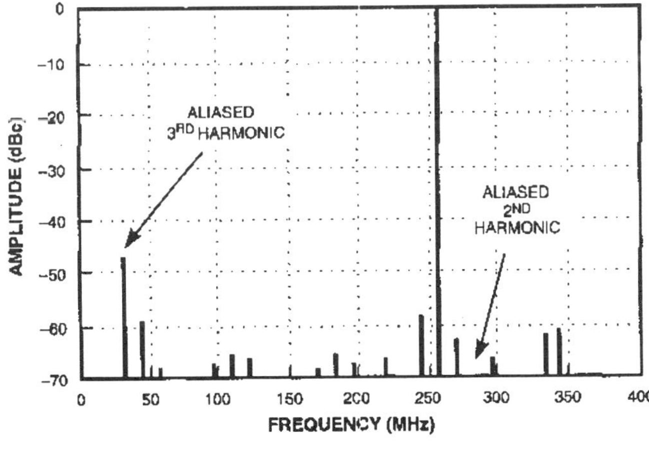

(a) $V_{EARLY} \approx 1.5V$.

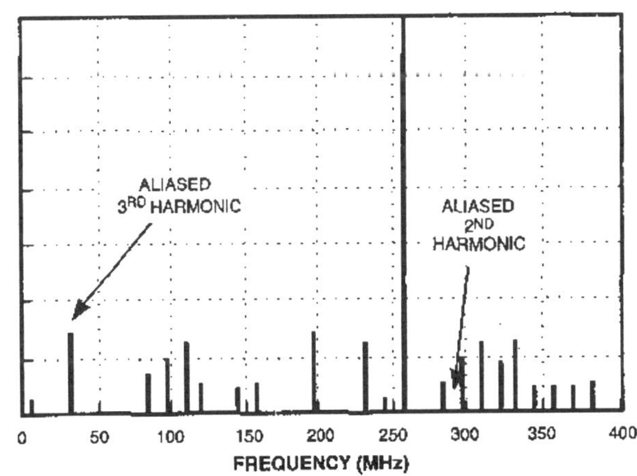

(a) $V_{EARLY} \approx 20V$.

Fig. 8. Balanced-DAC DDS output – frequency domain.

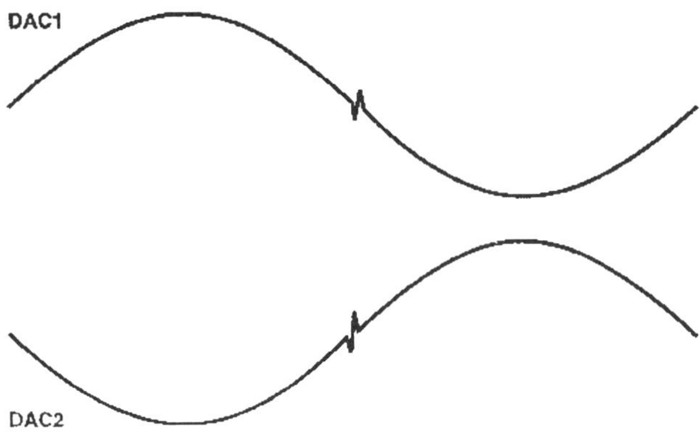

Fig. 10. Balanced-DAC DDS glitch doublet

While these qualitative arguments are appealing, the best proof of this technique is measurement. To this end, a breadboard 800 MHz balanced-DAC DDS has been built using a pair of single-ended monolithic GaAs HBT DDSs [15] synchronized to simulate the balanced-DAC DDS (Figure 11).

The two DDS outputs are subtracted via a surface-mount transformer. An on-chip phase-modulation port is used to program one DDS to be 180° out-of-phase with the other. The two DDSs are simultaneously reset then programmed to a given frequency. In order to ensure that the two accumulators are operating on the same 800 MHz clock cycle and contain the same phase value, the two DDS outputs are added together and detected. If the DDSs are synchronized properly, the sum of their two outputs should be zero. If the detector measures a signal above a pre-determined threshold, the DDSs are re-programmed. Since the ultimate balanced-DAC implementation will employ a single accumulator, this detector circuit is just required for this breadboard and will be eliminated in the final design.

As seen in Figures 12a-f, significant reductions in all of the spurious levels are obtained across the output band. Even- and odd-order, harmonics and aliased-harmonics, as well as the overall noise floor are all improved by at least 10 dB. No tuning was performed after the initial DAC setups. The cancellation is truly broadband and repeatable. A full statistical characterization was performed (Figure 13) in which the total spur power both close to the carrier (± 10 MHz) and in the full output band (2 to 380 MHz) was obtained at 138 different DDS output frequencies. The

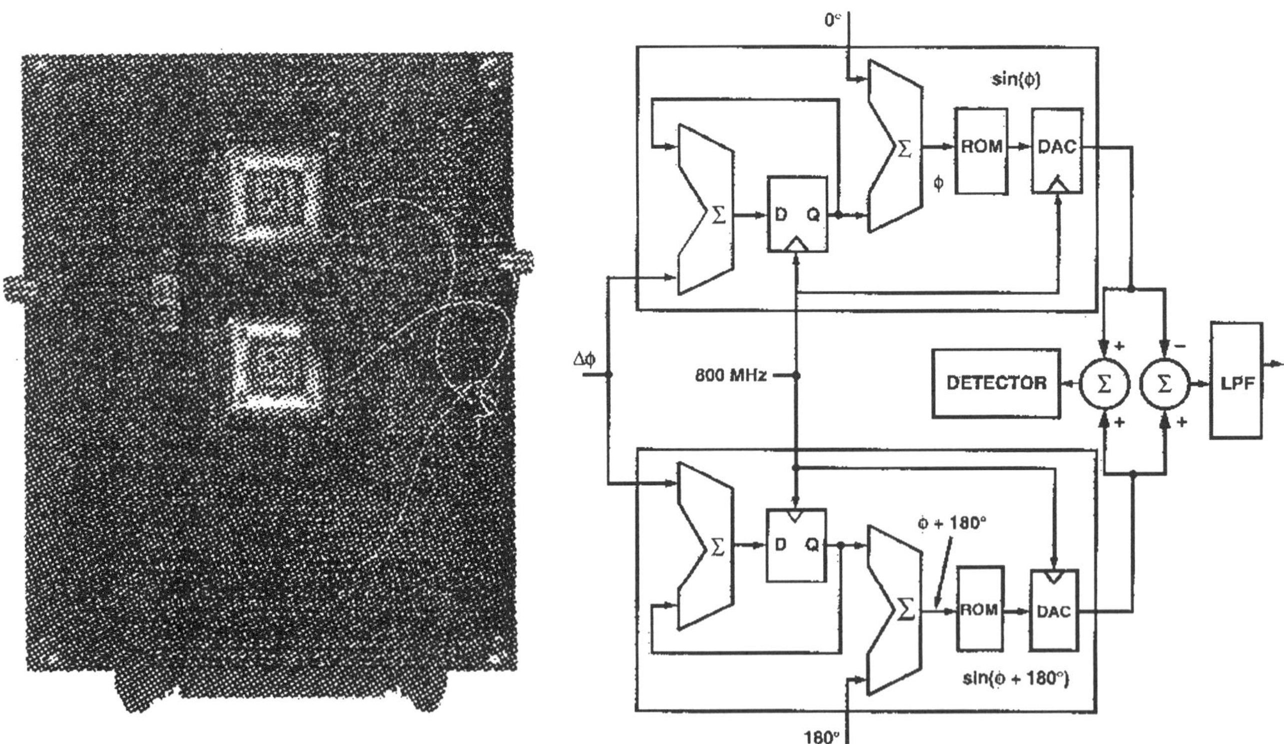

Fig. 11. Balanced-DAC DDS Breadboard

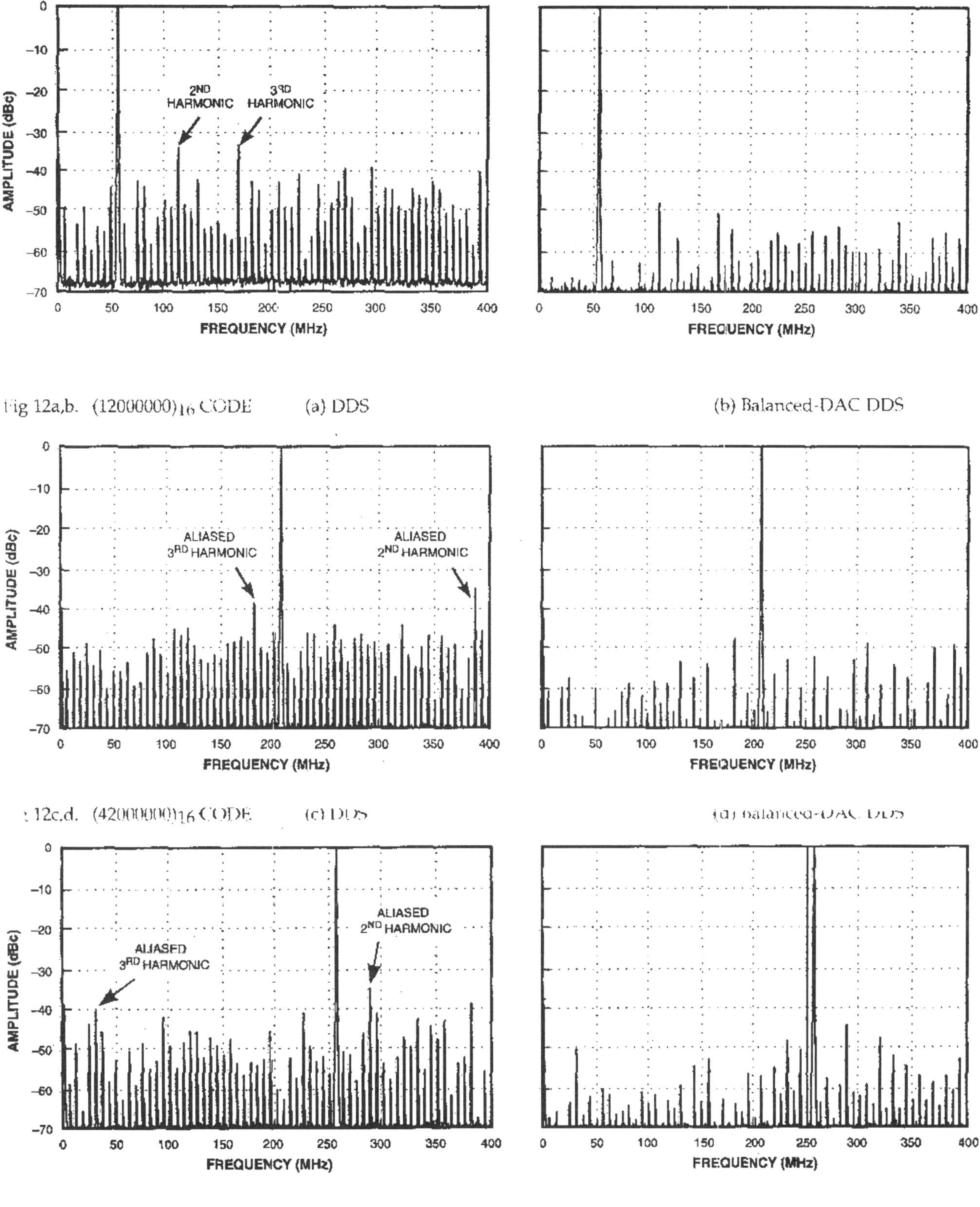

Fig 12a,b. (12000000)$_{16}$ CODE   (a) DDS   (b) Balanced-DAC DDS

Fig 12c,d. (42000000)$_{16}$ CODE   (c) DDS   (d) Balanced-DAC DDS

Fig 12e,f. (52000000)$_{16}$ CODE   (e) DDS   (f) Balanced-DAC DDS

mean spur power was improved by about 10 dB for both bandwidths. Further improvements are expected from a fully monolithic version, as matching should be improved with both DACs on the same die at the same temperature with the same common-mode noise.

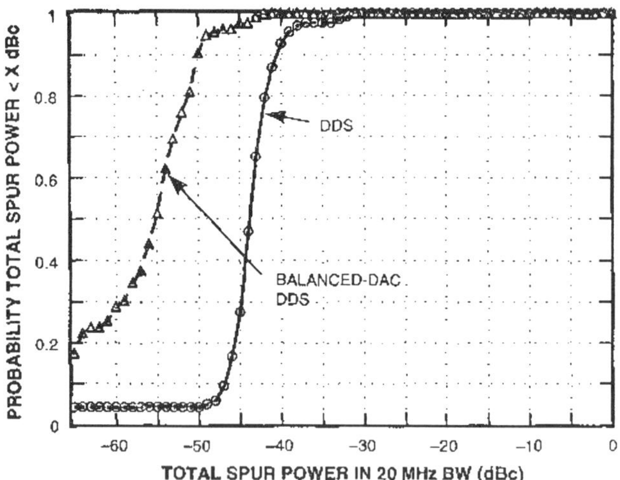

Fig. 13a. Cumulative distribution function of spurs in 20 MHz bandwidth

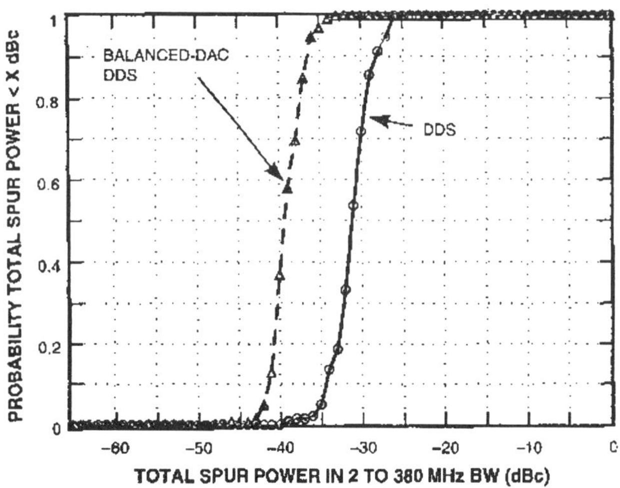

Fig. 13b. Cumulative distribution function of spurs in 378 MHz bandwidth.

III. MONOLITHIC BALANCED-DAC DDS

The architecture of Figure 6 has been implemented in monolithic form in Texas Instruments' GaAs HI^{2}L process [12], [13] in collaboration with TI's GaAs Signal Processing Group. Two versions have been implemented: one with an off-chip subtractor and a second with an on-chip subtractor. The version with the off-chip subtractor was very straight-forward to design, since all that is required is a second DAC and some inverters. This version has already been fabricated and is currently in test. Preliminary results are very encouraging, with spurious performance superior to the breadboard of section II.

The version with the on-chip subtractor requires a high-quality differential amplifier at the output. Since this HI^{2}L process has all of the emitters of all of the transistors grounded, conventional differential amplifiers cannot be used. Instead, a novel subtractor circuit, originally developed for a second-generation Composite DDS [16], [17], is employed. The design of this balanced-DAC DDS is complete (Figure 14) and is in fabrication. Detailed chip descriptions and measured performance of both versions will be the subject of a future paper.

Fig. 14. Monolithic Balanced-DAC DDS layout

IV. DISCUSSION & CONCLUSIONS

It has been known for decades that balanced designs result in reduced even-order distortion and provide common-mode rejection to noise. Power amplifiers and balanced mixers have capitalized on this fact for years. In fact, a number of prior DAC designs [18] have been developed that employ conventional differential amplifiers to steer the DAC current-source outputs into either of a pair of complemenatary (balanced) analog outputs (based on the input bit pattern), and could achieve most of these same advantages with the addition of an output subtractor.

The novel part of this newly proposed balanced-DAC DDS architecture is that it does not require the DAC itself to be balanced or differential in any way. Any pair of matched single-ended DACs can be employed in this new approach. Additionally, some excellent high-speed digital fabrication processes, notably GaAs HBT HI^{2}L, require all transistor emitters on the chip to be grounded, precluding a conventional differential-DAC design. The proposed balanced-DAC DDS architecture overcomes this process limitation, creating a balanced design in an inherently single-ended environment.

Simulations and breadboard measurements have been presented indicating that at least a 10 dB spurious performance improvement is possible with this new architecture. It is easy to implement, requiring only a second DAC, inverters, and an on- or off-chip subtractor. Two monolithic versions have been designed. Preliminary results are very encouraging, with measured spurious performance superior to the breadboard results presented here. If further improvements are desired, it may be advantageous to lay out the two single-ended DACs in an interlaced fashion, guaranteeing component matching and equal common-mode coupling, providing even better cancellation.

ACKNOWLEDGMENT

This work was sponsored by the US ARMY under contract F19628-95-C-0002. All of the 800 MHz DDS ICs used in this work were developed in collaboration with Van Andrew's GaAs Signal Processing Group at Texas Instruments.

The authors would like to thank David Snider, Ron Bauer, and Scott Sharp for their continued support and encouragement.

REFERENCES:

[1] J. Tierney, C. M. Rader, and B. Gold, "A digital frequency synthesizer," *IEEE Trans. Audio Electroacoust.*, vol. AU-19, pp. 43-57, Mar. 1971.

[2] C. E. Wheatley,III and D. E. Phillips, "Spurious suppression in direct digital dynthesizers," *Proc. 35th Annual Freq. Contr. Symp.*, May 1981, pp. 428-435.

[3] J. J. Olsen, "Phase truncation effects in direct digital frequency synthesis," *M. S. Thesis*, Massachusetts Institute of Technology, June 1985.

[4] D. A. Sunderland, et. al., "CMOS/SOS frequency synthesizer LSI circuit for spread spectrum communications," *IEEE Journal of Solid-State Ciruits*, vol. SC-19, no. 4, Aug. 1984, pp. 497-505.

[5] H. T. Nicholas III and H. Samueli, "An analysis of the output spectrum of direct digital frequency synthesizers in the presence of phase-accumulator truncation," *Proc. 41st Annual Freq. Contr. Symp.*, May 1987, pp. 495-502.

[6] H. T. Nicholas III, H. Samueli, B. Kim, "The optimization of direct digital frequency synthesizer performance in the presence of finite word length effects," *Proc. 42nd Annual Freq. Contr. Symp.*, May 1988, pp. 357-502.

[7] P. O'Leary and F. Maloberti "A Direct-Digital synthesizer with improved spectral performance," *IEEE Tran. on Communications*, vol. 39, no. 7, July 1991, pp. 1046-1048.

[8] J. A. Crawford, *Frequency Synthesizer Design Handbook*, Artech House, 1994, Chapt. 7.3.

[9] J. F. Garvey and D. Babitch, "An exact spectral analysis of a number controlled oscillator based synthesizer," *Proc. 44th Annual Freq. Contr. Symp.*, May 1990, pp. 511-521.

[10] V. S. Reinhardt, "Spur reduction techniques in direct digital synthesizers," *Proc. 1993 Int. Freq. Contr. Symp.*, June 1993, pp. 230-241.

[11] W. A. White, "Method and apparatus for digital to analog conversion with minimized distortion, *US Patent # 5,321,401*, June 14, 1994.

[12] H-T. Yuan, H-D Shih, J. Delaney, C. Fuller , "The development of heterojunction integrated injection logic," *IEEE Trans. on Electron Devices*, vol. 36, no. 10, October 1989, pp. 2083-2092.

[13] C. T. M. Chang and H-T. Yuan, "GaAs HBT's for high-speed digital integrated circuit applications," *Proc. IEEE* , vol. 81, no. 12, December 1993, pp. 1727-1743.

[14] J. M. Early, "Effects of space-charge layer widening in junction transistors," *Proc. IRE*, 40, p. 1401, 1952.

[15] G. Van Andrews et. al., "Recent progress in wideband direct digital synthesis," *1996 IEEE MTT-S Int. Microwave Symp. Dig.*, 20 June 1996, pp. 1347-1350.

[16] L. J. Kushner, "The composite DDS – A new direct digital synthesizer architecture," in *Proc. 1993 Int. Freq. Contr. Symp.*, June 2-4, 1993, pp. 255-260.

[17] L. J. Kushner, "An 800 MHz monolithic GaAs HBT serrodyne modulator," *IEEE J. Solid-State Ciruits*, vol. 30, no. 10, October 1995, pp. 1041-1050.

[18] Burr-Brown DAC650 datasheet, Jan. 1993.

A Direct-Digital Synthesizer with Improved Spectral Performance

Paul O'Leary and Franco Maloberti

Abstract—This paper presents a modified direct-digital synthesizer which uses noise shaping to reduce the effects of phase-accumulator truncation on the output spectrum. The discrete spectral disturbances associated with this truncation error are strongly reduced with the proposed method, making the synthesizer suitable for high performance signal processing. The proposed architecture uses first-order noise shaping. Higher order noise shaping can also be used if required.

I. Introduction

IN RECENT years direct-digital synthesis has been applied in many areas [1]–[8]. The spectral purity at the output is important in most applications, especially where the synthesizer is used in nonlinear processes.

In most applications the accumulator has a large bit width, typically 16–20 b. From the accumulator only a small number of bits are used as address for the Sine ROM, this corresponds to a phase truncation. The error due to this phase increment discretization (phase truncation) and due to the finite ROM data word are usually described in terms of a noise contribution. The power associated with that noise should have a white spectrum. However, with conventional direct-digital synthesizers (DDS), a white noise component is often accompanied by discrete lines. The presence of discrete spectral lines in the output spectrum is undesirable because they can result in intermodulation components. To avoid this undesirable effect, the conventional solutions dimension the sine ROM such that the output noise is dominated by the amplitude quantization and not by the phase truncation. Unfortunately, for high-precision sine waves the required ROM becomes prohibitively large. The use of trigonometrical decomposition [9] offers only a small improvement. The ROM size can also be reduced with the use of complex computation [10] but this is done at the cost of a more complex circuit.

This paper presents a direct-digital synthesizer where a noise shaping technique is applied to the phase truncation to overcome the problems of spurious spurs in the output spectrum. The modified circuit offers considerable improvement in the synthesizer performance when the generated frequency is low with respect to the clock frequency.

II. Conventional Direct-Digital Synthesizer

Fig. 1 shows the block diagram of a conventional direct digital synthesizer. It consists of an N bit accumulator the

Paper approved by the Editor for Synchronization Systems and Techniques of the IEEE Communications Society. Manuscript received July 16, 1990; revised October 16, 1990.

P. O'Leary is with Electronic Sensor Interfaces Group, Joanneum Research, Steyerergasse 17, A-8010 Graz Austria.

F. Maloberti is with the Department of Electronics, University of Pavia, 27100 Pavia, Italy.

IEEE Log Number 9100328.

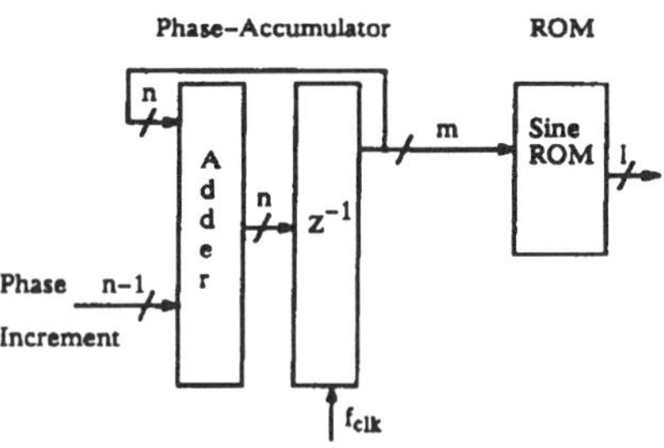

Fig. 1.   Conventional direct digital synthesizer.

output of which is truncated to m bits as the address of a Sine ROM with an L bit wide data bus.

The truncation of the phase word is equivalent to quantization. If it is assumed that quantization can be modelled as a linear additive process then the output of the synthesizer is given by

$$\text{Out}(t) = \sin\left(\frac{2\pi f_{\text{gen}}i}{2^m} + e_p(i)\right) + e_L(i) \qquad (1)$$

where

f_{gen} is the generated frequency.
$e_p(i)$ is the error associated with the phase truncation.
$e_l(i)$ is the quantization error due to the finite ROM data word.

Using a trigonometrical equality and assuming that the phase error is small relative to the phase, (1) becomes

$$\text{Out}(t) = \sin\left(\frac{2\pi f_{\text{gen}}i}{2^m}\right) + e_p(i)\cos\left(\frac{2\pi f_{\text{gen}}i}{2^m}\right) + e_L(i). \quad (2)$$

Thus, the phase error is amplitude modulated with the quadrature signal to the desired frequency. Consequently, discrete spectral lines in the phase error will be modulated to discrete spectral pairs in the synthesizer output spectrum.

The relevant contribution to the global error from the phase truncation error is demonstrated with the help of Figs. 2 and 3. Fig. 2 shows the phase truncation error spectrum from a conventional DDS running under the following conditions: $F_{\text{clk}} = 400$ KHz, $F_{\text{gen}} = 1080$ Hz, $N = 16$, $M = 6$, and $L = 14$.[1] Fig. 3 shows the synthesizer output spectrum of the same DDS system. The spectral pairs due to the phase truncation are easily identified.

A thorough analysis of phase truncation error effects on the output spectrum of conventional digital synthesizers has been presented by H. Nicholas *et al.* [11]. He shows that the worst

[1] All spectra presented in this paper were generated using a 4096 point FFT with Hanning window.

Reprinted from *IEEE Transactions on Communications*, Vol. 39, No. 7, pp. 1046-1048, July 1991.

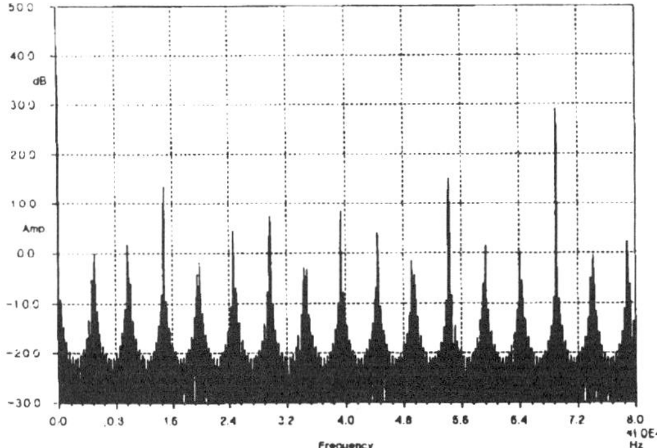

Fig. 2. Phase error spectrum for a conventional DDS.

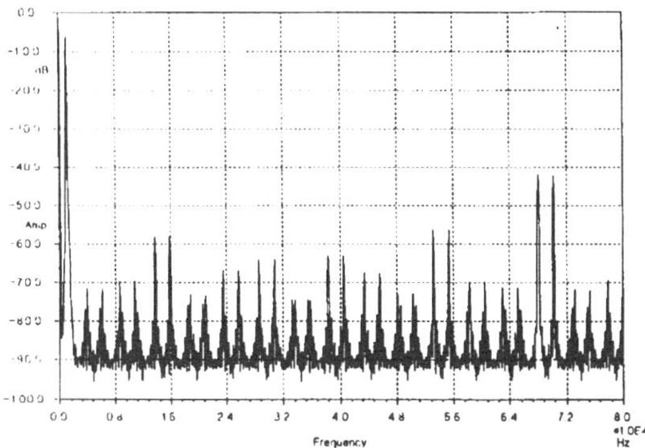

Fig. 3. Output spectrum for a conventional DDS.

case spur is given by

$$S_{\max} = \left[2^m \operatorname{sinc}\left(\frac{\pi\left(f_{\text{gen}} 2^{n-m} \right)}{2^{n-m}} \right) \right]^{-1} \qquad (3)$$

where $\left(f_{\text{gen}}, 2^{n-m} \right)$ denotes the greatest common divisor of f_{gen} and 2^{n-m}.

With (3), the peak output error signal due to the phase truncation error can be calculated. Evaluating the first term of (3) gives an output signal-to-noise ratio of

$$S_{\max} = -6.02m + 3.992 \text{ dB}. \qquad (4)$$

The finite bit width L of the ROM output word also results in a noise component, which can be approximated by:

$$\text{SNR} = -6.02L - 1.76 \text{ dB}. \qquad (5)$$

If m is chosen equal to L then, comparing (4) and (5) shows that the phase error can result in an output peak error which is 5.75 dB above the noise floor due to the output amplitude quantization.

Consequently, it should be chosen such that

$$m = L + 1 \qquad (6)$$

to ensure that the output error is dominated by the ROM data width and not by the address truncation. Unfortunately for high-precision sine waves this requires a very large ROM. (The ROM size increases exponentially in powers of 2 with the address width but only linearly in the number of data bits.)

III. MODIFIED ARCHITECTURE

The phase accumulator is often between 16 and 20 b wide. This corresponds to an almost ideal description of the phase. It has been noted that the truncation of this phase to m bits as ROM address corresponds to quantization. If the frequency being generated is low with respect to the used clock frequency, then there is an intrinsic oversampling. This oversampling can be used to apply a noise shaping technique [12] to reduce the effects of the phase truncation.

The use of an oversampling technique makes it possible to take into account the individual phase truncations by accumulating the resulting phase error. The total error can now be reduced by suitably correcting the ROM address with the accumulated error. The final result is equivalent to a linear (for first-order noise shaping) or a quadratic (for second order noise shaping) interpolation between two consecutive ROM addresses.

Fig. 4 shows the architecture of the modified direct digital synthesizer. A first-order noise shaper has been placed between the phase accumulator and the ROM. Using calculation methods similar to those used by J. Candy [12] it is possible to derive the following equation for synthesizer output signal:

$$\text{Out}(t) = \sin\left(\frac{2\pi f_{\text{gen}} i}{2^m} \right) + \{ e_p(i) - e_p(i-1) \}$$
$$\cos\left(\frac{2\pi f_{\text{gen}} i}{2^m} \right) + e_L(i). \qquad (7)$$

The additional processing of the phase error is equivalent to a first-order digital high-pass filtering of the phase "noise" before the amplitude modulation through the ROM. The improvement in the signal-to-noise performance for low-frequency signals is evident. More importantly the noise shaper also decorrelates the truncation error eliminating the discrete spectral lines in the output spectrum. Fig. 5 shows the output spectrum from the modified architecture with the same ROM and accumulator as for Figs. 2 and 3 but with the addition of the noise shaping (the synthesizer is running under the same conditions as the conventional synthesizer). It can be observed that a significant improvement in the output spectrum has been achieved. The output spectrum does not display strong discrete spectral disturbances, this makes the synthesizer suitable even for nonlinear applications.

IV. CONCLUSION

A modified direct digital synthesizer has been presented which uses noise shaping to reduce the effects of phase truncation on the output spectrum. It has been demonstrated that this can give a considerable improvement in the performance of the synthesizer, particularly for nonlinear applications. The

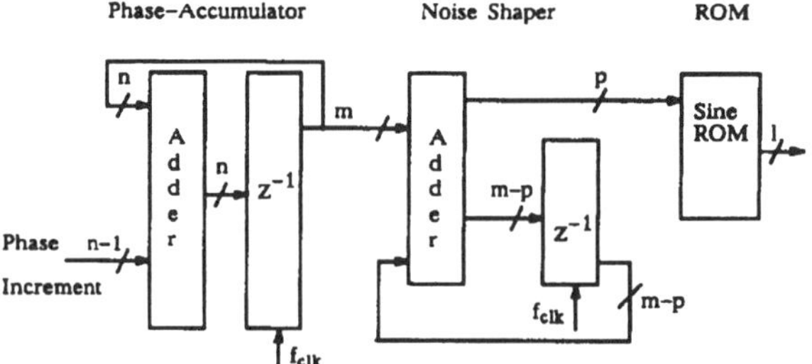

Fig. 4. New direct digital synthesis architecture.

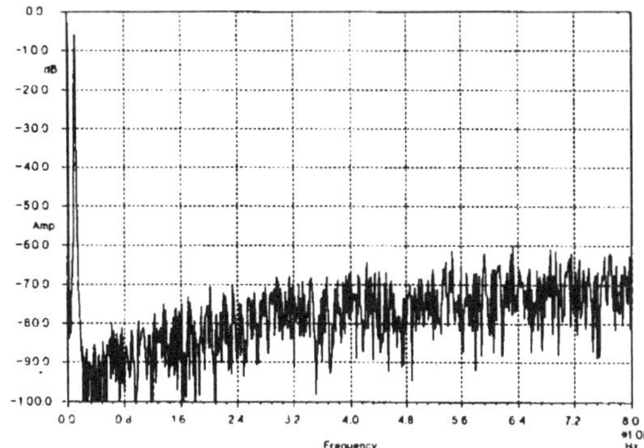

Fig. 5. Output spectrum for the new architecture.

proposed architecture used first-order noise shaping but the use of higher order noise shaping is possible.

REFERENCES

[1] T. A. C. M. Claasen and J. B. H. Peek, "A digital receiver for tone detection applications," *IEEE Trans. Commun.*, vol. COM-24, pp. 1291–1300, Dec. 1976.

[2] J. N. Denenberg, "Spectral moment estimators: A new approach to tone detection," *Bell Syst. Tech. J.*, vol. 55, no. 2, pp. 143–155, Feb. 1976.

[3] D. Shuterland *et al.*, "CMOS/SOS frequency systhesizer LSI circuit for spread spectrum communications," *IEEE J. Solid-State Circuits*, vol. SC-19, Aug. 1984.

[4] P. O'Leary, H. Horvat, and F. Maloberti, "Application of direct digital systhesis to modulators for data communications," in *Proc. 9th Int. Custom Microelectron. Conf.*, Apr. 24–26, 1990, London, England.

[5] P. O'Leary, M. Pauritsch, F. Maloberti, and G. Raschetti, "A DTMF generator based on interleaved oversampling," in *Proc. 15th European Solid-State Circuits Conf. 1989*, Vienna, pp. 200–203.

[6] P. Saul, D. Taylor, and T. Ward, "A 500 MHz output direct frequency systhesizer," in *Proc. 15th European Solid-State Circuits Conf. 1989*, Vienna, pp. 156–258.

[7] P. Saul and M. Mudd, "A direct digital systhesizer with 100 MHz output capability," *IEEE J. Solid-State Circuits*, vol. 23, June 1988.

[8] B. Giebel, J. Lutz, and P. O'Leary, "Digitally controlled oscillator," *IEEE J. Solid-State Circuits*, vol. 24, June 1989.

[9] C. R. Cole, "Design of a direct digital frequency systhesizer," Master's thesis, M.I.T., Cambridge, MA, 1982.

[10] J. Tierney, C. Rader, and B. Gold, "A digital frequency synthesizer," *IEEE Trans. Audio Electroacoust.*, vol. AU-19, Mar. 1971.

[11] H. T. Nicholas and H. Samueli, "An analysis of the output spectrum of direct digital frequency synthesizers in the presence of phase-accumulator truncation," *IEEE Proc. 41st Annu. Frequency Cont. Symp. 1987*, pp. 495–502.

[12] J. C. Candy and A. N. Huynh, "Double interpolation for digital to analog conversion," *IEEE Trans. Commun.*, vol. COM-34, pp. 77–81, Jan. 1986.

Spur-Reduced Digital Sinusoid Synthesis

Michael J. Flanagan and George A. Zimmerman

Abstract— This paper presents and analyzes a technique for reducing the spurious signal content in digital sinusoid synthesis. Spur reduction is accomplished through dithering amplitude and phase values prior to wordlength reduction. The analytical approach developed for analog quantization is used to produce new bounds on spur performance in these dithered systems. Amplitude dithering allows output wordlength reduction without introducing additional spurs. Effects of periodic dither similar to that produced by a pseudo-noise (PN) generator are analyzed. This phase dithering method provides a spur reduction of $6(M + 1)$ dB per phase bit when the dither consists of M uniform variates. While the spur reduction is at the expense of an increase in system noise, the noise can be made white, making the noise power spectral density small. This technique permits the use of a smaller number of phase bits addressing sinusoid look-up tables, resulting in an exponential decrease in system complexity. Amplitude dithering allows the use of less complicated multipliers and narrower data paths in purely digital applications, as well as the use of coarse-resolution, highly-linear digital-to-analog converters (DAC's) to obtain spur performance limited by the DAC linearity rather than its resolution.

I. Introduction

IT IS WELL KNOWN that adding a dither signal to a desired signal prior to quantization can render the quantizer error independent of the desired signal [1]–[3]. Classic examples of this work deal with the quantization of analog signals. Advances in digital signal processing speed and large scale integration have led to the development of all-digital receiver systems, direct digital frequency synthesizers and direct digital arbitrary waveform synthesizers. In all these applications, since finite wordlength effects are a major factor in system complexity, they may ultimately determine whether it is efficient to digitally implement a system with a particular set of specifications. Earlier work [4] has presented a technique for reducing the complexity of digital oscillators through phase dithering with the claim of increased frequency resolution. Recent research [5] has suggested mitigation of finite wordlength effects in the synthesis of oversampled sinusoids through noise shaping. This paper shows how the analysis techniques used for quantization of analog signals can be applied to overcome finite-wordlength effects in digital systems. The analysis in this paper shows how appropriate dither signals can be used

Paper approved by A. Duel-Hallen, the Editor for Communications of the IEEE Communications Society. Manuscript received June 8, 1992; revised January 26, 1993 and November 22, 1993. The work described in this paper was carried out at the Department of Electrical Engineering and at the Jet Propulsion Laboratory of the California Institute of Technology, and supported under a contract with the National Aeronautics and Space Administration. M. J. Flanagan was supported in part by a National Science Foundation Fellowship.

M. J. Flanagan is with AT&T Bell Laboratories, Whippany, NJ 07981 USA. G. A. Zimmerman is with PairGain Technologies, Tustin, CA 92608 USA.

IEEE Log Number 9411078.

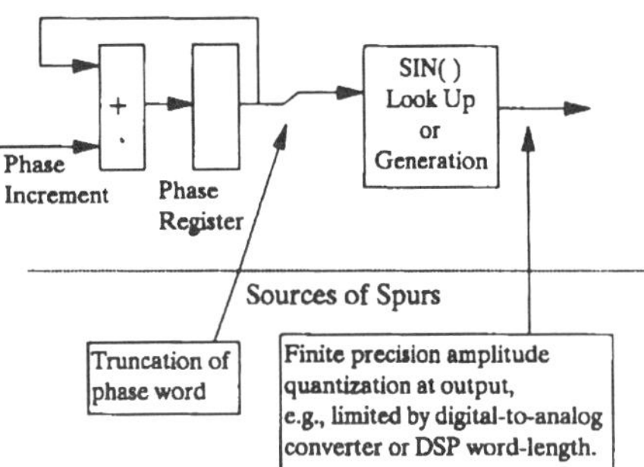

Fig. 1. Spur generation in conventional digital sinusoid generation.

to reduce word lengths in digital sinusoid synthesis without suffering the normal penalties in spurious signal performance. Furthermore, the dithering technique presented in this paper is not limited to the synthesis of oversampled signals.

Conventional methods of digital sinusoid generation [6], e.g., Fig. 1, result in spurious harmonics (spurs) which are caused by finite wordlength representations of both amplitude and phase samples [7]. Because both the phase and amplitude samples are periodic sequences, their finite wordlength representations contain periodic error sequences, which cause spurs. The spur signal levels are approximately 6 dB per bit of representation below the desired sinusoidal signal.

The technique presented in this paper reduces the representation word length without increasing spur magnitudes by first adding a low-level random noise, or dither, signal to the amplitude and/or the phase samples, which are originally expressed in a longer word length. The resulting sum, a dithered phase or amplitude value, is truncated or rounded to the smaller, desired word length. Of course, either the amplitude or the phase or both can be dithered. In phase dithering the spurious response is determined by the type of dithering signal employed. In amplitude dithering the spurious response is determined by the original, longer word length. While the amplitude-related spurious is generally related to the phase-related spurious, we will make the pre-dither amplitude word length long enough to satisfy spur power specifications. Then the exact relationship is unimportant, and since the phase dither signal is independent of the amplitude dither signal, the amplitude and phase dithering processes can be treated independently.

The next section describes the quantizer model. Amplitude and phase quantization effects are reviewed in Sections III and IV, and simple new bounds on spurious performance are presented. In contrast to bounds in the existing literature, the new bounds are straightforward and require little information

Reprinted from *IEEE Transactions on Communications*, Vol. 43, No. 7, pp. 2254–2262, July 1995.

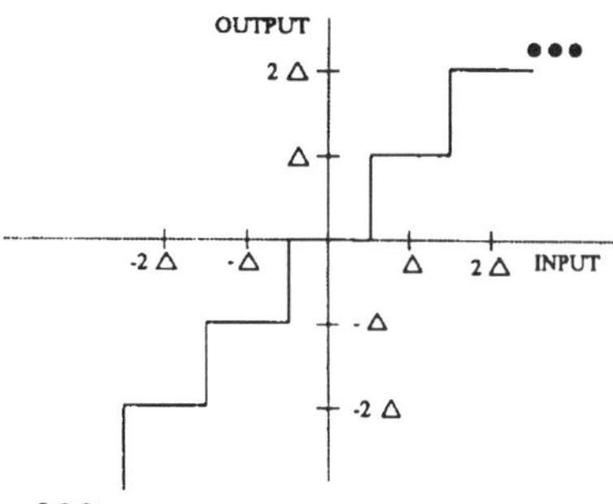

Fig. 2. Input–output relation of a midtread quantizer.

about the signal to be quantized. The derivations of the new bounds provide motivation for new analysis of dithered quantizer performance that occurs later in this paper. An analysis of dithering with a periodic noise source is presented in Section VI. The periodic noise source is considered because of its similarity to implementations involving linear feedback shift registers (LFSR's), or Pseudo-Noise (PN) generators. New analysis of phase dithering effects is presented in Sections VII and VIII, followed by simulation results and a design example.

II. QUANTIZER MODEL

When a discrete-time input signal, $x[n]$, is passed through an ideal uniform midtread quantizer [8], the output signal $y[n]$ can always be expressed as $y[n] = x[n] + e[n]$ where $e[n]$ is the quantization error, a deterministic function of $x[n]$. The input to the quantizer is mapped to one of 2^b levels where b is the number of bits which digitally represent the input sample. Output levels are separated by one quantizer step size, $\Delta = 2^{-b}$. Throughout this paper Δ_A will be used as the step size for amplitude quantization results, Δ_P will be used for phase quantization results, and Δ will be used if the result applies to both amplitude and phase quantization. Similar subscripting will be used on the quantization error.

The input/output relation of a mid-tread quantizer appears in Fig. 2. If the input does not saturate the quantizer then the quantizer error is [8]

$$e[n] = \sum_{\substack{k=-\infty \\ k \neq 0}}^{\infty} (-1)^k \frac{\Delta}{j2\pi k} \exp\left(\frac{j2\pi k x[n]}{\Delta}\right). \qquad (1)$$

Note that the above equation is not correct when the quantizer input is $\pm\Delta/2 \bmod \Delta$. This is not a problem since future analysis involving this expression will treat the input as having a piecewise continuous probability density function in which case the collection of points in question have probability measure zero. If the input signal is bounded so that $|x[n]| \leq A_Q$ where $A_Q = 1/2 - \Delta$, then the quantizer does not saturate and $|e[n]| \leq \Delta/2$. Throughout this paper, quantizers are always operating in nonsaturation mode.

III. AMPLITUDE QUANTIZATION EFFECTS

Let a discrete-time sinusoid with amplitude $A \leq A_Q$ and frequency ω_0 be the input to a mid-tread quantizer. If the sinusoid is generated in a synchronous discrete-time system, ω_0 can be expressed as 2π times the ratio of two integers. The input sequence is then periodic. Since the error sequence, $e_A[n]$, is a deterministic function of the input sequence, it is periodic as well. Therefore, the spectrum of the error sequence will consist of discrete frequency components (spurs) which contaminate the spectrum of $x[n]$.

The following argument leads to an upper bound on the size of the largest frequency component in the spectrum of $e_A[n]$. Assuming the quantizer is not saturated by the input signal $x[n]$, the maximum possible quantization error is $\Delta_A/2$ where Δ_A is the amplitude quantization step size. The total power in $e_A[n]$ is then bounded by $\Delta_A^2/4$. By Parseval's relation, the sum of the spur powers in the spectrum of $e_A[n]$ equals the power in $e_A[n]$. In order to maximize the power in a given spur, the total number of spurs must be minimized. Since $e_A[n]$ is real, the maximum power in a spur occurs when there are two frequency components at $+\omega_{\text{spur}}$ and $-\omega_{\text{spur}}$, with equal power.[1] With two frequency components the power in a single spur is $\leq \Delta_A^2/8$.

Since $x[n]$ is real, its spectrum consists of a positive and a negative frequency component, each having power $A^2/4$. Using the above bound on spur power, the Spurious-to-Signal Ratio (SpSR) is $\leq \Delta_A^2/(2A^2)$. If $A = A_Q \approx 1/2$ provided b is not small, then in deciBels with respect to the carrier (dBc), SpSR $\leq 3 - 6b$ dBc, where $\Delta_A = 2^{-b}$, and b is the wordlength in bits. In summary, this upper bound on power in a spur caused by amplitude quantization exhibits -6 dBc per bit behavior.

IV. PHASE QUANTIZATION EFFECTS

Let a phase waveform, $\phi[n]$, be the input to the mid-tread quantizer. The phase waveform, $\phi[n] = \langle fn + \Phi/2\pi \rangle$ is a sampled sawtooth with amplitude ranging from 0 to 1. The frequency to be generated, measured in samples/cycle is f and the phase, measured in radians is Φ. The fractional operator, $\langle x \rangle$, is defined so that $\langle x \rangle = x \bmod 1$, e.g., $\langle 1.3 \rangle = 0.3$. Since $\phi[n]$ is generated by a synchronous, finite-wordlength, discrete-time system, it has a finite period. The signal output from the quantizer can be expressed as $\phi[n] + e_P[n]$, where $e_P[n]$ represents the error introduced by quantization. Since $\phi[n]$ is periodic, $e_P[n]$ is periodic with a period less than or equal to the period of $\phi[n]$. After multiplication by 2π and passage through the ideal function generator, the output signal is $y[n] = A\cos(2\pi\phi[n] + 2\pi e_P[n])$. If the quantizer has many levels, i.e., >16, $e_P[n] \ll 1$, and the small angle approximation $y[n] \approx A\cos(2\pi\phi[n]) - 2\pi A e_P[n] \sin(2\pi\phi[n])$ may be used.

Since $e_P[n]$ and $\phi[n]$ are periodic, the total error $2\pi A e_P[n] \sin(2\pi\phi[n])$ is periodic. The total error power is bounded by $\pi^2 A^2 \Delta_P^2$ because $e_P[n]$ is bounded by $\Delta_P/2$ and the magnitude of a sinusoid is bounded by unity.

[1] DC offsets and half sampling rate spurs are excluded because they can be corrected by appropriate calibration and filtering.

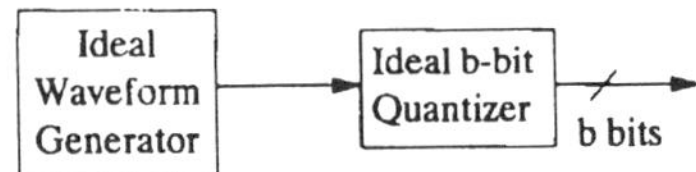

Fig. 3. Conceptual waveform generator model.

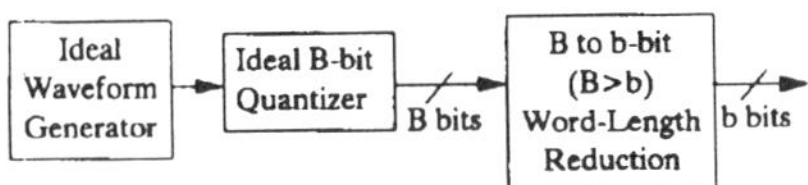

Fig. 4. Two-step waveform generator model.

Recalling the arguments in the previous section on amplitude quantization effects, the maximum spur power of the real error signal is bounded by placing the total error power into two spectral components. Therefore, the maximum spur power is $\pi^2 A^2 \Delta_P^2/2$, where $\Delta_P = 2^{-b}$ and b bits are used to represent phase samples. By the above approximation for $y[n]$ and the bound on the spur power, the Spurious-to-Signal Ratio bound is SpSR $\leq 2\pi^2 \Delta_P^2 = 13 - 6b$ dBc, independent of the signal amplitude, A. This new, simple derivation demonstrates the underlying -6 dBc per phase bit behavior, without the analytical complexity found in other existing bounds. More complicated arguments [7] improve the bound by about 9 dB.

V. AMPLITUDE DITHERING

In this section rounding the sum of an already quantized sinusoid and an appropriate dither signal is shown to cause spurious magnitudes which depend on the original (longer) wordlength, not the output (shorter) word length. This phenomenon occurs at the expense of increased system noise from the addition of the dithering signal. An important finite wordlength dithering system is subsequently shown to be equivalent to the continuous-amplitude uniformly-dithered system.

Consider the conceptual block diagram for a waveform generator shown in Fig. 3. The b-bit quantizer can be split into two parts as in Fig. 4: a high-resolution B-bit quantizer $(B > b)$ followed by truncation or rounding to b bits. Thus, the generation process consists of two separate steps: production of a high-resolution waveform and reduction of the word length. The number of bits used to represent the high-resolution samples should be sufficient to guarantee the desired spectral purity. Then, the word length should be reduced without creating excess signal-dependent quantization error.

The input in Fig. 5 is a B-bit representation of a sinusoid, $x[n] = A \sin(2\pi\phi[n]) + e_{A0}[n]$, where $e_{A0}[n]$ is the quantization error. The dither signal, $z_u[n]$, is white noise uniformly distributed in $[-\Delta_A/2, \Delta_A/2)$, where $\Delta_A = 2^{-b}$. The sum $z_u[n] + x[n]$ is rounded to retain only the b most significant bits. The rounding can be modeled as a uniform quantizer with step size Δ_A. The amplitude A is chosen to avoid saturating this quantizer when the dither signal is added, i.e., $A + \Delta_A/2 \leq A_Q$.

The output from the quantizer can be expressed as $y[n] = x[n] + z_u[n] + e_A[n]$. The characteristic function of the dither

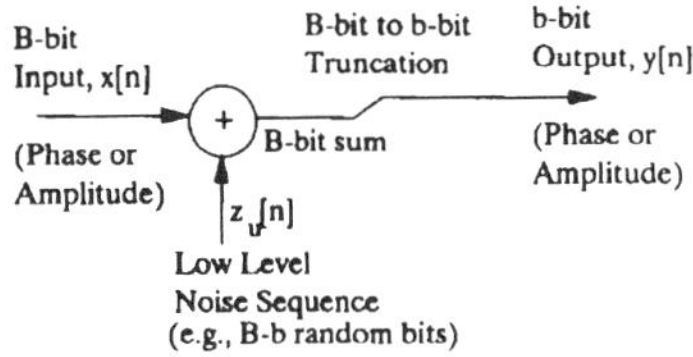

Fig. 5. Uniform dithered quantizer.

signal, $z_u[n]$, is

$$F_z(\alpha) = E\{\exp(j\alpha z[n])\} = \frac{2\sin(\alpha\Delta_A/2)}{\alpha\Delta_A}$$
$$= \text{sinc}\left(\frac{\alpha\Delta_A}{2\pi}\right) \tag{2}$$

which has zeros at nonzero integer multiples of $2\pi/\Delta_A$. Thus, as shown in [1], $e_A[n]$ will be a white, wide-sense stationary process, uniformly distributed over $[-\Delta_A/2, \Delta_A/2)$, and it will not contribute spurious harmonics to the output spectrum of $y[n]$. Any spurious components in $y[n]$ are therefore due to $e_{A0}[n]$, which are present in the B bit input $x[n]$.

It remains to comment on the noise power not isolated in discrete spurious frequency components. Gray and Stockham have shown [12] that the power in the signal $y[n] - x[n] = e_A[n] + z_u[n]$ is $\Delta^2/6$. This is approximately twice the error variance of a quantization system with no dithering signal. In summary $y[n]$ which is quantized to b bits, exhibits spurious performance as if it was quantized to B bits $(B > b)$ at the expense of doubling the white noise power.

Because the input $x[n]$ is expressed as a B bit value, an important system equivalent to continuous-amplitude, uniformly-dithered wordlength reduction can be constructed. Replace the uniformly distributed dither signal $z_u[n]$ by a finite wordlength representation of it, $z[n]$, which is said to be discretely and evenly distributed over the $(B - b)$ bit quantized values in the region $[-\Delta_A/2, \Delta_A/2]$. Heuristically, $z[n]$ randomizes the portion of the finite wordlength input, $x[n]$, that is about to be thrown away by the rounded truncation. This process is equivalent to continuous uniform dithering, since if $x[n]$ is padded out to an infinite number of bits by placing zeros beyond the least significant bit (LSB), then only the $B - b$ most significant bits of $z_u[n]$ will have an effect on the resulting sum, $x[n] + z_u[n]$. All of the bits below the most significant $B - b$ are added to zero, and cannot beget a carry. The output, $y[n]$, is identical in both systems. Therefore $z_u[n]$, continuously, uniformly distributed over $[-\Delta_A/2, \Delta_A/2)$ can be replaced by the discretely valued $z[n]$, and yield the same spurious response for $y[n]$.

It appears that the finite wordlength dither signal, $z[n]$, could be generated by a linear feedback shift register (LFSR), or PN generator. This will be strictly true only if the PN generator has an infinite period, since, at this time, the dither signal is required to be white. However, it is not surprising that ideal behavior is approached as the period of the PN generator gets longer. With a sufficiently long period, the case where spur magnitudes are limited by the original word length can be achieved. The following section gives a simple model for a system implementation using a periodic random sequence which can be approximated by a PN generator.

VI. Effect of Periodic Dither

This section analyzes the use of a periodic dither signal with a long period, L, for both amplitude and phase dithering. Since the dither signal is periodic, the discrete frequency components in its spectrum will contaminate the desired signal. It is shown that the period can be chosen to satisfy worst case spurious specifications. In this section, the case where the dither signal is generated using one uniform variate ($M = 1$) is given. When the dither signal is the sum of M independent uniform variates ($M > 1$), as in Section VIII, the analysis is the same because the resulting signal is an i.i.d. sequence of random variables.

Instead of using the white dither process, $z_u[n]$, described in the previous section, consider a substitute, $z_L[n]$, which is periodic with period L. Any two samples, $z_L[n]$ and $z_L[n + m]$, where $m \neq 0 \mod L$, are independent. Samples of $z_L[n]$ are uniformly distributed between $[-\Delta/2, \Delta/2)$, and the quantization step size is Δ.

When $z_L[n]$ is used as the dither signal, let the quantizer error be called $e_L[n]$. The autocorrelation of $z_L[n]$ when the lag m is an integer multiple of L is equal to $R_{z_L z_L}[0] = \Delta^2/12$. In the PN generator approximation to this noise source, $L = 2^l - 1$ where l is the length of the shift register in bits. At other lag values, the samples of $z_L[n]$ are independent, and since they have zero mean, the autocorrelation is zero. Therefore, expressing the autocorrelation as a discrete-time Fourier series

$$R_{z_L z_L}[m] = \frac{\Delta^2}{12}\delta[m \bmod L] = \sum_{l=0}^{L-1} \frac{\Delta^2}{12L}\exp\left(\frac{j2\pi ml}{L}\right)$$

where $\delta[m]$ is the Kronecker delta function ($\delta[0] = 1$, $\delta[m] = 0, m \neq 0$), and $z_L[n]$ contains L discrete frequency components, each with power $\Delta^2/(12L)$.

In the autocorrelation expression for $e_L[n]$, the expectation is taken over the random variables $z_L[n]$ and $z_L[n+m]$, using the definition of the autocorrelation and (1)

$$R_{e_L e_L}[n, n + m] = \sum_{\substack{k=-\infty \\ k \neq 0}}^{\infty} \sum_{\substack{l=-\infty \\ l \neq 0}}^{\infty} \alpha_k[n]\alpha_l^*[n + m]$$
$$\times E\left\{\exp\left(\frac{j2\pi}{\Delta}(kz_L[n] - lz_L[n + m])\right)\right\} \quad (3)$$

where

$$\alpha_k[n] = \frac{\Delta(-1)^k}{j2\pi k}\exp\left(\frac{j2\pi ks[n]}{\Delta}\right).$$

The desired signal to which the dither signal $z_L[n]$ is added is $s[n]$. Using the notation from earlier sections, in phase quantization, $s[n] = \phi[n]$ and in amplitude quantization $s[n] =$

$x[n]$. When the lag is not a nonzero integer multiple of L,

$$E\left\{\exp\left(\frac{j2\pi}{\Delta}(kz_L[n] - lz_L[n + m])\right)\right\}$$
$$= E\left\{\exp\left(\frac{j2\pi kz_L[n]}{\Delta}\right)\right\}$$
$$\times E\left\{\exp\left(\frac{-j2\pi lz_L[n + m]}{\Delta}\right)\right\}$$
$$= F_z\left(\frac{2\pi k}{\Delta}\right)F_z\left(\frac{-2\pi l}{\Delta}\right) = \delta[k]\delta[l].$$

This last fact is true because the characteristic function of $z_L[n]$ has zeros at all nonzero integer multiples of $2\pi/\Delta$ (2). But since k and l never assume the value 0 in (3), the autocorrelation function is zero when the lag is not 0 mod L. When the lag is 0 mod L

$$E\left\{\exp\left(\frac{j2\pi}{\Delta}(kz_L[n] - lz_L[n + m])\right)\right\}$$
$$= E\left\{\exp\left(\frac{j2\pi(k - l)z_L[n]}{\Delta}\right)\right\} = \delta[k - l].$$

This results in

$$R_{e_L e_L}[n, n + m]$$
$$= \frac{\Delta^2}{2\pi^2}\sum_{k=1}^{\infty} \frac{1}{k^2}\cos\left(\frac{2\pi k}{\Delta}(s[n] - s[n + m])\right). \quad (4)$$

Setting $m = 0$ in (4) and evaluating the resulting summation [9, p. 7] yields the power in $e_L[n]$: $R_{e_L e_L}[n, n] = \Delta^2/12$. From (4), $e_L[n]$ is a cyclostationary process because $s[n]$ has a finite period, N. Using the results of Ljung [10], spectral information is obtained when (4) is averaged over time. Note that when the lag m is not only an integer multiple of L, the period of the dither, but also an integer multiple of N, the autocorrelation function equals $\Delta^2/12$, independently of n. The smallest nonzero lag that satisfies these two conditions is the least common multiple of L and N, denoted by qL where q is an integer. Therefore, the period of the time-averaged autocorrelation function, $\bar{R}_{e_L}[m] = \text{Avg}_n(R_{e_L e_L}[n, n + m])$, is at least L and at most qL. Let the period equal cL, where c is an integer, $1 \leq c \leq q$. The function $\bar{R}_{e_L}[m]$ can be expressed as a sum of cL weighted complex exponentials

$$\bar{R}_{e_L}[m] = \sum_{l=0}^{cL-1} p_l\exp\left(\frac{j2\pi ml}{cL}\right),$$
$$m = \ldots, -1, 0, 1, 2, \ldots$$

where

$$p_l = \frac{1}{cL}\sum_{m=0}^{cL-1} \bar{R}_{e_L}[m]\exp\left(\frac{j2\pi ml}{cL}\right)$$
$$= \frac{1}{cL}\sum_{n=0}^{c} \bar{R}_{e_L}[nL]\exp\left(\frac{j2\pi nl}{c}\right).$$

The last equality is true since the autocorrelation function in (3) and its time-average, $\bar{R}_{e_L}[m]$ are zero for lags not equal to integer multiples of L. The weights, p_l, $l = 0, 1, \ldots cL - 1$, are the power magnitudes of the spurs. Since $\bar{R}_{e_L}[m] \leq$

$\Delta^2/12$, the spur power can be bounded: $p_l \leq \Delta^2/(12cL) \leq \Delta^2/(12L)$. Equality is achieved when the period of the time-averaged autocorrelation function is exactly L, the period of the dither.

As $L \to \infty$, the spacing between spurs goes to zero in the spectra of both $e_L[n]$ and $z_L[n]$. The power in an individual spur goes to zero, but the density (power per unit of frequency) tends to a constant. Thus, ideal white noise behavior is approached. While $z_L[n]$ and $e_L[n]$ are correlated in general, the worst case spur power scenario coherently adds the power spectra from both processes. For this reason, L should be chosen to satisfy $\Delta^2/(6L) < P_{\max}$, where $P_{\max}$ is the maximum acceptable spur power. When constructing a dither signal as the sum of $M \geq 1$ independent, uniform variates the noise autocorrelation becomes $R_{z_L z_L}[m] = (M\Delta^2/12)\delta[m \bmod L]$. The analysis follows closely to that for $M = 1$, and L should be chosen to satisfy $(M + 1)\Delta^2/(12L) < P_{\max}$.

As in the previous section, since the desired signal has finite word length, it is equivalent to round or truncate the dither signal to an appropriate word length. An implementation using a PN generator is an approximation to such a truncated periodic noise source which produces a periodic sequence of discretely and evenly distributed random numbers.

VII. PHASE DITHERING

In this section, phase dithering is analyzed using a continuous, zero-mean, wide-sense stationary sequence. As described in Section V on amplitude dithering, an evenly distributed discrete random sequence is equivalent to continuous uniform dithering when the initial phase word is quantized to a finite number of bits.

Let the digital sinusoid to be generated be

$$x[n] = \cos(2\pi(\phi[n] + \epsilon[n])) \tag{5}$$

so that the desired phase is $\phi[n]$ as defined in Section IV. The total quantization noise is $\epsilon[n] = e_P[n] + z[n]$, the sum of the dither signal and the quantizer error. Using small angle approximations

$$x[n] = \cos(2\pi fn + \Phi) - 2\pi\epsilon[n]\sin(2\pi fn + \Phi) + O((\max(\epsilon[n]))^2).$$

The total quantization noise will be examined by considering the first two terms above, and then the second-order, $O([\max(\epsilon[n])]^2)$, effect.

A. First-Order Analysis

Since the quantization error after dithering is independent of the input signal [2] $\epsilon[n]$ is uncorrelated with the desired sinusoids. Without loss of generality, and for ease of notation, let us shift the uniformly distributed dither random variate range to $[0, \Delta_P)$. The total phase quantization noise $\epsilon[n]$ will be $\epsilon[n] = -p[n]\Delta_P$ with probability $(1 - p[n])$, and $\epsilon[n] = (1 - p[n])\Delta_P$ with probability $p[n]$. The value $p[n]$ is the distance from the initial high-precision phase value, $\phi[n]$, to the nearest greater quantized value normalized by the

phase quantization step size Δ_P. The value of the probability sequence $p[n]$ varies periodically, since $p[n] = \phi[n] \bmod \Delta_P$, and $\phi[n]$ is periodic; however, at all sample times n the first moment of the total phase quantization noise, $E\{\epsilon[n]\}$, is zero.

Information about the spurs and noise in the power spectrum of $x[n]$ is obtained from the autocorrelation function. The autocorrelation of $x[n]$ is

$$\begin{aligned}
E\{x[n]x[n+m]\} &= \cos(2\pi fn + \Phi)\cos(2\pi f(n+m) + \Phi) \\
&\quad + 4\pi^2 \sin(2\pi fn + \Phi) \\
&\quad \times \sin(2\pi f(n+m) + \Phi) \\
&\quad \times E\{\epsilon[n]\epsilon[n+m]\} + O(\Delta_P^4).
\end{aligned}$$

Spectral information is obtained by averaging over time [10], resulting in

$$\bar{R}_{xx}[m] \approx \frac{1}{2}[1 + 4\pi^2 \bar{R}_{\epsilon\epsilon}[m]]\cos(2\pi fm)$$

where $\bar{R}_{\epsilon\epsilon}[m] = \text{Avg}_n(E\{\epsilon[n]\epsilon[n+m]\})$, the time-averaged autocorrelation of the total quantization noise.

The power spectrum of $x[n]$, the Fourier transform of the autocorrelation, is the power spectrum of the desired sinusoid of frequency f plus the total quantization noise amplitude modulated on the desired sinusoid. Note that since $\bar{R}_{\epsilon\epsilon}[m] = O(\Delta_P^2)$, and $\Delta_P \ll 1$, the modulation index is small.

To a first order, the AM signal produced by phase dithering is clear of spurious harmonics down to the level due to periodicities in the dither sequence. The next section will examine spur performance in more detail, but first it is important to consider the noise power spectral density resulting from the phase dithering process.

Recall that for any fixed time n, the probability distribution of $\epsilon[n]$, a function of $p[n]$, is determined by the input signal, but the outcome of $\epsilon[n]$ is determined entirely by the outcome of the dither signal $z[n]$. When $z[n]$ and $z[n+m]$ are independent random variables for nonzero lag m, $\epsilon[n]$ and $\epsilon[n+m]$ are also independent for $m \neq 0$, and hence $\epsilon[n]$ is spectrally white. In this case, the autocorrelation becomes

$$\bar{R}_{xx}[m] \approx \frac{1}{2}\cos(2\pi fm) + 2\pi^2\delta[m]\text{Var}(\epsilon)$$

where $\text{Var}(\epsilon)$ is the time-averaged variance of the total quantization noise. The resulting signal-to-noise ratio (SNR) is approximately $1/(4\pi^2\text{Var}(\epsilon))$.

When the dither signal is constructed from one uniform $[-\Delta_P/2, \Delta_P/2)$ random variate, the error $\epsilon[n]$ is bounded between $-\Delta_P$ to Δ_P where $\Delta_P = 2^{-b}$ and b is the number of bits in the phase representation after the wordlength is reduced. The number of bits b must be large enough to satisfy the small angle assumption earlier, e.g., $b \geq 4$. The time-averaged variance of $\epsilon[n]$ is less than or equal to $2^{-2b}/4$, and the SNR is $2^{2b}/\pi^2 = 6.02b - 9.94$ dB.

Since the sinusoid generated is a real signal, the signal power in the SNR will be equally divided between positive and negative frequency components. If the sinusoid is the result of a discrete-time random process with sampling frequency f_s, then the resulting noise power spectral density (NPSD) will

TABLE I
NOISE POWER SPECTRAL DENSITIES FOR 160 MHz SAMPLING RATE

b (bits/cycle)	Noise Power Spectral Density
5	-102.20 dBc/Hz
6	-108.22 dBc/Hz
7	-114.24 dBc/Hz
8	-120.26 dBc/Hz
9	-126.28 dBc/Hz
10	-132.30 dBc/Hz
11	-138.32 dBc/Hz
12	-144.35 dBc/Hz

be given by

$$\text{NPSD} \approx -[\text{SNR}/2 + 10\log_{10}(f_s/2)] \text{ dBc/Hz}$$
$$\leq 6.93 - 10\log_{10}(f_s/2) - 6.02b \text{ dBc/Hz}. \quad (6)$$

Table I gives noise power spectral densities as a function of the number of bits per cycle b at a 160 MHz sampling rate, calculated according to the above formula.

B. Second-Order Analysis: Residual Spurs

For a worst case analysis of second order effects, expand the initial cosine from (5) by the sum of angles formula

$$x[n] = \cos(2\pi\epsilon[n])\cos(2\pi fn + \Phi)$$
$$- \sin(2\pi\epsilon[n])\sin(2\pi fn + \Phi).$$

Information about the spurs in the power spectrum of $x[n]$ is obtained from the autocorrelation function at nonzero lags. When the dither sequence $z[n]$ is a sequence of i.i.d. variates, the autocorrelation function for $x[n]$, with lag m not equal to zero, is

$$R_{xx}[n, n+m] = E\{x[n]x[n+m]\}$$
$$= E\{x[n]\}E\{x[n+m]\}.$$

The expected value of $x[n]$ is a deterministic function of time. From the above expression, it follows that spectral information about the random process $x[n]$, with the exception of noise floor information, is contained in $E\{x[n]\}$, which we call the "expected waveform." Since $\epsilon[n]$ is zero mean at all sample times the expected waveform reduces to

$$E\{x[n]\} = (1 - 2\pi^2 E\{\epsilon^2[n]\})\cos(2\pi fn + \Phi) + O(\Delta_P^3).$$

The form of the expected waveform clearly shows that the spurious content of the signal will be derived from the dependence of second and higher order moments of the quantization noise. It is this fundamental principle that will ultimately lead to the -12 dBc per phase bit behavior for uniformly phase-dithered sinusoid generation.

It remains to consider the second moment of the total phase quantization noise, $E\{\epsilon^2[n]\}$, which we evaluate by using the probability sequence, $p[n]$, from the previous section as $E\{\epsilon^2[n]\} = \Delta_P^2(p[n] - p^2[n])$. As shown in the appendix, in the worst case this sequence produces a spur whose frequency is the reflection on the desired signal across 1/4 the sampling

rate, and this worst case is achieved for a large class of frequencies. The expected waveform is

$$E\{x[n]\} = ((1 - \pi^2\Delta_P^2/4) + (\pi^2\Delta_P^2/4)\cos(\pi n))$$
$$\times \cos(2\pi fn + \Phi) + O(\Delta_P^3)$$
$$= (1 - \pi^2\Delta_P^2/4)\cos(2\pi fn + \Phi) + (\pi^2\Delta_P^2/4)$$
$$\times \cos((2\pi f + \pi)n + \Phi) + O(\Delta_P^3),$$

clearly showing the desired signal and spur components. Thus, dropping the $O(\Delta_P^3)$ term, a -18 dB per bit power behavior, the worst-case spur level relative to the desired signal after truncating to b bits is (SpSR)

$$\text{SpSR} \approx \frac{\pi^4\Delta_P^4/16}{(1 - \pi^2\Delta_P^2/4)^2} \approx \frac{\pi^4\Delta_P^4}{16} = 7.84 - 12.04b \text{ dBc.}$$

In summary, if b bits of phase are output to a look-up table, and B bits of phase ($B > b$) are used prior to truncation, then the addition of an appropriate dithering signal using ($B - b$) bits will allow the word length reduction without introducing spurs governed by the usual $-6b$ dBc behavior. If a single random variate is added as a dither signal (first-order dithering), the spur suppression is accelerated to 12 dB per bit of phase representation. Since the table size is only effected linearly by the number of bits in a table entry, rather than exponentially as it is by the number of phase bits, the amplitude word length is of secondary importance to the phase wordlength, especially in all-digital systems. For example, -90 dBc spur performance would nominally require $b = 16$ bits of phase and a 65,536 entry table. With first-order dithering, this level of performance requires only $b \geq 8.1$ bits of phase per cycle in the look-up table addressing. Worst case spur performance of -100.5 dBc is achieved with 9 bits, a 512 entry table at most, and, at a 160 MHz sampling rate, Table I shows that with these realistic system parameters, the noise power spectral density is at a low -126 dBc/Hz.

VIII. ACCELERATED SPUR SUPPRESSION

Further analysis [11] based on an extension of results by Gray [12] indicates that the phase spur suppression rate can be increased in steps of 6 dBc per bit by adding multiple uniform random deviates to the phase value prior to truncation. The addition of M uniform random deviates produces a dither signal with Mth-order zeros in its characteristic function, thus making the Mth moment of the quantization error independent of the input sequence [12].

An example of this technique providing 18 dBc per phase bit spur performance is shown in Fig. 6. This technique involves adding two ($B - b$)-bit uniform deviates to produce a ($B - b + 1$) bit dither signal, which achieves the accelerated spur reduction due to second-order zeros in the dither characteristic function. Simulation results for when two uniform variates are added to the phase are presented in the next section. A straightforward extension of this technique to polynomial series allows spur-reduced synthesis of periodic digital signals with arbitrary waveforms.

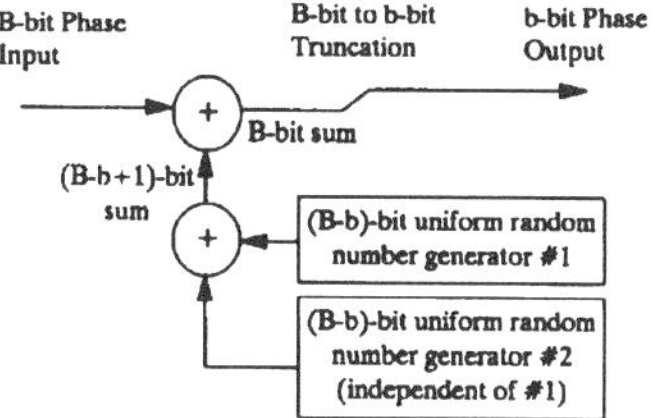

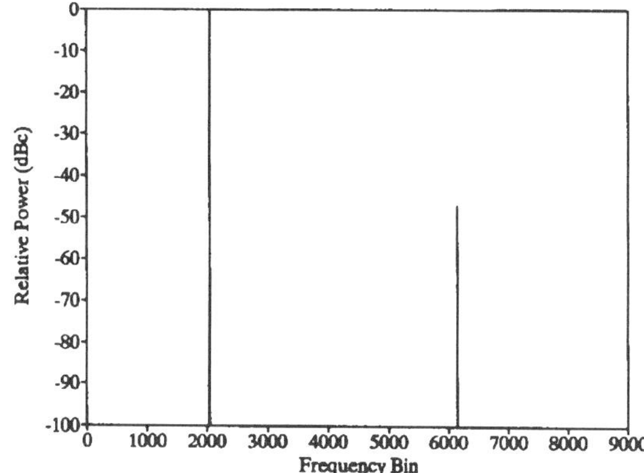

Fig. 6.   System for 18 dBc per phase bit spur reduction.

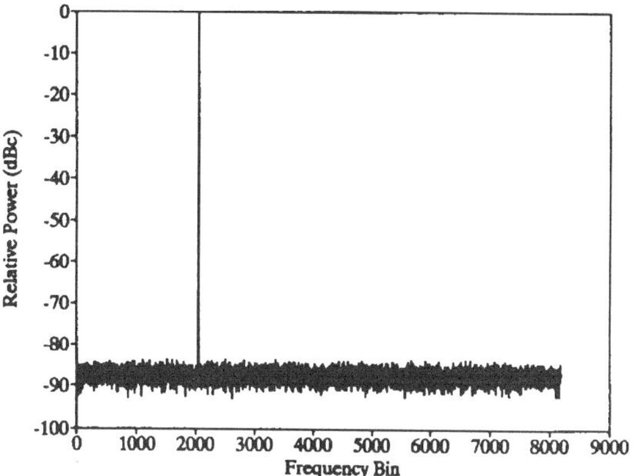

Fig. 8.   Power spectrum of 8 sample/cycle sine wave with amplitude dithering (8 bit amplitude quantization).

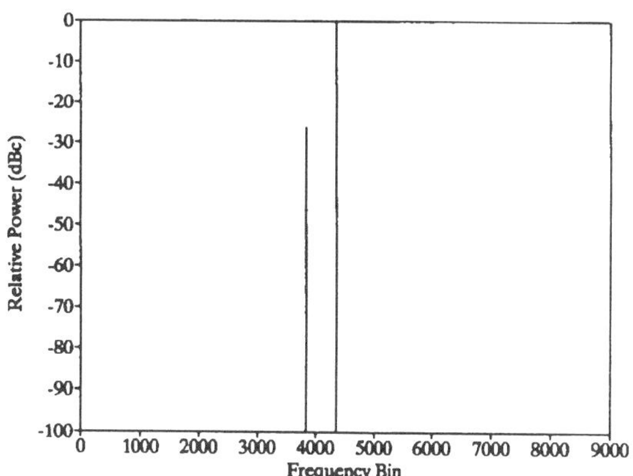

Fig. 9.   Power spectrum of 5 bit phase-truncated sine wave without phase dithering (high-precision amplitude).

Fig. 7.   Power spectrum of 8 sample/cycle sine wave without dithering (8 bit amplitude quantization).

IX. SIMULATION RESULTS

Simulations were performed to validate the results of this analysis. These results were obtained using 8192 point unwindowed FFT's, and the synthesized frequencies were chosen to represent worst-case amplitude and phase spur performance. In each of the figures, ten power spectra were averaged to better show the spurious content of the signals. Fig. 7 shows the power spectrum of a sine wave of one-eighth the sampling frequency truncated to 8 bits of amplitude without dithering. Fig. 8 shows the same spectrum with a 16 bit sinusoid amplitude dithered with one uniform variate prior to truncation to 8 bits. Note that the spurs have been eliminated to the levels consistent with those imposed by the initial sixteen bit quantization.

Fig. 9 shows the spectrum of a 5 bit phase-truncated sinusoid with high-precision amplitude values. A worst case example of first-order phase dithering is shown in Fig. 10. The measured noise power spectral density in Fig. 10 is -62.3 dBc per FFT bin, giving a noise density of $-23.2 - 10\log(f_s/2)$ dBc, in agreement with the upper bound derived in (6). The spur level is -52.3 dBc in the first-order dithered Fig. 10.

Fig. 11 shows the same example using second-order ($M = 2$) dithering using the sum of two uniform deviates. While the spectrum in Fig. 10 shows the residual spurs at -12 dBc per bit due to second-order effects, Fig. 11 shows no visible spurs, indicating better than -63 dBc spurious performance. Additional simulations involving Megapoint FFT's and not represented by figures confirm the -18 dBc per bit performance of the second-order phase-dithered system.

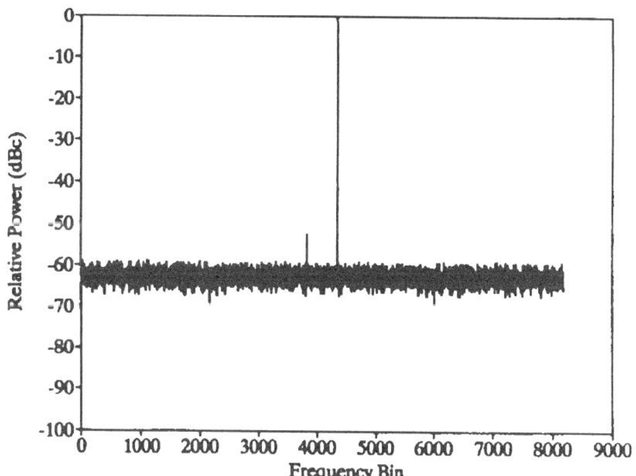

Fig. 10.   Power spectrum of 5 bit phase-truncated sine wave with first-order phase dithering (high-precision amplitude).

Finally, Fig. 12 shows a worst-case result for first-order phase dithering together with first-order amplitude dithering. The amplitude samples are truncated to 8 bits, as are the phase samples. Note that the spurs are not visible in the spectrum;

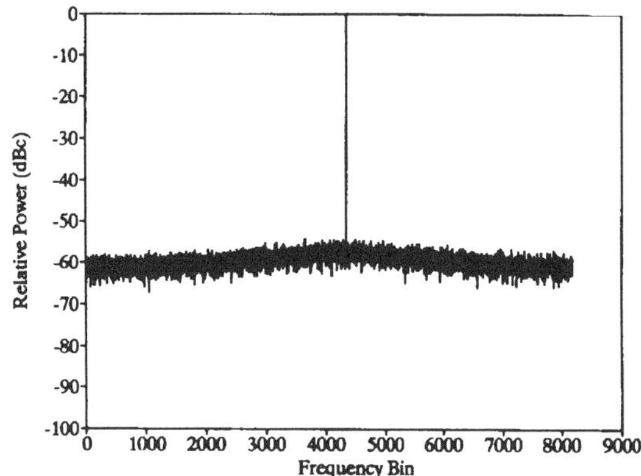

Fig. 11. Power spectrum of 5 bit phase-truncated sine wave with second-order phase dithering (high-precision amplitude).

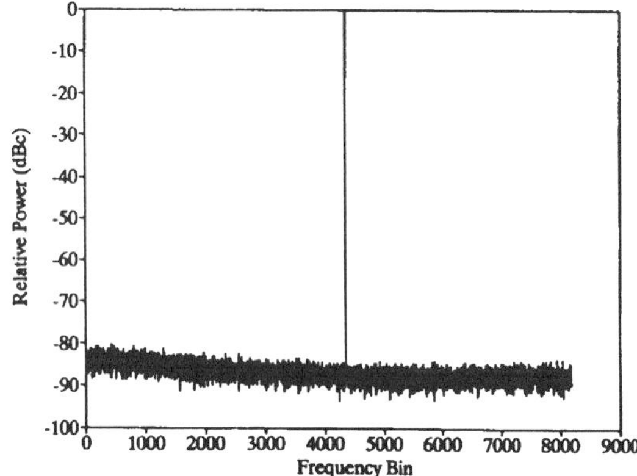

Fig. 12. Worst case power spectrum of sinusoid with first-order phase dithering and amplitude dithering (8 bits each).

however, close analysis has demonstrated that they are present at the -88 dBc level expected due to second-order effects.

X. A System Design Example

The block diagram of a direct digital frequency synthesizer based on the techniques presented here is shown in Fig. 13. The following system would perform at a sampling rate of 160 MHz, producing 8 bit digital sinusoids spur-free to -90 dBc with better than -120 dBc/Hz noise power spectral density. The system parameters are as follows

Phase bits are in unsigned fractional cycle representation with:

phase accumulator wordlength determined by frequency resolution, and ≥ 16 bits prior to addition of 1 uniform phase dither variate, with ≥ 9 bits after dither addition and truncation;

Amplitude look-up-table with:

$\geq 2^7 = 128$ entries (using quadrant symmetries) of ≥ 16 b each normalized so that the sinusoid amplitude equals

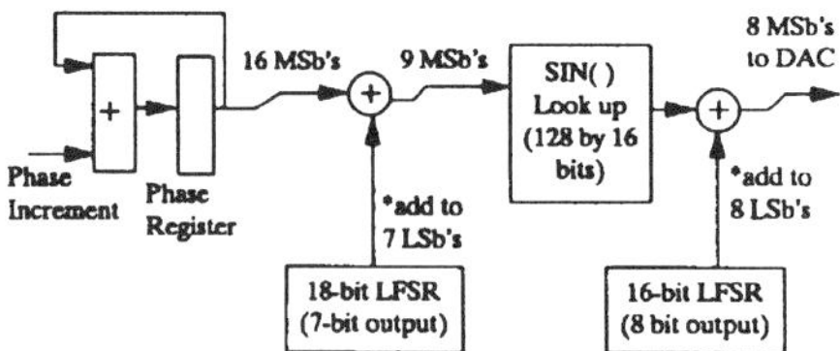

Fig. 13. Block diagram of spur-reduced direct digital frequency synthesizer.

512 16 bit quantization steps less than the full-scale value

Linear feedback shift register PN generator with ≥ 16 lags producing one 8 bit amplitude dither variate, and
One LFSR PN generator with ≥ 18 lags for generation of the 7 bit phase dither variate.

XI. Conclusion

A digital dithering approach to spur reduction in the generation of digital sinusoids has been presented. A class of periodic dithering signals has been analyzed because of its similarity to LFSR PN generators.

The advantage gained in amplitude dithering provides for spur performance at the original longer word length in an ideal system when the digital dithering signal is white noise distributed evenly, not uniformly, over one quantization interval. The reduced word length allows the use of less complicated multipliers and narrower data paths in purely digital applications. If the waveform is ultimately converted to an analog value, the reduced word length allows the use of fast, coarse-resolution, highly linear digital-to-analog converters (DAC's) to obtain sinusoids or other periodic waveforms whose spectral purity is limited by the DAC linearity, not its resolution. These results suggest that coarsely quantized, highly-linear techniques for digital-to-analog conversion such as delta-sigma modulation would be useful in direct digital frequency synthesis of analog waveforms.

The advantage gained in the proposed method of phase dithering provides for an acceleration beyond the normal 6 dB per bit spur reduction to a $6(M + 1)$ dB per bit spur reduction when the dithering signal consists of M uniform variates. Often the most convenient way to generate a periodic waveform is by table look-up with a phase index. Since the size of a look-up table is exponentially related to the number of phase bits, this can provide a dramatic reduction in the complexity of NCO's, frequency synthesizers, and other periodic waveform generators.

The advantages of dithering come at the expense of an increased noise content in the resulting waveform. However, the noise energy is spread throughout the sampling bandwidth. In high bandwidth applications, dithering imposes modest system degradation. It has been shown that high performance synthesizers with dramatically reduced complexity can be designed using the dithering method, without resulting in high noise power spectral density levels.

APPENDIX
WORST CASE PHASE SPUR ANALYSIS

Since the sequence $p[n]$ is bounded between 0 and 1, the function $u[n] = p[n] - p^2[n]$ is bounded between 0 and $1/4$, with its maximum value of $1/4$ at $p[n] = 1/2$. Therefore, $u[n]$ must have some nonzero DC (average) component. Any remaining components can be periodic in the worst case. Since all nonlinear operations have been performed, conservation of power (energy) arguments can be used to determine the total non-DC error power. The total power in the DC component of $u[n]$ is equal to the square of the average value of $u[n]$. Similarly, the total power in $u[n]$ is equal to the average value of $u^2[n]$. Thus, the power remaining for time-varying components of $u[n]$ is

$$\mathrm{Avg}(u^2[n]) - (\mathrm{Avg}(u[n]))^2 = \mathrm{Avg}((u[n] - \mathrm{Avg}(u[n]))^2).$$

This value is maximized by maximizing the dispersion of the samples about the mean. When the sample values are bounded, this maximization is achieved by placing half of the samples at each bound, so that the mean is equidistant from each bound. Since $0 \le u[n] \le 1/4$, the maximum power present in harmonic components is $1/64$.

To find a sequence achieving this bound, the sequence $E\{\epsilon^2[n]\}$ must be examined in more detail. The previous paragraph indicates that the bound will be achieved when half the samples of $E\{\epsilon^2[n]\}$ are 0 and half are $\Delta^2/4$. Since $\epsilon^2[n]$ is nonnegative, $E\{\epsilon^2[n]\} = 0$ implies that $\epsilon[n] = 0$. Note that the difference between $\epsilon[n]$ and $\epsilon[n + 1]$ is the phase increment modulo the quantization step size. If, for any n and $n + 1$, $\epsilon[n] = \epsilon[n + 1] = 0$, the phase increment can be exactly expressed in the new quantization step. By induction, $\epsilon[n]$ will be zero for all n if any two adjacent values $E\{\epsilon^2[n]\}$ and $E\{\epsilon^2[n + 1]\}$ are both zero. The only possible sequence $E\{\epsilon^2[n]\}$ achieving the worst case is therefore $0, 1/4, 0, 1/4, 0, 1/4 \ldots$, and thus, $u[n] = 1/8 - (1/8)\cos(\pi n)$. This sequence has a single sinusoidal component at the Nyquist frequency, half the sampling rate.

REFERENCES

[1] A. B. Sripad and D. L. Snyder, "A necessary and sufficient condition for quantization errors to be uniform and white," *IEEE Trans. Acoust. Speech and Signal Processing*, vol. ASSP-25, pp. 442–448, Oct. 1977.
[2] L. Schuchman, "Dither signals and their effects on quantization noise," *IEEE Trans. Commun. Technol.*, vol. COM-12, pp. 162–165, Dec. 1964.
[3] N. S. Jayant and L. R. Rabiner, "The application of dither to the quantization of speech signals," *Bell Syst. Tech. J.*, vol. 51, pp. 1293–1304, 1972.
[4] S. C. Jasper, "Frequency resolution in a digital oscillator," U.S. Pat. 4 652 832, Mar. 24. 1987.
[5] P. O'Leary and F. Maloberti, "A direct-digital synthesizer with improved spectral performance," *IEEE Trans. Commun.*, vol. 39, pp. 1046–1048, July 1991.
[6] J. Tierney, C. Rader, and B. Gold, "A digital frequency synthesizer," *IEEE Trans. Audio and Electroacoust.*, vol. AU-19, pp. 48–57, Mar. 1971.
[7] T. Nicholas and H. Samueli, "An analysis of the output spectrum of direct digital frequency synthesizers in the presence of phase accumulator truncation," in *41st Annu. Frequency Contr. Symp.*, USERACOM, Ft. Monmouth, NJ, 1987, pp. 495–502.
[8] L. E. Brennan and I. S. Reed, "Quantization noise in digital moving target indication systems," *IEEE Trans. Aerospace and Electron. Syst.*, vol. AES-2, pp. 655–658, Nov. 1966.
[9] I. S. Gradsteyn and I. M. Rizhk, *Table of Integrals, Series, and Products*. New York: Academic, 1980 (corrected and enlarged edition).
[10] L. Ljung, *System Identification: Theory for the User.* Englewood Cliffs, NJ: Prentice-Hall, 1987.
[11] M. J. Flanagan and G. A. Zimmerman, "Spur-reduced digital sinusoid generation using higher-order phase dithering," in *Conf. Rec.: 27th Annu. Asilomar Conf. Signals, Syst. and Comput.*, Pacific Grove, CA, IEEE Computer Society Press, Nov. 1993, pp. 826–830.
[12] R. M. Gray and T. G. Stockham, Jr., "Dithered quantizers," *IEEE Trans. Inform. Theory*, vol. 39, pp. 805–812, May 1993.

Spectral Properties of DDFS:
Computer Simulations
and Experimental Verifications

VĚNCESLAV F. KROUPA

Abstract: **In this paper we discuss the generation of spurious signals with computer simulations in direct digital frequency synthesizers (DDFSs) with different output waves.**

First, we deal with simple square-wave and triangular-wave outputs. Our main attention, however, is directed to the systems generating sine signals since they exhibit the lowest level attainable.

The major sources of spurious signals are phase truncation, finite wordlength in sine lookup tables, resolution of DAC, and intermodulation problems. The intermodulation problems are most difficult to appreciate.

I. INTRODUCTION

THE task of DDFS is to generate a desired output frequency f_x from the input standard or clock frequency f_i, generally, with the assistance of Modulo-N arithmetics. The governing equation of this system is

$$f_x = \frac{k}{N} * f_i \qquad [1]$$

where the denominator N is generally a power of two, that is,

$$N = 2^R \qquad [2]$$

For the examination of the DDFS properties, we must start from the normalized frequency

$$\xi_x = \frac{f_x}{f_i} = \frac{k}{2^R} = \frac{X}{2^L} = X/Y \quad L \le R \qquad [3]$$

where R is number of bits stored in the accumulator and X and Y are relatively prime integers (in instances where k is even, the effective accumulator bits are reduced to L). As a consequence, X is always odd, and Y is a power of two.

In recent years, we have studied the behavior of DDFS in respect to spectral purity [1 to 5]. First, we examined the simple system with the rectangular output waves. Then we investigated devices providing triangular output waves. However, most of our work was dedicated to sinusoidal frequency synthesizers.

Since the published data about spectral purity are often incomplete and contain too many spurious spectral lines, the origin of which is difficult to trace, we start with computer simulations.

II. SQUARE-WAVE OUTPUTS

Pulse and rectangular output waves are plagued with very large spurious signals [1,2,3] and are used only when we need to synthesize an exact value of the output frequency, for example, the PAL color subcarrier in [1]. However, there are also some special communication applications that require square-wave signals [6,7,8].

A. Quasiperiodic Omission of Pulses

In order to enhance the fundamental harmonic in the output signal, the rectangular output wave should have the space-mark ratio as close to 1:1 as possible. This requirement can be met by different means, the simplest of which is to put a flip-flop or binary divider at the end of the synthesizing chain [1,2].

For one single-output frequency, the author suggested as mathematical models:

Continued fraction expansion

Modified Engel series expansion

Combinations of both in some instances

Gear box approach (which provides output frequencies with very small errors [9]).

In a more general case we can use

Modulo-N arithmetics.

The advantage is the possibility of applying commercial IC chips and referring simply to the lows and highs of the MSB [7,8]. The disadvantage is a slow convergence of the binary systematic fractions in respect to the desired normalized frequency ξ_x.

All these theorems for approximating real numbers are discussed in detail by the editor in the introductory paper, and the shortened version of the continued fractions is in [1].

B. Spurious Signals

In the useful range of DDFS, extending ideally from zero to the Nyquist frequency, $f_c/2$, we encounter many discrete spurious signals that can be divided into several groups.

1. Harmonics of the Output Signal. Particularly when the normalized frequency from ξ_x is small compared with the

Nyquist "frequency" $Y/2$, we must be careful with rectangular wave harmonics (which are ideally only odd ones, though large since they decay inversely with their order). See Fig. 1.

2. *Harmonics of the Device Fundamental Frequency.* Most of the spurious signals in DDFS are harmonics of the device fundamental frequency

$$f_{\text{fund}} = \frac{1}{YT_c} \qquad [4]$$

However, it is generally easy to estimate power level and the frequency of the larger ones. The problem was discussed in depth by the author [1,2]. Here we will recall the results. The spurious phase-time modulation is given by

$$s(t_r) = T_c \left[r\frac{Y}{2X} - \text{integer}\left(r\frac{Y}{2X}\right) \right]$$

$$= T_c \left[r\left(A_o + \frac{a_1}{B_1} - \frac{a_1 a_2}{B_1 B_2} + \cdots \right) \right.$$

$$\left. - \text{integer}\left(r\frac{Y}{2X}\right) \right] \qquad [5]$$

where the series expansion of $Y/2X$ is based on the shortened continued fraction theorem [1]. At its first-order approximation by A_1/B_1, r in eq. (5) changes from 1 to B_1 ($r = 1, 2, \ldots B_1$) and we arrive at a sawtooth wave with the period $T_c A_1$ and the phase-time amplitude $1T_c$. The second-order approximation reveals an additional sawtooth wave with amplitude T_c/B_1 ($t = 1, 2, \ldots B_2$) and the new modulation period $T_c A_2$ which is also imposed on the first sawtooth wave. An example of such a modulation function is shown in Fig. 2 (see also Example 1).

To illustrate the problem, we will investigate the fundamental harmonic of the rectangular wave phase modulated by $\omega_x s(t_r)$, that is,

$$z(t) = 2\frac{A}{\pi} \sin\left[\omega_x t + \omega_x s(t_r)\right] \qquad [6]$$

which simplifies for $s(t_r)$ approximated with the largest sawtooth wave only to

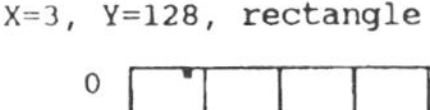

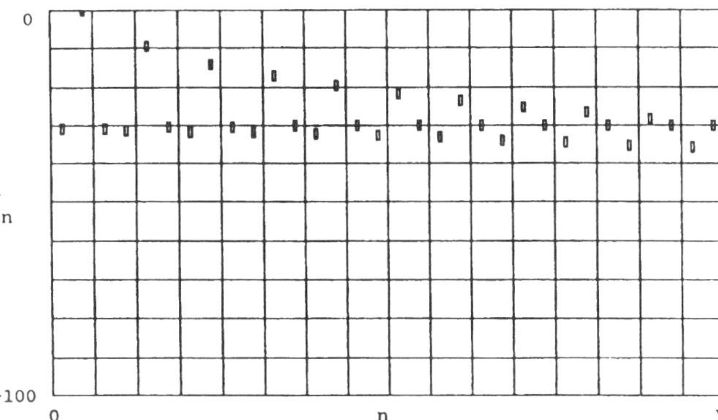

Fig. 1. Spurious signals in the rectangular wave with the normalized frequency $\xi_x = 3/128$.

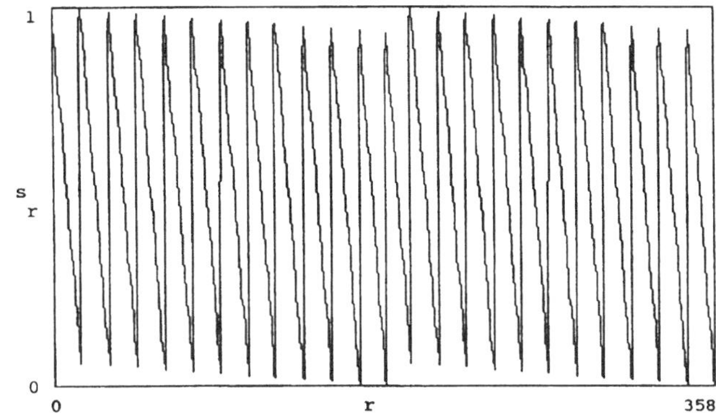

Fig. 2. Spurious phase-time modulation $s(t_r)$ for the normalized frequency $\xi_x = 179/10000$.

$$z'(t) = A \cos\left[\omega_x t + 2\frac{X}{YT_c}T_c \sum_{m=1,2,\ldots} \frac{(-1)}{m} \sin(m\nu t) \right] \qquad [7]$$

where the frequency of the sawtooth wave is the fundamental modulation

$$\nu = \frac{2\pi}{A_1 T_c} \qquad [8]$$

and amplitudes of individual harmonics are

$$a_m = \frac{2X}{mY} \qquad [9]$$

With the assistance of the theory of the phase or frequency-modulated signals we get for the level of the first spurious components

$$a_{1,1} \approx \frac{J_1(a_1)[J_o(a_2) \pm J_1(a_2)]}{J_o(a_1)J_o(a_2)} \qquad [10]$$

where the minus sign is for the lower and the plus sign for the upper side bands.

For Bessel functions with small indexes, the above equation simplifies to

$$a_{1,m} \approx J_1(a_m) \approx \frac{a_m}{2} \approx \frac{X}{Y} \qquad [11]$$

For the amplitudes of the next pair of spurious signals, we find

$$a_{2,m} \approx \frac{a_{1,m}}{4} \qquad [12]$$

and so on.

After introducing relation (9) into eq. (11), we get for the spurious signal levels "near" the carrier (in dB measure) for the first-order approximation, that is, for $r_{1,\max}$ in (5) being multiples of B_1,

$$a_{1,m_1} \approx 20\log\left(\frac{X}{m_1 Y}\right); \quad (m_1 = 1, 2, \ldots) \qquad [13]$$

with the modulation frequency

$$f_{m_1} \approx \frac{fc}{A_1} \qquad [14]$$

Relation (13) is often cited in the literature, but without warning that it gives useful results for very few m_1's.

For the second-order approximations, that is, for $r_{2,\max}$ being multiples of B_2, we have

$$a_{2,m_2} \approx 20 \log \left(\frac{X}{m_2 B_1 Y} \right); \quad (m_2 = 1, 2, \ldots) \qquad [15]$$

with the modulation frequency

$$f_{m_2} \approx \frac{f_c}{A_2} \qquad [16]$$

We can proceed in the same way until we arrive at the spurious signals closest to the carrier, and with the smallest modulation frequency

$$f_{m_n} = \frac{f_c}{A_n} = \frac{f_c}{Y} \qquad [17]$$

and amplitudes

$$a_{n,m_n} \approx 20 \log \left(\frac{X}{Y} \frac{1}{m_{n-1} B_{n-1}} \right) \qquad [18]$$

Note that these approximations pose some difficulties: relations for the level of the spurious signals are valid only for small values of m_1, m_2, and so on. Furthermore, they supply the better results the smaller is the ratio X/Y. However, in these instances we must be careful to remove harmonics of the output frequency f_x from the useful pass band (cf. Fig. 1). Another problem with the precision of the estimated amplitudes may be caused by aliasing.

Example 1. Estimate the major spurious signals in the square-wave output of the normalized frequency [10]

$$\xi_x = 179/10000$$

With the assistance of the modified continued fraction expansion of $Y/2X$, we get

TABLE 1

k	b_k	a_k	A_k	B_k
0	28	1	28	1
1	15	-1	419	15
2	12	-1	5000	179

from which we find for the normalized spurious phase-time modulation

$$\frac{s(t_r)}{T_c} = \left[r \left(28 - \frac{1}{15} - \frac{1}{15 * 179} \right) \right] \qquad [19]$$

Evidently, it is formed by two superimposed sawtooth waves as shown in Fig. 2.

The first sawtooth wave has the amplitude ≈ 1 and the modulation "frequency" $10000/419 = 23.87$.

Estimation of the first-order spurious signals according to (13) reveals

$$-34.94 \quad \text{dB}$$

Exact computation for normalized frequencies 179 ± 24 gives

$$r_{155} = -34.81 \quad \text{and} \quad r_{203} = -35.12 \quad \text{dB}$$

For the second-order approximation, we find for the first spurious side bands the modulation "frequency"

$$10000/5000 = 2$$

and for their level in accordance with (15)

$$-58.47 \quad \text{dB}$$

whereas the exact computation gives

$$r_{177} = -58.35 \quad \text{and} \quad r_{181} = -58.38 \quad \text{dB}$$

For actually computed spectra, see Fig. 3.

Example 2. In Fig. 4, reproduced from [1], we compare the simulated and actually measured spectra of the DDFS designed for the PAL color subcarrier.

$$\xi_x = 4436618.75/5.10^6$$

C. Limitation of the Sine Waves

Another way to generate the rectangular output way is by limitation of the output sine wave. The suggestion of Giebel, Lutz, and O'Leary [6, p. 641] is "to convert to a square wave,

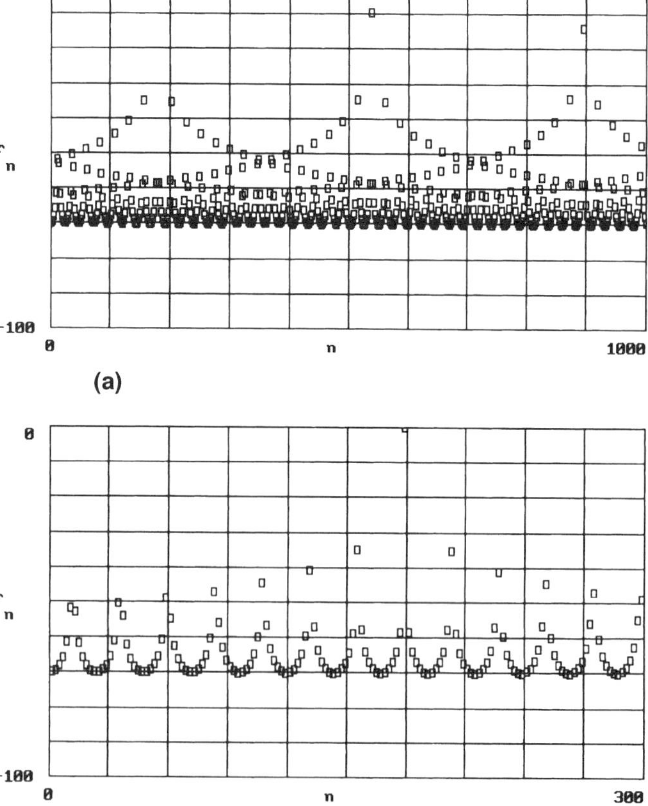

Fig. 3. Spurious signals in the rectangular wave with the normalized frequency $\xi_x = 179/10000$
(a) till $n = 1000$; (b) reveals more details around the carrier.

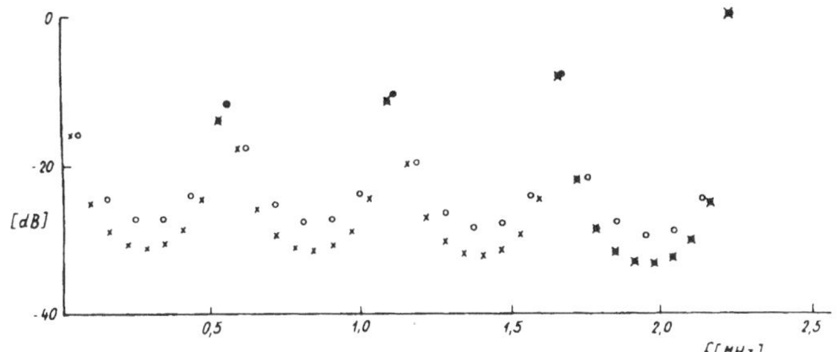

Fig. 4. DDFS for the PAL color subcarrier; expected first (.) and second (o) order spurious signal levels compared with actually measured (x) from [1].

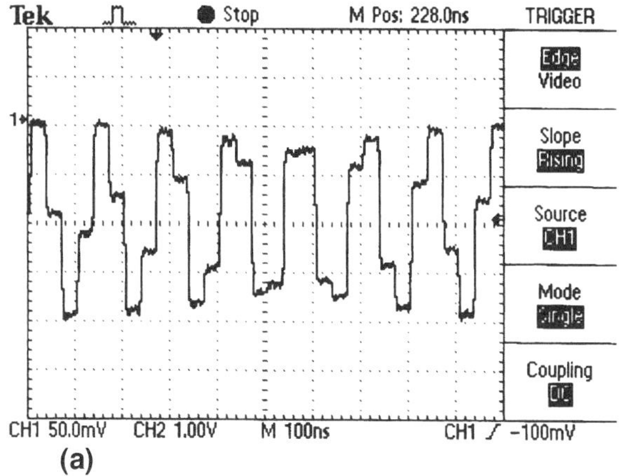

(a)

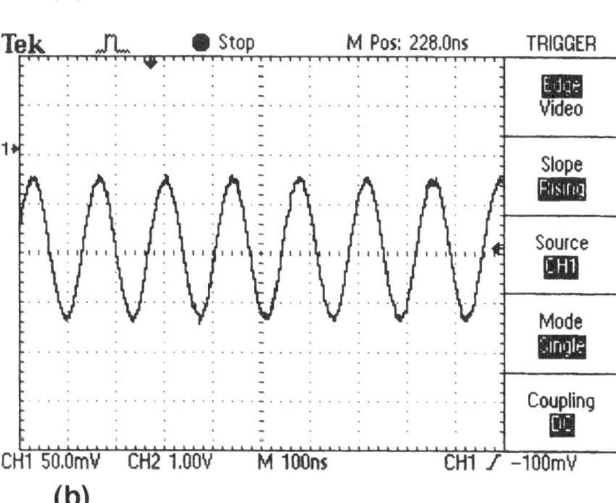

(b)

Fig. 5. (*a*) Output sine wave of the unfiltered DDSF output with the normalized frequency 31/128; (*b*) the same sine wave after passing the output filter.

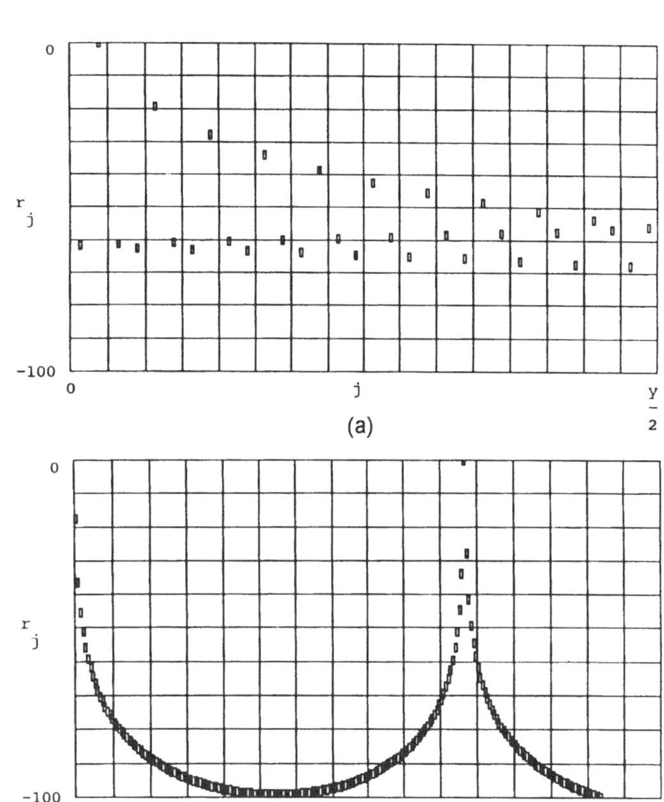

(a)

(b)

Fig. 6. Examples of the spectral properties of the triangular output wave signals.
(a) $\xi_x = 3/128$ and DAC $A = 10$
(b) $\xi_x = 341/1024$ with $A = 10$
(c) $\xi_x = 341/1024$ with $A = 5$

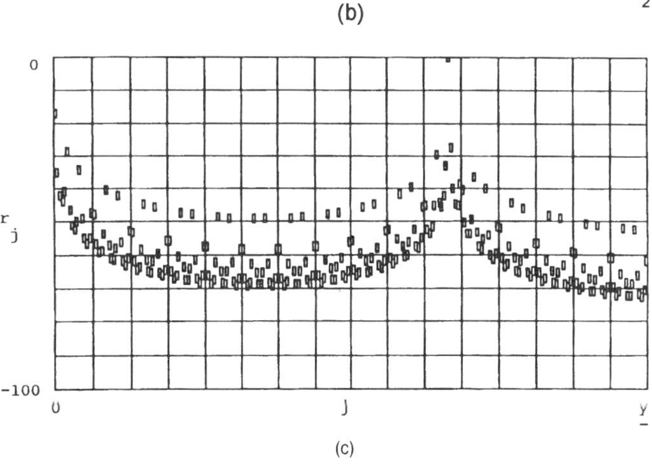

(c)

after low pass filtering, the 6 bit sine DAC output." In addition they state that "the synthesized output clock becomes asynchronous with respect to the reference clock."

The situation is illustrated in Fig. 5 by comparing the nonfiltered output sine wave with the same one passed by the output filter that suppresses all signals outside the Nyquist band (that is, all the "aliased signals"—cf. Fig. 16 in Part I).

III. TRIANGULAR-WAVE OUTPUTS

Another, rather simple, DDFS can be built with the triangular output wave. The advantage is that no sine lookup table is needed since the DAC is fed directly from the phase accumulator. Two MSBs are used for changing the sawtooth output from the accumulator into the triangular output.

In contradistinction to the rectangular output wave, the spectral properties are better since harmonic signals (ideally, only odd ones) of the f_x decay with the square of their order. Some examples are shown in Fig. 6.

IV. SINE-WAVE OUTPUT

The sine-wave output is the most desirable form of output signal since it exhibits the lowest level of inevitable spurious signals.

Two major spurious signal generators are the phase and the amplitude bit truncation. The level and locations of the spurs have been investigated in the past, for example in [11–14] and by the author in [2–5].

A. Fourier Analysis of the Quantized Output Sine Waves

The output wave of DDFSs is not an ideal sine but a staircase approximation, as shown with an example in Fig. 7. We can see that it repeats itself with the period YT_c. To compute the output spectrum, we apply the Fourier analysis on each rectangle in the output staircase curve; see the quantized sine wave in Fig. 7. With the assistance of the complex form, we get for the contribution to the nth harmonic from the mth rectangle

$$c_{n,m} = \frac{1}{YT_c} \int_{mT_c}^{(m+1)T_c} \sin\left(2\pi m \frac{X}{Y}\right) e^{-jn\left(\frac{2\pi}{YT_c}\right)t} dt$$
$$= \frac{1}{\pi n} \sin\left(\pi \frac{n}{Y}\right) e^{-j\pi \frac{n}{Y}} \sin\left(2\pi m \frac{X}{Y}\right) e^{-2j\pi \frac{nm}{Y}} \quad [20]$$

After performing summation over all m (from 0 to $Y-1$), we find that all c_n are equal to zero for all $n \neq X$ and only for $n = X$ do we arrive at a one-sided Fourier coefficient

$$c_X = \frac{\sin(\pi X/Y)}{\pi X/Y} \quad [21]$$

B. Spectral Analysis of Phase-truncated Sine Waves

As emphasized earlier, the high-frequency resolution of DDFS requires large accumulator capacity (32, 48, or even 64 bits). It is evident that the potential complexity of the hardware prevents us from referring to all these minute phase incre-

ments. Generally, we use only a small number of most significant bits (MSB), W, and disregard all the remaining least significant bits (LSB), $B = R - W$—cf. Fig. 8. Thus, from the original number $X(m)$ stored in the accumulator [2]

$$x(m) = mX - rY \leq Y \leq 2^R \quad [22]$$

we refer to

$$x'(m) = \text{integer}\left(\frac{mX - rY}{2^B}\right) \quad [23]$$

Since the information passed to the "sine lookup table" must be of the same order as the original $X(m)$, we have to multiply $X'(m)$ by 2^B. However, the phase readings are still performed at each mT_c $(m = 0, 1, \ldots, Y - 1)$. As a consequence, amplitudes of individual rectangles in the output sine wave change into (since $rY/2^B$ are always integers)

$$\sin\left[2\pi \frac{2^B}{Y} \text{integer}\left(\frac{mX}{2^B}\right)\right]$$
$$= \sin\left\{2\pi \frac{2^B}{Y}\left[\frac{mX}{2^B} - s(m)\right]\right\} \quad [24]$$

where

$$s(m) = \frac{mX}{2^B} - \text{integer}\left(\frac{mX}{2^B}\right) \leq 1 \quad [25]$$

Evidently,

$$\frac{2^B}{Y} s(m) \ll 1 \quad [26]$$

and relation (24) can be further simplified to

$$\sin\frac{2\pi mX}{Y} - 2\pi \frac{2^B}{Y} s(m) \cos\frac{2\pi mX}{Y} \quad [27]$$

The situation with eq. (27) is such that we face a nearly ideal sine wave with amplitude 1 and a disturbing cosine wave with a rather small amplitude; cf. eq. (28). We easily conclude that the largest spurious amplitude a_{sp} does not exceed the amplitude of the disturbing cosine wave, that is,

$$a_{sp} \leq 2\pi \frac{2^B}{Y} = 2\pi 2^{-W} \quad [28]$$

The exact computation of the output spectrum would require solution of the relation (24). But this is generally prohibitive because of the large $Y \leq 2^R$. The other way is to estimate some major disturbing components. To this end we will investigate properties of the phase modulation function $s(m)$. After

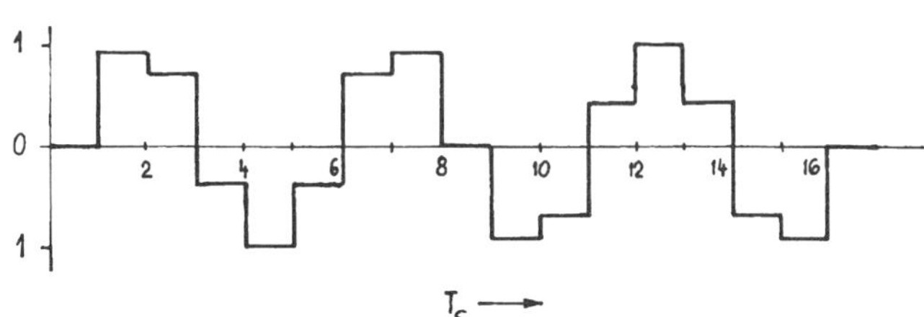

Fig. 7. Staircase approximation of the sine wave for the normalized frequency $\xi_x = 3/16$.

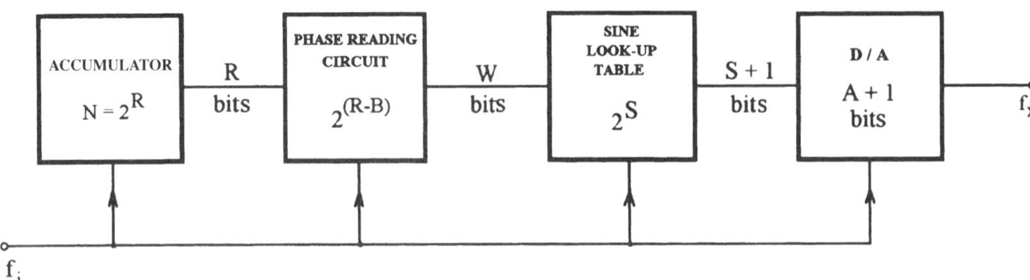

Fig. 8. The simplified block diagram showing sources of the spurious signals in DDFS due to the information loss in different system groups.

reverting to the earlier results found by the author for the quasiperiodic rectangular waves [1,3], we can expand $s(m)$ into a periodic series with the assistance of modified continued fraction expansion

$$s(m) = \frac{mX}{2^B} - \text{integer}\left(\frac{mX}{2^B}\right)$$

$$= m\left(A_0 + \frac{a_1}{B_1} - \frac{a_1 a_2}{B_1 B_2} + \cdots \right) \quad [29]$$

$$+ \frac{a_{n-1} a_n}{B_{n-1} B_n}\right) - \text{integer}\left(\frac{mX}{2^B}\right)$$

Examples of phase truncation functions $s(m)$ are shown in Fig. 9. Note that we again face a superposition of the sawtooth waves.

Examination of the plots reveals that the first-order approximation of $s(m)$ by A_1/B_1 is either a rectangular or a simple sawtooth wave with amplitude $1/2$ or 1 and the repetition period $B_1 T_c$. The second-order approximation adds another sawtooth wave with the amplitude $1/B_1$ and the repetition period $B_2 T_c$. The same happens with every other approximation.

Example 3. Find the phase modulation function $s(m)$ for the normalized frequency

$$\xi_x = 207/1024$$

with 5 LSB neglected. In accordance with (25), we get

$$s(m) = m(207/32) - \text{integer}(m.207/32)$$

The modified continued fraction expansion of 207/32 leads to

TABLE 2

k	b_k	a_k	A_k	B_k
0	6	1	6	1
1	2	1	13	2
2	7	1	97	15
3	2	1	207	32

and the respective series expansion reveals

$$s(m) = m\left\lfloor 6 + \frac{1}{2} - \frac{1}{2*15} + \frac{1}{15*32}\right\rfloor - \text{integer}\left(\frac{m207}{32}\right) \quad [30]$$

Note that since X and 2^B are relatively prime integers (X being odd), the latest divisor B_n is equal to 2^B. Series expansion (29) makes it possible to compute or estimate the levels and frequencies of spurious signals quite easily in comparison (to the author's knowledge) with earlier approaches [11,12,13]. In accordance with Appendix A, we will start with finding the remainder R_b for $s(m = 1)$

$$s(m = 1) = \frac{X}{2^B} - \text{integer}\left(\frac{X}{2^B}\right) = R_b \quad [31]$$

and then, taking into account that $s(m)$ is a periodic function, we arrive at a sum of sine waves, that is,

$$s(m) \doteq \frac{1}{\pi} \sum_{r=1,2,\ldots}^{2^{(B-1)}} \frac{1}{r} \sin(2\pi r m R_b) \quad [32]$$

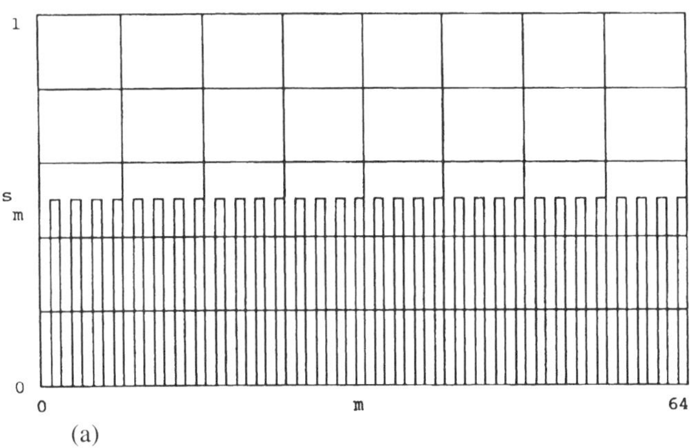

Fig. 9(*a*). Plot of the phase truncation function $s(m)$ for the normalized frequency $\xi_x = 207/1024$ for 1-bit LSB discarded:

$$s(m) = m(103 + 1/2) - \text{integer } 207(m/2).$$

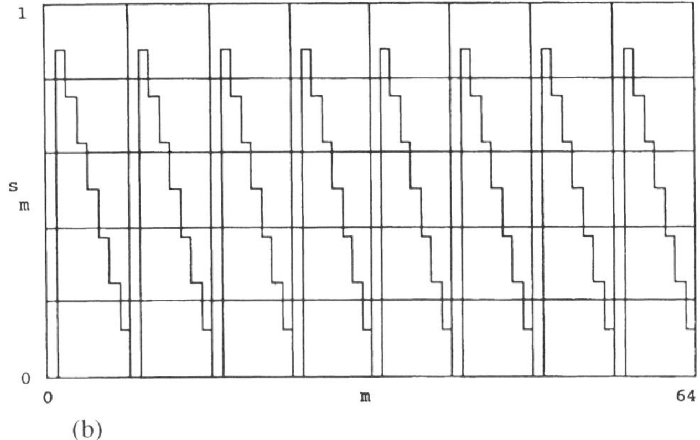

Fig. 9(*b*). Three bits discarded—one sawtooth wave predominates:

$$s(m) = m(26 - 1/8) - \text{integer}(m207/8).$$

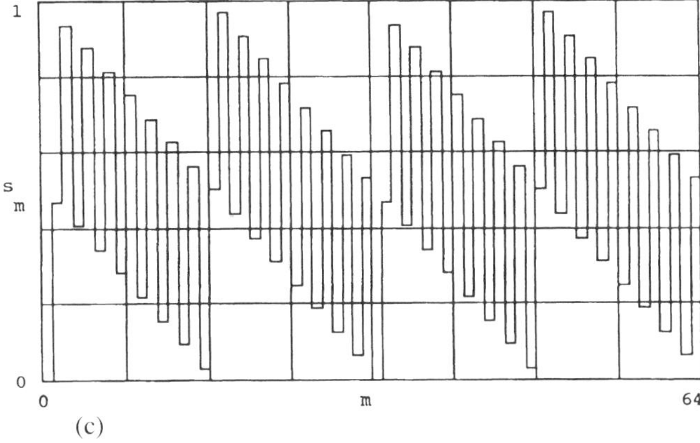

Fig. 9(*c*). Five bits discarded—superposition of rectangular and sawtooth waves:

$$s(m) = m(6 + 1/2 - 1/2.15 + 1/5.32) - \text{integer}(m207/32).$$

Note that we can always hold the value of the remainder in the range

$$0 \leq R_b = \frac{P}{Q} \leq \frac{1}{2} \quad (Q = 2^B) \quad [33]$$

which simplifies the location of spurious signals.

After introducing the rth harmonic from (32) into (27), we can expand the second term into

$$\frac{2^B}{rY}\left\{\sin\left[2\pi m\left(\frac{rP}{Q}+\frac{X}{Y}\right)\right]+\sin\left[2\pi m\left(\frac{rP}{Q}-\frac{X}{Y}\right)\right]\right\} \tag{34}$$

(34) represents two sine waves.

Evidently, the above expressions represent staircase values of spurious sine waves around the carrier. Consequently, we can compute their amplitudes with the assistance of (20), and in accordance with (21) we arrive at

$$c_{sr}=\frac{2^B}{rY}\frac{\sin\left[\pi\left(\frac{X}{Y}\pm\frac{rP}{Q}\right)\right]}{\pi\left(\frac{X}{Y}\pm\frac{rP}{Q}\right)} \tag{35}$$

After dividing the above relation by the carrier (21), we get for the rth order spurious signal level

$$\frac{c_{sr}}{c_X}=\frac{2^B}{rY}\cdot\frac{X/Y}{X/Y\pm rP/Q}\cdot\frac{\sin[\pi(X/Y\pm rP/Q)]}{\sin\left(\pi\frac{X}{Y}\right)}$$
$$\doteq\frac{2^B}{rY}=\frac{2^{-R+B}}{r}=\frac{2^{-W}}{r} \tag{36}$$

The approximation follows from the condition that the denominator $X/Y\pm rP/Q$ cannot exceed ±1. Consequently, in dB measure we get

$$\frac{c_{sr}}{c_X}\doteq-6W-20\log(r)\quad[\text{dB}] \tag{37}$$

The above relation is plotted in Fig. 10 for $r=1$ and reveals the worst-case spurious signal level as a function of the number W of the MSB used for phase reading from the accumulator of DDFS.

Several remarks need to be made, however. First, we would like to know the number of spurious signals caused by the phase truncation in the DDFS pass band, which in normalized notation is ideally equal to $Y/2$. The problem was already solved in [4]. We repeat it here briefly. From eq. (36) we can

deduce the spurious line numbers. (The carrier line number is X.) However, we must also take into account that some rP/Q exceed the pass band $Y/2$ but are returned as alias. As a consequence, the spurious spectral line numbers must meet condition

$$n_r=\left|X+\left(r\frac{PY}{2^B}+sY\right)\right|<Y/2$$
$$(s=\ldots,-2,-1,0,1,2,\ldots) \tag{38}$$

Investigation of the above inequality reveals different n_r only for $r<2^B/2$ and a double one for $r=2^{B-1}$. The result is that the lowest level of spurious signals caused by the phase truncation should not exceed

$$-6(W+B-1)=-6(R-1) \tag{39}$$

Since there are two side bands around the carrier for each r (i.e., both positive and negative), their number is given by (leaving out the carrier)

$$2^B-1 \tag{40}$$

that is, 31 in Fig. 11 and 7 in Fig. 12.

Second, with the assistance of relation (38), we easily arrive at actual spurious frequencies

$$f_{sr}=\frac{n_r}{Y}f_c=\left|f_x+\left(r\frac{P}{2^B}+s\right)\right|f_c \tag{41}$$

Example 4. With the assistance of FFT applied on eq. (24), we have computed the level of spurious side bands for the situation investigated in Example 3. The result is shown in Fig. 11 and reveals

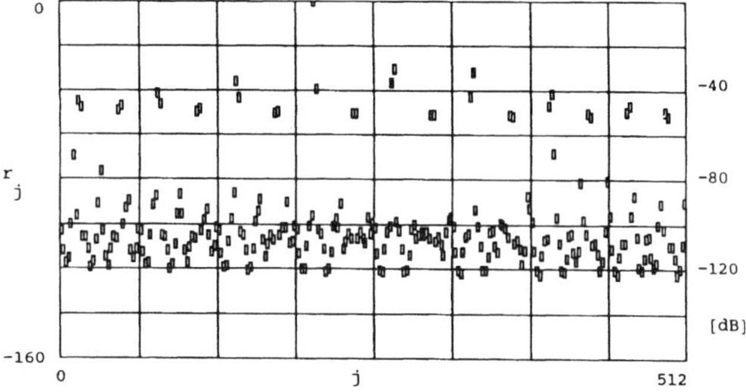

Fig. 11. An example of spurious side bands of $\xi_x=207/2^{10}$ with 5 LSB neglected $S=12$ [4].

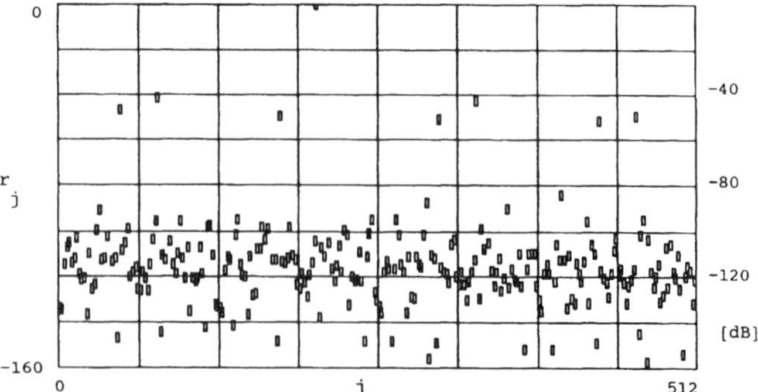

Fig. 12. An example of spurious side bands in DDFS with the normalized frequency $\xi_x=207/2^{10}$ with 3 LSB neglected, that is, $B=3$ and $W=7$ and $S=12$.

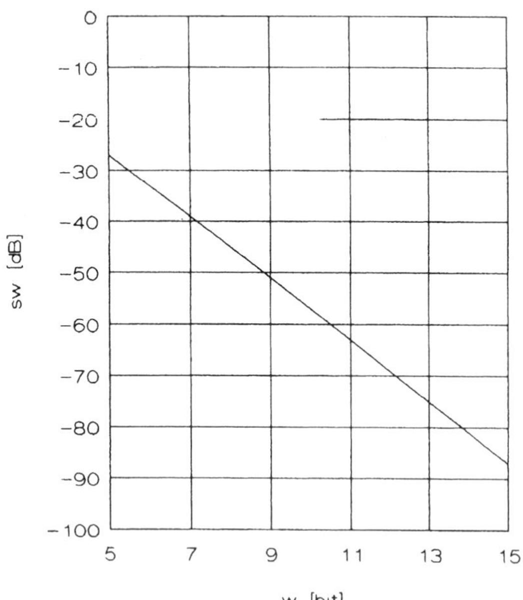

Fig. 10. Approximate spurious signal level s_w for W MSB retained.

that the largest spurious signals are about -30 dB as expected and extend down to about -50 dB. From eq. (39) we get -54 dB, which is a fairly good result.

B. Sine-Word Truncation

The ideal sine wave with staircase amplitudes

$$\sin\left(2\pi m \frac{X}{Y}\right) \qquad [42]$$

would reveal in the output pass band $Y/2$ no spurious components. However, the economy of actual DDFS requires that the information about the sine values in lookup tables is generally stored only as S-bit long words. The consequence is that besides those "high-level" spurious signals caused by the phase truncation, there is a large set of much smaller spurious signals the level of which depends on the number of bits used in lookup table words. As a result, truncation of stored sine values results in

$$\frac{1}{2^S}\text{integer}\left[2^S\sin\left(2\pi m\frac{X}{Y}\right)\right] \qquad [43]$$

or even in

$$\frac{1}{2^S}\text{integer}\left\{2^S\sin\left[2\pi\frac{2^B}{Y}\text{integer}\left(\frac{mX}{2^B}\right)\right]\right\} \qquad [44]$$

Example 5. With the assistance of the FFT algorithm and relation (44), we will investigate the DDFS spectra for different W and S. The results are shown in Figs. 11, 12, and 13. Everyone would expect that for smaller S the noise would be higher.

C. Estimation of Background Spurious Signal Level

When we first examine the above figures, we have a feeling that amplitudes of these background spurious signals are randomly distributed. However, this cannot be true since we have evaluated a periodic system. To get more insight, we will compute and plot the error signal

$$e(m) = \sin\left(2\pi m\frac{X}{Y}\right)$$
$$- \frac{1}{2^S}\text{integer}\left\{2^S\sin\left[2\pi\frac{2^B}{Y}\text{integer}\left(\frac{mX}{2^B}\right)\right]\right\} \qquad [45]$$

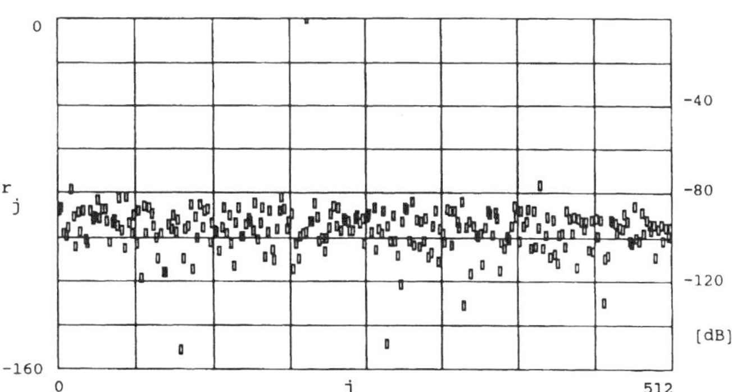

Fig. 13. Background spurious signal level for $\xi_x = 207/2^{10}$ with $B = 0$, $W = 10$, and $S = 10$.

From the above relation, we easily find out that

$$e(m)_{\max} < |1/2^S| \qquad [46]$$

A plot in Fig. 14 confirms the above inequality. Closer investigation reveals an odd symmetry and some other periodicities, that is, the process $e(m)$ is periodically stationary [15]. However, at first sight we have a feeling of randomness, and from this point of view we make further investigations.

For the amplitude distributions of $e(m)$'s, we will use the simplest one, namely, a uniform distribution between $e(m)_{\max,+}$ and $e(m)_{\max,-}$, that is,

$$p(e) = 1/(2*2^{-S}) \qquad \text{inside the interval}$$
$$p(e) = 0 \qquad \text{outside the interval} \qquad [47]$$

As a result, we find for the variance

$$\sigma^2(e) \doteq (1/3)*2^{-2S} \qquad [48]$$

By plotting the cumulative [16] distribution of $e(m)$ we find for larger m and S a straight line with two different slopes; see Fig. 15.

Evidently, the amplitude distributions of $e(m)$'s are constant in the ranges between $-1/2^S$ and 0 and from 0 to $+1/2^S$. This conclusion supports the assumption about the rectangular distribution of the error $e(m)$. Later on, we will find that the variances computed with the assistance of the relation (48) agree quite well with actually computed values as is shown in the following example.

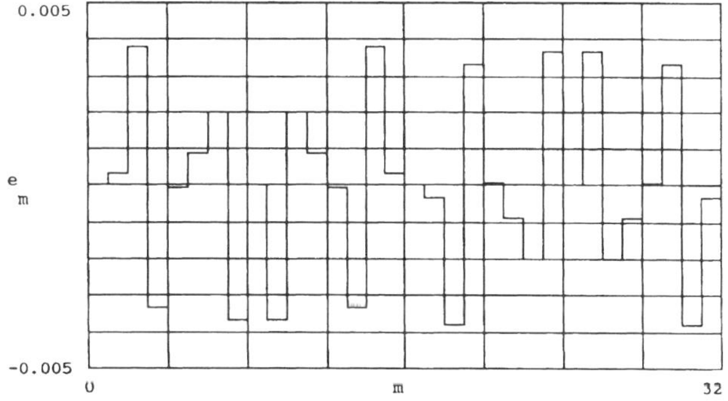

Fig. 14. The error signal $e(m)$ due to finite length of words, S, in sine lookup tables for the normalized frequency $\xi_x = 7/32$, $S = 8$, $B = 0$.

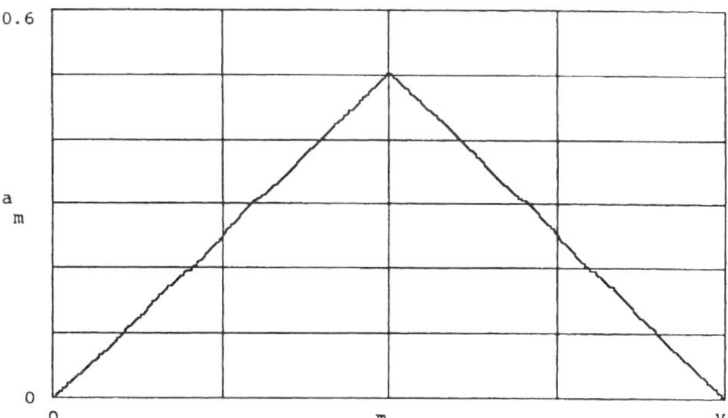

Fig. 15. Cumulative distribution of sine word errors: $m = Y = 1024$.

Example 6. Compare the computed variances of $\sigma^2(e)$ with the theoretical estimations by (48).

TABLE 3

S	$\sigma^2(e)$	var(e) $Y = 32$	var(e) $Y = 128$	var(e) $Y = 1024$
8	(5.09	5.33	5.18	5.48).10^{-6}
10	(3.18	2.85	3.44	3.15).10^{-7}
12	(1.99	0.96	1.90	2.00).10^{-8}
14	(1.24	1.16	1.40	1.27).10^{-9}

Investigation of Table 3 reveals rather small differences between estimated and actually computed variances. We can therefore conclude that we are justified in using relation (48) in our estimation of the "background" spurious signal levels.

Furthermore, investigation of Figs. 11, 12, and 13 reveals that there might exist a mean level $\langle r_i \rangle$ of the spurious spectral lines r_j with some dispersion. With the assistance of the δ-functions we can introduce the mean power spectral density (PSD) $\langle S(n) \rangle$ for discrete spurious signals.

With these encouraging findings, we can look for the closed-form solution of $\langle S(n) \rangle$ for different ξ_x and length S of the used sine words.

First, we will recall the relation between autocorrelation and the spectral density

$$R(0) = \int_0^\infty S(f)df \qquad [49]$$

where $S(f)$ is the one-sided power spectral density.

For evaluation of relation (49), we need to know their distribution. To this end, we have again plotted the cumulative distributions of the positive values of $|r_j|$ and arrived at straight lines, one of which is shown in Fig. 16.

Second, because of the odd symmetry of the error signal $e(m)$, its first moment is zero, and consequently the autocorrelation $R(0)$ is equal to the variance. Finally, by taking into account that the investigated PSD $S(f)$, extending from 0 to $Y/2$, is formed by spectral lines, the number of which is ideally

$$Y/4 - 1 \qquad [50]$$

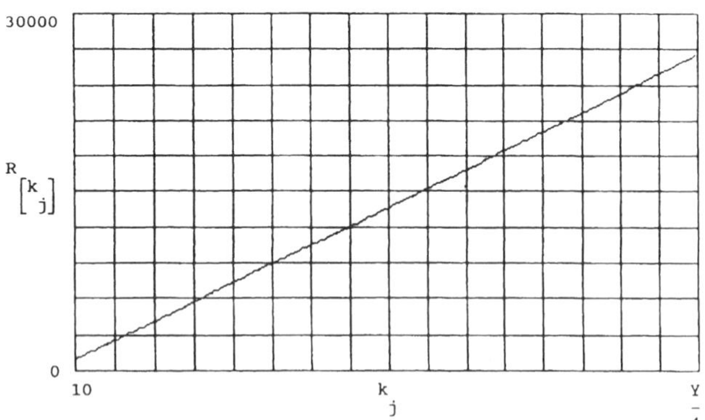

Fig. 16. Cumulative distribution of the level of the spectral lines $|r_j|$.

we must leave out the carrier. All spurious signals with even line numbers are negligibly small; however, for computation of the mean value $\langle S(n) \rangle$ of spectral lines $S(f = n_j \cdot f_c)$ they cannot be neglected. Consequently, relation (49) should be replaced with

$$\sigma^2(e) \approx (Y/2 - 1) * \langle S(n) \rangle \qquad [51]$$

After introducing (48) into the relation (51), we get for the mean value of the background spurious signals

$$\langle S(n) \rangle = \frac{\sigma^2(e)}{Y/2 - 1} = \frac{1}{3} * 2^{-2S} * \frac{2}{Y - 2} \qquad [52]$$

For large Y we can approximate $\langle S(n) \rangle$ in dB measure as

$$\langle S(n) \rangle \approx -1.75 - 6S - 10 * \log Y \quad [\text{dB}] \qquad [53]$$

Computer simulations summarized in Example 7 revealed values about -3 dB lower. To remove these discrepancies, we have tried a new approach. In a paper [5] we have computed the expected level of the background spurious signals with the assistance of the quantization noise spectra theory [15]. This time we have a rather good agreement (see Appendix B), that is,

$$\langle S(n) \rangle \approx \frac{1}{3Y} * 2^{-2S} \qquad [54]$$

Example 7. Compute the means $\langle S(n) \rangle$ and compare them with the means of actual values of the background spurious signals $\langle r_j \rangle$

$$\langle r_j \rangle = \langle 10 * \log[S(n_j)] \rangle$$
$$(j = 1, 3, \ldots j \neq X) \qquad [55]$$

The results are shown in Table 4 for the normalized frequencies $\xi_x = X/Y$ for different sine-lookup resolutions S; $\langle S(n) \rangle$ and $\langle r_j \rangle$ are means between $\xi_x = X/Y$ and $\xi_x \approx 1/3$.

TABLE 4

Y	S	$S(n)$	$\langle r_j \rangle$	σ
32	8	-68	-65	1.28
	10	-80	-78	3.61
	12	-92	-98.6	4.58
128	8	-94	-75	13.04
	10	-86	-87	8.12
	12	-98	-101	11.14
512	8	-80	-81	9.51
	10	-92	-93	9.19
	12	-104	-106	9.55
1024	8	-83	-85	10.95
	10	-95	-96	9.68
	12	-107	-109	10.18

Finally, in Example 7 we have also computed dispersions of the actual values of the background spurious signals around the mean (55) and have found that the respective standard deviation is in the range of 10 dB (see Example 7). Furthermore, our experimental evidence has taught us that only a few "background" signals exceeded the level given by

$$10 * \log[\langle S(n) \rangle] + 2\sigma \qquad [56]$$

This generally happens in the presence of phase truncation. A mere investigation of Figs. 11 and 12 reveals that spurious side bands due to this origin are much larger and rather few. Therefore, they can be omitted as outlayers [16], and the above arrived results are also valid in these instances.

D. Estimation of the Spurious Harmonics

Closer investigation of many simulations of background spurious signals in DDFS reveals that few of them exceed the level given by relation (56), even in instances where no phase truncation is present (see Fig. 17). We have looked for their origin and found it in higher harmonics of the output signal f_x.

Let us return to eq. (45) and compute the mean value of $e(m)$ for both the positive and negative half period of the output sine wave. We get a rectangular wave with the amplitude $2^{-S}/2$. The problem is illustrated with a simple example in Fig. 18. The Fourier series expansion reveals only odd harmonics

$$a_h = \frac{2 * 2^{-S}}{\pi} * \frac{1}{h} \quad (h = 3, 5, \ldots) \qquad [57]$$

For the level of the third spurious harmonic, we get in dB measure

$$20 * \log\left(\frac{c_{s,3x}}{c_X}\right) \approx -6S - 14 \quad [\text{dB}] \qquad [58]$$

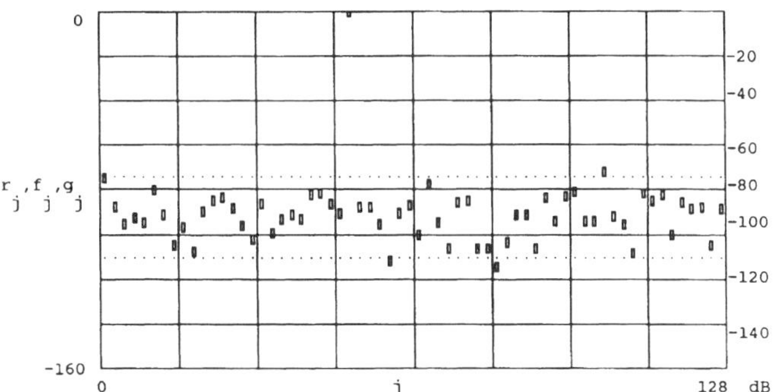

Fig. 17. FFT simulation of the background spurious signals for $S = 10$ and $\xi_x = 51/256$; point indicates 2σ dispersion around the mean (f_j and g_j).

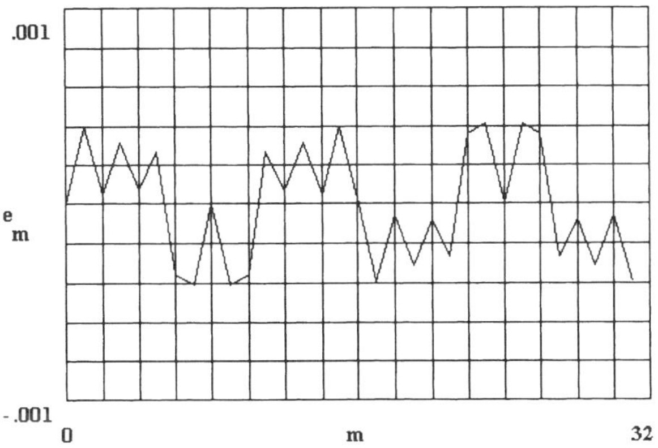

Fig. 18. Plot of the error function $e(m)$ for the normalized frequency $\xi_x = 3/32$ and $S = 11$.

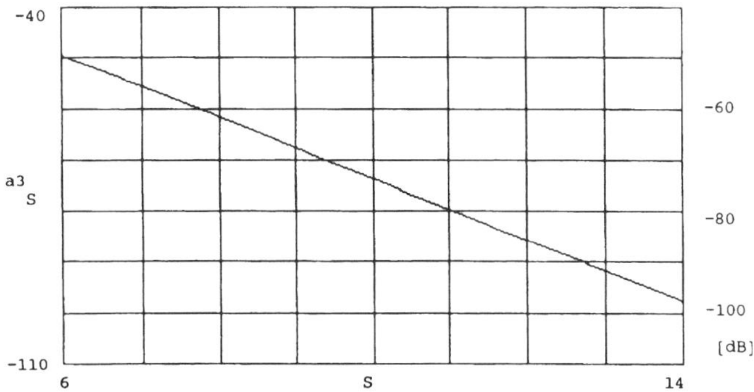

Fig. 19. The expected value of the third harmonic of the output frequency $(3f_x)$ as a function of the number of S bits used.

Note that we have again neglected the effect of sampling on amplitudes. Relation (58) is plotted in Fig. 19, and it is in a good agreement with many computer simulations. However, the fifth and higher harmonics are often masked by the dispersion of the background spurious signals.

E. Relation between W and S

In instances where we prefer that spurious signals due to the phase truncation not exceed the most prominent signals caused by the amplitude truncation, we must compare relation (58) with (37), and we arrive at the following inequality

$$W - 3 \leq S \leq W - 2 \qquad [59]$$

V. Digital-to-Analog Converter

The major causes of degradation of the output spectral purity by DAC are finite bit resolution, nonlinearity, and transient effects or glitches.

A. Finite Bit Resolution

With the present technology available, we are not able to build DAC with any arbitrary resolution. On the other hand, the higher the resolution, the more expensive DACs are. Consequently, we encounter in DDFS practice DAC with resolutions

$$D = 8 \text{ till } 12 \qquad [60]$$

Discussion of the DAC bit resolution reveals that from the S-bit-long sine word we use only A bits for conversion of digital information into its analog version. Note that when a D bit DAC is used in a binary version, the effective sine words are only A bit long, that is,

$$A = D - 1 \qquad [61]$$

As a result, in instances where $S > A$, we must replace in all relations for computation of background spurious signals the sine word length S by the bit resolution A (cf. paper by Garvey and Babitch [14]).

161

B. Background Level of Spurious Signals

After combining eqs. (61) and (54), for the background level of spurious signals we get

$$S(n_r) \doteq -4.77 - 6.02A - 10\log(Y) \qquad [62]$$

However, in instances where the denominator Y is large, particularly if the smallest frequency step Δf is much smaller than 1 Hz. That is, if

$$\Delta f_x \ll \frac{f_c}{Y} \ll 1 \qquad [63]$$

then the number of spurious spectral lines in 1 Hz bandwidth is

$$\frac{1}{\Delta f_x} = \frac{Y}{f_c} \qquad [64]$$

and the discrete background spectrum changes effectively into a continuous and white one. After dividing relation (54) with Δf_x, we get for the power spectral density

$$S(f) \doteq -4.77 - 6.02 * A - 10 * \log(f_c) \quad \text{dB} \qquad [65]$$

The conclusion is that the extent of the sine lookup table and the bit resolution of DAC are closely related in properly designed DDFS.

C. DAC Nonlinearity

Investigations of the real output spectra of DDFS also reveal even harmonics of the output signal in contradistinction to relation (57). Furthermore, we also encounter even harmonics of the fundamental spurious modulation frequency (63) (which theoretically should not be present) and the respective odd harmonics, often higher, than the values expected in accordance with (62) or (65). The reason is the nonlinearity of the DAC transfer characteristics, which generally is not larger than one LSB. Nevertheless, their influence is difficult to predict: as a rule of thumb, the background noise level increases to about 3 or 6 dB.

D. Glitches

An annoying difficulty associated with converting digital information into an analog signal is the transient effects generated by changes of digital states in DAC [17]. These undesired current or voltage spikes are called glitches or, incorrectly, "glitch energy." Every manufacturer introduces the means for reducing these transients just at the design stage. Nevertheless, application of an additional sample and hold circuit at the DAC output proves very useful [e.g., 17,18].

VI. INTERMODULATION SIGNALS IN DDFS

All frequency synthesizers are plagued with undesired signals accompanying the desired output signal. Very troublesome, particularly known from the theory of direct analog synthesizers [19,20], are the so-called intermodulation signals. They are generally defined as multiples of input frequencies generated and mixed in a nonlinear circuit to a frequency that may pass the output filter, that is,

$$|r * f_1 + s * f_2| = f_p \qquad [66]$$

where r and s are integers (either positive or negative) and f_p is the spurious frequency in the output filter pass band. Experience has taught us that the higher the intermodulation product is (i.e., the larger the integers r and s are), the smaller the respective intermodulation signal is.

In DDFS, the inevitable presence of the "switching" processes and the leakage of the clock frequency f_c are generators of the intermodulation signals [17].

Returning to eq. (66), we see that it can be solved in different ways. The "straight-line" intermodulation diagram in Fig. 20 adapted from [19] is very instructive.

Note that for

$$\xi_x = \frac{f_1}{f_2} \approx \frac{1}{4} \qquad [67]$$

the spectral components $f_2 - 3f_1$ and $5f_1 - f_2$ nearly coincide with the desired signal f_1. As a consequence, rather large spurious signals are present in the neighborhood of the desired signal. The remedy is to keep the leakage of the clock signal (f_2 in the above equations) to possible "mixers" as low as possible. In some instances, this difficulty can be alleviated by changing the desired output frequency or the clock frequency.

Today there exist many computer solutions of the intermodulation problem. One solution that is very simple and particularly suited for investigation of DDFS is given in Part XI of this book.

Since a DDFS is a highly nonlinear device, it is tempting to conclude that all spurious signals are intermodulation products of the type (66) [2,18]. However, pursuing this idea, we might arrive at very confusing results; see the table below taken from [16]. The intermodulation product in the last line is rather large, and at the same time it is of an extremely high order that is not consistent with the intermodulation theory. As a result, we always have to investigate all other possible sources of spurious signals in DDFS (see Fig. 8).

In this particular case, it was demonstrated that the intermodulation product in the last line in Table 5 can be traced

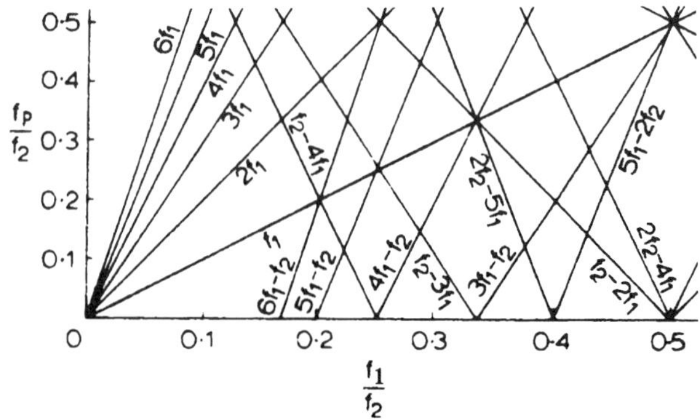

Fig. 20. The "straight-line" intermodulation diagram.

TABLE 5

Frequency (Hz)	Level (dBc)	Origin
263540	−72.5	$2Fs$
395310	−74.1	$3Fs$
472920	−78.2	$Fc\text{-}4Fs$
30710	−78.4	$23Fs\text{-}3Fc$
341150	−79.2	$Fc\text{-}5Fs$
449470	−79.7	$11Fs\text{-}Fc$
418760	−80.1	$2Fc\text{-}12Fs$
51450	−83.0	$(2^{14}+1)Fs\text{-}2159Fc$

to the accumulator phase truncation [5]; see the following example.

Example 8. From the Giffard et al. paper [18], we have

$$f_x = 131.77\,\text{kHz}, \quad f_c = 1\,\text{MHz}, \quad R = 32, \quad W = 14, \quad D = 12$$

from (31) we get for the remainder

$$R_b = 0.91968$$

To simplify computations, in accordance with (33) we use the value

$$1 - (R_b) = 0.08032$$

and with the assistance of relation (38), we get for the largest spurious signals

$$f_{s,1} = 51.450\,\text{kHz} \quad \text{and also} \quad f_{s,-1} = 212.09\,\text{kHz}$$

and for their level

$$-6*W = -84 \quad \text{dB}$$

Actual measurement performed by the authors revealed for $f_{s,1}$ the level

$$-83\,\text{dB}$$

To close the discussion about intermodulation problems, we have to note that often more than two signals are found at the input of the mixer. However, the solution in these instances is not a simple one [21].

VII. Conclusions

In this article, we have derived a simple closed-form solution for frequencies and power level of spurious signals in DDFS with square-wave, triangular, and sine-wave outputs.

Since DDFSs are generally plagued with spurious signals, sometimes even large ones, and often of unknown or dubious origin, we have presented a lot of computer simulations to provide the reader with information about the lowest level of spurious signals attainable.

VIII. Appendix

A. Investigation of the Modulation Function s(m)

Investigation of spurious signals generated by phase truncation starts by computing the remainder R_b. We easily change

eq. (31) into

$$R_b = \xi_x * 2^W - \text{integer}(\xi_x * 2^W) \tag{68}$$

with the assistance of the modified continued fraction expansion. Cf. eq. (29) where all a_1, a_2, and so on, are equal either to $+1$ or to -1 [1]. The first-order approximation of R_b is

$$R_b \approx (\pm)\frac{1}{B_1} \tag{69}$$

and $s(m)$ in the time domain reveals a sawtooth wave with the amplitude 1 and the repetition period $B_1 T_c$. Its Fourier series expansion reveals

$$\frac{1}{\pi}\sum \frac{1}{r}\sin\left(\frac{2\pi r}{B_1 T_c}t\right) \tag{70}$$

The second-order approximation of R_b is

$$R_b \approx \frac{a_1}{B_1} - \frac{a_2}{B_1 B_2} = \frac{A_2}{B_2} \tag{71}$$

Again $s(m)$ in the time domain would reveal a sawtooth wave with amplitude 1, but with the repetition period very close to that of the first-order approximation, namely, $(B_2/A_2)T_c$ (cf. Fig. 7 in [12]). With the assistance of the Fourier series expansion we get

$$\frac{1}{\pi}\sum \frac{1}{r}\sin\left(2\pi r\frac{A_2}{B_2}t\right) \tag{72}$$

Proceeding further in this way, we finally arrive at relation (32).

B. Computation of the Background Power Spectral Density

After replacing the variable m with t in relation (45), we can expand the random variable $\epsilon(t)$ as

$$\epsilon(t) = \sum_{-\infty}^{\infty} a_n e^{jn\omega_o T_i} \quad \omega_o = 2\pi/YT_i \tag{73}$$

Evidently, the nth spectral line of the kth rectangle $\epsilon_k(t)$ is

$$a_n = \frac{1}{YT_i}\int_{kT_i - T_i/2}^{kT_i + T_i/2} \epsilon(t)e^{-jn\omega_o t}dt$$
$$= \frac{1}{YT_i}*\epsilon(kT_i)*e^{-jn\omega_o kT_i}\frac{\sin(n\omega_o T_i/2)}{n\omega_o/2} \tag{74}$$

from which the variance is

$$a_n a_n^* = \frac{1}{Y^2}\sum \epsilon^2(kT_i)\,\text{Sin}^2(n\omega_o T_i/2)$$
$$= \frac{1}{Y}\frac{\sigma_e^2}{2}\,\text{Sin}^2(n\omega_o T_i/2) = \alpha_n \tag{75}$$

where we have introduced

$$\frac{\sin(x)}{x} = \text{Sin}(x) \tag{76}$$

Now we proceed with computation of the autocorrelation function $R(\tau)$

$$R(\tau) = \sum_{-\infty}^{\infty} \alpha_n e^{jn\omega_o \tau} \qquad [77]$$

Thus, we easily arrive at the double-sided power spectral density

$$\begin{aligned}
S(\omega) &= \int_{-\infty}^{\infty} R(\tau) e^{-j\omega\tau} d\tau \\
&= \int_{-\infty}^{\infty} \sum \alpha_n e^{jn\omega_o \tau} e^{-j\omega\tau} d\tau \qquad [78] \\
&= \sum \alpha_n \delta(\omega - n\omega_o)
\end{aligned}$$

Finally, for the mean value of one-sided power spectral density, we get with the assistance of (75) and (48)

$$S(n\omega_o) = \frac{1}{3Y} * 2^{-2S} \mathrm{Sin}(n\omega_o T_i / 2) \qquad [79]$$

Since the transfer function $\mathrm{Sin}(x)$ is rather flat in the pass band from zero to $Y/2$ (the maximum error is -4 dB), we feel justified with the simplification used in relation (54).

REFERENCES

[1] V. F. Kroupa. "Spectra of pulse rate frequency synthesizers." *Proceedings of the IEEE,* Vol. 67, p. 1680–82, December 1979. (Reprinted in Part IV.)

[2] V. F. Kroupa. "Spectral purity of direct digital frequency synthesizers." *Proceedings of the 44th Annual Symposium on Frequency Control,* pp. 498–510, May 1990.

[3] V. F. Kroupa. "Principles of direct digital frequency synthesizers." *Proceedings: Kleinheubacher Berichte, Band* 36, pp. 663–78, 1993.

[4] V. F. Kroupa. "Discrete spurious signals and background noise in direct digital frequency synthesizers," *Proceedings of the 1993 IEEE International Frequency Control Symposium,* pp. 242–249, June 1993.

[5] V. F. Kroupa. "Intermodulation signals in direct digital frequency synthesizers." *Proceedings of the 8th European Frequency and Time Forum,* pp. 627–45, March 9–11, 1994.

[6] B. Giebel, J. Lutz, and P. L. O'Leary. "Digitally controlled oscillator." *IEEE Journal of Solid-State Circuits,* vol. 24, pp. 640–45, June 1989.

[7] P. H. Saul and D. G. Taylor. "A high-speed direct frequency synthesizer." *IEEE Journal of Solid-State Circuits,* vol. 25, pp. 215–19, 1990. (Reprinted in Part III.)

[8] Fang Lu, H. Samueli, J. Yuan, and Ch. Svensson. "A 700-Mhz 24-b pipelined accumulator in 1.2 μm CMOS for application as a numerically controlled oscillator." *IEEE Journal of Solid State Circuits,* vol. 28, pp. 878–86, August 1993. (Reprinted in Part X.)

[9] G. W. Small. "A frequency synthesizer for $10/2\pi$ kHz." *IEEE Transactions on Instrumentation and Measurement,* Vol. IM-22, pp. 34–37, March 1973. (Reprinted in Part II.)

[10] V. F. Kroupa. "Useful computer programs for investigation of spurious signals in DDFS." Paper specially written for Part XI.

[11] S. Mehrgardt. "Noise spectra of digital sine-generators using the table-lookup method." *IEEE Trans. ASSP-31,* pp. 1037–39, August 1983. (Reprinted in Part IV.)

[12] H. T. Nicholas III and H. Samueli. "An analysis of the output spectrum of direct digital frequency synthesizers in the presence of phase-accumulator truncation." *Proceedings of the 41st Annual Frequency Control Symposium,* 1987, Proc. pp. 495–502. (Reprinted in Part IV.)

[13] H. T. Nicholas III, H. Samueli, and B. Kim. "The optimization of direct digital frequency synthesizer performance in the presence of finite word length effects." *Proc. 42nd AFCS,* pp. 357–63, 1988. (Reprinted in Part IV.)

[14] J. F. Garvey and D. Babitch. "An exact spectral analysis of a number controlled oscillator based synthesizer." *Proc. 44th AFCS,* pp. 511–21, May 23–25, 1990. (Reprinted in part IV.)

[15] M. Robert Gray. "Quantization noise spectra." *IEEE Transactions on Information Theory,* vol. IT-36 (1990), No. 6, pp. 1220/44.

[16] M. Gernot and R. Winkler. "Introduction to robust statistics and data filtering." Presented at the *1994 IEEE Annual Symposium on Frequency Control (Tutorials),* May 31, 1994.

[17] V. F. Kroupa. "Digital to analog converters." A specially written paper for Part VIII, 1995.

[18] P. Robin Giffard and S. Leonard Cutler. "A low frequency, high resolution digital synthesizer." *1992 IEEE Frequency Control Symposium,* pp. 188–92. (Reprinted in Part V.)

[19] V. F. Kroupa. *Frequency Synthesis: Theory, Design et Applications.* London: Ch. Griffin; New York: John Wiley, 1973.

[20] Vadim V. Manassewitch. *Frequency Synthesizers, Theory and Design.* New York: John Wiley, 1976, 1980.

[21] V. F. Kroupa. "Amplitude of the general intermodulation product." *Proc. IEEE,* vol. 58, pp. 851–52, May 1970.

Part VI

Combination of DDFS
with Phase-Locked Loops (PLLs)

THERE are two reasons for combining DDFS with PLL. The first one is in instances where we need to extend the synthesizer range to higher frequencies while retaining the fine step capabilities. In the second case, the PLL may provide the necessary filtering of spurious frequencies; cf. the generator of the PAL calor-subcarrier in the paper by the editor in Part XI [VI-1].

We begin this Section with a specially written paper by the editor recalling from the PLL theory all basic properties and equations important for the design of the combined PLL-DDFS systems.

The next paper, again by the editor (1983), is an extension of the PLL theory in respect to more complicated PLL devices. It considers expected noise properties of the output signal and introduces the phase noise power spectral densities (PSD) invariant to frequency multiplications or divisions—the so-called h-representation.

The resolution (the smallest frequency step) in one-loop PLL frequency synthesizers is equal to the reference frequency, which is often in kHz or MHz ranges. Before the practical applications of DDFS, smaller frequency steps required a system of PLLs with dividers in between (e.g. [VI-2]). As a consequence, frequency synthesizers of this type were bulky devices.

One of the early designs used the output frequency of a DDFS as the reference to the divided VCO carrier [VI-3]. The spurious signals due to the Modulo-N operated accumulator were corrected with the assistance of a DAC output voltage introduced in the feedback loop. In addition, the authors recommend application of two or three different clock frequencies to get larger spurious signals out of the PLL path band, since the level of the remaining spurious modulation after the digital-to-analog compensation is enhanced by the square of the division factor in the feedback path.

The difficulty with enhancing DDFS spurious signals, due to the divider in the feedback path of the PLL, can be overcome with the assistance of a translating PLL (e.g. [VI-4]).

A synthesizer of this type, described by M. V. Harris, has been designed for use in a spread-spectrum communication system. Two PLLs are used with a simple DDFS to secure 100-Hz resolution across an RF bandwidth of 2 GHz. The feature is a simple decimal, or base-10 direct digital synthesis.

A real revolution in DDFS applications in PLL systems was introduced with the so-called fractional-N loop at the end of the 1970s (e.g. [VI-5, VI-6, VI-7]). This arrangement made it possible to enhance the resolution of the output frequency to small fractions of the reference frequency and to retain, at the same time, a reasonable natural frequency, damping factor, and switching times of the PLL loop.

The problems of the fractional-N frequency synthesizers are discussed by the editor in a specially written paper. First, he repeats the principles; then he points to the origin of the spurious phase modulation with the expected modulation index. Finally, he investigates its suppression with the assistance of a digital-to-analog converter (DAC). Since pulses in the PLL feedback path should be removed in a quasiperiodic manner [VI-1], three pulse-removing or pulse-"swallowing" circuits are investigated.

A new solution to the problem of spurious signals in fractional-N PLL frequency synthesizers was suggested by B. Miller and B. Conley (1991) (originally 1990 [VI-8]). The idea is to push away from the carrier spurious signals introduced by noninteger division. This is accomplished with the assistance of oversampling A/D conversion technology incorporated into the fractional-N synthesis process. The idea was nearly simultaneously investigated by other researchers [VI-9].

A GaAs IC version of the PPL fractional frequency synthesizer, reported by Naber et al. (1992), exhibits a substantial reduction of the settling time, size, and dissipation.

Readers interested in the progress of digital and IC technology in PLL devices will find some information in the paper by Mijuskovic and others [VI-10] reporting about "a family of standard cells for PLL applications in frequency synthesis." Chips, using a special 1.5 μm CMOS process, contain loop filter, ring oscillator, and phase-frequency detector. More information about sampled PLL systems may be found in Crawford [11].

REFERENCES

[VI-1] V. F. Kroupa. "Quasiperiodic omission of pulses." (A specially written paper for Part XI of this volume.)

[VI-2] J. C. Shanahan. "Uniting signal generation and signal synthesis." *Hewlett-Packard Journal,* pp. 2–13, December 1971.

[VI-3] B. Bjerede and G. Fisher. "An efficient hardware implementation for high resolution synthesis." *Proceedings of the 30th Annual Frequency Control Symposium,* pp. 318–21, 1976.

[VI-4] D. P. O'Rourke. "SOS synthesizer development." *Proceedings of the 40th Annual Symposium on Frequency Control,* pp. 370–72, 1986.

[VI-5] J. Gibbs and R. Temple. "Frequency domain yields its data to phase-locked synthesizer." *Electronics,* pp. 107–13, April 27, 1978.

[VI-6] D. D. Danielson and S. E. Froseth. "A synthesized signal source with function generator capabilities." *Hewlett-Packard Journal,* pp. 18–26, January 1979.

[VI-7] U. L. Rohde. *Digital PLL Frequency Synthesizers.* Englewood Cliffs, N.J.: Prentice-Hall, 1983.

[VI-8] B. Miller and B. Conley. "A multiple modulator fractional divider." *Proceedings of the 44th Annual Symposium on Frequency Control,* pp. 559–68.

[VI-9] T. A. D. Riley, A. Copeland, and T. A. Kwasniewski. "Delta-sigma modulation in fractional-N frequency synthesis." *IEEE Journal of Solid-State Circuits,* pp. 553–59, May 1993.

[VI-10] D. Mijuskovic, M. Bayer, T. Chomicz, N. Garg, F. James, P. McEntarfer, and J. Porter. "Cell-based fully integrated CMOS frequency synthesizers." *IEEE Journal of Solid-State Circuits,* pp. 271–78, March 1994.

[VI-11] J. A. Crawford. *Frequency Synthesizer Design Handbook.* Boston/London: Artech House, pp. xiii + 435, 1994.

Principles of Phase-Locked Loops (PLL)

VĚNCESLAV F. KROUPA

I. Introduction

A major advantage of DDFS is the possibility of widespread use of off-the-shelf IC chips. Their application results in low-volume, low-weight, and often power-saving devices. At the same time, we also appreciate short switching times and very high-frequency resolution.

There are shortcomings too: a lot of spurious signals and the limited range of output frequencies. (Today commercial DDFSs exceed the 100-MHz bound, and laboratory devices work in ranges over 500 MHz.) These difficulties can be alleviated with the assistance of the phase-locked loops, particularly in frequency synthesizers designed for microwave range. The advantage of small tuning frequency steps can be retained with application of so-called fractional-N PLLs.

In the following paragraphs, we summarize the properties of PLL with some design-leading ideas.

II. Principles

The task of phase-locked loops is to maintain coherence between input (reference) signal frequency, f_i, and the respective output frequency, f_o, via phase comparison. The theory is explained in many textbooks (e.g. [1,2]) and in practically all books on frequency synthesis. Here, we shall repeat all major features with some new achievements [3,4].

A. Basic Equations

Each PLL loop works as a feedback system shown in Fig. 1a. Its forward gain can be written as

$$G(s) = \frac{K_d K_o F(s)}{s} \qquad [1]$$

and a feedback gain as $F_M(s)$. With the assistance of the block diagram identity, we easily come to Fig. 1b.

To explain quantities encountered in the above equation, we will investigate a bit extended Fig. 1c. We see that the output phase $\varphi_o(t)$ is compared with the input phase $\varphi_i(t)$ in phase detector (ring modulator, sampling circuit, etc.). At its output we get a voltage, $V_d(t)$, proportional to the phase difference of the respective input signals [1, and others]

$$V_d(t) = \left[\varphi_i(t) - \varphi_o(t)\right] K_d \qquad [2]$$

The proportionality factor, K_d [volt/2π], is called the "phase

(This paper was specially written for this volume.)

detector gain." Next, $v_d(t)$ passes the loop filter, $F(s)$ (a low-pass filter attenuating "carriers" with frequencies $\omega_i = \omega_o$, and ideally all undesired side bands note that the useful signal $v_d(t)$ is a slowly varying DC component), the output voltage of which is given by the following convolution

$$v_2(t) = v_d(t) \otimes h_f(t) \qquad [3]$$

where $h_f(t)$ is the time response of the loop filter. After applying $v_2(t)$ on the frequency control element of the voltage-controlled oscillator (VCO), we get the output phase

$$\varphi_o(t) = \int \omega_o(t)dt = \omega_c t + \int K_o v_2(t)dt \qquad [4]$$

with ω_c being the VCO free-running frequency. The proportionality factor, K_o [2π Hz/volt], is designated as the oscillator gain. Since, in most cases, K_d and K_o are voltage dependent, the general mathematical model of a PLL is a nonlinear differential equation. Its linearization, justified in small signal cases (steady-state working modes), provides good insight into the problem. With the assistance of the Laplace transform and Fig. 1, we can write

$$\left[\Phi_i(s) - \Phi_o(s)F_M(s)\right]K_d K_o F(s) = \frac{\Phi_o(s)}{s} \qquad [5]$$

The ratio, $\Phi_o(s)/\Phi_i(s)$, the PLL transfer function, is given by

$$H(s) = \frac{\dfrac{K F(s)}{s}}{1 + \dfrac{K F(s)F_M(s)}{s}} \qquad [6]$$

with the loop gain K,

$$K = K_d K_o \qquad [7]$$

By comparing this equation with Fig. 1b, we easily arrive at the conclusion that the transfer function $H(s)$ is a product of two factors, that is,

$$H(s) = H'(s)\frac{1}{F_M(s)} \qquad [8]$$

For frequency synthesis design, the knowledge of $H'(s)$ is very important since this is the effective transfer function governing basic parameters of the loop, that is, the damping factor ξ and the natural frequency ω_n, which will be discussed later. After computing and normalizing the phase error, we get

$$\frac{\Phi'_e(s)}{\Phi_i(s)} = 1 - H'(s) \qquad [9]$$

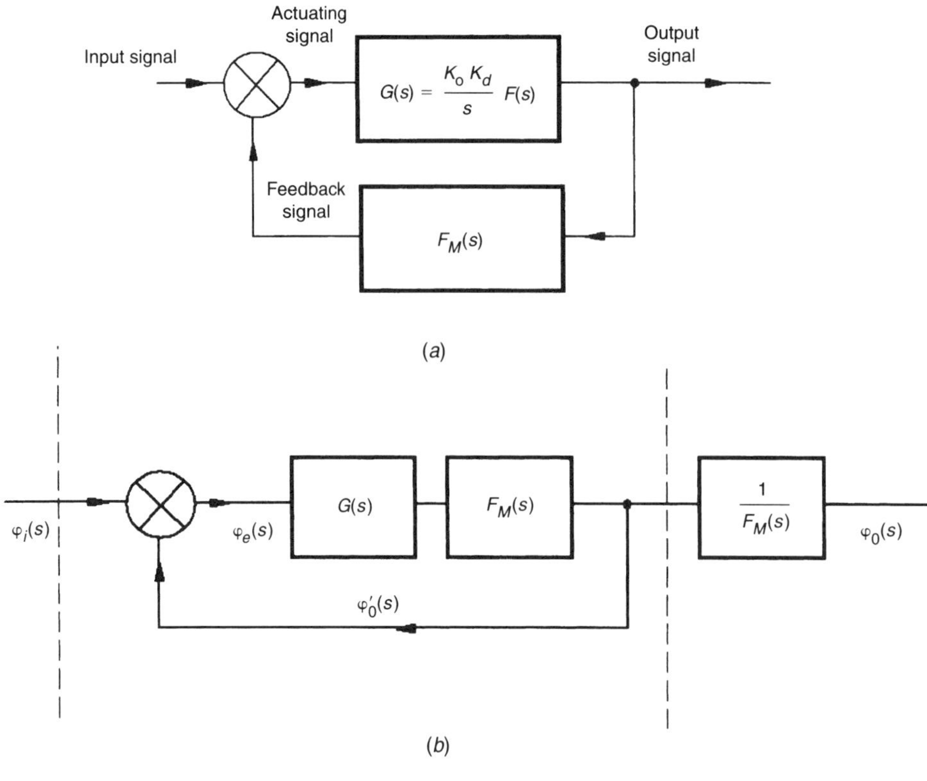

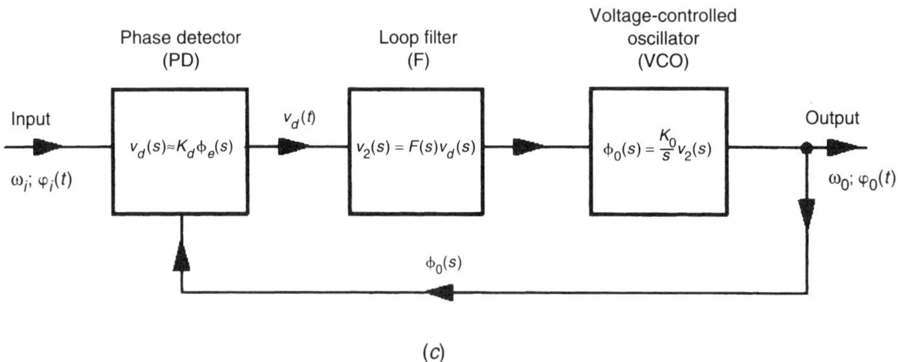

Fig. 1. Phase-locked loop: (*a*) basic feedback network; (*b*) simplification with the assistance of block diagram identity; (*c*) more extended block diagram $(FM(s) = 1)$.

B. Loop Filters

Before proceeding further, it is necessary to specify some basic loop filters (in the following we will silently assume that $F_M(s) = 1$ and consequently $H(s) = H'(s)$.

(*a*) *If no filter is used,* that is, $F(s) = 1$, the denominator in both eqs. (6) and (8) is a first-order polynomial in s and is therefore known as a first-order loop.

(*b*) *With the simple RC filter, we have* (see fig. 2*a*)

$$F(s) = \frac{1}{1 + sT_1} \qquad [10]$$

(*c*) *Or with a lag-lead filter* (Fig. 2*b*)

$$F(s) = \frac{1 + sT_2}{1 + sT_1} \qquad [11]$$

we arrive at the second-order loops (denominators in (6) and (8) are of the second-order).

In agreement with the servo technology [e.g. 5] we will use the "natural frequency" ω_n and the damping ξ factor as defining parameters

$$\omega_n = \sqrt{\frac{K}{T_1}}; \; \xi = \frac{1}{2}\sqrt{\frac{K}{T_1}}\left(T_2 + \frac{1}{K}\right) = \frac{\omega_n}{2}\left(T_2 + \frac{1}{K}\right) \quad [12]$$

After introducing the above relations into (6), we arrive at a general form for the transfer function $H(s)$

$$H(s) = \frac{s\omega_n\left(2\xi - \frac{\omega_n}{K}\right) + \omega_n^2}{s^2 + 2\xi\omega_n s + \omega_n^2} \qquad [13]$$

and similarly

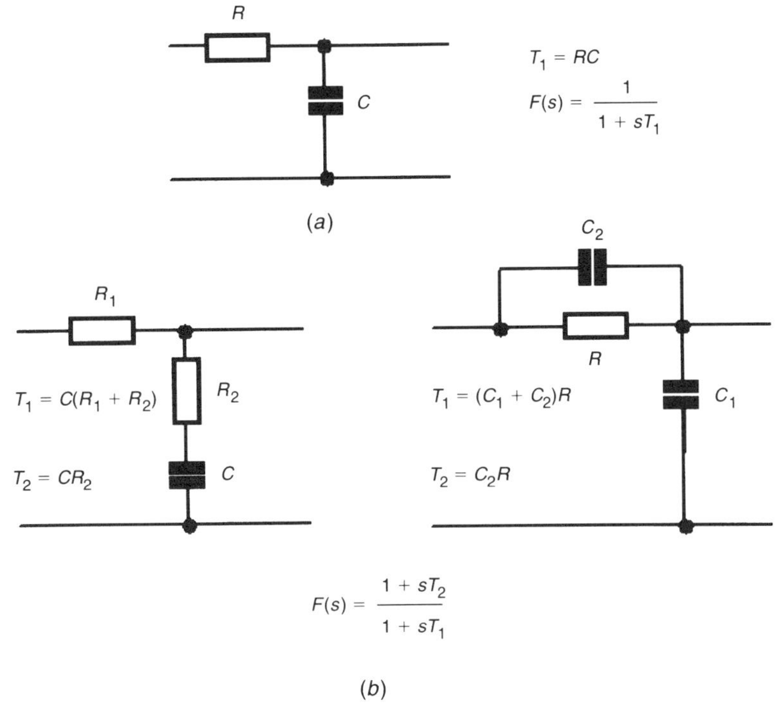

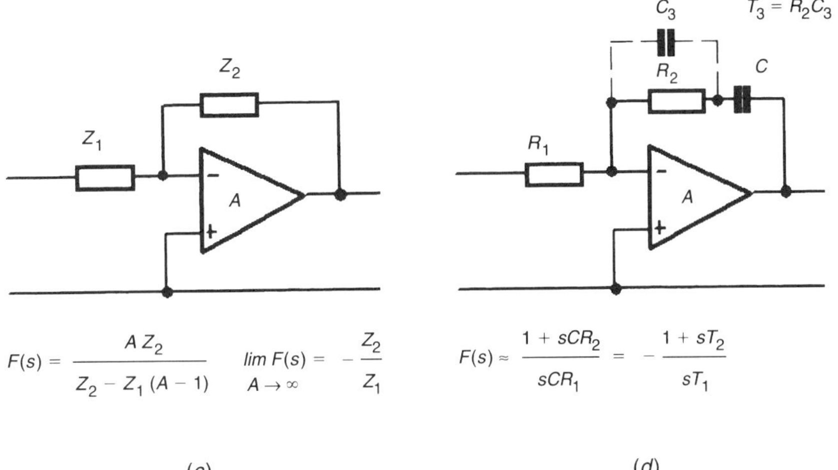

Fig. 2. Second-order phase-locked loop filters: (*a*) simple RC filter; (*b*) phase lag-lead or proportional plus integral networks (RRC filters); (*c*) general active filter configuration; (*d*) active phase lag-lead network (dashed is one of the third-order loop filter configurations).

$$1 - H(s) = \frac{s^2 - \frac{s\omega_n^2}{K}}{s^2 + 2\xi\omega_n s + \omega_n^2} \qquad [14]$$

The respective transfer functions for the so-called high-gain loop

$$\frac{\omega_n}{K} \ll 2\xi \qquad [15]$$

are plotted in the normalized form for $x = \omega/\omega_n$ in Fig. 3. (Observe that the above equations are valid for loops with simple RC filters if we introduce $\omega_n = 2\xi/K$); the respective transfer functions are plotted in Fig. 4. Note the change of the asymptotic slopes).

  (*d*) *Active lag-lead filter (Fig. 2d) with the transfer function*

$$F(s) = \frac{1 + sT_2}{sT_1} \qquad [16]$$

reduces (13) and (14) into

$$H(s) = \frac{2\xi\omega_n + \omega_n^2}{s^2 + 2\xi\omega_n s + \omega_n^2} \qquad [17]$$

$$1 - H(s) = \frac{s^2}{s^2 + 2\xi\omega_n s + \omega_n^2} \qquad [18]$$

Since in this case there are two integrators in the loop, we say that the loop is of the type 2. However, it still remains of the second-order, and its transfer functions are the same as for the high-gain loop. Cf. Fig. 3. Note that in instances where a di-

169

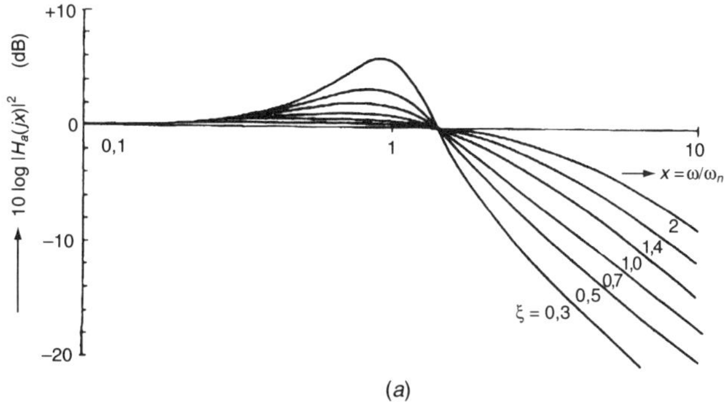

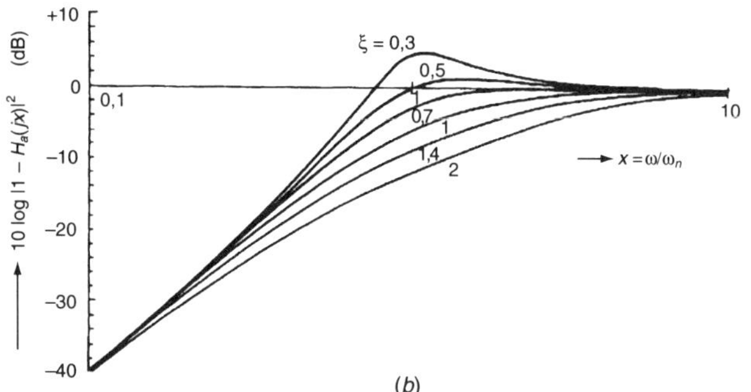

Fig. 3. Transfer functions of the second order "high-gain" phase-locked loops for six different damping factors; (a) $|H(jx)|^2$ and (b) $|1 - H(jx)|^2$.

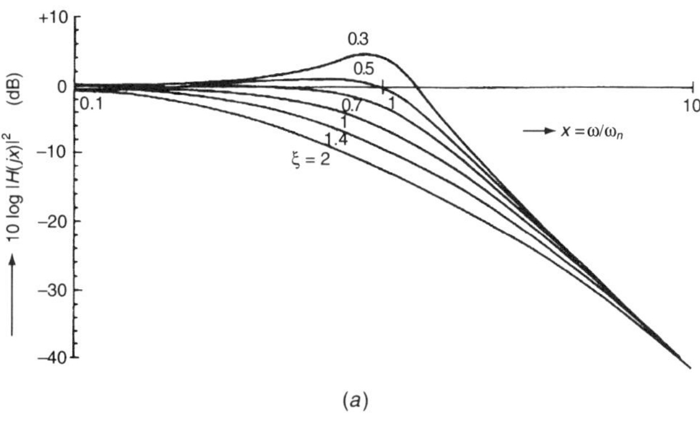

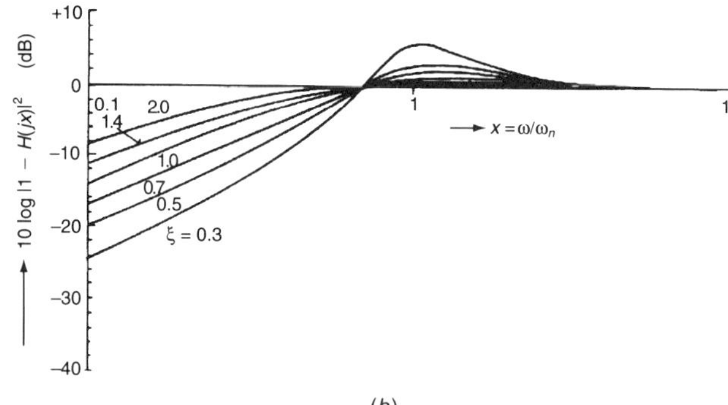

Fig. 4. Transfer function of the second-order phase-locked loop with simple RC filter; (a) $|H(jx)|^2$ and (b) $|1 - H(jx)|^2$.

vider by N or a multiplier by M is present in the feedback path we must introduce in the above equations $K' = K/N$ or $K' = KM$ instead of K in accordance with (6) and (8).

(e) Active lag-lead filter with additional RC section. Investigations of asymptotic approximation of eqs. (13) and (17) reveals that for a simple RC filter ($2\xi = \omega_n/K$) the slope of $|H(s)|^2$ outside of the pass band (i.e., for $\omega/\omega_n \gg 1$) is -40 dB/decade (cf. Fig. 4a), whereas for lag-lead filters it is only -20 dB/decade (cf. Fig. 3a). On the other hand, the slope of the phase error transfer function, $|1 - H(s)|^2$, in the stop band, that is, $\omega/\omega_n \ll 1$, is for all loops of the type 1 only 20 dB/decade (cf. Fig. 4b), but for loops of the type 2, 40 dB/decade (cf. Fig. 3b). As we will see later, in connection with noise properties of PLL frequency synthesizers, it is desirable to have both slopes in the respective stop band of $|H(s)|^2$ and $|1 - H(s)|^2$ 40 dB/decade. This task can be met only with the assistance of the type 2 loops modified by adding to the original loop filter another independent RC section. This can be easily done with a shunting capacitor C_3 shown dashed in Fig. 2d. As a consequence, the filter transfer function changes to

$$F(s) = \frac{1 + sCR_2}{sCR_1(1 + sC_3R_2)} = \frac{1 + sT_2}{sT_1(1 + sT_3)} \quad [19]$$

by putting

$$\frac{T_3}{T_2} = \frac{R_2C_3}{R_2C} = \frac{C_3}{C} = \kappa \quad [20]$$

We get for the PLL transfer function

$$H(jx) = \frac{j2\xi x + 1}{-j2\xi\kappa x^3 - x^2 + j2\xi x + 1}; \quad \left(x = \frac{\omega}{\omega_n}\right) \quad [21]$$

$$1 - H(s) = \frac{-j2\xi kx^3 - x^2}{-j2\xi\kappa x^3 - x^2 + j2\xi x + 1} \quad [22]$$

where parameters ξ are those defined for the original second-order PLL; see eq. (12) and ref. [6]. Squares of both transfer functions are shown in the normalized form in Fig. 5. Note that in the respective stop bands, they can be approximated with the following asymptotes

$$|H(j\omega)|^2_{\omega\gg\omega_n} \approx \left(\frac{\omega_n}{\omega}\right)^4 \frac{1}{\kappa^2} \quad [23]$$

$$|1 - H(j\omega)|^2_{\omega\ll\omega_n} \approx \left(\frac{\omega}{\omega_n}\right)^4 \quad [24]$$

(f) Fourth- and fifth-order loops. Very often we need to increase attenuation of the leaking "reference" signal or of some undesired mixing products. Solution of this problem may be provided with additional filtering elements. Recently, we have investigated properties of the fourth- and fifth-order loops [3,4]. They were formed by a parallel-T RC-network or a second-order active low-pass filter introduced in the second- or third-order loops of type 2; see Fig. 6.

In the first case (Fig. 6a), we appreciate the notch property of the transfer function at the desired frequency ω_{nf}.

By introducing the normalized frequency y as a function of the "notch" frequency (cf. Fig. 6a),

$$y = \frac{\omega}{\omega_{nf}}; \quad \omega_{nf} = \frac{1}{RC} \quad [25]$$

we get for the additive transfer function $F_{ad,T}$

$$F_{ad,T} \approx 1; \quad \varphi_{ad,T} \approx 230y \quad (°) \tag{27}$$

After introducing the normalized frequency x, we get

$$y = \frac{\omega}{\omega_n}\frac{\omega_n}{\omega_{nf}} = x\alpha \tag{28}$$

We see that for $x = 1$, and for $\omega_n/\omega_{nf} = \alpha \leq 0.03$, for example, the stability of the respective PLLs need not be degraded considerably ($\varphi_{ad,T} \approx 7°$) in contrast to a very large but selective attenuation at the notch frequency ω_{nf}.

In the second case of an additional low-pass filter (see Fig. 6b), we repeat only results important for the designer (the additive transfer function $F_{ad,A}$ and additive phase shift $\varphi_{ad,A}$)

$$F_{ad,T} = \frac{1}{1 + 2jdy - y^2} \tag{29}$$

where

$$y = \frac{\omega}{\omega_{cf}}; \quad d = \omega_{cf}T_2; \quad \omega_{cf} = \sqrt{\frac{1}{R^2 C_1 C_2}} \tag{30}$$

(ω_{cf} being the cutoff frequency and d the damping constant of the additional active filter). In the case where $y \ll 1$, we get form eq. (29) for the additive phase shift

$$\varphi_{ad,A} = 115dy = 115dx\alpha \quad (°) \tag{31}$$

and for reasonable values of d (≈ 0.5 to 0.7) and α (≈ 0.1)

$$\varphi_{ad,A} \approx 7° \tag{32}$$

However, the corner frequency where the original slope of $|H(jx)|^2$ acquires an additional slope of -40 dB/decade, that is, $\omega = \omega_n/\alpha$, is often too close to $\omega_r = \omega_i$. Another difficulty provides higher peaks of both $|H(jx)|^2$ and $|1 - H(jx)|^2$ in the neighborhood of $x \approx 1$. The last property is undesirable for low-noise design (see the transfer functions plotted in Figs. 7 and 8).

(g) A propagation time delay. A propagation time delay is generally introduced into phase locked loops with digital operation or sampling circuits. The proper solution would be with application of the z-transform. However, the zero order-sample and hold phase detector (PD) (cf. Fig. 9) used in PLL systems can be approximated with the following Laplace transform operations

$$\frac{1}{T}\left(\frac{1}{s} - \frac{1}{s}e^{-sT_r}\right) = e^{-sT_r}\frac{\sin\frac{sT_r}{2}}{\frac{sT_r}{2}} \approx e^{-\frac{sT_r}{2}} \tag{33}$$

Evidently, the additional phase shift is

$$\varphi \approx \frac{180}{\pi}\omega\frac{T_{ref}}{2} = 180\left(\frac{\omega_n}{\omega_r}\right)x \quad (°) \quad \left(x = \frac{\omega}{\omega_n}\right) \tag{34}$$

Since the third-order loops of the type 2, φ_{ad} cannot exceed about 10°, we get from the above relation that the reference frequency, ω_r, should be at least 20 times larger than the natural frequency, ω_n.

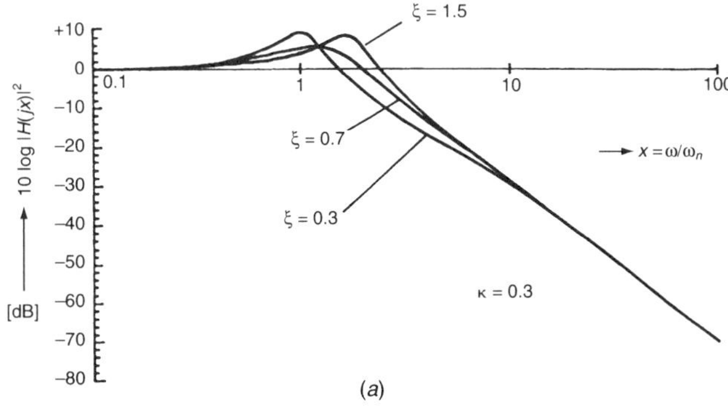

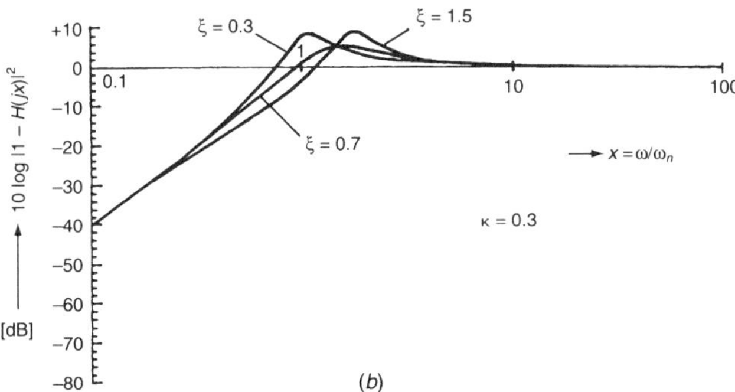

Fig. 5. Transfer function of the third-order phase-locked loops for three different damping factors, of the basic second-order loop; (a) $|H(jx)|^2$ and (b) $|1 - H(jx)|^2$.

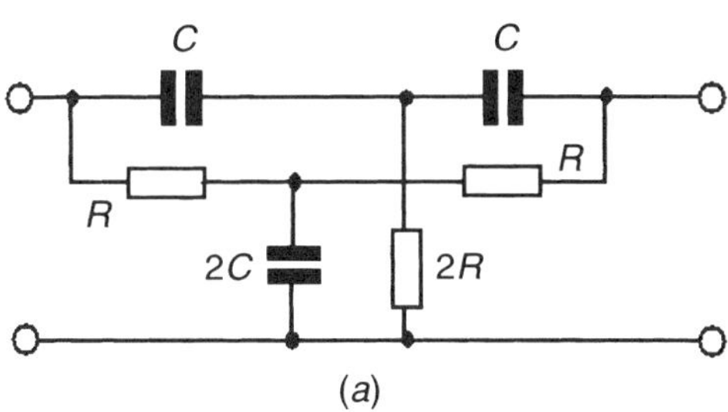

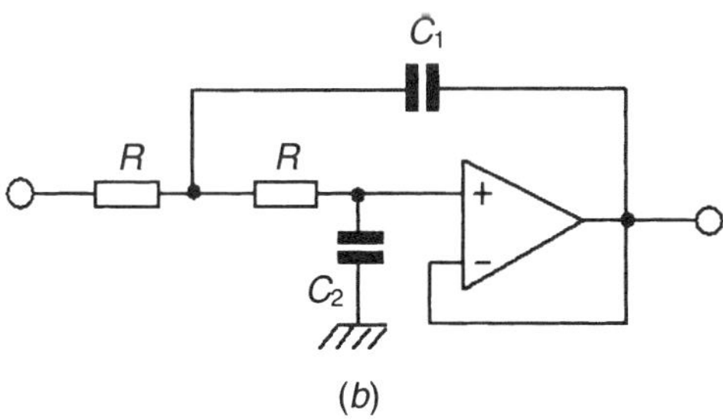

Fig. 6. (a) Additional parallel-T RC-network; (b) additional active second-order low-pass filter.

$$F_{ad,T} = \frac{1 - y^2}{1 - 4jy - y^2} \approx \begin{array}{ll} 1 & (y \ll 1) \\ 0 & (y = 1) \\ 1 & (y \gg 1) \end{array} \tag{26}$$

For small y ($y \ll 1$) the transfer function is $F_{ad,T}$, nearly frequency independent. However, the additive phase delay, $\varphi_{ad,T}$, should be respected because of the PLL stability:

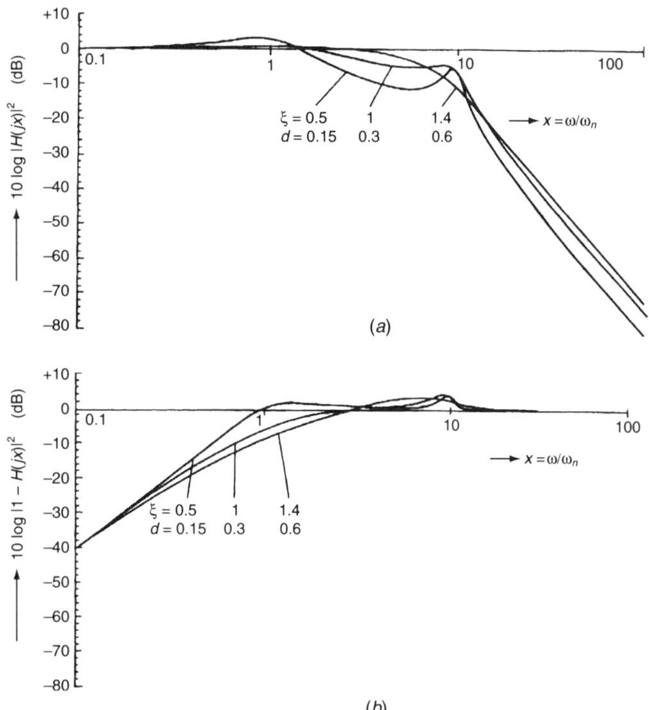

(a)

(b)

Fig. 7. Transfer function $|H(jx)|^2$ and $|1 - H(jx)|^2$ of the fourth-order phase-locked loop: second-order loop with an additional second-order active low-pass filter (ξ is that of the original second-order loop and $\alpha = 0.1$).

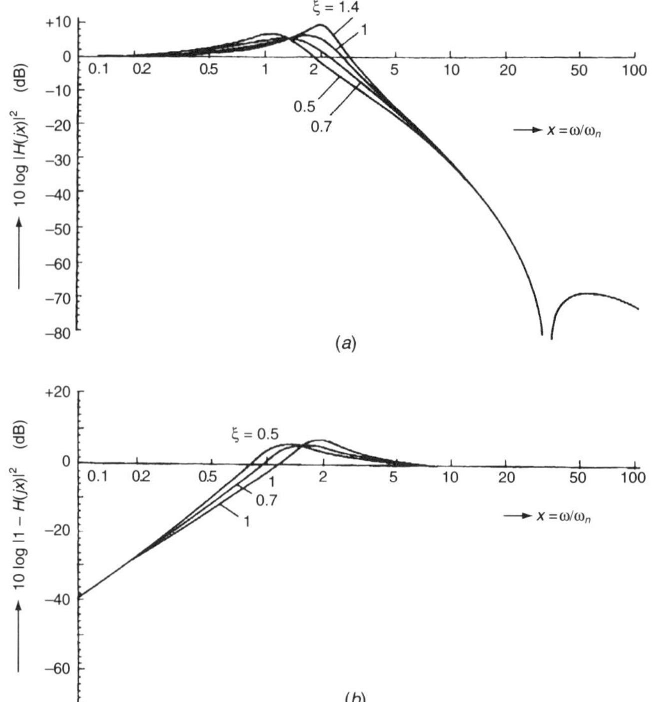

(a)

(b)

Fig. 8. Transfer function $|H(jx)|^2$ and $|1 - H(jx)|^2$ of the fifth-order phase-locked loop: third-order loop with $\kappa = 0.2$ and an additional parallel-T network with $\omega_n/\omega_{nf} = \alpha = .03$.

C. Stability of PLL

The phase-locked loops (PLL), as any feedback system, can oscillate if the denominator in (6) equals zero, that is,

$$1 + G(s)F_M(s) = 0 \qquad [35]$$

To this end, we solve the above equation and plot the locus of roots as a function of the gain K, in the s-plane. In the case where the respective curve enters the right-hand side of the plane (the real part of the root being positive), the feedback system is unstable. With this approach, we obtain a considerable degree of insight into the behavior of the respective PLL system.

We easily discover that all second- and third-order loops are unconditionally stable. However, in the presence of an excessive propagation delay or additional filtering sections, we recommend investigating the stability. With PLLs of type 1 and 2, the Bode plots provide a quick solution. After reverting to eq. (35), we get in the polar form

$$|G(j\varphi)|\,|F_m(j\varphi)|e^{j(\varphi_G+\varphi_M)} = -1 \qquad [36]$$

which is met if

$$|G(j\varphi)F_m(j\varphi)| = 1;$$
$$\varphi_G + \varphi_M = \pi(1 + 2k) \quad k = 0, 1, 2, \ldots \qquad [37]$$

For the open loop gain of the second-order loop type 2, we get with the assistance of Fig. 1 and eqs. (12) and (13)

$$G(j\omega) = K\frac{1 + j\omega T_2}{-\omega^2 T_1} = \frac{1 + 2jx\xi}{-x^2} \qquad [38]$$

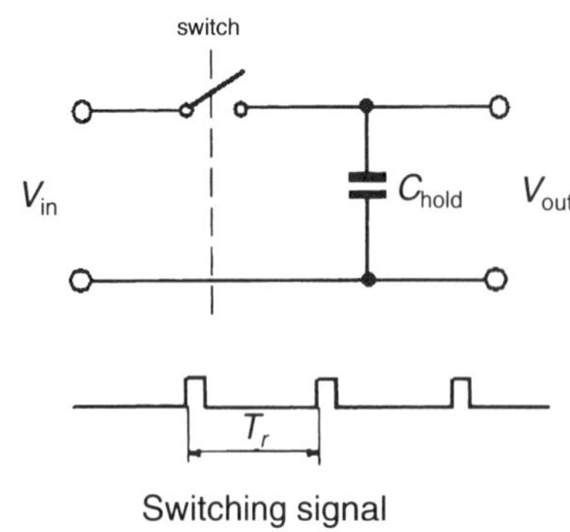

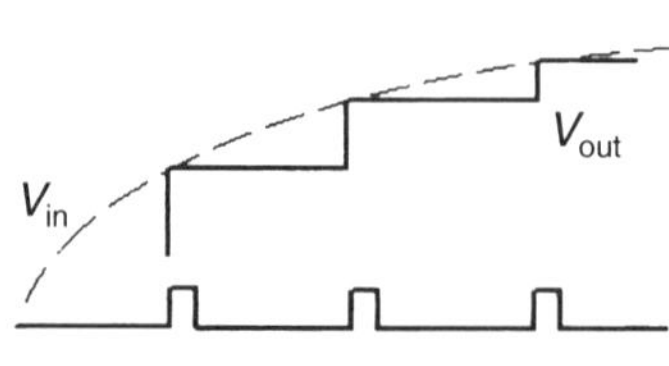

Fig. 9. Principle of the zero-order sample and hold circuit.

and for the third-order loop of type 2

$$G(j\omega) = -\frac{1}{x^2}\frac{1+2jx\xi}{1+2jx\xi\kappa} \qquad [39]$$

The respective phase shift around the loop is

$$\varphi = -180 + \frac{180}{\pi}\left[tg^{-1}(2x\xi) - tg^{-1}(2x\xi\kappa)\right] > -180 \quad [°] \qquad [40]$$

It is readily seen that for all x both PLL systems—eq. (38) and (39)—are unconditionally stable since the phase shift φ does not exceed $-180°$ all the time. This need not be true in instances where an excessive time delay is present. Cf. eq. (27), (31), or (34). In these cases, we have to investigate the so-called gain margin or phase margin, φ_{pm}, safety. The easiest way is to find out φ_{pm}. With the assistance of Fig. 10 we determine the normalized frequency x for which the magnitude of the $10\log|G(jx)|^2$ crosses the zero axis, and by using eq. (40) we compute the original phase margin from which we subtract the additional φ_{ad}. If the phase margin given by

$$\varphi_{pm} = 180 - \varphi - \varphi_{ad} \qquad [41]$$

is positive (generally about 20°), the PLL is stable.

D. Stability of the Higher Order System

Determine how stable the higher order systems are. Inspection of Figs. 5 and 10 together with eq. (40) reveals that the smaller the phase margin is, the higher the peaks of $10\log|H(jx)|^2$ are. By fitting the actual $|H(jx)|^2$ characteristic, at least in the pass band, to a suitable second-order loop characteristic, we can find an effective damping factor

$$\xi_{ef} = \frac{\varphi_{pm}}{100} \qquad [42]$$

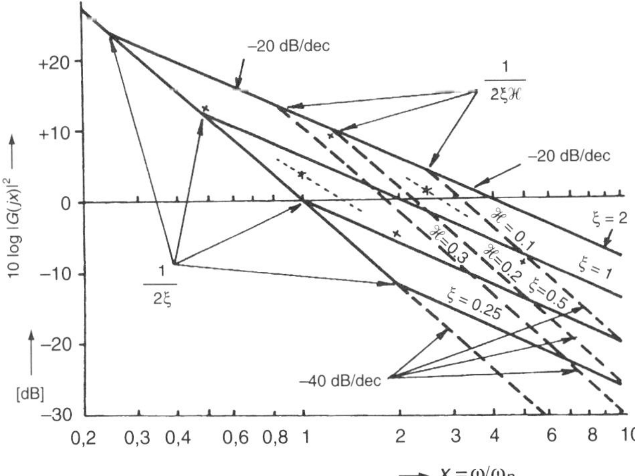

Fig. 10. Normalized open loop gains of second- and third-(dashed) order PLLs. Crosses indicate corrections needed to the asymptote approximations in the vicinity of the corner frequencies $1/2\xi$ and $1/2\xi\kappa$, respectively (3 dB) and (1 dB) at the half and two times of the corner frequencies.

E. Large-Signal Properties of PLLs

In the above discussion of PLL system properties, we have assumed "steady-state behavior." However, in instances where the loop is switched on sudden changes of input phases or frequencies occur (agile frequency synthesizers, etc.), the above theory fails to provide correct information. To remove this shortcoming, we will discuss the large-signal properties of PLLs as published in respective textbooks.

(a) Hold-in Range is the maximum frequency difference to be tolerated once the loop is locked (the phase lock being still hold)

$$\Delta\omega_H = \pm K_v \quad \text{(sine } PD) \qquad [43a]$$
$$\Delta\omega_H = \pm\pi K_v \quad \text{(sawtooth } PD) \qquad [43b]$$
$$\Delta\omega_H = \pm 2\pi K_v \quad \text{(phase-frequency detector)} \qquad [43c]$$

where K_v is the so-called velocity constant

$$K_v = K \cdot A \cdot F(0) \quad \text{(A is defined in Fig. 2}d\text{)} \qquad [44]$$

(b) Pull-out Frequency is the largest frequency step below which the loop does not skip cycles and remains locked; for high-gain second-order loops

$$\Delta\omega_{PO} \approx 1.8\omega_n(\xi + 1) \qquad [45]$$

(c) Pull-in Range is the maximum frequency difference $|\omega_c - \omega_i|$ for which the phase lock can always be acquired: for second-order loops

$$\Delta\omega_{p,2} \approx 2\sqrt{\omega_n K_v \xi} \qquad [46]$$

and for third-order loops

$$\Delta\omega_{p,3} \approx 1.24\sqrt[4]{\frac{K_v}{\omega_W \xi}\frac{1-\kappa}{\kappa^2}} \qquad [47]$$

For loops with a time delay, one generally finds

$$\Delta\omega_{P,\tau} \approx 2\sqrt{\omega_n K_v \xi(1 - \omega_n K_v \xi \tau^2)} \qquad [48]$$

More exact values for $\Delta\omega_{P,\tau}$ can be found in [8] where, in addition, $\Delta\omega_p$ for more complicated loop filters are considered.

(d) Lock-in Range is the maximum frequency difference for which neither discontinuity nor cycle skipping is expected. For first-order loops $\Delta\omega_L = \Delta\omega_H$, since second-order loops with lag-lead filters (RRC) can be considered for high frequencies as those of the first order, but with a reduced gain K_r

$$\Delta\omega_L \approx K\frac{T_2}{T_1} \approx 2\omega_n\xi \qquad [49]$$

On the contrary, for the second-order loops with a simple RC filter

$$\Delta\omega_L \approx \frac{\omega_n}{2\xi} \approx K \qquad [50]$$

Another very important design parameter, particularly for agile frequency synthesizers, is the whole settling time composed from pull-in time and lock-in time.

(e) Pull-in Time T_p is characterized by cycle skipping. For loops with RRC filters (cf. Fig. *2b,d*) and sine *PD*

$$2T_p\omega_n\xi \approx \left(\frac{\Delta\omega}{\omega_n}\right)^2 \qquad [51]$$

where $\Delta\omega$ is the initial detuning $|\omega_c - \omega_i|$. For sawtooth *PD*

$$2T_p\omega_n\xi \approx \frac{1.5}{\pi^2}\left(\frac{\Delta\omega}{\omega_n}\right)^2 \qquad [52]$$

and for modern phase-frequency detectors (PFD)

$$T_p\omega_n \approx \frac{\Delta\omega}{\omega_n} \qquad [53]$$

(f) Lock-in Time T_L. Investigations of transient phenomena in PLLs reveal that they do drop roughly in proportion to

$$e^{-\omega_n\xi t} \qquad [54]$$

As a consequence, we can estimate the lock time for an acceptable value of the phase error $\Delta\bar{\varphi}_{e,L}$ to be

$$T_L \approx \frac{-\ln\Delta\bar{\varphi}_{e,L}}{\omega_n\xi} \qquad [55]$$

By assuming $\Delta\bar{\varphi}_{e,L} \approx 0.02 \approx 1°$, we get for the second-order loops with linear phase detectors

$$T_L \approx \frac{4}{\omega_n\xi} \quad (\xi \leq 0.7) \qquad [56]$$

whereas for sine phase detectors [7]

$$T_L \approx \frac{10}{K_r} \qquad [57]$$

III. Conclusions

In the present paper we have summarized from the PLL theory all basic equations for the design of combined PLL-DDFS. Generally, some additional filter sections are introduced, either intentionally for improving PLL properties or unintentionally as spurious RC sections or mere time delays. We have also paid attention to these cases. (*Note:* all figures are reprinted from Kroupa [4]).

References

[1] F. M. Gardner. *Phase-Lock Techniques.* New York: John Wiley, 1966, 1979.

[2] W. C. Lindsey and C. M. Chie, eds. *Phase-Locked Loops.* New York: IEEE Press, 1985.

[3] V. F. Kroupa and L. Sojdr. "Phase-locked loops of higher orders." London, *IEE Conference Publication No. 303, Second International Conference on Frequency Control and Synthesis*, pp. 65–68, April 10–13, 1989.

[4] V. F. Kroupa. *Theory of Phase-Locked Loops and Their Applications in Electronics.* Prague: Academia, 1995 (in Czech).

[5] C. J. Savant, Jr. *Basic Feedback Control System Design.* New York, Toronto, London: McGraw-Hill, 1958.

[6] V. F. Kroupa. "Noise properties of PLL systems." *IEEE Transactions on Communications,* vol. COM-30, pp. 2244–53, October 1982.

[7] V. F. Kroupa. "Low-noise microwave-frequency synthesizers: Design principles." *IEE Proceedings,* vol. 130, Pt.H, pp. 483–88, December 1983.

[8] V. F. Kroupa. "Pull-in range of phase lock loops of the type two." *Archiv fuer Elektronik und Uebertragunstechnik,* vol. 39, pp. 37–44, January/February 1985 (in English).

Low-Noise Microwave-Frequency Synthesisers: Design Principles

V.F. KROUPA

Indexing terms: *Microwave systems, Synthesisers, Filters and filtering*

Abstract: In the present paper we have first recalled the basic phase-noise equations for a fairly general PLL system, and suggested their normalisation with respect to the square of the output frequency. In this way it is possible to proceed with a graphic solution for the noise spectrum, even in cases of complicated frequency synthesisers, using one and the same figure throughout, since the normalisation brings all sources to their frequency multiplication or division invariant noise power spectral densities. Further, we discuss the influence of the thermal noise, generated in low-pass PLL filters, and emphasise the importance of third-order type-2 loops for low-noise design. Theoretical findings are illustrated by two examples.

1 Introduction

Microwave generators with very low near-carrier phase noise and spurious signals are required for space communications and tracking systems and for modern radars, which use not only the amplitude of an echo signal, but also its phase, both as an aid to detection in the presence of clutter and to give instantaneous information about target velocity from the Doppler shift of the received signal.

A low-frequency crystal oscillator followed by a multiplier chain can meet the requirement, but provides an unacceptably high phase noise far from the carrier in communications systems. The last difficulty is usually solved with the assistance of phase-lock loop (PLL) systems. Unfortunately, in some ranges off the carrier, the multiplied noise of the phase detector and associated circuits often dominates the spectrum rather than the multiplied noise of the reference [1]. Resolution of the problem requires a trade-off between the complexity of the frequency synthesis system on the one hand and the price, size and weight, etc. on the other. In the following paragraphs we shall call attention to some design principles accompanied with practical examples.

2 Phase-lock loop

In this Section we rely on an earlier paper by the author [2] and therefore omit the tedious procedure of deriving the basic equations.

In Fig. 1 a general PLL arrangement is drawn with a phase detector (PD), a low-pass filter ($F_L(s)$), and a voltage-controlled oscillator (VCO) in the forward path. A mixer ($-$), an IF filter with the effective modulation transfer function ($F_M(s)$) [3, 4], and a divider ($\div N$) are in the feedback path. For completeness, we have placed a divider ($\div Q$) between the reference generator (RG) and the phase detector and a multiplier ($\times M$) between the RG and the second input to the mixer. However, we have to keep in mind that these two latter blocks, in an actual PLL system, are often replaced by more complicated frequency-synthesis circuits.

By assuming a locked loop, and by considering Fig. 1, we may write the following expression for the Laplace transform of the output-phase deviation:

$$\phi_{o,\,n} = \left[\left(\frac{\phi_{i,\,n}}{Q} + \phi_{DQ.\,n} - \phi_{DN.\,n} + \frac{V_{PD,\,n} + V_{F,\,n}}{K_d} \right) \frac{N}{F_M(s)} \right.$$

$$\left. + \phi_{MU.\,n} + M\phi_{i,\,n} - \phi_{MI.\,n} \right]$$

$$\times \frac{F_M(s)F_L(s)K_d K_o/Ns}{1 + F_M(s)F_L(s)K_d K_o/Ns}$$

$$+ \phi_{osc,\,n} \frac{1}{1 + \dfrac{F_M(s)F_L(s)K_d K_o}{Ns}} \tag{1}$$

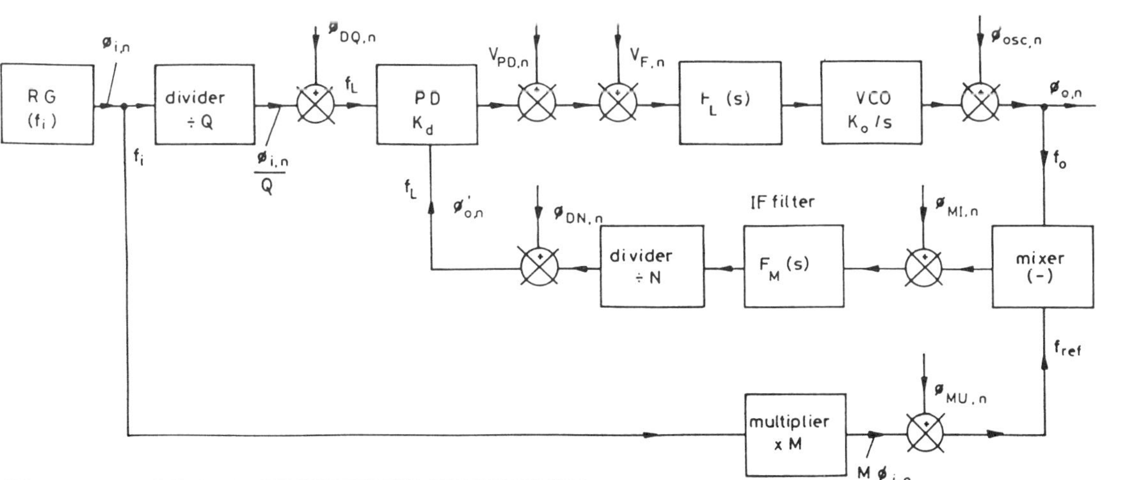

[Copyright *IEEE Trans.*, **COM-30**, Reference 2]

Fig. 1 *Block diagram of a general PLL system with additive noise sources*

Paper 2758H, first received 28th October 1982 and in revised form 29th July 1983

The author is with the Institute of Radio Engineering & Electronics, Czechoslovak Academy of Sciences, 182 51 Praha 8, Czechoslovakia

The noise components $\phi, \ldots, n$ and $V, \ldots, n$ are Laplace transformed quantities, i.e. $\phi, \ldots, n(s)$ etc. Eqn. 1 can be simplified with the assistance of the effective loop transfer function $H'(s)$ [5] into

$$
\phi_{o,\,n} = \left[\phi_{i,\,n}\left(M + \frac{N}{Q}\frac{1}{F_M(s)} \right) + \left(\phi_{DQ,\,n} - \phi_{DN,\,n} \right. \right.
$$
$$
\left. \left. + \frac{V_{PD,\,n} + V_{F,\,n}}{K_d} \right) \frac{N}{F_M(s)} + \phi_{MU,\,n} - \phi_{MI,\,n} \right]
$$
$$
\times H'(s) + \phi_{osc,\,n}[1 - H'(s)] \tag{2}
$$

where

$$
H'(s) = \frac{\dfrac{F_L(s)F_M(s)K_d K_o}{Ns}}{1 + \dfrac{F_L(s)F_M(s)K_d K_o}{Ns}} \tag{3}
$$

We see that the gain $K_d K_o = K$ is reduced in proportion to the division factor N to a new value

$$
K' = K/N = K_d K_o/N \tag{4}
$$

This is often compensated for by a DC amplifier incorporated into the $F_L(s)$ block.

Since all the noise components are random by nature and uncorrelated, we sum the respective power spectral densities $S_\phi, \ldots, (f)$ and we get the following expression for the overall phase spectral density: $F_m(s) \approx 1$

$$
S_{\phi_{o,\,n}}(f) = \left\{ S_{\phi_{i,\,n}}(f) \left[M + \frac{N}{Q} \right]^2 \right.
$$
$$
+ \left[S_{\phi_{DQ,\,n}}(f) + S_{\phi_{DN,\,n}}(f) + \frac{S_{V_{PD,\,n}}(f) + S_{V_{F,\,n}}(f)}{K_d^2} \right]
$$
$$
\left. \times N^2 + S_{\phi_{MU,\,n}}(f) + S_{\phi_{MI,\,n}}(f) \right\} |H'(\omega)|^2
$$
$$
+ S_{\phi_{osc,\,n}}(f)|1 - H'(\omega)|^2 \tag{5}
$$

By inspection of the above equation and Fig. 1 we see that the introduced noise deviations are associated with certain distinct 'carrier' frequencies, namely,

master oscillator frequency $= f_i$

output frequency $= f_o$ $\qquad\qquad\qquad$ (6)

phase detector locking frequency $= f_L$
$$
= f_i/Q
$$
$$
= (f_o - Mf_i)/N
$$

auxiliary reference frequency $= Mf_i$

which is often close to f_o, i.e. $Mf_i \approx f_o$.

We feel that some simplification together with a better insight might be gained by normalisation of eqn. 5 with respect to the square of the output frequency f_o^2, thus effectively deriving the phase-noise spectrum of a synthesiser invariant to frequency multiplication or division:

$$
h_o = \left\{ h_i\left(\frac{f_i}{f_o} \right)^2 \left(M + \frac{N}{Q} \right)^2 + S_{\phi L}(f)\left(\frac{N}{f_o} \right)^2 \right.
$$
$$
\left. + [S_{\phi_{MU,\,n}}(f) + S_{\phi_{MI,\,n}}(f)]\frac{1}{f_o^2} \right\} |H'(\omega)|^2
$$
$$
+ h_{osc}|1 - H'(\omega)|^2 \tag{7}
$$

where h_i stands for the sum

$$
\frac{S_{\phi i}}{f_i^2} \equiv h_i \equiv \frac{h_{i,\,-1}}{f^3} + \frac{h_{i,\,0}}{f^2} + \frac{h_{i,\,1}}{f} + h_{i,\,2} \tag{8}
$$

and similarly for h_{osc} and h_o.

The additive noise introduced by the frequency dividers, the phase detector itself and the circuits associated with the low-pass loop filter are summed up into a single term $S_{\phi L}(f)$:

$$
S_{\phi L}(f) = S_{\phi DQ}(f) + S_{\phi DN}(f) + \frac{S_{vPD}(f) + S_{vF}(f)}{K_d^2} \tag{9}
$$

In practice, we generally encounter three distinct PLL systems:

(a) no frequency mixing and multiplication is present, i.e.

$$
M = 0 \tag{10}
$$

In this case eqn. 7 reduces to

$$
h_o = \left[h_i + \frac{S_{\phi L}(f)}{f_L^2} \right] |H'(\omega)|^2 + h_{osc}|1 - H'(\omega)|^2 \tag{11}
$$

(b) the other extreme is in instances where

$$
M \gg \frac{N}{Q} \tag{12}
$$

By taking into account the above inequality we arrive at

$$
h_o \approx \left[h_i\left(\frac{Mf_i}{f_o} \right)^2 + \frac{N^2 S_{\phi L}(f) + S_{\phi MU}(f) + S_{\phi MI}(f)}{f_o^2} \right]
$$
$$
\times |H'(\omega)|^2 + h_{osc}|1 - H'(\omega)|^2 \tag{13}
$$

By considering that $Mf_i \approx f_o$ one can simplify eqn. 13 into the following form:

$$
h_o \approx \left[h_i + \frac{N^2 S_{\phi L}(f) + S_{\phi MU}(f) + S_{\phi MI}(f)}{f_o^2} \right] |H'(\omega)|^2
$$
$$
+ h_{osc}|1 - H'(\omega)|^2 \tag{14}
$$

(c) finally, there are frequency-synthesis systems where one of the input frequencies, f_{ref}, to the mixer is supplied by another PLL loop (cf. example 2 in Section 5). In that case we obtain

$$
h_o \approx \left[h_{ref}\left(\frac{f_{ref}}{f_o} \right)^2 + h_i\left(\frac{Nf_L}{f_o} \right)^2 \right.
$$
$$
\left. + \frac{N^2 S_{\phi L}(f) + S_{\phi MU}(f) + S_{\phi MI}(f)}{f_o^2} \right]
$$
$$
\times |H'(\omega)|^2 + h_{osc}|1 - H'(\omega)|^2 \tag{15}
$$

3 Additive noise in PLL systems

An earlier investigation of additive noises generated in frequency dividers, frequency multipliers, mixers and phase detectors [2] (see Appendix 8 for the results) has revealed that they are quite small and, in general, obey approximately the following experimental law:

$$
S_\phi(f) = \frac{10^{-14}}{f} + 10^{-16} \tag{16}
$$

However, there is a common experience that in actual PLL frequency synthesisers, the white-noise level of $S_{\phi L}(f)$ is generally higher than the -160 dB found for frequency dividers etc., and expected for phase detectors. Thus, the sole culprit must be the term

$$
\frac{S_{vF}(f)}{K_d^2} \tag{17}
$$

in eqn. 9. To obtain more insight into the problem, we have plotted in Fig. 2 the white-noise phase deviation $S_{\phi L}$

as a function of the natural frequency f_n of different phase-locked loops computed from published PLL noise characteristics. The inspection reveals an experimental relation

$$S_{\phi_L} \approx 10^{-11}/f_n \qquad (18)$$

for 'good cases', and a number of much 'worse cases'.

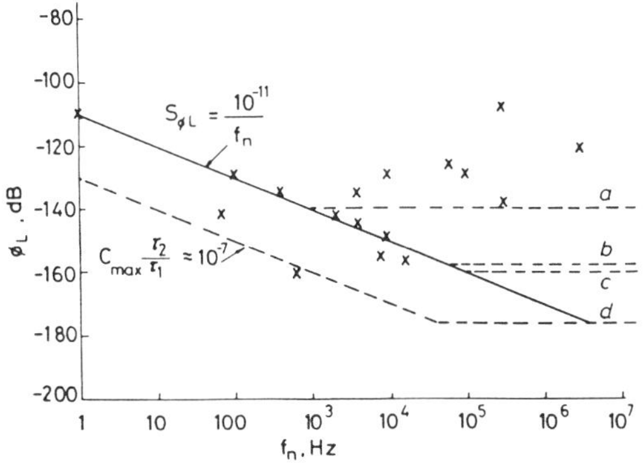

Fig. 2 *PSD S_{ϕ_L} of the additive white phase noise of different PLL systems ($\times$) together with practical and theoretical limits (– – –)*

a ECL level, *b* 10 μV level, *c* TTL level,
d Ring-modulator level

We have tried to find theoretical support for eqn. 18. With the assistance of the basic PLL equations (References 4 or 6) we have

$$\omega_n^2 = \frac{K_d K_o}{\tau_1} = \frac{K_d K_o}{RC} \qquad (19)$$

$$\omega_n = 2\zeta/\tau_2 \qquad (20)$$

and after dividing eqn. 19 by eqn. 20 we find another expression for ω_n, namely

$$\omega_n = \frac{K_d K_o}{2\zeta} \frac{\tau_2}{\tau_1} \qquad (21)$$

By assuming that the thermal noise of the loop-filter resistor R is responsible for S_{ϕ_L} we obtain

$$S_{\phi_L} = \frac{4kTR}{K_d^2} \qquad (22)$$

After substituting for R from eqn. 19 into eqn. 22

$$S_{\phi_L} = \frac{4kTK_o/\omega_n^2 C}{K_d} \qquad (23)$$

and for ω_n from eqn. 21 we have

$$S_{\phi_L} = \frac{4kT\zeta}{\pi f_n K_d^2 C} \frac{\tau_1}{\tau_2} \qquad (24)$$

Investigation of the above equation reveals that S_{ϕ_L} is indeed inversely proportional to f_n since neither the damping factor ζ and the phase detector gain K_d, nor the ratio of the time constants, τ_1/τ_2, change appreciably from loop to loop, and the capacitance C cannot exceed some C_{max} (for dimensional considerations on the one hand and increasing flicker noise on the other).

By considering some reasonable numerical values, i.e.

$$\zeta = 0.7, \; K_d = 5/2\pi \text{ V/rad}, \; C_{max} = 10^{-6} \text{ F}, \; \tau_1/\tau_2 \approx 10$$

we arrive, eventually, at

$$S_{\phi_L} \approx \frac{10^{-13}}{f_n} \qquad (25)$$

as a limit for S_{ϕ_L}. Note, that there is practically no mea-

surement point in Fig. 2 below this level, suggesting that for the loops considered C is less than a C_{max} of 1 μF.

4 Modification of output noise by loop filters

There is a general belief that in the PLL passband the output noise is that of the reference, and in the stop band that of the voltage-controlled oscillator (VCO). However, this is a simplification, which is true or nearly true in instances where the loop is of the third order and type 2 [2, 7]. The transfer functions of such a loop are given by

$$H(s) = \frac{s\omega_n(2\zeta - \omega_n/AK) + \omega_n^2}{s^2 + 2\zeta\omega_n s(1 + s^2 x/\omega_n^2) + \omega_n^2} \qquad (26)$$

and

$$1 - H(s) = \frac{s^2(1 + 2\zeta x s/\omega_n) + 4\theta s\omega_n^2/AK}{s^2 + 2\zeta\omega_n s(1 + s^2 x/\omega_n^2) + \omega_n^2} \qquad (27)$$

where ω_n and ζ have been defined by eqns. 19 and 20, and x is the ratio of the time constant τ_3 of an additional RC section to τ_2 (see Figs. 3a–c), i.e.

$$x = \tau_3/\tau_2 \qquad (28)$$

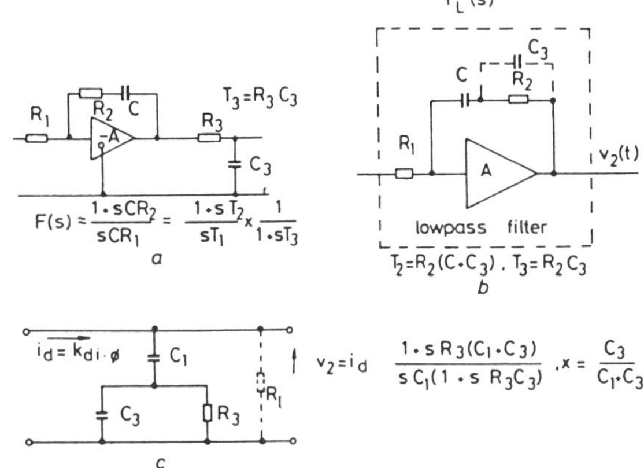

Fig. 3 *Three types of integrating phase lag-lead filters*

a Integrating lag-lead filter with additional RC section
b Integrating lag-lead filter with additional leaking capacitor C_3
c Current integration with additional leaking resistor R_3

The advantage of such a loop is illustrated in the following manner: for a conventional second-order loop, the noise within the loop passband, i.e. the region for which the condition

$$\omega \ll \omega_n \qquad (29)$$

holds, eqns. 11 or 14 approximate to

$$h_o \approx \left[h_i + \frac{N^2 S_{\phi_L}}{f_o^2} \right] + h_{osc} \frac{\omega^2}{\omega_n^4} [\omega^2 + (4\theta\omega_n^2/AK)^2] \qquad (30)$$

Evidently, the condition that the second term on the right-hand side in the above equation is negligible, particularly in instances where h_{osc} is proportional to $1/f^3$, requires a very large operational-amplifier gain.

For the second-order PLL stop band, i.e. if

$$\omega \gg \omega_n \qquad (31)$$

the output noise approximates to

$$h_o \approx \left[h_i + \frac{N^2 S_{\phi_L}}{f_o^2} \right] \omega_n^2 \frac{\omega^2(2\zeta - \omega_n/AK)^2 + \omega_n^2}{\omega^4[1 + (2\zeta\omega x/\omega_n)^2]} + h_{osc} \qquad (32)$$

To attenuate the white phase noise of $N^2 S_{\phi L}/f_o^2$ safely below h_{osc}, especially if the latter is proportional to $1/f^3$, the loop transfer function $H(\omega)^2$ must be proportional to $1/f^4$. This is the characteristic of the second-order loop with simple RC filter ($2\zeta = \omega_n/K$) or of the third-order type-2 loop which is achieved by adding extra filtering within a second-order loop. If simple RC filtering is used the transfer function approximates to

$$|H(\omega)|^2_{\omega \gg \omega_n} \approx \left(\frac{\omega_n}{\omega}\right)^4 \qquad (33)$$

or, where an integrating lag-lead filter with an additional RC section [7] is used (cf. Figs. 3a and b) the transfer function simplifies to

$$|H(\omega)|^2_{\omega \gg \omega_n} \approx \left(\frac{\omega_n}{\omega}\right)^4 \frac{1}{x^2} \qquad (34)$$

Note that an integrating loop filter is also preferable in the range $\omega \ll \omega_n$.

It has been shown elsewhere [7] that the safe phase margin (the loop stability) and a reasonably low overshoot are met with $x = 0.3$ and $\zeta = 0.7$ of the basic second-order loop (see Fig. 4). As a consequence, the attenuation of the

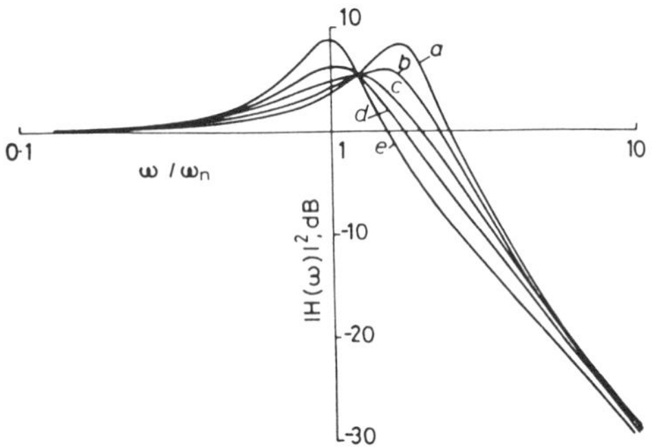

Fig. 4   Square of transfer function $|H(\omega)|^2$ as a function of normalised frequency ω/ω_n of the third-order type-2 PLL

$x = 0.3$
$a \quad \zeta = 1.5, \quad b \quad \zeta = 1.0, \quad c \quad \zeta = 0.7,$
$d \quad \zeta = 0.5, \quad e \quad \zeta = 0.3$

discrete spurious signals originating in the PLL loop is reduced about 10 dB compared with a simple RC system, cf. eqns. 33 and 34.

5   Examples

5.1   An ultra-low-noise microwave synthesiser
(According to G.D. Alley *et al.* [Reference 8])
A simplified block diagram is reproduced in Fig. 5. Evidently for its noise solution we shall apply eqn. 11 and

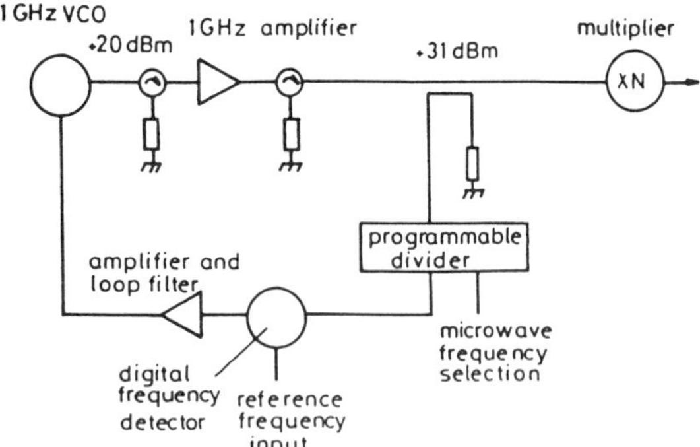

[Copyright *IEEE Trans.* **MTT-27**, Reference 8]

Fig. 5   *Block diagram of the microwave synthesiser according to G.D. Alley*

draw first the piecewise linearised noise characteristics h_i (of an ordinary ~ 5 MHz crystal oscillator) and h_{osc} (as computed for $Q = 7500$ from Fig. 8 in Reference 8) in Fig. 6; both noise spectra have been normalised in accordance

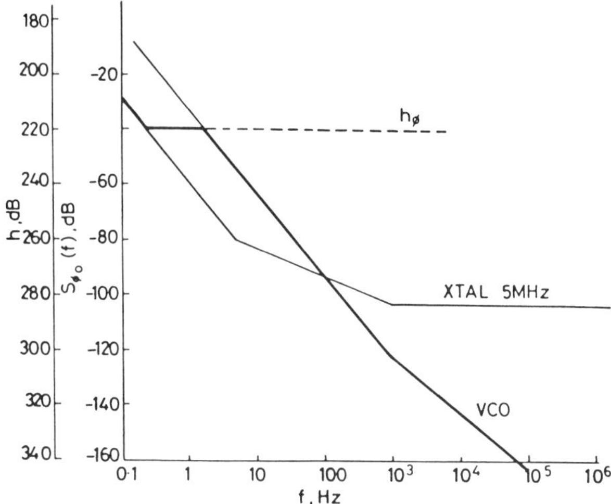

Fig. 6   *Asymptotic noise solution of an 'ultra-low-noise microwave synthesiser'* [8]

—— expected noise characteristics of VCO and crystal oscillators
- - - - expected level of noise in PLL loop filter systems
—— asymptotic solution of the output noise of the respective frequency synthesiser

with eqn. 8 and we see that the crossover point is in the neighbourhood of 100 Hz, and it would require the noise due to PD amplifiers etc. to be $S_{\phi L} \approx 10^{-27.3+11} = 10^{-16.3}$.

However, such a low level in accordance with Fig. 2 is nearly impossible to realise. By lowering f'_n we eventually arrive at $f'_n \approx 1$ Hz (as the authors did) with $S_{\phi L} \approx 10^{-11}$. Since the loop frequency, f_L, is approximately 3.10^5 Hz, eqn. 34 reveals a comfortable attenuation of -210 dB of the phase detector ripple components of frequency f_L. If the discrete spurious level of -120 dB at 1 GHz has been attained, the leakage of f_L signals through the phase detector is evidently not attenuated by the phase-lock feedback loop.

5.2   Low-noise synthesisers for radar and communications
(K.R. Slinn, P.A. Volckman and E. Scherer [9])
We shall start with loop 1 (see Fig. 1 in Reference 9) and assume a similar 5 MHz reference crystal oscillator as in the previous example and a VCO of $Q_L = 10$. We compute an approximate h_{osc} with the assistance of formulas summarised in the Appendix 8.

$$f_{01} \approx 440 \text{ MHz} \quad Q_{L1} \approx 10$$

$$h_{osc,1} = 10^{-15.6}/f^3 + 10^{-19.6}/f^2 + 10^{-32.3}$$

The loop frequency $f_{L,1}$ and the respective natural frequency $f'_{n,1}$ are

$$f_{L,1} = 10 \text{ KHz} \quad f'_{n,1} \approx 300 \text{ Hz}$$

Since the configuration of loop 1 is that of eqn. 11 we find with the assistance of Fig. 2 $S_{\phi L,1} \approx 10^{-13.5}$ from which $h_{\phi L,1} = S_{\phi L,1}/f_{L,1}^2 = 10^{-21.5}$. The asymptotic noise characteristics are plotted in Fig. 7.

Next we shall investigate loop 2. We shall start again with eqn. 11, using the same h_i (5 MHz crystal oscillator) and $h_{osc,2} = h_{osc,1}$ (VCO with the same Q_L). With the assistance of Fig. 2 we find for

$$f_{L,2} = 1 \text{ MHz} \quad \text{and} \quad f'_{n,2} \approx 10 \text{ KHz}$$

$$S_{\phi_{L,2}} \approx 10^{-15} \quad \text{and} \quad h_{\phi_{L,2}} \approx 10^{-27}$$

The asymptotic noise characteristics are plotted in Fig. 7.

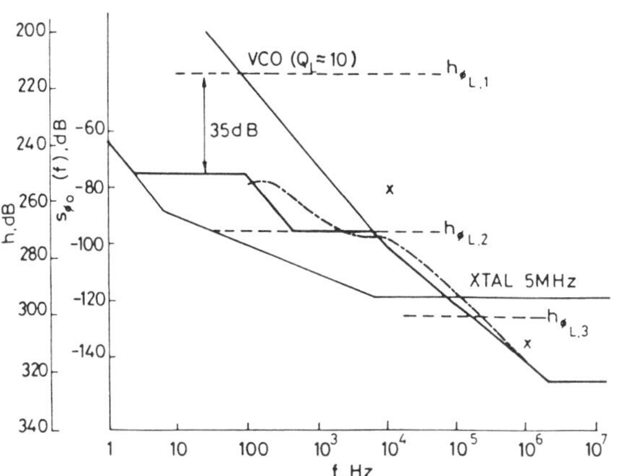

Fig. 7 *Asymptotic noise solution of 'low noise synthesisers for radar and communications' [9]*

——— expected noise characteristics of VCO and crystal oscillators
– – – expected level of noise in PLL loop filter systems
——— asymptotic solution of the output noise of the respective frequency synthesiser
–·–·–· taken from Fig. 2 in Reference 9
× estimated level of spurious signals at $f_o \simeq 570$ MHz

By solving the noise of the summing loop 3 we use eqn. 15 and put

$$h_{ref} = h_{0,2} \quad h_i = h_{0,1} \quad N = 1 \quad f_{L,3} \approx 11 \text{ MHz}$$

$$h_{osc,3} \approx h_{osc,2} \approx h_{osc,1}$$

Evidently, the natural frequency $f_{n,3}$ will be of the order

$$f_{n,3} \approx 10^5 \text{ Hz} \quad S_{\phi_{L,3}} \approx 10^{-16} \quad h_{\phi_{L,3}} \approx 10^{-30}$$

The noise equation is given by

$$h_{0,3} \approx \left[h_{0,2} + h_{0,1} \left(\frac{11 \times 10^6}{620 \times 10^6} \right)^2 + 1 \times 10^{-30} \right]$$

$$\times |H_3(\omega)|^2 + h_{osc,3} |1 - H_3(\omega)|^2$$

and the resulting noise characteristic is shown in Fig. 7 by thickened lines and compared with the result obtained by Slinn *et al.*

What remains is the evaluation of the level of spurious signals. Assuming a phase detector with a small leakage (sampling phase detector, phase detector followed by an integrator) of about -90 dB then

$$h_{sp,1} \approx -90 \text{ (phase detector leakage)}$$

$$-80 \, (-20 \log f_{L,1})$$

$$-35 \, (-20 \log (f_{L,3}/f_{0,3}))$$

$$-50 \, (-40 \log (f_{L,1}/f_{n,1}) - 20 \log x)$$

$$h_{sp,1} \approx -255 \text{ dB}, \ldots, S_{\phi_{sp,\, 10\,kHz}} = -79 \text{ dB}$$

$$h_{sp,2} \approx -90 - 120 - 70 - 30 = -310 \text{ dB}, \ldots, S_{\phi_{sp,\, 1\,MHz}}$$

$$= -134 \text{ dB}$$

6 Conclusions

In the present paper we have first recalled the basic phase-noise equation for a general second-order PLL system. In order to ease solution of noise problems of frequency synthesisers we have introduced noise power spectral densities invariant to frequency multiplication or division; this is the normalised h-representation. This procedure eases the graphic construction of the expected noise levels, and at the same time allows the contributions of individual sources to be examined.

It has been proved both theoretically and experimentally that the thermal noise generated in the low-pass loop filter dominates certain parts of the PLL noise characteristic. In this connection we have shown that the choice of the third-order type-2 PLL configurations is imperative for a low-noise design.

Coincidentally, the steep slope of $|H'(\omega)|^2$ (of 40 dB/decade) outside of the pass band is often necessary to keep some discrete spurious signals (the loop frequency) below a prescribed level.

Finally, the usefulness of the h-representations has been demonstrated with the assistance of two practical examples.

7 References

1 PAYNE, J.B.: 'Synthesiser designs depend on satcom uses', *Microwaves*, 1980, **19**, pp. 47–53, 99
2 KROUPA, V.F.: 'Noise properties of PLL systems', *IEEE Trans.*, 1982, **COM-30**, pp. 2244–2253
3 KROUPA, V.F.: 'Frequency synthesis: Theory, design and applications' (Griffin, London, 1973) p. 234
4 GARDNER, F.M.: 'Phase-lock techniques' (J. Wiley, New York, 1979, 2nd edn.) p. 147
5 KROUPA, V.F.: 'Frequency synthesis: Theory, design and applications' (Griffin, London, 1973) Chap. 7
6 KROUPA, V.F.: 'Frequency synthesis: Theory, design and applications' (Griffin, London, 1973) Chap. 6
7 KROUPA, V.F.: 'Spectral properties of third order phase-locked loop'. Proc. 5th summer symposium on circuit theory, Kladno, 1977, main lectures, pp. 201–215
8 ALLEY, G.D., and WANG, Hau-Chin: 'An ultra-low noise microwave synthesiser', *IEEE Trans.*, 1979, **MTT-27**, pp. 969–974
9 SLINN, K.R., VOLCKMAN, P.A., and SCHERER, E.: 'Low-noise synthesisers for radar and communications', *IEE Proc., Part H, Microwaves, Opt. & Antennas*, 1983, **130**, pp. 430–436
10 BURGOON, R., and WILSON, R.L.: 'Performance results of an oscillator using the SC cut crystal'. Proc. 33rd Annual Frequency Control Symposium, Atlantic City, 1979, pp. 406–410
11 ANDRICOS, C.: 'An L- and S-band radar exciter using agile low noise phase locked loop synthesisers'. Proc. 34th Annual Frequency Control Symposium, Philadelphia, 1980, pp. 202–212
12 VAN DER ZIEL, A.: 'Noise, sources, characterisation, measurement' (Prentice Hall, Englewood Cliffs, 1970) pp. 148–156
13 KERR, A.R.: 'Shot-noise in resistive-diode mixers and the attenuator noise model', *IEEE Trans.*, 1979, **MTT-27**, pp. 135–140
14 KEEN, N.J., VAN DER ZIEL, A., and SCHMIDT, R.R.: 'The effect of diode parameters on GaAs diode mixers in the 40–400 GHz range', *IEE Proc. H, Microwaves, Opt. & Antennas*, 1982, **129**, pp. 99–102
15 ALM, R.W.: 'Harmonic mixers elevate spectrum analysis to EHF', *Microwaves*, 1981, **20**, pp. 67–69
16 EGAN, W.F.: 'Frequency synthesis by phase lock' (Wiley, New York, 1981)
17 BLACHOVICZ, L.F.: 'Dial any channel to 500 MHz', *Electronics*, 1966, **39**, pp. 60–69
18 McALLISTER, P.A.: 'Phase-lock techniques for synthesis of microwave frequencies', *IEE Proc. H, Microwaves, Opt. & Antennas*, 1980, **127**, pp. 112–115
19 ŠOJDR, L.: Private communication
20 ROBINS, W.P.: 'Low-phase noise synthesisers for SHF satcom', *IEE Proc. H, Microwaves. Opt. & Antennas*, 1983, **130**, pp. 445–450
21 SCHERER, D.: 'Design principles and test methods for low phase noise R.F. and microwave sources'. Hewlett-Packard RF and Microwave Symposium, October 1978

8 Appendix

8.1 Noise properties of PLL building blocks

The problem has been discussed in depth by the author in Reference 2 and here only the results necessary for the evaluation of the phase-noise properties of PLL frequency synthesisers are summarised.

8.1.1 LC oscillator phase noise

$$h_{\mathrm{osc}} \equiv \frac{S_{\phi_{\mathrm{osc}}}(f)}{f_o^2} \equiv \frac{1}{f^3} \times \frac{10^{-11.6}}{Q_L^2} + \frac{1}{f^2} \times \frac{10^{-15.6}}{Q_L^2}$$
$$+ \frac{1}{f} \frac{10^{-11}}{f_o^2} + \frac{10^{-15}}{f_o^2} \tag{35}$$

8.1.2 Crystal oscillators

Since the product Qf_o is nearly constant in the whole range from 5 to 1000 MHz (it holds for SAW resonators too) the oscillator phase noise can be written as a function of the resonating frequency f_o only

$$h_{\mathrm{osc}} \equiv \frac{1}{f^3} \times 10^{-37.25} \times f_o^2 + \frac{1}{f^2} \times 10^{-39.4} \times f_o^2$$
$$+ \frac{1}{f} \times \frac{10^{-12.15}}{f_o^2} + \frac{10^{-15}}{f_o^2} \tag{36}$$

Note that both eqns. 35 and 36 are semi-theoretical and semi-experimental; actual h-coefficients may differ by -2 to $+1$-order (the latest 10 MHz crystal oscillators are reported to be better [10]).

8.1.3 Frequency dividers

Experimental findings revealed

$$S_{\phi_D}(f) \approx \frac{10^{-14.7}}{f} + 10^{-16.5} \tag{37}$$

where the white-noise level is caused by the shot noise in a TTL gate.

Larger loading resistors and smaller voltage differences between 'low' and 'high' states in ECL digital circuits are responsible for a higher white phase-noise level of about -140 dB/Hz. Andricos [11] published a better result

$$S_{\phi_D}(f) \approx \frac{10^{-11.3}}{f} + 10^{-14.7} \tag{38}$$

Also, Scherer's findings are of the same order [21].

8.1.4 Frequency multipliers

By referring the output noise of the best frequency multipliers to the input we have the following experimental relation:

$$S_{\phi_{\mathrm{MU.\,inp}}}(f) \approx \frac{10^{-14}}{f} + 10^{-16.5} \tag{39}$$

With step-recovery diodes we encounter a larger flicker noise, namely,

$$\frac{10^{-12.9}}{f} \tag{40}$$

8.1.5 Frequency mixers

There is both theoretical and experimental evidence [12–15] that the additive noise due to the mixers is quite small and of the order of the loading circuit noise. As a consequence we have for the white-noise region

$$S_{\phi_{\mathrm{MI}}}(f) \approx \frac{kT}{P_{\mathrm{in}} \times 10^{-L/10}} \tag{41}$$

where L is the conversion loss in dB and P_{in} is in watts or

$$S_{\phi_{\mathrm{MI}}}(f) \approx \frac{10^{-17.38}}{10^{(P_{\mathrm{in}} - L)/10}} \tag{42}$$

if P_{in} is in dBm.

8.1.6 Phase detectors

The experience gained by different authors with frequency-stability measurement systems revealed that the phase detectors with lowest noise are double-balanced mixers (with Schottky diodes) [2]

$$S_{\phi_{\mathrm{PD}}}(f) \approx \frac{10^{-14}}{f} + 10^{-17} \tag{43}$$

Unfortunately, these phase detectors are 'generators' of high-level discrete spurious signals [3, 16].

On the contrary, sampling or double-sampling types of phase detectors exhibit a small leakage of the 'loop-frequency' signals (inclusive of its harmonics) [3, 16, 17].

$$S_{\phi_{\mathrm{sp}}} \approx -90 \text{ to } -110 \text{ dB} \tag{44}$$

In addition, one would expect a very low background noise due mainly to the buffer amplifier. A similar behaviour in respect to spurious signals could be expected from popular phase-frequency detectors [16, 18] in the loops of type 2. The only difficulty is a larger flicker noise. Šojdr [19] has found, for phase frequency detectors based on TTL logic family,

$$S_{\phi_{\mathrm{PD}}}(f) \approx \frac{10^{-11}}{f} + 10^{-15.8} \tag{45}$$

and phase detector based on CMOS logic family

$$S_{\phi_{\mathrm{PD}}}(f) \approx \frac{10^{-12.7}}{f} + 10^{-16.2} \tag{46}$$

Note the increase of the flicker noise with respect to TTL frequency dividers. Similar results for the white phase-noise level of digital phase detectors has been published by Robins [20].

A J-Band Spread-Spectrum Synthesiser Using a Combination of DDS and Phaselock Techniques

M.V. HARRIS

Abstract.

The synthesiser described is for use in a spread-spectrum communication system where a high degree of immunity to interference is required. A frequency resolution of 100 Hz across an RF bandwidth of 2 GHz is achieved by the combination of two phaselocked loops (PLLs) with a relatively simple Direct Digital Synthesiser (DDS). The overall frequency resolution is determined by the frequency resolution of the DDS, which is potentially unlimited.

Decimal, or base-10 synthesis is obtained by using a decimal DDS, the design of which is outlined. By linearly ramping the DDS output frequency under digital control, it is possible to get sub-10 microsecond switching speeds with no cycle-slipping, with manageable PLL loop bandwidths (f_n of 700 kHz). The analysis is presented, together with detailed measurements of spurious levels from the DDS. Overall phase noise and spurious levels are presented.

A Description of the Synthesiser.

Figure 1 is a simplified schematic. PLL1 serves to upconvert the DDS output frequency to a microwave frequency (4020 ± 10 MHz). The absence of a frequency divider in this PLL means that DDS spurious levels are not enhanced between the DDS output and PLL1 output [1]. Similarly, the tuning range of PLL1 output is identical to that of the DDS output (20 MHz) and the frequency resolution is also as for the DDS (25 Hz). PLL2 has two input frequencies. PLL1 output forms one input to PLL2 via mixer M2. The other input to PLL2 is the fixed 10 MHz reference frequency, input to the Phase Frequency Comparator (PFC) of PLL2. PLL2 output frequency will change if either PLL1 output changes frequency or N (PLL2 programmable division ratio) changes. Both these changes can occur either independently or simultaneously. To summarise, the DDS/PLL1 circuit provides a vernier frequency control covering a 20 MHz bandwidth with 25 Hz resolution and PLL2 (by changing N) provides the coarse control. If N (variable between 26 and 51) is incremented by one, then PLL2 output changes frequency by 20 MHz. The total tuning range of PLL2 is 500 MHz, with 25 Hz resolution. After the final times-four frequency multiplier this becomes 2 GHz bandwidth with 100 Hz resolution.

The synthesis equation is:

$$f_{out} = 4 \left\{ 4 \text{ GHz} + f_{dds} - (20*N)\text{MHz} \right\} \qquad (1)$$

where f_{out} is the final output frequency, f_{dds} is that of the DDS output, and N is the division ratio in PLL2 (26 to 51).

Design Goals.

Listed overleaf is a summary design specification for the synthesiser.

This work was funded by Matra Marconi Space.

M.V.Harris is an independent consultant, formerly of GEC Plessey Semiconductors (Lincoln). This work was done while registered as a PhD student at the University of Bradford, UK.

<u>Design Specification.</u>

Output Frequency Range: 12 to 14 GHz
Minimum Frequency Step: 100 Hz
Fastest Hop Rate: 16 kHops per second
Output Power: 7 dBm
Mass: 1.5 kg
Dimensions: 160 x 80 x 150 mm
Power Consumption: 18 Watts
Switching Time: 6.25 us to within 1 kHz of final freq.
                               200 us to within 100 Hz of final freq.
Spurious Frequency Levels: -45 dBc max. at output.
Phase Noise (continuous):

          10 Hz: -58 dBc/Hz
          100 Hz: -69 dBc/Hz
          1 kHz: -90 dBc/Hz
          10 kHz: -95 dBc/Hz
          100 kHz: -95 dBc/Hz
          1 MHz: -95 dBc/Hz
          10 MHz: -101 dBc/Hz

The overall design aim is to produce a design which is easy to manufacture and low-cost. To this end, design margins are desirable. It will be shown that the DDS performance is well within the limits required. The main area of investigation is the dynamic performance of the phaselocked loops under switching conditions. There is, as usual, considerable tradeoff in the PLL design, between fast switching and low-noise performance. However it is possible to get a near-optimum design which meets the specification.

The justification for the DDS/PLL1 combination is that it is the most efficient way of upconverting the DDS output to a much higher frequency, without enhancing the spurious sidebands [1]. If a multiplier were placed on the DDS output, the spurious sidebands would be enhanced by $20 \log_{10}(P)$, where P is the multiplying factor. The power-efficiency of the PLL also applies to PLL2.

Philosophically, it is perhaps interesting to note that there are only two frequency multipliers in the synthesiser: the reference multiplier (output 4 GHz); and the final times-four multiplier. It is also apparent that all other frequency synthesis operations are achieved via the use of frequency dividers: the programmable divider in PLL2; and the DDS, which is in effect a frequency divider which divides by a programmable rational fraction [2].

<u>The Design of the Decimal DDS.</u>

Technological limitations are always a major consideration when incorporating a DDS into a subsystem design. The main limitations of the DDS are: limited clocking frequency; and slewing/glitching errors in the digital to analogue converter (DAC). Added to these considerations is the enormous increase in cost and power consumption as clock frequency is increased.

However, advances continue to be made and it is now possible to buy CMOS-based DDS chip sets which have input clock frequencies of hundreds of MHz. It is proposed that such a DDS is used in this synthesiser, although experimental results presented here are for an ECL-based DDS, clocking at 100 MHz. Because the digital logic in the DDS is clocked synchronously, the spurious levels at the DDS output should be dependent entirely upon the DAC performance, for a given DDS configuration (eg, number of Phase Accumulator bits), whether ECL or CMOS logic is used.

The design of the Decimal DDS is similar to that of a normal binary DDS, except that an extra memory circuit is used to force the overflow in the Phase Accumulator (PA) to occur at a decimal number, instead of occuring once in every 2^N pulses (where N is the PA word size).

Refer to Figure 2. The adder, latch, MEM1 memory and DAC are exactly the same as for a binary DDS. The adder and latch are 22-bits wide, to obtain the necessary frequency resolution, given that the clock frequency is 100 MHz and 25 Hz steps are required at the output. MEM1 is a look-up table for the DAC. MEM2 contents are mapped in Figure 2. MEM2 is transparent most of the time, because its digital output is equal to its digital input. Let's suppose that at count zero, the latch output is zero (all bits low). Because MEM2 is transparent, the DDS operates in the normal way (except that there is a small delay through MEM2), and the latch output increments upwards by one upon reciept of every clock pulse. This continues until the value of the latch output (MEM2 input) becomes D, whereupon the output of MEM2 becomes zero, and the adder output becomes 1. Therefore on the next clock pulse, the latch output goes to 1, MEM2 is now transparent again and so the latch output increments up by one every subsequent clock pulse. The cycle continues, and it can be seen that the overflow occurs when the latch output goes to D. As far as the PA is concerned, the effect of the overflow is the same as when, in the normal binary DDS, all the latch output bits become zero. To summarise, the count sequence at the PA output of the decimal DDS is:

0,1,2,3....D-1,D,1,2,3,...D-1,D,1,2,3,...

D can be any number, but to obtain exactly 25 Hz channel spacing from a 100 MHz clock, D must be:

$$D = (100 \times 10^6)/ 25 = 4 \times 10^6$$

$4 \times 10^6 < (2)^{22}$, therefore a 22-bit PA will suffice.

Why use a 100 MHz clock for the DDS when only a 20 MHz output frequency range is required? There were two reasons for this. Firstly, a 100 MHz DDS had been built and measured in detail. Secondly, this frequency plan avoids the DDS output passing through $f_{clock}/3$. At or near this output frequency, the maximum spurious levels can be as much as 23 dB higher than the typical figure. For this DDS, for an output frequency of up to 20 MHz, all spurs are below -60 dBc; but for output frequencies close to one third of the clock frequency, some spurs are as high as -47 dBc. The chosen frequency plan avoids this.

<u>Measured DDS Spurious Levels.</u>

Value of K	Largest Sideband (dBc)	Second Largest (dBc)	Mean Level of Sidebands (excl. largest) (dBc)
341	-47	-63	-71
343	-47	-65	-76
345	-73	-73	-76
254	-62	-66	-73
255	-60	-65	-74
257	-63	-66	-74
258	-63	-69	-75
259	-61	-68	-74
204	-62	-65	-70
205	-62	-68	-75
170	-73	-75	-77
171	-71	-72	-75

Levels of spurious single sidebands (see above) were measured for an ECL-based DDS of the normal binary type, with a 10-bit ECL DAC. The RAM look-up table was based on rounding down the digital sine values to the nearest integer. Spurious levels for the decimal DDS will be identical, except the frequencies of the spurs will be slightly offset. A HP3585A analyser was used, with a 1 MHz span selected. To check that the analyser input was not being saturated, the input attenuation was increased by 10 dB such that the measured sideband level (dBc) remained unaltered. This ensures that spurs are not enhanced by the measurement apparatus. The results are summarised:-The worst-case K-values shown above correspond to f_{dds} near $f_{clock}/3$, $f_{clock}/4$, $f_{clock}/5$ and $f_{clock}/6$, where f_{clock} is 100 MHz. On the basis of these results, a blanket specification of -60 dBc maximum for spurs (excluding the $f_{clock}/3$ case) seems reasonable.

<u>The Calculation of Synthesiser Output Spurious Noise Levels.</u>

There are three mechanisms for spurious frequency generation: intermodulation products generated by mixers M1 and M2; DDS spurious; and PLL reference frequency sidebands, produced by both PLLs. Harmonics of the final output frequency will be generated by the output multiplier, but these are not considered here.

<u>The Effect of the PLL on Phase Noise and Spurious.</u> Each of the two PLLs acts to reduce both continuous phase noise and discrete spurious sidebands from any signal which is injected into the PLL. However, whether the noise is reduced *inside* or *outside* the PLL loop bandwidth depends upon where the noise is injected. Looking at PLL1, noise on both input signals (DDS o/p and 4 GHz reference) will be transferred to PLL1 o/p without attenuation, for low offset frequencies which are well below the PLL natural frequency (f_n). But above f_n, these noise sidebands will be rolled-off at a rate of 6 dB per octave i.e 20 dB per decade. For noise added at the VCO input and for the free-running VCO noise, the converse applies: At frequencies *below* f_n, the noise sidebands will be attenuated at the rate of 12 dB per octave *reduction* in frequency (40 dB per decade) and for frequencies *above* f_n, the sidebands will be unattenuated. The same noise-reduction mechanisms apply for PLL2, except that, below f_n, the reference divider/PFC noise is enhanced by 20 log(2N) to the o/p of PLL2, due to the action of the programmable divider.

<u>Calculation of Overall Synthesiser Spurious Levels.</u> Using these principles it is relatively straightforward to calculate the overall phase noise and spurious levels at the synthesiser o/p, having calculated the levels of intermodulation products for the mixers, and knowing the phase noise performance of the VCO, logic, DDS, reference oscillator etc. For a mixer, relative phase noise powers of two inputs add linearly at the o/p. Thus if both inputs have the same noise power (in dBc or dBc/Hz), the output noise will be 3 dB higher.

Calculated maximum sideband level due to intermodulation in M1 and M2 is -70 dBc referred to mixer outputs . The output times-four multiplier enhances the sidebands by 12 dB. Therefore at the final o/p the sidebands due to mixer intermodulation are -58 dBc maximum, less any attenuation due to PLL filtering.

<u>PLL Reference Sidebands.</u> Another source of spurious is the PFC input frequency, which appears at the PFC output as short pulses. The resultant sidebands are symmetrical about carrier at offset frequencies equal to multiples of the reference frequency. As with most PLLs, the attenuation provided by the loop's active filter is insufficient and so extra passive L-C filtering is placed between the PFC and the active filter. The prediction of sideband level is based on estimating the rms voltage of the short pulses, then converting the voltage to a sideband level using the VCO sensitivity (K_o). This sideband level is reduced by any attenuation in the passive and active filters in the loop. The equation for single sideband level is:-

$$\left[\frac{N}{C}\right]_{ssb} = 20 \ log_{10}\left(\frac{K_o \ V_{rms}}{\omega_m}\right) + A \ (dB) \quad dBc/Hz \quad (2)$$

where K_o is the VCO gain (rad/SV); Vrms is the rms voltage of the PFC output pulses; ω_m is the PFC frequency (rad/S); and A is the gain (at ω_m) of the filtering between PFC and VCO. There is scope for error: the active loop filter can have wide variation of gain at ω_m of greater *or* less than unity, depending on the opamp response and circuit configuration. The required attenuation (-A) can thus be calculated for specified maximum spurious and for a given f_n.

<u>PLL Reference Filtering.</u> Calculations were done for a f_n of 700 kHz. To achieve the necessary spur level at either PLL output of -60 dBc (max), a passive 5-element Butterworth lowpass filter is inserted after the PFC, with f(3dB) of 4.8 MHz. This filter gives an attenuation at 10 MHz of 31 dB. Total attenuation required between the PFC and the VCO is 63.5 dB (a CMOS PFC is used in PLL2, so the output pulses are 5 Volts high). The remaining 32.5 dB is got by two means: the active filter provides 18 dB (measured) and 15 dB is from the use of a differential amplifier at the PFC output [3].

<u>Ramping the DDS to Reduce Design Conflict.</u> During the calculation of reference-filtering requirements, a strong conflict arose between minimising spurious sidebands and ensuring no loss of lock (cycle skipping). An optimum f_n is around 700 kHz for PLL2, but PLL1 (with f_n of 700 kHz) would lose lock if the DDS output were stepped across its full bandwidth instantaneously. This gave rise to the proposal to *linearly ramp* the DDS output frequency when changing channel. This can be done by continuously incrementing the digital frequency-control word by a fixed value at a constant rate. In this way, lock can be maintained for a smaller loop bandwidth, meaning in turn easier filtering of the reference sidebands.

<u>A Model of the PLL Dynamic Behaviour.</u>

When the PLL is tuned, it can be analysed as a second-order control system with a distubance. The transfer function of the loop filter approximates to:-

$$F(s) = \frac{1 + s\tau_2}{s\tau_1} \tag{3}$$

For PLL2, the frequency step, (due to N changing from N1 to N2) can be referred to the PFC input by scaling:-

$$\Delta f = \frac{(N1 - N2)}{N2} * f_{pfc} \tag{4}$$

When PLL1 is tuned by stepping the DDS o/p frequency, Δf is equal to Δf_{dds}. The phase error transient after a frequency-step input is:-

$$\psi_e \simeq \frac{\Delta\omega}{\omega_n \alpha} e^{-\zeta\omega_n t} \sin(\omega_n \alpha t) \tag{5}$$

where:-

$\psi_e(t) = \psi_i(t) - \psi_0(t) =$ loop phase error between PFC inputs, radians.

$\omega_n = 2\pi f_n =$ loop nat. fr.,   rad/S

$\zeta =$ loop damping ratio

$\alpha = \sqrt{(1 - \zeta^2)}$

$\Delta\omega = 2\pi\Delta f_n =$ equivalent input freq. step, rad/S

When differentiated with respect to time this gives:-

$$\frac{d(\psi e)}{dt} = \Delta\omega \; e^{-\zeta\omega_n t} \left[\cos(\omega_n\alpha t) - \frac{\zeta}{\alpha} \sin(\omega_n\alpha t) \right] \qquad (6)$$

which is zero when:-

$$\tan(\omega_n\alpha t) = \alpha/\zeta \qquad\qquad\qquad (7)$$

which gives the values of t for the maxima and minima of the phase error transient (equation 5). If we assume that the damping ratio is 0.707, then
$\zeta = \alpha = .707$, $\tan(\omega_n\alpha t)=1$ and

$$t = 1.11/\omega_n \qquad\qquad\qquad (8)$$

is the time of occurrence of the peak phase excursion (the first maximum). Substituting this into equation (5) and equating the phase error to 2π radians to ensure no loss of lock,

$$2\pi = \frac{\Delta\omega}{\omega_n\alpha} e^{-1.1\zeta} \sin(1.1\alpha) \qquad (9)$$

$$\Delta f/f_n = 13.78 \qquad\qquad\qquad (10)$$

so for a Δf of 9.61 MHz, the minimum f_n to avoid cycle-skipping is:-

$$f_n = 700 \text{ kHz} \qquad\qquad \text{for PLL2}$$

Repeating the calculation but for PLL1, the minimum f_n required for a 20 MHz step is:-

$$f_n = 1.45 \text{ MHz} \qquad\qquad \text{for PLL1}$$

Given the constraint that the minimum PFC frequency in PLL1 is 10 MHz, such a large value for f_n causes a major problem in the design of a lowpass filter for the PFC output. This led to the idea of ramping the DDS output frequency.

<u>Ramping the DDS Output Frequency.</u>

Assuming the same approximate loop filter transfer function as in equation (3) and an input ramp rate of D rad/S/S, the loop phase error (referred to the PFC) is:-

$$\psi_e(t) = \frac{D}{\omega_n^2} - \frac{D}{\omega_n^2} e^{-\zeta\omega_n t} \left[\cos(\omega_n\alpha t) + \frac{\zeta}{\alpha} \sin(\omega_n\alpha t) \right] \qquad (11)$$

$$\text{where } D = \frac{d\{\omega_i(t)\}}{dt} \qquad \text{and the ramp starts at } t=0$$

$$\frac{d}{dt}\,\psi_e(t) = \frac{D}{\omega_n\alpha} e^{-\zeta\omega_n t} \sin(\omega_n\alpha t) \qquad\qquad (12)$$

Equating this to zero gives the values of t for which the phase error is at a maximum or minimum on the transient waveform.

$$\sin(\omega_n\alpha t)=0 , \qquad\qquad \omega_n\alpha t=0,\pi,2\pi,...$$

$$\omega_n\alpha t=\pi \qquad\qquad \text{gives the value of t at the peak phase error}(\psi_{ep}).$$

Substituting $t = \pi/\omega_n\alpha$ into equation (11):-

$$\psi_{ep} = \frac{D}{\omega_n^2} \left[1 + e^{-\frac{\pi\zeta}{\alpha}} \right] \tag{13}$$

With a limiting maximum value of ψ_{ep} of 2π, and $\zeta = \alpha = 0.707$,

$$D/\omega_n^2 < 6 \qquad \text{rad}^{-1} \tag{14}$$

which fixes D, the maximum rate of change of frequency for a given ω_n. The next task is to find the total switching time for PLL1, for various values of f_n. The total time is divided into two intervals. The first is the duration of the ramp input to PLL1. The second interval is from the end of this ramp until the phase error transient has died away to an amplitude which is within the specification.

Assuming that the frequency ramp rate is D and the range of the ramp is Δf_i (Hz) or $\Delta \omega_i$ (rad/S), then the duration of the ramp is:-

$$\Delta t_r = \Delta \omega_i / D \tag{15}$$

Equation (11) describes the initial phase error transient during the ramp input (remember that the input is a *parabolic* phase input, i.e. linear frequency ramp). At the end of the ramp, the frequency instantaneously becomes constant again. This is equivalent to the linear superposition of a ramp input of gradient equal and opposite to the original ramp. In other words at t=0 a positive frequency ramp input occurs of gradient D (rad/S/S). Then when $t=\Delta t_r$, a frequency ramp input occurs which is of gradient -D. The second ramp input causes a transient of the same form as equation (11), but delayed by Δt_r and of opposite sign.

Because the duration (Δt_r) of the ramp input is close to the period of the oscillatory component ($1/f_n\alpha$) of the phase error transient, significant superposition of the two transients occurs. The worst-case superposition occurs when Δt_r is one half the oscillatory period

$$\Delta t_r = 1/(2f_n\alpha) \tag{16}$$

in which case it can be shown that the amplitude of the sum transient is 1.61 times greater than the second transient. The consequence of this is that the originally calculated value of D must be reduced by the factor 1.61. The envelope of the sum transient is (referring to equation 11) given by:-

$$\frac{1.61\ D}{\omega_n}\ e^{-\zeta\omega_n(t-\Delta t_r)} \tag{17}$$

where D is the reduced value:-

$$D/\omega_n^2 = 6/1.61$$

$$D = 3.73\ \omega_n^2 \tag{18}$$

It can be shown that the frequency error is given by:-

$$\frac{d}{dt}\ \psi_e(t) =$$

$$\frac{D}{\omega_n\alpha} \left[e^{-\zeta\omega_n t}.\sin \omega_n\alpha t - e^{-\zeta\omega_n(t-\Delta t_r)}.\sin \omega_n\alpha(t-\Delta t_r) \right] \tag{19}$$

and for the worst case, where $\Delta t_r = 1/2\alpha f_n$, and $t > \Delta t_r$,

$$\frac{d}{dt}\psi_e(t) = \left[1 + \frac{1}{\sqrt{e}}\right]\left[\frac{D}{\omega_n\alpha}e^{-\zeta\omega_n(t-\Delta t_r)}\sin \omega_n\alpha(t-\Delta t_r)\right] \quad (20)$$

So the total time for PLL1 to reach a specified frequency accuracy is obtained by calculating the value of t for which

$$1\cdot 61 \frac{D}{\omega_n\alpha}e^{-\zeta\omega_n t} = \text{spec. error} \qquad rad/s \qquad (21)$$

and adding $\Delta t_r = \Delta\omega_i/D$ S $\qquad\qquad\qquad (22)$

Choosing f_n of 700 kHz, from (18), (21) and (22):-

$D = 7.22 \times 10^{13}$ rad/S^2 $\qquad\qquad (23)$

$\Delta t_r = 1.74$ uS

$t = 3.24$ uS

$t + \Delta t_r = 5.0$ uS

This is the total switching time for PLL1 to be within 250 Hz of its final frequency, for a frequency ramp input to the DDS.

Consider PLL2 being switched separately by changing N from 51 to 26. The envelope of the *frequency* error (equation 6) is:-

$$\Delta f\, e^{-\zeta\omega_n t} \qquad\qquad\qquad\qquad (24)$$

The PLL2 switching time is got by equating this expression to the required frequency error (250 Hz referred to PLL2 o/p) when Δf is 9.61 MHz.

$t\,(\text{pll2}) = 3.4$ uS $\qquad\qquad (25)$

<u>Summary.</u>

A simple decimal DDS has been outlined. The DDS can be used in a low-spurious low-noise synthesiser with good noise design margins. This is possible by careful selection of the DDS output frequency range and by keeping the multiplication ratio low between the DDS output and the final output. By ramping the DDS a smaller loop bandwidth can be used in PLL1.

Independent switching times for PLL1 and PLL2 to be within 250 Hz of their final frequency (1 kHz referred to final o/p) for a worst-case switch are respectively 5.0 and 3.4 microseconds, implying a sub-ten microsecond switching time for the whole synthesiser.A phase noise estimate is shown in Figure 3.

Relatively high performance is achieved at low cost due to the small number of circuits and power-efficient design. Most of the logic including the DDS is realised in CMOS. Bipolar VCOs are used in the phaselocked loops. The VCOs each draw around 50 mA at 15 Volts and occupy a volume of typically 2.5 cubic centimetres. Due to the high proportion of MSI logic and analogue functions (eg.dividers, opamps) the whole synthesiser can be hybridised, with some RF screening of the mixers and multiplier circuits.

References.

[1] Robins W.P. 'Phase Noise in Signal Sources', Peter Peregrinus 1982, pp119-121

[2] Tierney J., Radar C.M., Gold B.: 'A Digital Frequency Synthesiser', IEEE Trans.Audio and Electroacoustics, Vol.AU-19, No.1, pp28-57, March 71.

[3] Harris M.V., 'A miniaturised 1.8 GHz synthesiser for use in mobile radio', IEE 6th Int.Conf.on Mobile Radio and Personal Communications, December 1991.

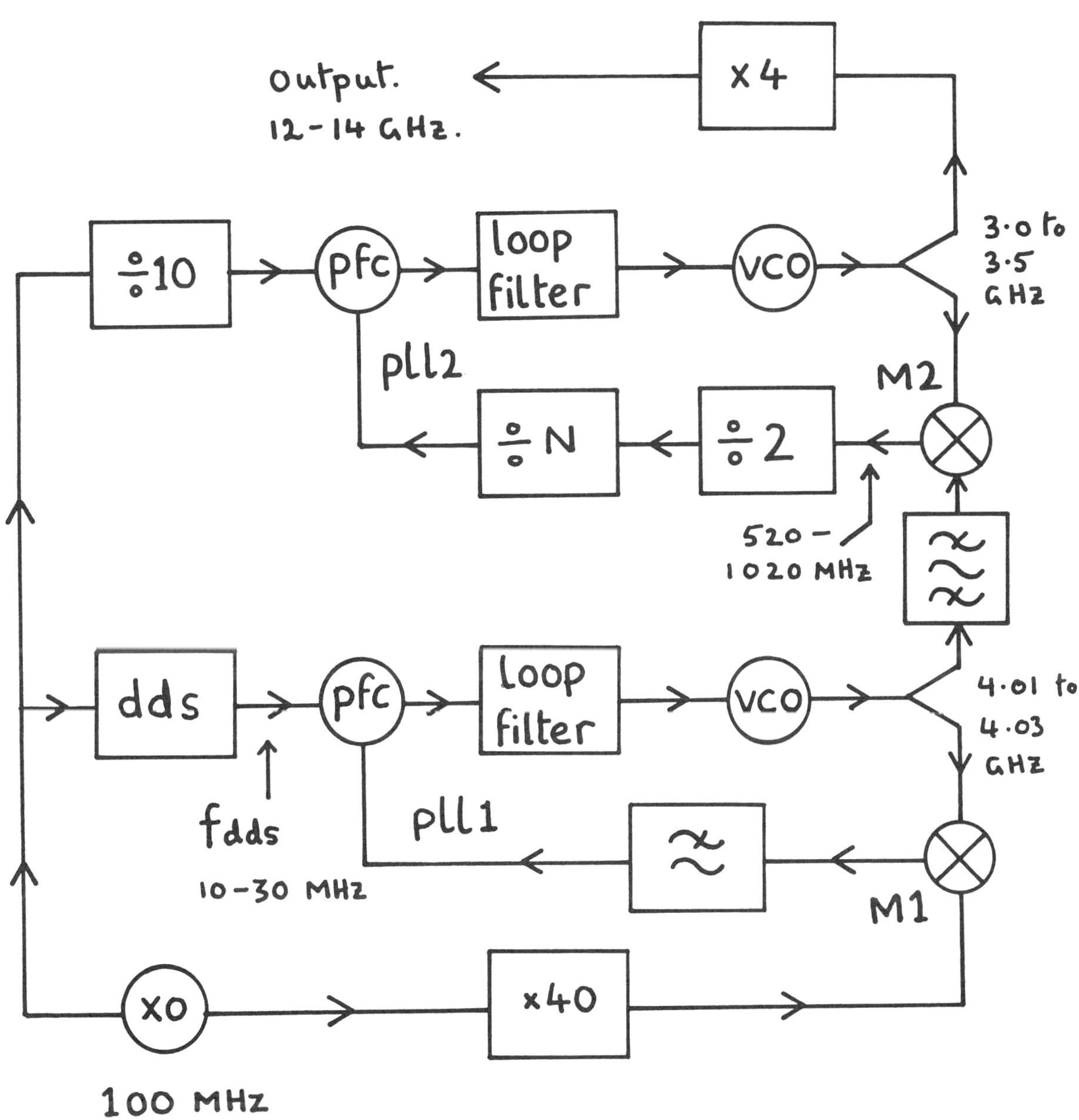

Figure 1. Synthesiser Schematic Diagram.

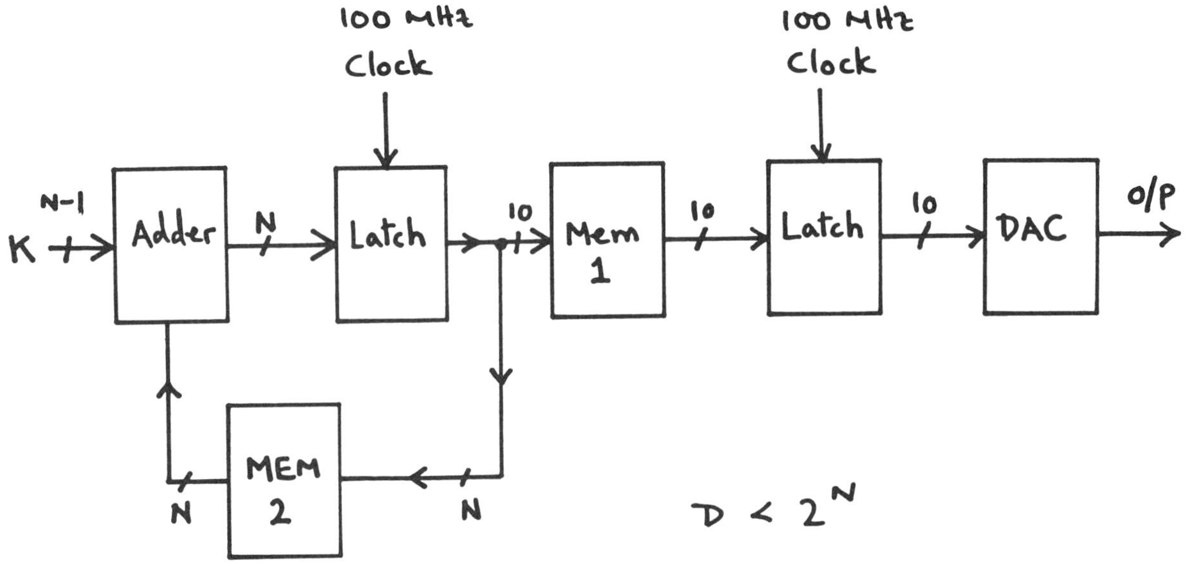

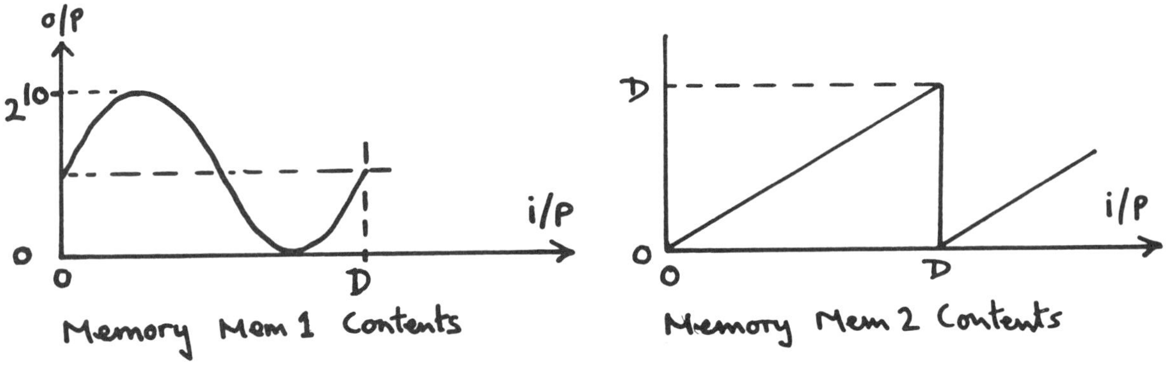

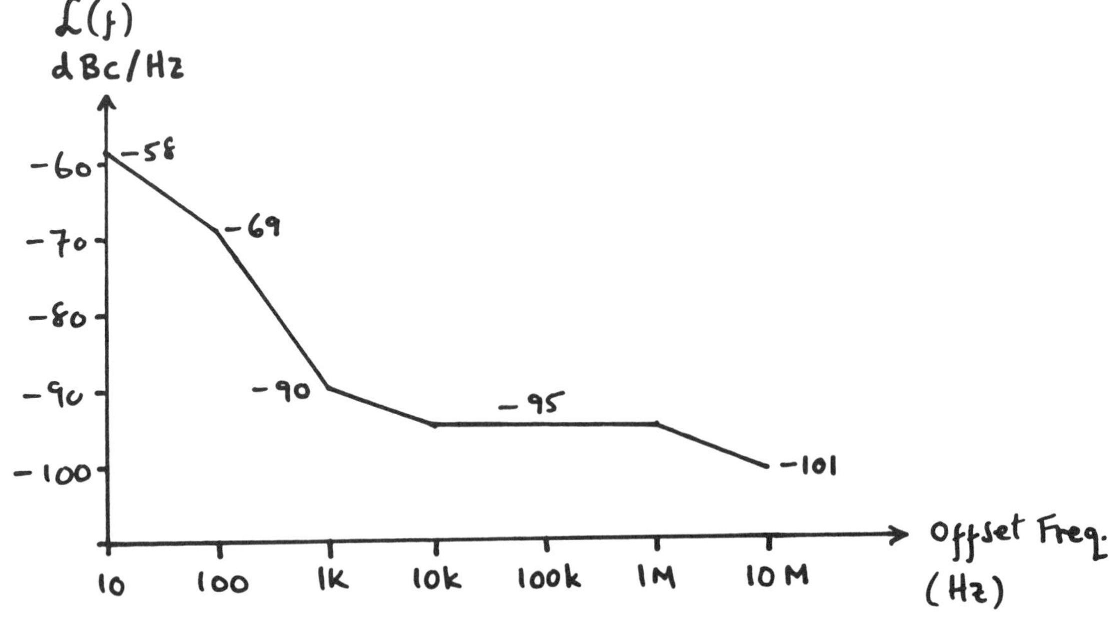

Figure 2. Principle of Operation of the Decimal DDS.

Figure 3. Worst-case Phase Noise at Final Output (predicted).

Principles of Fractional-*N* Frequency Synthesizers

VĚNCESLAV F. KROUPA

I. Introduction

THE fractional-*N* frequency synthesizer is similar to the divide-by-*N* PLL (divider in the feedback path). However, with the assistance of an auxiliary DDFS, the output frequency of the voltage-controlled oscillator (VCO) is not restricted to integral multiples of the reference signal only. Rather, it can also be locked to the fractional multiples, with the result of a substantial reduction of the frequency tuning steps in the useful bandwidth of the PLL without reduction of its pass band (i.e., without reduction of the loop natural frequency f_n). Consequently, these frequency synthesizers have both high-frequency resolution and short settling time, two essential requirements for modern applications, for example, in mobile radio sets. The major difficulties are spurious signals generated in the fractional-*N* divider systems.

To the editor's knowledge, this phase-locked loop system first became popular in connection with the HP 3325 Synthesizer/Function generator [1,2,3].

II. Principles

The idea and operation of the PLL fractional-*N* frequency synthesizer will be explained with the assistance of Fig. 1. The principle is in the quasiperiodic removing of pulses from the VCO pulse train. In actual devices, pulse suppression is provided with the assistance of two-mode dividers; this operation will be discussed in the Appendix.

For a better understanding of the problem, we assume that the DDFS is a simple divider of the reference frequency, f_r. By assuming the PLL loop without pulse swallowing, the period T_{rN}, at the divider-*N* output is equal to the *N* periods, T_o, of the VCO frequency

$$T_{rN} = T_o N \qquad [1]$$

However, after each overflowing of the divider-*F*, one pulse is missing in the VCO train T_o and the respective period of T_{rN} is prolonged by one T_o. Since the periods of the signals applied to the inputs of the phase detector (PD), T_r and T_{rN}, are very close to each other we can write

$$T_{rN} F \approx T_r F \approx T_o N F + T_o \qquad [2]$$

Finally, in the locked state of the respective PLL both T_r and T_{rN} are equal. Consequently, the fundamental repetition period of the signal supplied to the feedback port of the phase detector is exactly

$$T_o N F + T_o = T_r F \qquad [3]$$

From the above relation, the effective division ratio is easily computed as

$$N_{\text{eff}} = \frac{T_r}{T_o} = N + \frac{1}{F} \qquad [4]$$

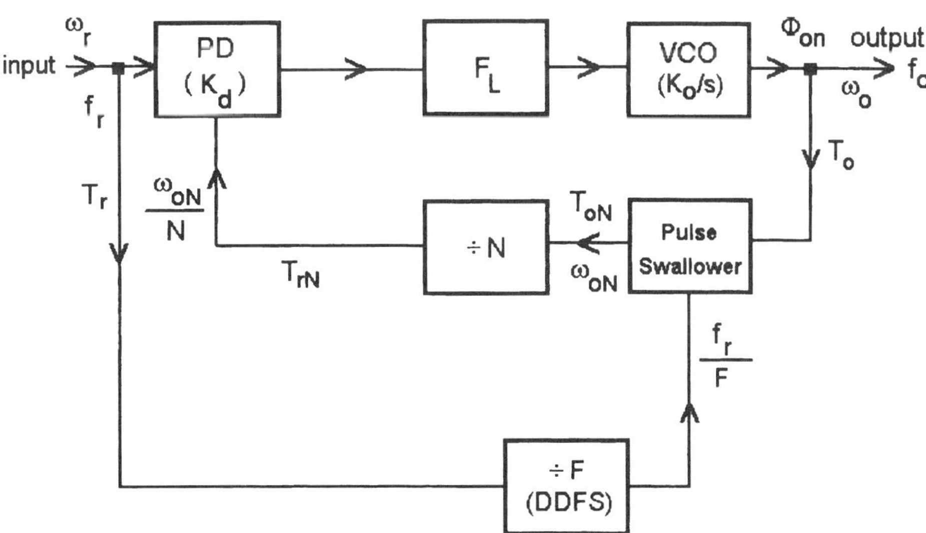

Fig. 1. Block diagram of the fractional *N* frequency synthesizer.

(This paper was specially written for this volume.)

For the later estimation of the spurious signals, introduced by the fractional-N behavior, we will also investigate solutions provided from the frequency domain point of view. The pulse swallower in Fig. 1 quasiperiodically removes pulses in the T_o train at each overflowing of the divider-F as follows [4, eq. (3)],

$$\omega_{oN}(t) = \omega_o - 2\pi\,\delta(t - FT_r) \qquad [5]$$

Evaluation of the steady-state input frequency ω_{oN} fed into the input of the divider-N is reduced to

$$\omega_{oN} = \omega_o - \frac{\omega_r}{F} \qquad [6]$$

where in the locked state

$$\omega_r = \frac{\omega_{oN}}{N} \qquad [7]$$

and finally

$$\omega_{oN} = \omega_o \frac{FN}{FN + 1} \qquad [8]$$

After the combination of these equations, we get for N_{eff} relation (4).

In the instance where F is a ratio of two integers

$$F = \frac{Y}{X} \qquad [9]$$

the theory of the quasiperiodic omission of pulses [4] reveals that the steady-state input frequency, ω_{oN}, to the divider-N will be reduced to

$$\omega_{oN} = \omega_o - \omega_r \frac{X}{Y} \qquad [10]$$

and the effective division ratio changes to

$$N_{\text{eff}} = N + \frac{X}{Y} \qquad [11]$$

III. Spurious Phase Modulation

Returning again to the theory of the quasiperiodic omission of pulses [4], we find that the system introduces a phase modulation. In the simplest case, the periodic suppression of a single pulse generates a sawtooth wave with the maximum amplitude of nearly 2π. In more complicated arrangements, where a simple divider by F in Fig. 1 is replaced by a DDFS, we encounter a superposition of several sawtooth waves.

We will start our investigation of the spurious phase modulation in the PLL with fractional-N division in the feedback path with the assumption that the loop is in the locked state. In that instance, we can replace T_r in eq. (5) with the assistance of eqs. (8) and (10) and get

$$\omega_{oN}(t) = \omega_o - 2\pi\,\delta(t - FN_{\text{eff}}T_o) \qquad [12]$$

Evidently, the signal fed to the divider-N is phase modulated and consequently also its output signal. The phase-time modulation, $s_r(t_k)$, of such a system has been investigated earlier

and with the assistance of eq. (11) in [4] we have arrived at

$$s_r(t_k) = T_o\left[k\frac{q_1 q_2}{q_1 - 1} - \text{integer}\left(k\frac{q_1 q_2}{q_1 - 1}\right)\right] \qquad [13]$$

After introduction of the above equation $q_1 = FN_{\text{eff}}$ and for $q_2 = N$, we find for the phase-time modulation of the signal at the output of the divider N

$$s_{rN}(t_k) = T_o\left[k\frac{NF + 1}{F} - \text{integer}\left(k\frac{NF + 1}{F}\right)\right] \qquad [14]$$

or in the general case where

$$q_1 = \frac{Y}{X}\frac{NY + X}{Y} = \frac{NY + X}{X}; \quad q_2 = N \qquad [15]$$

we have

$$s_r(t_k) = T_o\left[k\frac{NY + X}{Y} - \text{integer}\left(k\frac{NY + X}{Y}\right)\right] \qquad [16]$$

We find the shape and periods of the spurious signals with the assistance of the modified continued fraction expansion [5]. We will illustrate the problem with the following example.

Example 1. Let us examine the phase-time modulation of a fractional-N frequency synthesizer with

$$N = 32$$

and

$$X/Y = 45/128 \quad \text{(from which } N_{\text{eff}} = 4141/128)$$

Its modified continued fraction expansion is

$$\frac{4141}{128} = 32 + \frac{1}{3} + \frac{1}{3 * 17} - \frac{1}{17 * 37} + \frac{1}{37 * 128} \qquad [17]$$

and its plot is shown in Fig. 2 (with the mean value removed). Note the superposition of the sawtooth waves. The largest one lasts three clock periods T_r; the second sawtooth wave, with the amplitude 1/3, lasts nearly 17 clock periods; the third one (not easily identified), with the amplitude 1/17, has a duration of 37 clock periods, but with the opposite slope; and the last one with a rather small amplitude, 1/37, has the duration of $YT_r = 128T_r$.

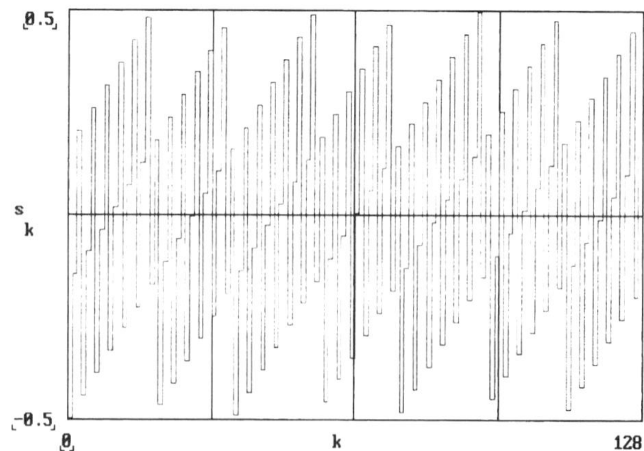

Fig. 2. Plot of the spurious modulation in fractional-N PLL with $N = 32$ and DDFS set to $X/Y = 45/128$.

By reverting to the general modified continued fraction expansion, we get from (17), $B_1 = 3$ (cf. also series (8) in [5]). We can write for the first-order approximation of the N_{eff}

$$N_{\text{eff}} \approx N + \frac{1}{B_1} \qquad [18]$$

Note that the above relation is exactly relation (4), only that F is replaced by B_1. Evidently, we can write for the period of the spurious modulation component, with the largest amplitude, approximately

$$T_{sp,1} \approx T_o B_1 N \approx B_1 T_r > T_r \qquad [19]$$

Since B_1 can be anywhere in the range

$$2 \leq B_1 \leq Y \qquad [20]$$

we find that $T_{sp,1}$ exceeds the reference period T_r, in some cases even substantially.

However, in accordance with the rule of thumb the natural loop frequency f_n of the PLL is generally in the range

$$0.01 * f_r < f_n < 0.1 * f_r \qquad [21]$$

We can conclude that the largest spurious modulation components, introduced by the fractional-N division, may be well in the PLL pass band and may appear, multiplied by N_{eff}, at the VCO output.

The respective spurious phase modulation at the PD input is

$$\varphi_r(t_k) = \omega_{ro} s_r(t_k)$$
$$\approx \frac{2\pi}{N_{\text{eff}}} \left[k\frac{X}{Y} - \text{integer}\left(k\frac{X}{Y}\right) \right] \qquad [22]$$

Due to the multiplication properties of PLLs with a divider in the feedback path, the spurious phase modulation of the output signal might be

$$\varphi_{\text{out}} \approx 2\pi \left[k\frac{X}{Y} - \text{integer}\left(k\frac{X}{Y}\right) \right] \qquad [23]$$

and reach an unacceptable amplitude of nearly 2π. At the same time, its frequency can be evaluated from the relation (19). Evidently, such an output signal is not acceptable for practical applications, and we must introduce the means for removing this difficulty.

Investigation of the information stored in the accumulator of the DDFS, due to its Modulo-Y operation, reveals

$$s_{\text{DDS}}(t_k) = T_r \left[k\frac{X}{Y} - \text{integer}\left(k\frac{X}{Y}\right) \right] \qquad [24]$$

which has the same form as the phase-time modulation fed to the feedback input of the PD; cf. relation (22). Consequently, the information stored in the DDFS accumulator can be used with the assistance of a DAC for compensation of the spurious modulation introduced by the fractional-N operation; see Fig. 3.

Investigation of the output phase noise with the assistance of Fig. 3 and the simplified eq. (1) in the preceding paper [6]

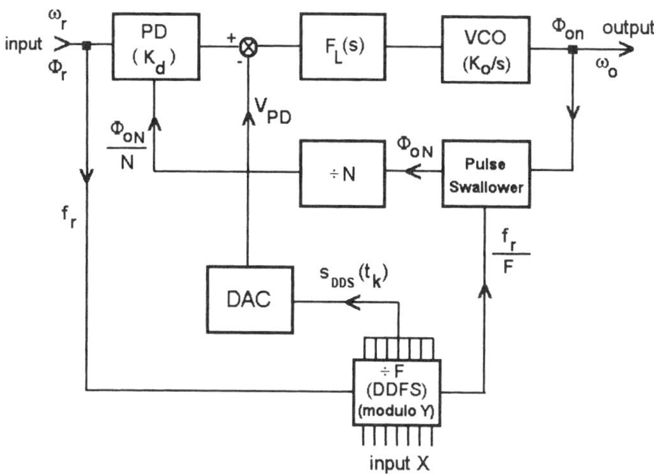

Fig. 3. Means for compensation of the spurious phase noise in fractional-N PLL.

leads to

$$\Phi_{o,n} = \left[\Phi_r - \frac{\Phi_{oN}}{N} + \frac{V_{PD}}{K_d} + \Phi_L \right]$$
$$* N_{\text{eff}} \frac{F_L(s)K/(N_{\text{eff}}s)}{1 + F_L(s)K/(N_{\text{eff}}s)} \qquad [25]$$

where ϕ_L is the noise generated in the PLL loop filter. For the spurious signals in the pass band, the above relation simplifies to

$$\Phi_{o,n} \approx \left[\Phi_r - \frac{\Phi_{o,N}}{N} + \frac{V_{PD}}{K_d} + \Phi_L \right] N_{\text{eff}} \qquad [26]$$

Clearly, by making the output voltage of the DDFS-DAC equal to

$$V_{\text{DAC}} = V_{PD} = K_d \frac{\Phi_{oN}}{N} \qquad [27]$$

we can compensate the spurious phase modulation in the loop to the level (eq. [11] in the paper [7])

$$N/S \approx -1.76 - 6.02D \quad [\text{dB}] \qquad [28]$$

with the assistance of a D-bit DAC. Because of the divider in the feedback path the spurious level at output of the VCO is again increased by about $20\log(N)$ dB, due to the multiplication properties of the PLL.

Example 2. Let us examine the spurious behavior of a fractional-N PLL. In instances where B_1 defined in eq. 18 is greater than 10, we can expect that the amplitude of the largest spurious signal will be approximately $2\pi/N$ and, in accordance with (21), the signal will pass the PLL pass band without any attenuation. However, if the compensation of the spurious phase ϕ_{oN}/N reaches the noise level ϕ_L, we can assume that all discrete disturbing signals will disappear from the VCO output.

Now returning to the preceding paper [6], to eq. (18) and the accompanying Fig. 2, we get for the natural loop frequency $f_n \geq 10$ kHz (which is a reasonable assumption in fractional-N PLL) for the power spectral density $S_{\phi L}$

$$S_{\phi L} \approx 10^{-15}$$

By assuming a perfect compensation, assisted, for example, with a feedback circuit [8], via a 16-bit DAC, we can hardly achieve such a low-noise level. Let us consider the case where $N = 1000$ and a 16-bit DAC is used. Computation of the spurious level

$$(2\pi * 2^{-16} * 10^{-3})^2 \approx 10^{-14}$$

reveals a value about one order higher than required.

In addition DACs are not perfect devices (nonlinearities, glitches, etc.). Therefore, the Hewlett-Packard Co. introduced a specially designed analog phase interpolating circuit [2,3] instead of a DAC by using the top five decimal digits from the DDFS for switching compensation currents into the loop. All spurious signals were reduced below the level of -90 dB in the output signal—actually below the loop noise level.

Recently, the newly introduced concept stating that the interpolating DDFS behaves as a sigma-delta modulator was used [9,10] for pushing spurious signals out of the PLL pass band. The principle is discussed in detail in the next paper [9]. Some perfection of the system has been suggested by Chodora [11] by changing the division factor N in a random manner between $N - 3$ and $N + 4$. These digital corrections take place every reference cycle. In this way, the noise starts to be "white" from the start.

IV. Conclusions

The major advantage of the fractional-N PLL frequency synthesizers is that they extend application of the low-frequency DDFS well into the microwave frequency ranges. They increase substantially the output frequency resolution without reduction of natural frequency, f_n, of the PLL, that is, without prolonging the switching times. The difficulty is that the output signal has rather poor spectral properties that can be alleviated only with complicated compensation circuits.

V. Appendix

A. Pulse-Inhibiting Circuits

a. Simple J-K Flip-Flops. These pulse-inhibiting or pulse-subtracting circuits are very simple as proven in Fig. 4 in the paper by Small [12]. The difficulty might be with time delays in the dividers.

b. One Pulse-Removing Circuit. The above difficulty can be removed with the circuit shown in Fig. 4 [13]. It contains two D flip-flops and six gates. The original pulse train is supplied to the input "A" and the "pulse remove command" to the input "B." Its positive edge activates the first flip-flop and via $Q_1 - D_2$ prepares activation of the second flip-flop with the sequence of the inverted A (Q_2 high). After nearly a half period later, A is high and the output of G_2 low and that of G_3 high. The next change of the state of A, besides G_1 and G_2, also activates G_4. Due to the delay in the interconnecting RC

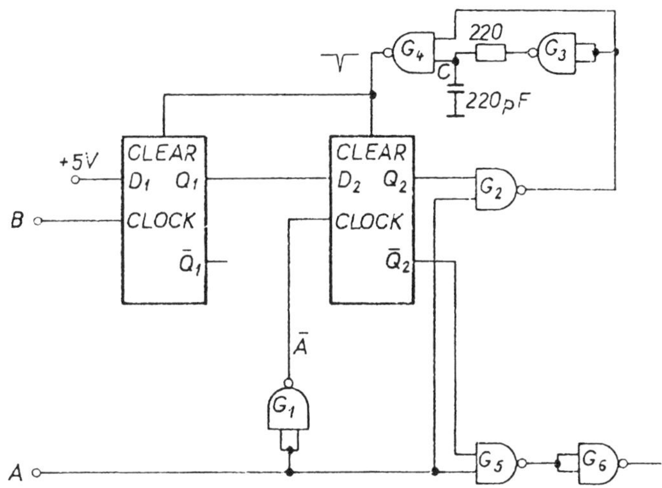
Fig. 4. Block diagram of a pulse-removing circuit with a delaying memory.

circuit, both inputs to G_4 are simultaneously high for a very short time; the negative pulse at its input clears both flip-flops and prepares the next subtracting operation. In the meantime, when the inverted Q_2 was low, it blocked the passage of sequence A through the gate G_5. The following inversion by G_6 supplies the original pulse train with one pulse missing (see Fig. 5).

c. Dual-Modulus Dividers. In most instances, the general practice for the pulse-removing or "swallowing" process is application of dual-modulus dividers. The functional block diagram of a 16/17 prescaler is shown in Fig. 6 [14]. The design uses a 2/3 divider as input stage. If the prescaler is intended for 16/17 operation, for example, the 2/3 divider divides by two continuously so that the output of the fifth flip-flop (TFF-5) is 1/16 of the VCO frequency. If the prescaler is to divide by 17, NAND gates G_3 and G_4 generate a signal that changes the modulus of the 2/3 to three for one out of four cycles, resulting in a total division factor of 17.

In Fig. 7 we reproduce another type of dual-modulus frequency divider. The operation is as follows: First, the input di-

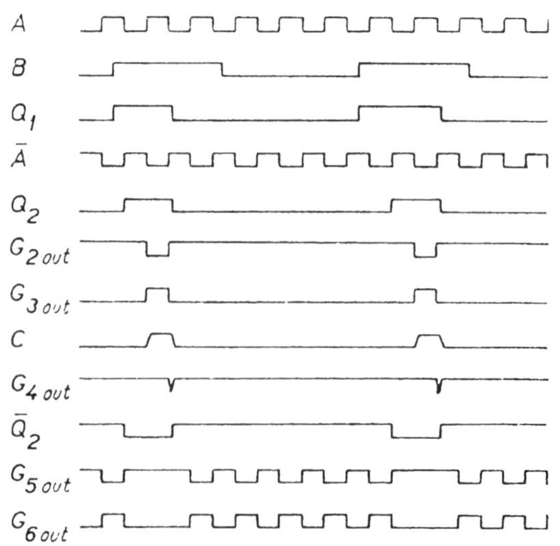
Fig. 5. Time diagram of the pulse-removing circuit with a delaying memory.

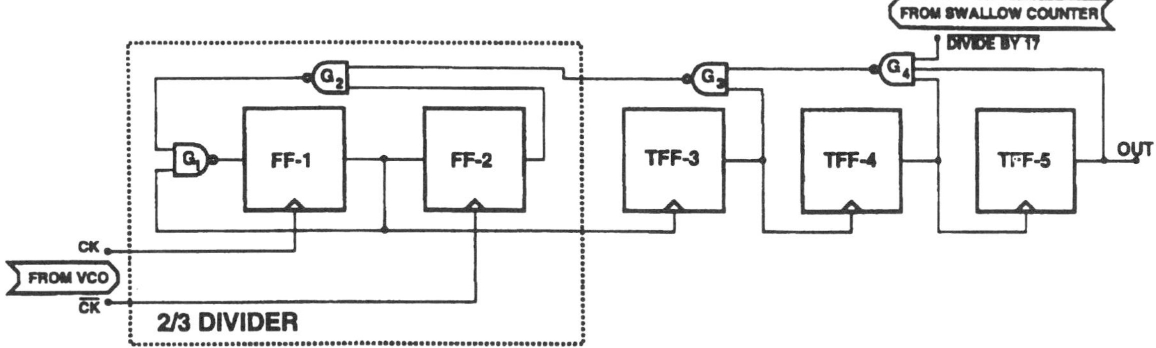

Fig. 6. Dual-modulus 16/17 divider-prescaler (from [14]).

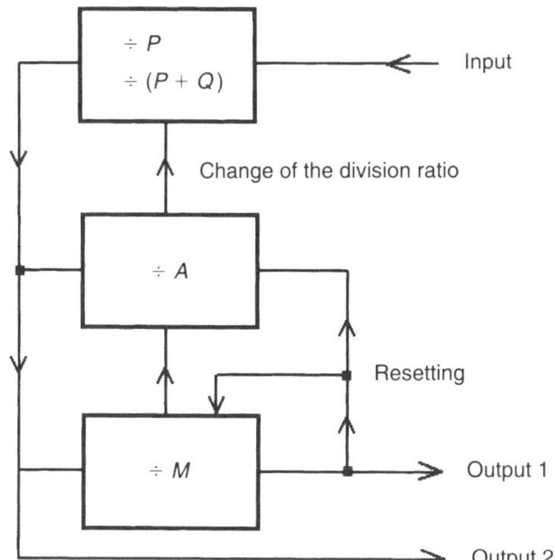

Fig. 7. Block diagram of another type of dual-modulus divider.

vider operates with the division factor $P + Q$. Its output is fed simultaneously into both auxiliary dividers with the division factors A and M. As soon as the divider A overflows, the output signal changes the division factor $P + Q$ to P and blocks the input to itself. This state remains unchanged until the divider M is full. Its output signal resets the main divider to the $P + Q$ state again and the auxiliary divider A to zero, and the cycle starts to repeat. The time available for the change is ideally P periods. The final division ratio from the input to the output 1 is computed from the following relation:

$$N = (P + Q)A + P(M - A) = PM + AQ \qquad [29]$$

The number of pulses stored in M is at least

$$M_{min} \geq A_{min} + 1 \qquad [30]$$

In the case where only one pulse should be removed, that is,

$$Q = 1 \qquad [31]$$

the division ratio N would comprise the whole set of integers from PM_{min} till $(PM_{max} + A_{max})$ if the division ratio A would be in the range

$$0 \leq A \leq P - 1 \qquad [32]$$

From the above discussion, we find the minimum N

$$N_{min} = PM_{min} + A_{min} = P^2 \qquad [33]$$

Evidently, the smallest possible dual-modulus divider is $P/Q = 2/3$.

These dual-modulus frequency dividers make it possible to extend the range of the fractional-N frequency synthesizer into GHz ranges [e.g. 14,15,16].

B. Modifications of Fractional-N Circuitry

To increase the reference frequency in an ordinary PLL frequency synthesizer above the desired frequency step, Nakagawa and Ohira [17] suggest the addition of M pulses into the feedback path. As a consequence, the relationship between the VCO output frequency and the reference frequency is given by

$$\frac{f_{VCO}}{N} \cdot (M + 1) = f_r \qquad [34]$$

A 2-GHz PLL frequency synthesizer with dual-modulus divider (16/17) in the feedback path with the output 1 (see Fig. 7) connected to the phase detector was suggested by Aytur and Razavi [14]. The final division ratio is

$$N = PM + A \qquad [35]$$

The advantage of this arrangement is that it is simple and has relatively few blocks.

Another arrangement of the fractional-N frequency synthesizer was suggested by Nakagawa and Tsukahara [18]. Again, a dual-modulus frequency divider of the type shown in Fig. 7 is put in the PLL feedback path. However, this time the output 2 is fed to the PLL phase detector, and the effective division ratio is

$$N_{eff} = P + \frac{A}{M}Q \qquad [36]$$

In instances where $Q = 1$ the advantage is that the division ratio may have an arbitrary numerator as well as an arbitrary denominator, since

$$N_{eff} = P + \frac{A}{M} \qquad [37]$$

*C. The Smallest Frequency Steps
at the Output of the Fractional-N
Frequency Synthesizers*

In the systems where an ordinary DDFS is used in the pulse-swallowing process (see Fig. 1), we get from the relation (11)

$$\Delta f_{o,\min} = f_r \frac{1}{Y} \qquad [38]$$

In instances where a programmable divider F [cf. eq. (4)] or M [Cf. eq. (36)] is changed

$$\Delta f_{o,\min} \approx f_r \frac{1}{F^2} \qquad [39]$$

or

$$\Delta f_{o,\min} \approx f_r \frac{A}{M^2} \qquad [40]$$

REFERENCES

[1] J. Gibbs and R. Temple. "Frequency domain yields its data to phase-locked synthesizer." *Electronics,* pp. 107–13, April 27, 1978.

[2] D. D. Danielson and S. E. Froseth. "A synthesized signal source with function generator capabilities." *Hewlett-Packard Journal,* pp. 18–26, January 1979.

[3] U. L. Rohde. *Digital PLL Frequency Synthesizers.* Englewood Cliffs, N.J.: Prentice-Hall, 1983.

[4] V. F. Kroupa. "Periodic omission of pulses." (A specially written paper for Part XI in this volume), 1995.

[5] V. F. Kroupa. "Spectra of pulse rate frequency synthesizers." *Proceedings of the IEEE,* pp. 1680–82, December 1979. (Reprinted in Part IV.)

[6] V. F. Kroupa. "Low-noise microwave-frequency synthesizers: Design principles." *IEE Proceeding-H,* pp. 483–88, December 1983. (Reprinted in this Section.)

[7] V. F. Kroupa. "Digital to analog converters." (A specially written paper for Part VIII of this volume.)

[8] C. Little. "Fractional-*N* synthesis." *Electronics World,* pp. 130–35, February 1996, pp. 196–98, March 1996.

[9] B. Miller and R. J. Conley. "A multiple modulator fractional divider." *IEEE Transactions IM-40,* pp. 578–83, June 1991. (Reprinted in this Section.)

[10] T. A. D. Riley, A. Copeland, and T. A. Kwasniewski. "Delta-sigma modulation in fractional-N frequency synthesis." *IEEE Journal of Solid-State Circuits,* pp. 553–59, May 1993.

[11] J. Chodora. "A digitally corrected fractional-*N* synthesizer." *Hewlett-Packard Journal,* p. 44, April 1993.

[12] G. W. Small. "A frequency synthesizer for $10/2\pi$ kHz." *IEEE Transactions on Instrumentation and Measurement,* pp. 34–37, March 1973. (Reprinted in Part II.)

[13] V. F. Kroupa. "Pulse subtracter for frequency synthesis." *Electronic Engineering,* p. 25, January 1977. (Discussion in this Appendix is also based on the original author's notes and figures.)

[14] T. S. Aytur and B. Razavi. "A 2-Ghz, 6-mW BiCMOS frequency synthesizer." *IEEE Journal of Solid-State Circuits,* pp. 1457–62, December 1995.

[15] M. Rochi et al. "A 1.2-GHz frequency synthesizer using a custom-design divide-by-20/21/22/23/24." *IEEE Journal of Solid-State Circuits,* pp. 1194–99, December 1986.

[16] N. Foroudi and T. A. Kwasniewski. "CMOS high-speed dual-modulus frequency divider for RF frequency synthesis." *IEEE Journal of Solid-State Circuits,* vol. 30, pp. 93–100, February 1995.

[17] T. Nakagawa and T. Ohira. "A phase noise reduction technique for MMIC frequency synthesizers that uses a new pulse generator LSI." *IEEE Transactions on Microwave Theory and Technique,* vol. 42, pp. 2579–82, December 1994.

[18] T. Nakagawa and T. Tsukahara. "A low phase noise C-band frequency synthesizer using a new fractional-N PLL with programmable fractionality." *IEEE Transactions on Microwave Theory and Technique,* vol. 44, pp. 344–46, February 1996.

A Multiple Modulator Fractional Divider

Brian Miller, *Member, IEEE*, and Robert J. Conley

Abstract—Fractional-N synthesis allows a PLL to achieve arbitrarily fine frequency resolution. Because the technique modulates the instantaneous divide ratio, fractional-N synthesizers suffer from fractional spurs. Various cancellation schemes allow fractional spur reduction to about -70 dBc at the expense of hardware cost and complexity.

Recent advances in oversampling A/D conversion technology can be incorporated into fractional-N synthesis, allowing the spectrum of error energy to be shaped so that fractional synthesis error energy is pushed away from the carrier. Based on this new technology, a CMOS integrated fractional-N divider was successfully developed. A complete fractional-N PPL was constructed utilizing only the CMOS divider, a dual modulus prescaler, a simple loop filter, and VCO. The resulting PLL exhibits no fractional spurs.

I. Review of Fractional-N Synthesis

IN a conventional PLL (Fig. 1) $f_{vco} = N * f_{ref}$. The divide ratio N must be an integer and therefore the frequency resolution of the locked loop is f_{ref}. Fine frequency resolution requires a small f_{ref} and a correspondingly small loop bandwidth. Narrow loop bandwidths are undesirable because of long switching times, inadequate suppression of vco phase noise, and susceptibility to hum and noise.

Fractional-N synthesis was developed to allow a phase-locked loop to have frequency resolution finer than f_{ref} [1], [2]. In a fractional-N divider, the integer divide ratio is periodically altered from N, to $N + 1$ (Fig. 2). The resulting average divide ratio will be increased from N by the duty cycle of the $N + 1$ division.

$$
\begin{aligned}
\text{average } f_{vco} &= \frac{1}{(T_N + T_{N+1})} [T_N * N * f_{ref} \\
&\quad + T_{N+1} * (N + 1) * f_{ref}] \\
&= [N + T_{N+1}/(T_N + T_{N+1})] * f_{ref} \\
&= (N.f) * f_{ref}
\end{aligned}
$$

where N denotes the integer portion of the divide ratio and $.f$ is the fractional component of the average divide ratio.

Typically, the overflow from an accumulator is used to modulate the instantaneous divide ratio (Fig. 3) [3]. Given that

a) VCO frequency = $N.f * f_{ref}$,
b) Accumulator maximum capacity = $C - 1$,
c) $X = .f * C$ (so that $X/C = .f$),

Manuscript received May 3, 1990; revised December 31, 1990. This paper was presented at the 1990 Frequency Control Symposium.

The authors are with Research & Development Laboratory, Hewlett-Packard, TAF C-34, Spokane, WA 99220.

IEEE Log Number 9143071.

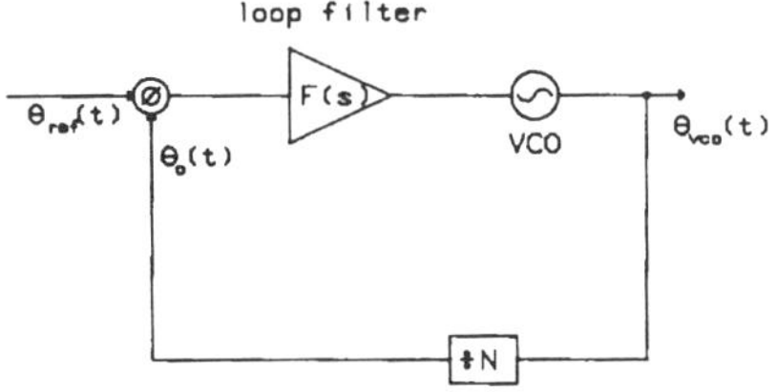

Fig. 1. Basic PLL.

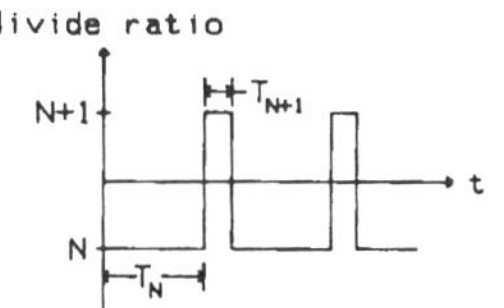

Fig. 2. Alternating divide ratio of fractional-N PLL.

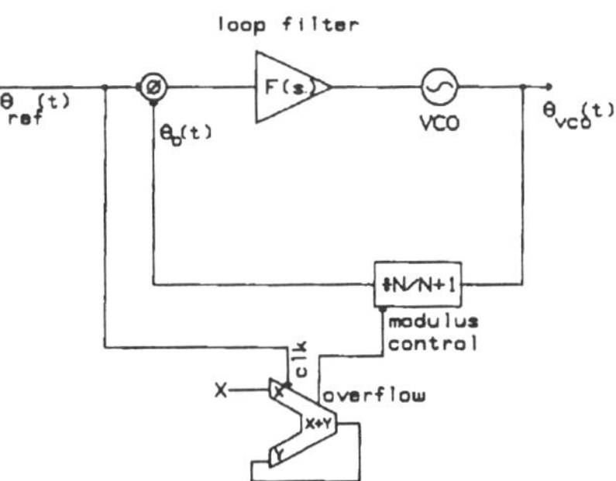

Fig. 3. Fractional-N PLL showing accumulator control of divide ratio.

then, while the divider is programmed to divide by N, the VCO signal at the phase detector will be at a frequency of $f_{ref} + (.f/N) * f_{ref}$ and the loop phase error will begin to advance at a rate of $2\pi * (.f/N) * f_{ref}$ rad/s. The phase error, referred to the VCO, advances at a rate of $2\pi * .f * f_{ref}$ rad/s. Because the accumulator is summing the same fraction as the VCO phase-error/reference-cycle, an accumulator overflow corresponds to a VCO phase error exceeding 2π radians and indicates the need to remove 2π of phase from the VCO output; accomplished by changing the divider modulus to $N + 1$ for a single reference cycle.

This periodic modification of the divider modulus gives rise to a sawtooth phase error (Fig. 4). If unfiltered, the phase error causes severe spurious tones (fractional spurs) at all multiples of the offset frequency ($.f * f_{ref}$).

Reprinted from *IEEE Transactions on Instrumentation and Measurement*, Vol. 40, No. 3, pp. 578–583, June 1991.

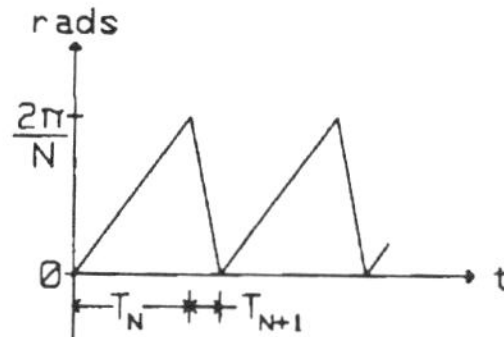

Fig. 4. Sawtooth phase error of conventional fractional-*N* synthesis.

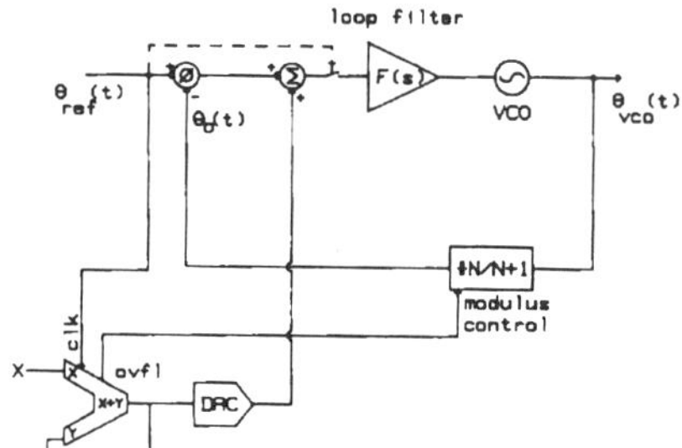

Fig. 5. Fractional-*N* PLL incorporating phase interpolation.

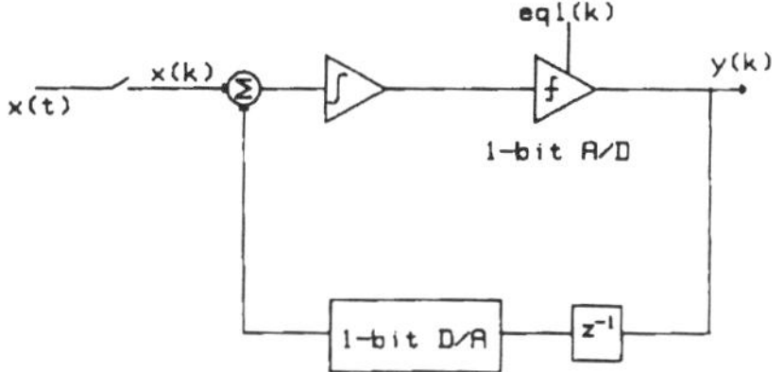

Fig. 6. Sigma–delta modulator.

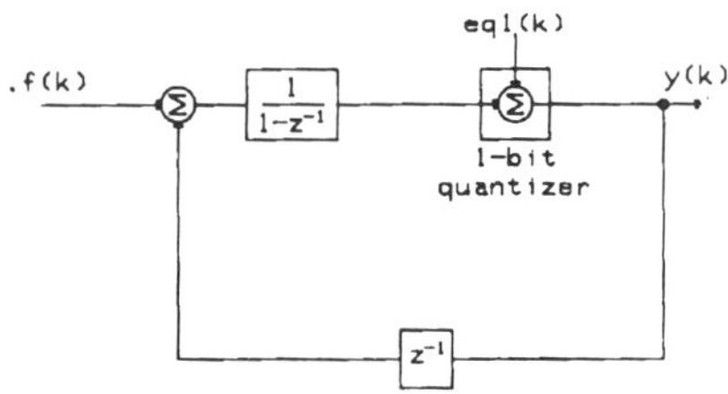

Fig. 7. Sigma–delta modulator suitable for fractional-*N* synthesis.

Fortunately, it is easy to predict what the phase error will be at any given time and a compensating signal can be summed into the PLL to cancel the error signal (Fig. 5) [4]. The phase detector output is sampled to allow settling of the interpolation DAC every reference cycle. This form of correction is called phase interpolation and fractional spurs are reduced to the extent that the phase interpolation signal exactly matches the phase error. Through the use of precision DAC's and carefully designed phase detector and sampler circuitry, fractional spurs of −70 dBc can be achieved. The complexity and expense of the interpolation circuitry make this form of fractional-*N* synthesis unsuitable in many applications.

II. A New Approach to Fractional Division

Interpolative A/D converters primarily based on sigma-delta modulators, have recently been developed into a viable technology for low frequency measurement and audio systems [5]–[7]. Interpolative A–D converters operate by greatly oversampling the input with a coarse (usually 1-bit) converter and then digitally filtering the 1-bit output stream to eliminate out-of-band quantization noise. S/N is further enhanced by embedding the 1-bit converter in a recursive filter structure which shapes the quantization noise present at the converter output so that most of the noise energy lies outside the band of interest and will be removed during filtering.

The same concept may be applied to fractional-*N* synthesis. Fractional-*N* synthesis attempts to achieve fine frequency resolution through manipulation of a coarse, integer divider. The desired fractional frequency is analogous to the analog input of an A–D. The integer-restricted divider is analogous to the 1-bit converter utilized in interpolative A–D converters. Consequently, most of the recent developments in interpolative A–D converter technology are also applicable to fractional-*N* synthesis.

A multistage, sigma-delta modulator architecture based on the work of Matsuya *et al.* [5] was chosen to implement the new fractional-*N* divider. Fig. 6 shows the basic modulator used in sigma-delta A–D converters. $x(k)$ is the modulator input, $y(k)$ is the modulator output, and $e_{q1}(k)$ is the quantization error added by the 1-bit A/D. In fractional-*N* synthesis applications, the input to the sigma-delta modulator is the desired fractional offset, which is a digital word. Consequently, the integrator may be digitally implemented and the 1-bit D–A is not required. Fig. 7 shows a sigma–delta modulator suitable for fractional-*N* synthesis.

$$Y(z) = \frac{1/(1 - z^{-1})}{1 + z^{-1}/(1 - z^{-1})} (.F(z))$$

$$+ \frac{1}{1 + z^{-1}/(1 - z^{-1})} E_{q1}(z)$$

$$= .F(z) + (1 - z^{-1})E_{q1}(z) \qquad (1)$$

where $Y(z)$, $.F(z)$ and $E_{q1}(z)$ are the Z-transforms of $y(k)$, $.f(k)$, and $e_{q1}(k)$, respectively.

A block diagram of a three-modulator fractional divider is shown in Fig. 8. System behavior is assumed sufficiently random to justify modeling each 1-bit quantizer as a unity gain element with added quantization noise [8]. $N.F$ is the desired rational divide ratio and $N_{\text{div}}(k)$ is the actual sequence presented to the integer restricted divider. Using

$$N_1(z) = (1 - z^{-1})E_{q1}(z) + .F(z) \qquad (1)$$

$$N_2(z) = -E_{q1}(z) + (1 - z^{-1})E_{q2}(z) \qquad (2)$$

$$N_2'(z) = -(1 - z^{-1})E_{q1}(z) + (1 - z^{-1})^2 E_{q2}(z) \qquad (3)$$

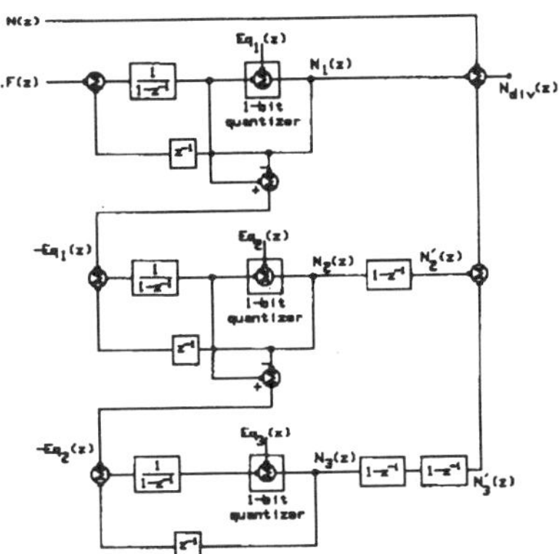

Fig. 8. 3-modulator fractional-N divider.

$$N_3(z) = -E_{q2}(z) + (1 - z^{-1})E_{q3}(z) \qquad (4)$$

$$N_3'(z) = -(1 - z^{-1})^2 E_{q2}(z) + (1 - z^{-1})^3 E_{q3}(z) \qquad (5)$$

$$N_{\text{div}}(z) = N(z) + N_1(z) + N_2'(z) + N_3'(z)$$

$$= N.f(z) + (1 - z^{-1})^3 E_{q3}(z) \qquad (6)$$

In a locked PPL,

$$f_{\text{out}}(k) = N_{\text{div}}(k)f_{\text{ref}}. \qquad (7)$$

Using (7), we can then write

$$F_{\text{out}}(z) = N.F(z)f_{\text{ref}} + (1 - z^{-1})^3 f_{\text{ref}}E_{q3}(z) \qquad (8)$$

where the first term of (8) is the desired frequency, and the second term represents frequency noise due to fractional division. This form is not useful for frequency synthesis applications and needs to be converted into single-sideband phase noise, $\mathcal{L}(f)$.

Each of the 1-bit quantizers is assumed to have uniform quantization error. The error power is then $(\delta)^2/12$; where δ is the minimum stepsize of the quantizer. Because we are quantizing to integers, $\delta = 1$ and the quantization error power $= 1/12$. This power is spread over a bandwidth of f_{ref}. Consequently, the power spectral density (PSD) of $e_{qj} = 1/(12f_{\text{ref}})$. Defining, $v(z) = $ frequency fluctuations of $F_{\text{out}}(z)$

$$S_v(z) = |(1 - z^{-1})^3 f_{\text{ref}}|^2 (1/12f_{\text{ref}}) \qquad (9)$$

$$= |1 - z^{-1}|^6 (f_{\text{ref}}/12). \qquad (10)$$

We want phase fluctuations, not frequency fluctuations.

$$\phi(t) = \int \omega(t)\,dt = 2\pi \int vt\,dt.$$

Employing a simple rectangular integration to represent $\int dt$ in the z-domain,

$$\Phi(z) = \frac{T_s W(z)}{1 - z^{-1}} = \frac{2\pi T_s v(z)}{1 - z^{-1}}, \qquad (11)$$

where T_s is the sample period.

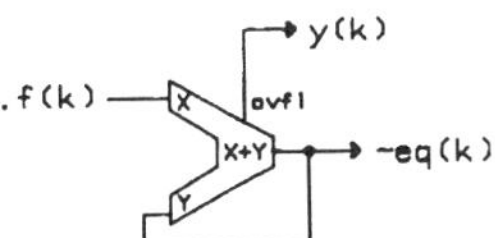

Fig. 9. An accumulator regarded as a sigma–delta modulator.

Using $T_s = 1/f_{\text{ref}}$ and (10) in (11), we obtain

$$S_\Phi(z) = \frac{(2\pi)^2}{|1 - z^{-1}|^2 f_{\text{ref}}^2} * \frac{|1 - z^{-1}|^6 f_{\text{ref}}}{12}$$

$$= \frac{(2\pi)^2}{12 f_{\text{ref}}} |1 - z^{-1}|^4 \ \text{rad}^2/\text{Hz}. \qquad (12)$$

If $S_\phi(f)$ is a two-sided PSD, then; $\mathcal{L}(f) = S_\phi(f)$. Therefore,

$$\mathcal{L}(z) = \frac{(2\pi)^2}{12 f_{\text{ref}}} |1 - z^{-1}|^4 \ \text{rad}^2/\text{Hz}. \qquad (13)$$

Converting to the frequency domain and generalizing to any number of modulator sections,

$$\mathcal{L}(f) = \frac{(2\pi)^2}{12 f_{\text{ref}}} [2 \sin (\pi f/f_{\text{ref}})]^{2(m-1)} \ \text{rad}^2/\text{Hz}, \qquad (14)$$

where m is the number of modulator sections.

Typically we are concerned with offset ranges small compared to the reference frequency allowing;

$$\mathcal{L}(f) \approx \frac{(2\pi)^2}{12 f_{\text{ref}}} \left[\frac{f}{f_{\text{ref}}/2\pi} \right]^{2(m-1)} \ \text{rad}^2/\text{Hz}. \qquad (15)$$

Equation (15) gives the colored quantization noise produced by the multiple modulator synthesis technique. Instead of discrete spurs, error energy produced by the fractional division will be manifested as noise. This noise must be filtered prior to the VCO to prevent unacceptable degradation of spectral purity. Interpolative A–D converters utilize digital filters to remove out-of-band quantization noise. In a fractional-N synthesis application, the PLL lowpass characteristic may be utilized to filter the quantization noise. A circuit example is given in Section IV.

The system of Fig. 8 can be simplified when it is recognized that an accumulator is a compact realization of the sigma–delta modulator. Fig. 9 shows an accumulator based sigma–delta modulator. The feedback of Fig. 7 occurs implicitly in the internal logic of the accumulator. Incorporating accumulators into the three-stage sigma–delta modulator yields Fig. 10. The topology of Fig. 10 is readily amenable to integration. The multiple modulator fractional-N control system reduces to a forward path of accumulators and a reverse path of differentiators.

III. Fractional Divider Implementation

An IC was conceived to implement all the digital functions of a PLL incorporating a multiple modulator fractional divider. The part is fabricated in a 1.5-μm, 5-V

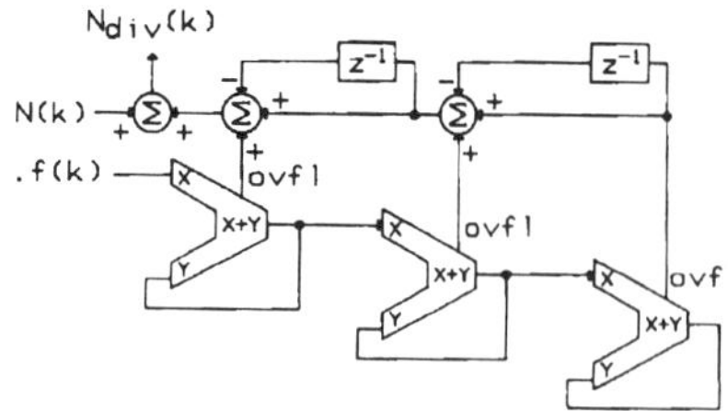

Fig. 10. Multiple modulator divider implemented with accumulators.

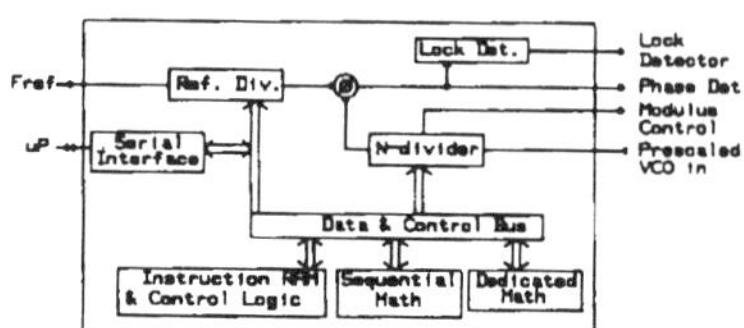

Fig. 11. Block diagram of the fractional-N IC.

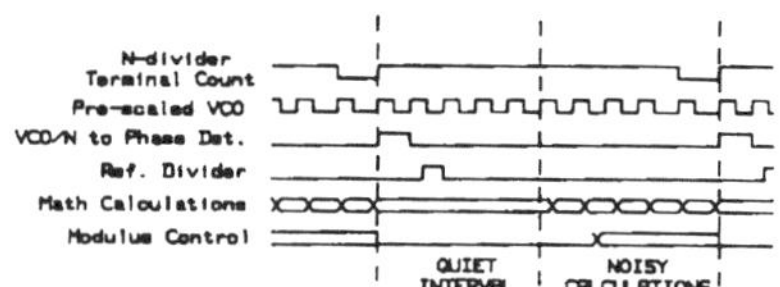

Fig. 12. Timing diagram of the basic PLL cycle.

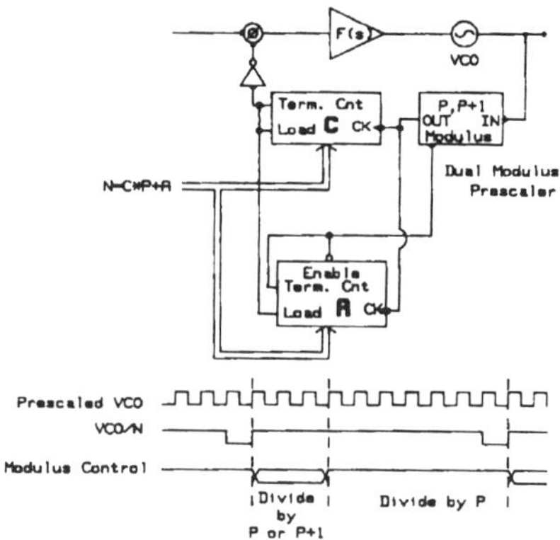

Fig. 13. Prescalar controller with modulus changes referenced to the beginning of the reference cycle.

CMOS process and packaged in a 44-pin plastic leaded chip carrier. The part dissipates 75 mW when clocked at 15 MHz.

Fig. 11 illustrates the overall block diagram of the IC. This diagram can be broken down into two distinct sections:

1) Basic Synthesizer	2) Computation and Control
a) N divider	a) Sequential Math Section
b) Phase Frequency Detector	b) Dedicated Math Section
c) Reference Divider	c) Data Flow Control
d) Out of Lock Detector	
e) Serial Interface	

Not shown in Fig. 11 but included on the IC, is circuitry to support FM inside the PLL bandwidth, synthesized sweep, and a general purpose 12-bit parallel data bus for control of external peripheral devices.

A. Basic Synthesizer

Although the basic synthesizer looks very much like the parts available from Motorola [11], Fujitsu [12], and others, significant differences exist, particularly in the N divider. Those who have built and tested fractional N PLL's know the importance of avoiding spurious coupling from the N divider, reference divider, and the fraction computation logic onto the edges presented to the phase detector. With this in mind, the following basic premise governed the architecture of the N divider and mathematical sections of the IC.

"Avoid spurious coupling of divider and computation events to the phase detector edges by restricting both to separate times within the PLL cycle."

This, coupled with the following two observations, resulted in the general PLL cycle requirements illustrated in Fig. 12.

a) To reduce the delay of FM, sweep, and other external requests, the requests should be sampled and their associated calculations executed at the last possible moment before the N-divider is reloaded.
b) The VCO$/N$ edge placement must represent exactly the number of VCO cycles calculated by the sigma-delta modulators. No incidental PM can be allowed to modulate the VCO$/N$ edge.

The start of the reference cycle is "quiet" and reserved for the phase detector edges. The last portion of the cycle is reserved for the noisy calculations and prescalar modulus control changes.

The N-divider controller is based on the well known dual modulus prescalar [13] technique. In a typical prescaler based divider, prescaler modulus changes are referred to the beginning of the reference cycle as illustrated in Fig. 13. However, unlike most dividers found in single chip synthesizers, the new IC references changes of the prescalar modulus to the end of the PLL reference cycle (Fig. 14) to facilitate time separation of phase detector edges and noisy computations. Instead of using two counters to generate prescalar modulus control, a counter and comparator are used. The Q outputs of the down counter are compared to the value stored in the modulus control register. Although other dual modulus dividers have been built, based on a single counter and comparators [14], they do not function to reference the modulus control changes to the end of the PLL cycle.

B. Computation and Control

Consider the implementation of three-modulator divider illustrated in Fig. 10. In addition let the desired

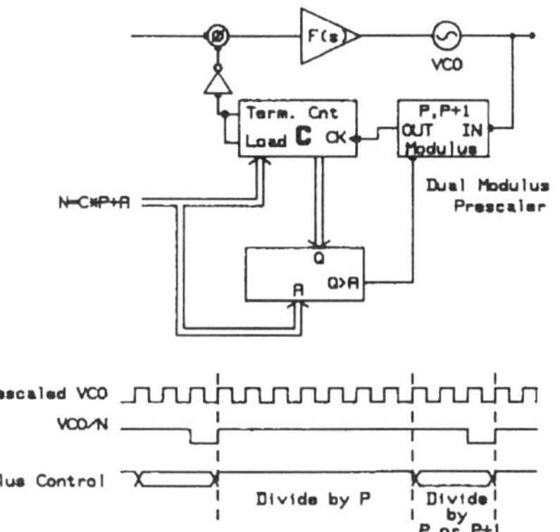

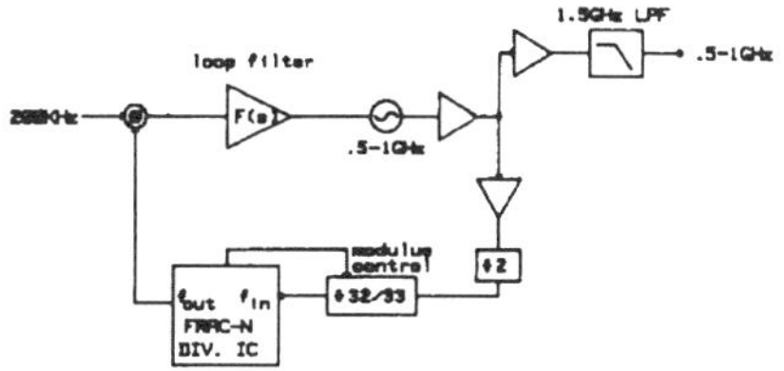

Fig. 15. 0.5–1-GHz fractional-*N* PLL.

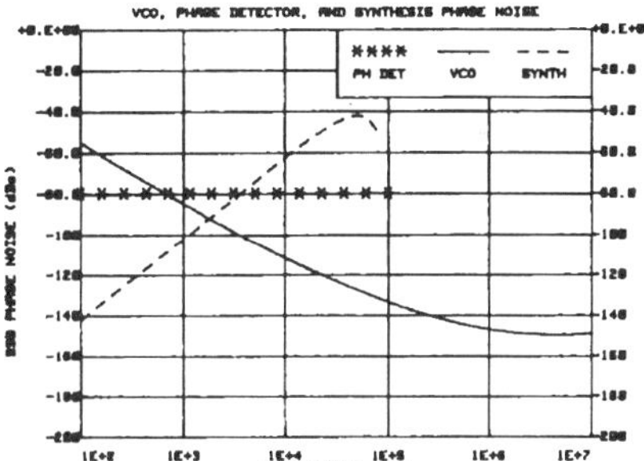

Fig. 14. Prescalar controller with modulus changes referenced to the end of the reference cycle.

Fig. 16. Phase noise sources in the 0.5–1-GHz PLL.

fractional resolution require a 24-bit word width in each modulator and an *N*-value less than 4096. The major logic required to implement this is cataloged as follows:

1) 3 24-bit adders,
2) 4 24-bit registers,
3) 6 24-bit data paths,
4) 2 $1 - z^{-1}$ operators,
5) 1 12-bit adder,
6) 3 12-bit data paths.

This could be a large integration problem. Since multiple prescalar output cycles are available every PLL cycle, a sequential approach to the accumulations was chosen. This approach complicates IC control but saves considerable die area. The number of prescalar output cycles per PLL cycle are limited, so the small valued computations such as the $1 - z^{-1}$ operators are implemented in dedicated logic. The $1 - z^{-1}$ operators can be scaled, since their peak numeric value is bounded (i.e., $+1, -1$ for the third modulator section). This further reduces die requirements.

The control section, in addition to scheduling accumulations, is required to fetch dedicated computations into the sequential section and pass new values off to the *N* divider and peripherals used in FM [15] and sweep functions. The state machine required to implement the sequential math and control could have been implemented in PLA, ROM, or RAM. Cell-based RAM was selected for its obvious versatility as well as overall speed when compared to available cell-based ROM. The IC has a 16-instruction control RAM which must be programmed upon application of power to the part. RAM-based control hastened development of the part and its intended application because it was easy to reconfigure the IC for experimentation. The RAM-based control also provides flexibility to PLL designers who, with the IC, can program unique combinations of correction, specialized sweep, and FM or PM.

IV. Complete Fractional-*N* Synthesis PLL

The new fractional-*N* divider was used to construct a 0.15–1-GHz PLL (Fig. 15). The divider IC was programmed to implement a 3-modulator interpolative divider. An external 32/33 prescalar was utilized in conjunction with the fractional-divider IC to implement the *N*-divider function. The prescaler is preceeded by a divide-by-2 because the 32/33 prescalar is not able to toggle at 1 GHz.

The VCO is a single band, 0.5–1-GHz, varactor tuned transistor oscillator. The loop f_{ref} is 200 KHz. In addition to the normal phase control objectives, the loop filter must also provide adequate rejection of the synthesis noise. The predicted synthesis noise is overlaid on a plot of the free-running VCO phase noise (Fig. 16) to determine the loop filter design parameters. The synthesis noise plotted in Fig. 16 is increased 6 dB over (14) because of the divide-by-2 prescaler between the VCO and the fractional divider circuit. Note how the fractional division error energy has been pushed away from the carrier. In this case, the error energy intersects the VCO phase noise at an offset greater than where the phase detector noise intersects the VCO phase noise. Notice also that the error energy rises at 40-dB/decade. For this application, the servo requirements were satisfied with a type II, second-order loop of approximately 750 Hz bandwidth. The filtering requirements were met by adding two additional real-axis poles (1.8 kHz, 3 kHz) to the loop filter.

The results of a phase noise measurement on the locked loop are given in Fig. 17. The fractional divider was programmed for a frequency offset of 2 kHz. None of the traditional 2 kHz (and multiples thereof) spurs are present.

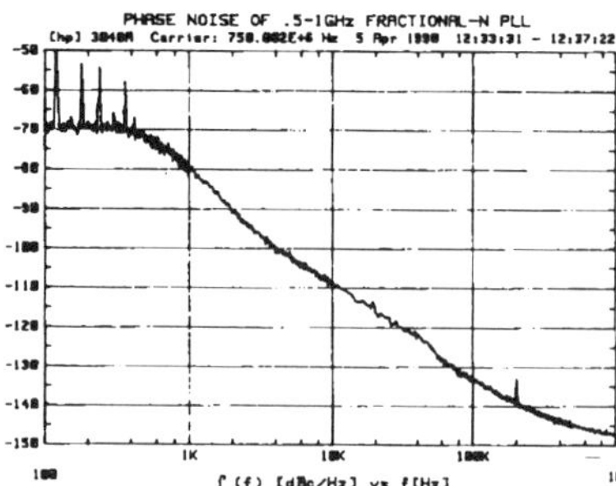

Fig. 17. Measured phase noise of the PLL.

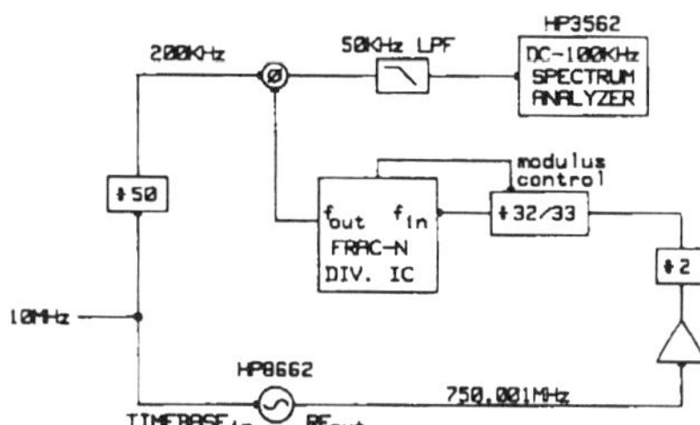

Fig. 18. Apparatus to measure synthesis noise of new fractional-*N* technique.

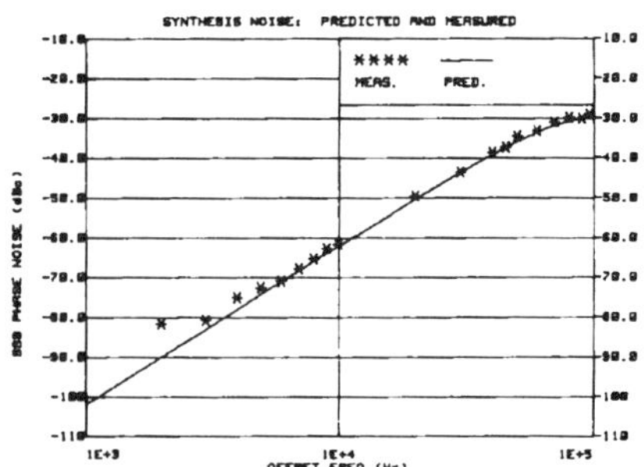

Fig. 19. Theoretical and measured synthesis noise.

Equation (14) was verified with the apparatus of Fig. 18. $N.f$ was chosen to divide the applied RF (750.001 MHz) down to the reference frequency (200 KHz). The baseband noise recorded on the spectrum analyzer was adjusted by the reciprocal of the low-pass filter gain, which was measured separately. Results are given in Fig. 19. The noise floor of the measurement system was approximately -80 dBc.

The mathematical prediction of synthesis error energy (14) relies on a uniform quantization noise model. The reader may rightly wonder if this is justifiable. Several authors [9], [10] have addressed the exact output spectra of a sigma–delta modulator but have limited their analyses to single-stage modulators. For single-stage modulators, the output spectra is strongly dependent on the dc input to the modulator and a quantization noise model is not appropriate [9]. Heuristically, higher order coders decrease the correlation between the input signal and the quantization error [10], permitting a quantization noise model.

We found the three-stage ΣDM to comply with the uniform noise model if the first accumulator experiences activity in, or near, the LSB bit position. Inputs which excite only bits near the MSB position, such as $.f = 0.5$, 0.75, 0.25, etc., result in a limit cycle of short duration and insufficient randomness to decorrelate the quantization error. Fortunately, it is easy to always add 1 LSB to the desired frequency offset if the desired $.f$ does not already have the LSB set. Because the accumulator length is so large, a 1-LSB frequency error is easily tolerable. In our test circuit utilizing a 24-bit accumulator, 1 LSB corresponds to 0.0238 Hz, or 4.77×10^{-5} ppm worst case.

V. Summary

A new technique for fractional-*N* synthesis has been developed. The technique causes error energy to be converted into colored noise instead of discrete spurious tones. The noise energy is suppressed at small offsets, allowing the remaining error to be rejected by simple filtering. The technique provides a tremendous advantage in cost, size, and complexity over traditional fractional-*N* synthesis techniques which utilize error correction. The prototype circuit achieved a 2.5 times reduction in size with respect to previous generation fractional-*N* synthesis circuits.

Acknowledgment

The authors would like to thank Scott Grimmett, Marcus DaSilva, Mark Talbot, Ben Flugstad, and Jim Catlin for their contribution to the successful development of the IC.

References

[1] V. Manassewitsch, *Frequency Synthesizers.* New York: Wiley, 1987, pp. 43–48.

[2] W. F. Egan, *Frequency Synthesis by Phase Lock.* New York: Wiley, 1981, pp. 196–202.

[3] C. A. Kingsford-Smith, U.S. Patent 3 928 813, Dec. 23, 1975.

[4] R. G. Cox, U.S. Patent 3 976 945, Aug. 24, 1976.

[5] Y. Matsuya *et al.*, "A 16-bit oversampling A/D conversion technology using triple integration noise shaping," *IEEE J. Solid-State Circuits*, vol. SC-22, pp. 921–929, Dec. 1987.

[6] J. C. Candy, "A use of double-integration in sigma-delta modulation," *IEEE Trans. Commun.*, vol. COM-33, pp. 249–258, Mar. 1985.

[7] D. R. Welland, *et al.*, "A stereo 16-bit delta–sigma A/D converter for digital audio," *J. Audio Eng. Soc.*, vol. 37, no. 6, pp. 476–484, June 1989.

[8] B. Agrawal and K. Shenoi, "Design methodology for ΣDM," *IEEE Trans. Commun.*, vol. COM-31, pp. 360–370, Mar. 1983.

[9] J. C. Candy and O. J. Benjamin, "The structure of quantization noise from sigma–delta modulation," *IEEE Trans. Commun.*, vol. COM-29, pp. 1316–1323, Sept. 1981.

[10] R. M. Gray, "Oversampled sigma–delta modulation," *IEEE Trans. Commun.*, vol. COM-35, pp. 481–489, May 1987.

[11] *Motorola CMOS Special Functions Data Book.* Q2, 1988. Motorola Semiconductor Products, P.O. Box 20912, Phoenix, AZ 85036.

[12] *Fujitsu Telecommunications Data Book*, 1987. Fujitsu Limited, Integrated Circuits and Semiconductor Marketing, Furukawa Soyo Bldg., 6-1, Marunouchi, 2-Chrome Chiyodu-ku, Tokyo 100, Japan.

[13] Motorola Application Note AN-827.

[14] J. A. Borras, U.S. Patent 4 427 820, Sept. 18, 1984.

[15] M. K. DaSilva, *et al.*, U.S. Patent 4 546 331, Oct. 8, 1985.

A Fast-Settling GaAs-Enhanced Frequency Synthesizer

John F. Naber, *Member, IEEE,* Hausila P. Singh, *Senior Member, IEEE,* William J. Tanis, *Member, IEEE,* Andrew J. Koshar, and Gregory L. Segalla

Abstract—An indirect, phase-locked loop (PLL), GaAs-enhanced frequency synthesizer with 700-ns loop settling time has been developed. The two-chip GaAs insertion reduced the size of an existing synthesizer from 90 in³ to only 30 in³. The 6.0 × 5.5 × 0.9-in module contains a 400-gate GaAs programmable divider and a sample and hold (S/H) which improved the settling time 77% and reduced the size 67% over the current state-of-the-art synthesizer. Futhermore, the divider reduced power dissipation by 9.7 W and the S/H reduced power dissipation by 1.3 W.

I. Introduction

THE direct digital synthesizer (DDS) is almost invariably considered for requirements that call for finely stepped submicrosecond settling, ultracompact, low-power, and low-cost frequency synthesis [1], [2]. However, when requirements call for wide frequency range signals above a few hundred megahertz as well, the DDS output must be either multiplied or up-converted using costly, high-power, and space-consuming RF-microwave hardware. The GaAs-enhanced fractional-division (FD) frequency synthesizer described here essentially has a low-frequency DDS (without the sine-wave look-up table) embedded within its phase-locked loop (PLL) to provide fine steps at microwave frequencies [3]. The FD synthesizer provides the advantages of the DDS without the disadvantages of frequency multiplication or up-conversion at microwave output frequencies. In addition, the FD synthesizer loop provides inherent low-pass filtering for attenuating wide-band spurs. For some applications a second PLL may be required to further reduce close-to-the-carrier FM sidebands. The intended use of this synthesizer requires the highest efficiency and reliability with the lightest weight in the smallest possible size. Frequency synthesizers have been previously demonstrated incorporating GaAs to enhance performance, but with much smaller complexity [4]. The objective of this work is to develop GaAs IC's that can enhance the capabilities of an existing frequency synthesizer by significant performance improvements. Prescalers, counters, and sam-

Manuscript received April 15, 1992; revised June 10, 1992.

J. F. Naber and H. P. Singh are with ITT Gallium Arsenide Technology Center, Roanoke, VA 24019.

W. J. Tanis, A. J. Koshar, and G. L. Segalla are with ITT Avionics, Nutley, NJ 07110.

IEEE Log Number 9202432.

ple and holds (S/H's) have been described previously and used as building blocks for the GaAs IC's embodied in this current synthesizer [5], [6].

II. Synthesizer Design

Fig. 1 shows a block diagram of the fast-settling frequency synthesizer with the digital and analog GaAs IC's highlighted. The GaAs IC's include a S/H and a programmable divider. The operation of the synthesizer commences after a reset from the control block. The output of the voltage-controlled oscillator (VCO) is fed to the GaAs divider, which in turn provides the input to the GaAs S/H phase detector. The programmable divider can provide division ratios from 12 to 2047 dependent upon the application. The phase detector compares the phase relationship of the divided VCO signal to the reference and generates an error voltage. The voltage is amplified and integrated by the loop integrator and fed back to the VCO via the summing amplifier to correct for variation in the VCO phase, thereby completing the PLL. The VCO is initially tuned open-loop and takes less than 50 ns to be within the capture or lock range of the loop, which minimizes the settling time. This is accomplished by the digital tuning network consisting of a programmable read-only memory (PROM) and a 12-b DAC. Fractional division techniques are used in the synthesizer to obtain resolutions as fine as 10 Hz with a 32-MHz reference frequency. The fractional division controller enables the programmable divider to divide by the fractional numbers.

III. IC Circuit Design

The programmable divider is a 400-gate IC that is capable of dividing inputs from 10 MHz to 2.0 GHz by an integer from 12 to 2047. A block diagram of the divider is shown in Fig. 2. The divider has a total of 11 b of digital control, with 2 b used to control the modulus control counter and 9 b for the down counter. Direct-coupled-FET logic (DCFL) is used throughout the design based on a well-characterized standard-cell library. DCFL offers very high speed at very low power. The divider interfaces with ECL control levels while requiring only a single −2-V power supply. Moreover, a second design was completed as well using TTL control in place of the

Reprinted from *IEEE Journal of Solid-State Circuits,* Vol. 27, No. 10, pp. 1327-1330, October 1992.

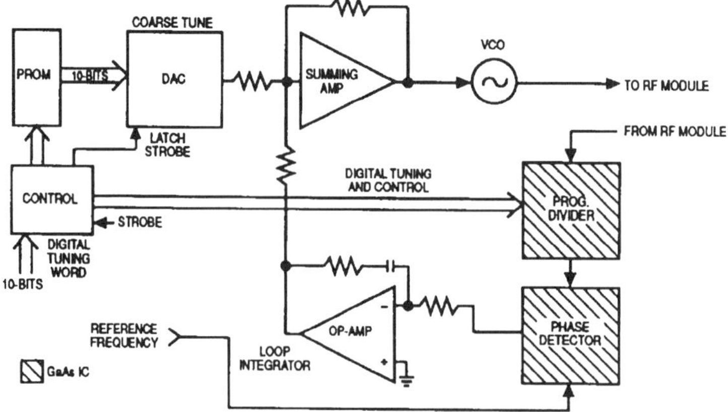

Fig. 1. Block diagram of a 700-ns settling frequency synthesizer.

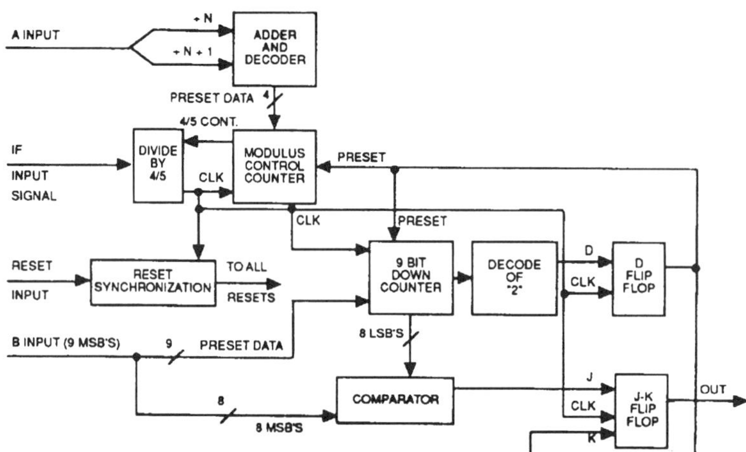

Fig. 2. Block diagram of the GaAs 400-gate programmable divider.

ECL levels. All divide conditions must function over the military temperature range of -55 to $+125°C$. The most significant feature of the GaAs divider is that it replaces 35 ECL chips in an existing synthesizer. This single-chip GaAs replacement results in a 97% power reduction from 10 W down to 0.3 W. A block diagram of the S/H is shown in Fig. 3. The S/H phase comparator is used instead of the more common digital phase comparator because it gives an order of magnitude increase in gain with reduced phase noise and sampling spikes. Furthermore, the reduced phase noise and spikes allow a wider loop bandwidth for faster response. The S/H uses a FET switch design that consists of over 50 active devices. The FET switch was used in place of a bridge design to minimize the distortion. The S/H is controlled by an ECL-driven clock source to toggle between the sample and hold modes. Futhermore, bootstrapped cascoded input and output amplifiers were used to limit the effects of loading on the amplifiers. The S/H uses ± 5-V power supplies and can accept signals as large as ± 2 V peak to peak without

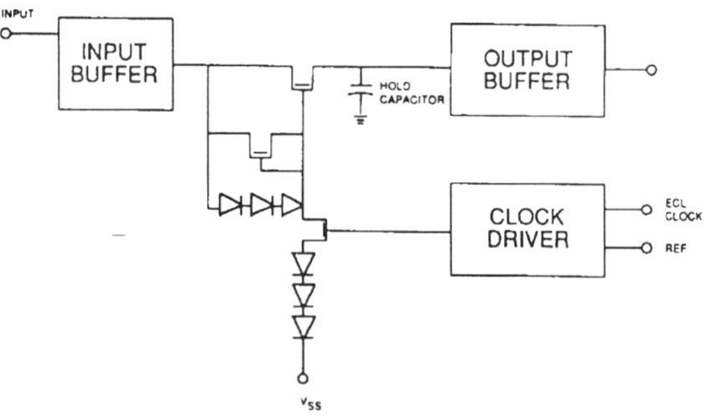

Fig. 3. Block diagram of the GaAs S/H.

saturating the amplifiers. The most significant benefit of the GaAs S/H is the reduction in power and size. The S/H replaces a silicon hybrid S/H measuring 1.5×0.75 in and dissipating 1.5 W with a monolithic IC measuring 1 mm $\times$ 1 mm and dissipating 200 mW of power. This results in an 87% reduction in power.

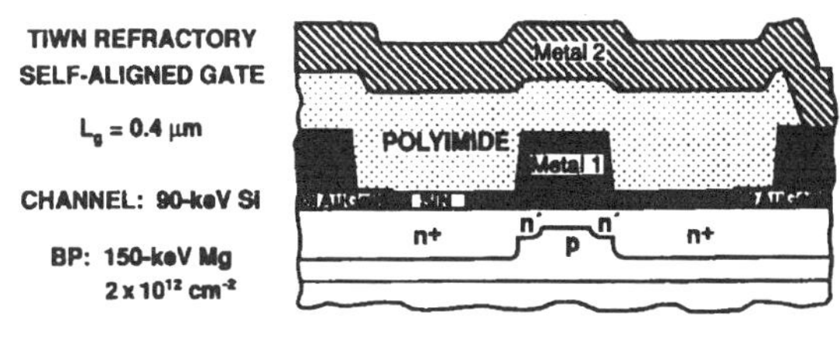

Fig. 4. Cross section of a MSAG 0.4-μm FET.

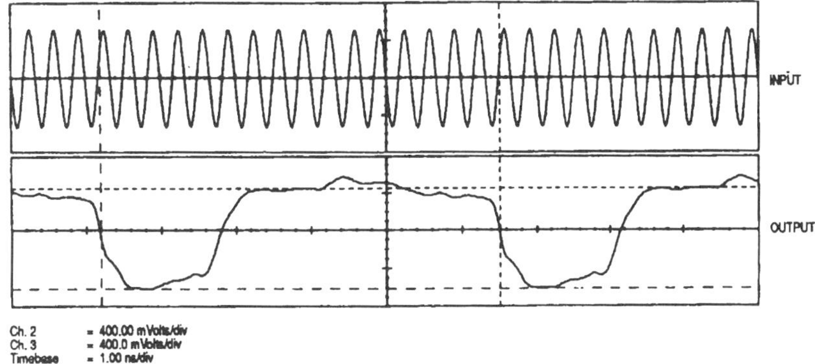

Fig. 5. High-frequency performance of the divider, with the input operating at 3.0 GHz and the output at 187.5 MHz in the divide-by-16 mode.

IV. Fabrication Process

The IC's were fabricated with a titanium tungsten nitride 0.4-μm multifunction self-aligned gate (MSAG) process that provides excellent performance and yield [7]. A cross section of a MSAG FET is shown in Fig. 4. On 3-in wafers, the E-FET's have an average threshold voltage of +0.20 V, with a standard deviation of 15 mV. DFET's have a typical theshold voltage of -0.40 V and a standard deviation of 25 mV. The average transconductance over a wafer is typically 270 mS/mm for E-FET's and 200 ms/mm for D-FET's. Typical inverter propagation delay is 40 ps (FI = FO = 1) with a power dissipation of 0.5 mW per gate at $V_{DD} = 2.0$ V.

V. Test Results

Fig. 5 shows the divider operating in the divide-by-16 mode, with the clock operating at a maximum frequency of 3.0 GHz and the output operating at 187.5 MHz. The average on-wafer speed of these dividers is 2.3 GHz, with an average power dissipation of 300 mW and an average yield of over 80%. These dividers are tested at 20 MHz for full functionality with an array of test vectors providing 98% fault coverage. After passing this test the dividers are tested at high speed and over temperature while in a self-test mode utilizing 60% of the gates. The functionality of the S/H is shown in Fig. 6 in an undersampling

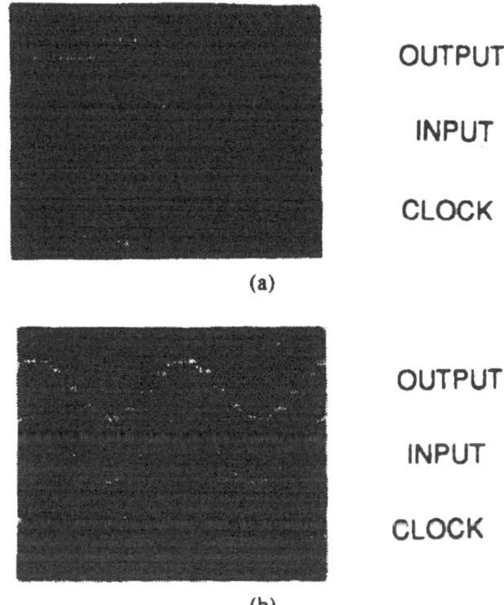

Fig. 6. Functionality of the S/H in the oversampling as well as undersampling mode of operation. (a) Undersampled, input = 1.0 MHz clock = 100 kHz. (b) Oversampled, input = 100 kHz, clock = 1.0 MHz.

as well as in an oversampling mode. A low-speed on-wafer test was done to verify that the circuit was functional. The high-speed testing was done on packaged devices up to the 400-MHz limit of the test equipment.

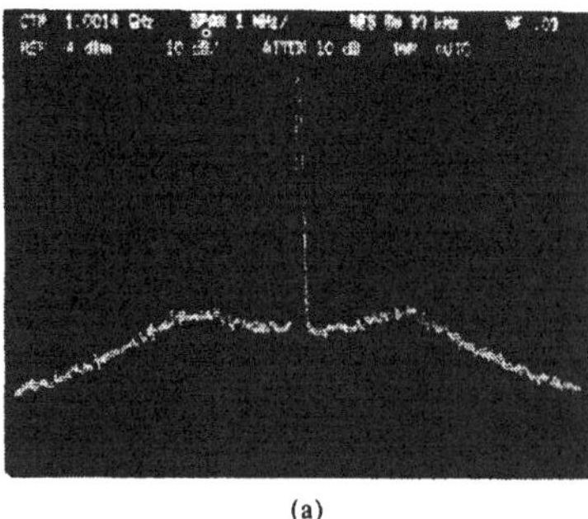

(a)

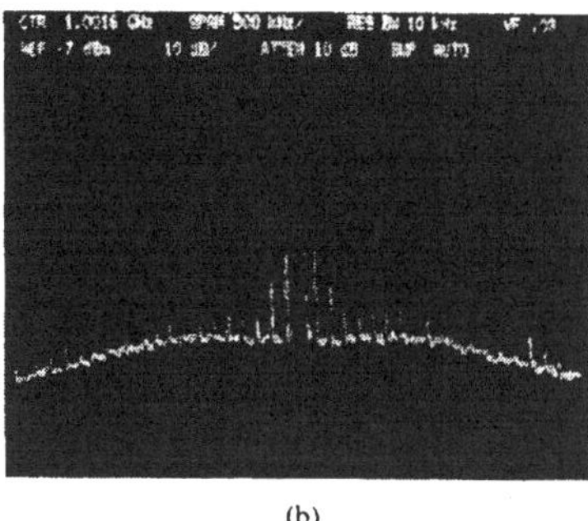

(b)

Fig. 7. Output spectrum of the frequency synthesizer with and without fractional division enabled. (a) Without fractional division, 10-MHz span, integer steps. (b) With fractional division, 5-MHz span, −36-dBC FD sidebands, fractional steps.

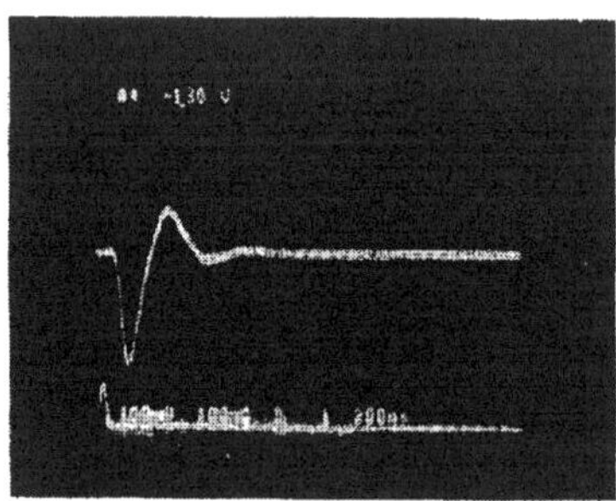

Fig. 8. The PLL settling to within 700 ns after a disturbance to the loop.

Moreover, the S/H had an average power dissipation of 200 mW with ±5-V supplies and an average yield of 85%. Fig. 7 shows the output spectrum of the frequency synthesizer with the fractional division enabled as well as without the fractional division enabled. Fig. 7(b) shows the fractional division sidebands, whereas Fig. 7(a) has no sidebands because of the integer steps. Fig. 8 shows the PLL settling time to be 700 ns after a disturbance to the loop. The bottom trace of Fig. 8 shows the blocking pulse to the programmable divider that was used to create the disturbance to the loop. A summary of the overall breadboard performance of the frequency synthesizer is shown in Table I for the basic loop as well as the FD loop.

TABLE I
FREQUENCY SYNTHESIZER PERFORMANCE SUMMARY

Parameter	Breadboard Performance	
	Basic Loop	FD Loop
Output frequency	1–2 GHz	1–2 GHz
Resolution	32 MHz	10 Hz
Settling	700 ns	700 ns
Spurious		
within 10 MHz	< −60 dBc	< −35 dBc
outside of 10 MHz	< −60 dBc	< −60 dBc
SSB Phase noise		
100-kHz offset*	98 dBc/Hz	98 dBc/Hz

*Reference dependent within 2-MHz loop bandwidth.

VI. CONCLUSIONS

Significant space and power savings, in addition to higher reliability and performance, have been realized using GaAs IC technology for next-generation frequency synthesizers. Furthermore, a significant cost reduction is achieved by eliminating a hybrid containing over 35 ECL integrated circuits and an expensive high-performance silicon S/H hybrid from today's existing frequency synthesizer. The GaAs programmable divider reduced power dissipation by 9.7 W, while the GaAs S/H reduced power dissipation by 1.3 W. These GaAs insertions resulted in a significant power reduction of 67% over the currently used synthesizer as well as 77% improvement in settling time.

REFERENCES

[1] L. Jackson, "Digital frequency sysnthesizer," U.S. Patent 3 735 269, May 1973.

[2] V. Manassewitch, *Frequency Synthesizer Theory and Design.* New York: Wiley, 1980.

[3] W. Tanis, "Frequency synthesizer having fractional frequency divider in phase-locked loop," U.S. Patent 3 959 737, May 1976.

[4] M. Rocchi *et al.*, "A 1.2-GHz frequency synthesizer using a custom divide-by-20/21/22/23/24 GaAs circuit," *IEEE J. Solid-State Circuits*, vol. SC-20, no. 6, pp. 1194–1199, Dec. 1985.

[5] H. Singh *et al.*, "GaAs prescalers and counters for fast-settling frequency synthesizers," *IEEE J. Solid-State Circuits*, vol. 25, no. 1, pp. 239–245, Feb. 1990.

[6] J. Naber, H. Singh, and R. Sadler, "A 2.5-GHz GaAs sample and hold," in *Proc. IEEE Int. Symp. Circuits Syst.*, vol. 4, pp. 3077–3080, May 1990.

[7] R. Sadler *et al.*, "A manufacturable 0.4 μm process for high-performance LSI circuits," in *Proc. IEEE GaAs IC Symp.*, Oct. 1990, pp. 63–66.

Part VII

Phase and Background Noise
in DDFS

I N Parts IV and V we have focused on discrete spurious signals in DDFS. Here, we return to the problem of the omnipresent background noise.

Every physical phenomenon is accompanied by smaller or larger fluctuations. In instances where these fluctuations are random, we speak about noise. The same is also true for all frequency generators where we encounter both amplitude and phase or frequency variations that cause the so-called frequency instability of oscillators, frequency synthesizers, and so on.

The problem is not fully understood in such complicated devices as DDFS. The frequency stability and frequency purity (due to discrete spurious components) depend on many factors: on the clock oscillators that are used, on the chosen frequency synthesis procedure, and on the acceptable dimensions, weight, and cost.

We often encounter statements that the output phase noise of DDFS is that of the reference oscillator reduced by the square of the normalized frequency, that is, by $(f_x/f_c)^2 = \xi_x^2$. However, generally this need not be true because nearly all building blocks may be noise generators; see Fig. 11 in the introductory paper by the editor in Part I.

We start Part VII with a concise introduction to the art of frequency stability written by the editor. (The problem has been investigated by many authors; here we refer only to [VII-1 and VII-2]). This paper also contains some guidelines for estimating the crystal oscillator and frequency divider background noises.

To estimate the DDFS noise based on the published results is difficult for two reasons: (1) one encounters the tendency to reproduce spectra of only a few normalized frequencies, often of the good ones; and (2) we usually find spectrograms measured with frequency analyzers. In these instances, it is difficult to distinguish between the background noise of the frequency synthesizer and that of the analyzer. Actual power spectra densities, for only a few normalized frequencies, are published by some producers (e.g., [VII-3, VII-4]). The integrated spurious or the worst case spurious plots are rare [VII-5].

Information about the close-to-the-carrier-noise is provided with the second paper by the editor. This noise is significant when zero crossings of the output wave are of importance.

Mattison and Coyle (1988) provide valuable information by comparing noise for both sine- and square-wave DDFS. We chose to reprint the paper since it is still frequently referenced.

The reprinted paper by Saul and Mudd (1988) contains two interesting spectra of DDFSs with 100-MHz capability. A deeper discussion of the results is provided by the editor in his first paper in this section.

In the last reprinted paper, by J. A. Connelly and K. P. Taylor (1991), the authors consider thermal noises in resistors combined with effective input noises of operational amplifiers in D/A and A/D converters.

REFERENCES

[VII-1] V. F. Kroupa, ed. *Frequency Stability: Fundamentals and Measurement.* New York: IEEE Press, 1983.

[VII-2] D. B. Sullivan, D. W. Allan, D. A. Howe, and F. L. Walls, eds. *Characterization of Clocks and Oscillators.* Boulder, Colo.: NIST 1337 Note, 1990.

[VII-3] A. Hill and J. Surber. "Using aliased-imaging techniques in DDS to generate RF signals." *RF Design,* p. 31, September 1993.

[VII-4] Bar-Giora Goldberg. "Linear frequency modulation—theory and practice." *RF Design,* pp. 39–46, September 1993.

[VII-5] G. W. Kent and N.-H. Sheng. "A high purity, high speed direct digital synthesizer." *IEEE International Frequency Control Symposium Proceedings,* 1993, pp. 207–11. (Reprinted in Part IX.)

Introduction to the Noise Properties
of Frequency Sources

VĚNCESLAV F. KROUPA

I. INTRODUCTION

IN the present paper, we first recall basic definitions of the frequency stability in both frequency and time domains. Then we discuss oscillator noise equations with the estimated noise coefficients, which may be useful when evaluating the correctness of measured results. Finally, we focus on the noise properties of all major building blocks used in DDFS.

II. NOISY SIGNALS

No generator has an output signal that can be represented, in the frequency domain, with a δ-function-like carrier frequency. In the most general case, we must expect that both spurious amplitude and phase modulation would be present, that is,

$$v(t) = V_o\left[1 + A(t)\right]\cos\left[\omega_o + \varphi(t)\right] \tag{1}$$

However, the limiting behavior of the oscillating loops and frequency synthesis devices reduces spurious side bands owing to the amplitude fluctuations. As a consequence, we can simplify eq. (1) into

$$\begin{aligned} v(t) &= V_o\cos\left[\omega_o t + \varphi(t)\right] \\ &= V_o\cos\left[\omega_o t + \int \dot{\varphi}(t)dt\right] \end{aligned} \tag{2}$$

where

$$\dot{\varphi}(t) \approx \Delta v(t) + \text{discrete spurious signals} + \text{secular terms} \tag{3}$$

Discrete spurious signals are of concern in frequency synthesizers, particularly in DDFS, whereas "secular" terms generally include aging and frequency variations caused by changing environmental conditions (temperature, humidity etc.).

III. BASIC FREQUENCY INSTABILITY MEASURES

In the following discussions, we consider only the first term on the right-hand side of eq. (3).

A. Frequency Domain Measures

Short-term fluctuations of the carrier, $\Delta v(t)$, or of its phase, $\Delta\varphi(t)$, are used to define the frequency instability. The first parameter should be preferred since its first moment, m_1, is

(This paper was specially written for this volume.)

very close to zero:

$$m_1 = \langle\Delta v\rangle \approx 0 \tag{4}$$

This is generally not true for the mean value of the phase fluctuations. Furthermore, by considering the second moment, m_2, or the autocorrelation of the frequency fluctuations, we get

$$m_2 = \left\langle\left[\Delta v(t)\right]^2\right\rangle = \int_0^\infty S_{\Delta v}(f)df \tag{5}$$

where $S_{\Delta v}(f)$ is the so-called power spectral density (PSD) of the frequency fluctuations, with the dimensions Hz^2/Hz, and is used as the frequency stability measure in the frequency domain.

Since the PSD of the time derivatives of a random variable, $x(t)$, is defined as $\omega^2 S_x(\omega)$, we often use this relation the other way and write for PSD of phase fluctuations

$$S_\varphi(f) = \frac{1}{f^2}S_{\Delta v}(f) \tag{6}$$

By introducing fractional frequency fluctuations

$$y(t) = \frac{\Delta v(t)}{\omega_o} \tag{7}$$

we arrive at the basic frequency instability measure independent of the carrier frequency ω_o. The advantage of the above representation is that its PSD $S_y(f)$ is invariable with respect to the ideal frequency multiplications and divisions. As a consequence, we can compare noise properties of generators with different input or output frequencies in one and the same figure. Finally, we can introduce the normalized phase measure

$$\frac{S_\varphi(f)}{f_o^2} = \frac{S_y(f)}{f^2} \tag{8}$$

The advantage is that this normalized phase measure retains all the original slopes of $S_\varphi(f)$ and the frequency invariability of $S_y(f)$.

B. Time Domain Measures

Before the advent of the FFT (fast Fourier transform), the direct measurement of $S_\varphi(f)$ had a lower bound, in the neighborhood of 10 Hz. This difficulty was solved with the time domain approach. The workhorse is the two-sample variance made from the mean differences of fractional frequency fluctuations during the successive time intervals, τ, that is,

$$\sigma_y^2(\tau) = \frac{1}{2}\langle(\bar{y}_{k+1} - \bar{y}_k)^2\rangle \qquad [9]$$

with

$$\bar{y}_k = \frac{1}{\tau}\int_{t_k}^{t_k+\tau} y(t)dt; \quad t_k + \tau = t_{k+1} \qquad [10]$$

Note that this "two-sample variance" is generally called the Allan variance [1].

Relation of the $\sigma_y^2(\tau)$ with the above introduced PSD $S_y(f)$ is similar to the definition of the second moment in (5), though with the assistance of the frequency-dependent transfer function, $H_A(jf)$, that is,

$$\sigma_y^2(\tau) \approx \int_0^\infty S_y(f)|H_A(jf)|^2 df \qquad [11]$$

where

$$|H_A(x)|^2 = 2\frac{\sin^4(x/2)}{(x/2)^2}; \quad x = 2\pi f\tau \qquad [12]$$

and the frequency characteristic is shown in Fig. 1. Note its behavior as a bandpass filter with a very low Q. Another consequence is that we can evaluate the integral in (11), in the closed form, but only for a very particular form of $S_y(f)$, namely, a piecewise linearized expressed as

$$S_y(f) \approx \frac{h_{-2}}{f^2} + \frac{h_{-1}}{f} + h_0 + h_1 \cdot f + h_2 \cdot f^2 \qquad [13]$$

Where all important noise processes, generally encountered by evaluating the frequency instability, are considered

the random walk of frequency with the noise constant $\quad h_{-2}$

the flicker frequency noise with the noise constant $\quad h_{-1}$

the white frequency noise with the noise constant $\quad h_0$

the flicker phase noise with the noise constant $\quad h_1$

the white phase noise with the noise constant $\quad h_2$

Furthermore, the periodicity of the transfer function $H_A(x)$ would cause divergence of the integral (11) for white and flicker phase noises. Therefore, we have to introduce some low-pass filtering with a cutoff frequency f_H, the magnitude of

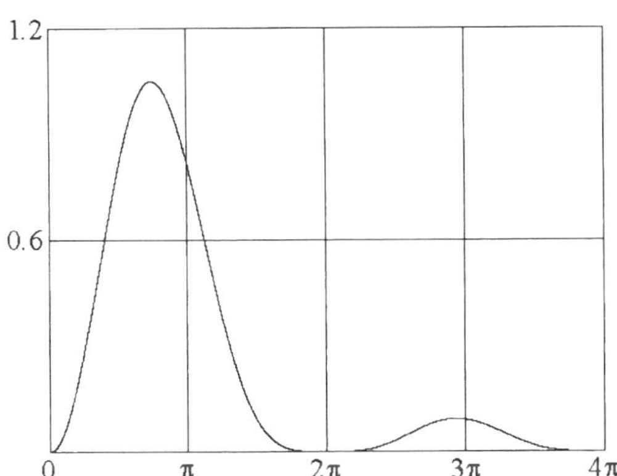

Fig. 1. The filter function of the Allan variance transfer function $|H_A(x)|^2$.

TABLE 1

$S_y(f)$	h_{-2}/f^2	h_{-1}/f	h_o	$h_1 f$	$h_2 f^2$
$\sigma_y^2(\tau)$	$(2\pi)^2\tau h_{-2}/6$	$2h_{-1}$	$h_o/2\tau$	$h_1(2\pi\tau)^{-2} \times(1.38+3\ln\omega_H\tau)$	$3h_2 f_H(2\pi\tau)^{-2}$

which should always accompany the respective measurement data. Another important question has to do with the relation between the time domain and frequency domain measurements. The conversion table (Table 1) reprinted from [1] may solve the problem.

Investigation of the Allan variance reveals that $\sigma_y^2(\tau)$ has the same slope for white and flicker frequency noises. In this case, only change of the f_H can resolve the problem. Another possibility is application of the "modified Allan variance." For more information, see, for example, [1]. The difficulty with the poor selectivity—cf. Fig. 1—can be removed with more complicated variances such as the Hadamard variance.

IV. Noise Properties of Different Circuits

All building blocks in DDFS might be noise generators. In the following paragraphs we will summarize properties of the most important circuits.

A. Amplifiers

The noise in transistors and integrated circuits has been discussed by many authors in the past. Here, we feel it is sufficient to mention only books by Van der Ziel [2,3]. Some information about noise in LF operational amplifiers contains the reprinted paper by Connelly and Taylor [4]. Kroupa [5] summarized earlier measurements on the best LF amplifiers with the results

$$S_\varphi(f) \approx \frac{10^{-16(+2)}}{f} + 10^{-18(+2)}; \quad [V^2/\text{Hz}] \qquad [14]$$

Recently published measurements [6] reveal for bipolar transistor (BFR92)

$$S_\varphi(f) \approx \frac{10^{-12.5}}{f} + 10^{-16.5} \qquad [15]$$

and for GaAs-MESFET CF300

$$S_\varphi(f) \approx \frac{10^{-10}}{f} + 10^{-17} \qquad [16]$$

However, GaAs/GaAlAs heterojunction bipolar transistors (HBTs) exhibit much lower $1/f$ noise [7] and in addition high f_T and high $f_{\max}$.

$$S_\varphi(f) \approx \frac{10^{-13}}{f} \qquad [17]$$

The HBT process is also suitable for high-speed digital IC

209

used—for example, as dual mode prescalers in fractional-N PLL frequency synthesizers.

For RF amplifier noise, where we only consider a narrow bandwidth around the carrier, one-half of the thermal white noise contributes to the amplitude noise modulation and the other half to the phase noise modulation. In the following we will investigate the theoretical limit of the phase noise at the RF amplifier output

$$\mathcal{L}(f) = 2\frac{kTR}{V_{rms}^2} \qquad [18]$$

For the quantitative information of the readers, after introducing $R = 50\Omega$ and $V_{rms} = 1V$, we find for $\mathcal{L}(f)$

$$\mathcal{L}(f) \geq -187 \quad \text{dB} \qquad [19]$$

However, generally a noise factor of about $+10$ to $+20$ dB, due to the electronics, must be added to the theoretical white noise level (see Fig. 2). Note also the approximate relation

$$\mathcal{L}(f) \approx \frac{1}{2}S_\varphi(f) \qquad [20]$$

B. Frequency Multipliers

The PSD of the phase noise at the output of a frequency multiplier is equal to the input $S_\varphi(f)$ multiplied by the square of the multiplication factor N plus an additive term

$$S_{\varphi,\text{out}} \approx S_{\varphi,\text{inp}} * N^2 + S_{\varphi,\text{add}} \qquad [21]$$

where the additive term has been found as follows:

$$S_{\varphi,\text{add}} \approx \frac{10^{-13\pm2}}{f} + 10^{-16\pm1} \qquad [22]$$

However, the flicker phase noise contribution can be reduced

to the indicated lower level only with the assistance of the local RF negative feedback (emitter degradation) applied to the respective transistors [1, papers by Halford et al., by Baugh, and by Andresen et al., pp. 322–42]; see also Fig. 2. The same is also true for oscillators.

C. Frequency Dividers

Theoretically, the division process reduces the input PSD in proportion to the square of the division factor N^2. However, investigation of the divider output phase noise performed by different authors [9–12] revealed for lower frequencies the predominance of an additive noise given practically by relation (22). For ECL gate-systems, the additive divider noise increases approximately to

$$S_\varphi(f) \approx \frac{10^{-11.5}}{f} + 10^{-15\pm1} \qquad [23]$$

whereas for GaAs gate-systems, intended for higher output frequencies, the flicker phase noise increases to

$$S_\varphi(f) \approx \frac{10^{-8.5\pm.5}}{f} \qquad [24]$$

D. Noise Properties of Oscillators

In the past, many authors investigated the problem of the noise in oscillators [e.g., 13,14] and in the resonators as well [15,16, and others]. The editor of this volume summarized all the available experimental data [17]. He proceeded with the solution of the noise behavior of the basic oscillator circuit shown in Fig. 3 by investigating the influence of the additive phase shift introduced via maintaining electronic circuits (the well-known Leeson solution [18]) and introduced the phase

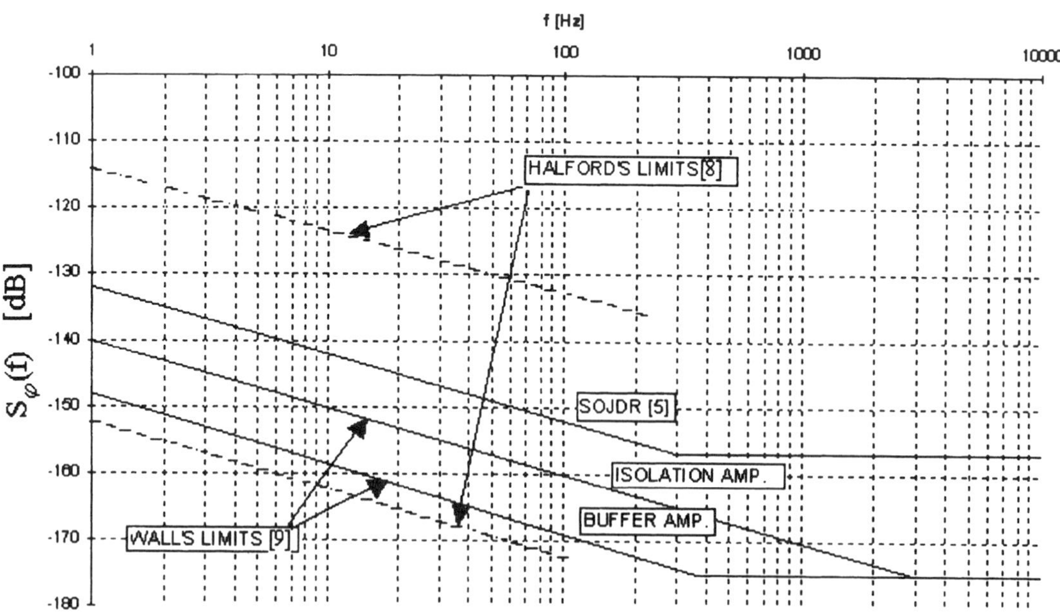

Fig. 2. Noise in RF and isolation amplifiers: limits in accordance with Halford [8] and Walls [9].

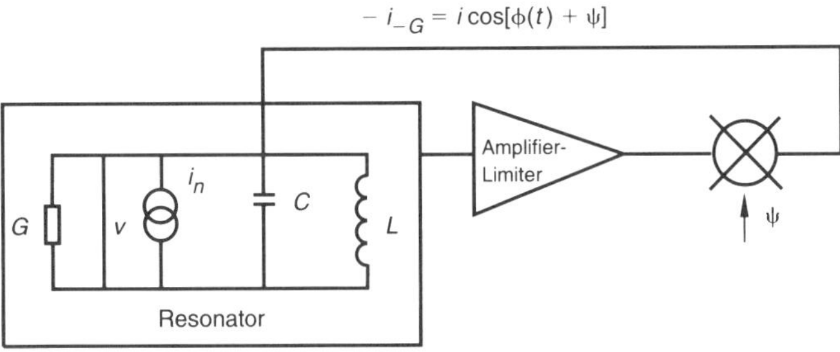

Fig. 3. Simplified block diagram for the solution of the noise behavior of oscillators.

shift compensation resonator fluctuations [17]. Finally, with the assistance of the superposition principle, he arrived at the normalized phase noise PSD of the crystal oscillator

$$\frac{S_{\varphi,\text{out}}(f)}{f_{o2}} \approx \frac{1}{f^3}\frac{a_{-1R}+a_{-1E}}{Q_U^2} + \frac{1}{f^2}\left(\frac{a_{-1R}}{Q_U f_o} + \frac{a_{oE}}{Q_U^2}\right)$$
$$+ \frac{1}{f}\frac{a_{-1E}}{f_o^2} + \frac{a_{oE}}{f_o^2} \qquad [25]$$

where Q_U is the unloaded Q-factor of the resonant circuit (by silently assuming, in accordance with [19], that $Q_U \approx 2Q_L$; that is, the loaded Q-factor), a_{-1E} and a_{oE} are the flicker and noise constants of the electronic circuits, and a_{-1R} is the flicker frequency noise constant of the resonator—all at the Fourier frequency 1 Hz.

From the investigation of the noise in amplifiers, we suggest practical values of noise constants that are of interest for designers and researchers

$$a_{oE} \approx \frac{4kT}{P_{\text{out}}}F_N \approx 10^{-14}\ldots 10^{-17} \qquad [26]$$

and

$$a_{-1E} \approx 10^{-11}\ldots 10^{-14} \qquad [27]$$

The lower values are arrived at in instances where a RF feedback (an unbiased resistor) is applied into the emitter circuit and oscillator transistors are carefully selected.

A very large number of crystal resonators exhibit a noise constant (the origin of which is not yet fully understood), approximately

$$a_{-1R} \approx 10^{-12.75\pm1} \qquad [28]$$

and a nearly constant product between unloaded-Q and the resonant frequency in the whole range from 2.5 MHz to 500 MHz [17]

$$Q_u f_o \approx 10^{13} \qquad [29]$$

For readers interested in the noise properties of the investigated clock oscillators, we have composed a computer program for solving eq. (25) and enclosed it into a specially written paper in Part XI.

E. Noise Properties of Phase-Locked Loops (PLL) and Voltage-Controlled Oscillators (VCOs)

Major properties of PLL have been discussed in Part VI. In a properly designed PLL, its natural frequency f_n is generally chosen in the neighborhood of where the reference and the respective VCO noise characteristics cross. As the rule of thumb for PSD of VCOs, we recommend using eq. (35) from the second reprint paper in Part VI.

$$\frac{S_{\varphi\text{osc}}(f)}{f_o^2} \approx \frac{1}{f^3}\frac{10^{-11.6}}{Q_L^2} + \frac{1}{f^2}\frac{10^{-15.6}}{Q_L^2} + \frac{1}{f}\frac{10^{-11}}{f_o^2} + \frac{10^{-15}}{f_o^2} \qquad [30]$$

F. Noise Properties of DAC

In the last reprinted paper in this Part, J. A. Connelly and K. P. Taylor [4] consider thermal noises in resistors combined with effective input noises of operational amplifiers in D/A and A/D converters and arrive at the following noise characteristic

$$S(f) \approx \frac{10^{-13\pm.3}}{f} + 10^{-15\pm.3} \qquad [31]$$

By comparing this relation with other noise sources in DDFS, we conclude that its contribution to the overall output noise is not too important.

V. NOISE PROPERTIES OF DDFS

The difficulty with estimating the DDFS output noise is that, to the editor's knowledge, only very few reliable measurements have been published [20–23].

We often encounter the statement that DDFS output phase noise is that of the reference or clock oscillator reduced by the square of the normalized frequency. The estimated power spectral density would then be given by

$$S_{\varphi,\text{DDS}} \approx S_{\varphi,\text{clock}} * \xi_x^2 \qquad [32]$$

As an example, we reproduce in Fig. 4 the phase noise performance of a commercial DDFS-type AD9955 [20]. Note the decrease of the phase noise with the reduction of the output frequency in accordance with the relation (32). Another peculiarity with this measurement is a very large slope of the nearly $1/f^5$ for Fourier frequencies in the range from 1 Hz to 10 Hz. A similar phenomenon also exhibits the characteristics reproduced in the paper by Olbrich et al. [6]. From our own measurements we conclude that the natural frequency of the PLL in the measurement device has probably been too large; cf. eq. (7) in the following paper by the editor.

The validity of the relation of (32) has also been verified in instances where the output frequency f_x was a small integer fraction of the clock frequency f_c such as ξ_x in Fig. 5a (or in Fig. 5 in [21]). Note the power mains spurious at 50 Hz and its harmonics. However, the phase noise in DDFS is generally corrupted with the background spurious signals, and thus its

211

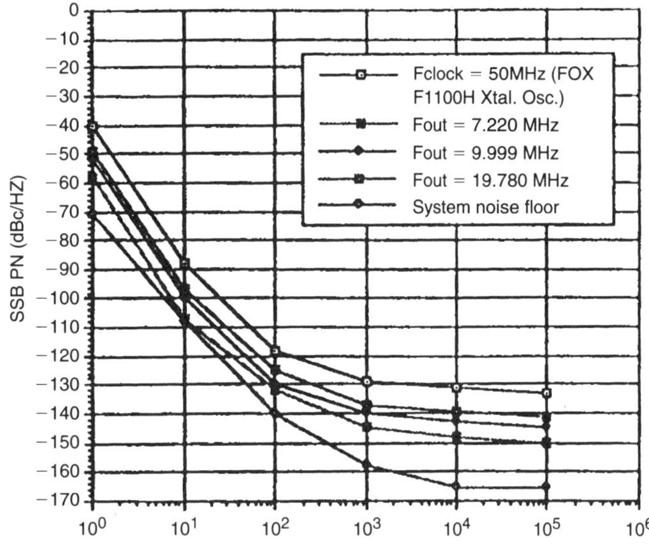

Fig. 4. Phase noise performance of a commercial DDFS-type AD9955 [from 20]. © Intertec Publishing. Used by permission.

definition starts to be vague. The problem is illustrated with the assistance of Fig. 5*b* where we have evaluated a signal with nearly the same frequency as in Fig. 5*a,* but with the LSB of the 24-bit accumulator turned on. We witness a comb of spurious frequencies about 2.38 Hz apart.

The problem is discussed in depth by the editor in papers [24 and 25]. Here, for illustration we provide its solution for the situation in Fig. 5*b.*

Example 1. Let us perform evaluation of the comb frequencies in Fig. 5*b*. The fundamental frequency of the comb $f_{1,sp}$ follows from the expansion of $\xi_x = (2^{20} + 1)/2^{24}$ into a continued fraction [26].

$$\xi_x = \frac{2^{20} + 1}{2^{24}} = \frac{1}{B_1} \pm \frac{1}{B_1 B_2} \pm \cdots$$

$$= \frac{1}{16} + \frac{1}{16 * 1048575} + \cdots \quad [33]$$

from which we get for $B_1 = 16$ and $B_2 = 1048575$ and for $f_{1,sp}$ with the clock frequency 2.5 MHz,

$$f_{1,sp} = \frac{f_c}{B_2} = \frac{2500000}{1048575} \approx 2.38 \quad \text{Hz} \quad [34]$$

In instances where B_2 is very large, the individual spectral lines are so close to each other that they form nearly a continuum, with PSD close to the carrier, exhibiting the slope $1/f^2$. See Fig. 2 in the following paper [25].

At higher output frequencies, particularly where ECL or GaAs IC-gate circuits are used, the "frequency divider noise" must be taken into account and the above relation (32) changes into

$$S_{\varphi,\text{DDS}} \approx S_{\varphi,\text{clock}} * \xi_x^2 + S_{\varphi,\text{divider}}(f) \quad [35]$$

where for $S_{\varphi,\text{divider}}(f)$ we introduce either relation (23) or (24); cf. Fig. 6 in the paper by Saul and Mudd [21]. Examina-

tion of this figure reveals a lot of spurious signals. Since the fundamental repetition period of the DDFS output wave lasts YT_c, the expected frequencies of the spurious signals (inclusive the carrier) will be

$$f_{sp} = n \frac{f_c}{Y} = n * \Delta f_x; \quad (n = 1, 2, \ldots, Y/2) \quad [36]$$

However, the DDFS output wave exhibits an odd symmetry, that is,

$$f_{\text{out}}(x + \pi) = -f_{\text{out}}(x) \quad [37]$$

Consequently, all even spurious signals are missing, and we should find around the carrier only spectral lines of the type

$$n_{sp} * 2\Delta f_x = n_{sp} * 2\left(\frac{f_c}{Y}\right); \quad (n_{sp} = 1, 2, \ldots, Y/4) \quad [38]$$

Example 2. We will examine the situation illustrated in Fig. 6 in the paper [21].

$$f_c = 327.68 \, \text{MHz}; \quad f_x = 81.925 \, \text{MHz};$$

$$\xi_x = 1/4 + 1/2^{16}; \quad \Delta f_x = 5 \, \text{kHz}$$

The first larger Fourier spectral line is at 10 kHz as expected from the relation (38). It is part of the "background" spurious signals the level of which for a DAC of D and Y bits, that is,

$$D = 8 \quad \text{and} \quad Y = 2^{16}$$

is in accordance with eq. (54), in the paper by the editor in Part V [26],

$$\langle S(n) \rangle \approx -5 - 6(D - 1) - 10\log(2^{16})$$
$$\approx -5 - 42 - 48 = -95 \, \text{dB} \quad [39]$$

However, the second and other large spurious signals are 20 kHz apart and have a level orders higher. Their origin must be different. There are two possibilities: either they are intermodulation products between harmonics of f_x and the clock frequency, for example,

$$3f_x - f_s = 245775000 - 327680000 = -81905000$$
$$5f_x - f_s = 409625000 - 327680000 = 81945000$$
$$7f_x - 2f_s = 573475000 - 655360000 = -81885000$$

and so on—see the program for computation of intermodulation signals in Part XI [27]—or they are the first of the comb frequencies of the type (34) since the expansion of the normalized frequency $\xi_x = (2^{14} + 1)/2^{16}$ into the continued fraction expansion reveals

$$B_1 = 4 \quad \text{and} \quad B_2 = 16383$$

from which

$$f_{1,sp} = 327680000/16383 = 20001.22 \quad \text{Hz}$$

What remains is to explain the presence of the 10-kHz spurious signal close to the carrier. With the assistance of (39) and (41) from [26], and the largest $r = 2^{B-1}$ the respective spur frequency is

$$f_{\text{spur}} = 81.925 - 168.84 = -81.915 \quad \text{MHz}$$

The difference between the carrier and the first spurious signal is 10 kHz. Note that this spur is a doublet. Since the accumulator has

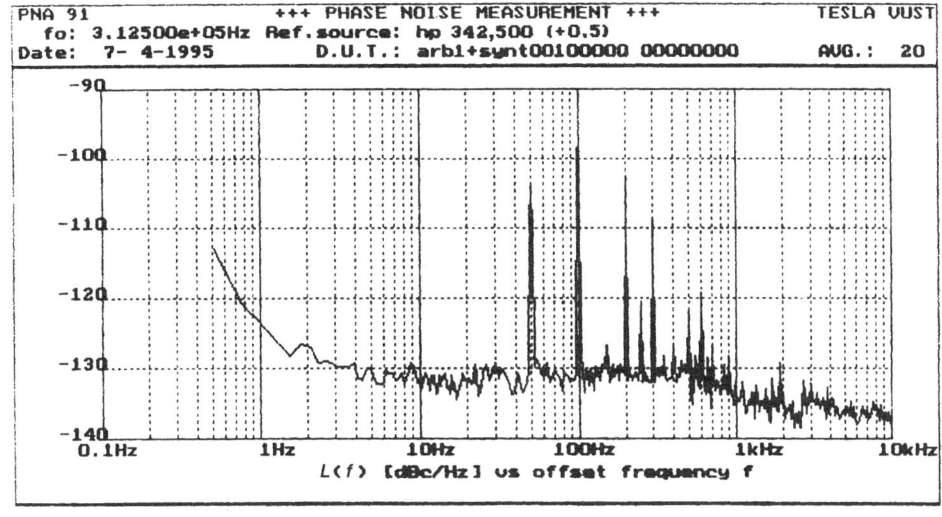

(a)

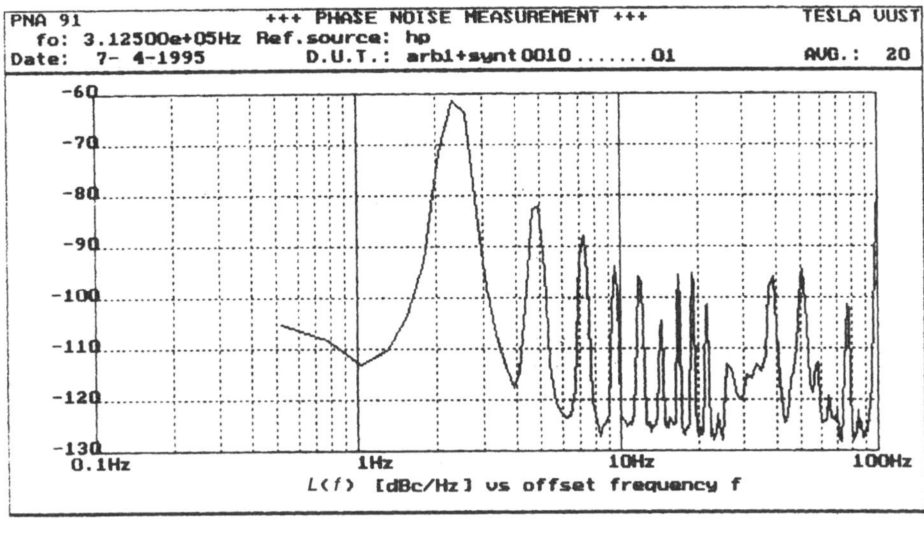

(b)

Fig. 5. The background noise of a DDS with $f_c = 5$ MHz; (a) $\xi_x = 1/16$; (b) $\xi_x = 1/16 + 1/2^{24}$.

---

only $R = 16$ active bits, the expected spurious level—from eq. (39) in [26]—should be in the range of

$$-6 * 15 + 3 = -87 \quad \text{dB}$$

Figure 6 in the paper by Saul and Mudd in this part [21] confirms both these results.

An earlier discussion of $1/f$ noise in DDFS was provided nearly a decade ago by Mattison and Coyle (1988). We chose to reprint this paper since it is still often mentioned in references [22]. It deserves attention particularly for comparing noise for both sine- and square-wave DDFS outputs with no appreciable difference; see their Table 1. Investigation of their Fig. 8 reveals for the flicker phase noise

$$S(f) \approx \frac{10^{-11}}{f} \tag{40}$$

which is much higher than one would expect for an ordinary digital divider.

Note also three background noise spurious signals in the range from 100 to 1000 Hz in Mattison's Fig. 8. Their origin can be traced to the accumulator sawtooth-wave modulation. See the following example.

Example 3. Provide the modified continued fraction expansion of the normalized frequency examined by Mattisson et al. and set with a DDFS (the accumulator has 64 bits). With the assistance of continued fraction expansion [28,29], we get

TABLE 2

$$\xi_x = 405750/5000000$$

k	b_k	a_k	Ak	B_k
0	0	1	0	1
1	12	1	1	12
2	3	1	3	37
3	10	1	31	382
4	4	1	127	1565
5	3	-1	350	4313
6	5	-1	1623	20000

For the lowest spurious signal we get

$$f_{1,sp} = f_c/B_6 = 5000000/20000 = 250 \text{ Hz}$$

in agreement with Fig. 8 in [22].

Nearly the same value for the DDFS flicker phase noise follows from a set of our own measurements [25] and from the measurements performed by F. L. Walls [30].

Another DDFS $1/f$ data point is provided by Bar-Giora Goldberg [23]. From it we deduce, after eliminating clock oscillator noise, for $f_{\text{out}} \leq 230$ MHz

$$S_{\varphi,\text{DDS}}(f) \approx \frac{10^{-8}}{f} \qquad [41]$$

which agrees with the Saul and Mudd results [21].

The problem of the "white phase noise" has been discussed by the editor in Part V [28]. As a consequence, we can write for the additive DDFS noise characteristic the following relation ($Y = 2^R$ being larger than f_c; otherwise Y and f_c must be interchanged)

$$S_{\varphi,\text{DDS}}(f) \approx \frac{10^{-10\pm2}}{f} + \frac{1}{3 \cdot 2^{2S} \cdot f_c} \qquad [42]$$

Evidently, both the flicker and white noise levels at the DDFS outputs are often higher than one would expect from a simple digital divider behavior. As long as we distinguish individual spectral lines one from the other, we can estimate the DDFS background spurious noise as

$$S_{\varphi,\text{DDS}}(f) \approx S_{\varphi,\text{clock}}(f) \cdot \xi_x^2 + \frac{10^{-10\pm2}}{f} + 10^{-15\pm1} \qquad [43]$$

V. Conclusions

We have seen that both the flicker and white phase noise at the DDFS outputs may often be larger than one would expect from a simple digital divider behavior [e.g., 11,12]. In addition, there are many spurious signals close to the carrier whose amplitude and frequency change significantly with the change of the normalized frequency ξ_x. The problem is discussed in the following paper [25].

References

[1] V. F. Kroupa, ed. *Frequency Stability: Fundamentals and Measurement.* New York: IEEE Press, 1983.

[2] A. Van der Ziel. *Noise: Sources, Characterization, & Measurement.* Englewood Cliffs, N.J.: Prentice-Hall, 1970.

[3] A. Van der Ziel. *Noise in Solid State Devices and Circuits.* New York: John Wiley, 1986.

[4] J. A. Connelly and K. P. Taylor. "An analysis methodology to identify dominant noise sources in D/A and A/D converters." *IEEE Transactions on Circuits and Systems,* pp. 1133–44, October 1991. (Reprinted in this Section.)

[5] V. F. Kroupa. *Theory of Phase-Locked Loops and Their Applications in Electronics.* Prague: Academia, publishing house of the Academy of the Czech Republic, 1995 (in Czech).

[6] G. R. Olbrich, P. Russer, B. Scheffler, and P. Kraus. "Phasenrauscharmer 80–110 MHz Oszillator mit DDS-frequenzquelle." *Kleinheubacher Berichte,* vol. 36, pp. 621–27, 1992–1993.

[7] P. J. Topham. "GaAs bipolar transistors for microwave and digital circuits." *GEC Journal of Research,* vol. 9, No. 2, 1991.

[8] D. Halford, A. E. Wainwright, and J. A. Barns. "Flicker noise in RF amplifiers and frequency multipliers: characterization, cause, and cure." *Proceedings of the 22nd Annual Symposium on Frequency Control,* pp. 340–41, 1968.

[9] F. L. Walls. "Low noise frequency synthesis." *Proceedings of the 41st Annual Symposium on Frequency Control,* pp. 512–18, 1987.

[10] V. F. Kroupa. "Electromagnetic compatibility of frequency synthesizers." *Proceedings of the 3rd Symposium on Electromagnetic Compatibility,* pp. 535–40, Rotterdam, May 1979.

[11] W. F. Egan, "Phase noise modelling in frequency dividers." *Proceedings of the 45th Annual Symposium on Frequency Control,* pp. 629–35, May 1991.

[12] M. M. Driscoll and T. D. Merrell. "Spectral performance of frequency multipliers and dividers." *Proceedings of the 1992 IEEE Frequency Control Symposium,* pp. 193–200, May 1992.

[13] W. A. Edson. "Noise in oscillators." *Proceedings IRE,* pp. 1454–66, August 1960.

[14] K. Kurokawa. "Noise in synchronized oscillators." *IEEE Transactions* vol. MTT-16, pp. 234–40, April 1968.

[15] F. L. Walls and A. E. Wainwright. "Measurement of the short-term stability of quartz crystal resonators and the implications for crystal oscillator design and applications." *IEEE Transactions,* vol. IM-24, pp. 15–20, March 1975.

[16] G. Marianneau, J.J. Gagnepain, and J. Uebersfeld. "Bruit de phase et d'amplitude des resonateurs et oscillateurs a quartz." *Proceedings International Symposium in Telecommunications,* Lanion, France, pp. 240–44, October 1977.

[17] V. F. Kroupa. "Flicker frequency noise in BAW and SAW quartz resonators." *IEEE Transactions,* UFFC-35, pp. 406–20, 1988.

[18] D. B. Leeson. "A simple model of feedback oscillator noise spectrum." *Proceedings of the IEEE,* pp. 329–30, February 1966. (Reprinted in [12].)

[19] J. K. A. Everard. "Minimum sideband noise in oscillators." *Proceedings of the 40th Annual Symposium on Frequency Control,* 1986.

[20] A. Hill and J. Surber. "Using aliased-imaging techniques in DDS to generate RF signals." *RF Design,* p. 31, September 1993.

[21] P. H. Saul and M. S. J. Mudd. "A direct digital synthesizer with 100 MHz output capability." *IEEE Journal of Solid-State Circuits,* June 1988. (Reprinted in this Section.)

[22] E. M. Mattison and L. M. L. Coyle. "Phase noise in direct digital synthesizers." *Proceedings of the 42nd Annual Frequency Control Symposium,* 1988. (Reprinted in this Section.)

[23] Bar-Giora Goldberg. "Linear frequency modulation—theory and practice." *RF Design,* pp. 39–46, September 1993.

[24] V. F. Kroupa. "Synthesis techniques." *1995 IEEE Frequency Control Symposium, Tutorials.*

[25] V. F. Kroupa. "Close-to-the-carrier noise in DDS." *1996 IEEE Frequency Control Symposium.* (Reprinted in this Part.)

[26] V. F. Kroupa. "Spectral properties of DDFS: Computer simulations and Experimental verifications." (Specially written paper for Part V.)

[27] V. F. Kroupa. "Useful computer programs for investigations of spurious signals in DDFS." (See Part XI.)

[28] V. F. Kroupa. "Approximating frequency synthesizers." *IEEE Transactions on Instrumentation and Measurement,* Vol. IM-23, No. 4, pp. 521–24, December 1974. (Reprinted in Part II.)

[29] V. F. Kroupa. "Spectra of pulse rate frequency synthesizers." *Proceedings of the IEEE,* pp. 1680–82, December 1979. (Reprinted in Part IV.)

[30] F. L. Walls, private communication, 1994.

Close-to-the-Carrier Noise in DDFS

VĚNCESLAV F. KROUPA

Abstract: **In this paper we will discuss the close-to-the-carrier noise in direct digital frequency synthesizers (DDFS). Large spurious signals due to the accumulator truncation are well understood. The same might be said about the background noise due to the finite bit length of the sine words in lookup tables and to some extent also about the noise caused by DA converters [1,2]. However, the same cannot be said about the close-to-the-carrier noise. We often encounter the statement that the output noise of DDFS is that of the reference oscillator reduced by the square of the normalized frequency. But our investigations proved that this need not be generally true.**

In the following sections, we will provide new insight both from the theoretical and experimental point of view, with the result that we often face a "continuum" of closely spaced discrete components.

I. INTRODUCTION

A general belief is that the phase noise power spectral density (PSD) $S_\varphi(f)_{out}$ of the DDFS output signal f_x is that of the reference or clock (f_c) phase noise reduced by the square of the normalized frequency ξ_x [3,4,5, and others]

$$\xi_x = \frac{f_{out}}{f_{clock}} = \frac{f_x}{f_c} = \frac{X}{Y} \qquad [1]$$

that is,

$$S_{\varphi,out}(f) = S_{\varphi,clock}(f) \cdot \xi_x^2 \qquad [2]$$

This is true in those cases where the effective division ratio is equal to the small integer fractions exactly (see, e.g., the paper by Saul and Mudd in this section). An additional condition is that the denominator in (1) is a power of two (in the binary systems).

In other instances, the experimental measurements provided a comb of spurious signals [e.g. 6] sometimes forming nearly a continuum [7]. The reason is the sampled phase generated in the accumulator and converted to the desired sine wave in some instances and intermodulation between the output and clock frequencies in other cases. In the following sections we will investigate possible contributions to the close-to-the-carrier noise in DDFS.

II. FREQUENCY DIVIDER NOISE

Investigation of the divider output phase noise, with the division factor N, revealed an experimental formula for lower fre-

Institute of Radio Engineering and Electronics, Academy of Sciences of the Czech Republic, 182 51 Praha, Czech Republic

quencies [8,9,10]

$$S(f) \approx \frac{10^{-12}}{f} + 10^{-14} + \frac{S(f)_{clock}}{N^2} \qquad [3]$$

and

$$S(f) \approx \frac{10^{-8}}{f} + 10^{-14} + \frac{S(f)_{clock}}{N^2} \qquad [4]$$

for systems based on GaAs gates intended for division at higher output frequencies [10,11,12].

III. EXAMPLES OF THE OUTPUT NOISE IN DDFS

In Fig. 1 we reproduce a block diagram for evaluation of the phase noise generated by DDFS. To simplify the discussion, we assume noiseless clock generators and the disturbing phases being added to the output of the investigated synthesizers. In such a case, the PLL theory [13 and others] reveals that the phase φ_e at the phase discriminator output is

$$\varphi_e = \varphi_2 - (\varphi_1 + \varphi_{add}) \qquad [5]$$

where the additive phase disturbing component φ_{add} is equal to (if PLL loop filter $F(s)$ is considered)

$$\varphi_{add} = \varphi_e \frac{K_d K_o F(s)}{s} \xi_x \qquad [6]$$

After introduction of the above equation into (5), we get for the relation between the DDFS noises and the output phase noise φ_e

$$\varphi_2 - \varphi_1 = \varphi_e \left(1 + \frac{K F(s)\xi_x}{s} \right); \quad (K = K_d K_o) \qquad [7]$$

In instances where the open loop gain of the PLL,

$$G(s) = \frac{K F(s)\xi_x}{s} \qquad [8]$$

is smaller than "1," the output phase φ_e is approximately

$$\varphi_e \approx \varphi_2 - \varphi_1 \qquad [9]$$

In the majority of cases, the DDFSs noises are uncorrelated, and the measured output power spectral density, PSD, is equal to (by assuming that $K F(s)/s \ll 1$)

$$S_{\varphi_e}(f) \approx S_{\varphi_1}(f) + S_{\varphi_2}(f) \qquad [10]$$

The measured PSD of a DDFS, set in accordance with the numerical arrangement in Fig. 1, is shown in Fig. 2.

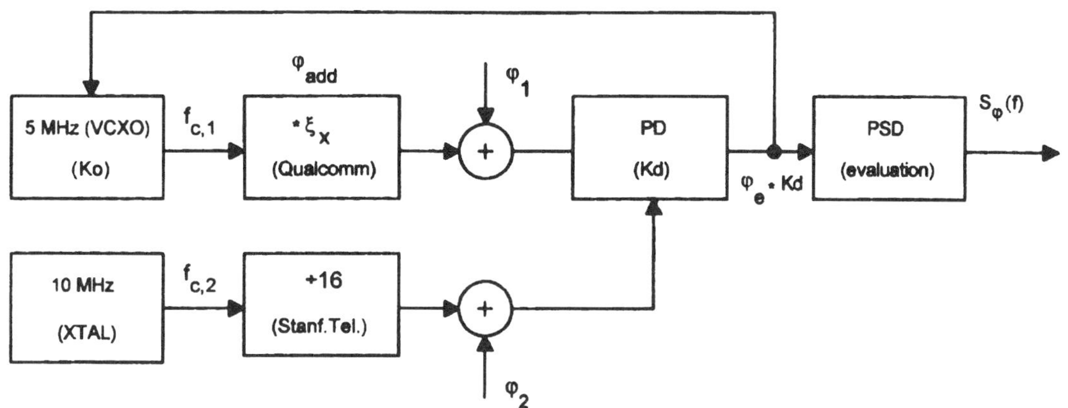

Fig. 1. Block diagram for the evaluation of the phase noise generated in DDFS.

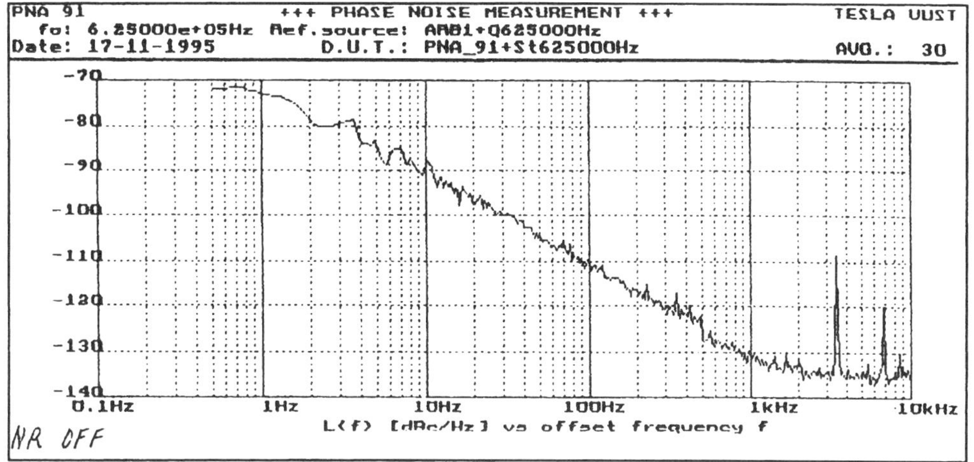

Fig. 2. Phase noise characteristic of a DDFS with $\xi_x = (2^{29} - 1)/2^{32}$.

We find that the slope of the measured PSD, close to the carrier, is proportional to the square of the respective Fourier frequency, that is,

$$S_\varphi(f) \sim \frac{1}{f^2} \qquad [11]$$

Evidently, the expected PSDs of the crystal oscillator noises, multiplied by ξ_{x2}, with the frequency flicker noise slope $1/f^3$, are masked by noises generated in DDFS.

However, a small detuning revealed a comb of spurious frequencies very close apart (see Fig. 3).

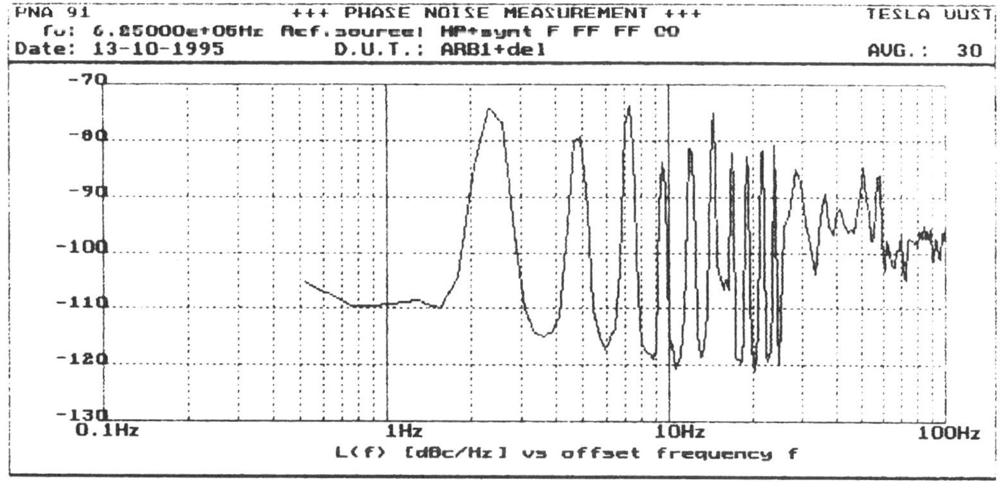

Fig. 3. Phase noise characteristic of a DDFS with $\xi_x = (2^{28} - 64)/2^{32}$
$= (2^{22} - 1)/2^{26}$.

IV. Theoretical Background

Without any loss of generality, we will first consider the case where all the accumulator bits, that is, the phase bits, are used for generation of the output wave.

We have to start with investigation of the accumulator behavior in DDFS. Its overflowing assures the Modulo-N operation on the clock pulse rate and as a consequence the quasi-periodic operation of the system [e.g. 7,14,15]. By DDFS setting, the ideal output frequency is

$$\omega_x = \frac{X}{Y}\omega_c \qquad [12]$$

Due to the clocked operation, the lowest modulation frequency is equal to the repetition period, that is,

$$f_{m,\min} = \frac{1}{Y \cdot T_c} \qquad [13]$$

In addition, one encounters all its multiples. However, there are also other modulation frequencies which are closely related to the quasiperodic theorem resulting from the step-by-step approximation of the normalized frequency ξ_x in DDFS. The effective mathematical model is the modified continued fraction expansion [14]. In this way we get

$$\frac{X}{Y} = \frac{1}{B_1} - \frac{a_1 a_2}{B_1 B_2} + \frac{a_1 a_2 a_3}{B_2 B_3} - \cdots + (-1)^{n-1}\frac{a_1 \cdots a_n}{B_{n-1} \cdots B_n} \qquad [14]$$

where B_n is equal to the denominator Y.

The actual output phase, supplied by the accumulator, is

$$\varphi(mT_c) = 2\pi f_c T_c \left[m\frac{X}{Y} - \text{integer}\left(m\frac{X}{Y}\right) \right] \qquad [15]$$

Evidently, we face a set of sawtooth waves. A simple example for the generated normalized frequency $\xi_x = 17/64$ with its continued fraction expansion

$$\frac{17}{64} = \frac{1}{4} + \frac{1}{4*15} - \frac{1}{15*64} \qquad [16]$$

is shown in Fig. 4 (note the 17 divisions on the x-axes).

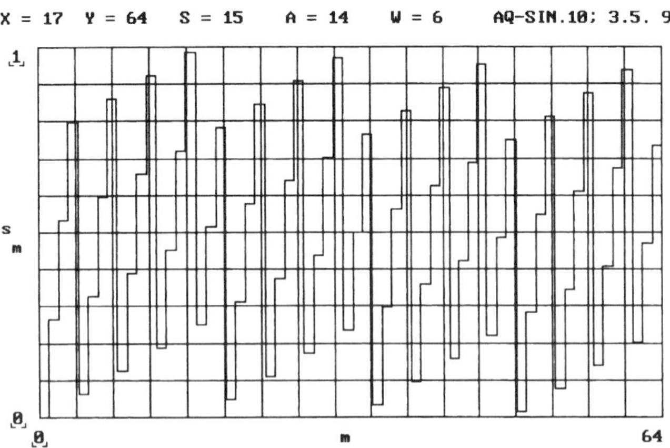

Fig. 4. An example of the output phase supplied by the accumulator for the normalized frequency $\xi_x = 17/64$.

In the first-order approximation, the output frequency ω_{x1} is given by

$$\omega_{x1} = \frac{\omega_c}{B_1} \qquad [17]$$

The first-order approximation of ξ_x is simple, that is, $\xi_{x1} = 1/B_1$, and the accumulator will overflow for $m = B_1$, $2B_1$, etc., and the output phase will reach nearly 2π and return effectively to zero (cf. Fig. 4). Furthermore, after each B_2th pulse the output phase would experience a smaller step, namely,

$$\Delta\varphi_2 \approx \frac{2\pi}{B_1} \qquad [18]$$

Its Fourier series expansion results in a comb of spurious frequencies, decreasing in proportion to the respective harmonic order

$$\Delta\varphi_2(t) = 2\frac{1}{B_1}\sum_{r=1}^{\infty}\frac{1}{r}\sin\left(r\frac{\omega_c}{B_2}t\right) \qquad [19]$$

We can proceed this way and investigate spurious modulation introduced by further terms in the expansion (14). However, their influence is generally negligible due to the other sources of spurious signals in DDFS.

V. Estimation of the Spurious Spectra

Estimation of the spurious spectra, due to the accumulator phase modulation, at the DDFS output is complicated by the sine-wave transformation process. Investigation of the final output waves reveals additional amplitude and phase or frequency modulation. Only zero crossings of the output wave copy the original accumulator phase (see Fig. 5).

Since the zero crossings of the output wave are sometimes of importance, we shall proceed with the investigation or more exactly with the estimation of the spurious signals caused by the Modulo-Y process in the DDFS accumulator.

1. The Frequency Domain Approach. From relation (19) it follows that the individual amplitudes after the second-order approximation are

$$\Delta\varphi_{2,r} = \frac{2}{B_1}\cdot\frac{1}{r} \qquad [20]$$

and frequencies are

$$v_r = \frac{r}{B_2 T_c} \qquad [21]$$

In the instance where the fundamental modulation frequency v_1 is very small, particularly when

$$v_r = \frac{r}{B_2 Tc} \ll 1 \text{ Hz} \qquad [22]$$

the spurious components are so close one to the other that they form nearly a continuum. In such a case, the spurious power spectral densities (PSD) in a narrow frequency band from v_{r1} to v_{r2} can be estimated with the assistance of (17) as

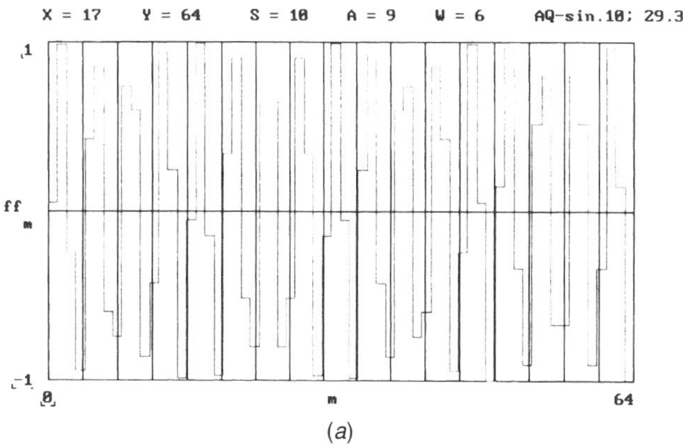

(a)

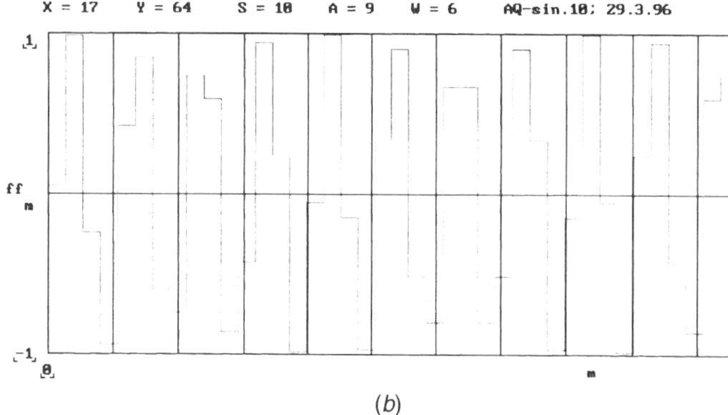

(b)

Fig. 5. DDFS output wave for $\xi_x = 17/64$: (*a*) full period, (*b*) enlarged portion.

$$S_\varphi(f) \approx \frac{1}{2}\left(\frac{2}{B_1}\right)^2 \int_{\nu_{r_1}}^{\nu_{r_2}} \left(\frac{\nu_c}{\nu_r}\right)^2 d\nu_r$$

$$\approx \frac{1}{2}\left(\frac{2}{B_1}\right)^2 \left(\frac{\nu_c}{\nu_r}\right)^2 \approx \frac{2}{B_2^2}\left(\frac{f_{x1}}{f}\right)^2 \qquad [23]$$

The estimation reveals that the PSD $S_\varphi(f)$ is inversely proportional to the square of the Fourier frequency, f, which is in good agreement with Fig. 2. The device exhibits a "white" frequency noise behavior.

From the theory of the fractional frequency noise PSD (defined in the preceding paper) we compute $S_y(f)$ as

$$S_y(f) = S_\varphi\left(\frac{f}{f_x}\right)^2 = \frac{2}{B_2^2} = h_o \qquad [24]$$

2. The Time Domain Investigation. To get qualitative information about the spectral properties of DDFS in the neighborhood of the carrier, we will also investigate time variances. First, we will repeat relations for the Allan variance in the presence of discrete spurious signals [16, p. 79].

$$\sigma_y^2(\tau) = \left(\frac{\Delta\nu_o}{\nu_o}\right)^2 \frac{\sin^4 \pi f_m \tau}{(\pi f_m \tau)^2} \qquad [25]$$

In our case ν_o is approximately equal to the output frequency $f_x \approx f_c/B_1$, and $\Delta\nu_o$, the frequency modulation amplitude, is found by derivation of relation (19)

$$\frac{\Delta\nu_o}{\nu_o} \approx \frac{2}{B_2} \qquad [26]$$

However, in relation (25) we have to perform summation over all f_m, that is, over all ν_r. Again we will approximate summation with integration, and with the assistance of integral tables we get

$$\sum_{\nu_1}^{\infty} \frac{\sin^4 \pi \nu_r \tau}{(\pi \nu_r \tau)^2} \approx \int_0^\infty \frac{\sin^4 z}{z^2} \frac{dz}{\pi \tau} \approx \frac{1}{4\tau} \qquad [27]$$

After combination of relations (25), (26), and (27), we get for the Allan variance

$$\sigma_y^2(\tau) = \frac{1}{\tau B_2^2} = \frac{h_o}{2\tau} \qquad [28]$$

The noise coefficient h_o computed via the time domain approach is the same as that found in frequency domain investigations, namely,

$$h_o \approx \frac{2}{B_2^2} \qquad [29]$$

We again arrive at the "white" frequency noise behavior.

VI. Experimental Results

Our earlier computer simulations of DDFS spectral behavior did not reveal the close-to-the-carrier spurious shown in Figs. 2 and 3. Therefore, we started a series of experimental investigations with the assistance of our bread board device and three commercial DDFS synthesizers.

1. PSD Measurements. In all instances where the normalized frequency was close to the ratio of small integers (e.g. $\approx 1/5$, 1/8, 1/16), we have found the validity of the relation (21) for the comb of spurious signals. In addition, the $1/f^2$ law for the noise was confirmed where condition (22) was met.

On the other hand, the difficulty was with the expected noise levels in accordance with (23). The situation will be illustrated with two examples.

Example 1. In Fig. 2 we have plotted the PDS close to the carrier in the instance where one normalized frequency ξ_x has been set from $f_c = 10$ MHz to

$$\xi_x = 1/16$$

exactly and phase-locked to the other DDFS also set to $\xi_x \doteq 1/16$ ($f_c = 5$ MHz) in accordance with Fig. 1. However, its adjusting from the PC keyboard has been performed in the hexadecimal form; consequently, the actual ξ_x has been

$$\xi_x = (2^{29} - 1)/2^{32}$$

The continued fraction expansion of the above normalized frequency reveals

$$B_1 = 8 \qquad B_2 = 536870905$$

With the assistance of (21) we get for

$$f_{m1} = v_1 = .0186 \text{ Hz}$$

and for the PSD we would get from (23)

$$S_\varphi(1) = -55.67 \text{ dB}$$

However, our measurement reproduced in Fig. 2 reveals a value of about

$$S_\varphi(1) = -70 \text{ dB}$$

Example 2. Investigation of Fig. 3 reveals the first two denominators of the continued reaction approximation to be

$$B_1 = 16 \qquad B_2 = 4194289$$

and for

$$f_c = 10 \text{ MHz} \qquad f_{m1} = 2.38 \text{ Hz}$$

Note that the theoretical comb of the close-to-the carrier spurious signals is in good agreement with that shown in Fig. 3. However, this is not true of the spurious level, since with the assistance of the relation (23) we would get orders of smaller values than

$$S_\varphi(2.38) = -75 \text{ dB}$$

read from the figure.

---

2. Time Variance Measurement. The difficulty with the noise level was so puzzling that we have started time domain investigation. One of the results is presented in the following example.

Example 3. In the third example, we have investigated the Allan variance of the 1-MHz output signal generated in a commercial DDFS (Qualcomm 2334) clocked by a 5-MHz reference. The actual normalized frequency in the hexadecimal notation is

$$\xi_x = 19\ 99\ 99\ 99/\ 1\ 00\ 00\ 00\ 00$$

The measured Allan variance is reproduced in Fig. 6. Note that in the range from 0.2 to 0.8 s the slope is proportional to the $1/\sqrt{\tau}$, which indicates a white frequency noise behavior (for larger τ the steeper slope is probably caused by a systematic error). With the assistance of the plot in Fig. 6 and the respective conversion formula, we get for the noise coefficient h_o,

$$h_o \approx 10^{-17.74}$$

When using eq. (30), we find

$$h_o = 10^{-17.4}$$

After comparing experimental and theoretical values for h_o, we encounter only a very small difference.

The curve ($\square$) in Fig. 6 was measured with the "noise reduction circuit OFF," whereas the curve ($\Diamond$) was gained with the "noise reduction circuit ON." Its investigation reveals a nearly flat part in the interval from 0.2 to 0.8 s, which indicates that the $1/f^2$ behavior was destroyed.

3. Spectral Analysis. The third experimental approach was the measurement with the assistance of a spectral analyzer. In the first place, we focused on the close-to-the-carrier noise.

Example 4. Investigation of the situation where

$$f_x \doteq 625 \quad \text{kHz} \qquad f_c = 5 \quad \text{MHz} \qquad \xi_x = (2^{23} - 3)/2^{26}$$

resulted in the first two denominators in the continued fraction expansion

$$B_1 = 8 \qquad B_2 = 2796201$$

The expected f_{m1} was

$$f_{m1} \doteq 1.79 \quad \text{Hz}$$

The result is shown in Fig. 7. Note the comb of spurious signals about 1.8 Hz apart.

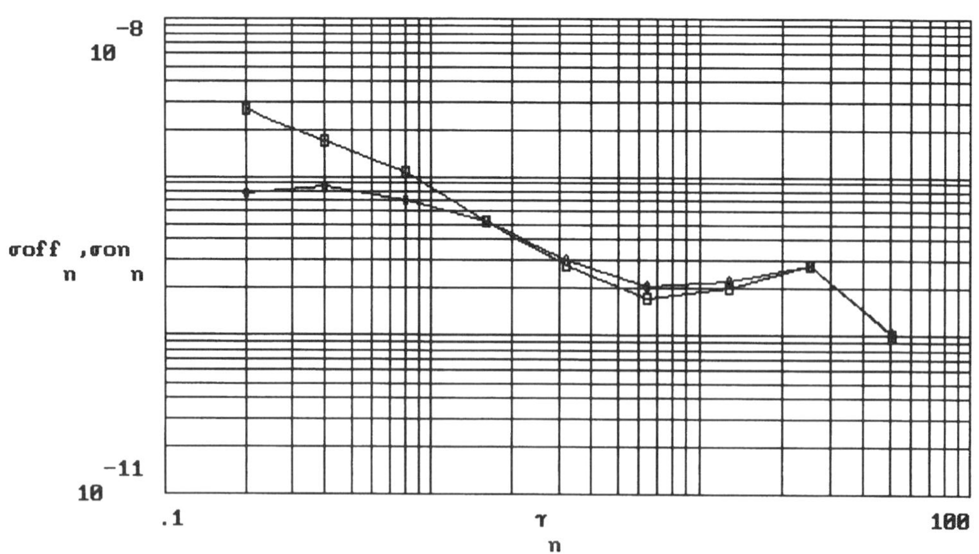

Fig. 6. The Allan variance for DDFS set to $f_x \doteq 1$ MHz from $f_c = 10$ MHz ($\square$) Noise reduction circuit OFF ($\Diamond$) Noise reduction circuit ON.

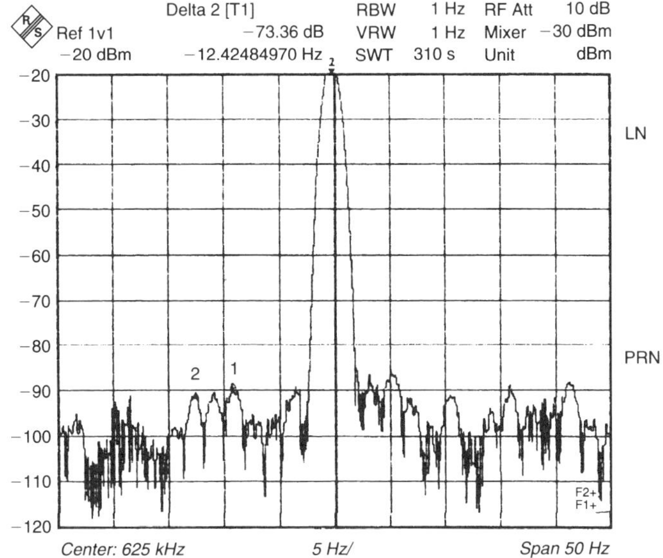

Fig. 7. Spectral analysis of a DDFS output with $f_x \doteq 625$ kHz, $f_c = 5$ MHz, i.e., $\xi_x = (2^{23} - 3)/2^{26}$.

Example 5. In the case where

$$\xi_x = 17/64$$

that is,

$$f_x \doteq 1.328 \quad \text{MHz} \qquad f_c = 5 \quad \text{MHz}$$

we have looked for $f_{m1} = 3.33$ kHz, that is, $f_x = 1.661$ MHz. However, we have spotted the twenty-first spurious line with frequency

$$f_{sp} = 1.640 \quad \text{MHz}$$

VII. DISCUSSION OF THE RESULTS

Examination of the output sine wave generated by a DDFS reveals spurious modulations. Computer simulation (e.g. in Fig. 5) agrees quite well with observation on the scope (see Fig. 8). We clearly note some sort of amplitude modulation with the frequency

$$f_{a,\mathrm{mod}} \approx f_c/15 \quad \text{or} \quad f_{a,\mathrm{mod}} \approx Y/15$$

At the same time the zero crossings are shifted in accordance with the sawtooth-wave expansion (14) (see Fig. 5).

Another example is shown for $\xi_x = 207/1024$ in Figs. 9 and 10.

$$\frac{207}{1024} = \frac{1}{5} + \frac{1}{5 \cdot 94} + \frac{1}{94 \cdot 465} - \frac{1}{465 \cdot 1024}$$

This time the output sine wave seems superimposed on a low-frequency wave with the fundamental component

$$W_{SP} = 0.1 \sin\left(2\pi m \frac{1}{94}\right)$$

Numerical investigation of the zero crossings in Fig. 9 reveals their sawtooth wave shifting till the 94th sample where they experience a step change with a loss of one clock period.

Note that "sines waves" in Fig. 9*a* have only five levels that do not depend much on the number of sine bits or DAC bits. Another interesting finding is that the shortened version of $\xi_{xo} = 207/1024$ to $\xi_{x2} = 19/94$ exhibits nearly the same output wave (see Fig. 10).

Evidently, the first two terms in the expansion (14) are very important for the shape of the "modulation" of the DDFS output wave.

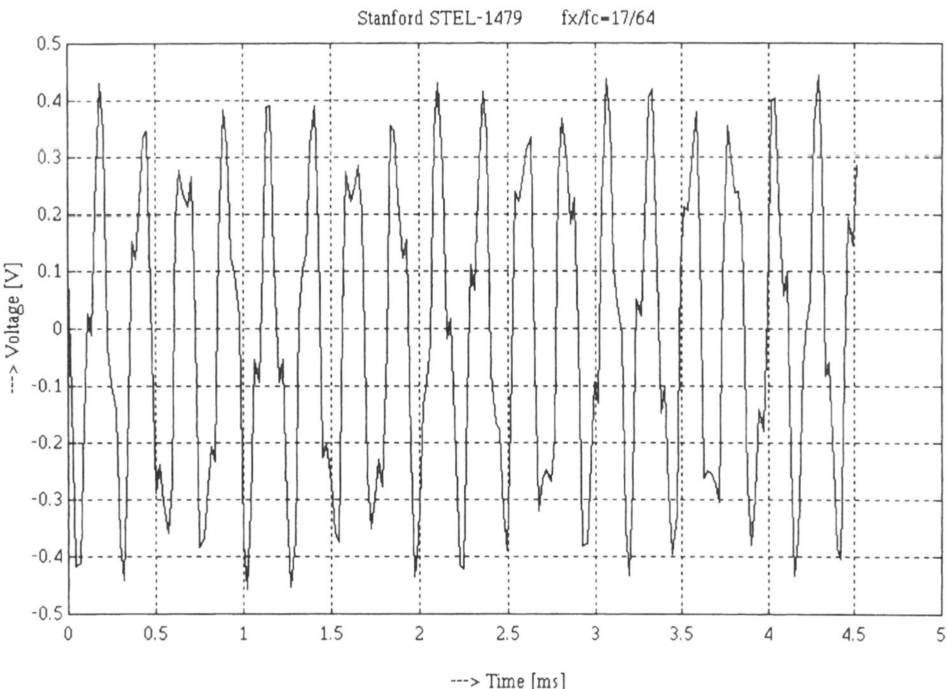

Fig. 8. DDFS output wave for $\xi_x = 17/64$ taken from the scope.

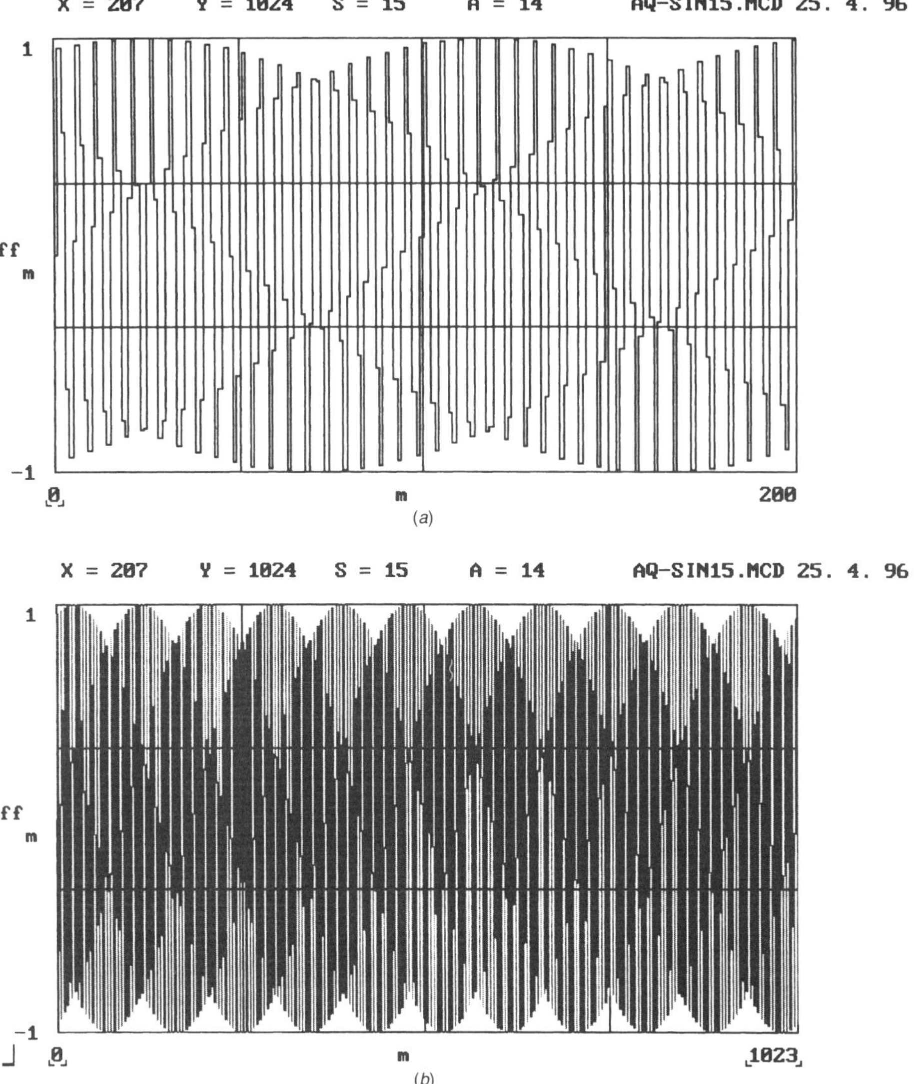

Fig. 9. Computer-simulated DDFS output waves for $\xi_x = 207/1024$.

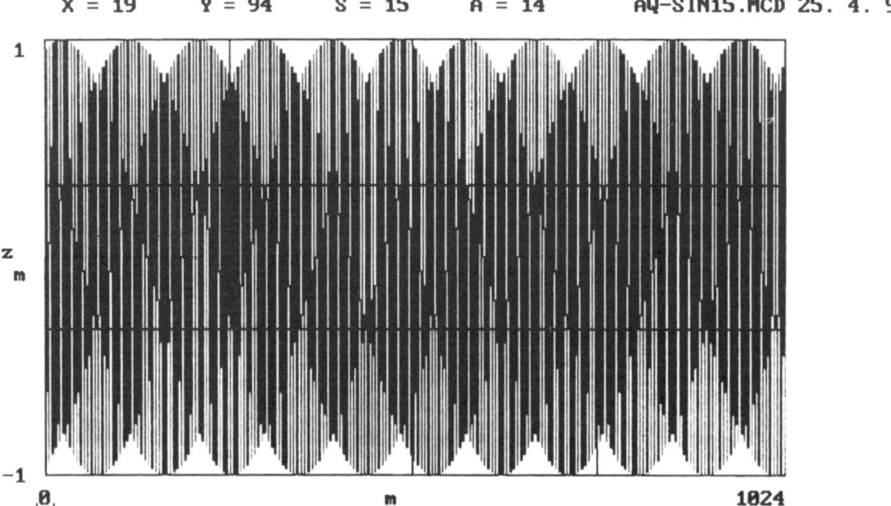

Fig. 10. Computer-simulated DDFS output wave for the second-order approximation of $\xi_{xo} = 207/1024$ to $\xi_{x2} = 19/24$.

VIII. Conclusions

In the ideal case, the output wave is a staircase sine wave as is often shown in the literature about DDFS. However, this is not true in instances where the normalized frequency ξ_x is nearly equal to the ratio of small integers of Fig. 5 or Fig. 9a. In these cases, the close-to-the carrier spurious signals originating in the accumulator may appear, particularly where the denominator in the second-order approximation B_2 is much larger than B_1. This phenomenon is significant in those instances where zero crossings of the output wave are of importance for further DDFS applications.

IX. Acknowledgment

The assistance of the staff of the several departments of the Institute of Radio Engineering and Electronics of the Academy of Sciences of the Czech Republic and of Mr. R. Gilmore from the Qualcomm Co. is appreciated.

References

[1] H. T. Nicholas III and H. Samueli. "An analysis of the output spectrum of direct frequency synthesizers in the presence of phase-accumulator truncation." *Proceedings of the 41st Annual Frequency Control Symposium,* pp. 495–502, 1987.

[2] V. F. Kroupa. "Spectral properties of DDFS: Computer simulation and experimental verifications." *1994 IEEE International Frequency Control Symposium,* Proceedings, pp. 613–23.

[3] Stanford Telecom. "Alias and spurious responses in DDS systems." Application Note 102, Rev. 4, 1994.

[4] Qualcomm. "Hybrid PLL/DDS frequency synthesizers." Application Note AN 2334-4, March 1992.

[5] M. J. Underhill, M. J. Blewett, and P. Jenkins. "Spectral improvement of a direct digital frequency synthesizer." *10th European Frequency and Time Forum,* pp. 452–60, March 1996.

[6] F. L. Walls, private communication, 1994.

[7] V. F. Kroupa. "Synthesis techniques (DDS)." *1995 IEEE International Frequency Control Symposium,* Tutorials, Session 2C.

[8] V. F. Kroupa. "Electromagnetic compatibility of frequency synthesizers." *Proceedings of the 3rd Symposium on Electromagnetic Compatibility,* pp. 535–40, Rotterdam, May 1979.

[9] F. L. Walls. "Low noise frequency synthesis." *Proceedings of the 41st Annual Symposium on Frequency Control,* pp. 512–518, 1987.

[10] W. F. Egan. "Modelling phase noise in frequency dividers." *IEEE Transactions on Ultrasonics, Ferroelectrics and Frequency Control,* vol. 37, 307–15, July 1990.

[11] W. F. Egan. "Phase noise modelling in frequency dividers." *Proceedings of the 45th Annual Symposium on Frequency Control,* pp. 629 35, May 1991.

[12] M. M. Driscoll and T. D. Merrell. "Spectral performance of frequency multipliers and dividers." *IEEE Frequency Control Symposium, Proceedings,* 1992, pp. 193–200.

[13] F. M. Gardner. *Phase-Lock Techniques.* New York: John Wiley, 1969, 1979.

[14] V. F. Kroupa. "Spectra of pulse rate frequency synthesizers." *Proceedings of the IEEE,* vol. 67, pp. 1680–82, December 1979.

[15] V. F. Kroupa. "Spectral purity of direct digital frequency synthesizers." *Proceedings of the 44th Symposium on Frequency Control,* pp. 498–510, 1990.

[16] V. F. Kroupa, ed. *Frequency Stability: Fundamentals and Measurement.* New York: IEEE Press, 1983.

Phase Noise in Direct Digital Synthesizers

EDWARD M. MATTISON AND LAURENCE M. COYLE
SMITHSONIAN ASTROPHYSICAL OBSERVATORY
CAMBRIDGE, MASSACHUSETTS, 02138

Introduction

The receiver systems of most atomic clocks include a frequency synthesizer to adjust the frequency of a voltage controlled crystal oscillator that is phase locked to the atomic transition frequency. In the case of hydrogen maser receivers constructed by the Smithsonian Astrophysical Observatory (SAO), as with many other such systems, synthesizer phase variations contribute to the output of the voltage controlled crystal oscillator (VCXO) at the same level as phase noise from the maser signal; therefore a knowledge of the synthesizer phase noise is important for an understanding of the phase limitations of the instrument. We have designed and constructed a high resolution direct digital synthesizer (DDS) intended for hydrogen masers. We have investigated its phase noise by simulating its output signal and calculating its phase spectrum, and by measuring its phase spectrum directly. In addition we have compared its phase noise with that of other synthesizers used in hydrogen masers.

Direct Digital Synthesizer Fundamentals

The basic components of the DDS[1,2] are shown in Fig. 1. At each cycle or pulse of a clock signal, a number that

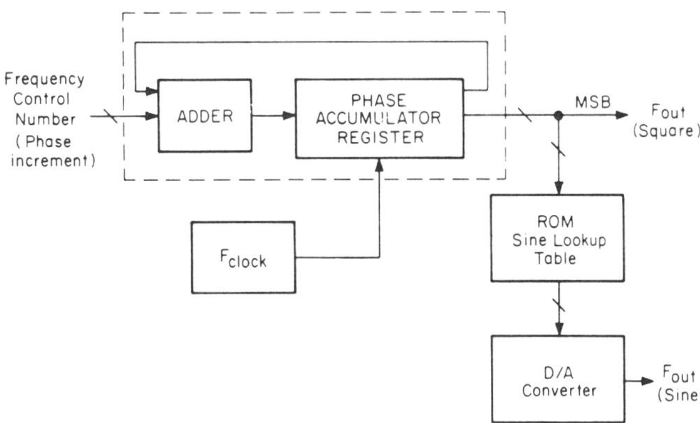

Figure 1. Block Diagram of Basic DDS Components

represents a phase increment is added to a digital accumulator whose content is proportional to the phase of the output signal. At each clock cycle the accumulator value is passed to a read-only sine lookup table (ROM) whose output goes to a digital-to-analog converter (DAC). The DAC produces a series of voltage steps that represent a sine wave sampled at the rate of the clock pulses. When the accumulator overflows as the result of the repeated additions, its most significant bit (MSB) changes from 1 to 0, and a new output cycle has begun. If only the MSB is observed, the DDS produces a square wave that changes sign coincidentally with the zero crossing of the sampled sine wave. Because the square wave can change sign only at a clock cycle, its phase is in general delayed by a varying amount relative to the ideal phase of the signal to be produced. The phase delay is proportional to the number remaining in the accumulator after an overflow.

If the accumulator capacity is L bits, and the phase increment number added at each clock cycle is P, the DDS's output frequency is given by

$$F_{out} = F_{clock}\frac{P}{2^L} \tag{1}$$

If P is divisible by an integral power of 2, the ratio $P/2^L$ can be reduced to lowest terms, $M/2^N$, where the reduced phase increment, M, is odd. If, for example, $L = 8$ bits and $P = 00111000$ binary, P and 2^L are each divisible by 2^3 to give $M = 00111$ (= 7 decimal) and $N = 5$. Then Eq.1 can be written

$$F_{out}/F_{clock} = \frac{00111000}{100000000} = \frac{00111}{100000} = \frac{7}{32}(\text{decimal}) \tag{2}$$

On average, one output period for this example corresponds to $32/7$ clock periods. Because in general a non-zero value remains in the accumulator after each overflow, the accumulator sequence is not the same for every output cycle; rather, the sequence repeats when the accumulator contains zero, which occurs after M output cycles, or 2^N clock cycles. Thus the lowest Fourier frequency to be expected in the phase spectrum is

$$F_{min} = \frac{F_{out}}{M} = \frac{F_{clock}}{2^N} \tag{3}$$

In the example above, the repetition time for the accumulator sequence, which represents the longest variation period of the output waveform, corresponds to 7 output cycles, or 32 clock cycles.

SAO Direct Digital Synthesizer

The SAO maser synthesizer is part of a receiver system that phase locks a 100 MHz VCXO to the 1420.4 MHz hydrogen maser signal. Using a 5 MHz clock signal derived from the VCXO, the synthesizer produces an output signal of approximately 405 kHz. The DDS is controlled by a Z-80 microprocessor that is the central component of a digital control and monitoring system. This system permits local or remote control of the maser's tuning diode voltage, hydrogen pressure, and synthesizer frequency. In addition, it samples and stores the maser's analog operating parameters at selectable intervals, and transfers the data to an external computer upon request. The synthesizer, which is constructed on a single STD bus wire-wrap card (Fig. 2), is based upon a pair of 32-bit number controlled oscillators[3]. It has an accumulator length of 64 bits, and thus a maximum resolution of 2.7×10^{-13} Hz, corresponding to a fractional VCXO frequency variation of 1.9×10^{-22}. While the synthesizer is capable of producing frequencies up to approximately 40% of the clock frequency, the microprocessor software has been written to allow synthesizer frequencies in the range from 405750 Hz to 405760 Hz (the frequency interval relevant for the maser) with a resolution of 10^{-8} Hz, corresponding to a VCXO fractional frequency variation of 7×10^{-18}.

Simulation of DDS Phase Variation

Because the DDS can make output transitions only coincidentally with clock pulses, the zero crossings of its output do not in general coincide with the theoretical zero crossings of the ideal waveform to be approximated. (Only when the phase increment P is an integral power of 2 is there no accumulator overflow, and the output is exactly equivalent to the ideal waveform.) The accumulator remainder following each overflow is proportional to the phase delay between the ideal

Reprinted from *Proceedings of the 42nd Annual Frequency Control Symposium*, pp. 352-356, 1988.

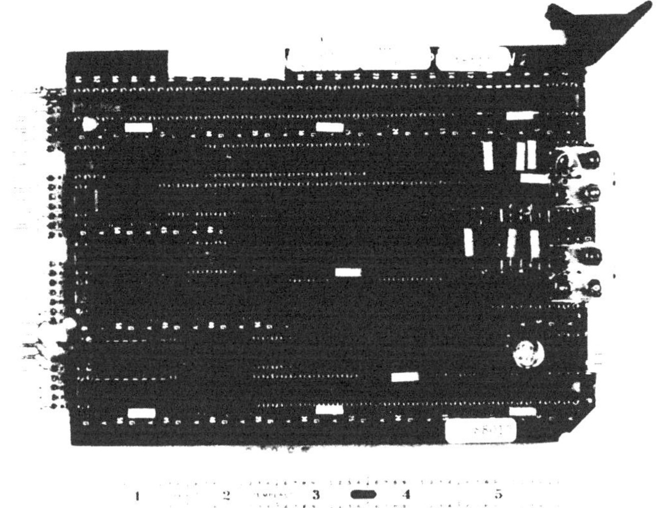

Fig. 2.   SAO 64-bit Direct Digital Frequency Synthesizer

waveform and the actual zero crossing. The series of remainders, then, represents the phase variation of the square wave DDS output, sampled once each output cycle. We simulated the DDS phase spectrum by calculating the remainder series for particular phase increment numbers, and then calculating the discrete Fourier spectrum of the series by means of a fast Fourier transform software package.

We simulated the phase spectrum for two ranges of phase increment numbers: "simple" numbers with few non-zero bits, all within the eight most-significant bits; and the phase increment that produces the frequency 405750 Hz for a clock frequency of 5 MHz. Simple numbers produce phase spectra with a small number of well separated lines, making comparison with measured spectra relatively straightforward, while 405750 kHz is within the frequency range used in hydrogen masers. The simulated phase spectra for simple phase increment numbers display M equally spaced spectral lines, separated by (F_{out}/M) Hz, for Fourier frequencies f in the range $0 \le f < F_{out}$.

It has been pointed out[2] that a source of DDS phase variation is the phenomenon of "phase truncation". Although DDS accumulators generally are designed to store a large number of bits, hardware limitations make it practical to pass only a small number of the most significant bits to the sine look-up ROM. (In the STI 1172A, 8 bits are passed to the ROM.) If the number of bits in the reduced phase increment M is larger than the number of bits sent to the ROM, the consequent truncation of the accumulator contents produces an error in the output phase. For the "simple" values of M considered here, which contain at most 8 bits, phase truncation does not occur, and thus does not affect the correspondence between the simulated and measured spectra.

Measurement of DDS Phase Spectra

The DDS phase spectra were measured by comparing the outputs of two synthesizers operated from a common clock reference. The signals were combined in a double-balanced mixer whose output was amplified and filtered, and the resulting phase spectrum was measured with a digital spectrum analyzer[4]. The mixer's output is proportional to the phase variation of the input signals when the average phase difference between the inputs is 90°.

Because the DDS signal is produced by a deterministic sequence of accumulator values beginning with zero, one expects the observed relative phase variation between two DDS's to depend upon the delay between the start of one DDS sequence and the start of the second. (If the DDS's are started simultaneously, no relative phase variation should be observed.) Using a circuit that reset the accumulators of two DDS's with a selectable time delay, we observed variations in the measured phase spectrum with initial delay. To eliminate this source of variability, we measured the phase spectra of a single DDS relative to a sine-wave analog frequency synthesizer (HP5100A), whose output is uniform in time aside from random phase noise. For Fourier frequencies less than 100 Hz the DDS phase noise is considerably lower than that of the HP5100A, and in that frequency range we measured the phase spectral densities of two DDS's relative to one another.

The DDS synthesizers used for these measurements were capable of producing either a square wave output from the accumulator MSB, or a sine wave from the DAC. The controlling software was modified to allow any frequency to be selected.

Results

Simulated and measured phase spectra for phase increment $M = 00111$ (=7 decimal) are shown in Fig. 3. The simulated square-wave spectrum, shown by triangles, was calculated for Fourier frequencies up to the Nyquist limit, or half the sampling frequency. (The sampling frequency equals F_{out} since samples are calculated once each output cycle.) The open triangles represent the calculated lines reflected about the Nyquist frequency. For the measured spectra $F_{clock} = 400$ kHz,

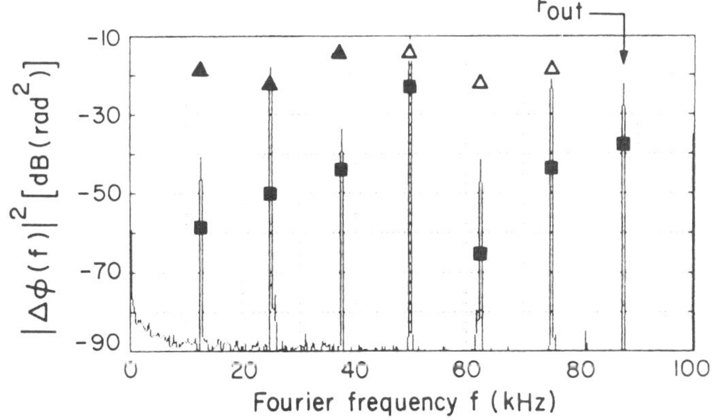

Fig. 3.   DDS Phase Spectra for $M = 00111$

giving $F_{out} = (7/32)F_{clock} = 87.5$ kHz. The lines in the measured spectra correspond in number — 7 — and position to those predicted by the simulation. The magnitudes of the even lines in the measured square wave spectrum $(2, 4, 6...\times F_{min})$ agree within a few dB with those of the simulated spectrum, while the odd spectral lines $(1, 3, 5...\times F_{min})$ are lower than the predicted levels by as much as 30 dB. The levels of the sine wave phase spectrum are considerably below both the simulated and the measured square wave spectra.

Spectra for $M = 000111001$ (=57 decimal) are shown in Figs. 4 and 5. Here $F_{clock} = 400$ kHz and $F_{out} = (57/256)F_{clock} = 89062.5$ Hz. The square wave spectrum (Fig. 4) clearly shows the predicted 57 lines between 0 Hz and F_{out}, and suggests the rapid increase in spectral complexity with increasing length of the phase increment. The dots represent the simulated spectral line strengths. This spectrum also has

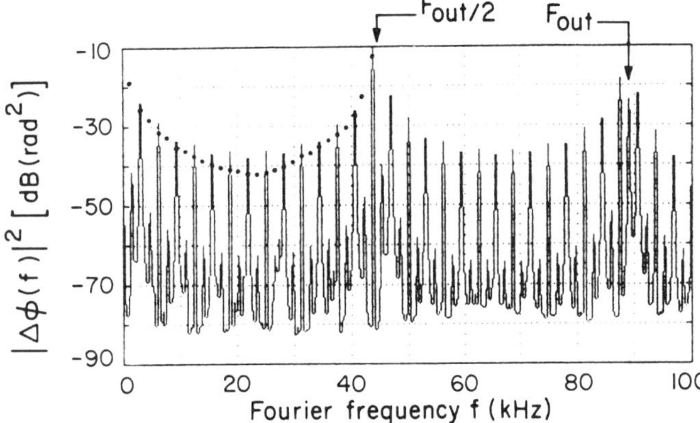

Fig. 4. DDS Square Wave Phase Spectra
for M = 00111001

even spectral components that agree in magnitude with the simulated spectrum, and odd components that are considerably lower. The sine wave phase spectrum (Fig. 5) is substantially lower in magnitude than the square wave spectrum.

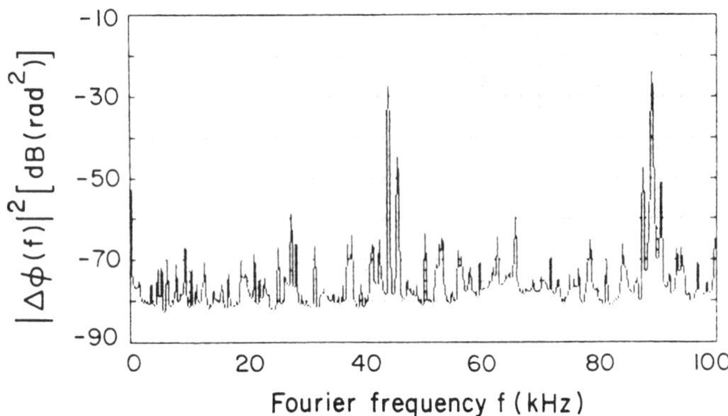

Fig. 5. Measured DDS Sine Wave Phase Spectrum
for M = 00111001

The improvement of the sine wave phase spectrum over the square wave spectrum can be understood by the following heuristic argument. Because the sine wave DAC changes its output coincidentally with clock pulses, and thus theoretically makes zero crossings at the same instant as the square wave output, one expects the phase variations of the sine wave to be identical to those of the square wave. However, the DAC circuitry has associated capacitances that cause slew-rate limiting, which is equivalent to low-pass filtering. Fig. 6 shows the effect of a low pass filter (LPF) on the DAC output. In both 6a and 6b the theoretical DAC zero transition, shown by a dashed line, coincide with the square wave zero crossing. However, the LPF causes the DAC output to decay exponentially from the previous level to the new one. In 6a the phase delay $\Delta\phi$ between the theoretical sine wave and the DAC transition is small, the DAC voltage goes from a relatively large positive value to a small negative value, and the exponential decay delays the actual zero crossing for a relatively long time, increasing the phase error by a large fraction. In 6b, by contrast, $\Delta\phi$ is large, the DAC voltage goes from a small positive to a large negative value, and the exponential decay delays the zero crossing by a relatively short time, making a small increase in the phase error. Therefore the net phase errors – theoretical phase error plus LPF delay – for the two cases tend toward equality, reducing the net phase variation from one cycle to the next.

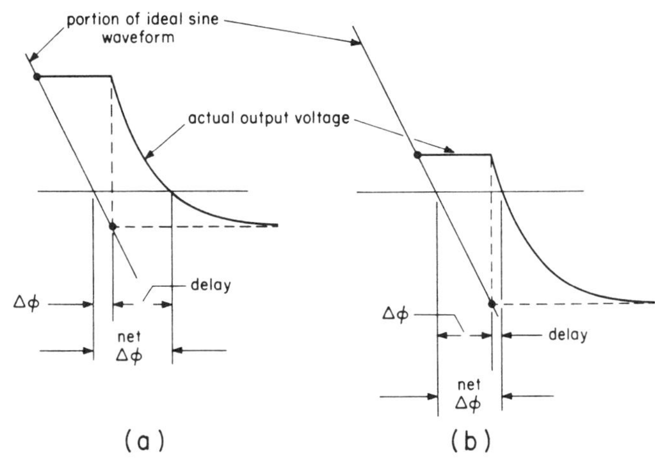

Fig. 6. Effect of Filtering on DDS Stepped Sine Output

The measured and simulated square wave phase spectra for F_{out} = 405750 kHz are shown in Fig. 7. (For the measurement, values of F_{clock} = 1 MHz and F_{out} = 81150 Hz were used, in order to be able to display the spectum up to F_{out} on the 100 kHz spectrum analyzer.) The number and positions of major lines in the simulated and measured spectra agree, although the measured spectrum has lower resolution than the simulation. For convenience, the major lines of the simulated spectrum are indicated on the measured spectrum by dots.

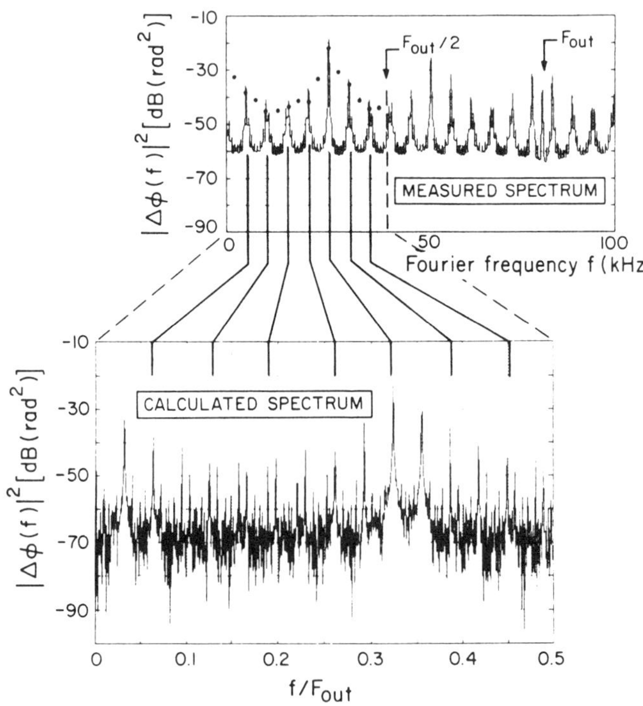

Fig. 7. Simulated and Measured DDS Square Wave Spectra
F_{out}/F_{clock} = 405750/5 000 000

(The phase increment for this output frequency has a 1 in its 64th, or least signficant, bit. For a 5 MHz clock clock signal, the longest repetition time of the accumulator sequence is 2^{64} clock cycles, or 1.2×10^5 years. It is clearly impossible to integrate a measurement long enough to observe this period, or to simulate the spectrum with enough accumulator values to resolve the minimum line separation.) As with the other measurements, the odd major components of the measured spectrum are suppressed relative to the predicted levels.

For evaluating synthesizer operation in the maser, Fourier frequencies lower than roughly 100 Hz are of interest. Because the lock-loop time constant for the maser receiver system is on the order of 10 Hz, higher frequency synthesizer phase variations are attenuated by the phaselock LPF. Fig. 8 shows the phase spectral density of two sine-wave DDS's for Fourier frequencies below 100 Hz, for F_{clock} = 5 MHz and F_{out} = 405750 Hz. (Because the synthesizers are equivalent, the spectral density of a single synthesizer is 3 dB lower than the levels shown.) Because the low-frequency spectrum is free of discrete lines, the spectral density (with units of dB relative to 1 radian^2/Hz) is the appropriate measure of phase noise. The spectral density at these Fourier frequencies was not noticeably affected by the starting delay between the two synthesizers.

In order to establish a basis for comparison of the DDS phase performance, we measured the phase spectra of two other types of frequency synthesizers used in hydrogen masers, a phase-lock loop (PLL) synthesizer made by SAO, and a Dana Laboratories Corp. model 7010 synthesizer. The results, shown in Table 1, indicate that the phase noise of the DDS synthesizer is less than or at most equal to the other synthesizers for Fourier frequencies below 1 kHz.

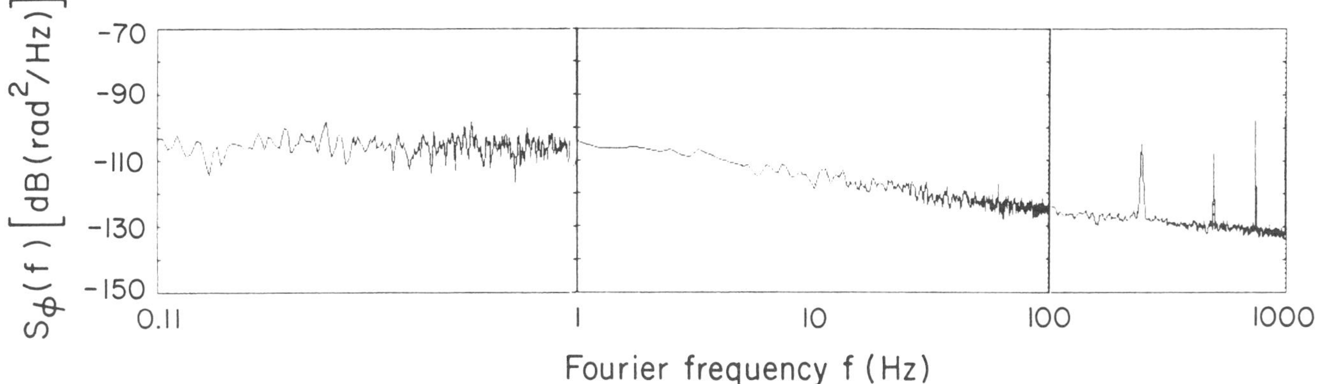

Fig. 8.   Phase Spectral Density of DDS Sine Wave
F_{clock} = 5 MHz; F_{out} = 405750 kHz

Table 1
Comparison of Phase Noise Among Synthesizers
F_{clock} = 5 MHz; F_{out} = 405750 Hz
Units = dB(rad^2/Hz)

Freq (Hz)	0.1	1.0	10	100	1000
Oscillators:					
DDS Sq/Sq	—	−103	−113	−120	−127
DDS Sin/Sin	−105	−104	−115	−124	−132
PLL	—	−95	−112	−120	−135
		to −95	to −110	to −120	
Dana	−75	−90	−110	−118	−115

Summary and Discussion

With increasing use of digital frequency synthesizers in electronic systems, an understanding of their phase noise performance is increasingly important. We have demonstrated the ability to calculate accurately the locations of phase spectral lines, and have compared simulated spectra with measured values. The measured line magnitudes agree with the calculated values for the square wave spectra, with the exception of the suppression of alternate (odd) lines. Because we observed this suppression with an analog (swept-filter) spectrum analyzer as well as with the digital analyzer, we conclude that it is not related to spectrum analyzer performance. We are currently investigating the phenomenon by means of high-resolution calculations of the process of multiplying a DDS square wave with a square wave of uniform period.

Our spectral measurements clearly show that the use of a sine-wave output with low pass filtering improves the DDS phase spectrum. It would be valuable to investigate quantitatively the effects of LPF cutoff frequency and slope on the phase spectrum.

It has been shown[2] that the power spectrum of the DDS can be very sensitive to specific values of the phase increment number. A similar sensitivity is to be expected for the phase spectrum. In any particular application, it is important to investigate the phase behavior of a DDS in the frequency range of interest. A large accumulator capacity makes possible an extremely large number of discrete output frequencies within any macroscope frequency range; thus it is a practical impossibility to examine the phase spectrum for every possible frequency. However, useful conclusions may be drawn by studying the trend of the phase spectrum with frequency and by identifying frequency ranges associated with particularly large values of phase variation. Further work along these lines is in process at SAO.

Acknowledgements

We are pleased acknowledge the support of the Jet Propulsion Laboratory for providing the impetus and funding for this work. For interesting and helpful discussions we are pleased to thank Leonard Cutler, Paul Kuhnle, Henry Nicholas, III, Donald Percival, Richard Sydnor, Robert Vessot, and Fred L. Walls.

References

1.   V. Reinhardt, K. Gould, K. McNab, and M. Bustamante, "A short survey of frequency synthesizer techniques." Proc. 40th Annual Frequency Control Symp., 1986, pp. 355-365.

2. Henry T. Nicholas, III, and Henry Samueli, "An analysis of the output spectrum of direct digital frequency synthesizers in the presence of phase-accumulator truncation." Proc. 41st Annual Frequency Control Symp., 1987, pp. 495-502.

3. Model 1172A number controlled oscillator, Stanford Telecommunications, Inc., Santa Clara, CA 95054

4. Mixing and amplification were performed by a Femtosecond Systems, Inc., model 600 phase noise detection unit. The spectrum analyzer was an HP model 3561A.

A Direct Digital Synthesizer with 100-MHz Output Capability

P.H. SAUL AND M.S.J. MUDD

Abstract —A bipolar direct digital synthesizer with 5-kHz to 100-MHz output capability is described. Channel spacing is 5 kHz with a "hop" time between discrete output frequencies of less than 17 ns. Spurious output rejection is better than −32 dBc, and close-to-carrier measurements indicate a noise floor lower than −138 dBc at ±25 kHz.

I. Introduction

There is an increasing trend towards "analog" circuits which are primarily digital in operation, but contain data conversion components at inputs and outputs. The device described in the paper uses such a technique, so that digital manipulation is used to provide a drive code for an output digital-to-analog converter (DAC). The first generation of such a device is described. The limitations of the device are seen to be in the spurious output levels; these will be improved with the incorporation of newer and faster processes.

Direct digital synthesis has been described in the literature [1], but has been considered as a low-speed technique only, since the achievable performance is limited by the DAC's and high-speed logic available. The major difference between direct digital frequency synthesizers (DDFS's) and phase-locked-loop (PLL) types is that the former do not contain feedback loops. This can be a major advantage in terms of settling time to a new frequency; a good PLL has acquisition times of the order of 1 ms, whereas a DDFS can acquire in a time ultimately limited only by the DAC settling time. The DDFS designs contain no loops or indeed VCO's. The primary source of stability is the clock oscillator, so that, in the limit, since the clock is always at a higher frequency than the output, the output phase noise due to integration is better than the clock itself. This assertion, while correct, must be carefully interpreted; since the clock is likely to be several hundred megahertz, the phase noise will not be as good as the source for a PLL at, say, 4 MHz. Also, this very high-frequency clock will usually be derived by phase locking to a lower frequency, so the gains are less great than the initial statement may imply. Finally, the phase noise from the clock is integrated over a number of cycles depending on the ratio of the clock to output frequency, but the improvement in phase noise can never be as great as this ratio implies. Practically, as will be seen below, phase noise is rarely a limitation; the presence in the output of spurious lines due to finite clock pulse width and DAC resolution imposes much more fundamental operational limits for the general case.

Manuscript received October 12, 1987; revised January 26, 1988. This work was supported by Plessey Semiconductor.

The authors are with the Allen Clark Research Centre, Plessey Research Caswell Limited, Caswell, Towcester, Northamptonshire NN12 8EQ, England.

IEEE Log Number 8820722.

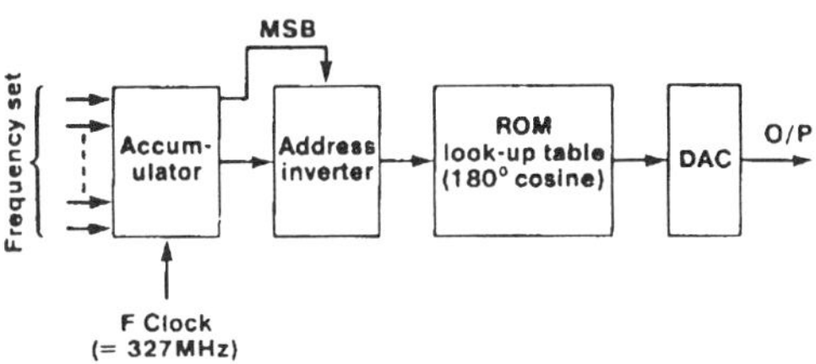

Fig. 1.   Direct digital synthesizer block diagram.

II. System Description

The block diagram of a system is shown in Fig. 1. A frequency setting word is applied to an accumulator, which is clocked at the system clock frequency of, in this case, 327 MHz. The origin of this frequency is the required channel spacing (5-kHz basic increment) times the accumulator length (2^{16}). Hence, for a 5-kHz increment the exact clock frequency is 327.680 MHz. Obviously different clock frequencies can be chosen for alternative increments; a particular example is 3.125-kHz increments from a 204.800-MHz clock. This example would fit the European 12.5-kHz based channel spacings, while 5 kHz would fit U.S./Japanese 10-kHz channels. The accumulator could be of arbitrary length, depending on the channel spacing required; the example constructed had 16 bits.

To avoid the need for storage of a full 360° in the ROM, the MSB output of the accumulator is used as a sign inverter, which, with the LSB's, forms a digitized triangular number sequence. The ROM contains all the data for 180°, stored in a cosine sequence in this instance. ROM size is 1 kbit, i.e., 128×8 bits. With the reflection about zero implicit in the 180° storage, this is equivalent to 256 words of data, i.e., the storage density is equal to the word length. The minimum ROM would have stored only 90°, but would have entailed both amplitude and phase inversion of data. The system used here was chosen partly for simplicity, and partly because a ROM cell of this size already existed in the library.

The data pass through retiming latches at each stage including the output in order to provide accurate data at the high clock rate; pipeline delays are unimportant in a nonlooped system. Finally, the DAC constructs the output waveform, which consists of discrete points on the cosinusoid. Interpolation could be carried out by low-pass filtering; in practice, for these tests, no filters were used except the inherent low-pass action of the DAC. Clearly from Fig. 2 it would have been possible to increase the ROM size. However, increase in DAC word length implies a new DAC design. The important feature of the DAC is the ability to settle fast to a given accuracy. Increasing the DAC from 8 bits as used here, to say 10 bits, but slowing the response time, would have a negative effect on the output spurious levels. Increasing the DAC size would also require an increase in ROM size with a

Reprinted from *IEEE Journal of Solid-State Circuits*, Vol. 23, No. 3, pp. 819-821, June 1988.

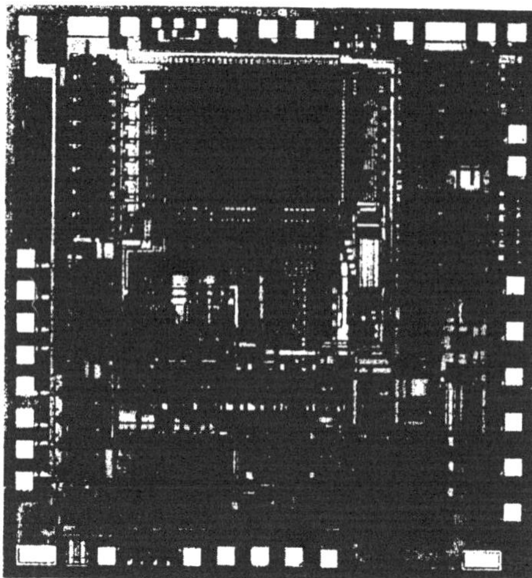

Fig. 2. Direct digital synthesizer chip microphotograph.

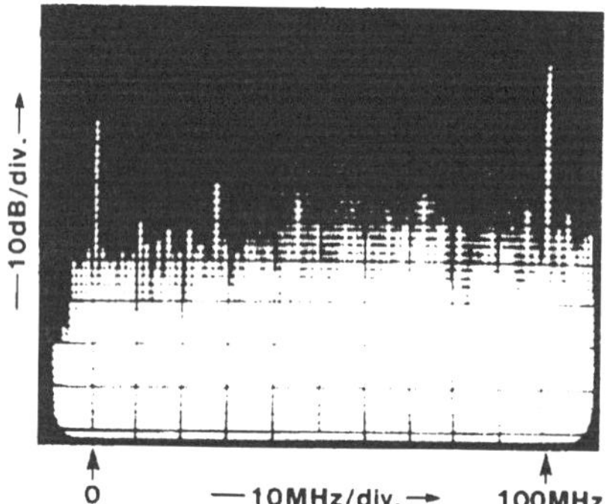

Fig. 3. Frequency spectrum for 100-MHz output with F_{clock} = 327.68 MHz.

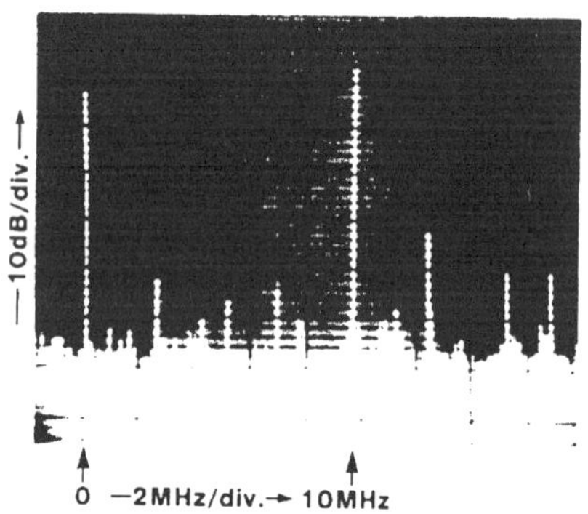

Fig. 4. Frequency spectrum for 10-MHz output with F_{clock} = 32.768 MHz.

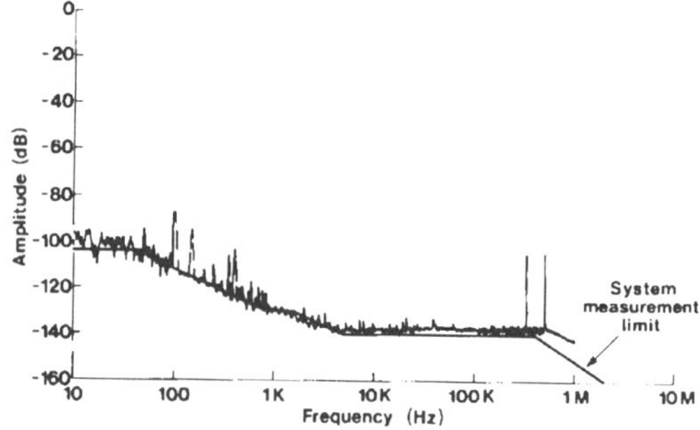

Fig. 5. Spectrum at 81.92 MHz.

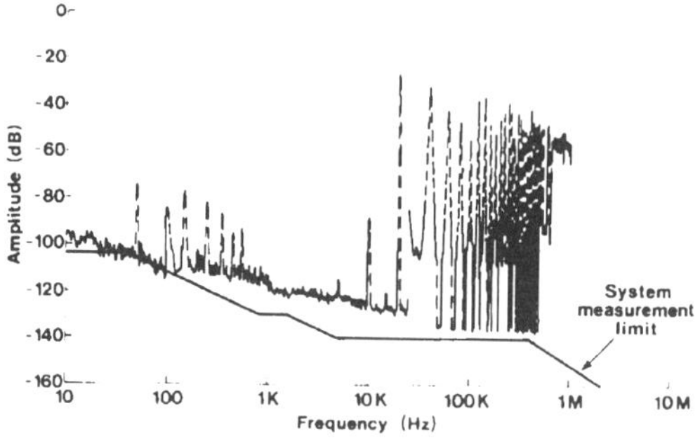

Fig. 6. Spectrum at 81.925 MHz.

corresponding increase in access time and hence reduction in maximum output frequency.

The circuit design was based around a provisional specification calling for 100-MHz output at 5-kHz spacing with 40-dBc/Hz spurious rejection. With a 16-bit binary channel select number, this leads to a clock frequency of 327.68 MHz. The requirement for 40-dB spurious rejection is within the reach of an 8-bit DAC drive word, although the DAC used cannot in practice achieve settling to 8 bits at this speed. Figs. 3 and 4 illustrate this point.

In Fig. 3, the output is 100 MHz, with a spurious rejection of 32 dBc. The largest spurs are at $(327.68 - 3 \times 100)$ MHz and $(4 \times 100 - 327.68)$ MHz. At 10-MHz output, using the same program code, with 32.768-MHz clock (Fig. 4), the largest spur is at -38 dBc, although this is at a frequency which could be removed by low-pass filtering. The true spur level is -46 dBc, close to the theoretical limit for an 8-bit system.

Fig. 5 shows the output close-to-carrier spectrum at 81.92 MHz (0100 0000 0000 0000). Spurious level is excellent, closely following the system noise level represented by the solid line. At 81.925 MHz, Fig. 6, the spurious level is of course degraded. Spurious lines occur at harmonics of 20 kHz, i.e., the ratio of the clock to output frequency (4) times the channel increment (5-kHz).

III. PROCESS

The device described was designed on a 3-μm minimum feature size oxide isolated process (Plessey Process HA) using three-layer interconnect. The upper layer is a 12-μm minimum pitch, used for power buses, while the first two layers are at 8-μm pitch for interconnect. The circuit uses a restricted library of standard cells, and so is midway in design concept between a full-custom design and a truly cell-based automatic design approach. This was in effect a compromise to realize a quick turnaround time with the design tools readily available; the compromise was successful in that the devices were fully operational at the first attempt. Minimum gate delays predicted on SPICE are 250 ps, with 2-ns ROM access time. These delays have been confirmed by measured results.

All measurements reported here were carried out using a standard 8-bit DAC (SP9768) based on a design reported in 1980 [2]. This device is the most limiting aspect of the present system, and a faster DAC should give at least a 6-dB improvement in spurious output levels.

230

TABLE I
SUMMARY OF SYNTHESIZER

Frequency range	0 → 100MHz
Channel spacing	5KHz
Programming	16 BIT parallel
Clock Frequency	327.68MHz
Spurious output level	32dB at 100MHz O/P 38dB at 30MHz O/P
Close to carrier noise floor	Better than − 138dBc at ± 25KHz
I/O Interface	F100K
Power consumption	1.8W
Supply voltage	−5.2V
Chip size	4.65mm x 4.80mm
Complexity	2700 transistors

IV. CONCLUSIONS

Because the DDFS system is not looped, the acquisition of new frequencies is very fast, limited only by the DAC settling time (5 ns) and in this case a further four clock periods of delay incurred in the accumulator, which is not fully relatched between stages. This leads to a theoretical frequency hop time of 17 ns. This has not yet been measured, but is of course practically instantaneous; it is at least 5 orders of magnitude faster than loop synthesizers.

Table I is a summary of the primary characteristics of the device. The design is all-digital and contains no analog loops or indeed any reactive components. The close-to-carrier noise is excellent, and arbitrarily narrow channel spacing could be accommodated by the use of a larger accumulator without compromise in performance. A faster DAC will achieve the 40-dB spur target, while a fully integrated synthesizer/DAC would further stretch the performance up in frequency, possibly to 500 MHz on a 1-μm process.

ACKNOWLEDGMENT

The authors gratefully acknowledge the assistance of T. Hinch and P. Moore (Plessey Radar) in preparation of Figs. 5 and 6.

REFERENCES

[1] V. Manassewitsch, *Frequency Synthesis and Design.* New York: Wiley, 1980, pp. 37–49, pp. 494–501.
[2] P. H. Saul, P. J. Ward, and A. J. Fryers, "An 8-bit, 5-ns monolithic D/A converter subsystem," *IEEE J. Solid-State Circuits*, vol. SC-15, no. 6, pp. 1033–1039, Dec. 1980.

An Analysis Methodology to Identify Dominant Noise Sources in D/A and A/D Converters

J. Alvin Connelly, *Senior Member, IEEE*, and Katherine P. Taylor, *Member, IEEE*

Abstract —This paper presents a methodology of noise analysis to identify dominant noise sources in D/A and A/D converters that is applicable to any converter topology and integrated circuit technology (e.g., bipolar, MOS, GaAs) provided that the noise contribution from individual converter elements is known. The methodology is demonstrated using a simple voltage—referenced D/A topology and a flash A/D converter. The thermal noises in an R-2R network and in a binary-weighted network are compared first as part of a theoretical analysis of the D/A converter. Then the output noise from an R-2R network with three different operational amplifier (op amp) configurations is compared to the output noise of a binary-weighted network with an op amp in an inverting summer configuration. The white and 1/f noise contributions from the op amp are included in this part of the analysis. For the flash A/D converter, the noise from a voltage reference, a resistive divider, and comparators are examined in terms of equivalent noise sources placed at the inputs of the comparators. SPICE input files constitute a noise model of a particular converter topology whose noise parameters may be changed to match actual converter noise behavior more closely. SPICE simulations are used to verify the analytical results and to examine typical noise levels in all of the circuits. The SPICE noise models allow easy identification of dominant noise sources within a given topology for a given set of noise parameters. The effects of changing the parameters of the dominant noise sources (e.g., using components designed for low noise) are examined to address design implications of the noise analysis. Finally, the effects of all noise sources on an analog-to-digital-to-analog (A-D-A) converter system are examined.

I. Introduction

SIGNAL processing applications of analog-to-digital (A/D) and digital-to-analog (D/A) converters continue to require finer and finer resolution. As resolution starts to approach the noise limits of a converter, the usefulness of the less significant bits becomes questionable since the probability of bit errors significantly increases. Effective bit resolution is determined by both interference-type noise as well as inherent fundamental noise within the converter circuitry.

Most interference-type noise can be reduced or eliminated through proper shielding and grounding tech-

Manuscript received October 6, 1989; revised June 10, 1990. This paper was recommended by Associate Editor T. T. Vu.

The authors are with the School of Electrical Engineering, Georgia Institute of Technology, Atlanta, GA 30332-0250.

IEEE Log Number 9102160.

niques. At some point, however, the noise of a converter will be dominated by fundamental device noise. Reduction of fundamental noise requires converter operation at low temperatures, which is often an impractical solution, or converter design, which identifies dominant noise sources and seeks to reduce their contributions by changing transistor operating points or transistor types, lowering resistances, and optimizing the converter topology from a noise perspective.

Numerous articles and publications address noise models and noise analysis in integrated circuit devices, i.e., resistors, diodes, bipolar junction transistors (BJT's), MOSFET's, and operational amplifiers (op amps) [1]–[9]. Relatively few references [10]–[12] consider noise in more complex circuits such as A/D and D/A converters, and none of these seeks to model converters with anything more than a single comparator and one noise source.

Consequently, there is a need for noise analysis and noise modeling methodologies in A/D and D/A converters using more complete circuit models. This paper describes a methodology of noise analysis to identify dominant noise sources in D/A and A/D converters. The noise modeling methodology can be adapted to other converter topologies with little modification provided that the noise contributions from individual converter elements are known. In addition, the modeling methodology is not technology specific and is therefore applicable to noise simulation of bipolar, MOS, and GaAs converters. The methodology is demonstrated using a simple voltage-referenced D/A topology and a flash A/D converter.

While the voltage-referenced D/A is not often implemented by integrated circuit manufacturers, it is chosen here because this topology is often used to explain D/A converter concepts and is well defined and understood. The flash A/D converter is chosen because of its simple topology with no D/A converter to compound the noise analysis.

This paper first presents a noise analysis of binary weighted and R-2R resistor networks commonly found in D/A converters. Four different op amp summation schemes for producing the analog output of the D/A from the resistor networks are then considered. The noise from a voltage reference, a resistive divider, and compara-

Reprinted from *IEEE Transactions on Circuits and Systems*, Vol. 38, No. 10, pp. 1133-1144, October 1991.

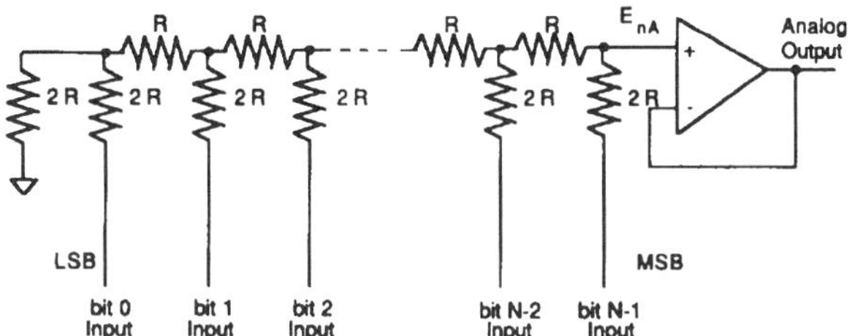

Fig. 1. Simplified *N*-bit D/A converter with R-2R network.

tors of a flash A/D are examined in terms of equivalent noise sources placed at the inputs of the comparators.

SPICE input files developed in conjunction with this methodology constitute a noise model of a particular converter topology whose noise parameters may be changed to match actual converter noise behavior more closely. SPICE simulations are used to verify the analytical results and to determine typical noise levels in the circuits considered here. The SPICE noise models allow easy identification of dominant noise sources within a given topology for a given set of noise parameters. The effects of changing the parameters of the dominant noise sources (e.g., using components designed for low noise) are examined to address design implications of the noise analysis. Finally, the noise in a complete analog-to-digital-to-analog (A-D-A) converter system is examined.

II. Resistor Networks

Figs. 1 and 2 show a simplified *N*-bit D/A converter using an R-2R network and a binary-weighted network, respectively. The first part of the analysis considers only the resistors in the networks and excludes op amp noise sources and noise from R_F in the binary-weighted case.

Noise analysis of both resistor networks entails grounding the digital input bits and placing thermal noise voltage sources in series with each resistor. The value of the spectral density noise voltage squared of each source is $4kT$ times the corresponding resistance, where k is Boltzmann's constant (1.38×10^{-23} J° K^{-1}) and T is absolute temperature in degrees Kelvin. The equivalent network noise output voltage squared per Hertz, E_{nA}^2 is the sum of the individual resistor noise contributions evaluated through the appropriate voltage divider for each noise source.

The spectral density of E_{nA}^2 of the R-2R network is independent of the number of bits, N, and has a value of $4kTR$. E_{nA}^2 of the binary-weighted network is given by (1).

$$E_{nA}^2 = \frac{4kTR}{1 + \left[\dfrac{2^{N-1} - 1}{2^{N-1}} \right]} . \tag{1}$$

As N increases, E_{nA}^2 rapidly approaches a value of $2kTR$. For $N = 6$, this approximation results in an E_{nA}^2 error of 1.4%.

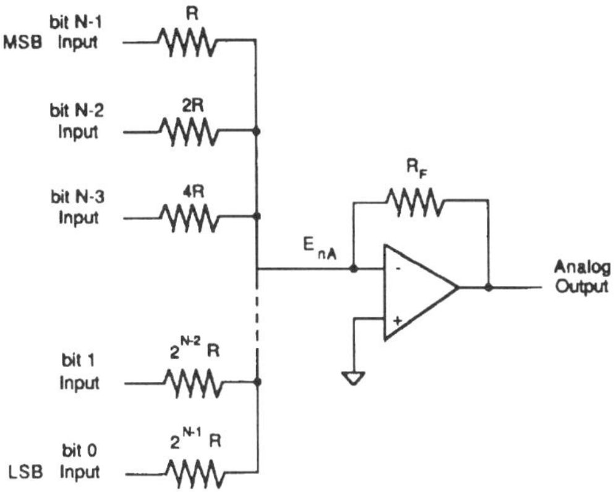

Fig. 2. Simplified *N*-bit D/A converter with binary-weighted network.

III. Resistor Networks and Summation Topologies

Four different op amp topologies which could be used in the simplified models of Figs. 1 and 2 are shown in Fig. 3. For the noise analysis of these topologies, the op amp noise was modeled with a noise voltage source, E_n (V/√Hz), in series with the noninverting input terminal, and a noise current source, I_n (A/√Hz), located between the two input terminals as shown in Fig. 4. Both E_n and I_n may be modeled as having flat noise spectra, 1/f noise spectra, or a combination of these two spectra such that the spectrum is flat for frequencies greater than a noise corner frequency, f_{nc}, and has a 1/f characteristic for frequencies below f_{nc}.

The contributions to E_{no}^2 (i.e., the square of the noise output voltage) from each noise voltage source are calculated by squaring the source noise voltage and multiplying by the square of the voltage gain from the source's location to the output. The I_n contribution to E_{no}^2 is calculated by squaring the value of I_n, multiplying by the square of the equivalent source resistance seen by I_n, and then multiplying by the square of the noninverting forward gain of the circuit.

For example, consider the R-2R network with a noninverting amplifier shown in Fig. 3(c). The resistor network has an equivalent resistance, R_{eq}, equal to R seen looking back into the network. The network also introduces a bit factor (BF) which is the voltage gain of the network if all

233

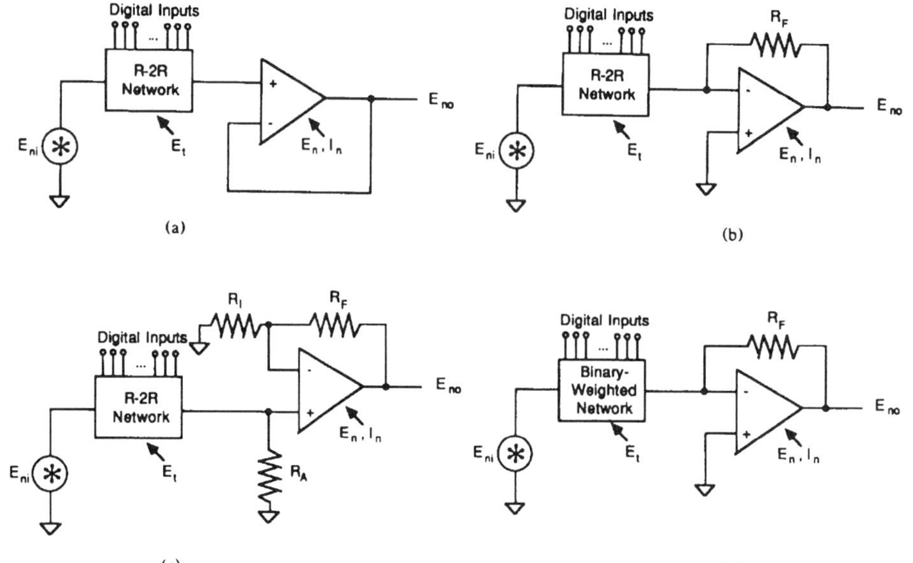

Fig. 3. Four summation schemes. (a) R-2R network with voltage follower. (b) R-2R network with inverting amplifier. (c) R-2R network with noninverting amplifier. (d) Binary-weighted network with inverting summer.

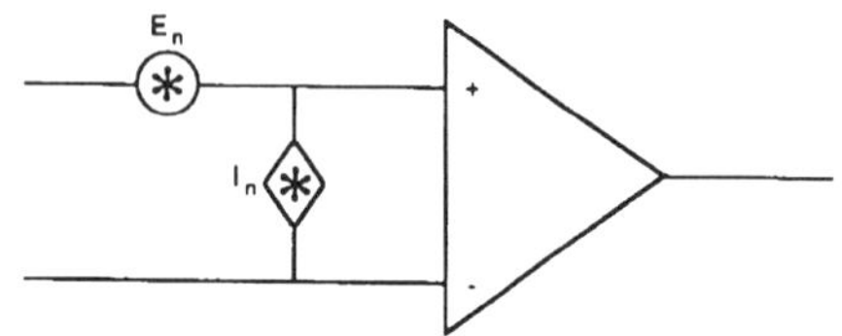

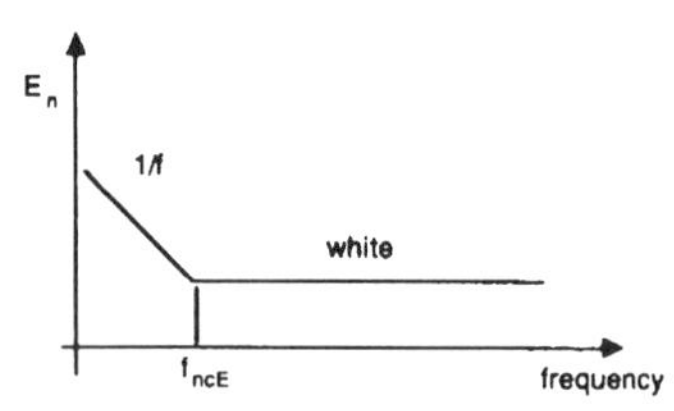

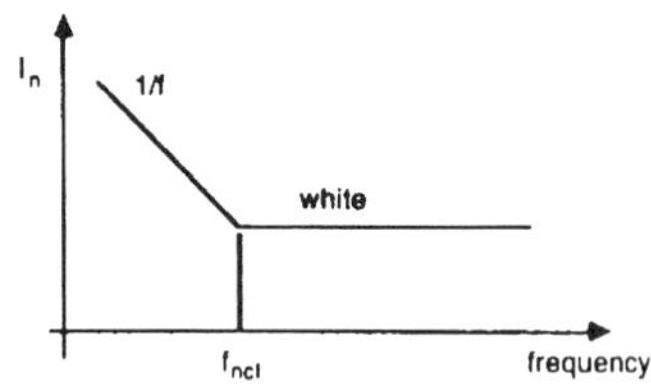

Fig. 4. Operational amplifier noise sources.

of the inputs are connected together. This bit factor is given by

$$BF = \frac{2^N - 1}{2^N}$$ (2)

and is approximately equal to 1.

Ignoring the input and output resistances of the op amp, the other circuit elements make the following contributions in units of V^2/Hz to E_{no}^2.

R-2R network:

$$4kTR_{eq}(BF)^2\left[\frac{R_A}{R_A + R_{eq}}\right]^2\left[\frac{R_F + R_I}{R_I}\right]^2$$ (3)

$$E_n: \quad E_n^2\left[\frac{R_F + R_I}{R_I}\right]^2$$ (4)

$$R_A: \quad 4kTR_A$$ (5)

$$R_I: \quad 4kTR_I\left(\frac{R_F}{R_I}\right)^2$$ (6)

$$R_F: \quad 4kTR_F$$ (7)

$$I_n: \quad I_n^2\left[\frac{R_A R_{eq}}{R_A + R_{eq}} + \frac{R_F R_I}{R_F + R_I}\right]^2\left[\frac{R_F + R_I}{R_I}\right]^2.$$ (8)

TABLE I
E_{no}^2 TERMS

Topology	Bit factor	R_{eq}	E_n^2	I_n^2	Network E_t^2	E_t^2 of R_F	E_t^2 of R_I	E_t^2 of R_A
R-2R with voltage follower	$\dfrac{2^N-1}{2^N}$	R	1	$(R_{eq})^2$	$(BF)^2$	—	—	—
R-2R with inverting amplifier	$\dfrac{2^N-1}{2^N}$	R	$\left(\dfrac{R_F+R_{eq}}{R_{eq}}\right)^2$	R_F^2	$(BF)^2\dfrac{R_F^2}{R_{eq}^2}$	1	—	—
R-2R with noninverting amplifier	$\dfrac{2^N-1}{2^N}$	R	$\left(\dfrac{R_F+R_I}{R_I}\right)^2$	$\left(\dfrac{R_F+R_I}{R_I}\right)^2 \cdot \left(\dfrac{R_A R_{eq}}{R_A+R_{eq}}+\dfrac{R_F R_I}{R_F+R_I}\right)^2$	$\left(\dfrac{R_F+R_I}{R_I}\right)^2 (BF)^2 \cdot \left(\dfrac{R_A}{R_A+R_{eq}}\right)^2$	1	$\left(\dfrac{R_F}{R_I}\right)^2$	1
Binary-weighted with inverting summer	1	$\dfrac{R}{1+\dfrac{2^{N-1}-1}{2^{N-1}}}$	$\left(\dfrac{R_F+R_{eq}}{R_{eq}}\right)^2$	R_F^2	$\dfrac{R_F^2}{R_{eq}^2}$	1	—	—

TABLE II
E_{ni}^2 TERMS

Topology	E_{no}^2/E_{ni}^2	E_n^2	I_n^2	Network E_t^2	E_t^2 of R_F	E_t^2 of R_I	E_t^2 of R_A
R-2R with voltage follower	$(BF)^2$	$\dfrac{1}{(BF)^2}$	$\dfrac{R_{eq}^2}{(BF)^2}$	1	—	—	—
R-2R with inverting amplifier	$(BF)^2\dfrac{R_F^2}{R_{eq}^2}$	$\left(\dfrac{R_F+R_{eq}}{(BF)R_F}\right)^2$	$\dfrac{R_{eq}^2}{(BF)^2}$	1	$\left(\dfrac{R_{eq}}{(BF)R_F}\right)^2$	—	—
R-2R with noninverting amplifier	$\left(\dfrac{(BF)R_A}{R_A+R_{eq}}\right)^2 \cdot \left(\dfrac{R_F+R_I}{R_I}\right)^2$	$\left(\dfrac{R_A+R_{eq}}{(BF)R_A}\right)$	$\left(\dfrac{R_A R_{eq}}{R_A+R_{eq}}+\dfrac{R_F R_I}{R_F+R_I}\right)^2 \cdot \left(\dfrac{R_A+R_{eq}}{(BF)R_A}\right)^2$	1	$\left(\dfrac{R_A+R_{eq}}{(BF)R_A}\right)^2 \cdot \left(\dfrac{R_I}{R_F+R_I}\right)^2$	$\left(\dfrac{R_A+R_{eq}}{(BF)R_A}\right)^2 \cdot \left(\dfrac{R_F}{R_F+R_I}\right)^2$	$\left(\dfrac{R_A+R_{eq}}{(BF)R_A}\right)^2 \cdot \left(\dfrac{R_I}{R_F+R_I}\right)^2$
Binary-weighted with inverting summer	$\dfrac{R_F^2}{R_{eq}^2}$	$\left(\dfrac{R_F+R_{eq}}{R_F}\right)^2$	R_{eq}^2	1	$\left(\dfrac{R_{eq}}{R_F}\right)^2$	—	—

Table I lists the multiplication factor for each noise source's contribution to E_{no}^2 in the four topologies. The bit factors and R_{eq} for each topology are listed as well. Table II lists the multiplication factor for each noise source's contribution to E_{ni}^2 in the four topologies. E_{ni} is an equivalent noise voltage source located at the input to the resistor network whose squared value is equal to E_{no}^2 divided by the square of the voltage gain from the location of E_{ni} to the output. E_{no}^2/E_{ni}^2 for each topology is also given.

The analysis is simplified by considering that for equal digital inputs to the four topologies, the four analog outputs should be equal. Therefore, the following assumptions are made.

$$BF = 1. \tag{9}$$

R-2R networks:

$$R_F = R_I = R_A = R. \tag{10}$$

Binary-weighted network:

$$R_F = R_{eq} = \frac{R}{2}. \tag{11}$$

These assumptions make the transfer function from E_{ni} to E_{no} equal to 1 in all four topologies. Thus $E_{no}^2 = E_{ni}^2$. Equations (12)–(15) define E_{no}^2 for each topology. E_t^2 is equal to $4kTR$ in V^2/Hz.

R-2R with a voltage follower:

$$E_{no}^2 = E_n^2 + E_t^2 + I_n^2 R^2. \tag{12}$$

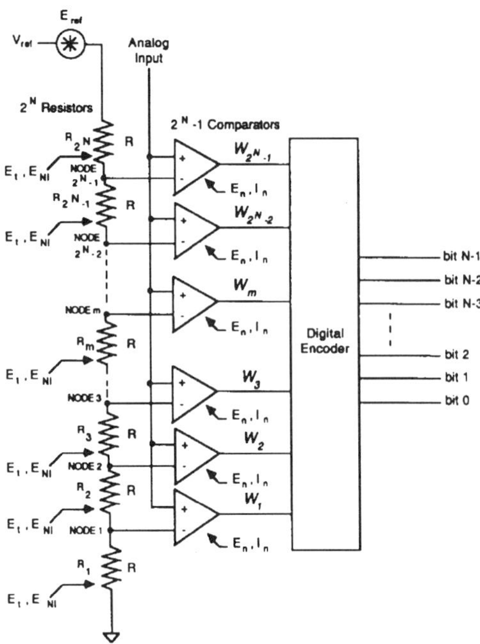

Fig. 5. Simplified N-bit flash A/D converter.

R-2R with an inverting amplifier:

$$E_{no}^2 = 4E_n^2 + 2E_t^2 + I_n^2 R^2. \tag{13}$$

R-2R with a noninverting amplifier:

$$E_{no}^2 = 4E_n^2 + 4E_t^2 + 4I_n^2 R^2. \tag{14}$$

Binary-weighted with an inverting summer:

$$E_{no}^2 = 4E_n^2 + \frac{3}{2}E_t^2 + \frac{1}{4}I_n^2 R^2. \tag{15}$$

These equations show that the R-2R network with a voltage follower produces the lowest noise level. The largest noise level comes from the R-2R network with a noninverting amplifier. The voltage follower topology produces the least noise not only because it has the fewest components but because its forward noninverting gain is equal to one. Thus there is no multiplication of the E_n and I_n terms contributing to E_{no}^2 as there are with the other topologies. These equations also show that if the amplifier's noise is predominantly due to I_n, then a binary-weighted network could possibly be better than the voltage follower topology because the I_n term is divided by four.

IV. Noise in a Flash A/D Converter

The flash A/D converter of Fig. 5 has a labeled positional index, m, to designate the mth node of the circuit. A voltage reference V_{REF} is divided into equally spaced threshold voltages at the inverting inputs of the comparators. The analog input voltage is simultaneously compared

to each of the threshold voltages. The voltage reference noise and the comparators' noise sources exhibit voltage and current noise with white noise spectra beyond a lower noise corner frequency and $1/f$ spectra at frequencies below the noise corner frequency. The resistors in the divider generate thermal noise as well as excess noise with a $1/f$ spectrum. A bit error will occur if the sum of the instantaneous noise and the analog input voltage exceed the quantization interval in which the analog input occurs. It is possible to include a noise contribution from the digital encoder, E_{DNL}, by adding another noise source whose value corresponds to the encoder noise at the output of the comparator of interest multiplied by the reverse transmission gain of the comparator. Since the reverse transmission gain of a comparator is quite small, E_{DNL} is usually not significant in determining whether or not an error will occur in a comparator's decision. Thus E_{DNL} is excluded from this analysis. E_{DNL} becomes significant at the output of the converter only if it exceeds the noise margin of the decoder circuitry.

The noise generated by a comparator operating in its transition region can be modeled in terms of an E_n noise voltage source and an I_n noise current source in the same way that the noise of an op amp is described. Noise at the input to any one comparator has contributions from that comparator's E_n source, all I_n sources, each resistor in the divider, and the voltage reference. The other comparators' E_n sources do not contribute to the equivalent noise at the input of the comparator of interest because of the high input impedance of the comparators associated with the other E_n sources. Using the positional index m to designate the mth node at the input to the mth comparator, (16) gives the total spectral density noise voltage pesent at the input to the mth comparator, $E_{ni_m}^2$.

$$E_{ni_m}^2 = \left[E_{n_m}^2 + \frac{E_{n_m}^2(f_{nc_E})}{f} \right] + E_{t_m}^2 + \sum_{j=1}^{2^N-1} E_m^2(I_{n_j})$$

$$+ \left(\frac{m}{2^N} \right)^2 \left[E_{REF}^2 + \frac{E_{REF}^2(f_{nc_{REF}})}{f} \right] + E_{ex_m}^2,$$

$$\text{for } m = 1,2,3,\cdots,2^N-1. \tag{16}$$

$E_{t_m}^2$ and $E_{ex_m}^2$ are the total noise contributions at the mth node caused by thermal noise and resistor current noise, respectively. They are described by (17) and (18) where E_t^2 is the thermal noise of a single resistor of resistance R, and E_{ex}^2 is the excess noise of a single resistor in the voltage divider.

$$E_{t_m}^2 = mE_t^2 \left(\frac{2^N - m}{2^N} \right)^2 + (2^N - m) E_t^2 \left(\frac{m}{2^N} \right)^2 \tag{17}$$

$$E_{ex_m}^2 = mE_{ex}^2 \left(\frac{2^N - m}{2^N} \right)^2 + (2^N - m) E_{ex}^2 \left(\frac{m}{2^N} \right)^2. \tag{18}$$

$E_m^2(I_{n_j})$ is the noise voltage squared at the mth node caused by the jth I_n noise current source. The I_n contri-

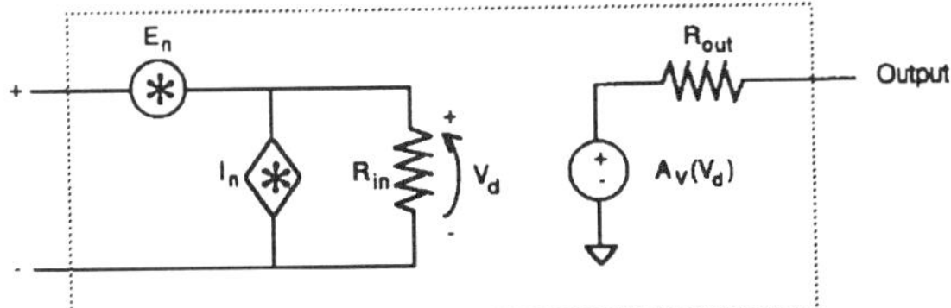

Fig. 6. Operational amplifier model.

butions to the noise at node m are given by

$$E_m^2(I_{n_j}) = \left[E_j^2(I_{n_j})\right]\left(\frac{m}{j}\right)^2 \qquad (19)$$

if $m \leq j$, or

$$E_m^2(I_{n_j}) = \left[E_j^2(I_{n_j})\right]\left[\frac{2^N - m}{2^N - j}\right]^2 \qquad (20)$$

if $m \geq j$. $E_j^2(I_{n_j})$ is the noise at the jth node caused by the jth I_n noise current source and is equal to

$$E_j^2(I_{n_j}) = \left[\frac{j(2^N - j)R}{2^N}\right]^2\left[I_{n_j}^2 + \frac{I_{n_j}^2(f_{nc_I})}{f}\right]. \qquad (21)$$

Determining which comparator has the greatest amount of equivalent noise at its input requires analysis of a complex function of m, N, frequency, and the spectra of the noise sources. As far as individual sources are concerned, the voltage reference contributes the most noise at the input to the 2^N-1th (i.e., the top) comparator. All other sources of noise contribute the most noise at the 2^{N-1}th (i.e., the middle) comparator.

V. D/A Spice Simulations

The circuits of Fig. 3 were simulated in PSpice [12] with $N = 8$ and $R = 10$ kΩ. For Fig. 3(b) and (d), R_F was equal to R. For Fig. 3(c), $R_A = R$, and $R_F = R_I = 1$ kΩ. The model of Fig. 6 was used for the op amps. E_n and I_n were modeled as white noise sources using the technique outlined in [13]. E_n was set to equal to 20 nV/$\sqrt{\text{Hz}}$, and I_n was set equal to 0.5 pA/$\sqrt{\text{Hz}}$; these values are typical for a 741 op amp [1], [14]. Table III compares the PSpice E_{no}^2 (V^2/Hz) values and the theoretical E_{no}^2 (V^2/Hz) values derived from the results in Table I for each of the four topologies. Table IV compares the SPICE E_{ni} (V/$\sqrt{\text{Hz}}$) values with those derived from the results in Table II.

Clearly, the correspondence between the theoretical results and the PSpice results is good ($< 0.51\%$ difference). Again, the R-2R network with a voltage follower was found to produce the lowest noise. The R-2R network with a noninverting amplifier has an E_{no}^2 that is nearly 3.4 times greater than the E_{no}^2 of the voltage follower configuration for the component values chosen. It should also be noted that the binary-weighted network with an inverting summer produces the second lowest

noise level for the component values chosen because of the reduction of the I_n term by a factor of four.

Table V lists the contributions to E_{no}^2 made by each of the terms for $N = 8$, $R = 10$ kΩ, $E_n = 20$ nV/$\sqrt{\text{Hz}}$, and $I_n = 0.5$ pA/$\sqrt{\text{Hz}}$. The E_n term is the dominant noise source in all four topologies. If a low noise amplifier such as the OP-37 [15], [16] with a much lower E_n is used, the noise is significantly reduced. The values in Table VI, calculated with an E_n of 3 nV/$\sqrt{\text{Hz}}$ and an I_n of 0.4 pA/$\sqrt{\text{Hz}}$, reflect this fact.

The low noise amplifier significantly reduces the noise figure of the circuits because the optimum source resistance (R_o) is closer to the equivalent resistance (R_s) seen by the I_n noise source. For R_s to be equal to R_o, R_s must equal E_n / I_n which is 7.5 kΩ for the low noise amplifier. For the voltage follower topology and $R = 10$ kΩ, the noisier op amp has a noise figure of 5.68 dB, while the low noise op amp has a noise figure of 0.63 dB. If the optimum source resistance is used (i.e., $R = 7.5$ kΩ in the voltage follower topology) with the low noise op amp, the noise figure is further reduced to 0.607 dB.

VI. A/D Spice Simulation

This A/D analysis is concerned only with finding the equivalent input noise at the input to each comparator and not with modeling of the A/D function. The op amp model of Fig. 6 was used in the PSpice simulation connected as a voltage follower in place of each comparator. To simulate the equivalent noise at the input to a comparator, it is only necessary that the E_n and I_n noise

TABLE III
E_{no}^2 Comparison

Topology	SPICE (V^2/Hz)	Theoretical (V^2/Hz)
R-2R with voltage follower	5.92E-16	5.89E-16
R-2R with inverting amplifier	1.96E-15	1.96E-15
R-2R with noninverting amplifier	2.00E-15	1.99E-15
Binary-weighted with inverting summer	1.77E-15	1.76E-15

TABLE IV
E_{ni} Comparison

Topology	SPICE (V/$\sqrt{\text{Hz}}$)	Theoretical (V/$\sqrt{\text{Hz}}$)
R-2R with voltage follower	2.44E-8	2.44E-8
R-2R with inverting amplifier	4.44E-8	4.44E-8
R-2R with noninverting amplifier	4.49E-8	4.49E-8
Binary-weighted with inverting summer	4.22E-8	4.22E-8

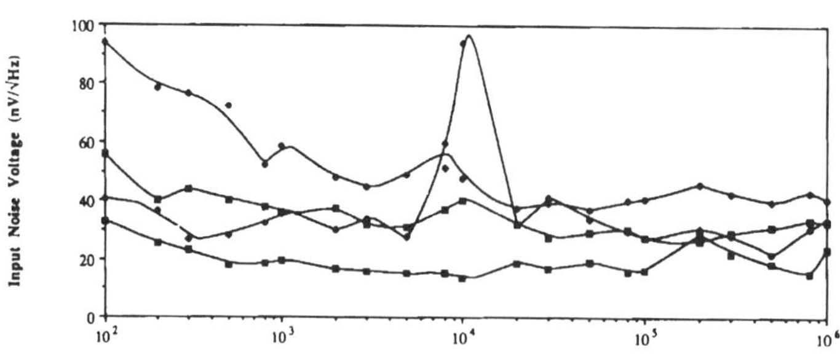

Fig. 7. Measured LM-339 E_n noise data.

TABLE V
INDIVIDUAL CONTRIBUTIONS TO E_{no}^2 (V^2/Hz)

Topology	E_n^2	I_n^2	Network E_t^2	E_t^2 of R_F	E_t^2 of R_I	E_t^2 of R_A
R-2R with voltage follower	4.0×10^{-16}	2.5×10^{-17}	1.64×10^{-16}	—	—	—
R-2R with inverting amplifier	1.6×10^{-15}	2.5×10^{-17}	1.64×10^{-16}	1.66×10^{-16}	—	—
R-2R with noninverting amplifier	1.6×10^{-15}	3.03×10^{-17}	1.64×10^{-16}	1.66×10^{-17}	1.66×10^{-17}	1.66×10^{-16}
Binary-weighted with inverting summer	1.59×10^{-15}	6.25×10^{-18}	8.25×10^{-17}	8.28×10^{-17}	—	—

TABLE VI
INDIVIDUAL CONTRIBUTIONS TO E_{no}^2 USING LOW NOISE AMP (V^2/Hz)

Topology	E_n^2	I_n^2	Network E_t^2	E_t^2 of R_F	E_t^2 of R_I	E_t^2 of R_A
R-2R with voltage follower	9.0×10^{-18}	1.6×10^{-17}	1.64×10^{-16}	—	—	—
R-2R with inverting amplifier	3.6×10^{-17}	1.6×10^{-17}	1.64×10^{-16}	1.66×10^{-16}	—	—
R-2R with noninverting amplifier	3.6×10^{-17}	1.94×10^{-17}	1.64×10^{-16}	1.66×10^{-17}	1.66×10^{-17}	1.66×10^{-16}
Binary-weighted with inverting summer	3.59×10^{-17}	4.0×10^{-18}	8.25×10^{-17}	8.28×10^{-17}	—	—

spectra of the comparator be the same as the E_n and I_n noise spectra of the op amp used in the PSpice simulations. Then, the noise at the output of each op amp in a voltage follower configuration is equal to the noise at the input of each op amp, and therefore *represents* the noise that would be present at the input to each comparator.

For the PSpice simulation, R was set equal to 100 Ω. E_{REF} was set equal to 100 nV/$\sqrt{Hz}$ with a noise corner frequency, $f_{nc_{REF}}$ of 10 Hz. These values are typical of manufacturers' voltage reference data [14]–[16]. Measured E_n data (shown in Figs. 7 and 8) on four units of two commercially available comparators, LM-339's and L-161's, yielded values from approximately 10 nV/$\sqrt{Hz}$ to 30 nV/$\sqrt{Hz}$, which are in the same general range of op amp E_n data. Thus a value of 20 nV/$\sqrt{Hz}$ was chosen for E_n in the simulation. E_n noise corner frequencies ranged from approximately 1 kHz for the LM-339's to frequencies below 100 Hz for the L-161's. Consequently, a value of 100 Hz was chosen for the E_n noise corner frequency f_{nc_E}. Since it was impossible to stabilize the comparator to measure I_n, I_n was set equal to 0.5 pA/$\sqrt{Hz}$, and the I_n noise corner frequency, f_{nc_I}, was set equal to 100 Hz.

The op amp was modeled as a subcircuit in the PSpice listing, and the subcircuit was called seven times in simu-

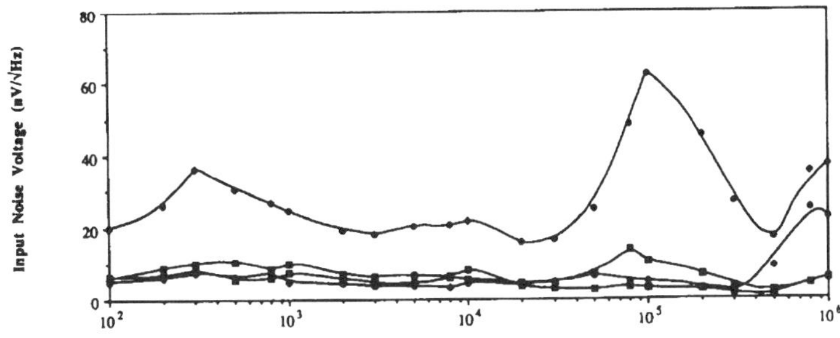

Fig. 8.   Measured L-161 E_n noise data.

TABLE VII
TOTAL E_{ni}^2 (10^{-16} V^2/Hz) AT EACH NODE VERSUS FREQUENCY

Frequency (Hz)	Node 1	Node 2	Node 3	Node 4	Node 5	Node 6	Node 7
1	1015	1491	1831	2036	2106	2041	1840
10	106.5	158.3	199.4	229.8	249.5	258.4	256.6
100	15.68	25.09	36.25	49.14	63.78	80.16	98.28
1000	6.594	11.77	19.93	31.08	45.22	62.34	82.45
10000	5.686	10.44	18.30	29.27	43.66	60.56	80.86
100000	5.595	10.30	18.14	29.09	43.17	60.38	80.70
1000000	5.586	10.29	18.12	29.07	43.15	60.36	80.69

TABLE VIII
CONTRIBUTIONS TO THE TOTAL E_{ni}^2 (10^{-16} V^2/Hz) AT THE MIDDLE
NODE FOR A 3-BIT FLASH A/D CONVERTER

Frequency (Hz)	Noise Source					Total E_{ni}^2
	E_{REF}	Thermal*	Excess	E_n	I_n^*	
1	275.0	—	1357	404.0	—	2036
10	50.03	—	135.7	44.01	—	229.8
100	27.53	—	13.57	8.005	—	49.14
1000	25.28	—	1.357	4.405	—	31.08
10000	25.06	—	—	4.045	—	29.27
100000	25.04	—	—	4.009	—	29.09
1000000	25.03	—	—	4.006	—	29.07

*Contributions from these noise sources were always less than 0.14×10^{-16} V^2/Hz and consequently represent insignificant additions to the total E_{ni}^2.

lating the case of $N = 3$. The noise sources were modeled using diodes and dependent sources as outlined in [13]. The subcircuit approach is a much simpler way to simulate the numerous noise sources while keeping E_n and I_n sources uncorrelated. The same result would have occurred if fourteen separate diode noise references were used for the E_n and I_n noise sources.

Table VII shows the square of the total amount of equivalent input noise, $(E_{ni})^2$, which is present at the inputs of all comparators as a function of m and frequency for a 3-bit flash A/D converter. The data in Table VII show that the node with the most noise is a function of frequency. However, as N increases, the function which

TABLE IX
CONTRIBUTIONS TO THE TOTAL E_{ni}^2 (10^{-16} V^2/Hz) AT THE TOP
NODE FOR A 5-BIT FLASH A/D CONVERTER

Frequency (Hz)	Noise Source					Total E_{ni}^2
	E_{REF}	Thermal*	Excess	E_n	I_n^*	
1	1032	—	41.08	404.0	—	1478
10	187.8	—	4.108	44.01	—	235.9
100	103.4	—	—	8.005	—	111.8
1000	94.92	—	—	4.405	—	99.38
10000	94.06	—	—	4.045	—	98.13
100000	93.98	—	—	4.009	—	98.01
1000000	93.98	—	—	4.006	—	98.00

*Contributions from these noise sources were always less than 0.42×10^{-16} V^2/Hz and consequently represent insignificant additions to the total E_{ni}^2.

TABLE X
CONTRIBUTIONS TO THE TOTAL E_{ni}^2 (10^{-16} V^2/Hz) AT THE MIDDLE NODE FOR A 5-BIT FLASH A/D CONVERTER

Frequency (Hz)	Noise Source					Total E_{ni}^2
	E_{REF}	Thermal	Excess	E_n	I_n*	
1	275.0	—	339.2	404.0	—	1020
10	50.03	—	33.92	44.01	—	128.3
100	27.53	—	3.392	8.005	—	39.10
1000	25.28	—	0.339	4.405	—	30.18
10000	25.06	0.133	—	4.045	—	29.29
100000	25.04	0.133	—	4.009	—	29.20
1000000	25.03	0.133	—	4.006	—	29.19

*Contributions from these noise sources were always less than 1.05% of the dominant noise source at a particular frequency and consequently represent insignificant additions to the total E_{ni}^2.

TABLE XI
CONTRIBUTIONS TO THE TOTAL E_{ni}^2 (10^{-16} V^2/Hz) AT THE TOP NODE FOR AN 8-BIT FLASH A/D CONVERTER

Frequency (Hz)	Noise Source					Total E_{ni}^2
	E_{REF}	Thermal*	Excess*	E_n	I_n*	
1	1092	—	—	404.0	—	1497
10	198.5	—	—	44.01	—	242.7
100	109.3	—	—	8.005	—	117.3
1000	100.3	—	—	4.405	—	104.8
10000	99.46	—	—	4.045	—	103.5
100000	99.36	—	—	4.009	—	103.4
1000000	99.36	—	—	4.006	—	103.4

*Contributions from these noise sources were always less than 0.66×10^{-16} V^2/Hz and consequently represent insignificant additions to the total E_{ni}^2.

TABLE XII
CONTRIBUTIONS TO THE TOTAL E_{ni}^2 (10^{-16} V^2/Hz) AT THE MIDDLE NODE FOR AN 8-BIT FLASH A/D CONVERTER

Frequency (Hz)	Noise Source					Total E_{ni}^2
	E_{REF}	Thermal*	Excess*	E_n	I_n	
1	275.0	—	—	404.0	865.0	1588
10	50.03	—	—	44.01	94.22	193.8
100	27.53	—	—	8.005	17.14	54.45
1000	25.28	—	—	4.405	9.432	40.51
10000	25.06	—	—	4.045	8.662	39.12
100000	25.04	—	—	4.009	8.585	38.98
1000000	25.03	—	—	4.006	8.577	38.96

*Contributions from these noise sources were always less than 4.3% of the dominant noise source at a particular frequency and consequently represent insignificant additions to the total E_{ni}^2.

determines the node with the most noise will depend less on frequency since E_{REF} which is greatest at the top node will have a greater contribution to E_{ni}^2. Table VIII shows the individual noise source contributions to $(E_{ni})^2$ at the middle comparator ($m = 4$) at node 4 as a function of frequency for a 3-bit flash A/D converter. Notice that the voltage reference noise E_{REF} is dominant at frequencies of 100 Hz and above even at the middle comparator where all of the other noise sources have their maximum contribution. This dominance by E_{REF} explains why the greatest equivalent input noise is found at the top comparator for frequencies above 100 Hz in Table VII. The smallest equivalent noise is found at the first ($m = 1$) comparator since the contributions from all of the noise sources are at a minimum for $m = 1$.

Tables IX and X, respectively, show the individual noise contributions to $(E_{ni})^2$ versus frequency for a 5-bit flash converter at the top ($2^N - 1$th = 31st) comparator where E_{REF} is most significant, and at the middle (2^{N-1}th = 16th) comparator where all other noise sources are most significant. Tables XI and XII show the individual noise contributions to $(E_{ni})^2$ versus frequency for an 8-bit flash converter.

The total amount of equivalent input noise increases as N increases, but the percentage increase in E_{ni} per 1 bit increase in N decreases as N becomes large. The dominant noise source in all cases in the white noise region of

E_{ni} is the voltage reference. E_n is also a significant noise contributor with respect to the other noise sources, particularly at the top comparator. At the middle comparator, the E_n contribution becomes less significant as N increases. Increasing N to 8 or more bits causes the I_n contribution to become more significant than the E_n contribution. Resistor excess noise is greatest when N is small (e.g., $N = 3$) because the dc voltage drop across the resistors in the divider is greater when N is small.

Since the largest noise is typically at the $2^N - 1$th (top) comparator, the greatest probability of a bit error occurs when the analog input voltage (V) is in the range of $[V_{REF}(2^N - 2)/2^N]$ to $[V_{REF}(2^N - 1)/2^N]$ where V_{REF} is the reference voltage. These are the lower and upper limits of the quantization interval at the top comparator, respectively. $[V_{REF}(2^N - 2)/2^N] + 1/2$ LSB is the center of the quantization interval. This is illustrated in Fig. 9. It is assumed that the noise has a Gaussian amplitude distribution which is centered at the analog input voltage, and that there is a uniform distribution of shift (E) of V from $[V_{REF}(2^N - 2)/2^N] + 1/2$ LSB. Fig. 10 shows the probability of a bit error for several shifts of V from the center of the quantization interval. The shaded areas of Fig. 10 correspond to the probability of a bit error for an associated value of E since the noise plus the analog input in the shaded regions exceed the boundaries of the quantization interval. Following Gordon's analysis [11],

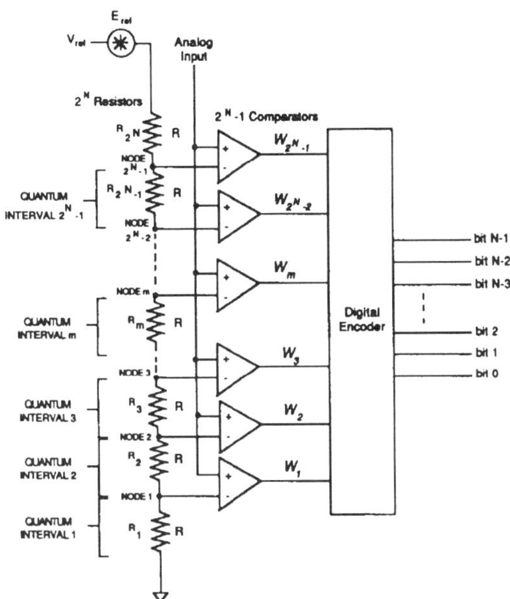

Fig. 9. Quantization interval boundaries for a flash A/D converter.

the probability of a bit error for an analog input anywhere in the quantization interval is given by

$$P_E = \int_{-(LSB)/2}^{+(LSB)/2} P(X|E) P(E)\, dE$$

$$= \frac{1}{(LSB)} \int_{-(LSB/2)}^{+(LSB/2)} \left[1 - \frac{1}{\sqrt{2\pi}} \int_{-LSB/2-E}^{LSB/2-E} \exp\left(\frac{-x^2}{2}\right) dx \right] dE. \quad (22)$$

P_E is the probability of a bit error, $P(E)\,dE$ is the probability of the analog input being shifted from half-scale $+ 1/2$ LSB by an amount between E and $E + dE$, and $P(X|E)$ is the probability that the Gaussian noise $P(X)$ will exceed the quantization interval when it is superimposed on V with its mean at E.

The integrand of the outer integral represents the area under the Gaussian curve that exceeds the quantization interval boundaries for a fixed value of E. For a particular value of E expressed as a fraction of an LSB, this area can be found using a look-up table of areas under a Gaussian curve. The outer integral is then evaluated numerically to give the total probability of a bit error for an analog input occurring anywhere within the quantization interval (i.e., $|E| \leq 1/2$ LSB). For the examples that follow, (22) was evaluated in a spreadsheet program using a rectangular approximation by dividing the $\pm 1/2$ LSB interval into 1000 equally spaced segments ($\Delta E = 0.001$).

For the case of $N = 3$ with a comparator having an E_n of 20 nV/$\sqrt{\text{Hz}}$, E_{ni} equals σ, which equals 89.83 nV/$\sqrt{\text{Hz}}$

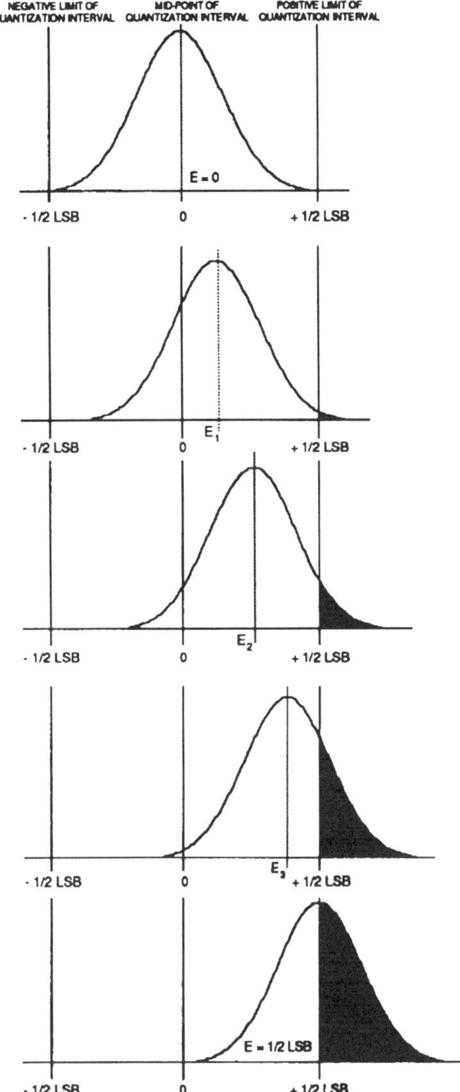

Fig. 10. Probability of bit errors for several shifts of analog input.

at the top comparator in the white noise portion of the E_{ni} spectrum. Assuming a rise time of 0.2 μs, which is typical of commercially available comparators, leads to a -3 dB-bandwidth of 1.7 MHz. Assuming a single pole response for the comparator gives a noise bandwidth of approximately 2.75 MHz. The broadband noise at the top comparator is then 148.97 μV. For $N = 5$, the broadband noise is 164.17 μV, and for $N = 8$, the broadband noise is 168.65 μV.

For a full-scale voltage (V_{FS}) of 10 V, (22) gives a probability of a bit error of 6.55% for $N = 3$, 7.32% for $N = 5$, and 7.60% for $N = 8$. If the full-scale voltage is reduced to 5 V, these probabilities become 28.7%, 31.8%, and 32.8%, respectively. If the dominant noise source,

E_{REF}, is reduced by 10% to 90 nV/$\sqrt{Hz}$, the spectral density input noises become 81.32 nV/$\sqrt{Hz}$ for $N = 3$, 89.52 nV/$\sqrt{Hz}$ for $N = 5$, and 91.93 nV/$\sqrt{Hz}$ for $N = 8$. The probabilities of a bit error for $V_{FS} = 10$-V range from approximately 5.8% to 7%.

For $E_{REF} = 100$ nV/$\sqrt{Hz}$, $R = 1$ kΩ, and $N = 8$, the contribution of the I_n noise sources becomes much more important. At the top comparator, however, the equivalent input spectral density noise only increases by 0.2 nV/Hz or 0.2%. The I_n contribution is most significant at the middle comparator where it is the dominant noise source for the circuit values chosen. At the middle comparator, E_{ni} is 306.9 nV/$\sqrt{Hz}$, which leads to a probability of a bit error of 25.4% at the middle comparator. Thus for large values of N and moderate values of R, it is important to examine the dominant noise sources at the top ($2^N - 1$th) comparator and at the middle (2^{N-1}th) comparator.

The probability of a bit error is reduced by reducing the ratio of the noise present at the comparator input to the resolution of the converter. This is done either by increasing V_{FS} at the expense of the converter's resolution, or by reducing the input noise level. Reduction of the noise entails using low values of R, using comparators with low E_n values if N is small (e.g., $N = 3$), using comparators with low I_n values if N is large (e.g., $N = 8$), and using a reference with the lowest possible noise.

VII. A-D-A Noise Analysis

The effect of noise on an 8-bit A-D-A system is examined for a full-scale voltage of 10 V, and an input voltage of 10 V minus 1.5 LSB, or 9.9414 V. The comparators of the flash A/D section are assumed to have a noise bandwidth of 2.75 MHz, and R is 100 Ω. The D/A converter uses a voltage follower and an R-2R network with $R = 10$ kΩ. The op amp is assumed to have a noise bandwidth of 1.57 MHz. The comparators and the op amp have $E_n = 20$ nV/$\sqrt{Hz}$ and $I_n = 0.5$ pA/$\sqrt{Hz}$ and noise corner frequencies of 100 Hz.

From the SPICE simulation, the rms noise present at the input to the top comparator is 101.7 nV/$\sqrt{Hz}$ or 168.65 μV over the noise bandwidth of the comparator. This magnitude of noise leads to a digital output of $(254)_{10}$ 99.7% of the time. Excluding noise in the D/A converter, this digital output corresponds to an analog voltage of 9.9219 V.

The rms noise voltage from the R-2R network and the voltage follower is equal to 30.6 μV over the 1.57 MHz noise bandwidth. Thus, 99.7% of the time, the D/A converter will add an amount of noise whose magnitude is less than or equal to 91.7 μV to the output of the A-D-A system. Thus, the sum effect of noise on the overall system will cause the output to be approximately 9.9219 V, and the noise of the D/A converter is negligible.

If the analog input is allowed to be anywhere within the LSB interval (i.e., between 9.9219 V and 9.9609 V), the digital outputs will be between $(253)_{10}$ and $(255)_{10}$ 99.7% of the time. The analog output range will be between 9.8828–9.9609 V. The maximum error caused by noise is ± 1LSB.

The error caused by fundamental noise in a D/A or A/D converter can be predicted using the methodology described in this paper. This error can then be included in a converter's error budget along with errors due to offsets, nonlinearities, etc., and a more accurate determination of a converter's true resolution can be made, which includes the effects of fundamental noise. Other noise sources (perhaps man-made noise) can easily be added to the noise model if their values at some point in the converter are known.

VIII. Conclusions

This paper describes a methodology by which noise sources in converters can be modeled using SPICE. This methodology will allow circuit designers to examine the effects of various noise sources on converter performance, determine dominant noise sources, compare different converter topologies with respect to noise, and suggest possible design changes that will reduce fundamental noise in a converter. The modeling methodology used herein is applicable to bipolar, MOS and GaAs converters and is easily adaptable to other converter topologies.

Two converter topologies were used to illustrate the noise analysis methodology. A noise analysis of an R-2R network and a binary-weighted network has been presented. Four different op amp summation schemes for voltage-referenced D/A converters were considered for the networks and noise contributions at the output of the converters compared. The noise contributions at the inputs to the comparators of a flash A/D converter were also analyzed.

The SPICE files used in the noise analysis constitute noise models of the particular converter topologies considered here. The noise parameters of the models are easily changed to adapt the model for more involved noise analysis of converters (e.g., to add other noise sources than those considered here) or to match the models' predicted noise behavior to actual converter noise performance. The SPICE noise models were used to identify dominant noise sources within a given topology and the effects of changing noise parameters of various noise sources in the model were examined to predict and evaluate the converter noise performance improvement that could be obtained with low-noise design techniques.

The results of D/A noise model simulations with typical component values and noise characteristics show that the R-2R network with a voltage follower produces the lowest noise. The R-2R network with a noninverting amplifier has an E_{no}^2 which is nearly 3.4 times greater than the E_{no}^2 of the voltage follower configuration for the component values chosen. The binary-weighted network with an inverting summer produces the second lowest noise level for the component values chosen because of the reduction of the I_n term by a factor of four. Using a low-noise op amp significantly reduces the output noise of a D/A converter since op amp E_n is by far the dominant noise source among the sources considered.

In the flash A/D converter, fundamental noise is dominated by voltage reference noise while the importance of other noise sources is dependent on the number of bits, frequency, the node of interest, and the converter topology.

For example, for the 3-bit flash converter, it was found that thermal noise and comparator I_n noise were not significant. For an 8-bit flash converter, I_n becomes very significant along with the reference noise while E_n becomes insignificant, and thermal noise remains insignificant. For any number of bits, E_n contributes noise only at the node of interest, while the other noise sources contribute noise at all nodes. Furthermore, reference noise is most significant at the top node and least significant at node 1. All other noise sources are most significant at the middle node of the converter.

The results of this analysis show that inherent noise is an important consideration in the design of A/D and D/A converters. The methodology and models described in this paper can be used in conjunction with Gordon's probabilistic error analysis [10], [11] to predict more accurately the probability of a bit error in a converter. Thus the effects of noise can be taken into account in determining the true resolution of a converter.

Acknowledgment

The authors wish to thank Professor P. E. Allen, Greg Fisher, and the other reviewers for their comments on this paper, and Dr. Tom Brewer for his help in making the laboratory measurements. Also, the authors gratefully acknowledge the support of AT&T Bell Laboratories for the purchase of test equipment for noise characterization of op amps and comparators.

References

[1] C. D. Motchenbacher and F. C. Fitchen, *Low-Noise Electronic Design*. New York: Wiley, 1973.

[2] M. J. Buckingham, *Noise in Electronic Devices and Systems*. New York: Halstead Press, 1983.

[3] R. Rohrer, L. Nagel, R. Meyer, and L. Weber, "Computationally efficient electronic-circuit noise calculations," *IEEE J. Solid-State Circuits*, vol. SC-6, pp. 204–213, Aug. 1971.

[4] R. C. Jaeger and J. Brodersen, "Low-frequency noise sources in bipolar junction transistors," *IEEE Trans. Electron Devices*, vol. ED-17, pp. 128–134, Feb. 1970.

[5] A. J. Brodersen, E. R. Chenette, and R. C. Jaeger, "Noise in integrated-circuit transistors," *IEEE J. Solid-State Circuits*, vol. SC-5, pp. 63–66, Apr. 1970.

[6] R. G. Meyer, L. Nagel, and S. K. Lui, "Computer simulation of 1/f noise performance of electronic circuits," *IEEE J. Solid-State Circuits*, vol. SC-8, pp. 237–240, June 1973.

[7] A. Bilotti and E. Mariani, "Noise characteristics of current mirror sinks/sources," *IEEE J. Solid-State Circuits*, vol. SC-10, pp. 516–523, Dec. 1975.

[8] G. Nicollini, D. Pancini, and S. Pernici, "Simulation-oriented noise model for MOS devices," *IEEE J. Solid-State Circuits*, vol. SC-22, pp. 1209–1212, Dec. 1987.

[9] F. N. Trofimenkoff and O. A. Onwuachi, "Noise performance of operational amplifier circuits," *IEEE Trans. Education*, vol. E-32, pp. 12–16, Feb. 1989.

[10] B. M. Gordon, "Noise-effects on analog to digital conversion accuracy—Part 1," *Computer Design*, pp. 65–76, Mar. 1974.

[11] ____, "Noise-effects on analog to digital conversion accuracy—Part 2," *Computer Design*, pp. 137–145, Apr. 1974.

[12] P. W. Tuinenga, *SPICE: A Guide to Circuit Simulation and Analysis Using PSPICE*. Englewood Cliffs, NJ: New Jersey: Prentice-Hall, 1988.

[13] G. J. Scott and T. M. Chen, "Addition of excess noise in SPICE circuit simulations," in *Proc. IEEE Southeastcon Conf.*, vol. 1, pp. 186–190, 1987.

[14] National Semiconductor Corp., *Linear Databook*, vol. 2, 1988.

[15] Analog Devices, *Integrated Circuits Databook*, vol. 1, 1988.

[16] Precision Monolithics, Inc., *Analog IC Databook*, 1988.

Part VIII

Digital-to-Analog Converters

DIGITAL-TO-ANALOG converters (DACs) have almost unlimited applications—for example, in closed-loop process control systems, space telemetry systems, high-definition digital TV, computer graphics, digital oscilloscopes, and many others. As a result, their development and perfection proceeded for nearly half a century. However, because of their complexity, widespread use did not occur until their realization in the form of IC chips.

Since the early DAC chips were rather slow, their applications in DDFS were exceptional. Nevertheless, technological progress soon overcame this difficulty. If Dooley [VIII-1] indicated for a 10-b DAC the settling time 1.5 μs, seven years later Saul et al. [VIII-2] reported for an 8-b device a 5 ns value. In the mid 1980s, the range of clock frequencies was pushed to 500 MHz [VIII-3], and soon the threshold of 1 GHz was crossed [VIII-4, VIII-5].

The first survey paper is specially written by the editor for this volume. This article deals with such practical problems as codes, signal-to-noise ratios, linearity, glitches, and deglitching circuits, and presents a wealth of references.

As an introduction to the problem of monolithic DAC (and for those who want to get more in-depth information), we reprint one of the first papers by Kelson, Stellrecht, and Perloff (1973), written more than 20 years ago. Its major advantage is a lucid discussion of nearly all background circuits.

The tutorial introduction proceeds with the paper by Schoeff (1979). After a short review of technology, particularly of diffused resistors, the author suggests building a 12-bit DAC segmented from three different types of current sources. In this way, he alleviates the linearity problem. With the assistance of the offset binary operation, an output swing of ± 10 V was achieved. The settling time to ± 0.01 percent is 0.25 μs.

Microprocessor-compatible 8-bit DAC by Amazeen et al. [VIII-6] operating on a single $+5$ V supply (not reprinted here) is nevertheless suggested as continuation of the above introductory papers for interested readers.

An 8-bit, 5 ns was described by Saul, Ward, and Fryers (1980). The suggested multiple-current source approach uses for MSBs four, two, and one equal value and a half-value current sources. Four LSBs are generated in a second current source. By careful optimization, a respectably high speed (500 MHz) was achieved [VIII-5].

Maio and coauthors (1985) realized a 500-MHz, 8-bit DAC (3 MSB segmented and the remaining 5 LSB equal current sources with the R-2R ladder network), with the assistance of a new high-speed conversion technique—the data multiplexing method.

The reprinted paper by Kamoto and Shinagawa (1988) deals with further progress of 8-bit monolithic DAC. They report the maximum conversion rate of 1 GHz (by using transistors with f_T approximately 18 GHz), glitch "energy" 2 psV, and 10-bit linearity accuracy without trimming.

The paper by Goodenough (1994) [VIII-7] presents the state of the art: a 12-bit DAC that runs at 1 GHz and puts 20 mA into 50 Ω. Special SiGe heterojunction bipolar transistors combine f_T beyond 100 GHz with beta-Early-voltage products exceeding 48000. The larger the Early voltage is, the smaller the transistor output conductance [VIII-6, VIII-8].

REFERENCES

[VIII-1] D. J. Dooley. "A complete monolithic 10-b D/A converter." *IEEE Journal of Solid-State Circuits,* SC-8, pp. 404–8, December 1973.

[VIII-2] P. H. Saul, P. J. Ward, and A. J. Fryers. "An 8-bit, 5 ns monolithic D/A converter subsystem." *IEEE Journal of Solid-State Circuits,* pp. 1033–39, December 1980 (Reprinted in this Part).

[VIII-3] K. Maio, S.-I. Hayashi, M. Hotta, T. Watanabe, S. Ueda, and N. Yokozawa. "A 500-MHz 8-bit D/A converter." *IEEE Journal of Solid-State Circuits,* pp. 1133–36, December 1985.

[VIII-4] K.-C. Hsieh, T. A. Knotts, G. L. Baldwin, and T. Horak. "A 12-bit 1-Gword/s GaAs digital-to-analog converter system." *IEEE Journal of Solid-State Circuits,* SC-22, pp. 1048–54, December 1987.

[VIII-5] P. H. Saul and D. G. Taylor. "A high speed direct frequency synthesizer." *IEEE Journal of Solid-State Circuits,* SC-25, pp. 215–19, February 1991. (Reprinted in Part III.)

[VIII-6] B. E. Amazeen, P. R. Hollway, and D. G. Mercer. "A complete single-supply microprocessor-compatible 8-bit DAC." *IEEE Journal of Solid-State Circuits,* SC-14, pp. 1059–70, December 1980.

[VIII-7] F. Goodenough. "12-bit DAC runs at 1 GHz, puts 20 mA into 50 Ω." *Electronic Design,* pp. 47–48, 51, 52, February 1994.

[VIII-8] F. A. Lindholm and D. J. Hamilton. "Incorporation of the Early effect in the Ebers-Moll model." *Proceedings of the IEEE,* vol. 59, pp. 1377–78, September 1971.

Digital-to-Analog Converters

VĔNCESLAV F. KROUPA

I. Introduction

WITH the advent of digital technologies, the analog-to-digital and digital-to-analog converters made their way into practical applications [1,2]. However, early devices made as networks, consisting of individual components, were too bulky and too slow for widespread use. Their victorious way from the early 1970s is connected with the introduction of integrated circuits (ICs). For selection of a good DAC for DDFS application, we have to consider three major criteria: bit resolution, accuracy, and speed. In addition, linearity, temperature stability, coding, and power consumption must be considered in accordance with intended applications. Speed and power consumption depend to a very high degree on the digital logic used.

II. Digital Codes

In digital-to-analog converters (DACs), as well as in analog-to-digital converters (ADCs), binary coding systems are prevailing. Any real number can be expressed as a systematic fraction with the base 2 (in contrast to our everyday calculations that are based on systematic fractions with the base 10), that is,

$$\xi_0 = a_n 2^n + a_{n-1} 2^{n-1} + \cdots + a_1 2 + a_0 2^0 + a_{-1} 2^{-1} + \cdots \tag{1}$$

where $a_n, a_{n-1}, \ldots$ are equal either to 1 or zero. In actual digital systems, the number of useful places or bits is limited; for example, in Table 1 representing the unipolar binary code we have chosen, $n = 4$.

A. Unipolar Straight Binary Code

Unipolar binary coding is the simplest one for understanding. In Table 1 we compare 4-bit binary numbers with decimal equivalents and analog output voltage for the full-scale swing of VFS = V_{pp} = 1 V. Note that the largest output voltage is by one LSB (least significant bit) smaller than V_{pp} (peak to peak). The output wave has a staircase sawtooth wave form.

B. Bipolar Offset and Complementary Offset Binary Codes

In DDFS applications, the desired output voltage of the DAC should be bipolar. This requirement is met with the offset binary codes. Examination of Table 2 reveals that the MSB is responsible for the sign of the output voltage. As long as the

(This paper was specially prepared for this volume.)

TABLE 1. UNIPOLAR STRAIGHT BINARY CODE (USB).

Binary Code Setting				Decimal Value	Output Voltage
a_3	a_2	a_1	a_0	$X/2^n$	$V_{pp} = 1V$
0	0	0	0	0	.0000
0	0	0	1	1/16	.0625
0	0	1	0	2/16	.1250
0	0	1	1	3/16	.1875
0	1	0	0	4/16	.2500
0	1	0	1	5/16	.3125
0	1	1	0	6/16	.3750
0	1	1	1	7/16	.4375
1	0	0	0	8/16	.5000
1	0	0	1	9/16	.5625
1	0	1	0	10/16	.6250
1	0	1	1	11/16	.6875
1	1	0	0	12/16	.7500
1	1	0	1	13/16	.8125
1	1	1	0	14/16	.8750
1	1	1	1	15/16	.9375
0	0	0	0	0	.0000

TABLE 2. BIPOLAR OFFSET BINARY CODE (BOB).

a_3	a_2	a_1	a_0	$V_{pp} = 1$ V
0	0	0	0	-0.5
0	0	0	1	-0.4375
0	0	1	0	-0.375
0	0	1	1	-0.3125
0	1	0	0	-0.25
0	1	0	1	-0.1875
0	1	1	0	-0.125
0	1	1	1	-0.0625
1	0	0	0	0.0
1	0	0	1	0.0625
1	0	1	0	0.125
1	0	1	1	0.1875
1	1	0	1	0.25
1	1	0	1	0.3125
1	1	1	0	0.375
1	1	1	1	0.4375

MSB is zero, the output voltage is negative; after changing the MSB into the logic 1 (all other bits being at logic 0), the output voltage is zero and positive for all other settings of smaller bits. The consequence is that the peak positive voltage is by one LSB smaller than the ideal value $V_{pp}/2$. Comparison of Tables 2 and 3 reveals that the relationship between bipolar offset binary code (BOB) and complementary offset binary code (COB) coding schemes is one's complement (all bits inverted) of the other.

Application of offset binary codes generates a staircase output sine wave that exhibits a DC shift (cf. Fig. 1). However, with the assistance of additional output circuits, a true zero

TABLE 3. COMPLEMENTARY OFFSET BINARY CODE (COB).

a_3	a_2	a_1	a_0	$V_{pp} = 1$ V
1	1	1	1	-0.5
1	1	1	0	-0.4375
1	1	0	1	-0.375
1	1	0	0	-0.3125
1	0	1	1	-0.25
1	0	1	0	-0.1875
1	0	0	1	-0.125
1	0	0	0	-0.625
0	1	1	1	0.0
0	1	1	0	0.0625
0	1	0	1	0.125
0	1	0	0	0.1875
0	0	1	1	0.25
0	0	1	0	0.3125
0	0	0	1	0.375
0	0	0	0	0.4375

TABLE 4. BINARY TWO'S COMPLEMENT CODE (BTC).

a_3	a_2	a_1	a_0	$V_{pp} = 1$ V
1	0	0	0	-5.000
1	0	0	1	-4.375
1	0	1	0	-3.750
1	0	1	1	-3.125
1	1	0	0	-2.500
1	1	0	1	-1.875
1	1	1	0	-1.250
1	1	1	1	-0.625
0	0	0	0	0.0
0	0	0	1	0.0625
0	0	1	0	0.1250
0	0	1	1	1.875
0	1	0	0	2.500
0	1	0	1	3.125
0	1	1	0	3.750
0	1	1	1	4.375

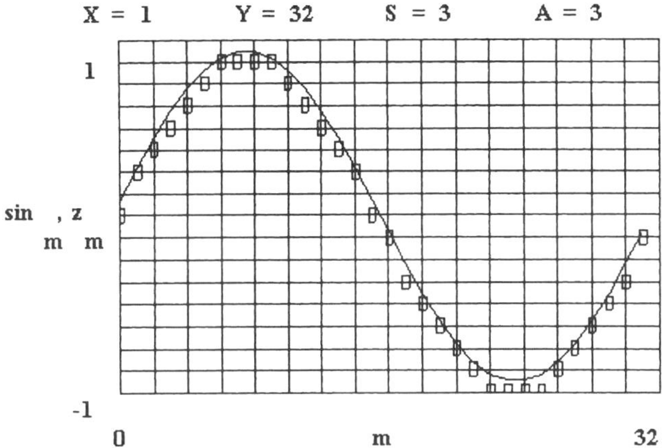

Fig. 1. Idealized and actual sine-wave output waveform.

TABLE 5. COMPLEMENTARY TWO'S COMPLEMENT CODE (CTC).

a_3	a_2	a_1	a_0	$V_{pp} = 1$ V
0	1	1	1	-5.000
0	1	1	0	-4.375
0	1	0	1	-3.125
0	0	1	1	-2.500
0	0	1	0	-1.875
0	0	0	1	-1.250
0	0	0	0	-0.625
1	1	1	1	0.000
1	1	1	0	0.625
1	1	0	1	1.250
1	1	0	0	1.875
1	0	1	1	2.5
1	0	1	0	3.125
1	0	0	1	3.750
1	0	0	0	4.375

TABLE 6. COMPARISON OF DIFFERENT DIGITAL CODES
AND ANALOG OUTPUT VOLTAGES OF DAC.

Digital input		Analog output		
MSB LSB		CSB	COB	CTC
000000000000		+Full Scale	+Full Scale	-1LSB
011111111111		+1/2 Full Scale	Zero	$-$Full Scale
100000000000		+1/2 Full Scale-1LSB	-1LSB	$-$Full Scale
111111111111		Zero	$-$Full Scale	Zero

output can be realized; in individual cases, we recommend consulting the respective company's "Application Notes" or DAC "Data Books."

C. Two's Complement Codes

In instances where computational operations are needed or microprocessors are used, the two's complement code is very useful (see Tables 4 and 5). Generally, when used, a logic inverter is connected ahead of the most significant bit (MSB).

A comparison of the different digital codes and the corresponding analog output voltage of DAC is shown in Table 6.

Examination of Tables 2 and 3 or 4 and 5 reveals that the MSB (most significant bit) is responsible for the sign of the output voltage. As long as the MSB is zero, the output voltage is negative. Its change into 1 (all other bits being zero) indicates the bipolar zero and positive output voltage for all other settings of smaller bits. The consequence is that the peak positive voltage is by one LSB smaller than the ideal value V_{pp}.

Without additional measures, however, bipolar outputs are difficult to obtain. Dooley [3] has suggested one solution for the sign magnitude codes by using the MSB as the sign bit (see

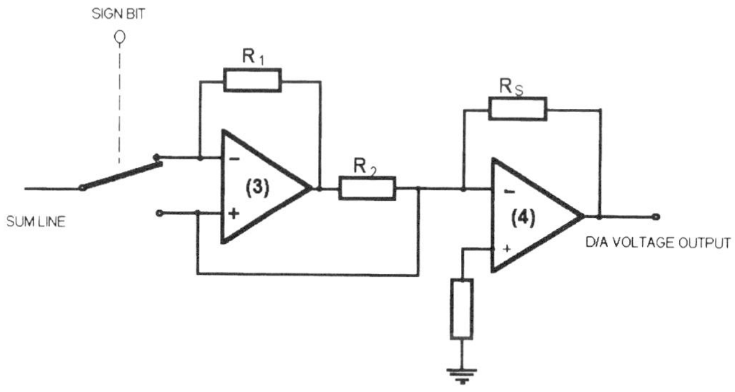

Fig. 2. Block diagram of the sign bit current-inverting amplifier.

Fig. 2). The current sum line is switched under logic control either to inverting or noninverting input of the sign bit OP amp (3). When the switch is connected to the noninverting terminal, the sum line current is drawn directly from the output OP amp (4) through the sum resistor Rs. The DAC will now be in the positive sign condition. The state change of MSB switches the sum line current to the inverting terminal of OP-3, and in the instance that R1 is equal to R2 the DAC will be in the negative sign condition. The difficulty is the accuracy problem. Dooley [3] has suggested a solution by changing the voltage switching to current steering.

Another approach for the sign inversion is shown in Fig. 3; with all current switches off, V_{out} is equal to $+1$ V. In the case that all current switches are on, $I_{\text{sum}} = -40$ mA, and the output voltage is -1 V.

In DDFS we generally require a sine-wave (or triangular-wave) output. To achieve this goal, we must combine, for example, the bipolar offset binary code (BOB) with the complementary offset binary code (COB). The relationship between COB and BOB is that each coding scheme is the one's complement (all bits inverted) of the other; compare Table 2 with Table 3.

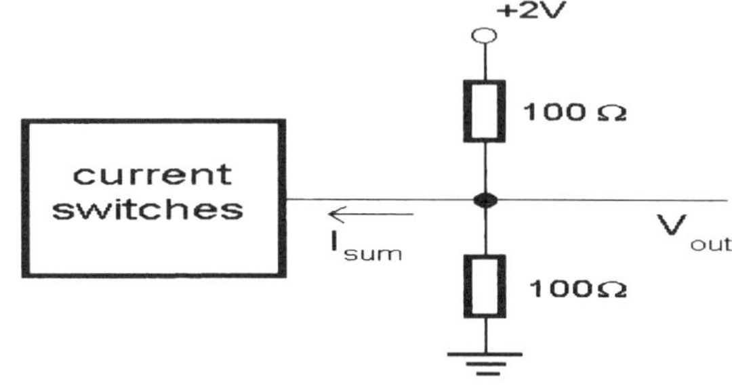

Fig. 3. Another solution for bipolar offsetting.

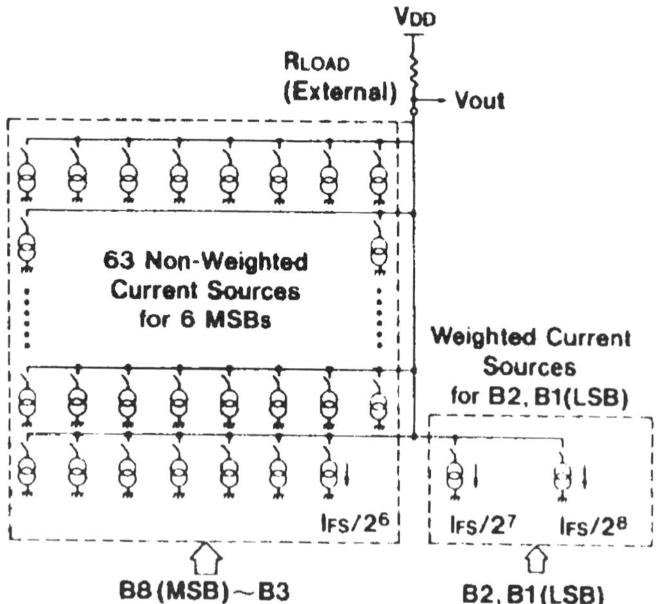

Fig. 4. A two-stage segmented DAC (from [4]).

D. Thermometer Code

In segmented DACs, where the output current for several MSBs is generated in nonweighted current cells, special decoding circuits are necessary. One implementation suggested by Nojima and Gendai [4] is shown in Fig. 4. The input binary code is changed into a steadily increasing one, the so-called thermometer code (see Fig. 5).

E. Johnson Code

Application of the Johnson counter in DDFS is unique [5]. Nevertheless, we record this code in Table 7 together with the effective output. The principle block diagram is shown in Fig. 6.

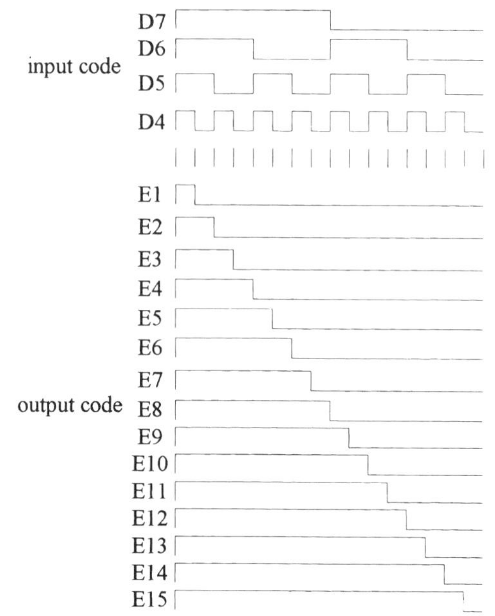

Fig. 5. Four-bit decoder for the MSB part. D4–D7, are the input binary bits; E1–E15 are the output thermometer code.

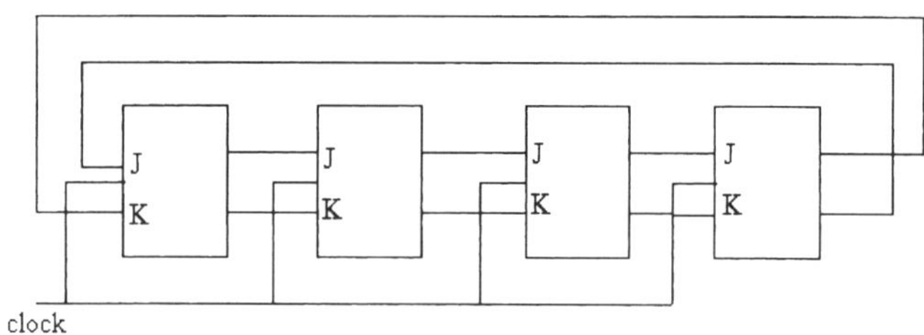

Fig. 6. Principle block diagram of the Johnson counter.

TABLE 7. JOHNSON CODE

States of the counter				Linear output
0	0	0	0	0
1	0	0	0	1
1	1	0	0	2
1	1	1	0	3
1	1	1	1	4
0	1	1	1	3
0	0	1	1	2
0	0	0	1	1
0	0	0	0	0

III. OUTPUT WAVES

All the binary codes discussed until now supply at the output of DAC sawtooth waves, either with a DC component (unipolar binary code) or with an odd symmetry around zero (offset binary codes). However, the most desired output of DDFS is periodic and symmetrical sinusoidal signal; in some instances, a triangular one is acceptable.

A. The Triangular-Wave Output

We will start with the triangular-wave output, since its generation does not require a special memory circuit. For the demonstration we chose a 4-bit DAC and a 5-phase bit accumulator. By using the bipolar offset code, we start with the most negative output voltage -5.0 V for all zeros (cf. Table 2). After the fifteenth clock pulse, the output voltage is the most positive $+5.0$ V $-$ LSB $= +4.375$ V; the sixteenth pulse changes the MSB to "1" and the output code to the "complementary offset binary" one. The result is that the slope of the output signal is changed and the output voltage decreases till -5.0 V—where all bits are in the "1" state. The next pulse, the 32nd, changes all 5 bits to "0," and consequently the complementary offset binary code, back to the ordinary bipolar offset binary. The output wave is shown in Fig. 7. Its investigation reveals a small DC shift ($-1/2$ LSB) and shows that the spurious modulation due to the quantization consists of sawtooth waves with periods equal to the clock period.

B. The Sine-Wave Output

Generation of the sine-wave output requires a lookup table where sine values, corresponding to the phase arguments, are stored. Generally, we refer only to the first quadrant in order to keep the ROM memory as small as possible, particularly when it forms part of the IC chip. The second MSB serves for quadrant discrimination, and the first MSB, for changing the bipolar offset binary code into the complementary one, is used as a sign bit. See, for example, the block diagram in the paper by Sunderland et al. [6]. This arrangement introduces an additional shift caused by the asymmetry of the bipolar binary codes used; see Fig. 1.

C. Arbitrary Waveform Output

By replacing the sine lookup table with a programmable random access memory (RAM), we can generate arbitrary output waveforms, for example, [7]. Setting from the keyboard would be a tedious task. Use of special software, provided by manufacturers, as well as personal computers, solves the problem.

IV. BASIC CIRCUITS

In this section we will discuss digital-to-analog converters suitable for the ICs production.

1. Voltage-Switching Converters. The converter reproduced in Fig. 14 in the introductory part is of the voltage-switching type and uses an R-2R resistor ladder network.

This 8-bit DAC (an early production by Plessey-Semiconductors type ZN426-8 [8]) is simple but rather slow, with the settling time of 1 μs.

2. Current-Switching Converters. Since, for several reasons, currents may be switched more rapidly (at least by one order) than voltages today, commercial DACs are produced in this way. More detailed discussion was provided by Kelson et al. 20 years ago [9]. Also, in these integrated D/A converters, an R-2R ladder network is used to generate binary-weighted currents.

A. Summing Binary-Weighted Currents

The binary weighting approach is shown in Fig. 8. This type of switching is considered one of the fastest current-switching methods. The difficulty is the necessity of proper scaling transistors in accordance with weighted currents. A remedy was suggested by Schultz [10], and another by Shoji and Saari [11].

The block diagram of commercial DAC of this type is reproduced in Fig. 12 in the introductory paper.

3. Binary Attenuation of Equal-Valued Currents. Binary attenuation of equal-valued current removes the necessity of transistor scaling. The principle is shown in Fig. 9. This technique is widely used; see, for example, Fig. 13 in Part I [12].

4. Segmentation of DAC. Another approach to constructing DAC is segmentation, which makes it possible to increase either speed or resolution and reduce spurious glitches.

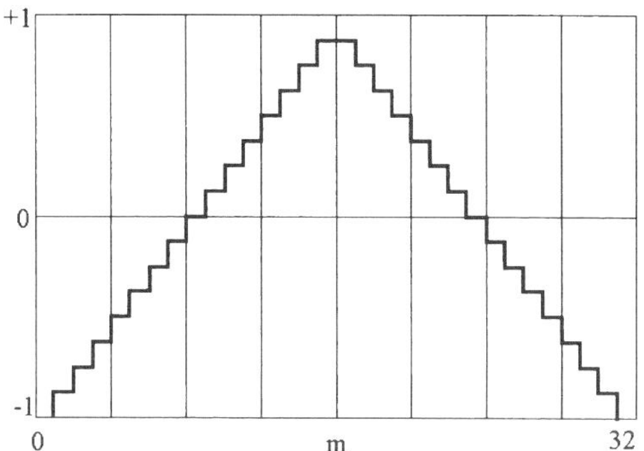

Fig. 7. The triangular wave generated with the assistance of the bipolar offset binary and complementary offset binary code.

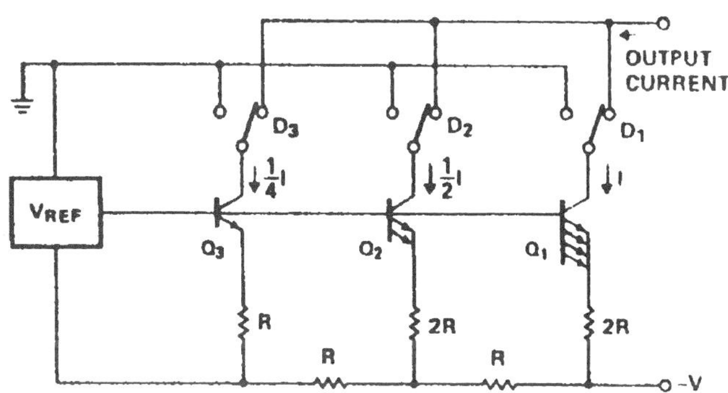

Fig. 8. Block diagram of a 3-bit DAC with binary-weighted currents.

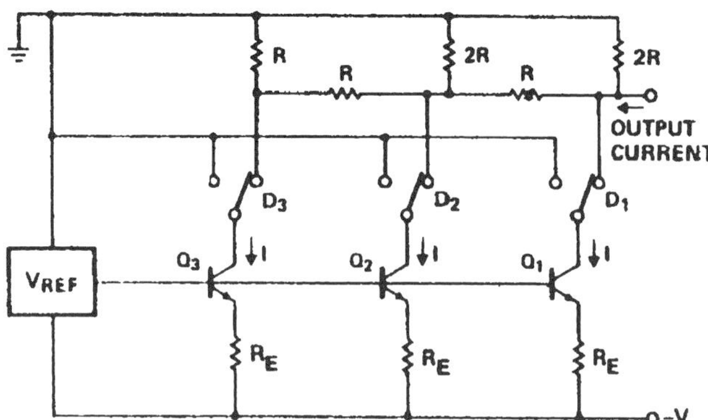

Fig. 9. Block diagram of a 3-bits DAC based on binary-attenuated, equal-valued currents (from [9]).

A two-stage segmented DAC is shown in Fig. 4. It is composed of a set of nonweighted current cells and an ordinary part of weighted current sources. The 6 MSB are decided in accordance with the thermometer code; for example, if they are set to the decimal number of 30, 30 current sources are switched to the sum line. The output is obtained from the voltage drop across the external load resistor. The remaining current proportional to the 2 LSB is supplied by two weighted current sources. The advantage is that the mismatch problem is relaxed and glitches are reduced.

The principle is used, for example, by Sony Corporation on an 8-bit 40-MHz DAC-type CXD 1171M [13]; see also Fig. 14 in the introductory paper [8]. These segmentations were also discussed in more detail by Miki et al. [14].

A three-stage segmented DAC was suggested by Schoeff in a paper reprinted in this part [15]. The principle was applied by Burr-Brown Company in its 12-bit 500-MHz DAC-type DAC 650 [16]. There are seven current sources for the 3 MSB. The following current sources for bits 4 to 8 are scaled down in binary fashion and switched directly to the output of the R-2R ladder. Bits 9-12 are fed to the laser-trimmed R-2R ladder for proper scaledown. In this way, further minimization of output glitches was achieved.

V. Properties Important for DDFS Applications

The aim of both designers and users of DDFS is to have an undisturbed sinusoidal output signal with the desired frequency only. Such an aim is not realistic. We have already seen the problem between the input binary codes (cf. Tables 2 and 3) and the output voltage. However, there are other difficulties too.

A. Signal-to-Noise Ratio (S/N) Due to Quantization

The change of the steadily increasing phase into a periodic sine wave requires lookup tables with stored sine values in finite binary words. However, generally $\sin x$ is an irrational number, and the respective binary fraction is an infinite one.

Practically all DDFSs are working in a binary number system, that is, using systematic fraction series expansion with the base 2. Thus, the sine waves supplied by lookup tables are of the type

$$\sin x = \frac{a_1}{2} + \frac{a_2}{2^2} + \cdots + \frac{a_{S-1}}{2^{S-1}} + \frac{a_S}{2^S} + \frac{a_{S+1}}{2^{S+1}} + \cdots + \text{in inf} \qquad [2]$$

where a_i ($i = 1, 2, \ldots,$ in inf.) have value of either 1 or 0.

To keep the memory in DDFS needed for ROM lookup tables in reasonable dimensions and cost, we must cut the series (2), say at the member, $a_S/2^S$. The resulting error can be nearly as large as

$$\frac{1}{2^{S+1}} + \frac{1}{2^{S+2}} + \cdots + \text{in inf} = \frac{1}{2^{S+1}} \frac{1}{1 - \frac{1}{2}} = \frac{1}{2^S} \qquad [3]$$

or zero in instances where all $a_{s+1}, a_{s+2}, \ldots$ are equal to zero. (In that case, the number in eq. (2) would be a rational one, for example, $\sin 30° = 1/2$.)

After rounding off the expansion (2) to "S" places, we can reduce the error to

$$\pm \frac{1}{2} \frac{1}{2^S} = \pm \frac{1}{2^{S+1}} \qquad [4]$$

By assuming a linear distribution of the truncation error in the interval error from

$$-\frac{1}{2} 2^{-S} \quad \text{to} \quad +\frac{1}{2} 2^{-S} \qquad [5]$$

the probability, $p(e)$, is

$$p(e) = 2^S \quad \text{inside and} \quad p(e) = 0 \quad \text{outside of the interval} \qquad [6]$$

In this case the variance of the truncation error is

$$\sigma_e^2 = \frac{1}{12} 2^{-2S} \qquad [7]$$

Now we face the problem of changing the digital information supplied by the sine lookup table into a respective analog value. By assuming an ideal DAC with the resolution D-bits in the range of V_{pp} (peak-to-peak) volts, the LSB is

$$\text{LSB} = \frac{V_{pp}}{2^D} \qquad [8]$$

The random voltage error will occupy the range from $-1/2$ LSB to $+1/2$ LSB. Consequently, its variance will be

$$\sigma_{e,\text{DAC}^2} = \frac{1}{12}(\text{LSB})^2 = \frac{1}{12} \left(\frac{V_{pp}}{2^D} \right)^2 \qquad [9]$$

and the ideal signal-to-noise ratio is

$$\frac{S}{N} = \frac{\frac{1}{2} \left(\frac{V_{pp}}{2} \right)^2}{\frac{1}{12} \left(\frac{V_{pp}}{2^D} \right)^2} = \frac{3}{2} 2^{2D} \qquad [10]$$

or in dB measure

$$\frac{S}{N} = 1.76 + 6.02D \quad \text{dB} \tag{11}$$

Note that generally only one quarter of the sine values is stored in the ROM lookup table (the two MSBs supply information about the sign and slope of the sine). Consequently, matching would require that

$$S \geq D - 1 \tag{12}$$

Example 1. For the matched arrangement, an ideal D-bit DAC, $D = 12$, will generate S/N

$$\frac{S}{N} = 1.76 + 6.02D = 74 \text{ dB} \tag{13}$$

which is about 2 dB more than one gets from the rule of thumb formula

$$\frac{S}{N} \approx 6D \text{ dB} \tag{14}$$

Besides the matched devices, we can encounter a DAC the resolution of which is smaller than required by (12). In that case, the expression (2) will be cut at the Ath place

$$A = D - 1 < S \tag{15}$$

The sine value will not be rounded; that is, the error will be in the range from 0 to $1/2^A$. By assuming the linear distribution, the probability will be $p(e) = 2^A$ in the respective range we find for the mean level

$$\eta = \frac{1}{2}2^{-A} \tag{16}$$

and the variance remains unchanged.

However, a spurious rectangular wave modulation with an amplitude $2^{-A}/2$ is present. The editor provides more detailed discussion in his paper in Part V.

$$\sigma_{eA}^2 = \frac{1}{3}2^{-2A} - \frac{1}{4}2^{-2A} = \frac{1}{12}2^{-2A} \tag{17}$$

On the other hand, in the case where the DAC resolution is higher than the matching requires, that is, in accordance with the relation (12), the S/N is controlled by the sine lookup table. However, the mean spurious level of the individual spectral lines caused by this process is dependent on the number of bits stored in the accumulator or on the clock frequencies. See the paper by the editor in Part V.

B. Spurious Signals Due to DAC Nonlinearities

The manufacturing processes are plagued with inaccuracies discussed in detail in the papers reprinted in this Section. Together with environmental influences, particularly temperature fluctuations of the current weighting circuits in the DAC, such as scaling resistors, base emitter voltages, and beta drift of current switches, DACs exhibit additional nonlinearities that further decrease the output signal purity.

The difficulty one encounters with DAC evaluation centers on definitions and terminology as used by different authors [e.g. 2,15,17] and producers [e.g. 12,16,18,19]. For this reason we will first introduce a vocabulary of definitions and then perform a deeper investigation.

1. Vocabulary.

Absolute accuracy: The difference between the actual output voltage and the ideal for a given digital code. It includes all the errors, and it is usually expressed as a percentage of the full-scale range.

A.C. feedthrough: [19,20] The ratio of the amplitude of the DAC output signal to the reference input with all DAC switches off.

Differential nonlinearity (DNL): [16,19] The worst-case deviation from one LSB output change from one adjacent state to the next.

Digital feedthrough: [19] A glitch impulse transferred from the digital DAC input to DAC output. It is measured with $V_{\mathrm{REF}} = 0$ and specified in psV.

Gain drift: [16,19] A measure of the change in the full-scale range (FSR) output over the specification temperature range. This parameter has units of % FSR, or ppm FSR, or % FSR/°C, or ppm FSR/°C.

Gain error: [12] The difference between the actual and the ideal analog output. It is expressed in percentage measure.

Integral nonlinearity: [2,12,19,20] The maximum deviation from a straight-line approximation of the DAC transfer function.

Monotonicity: [2,12,16,19] A DAC is monotonic if the output either increases or remains the same for increasing digital input values. If the DNL is less than or equal to ± 1, LSB monotonicity is guaranteed. If the converter is monotonic, its output will yield the full number of levels indicated by the number of bits of resolution.

Offset error: [12,16,17,19] Error encountered in DACs, the transfer function of which is either to the right or left of the ideal one but parallel to it (see Fig. 10).

Power supply rejection ratio: [12] The ratio of fractional output voltage (DVout/Vout) to the fractional power supply voltage (DVdd/Vdd).

Temperature dependence: [2] Monotonicity and linearity must be maintained on a large temperature range (e.g. from $-20°C$ to $+85°C$) to keep distortion and the S/N ratio within the specified range; see also *gain drift.*

2. Major Sources of DAC Nonlinearities. Investigation of the constant offset of the output voltage, illustrated in Fig. 10, reveals that its major source is the shift of the reference voltage [6,17] and is given by

$$V_{pp} = (V_{+FS} - V_{-FS})\frac{2^n}{2^{n-1}} \tag{18}$$

$$V_{\mathrm{out}} = V_{pp}\left(\frac{a_1}{2} + \frac{a_2}{4} + \cdots + \frac{a_N}{2^N}\right) + V_{-FS} + V_{\mathrm{offset}} \tag{19}$$

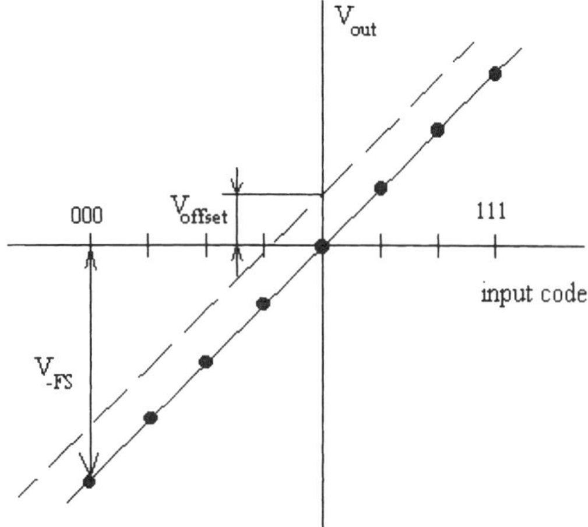

Fig. 10. The offset drift: all V_{out} are shift up or down regardless of the output code.

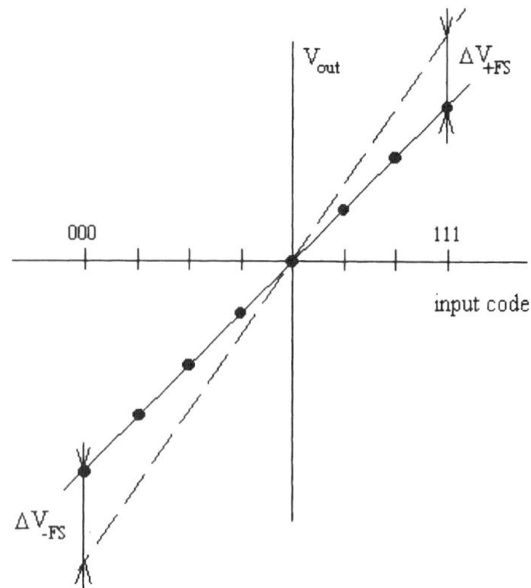

Fig. 11. The gain drift rotates the V_{out} about the midpoint.

Similarly, the varying temperature coefficients of the DAC resistors, base emitter voltages, and transistor betas cause additional nonlinearities.

In most instances, the differential error is in the range of $\pm 1/2$ LSB, and its variance is given by the relation (7). Since in the matched case [see relation (12)] both errors, that of the sine ROM and the DAC, are mutually independent processes, the denominator in (10) is two times larger and S/N decreases by 3 dB.

Example 2. Consider that the DAC error would be reduced to the range S; then the error variance would be

$$\sigma_{e,\text{DAC}^2} \approx \frac{1}{12}\left(\frac{V_{pp}}{2^{S+1}}\right)^2$$

After its introduction of the above variance into the denominator of (10), we find out that S/N decreases only by 1 dB.

The gain drift rotates the output characteristic about the midpoint as is illustrated in Fig. 11. This error is caused mainly by the drifting reference voltage.

However, individual output voltage errors, εD_{out}, often exhibit a systematic component as is schematically illustrated in Fig. 12. Note that for zero errors at the end points, there exists for every possible input code an equal and opposite error associated with one's complement of that code [18,21]. The consequence is generation of higher harmonics of the output signal.

C. Glitches

1. Generalities. DACs require a lot of switching operations that should ideally be performed simultaneously. However, this is not the case in large real-world networks. The consequence is that spurious voltage or current steps are generated and decay with transient effects called glitches. A particularly large glitch is encountered at midrange change of binary codes, for example, from

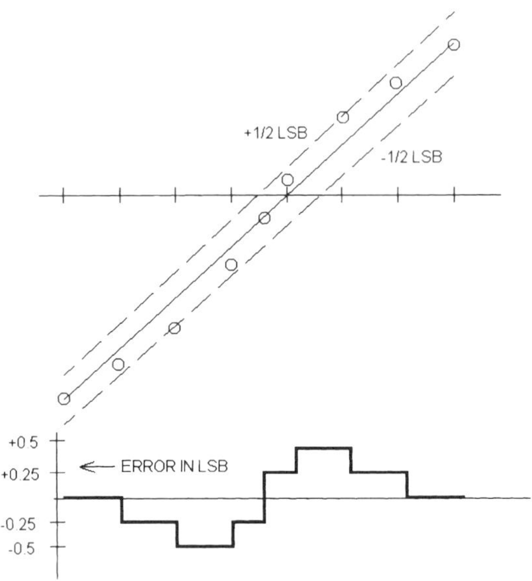

Fig. 12. The systematic differential error.

$$0111 \quad \text{to} \quad 1000 \qquad [20]$$

Before the final transition is realized, some intermediate states may be formed [22]. In most unfavorable instances, we should encounter for a 4-bit DAC transition states

$$0000 \quad \text{or} \quad 1111 \qquad [21]$$

for a while during the settling time. The consequence is a large voltage step, equal to half of the output swing V_{pp} lasting a very short time as shown schematically in Fig. 13.

The actual shape and magnitude (in pVs) of the generated spurious signal, the so-called glitch, depends on the effective time skew, the duration of the transient effect T_p, and the facing impedance. By assuming that the output circuit behaves as a simple RC network (see Fig. 14), the time integral of the spurious output voltage is

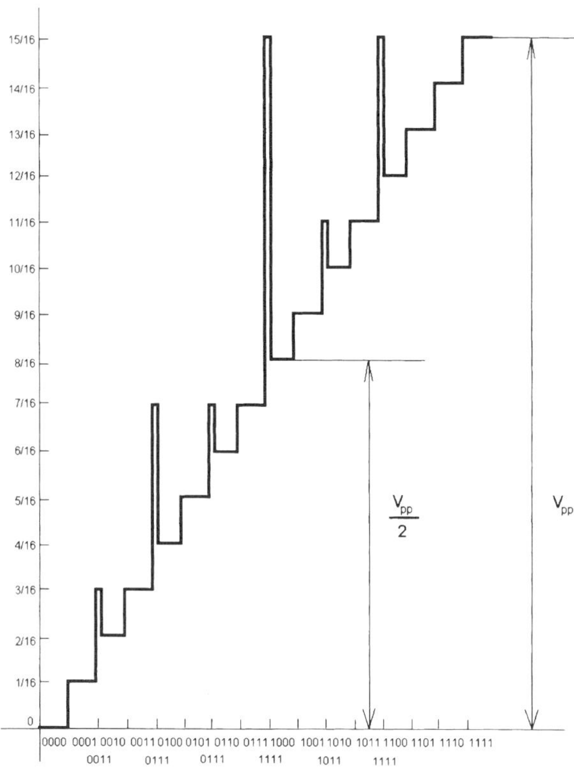

Fig. 13. Principle of glitch generation; note the large midrange excursion (adopted from [22]).

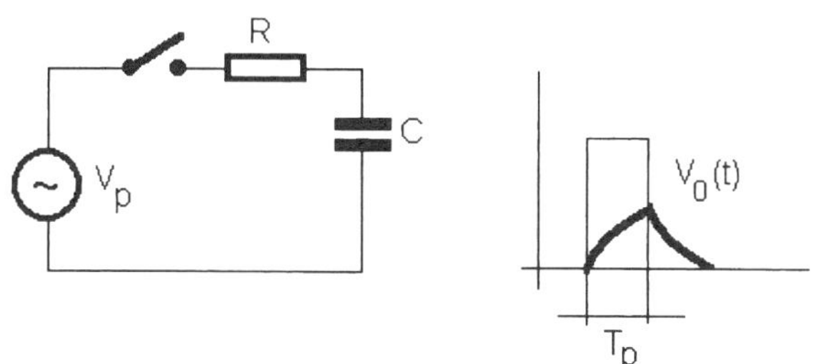

Fig. 14. Reshaping of the glitch after transition through a spurious RC circuit.

$$v_{g\tau} = \int_0^\infty v_o(t)dt$$

$$= \int_0^{T_p} \frac{v_{pp}}{2}\left(1 - e^{-\frac{t}{RC}}\right)dt + \int_{T_p}^\infty \frac{v_{pp}}{2}\left(1 - e^{-\frac{T_p}{RC}}\right)e^{-\frac{(t-T_p)}{RC}}dt$$

$$= \frac{v_{pp}}{2}T_p \qquad [22]$$

T_p is generally unknown. Nevertheless, we can estimate the magnitude of $V_{g\tau}$ from the effective glitch by taking half of its maximum voltage $V_o(t)$ and multiplying it by the pulse width, τ, at its middle.

In Tables 8 and 9 we have summarized some published data for 8-bit DACs.

2. Signal-to-Noise Ratio Due to Glitches. The next question is about the relationship between glitch magnitude and the degradation of the output spectrum. To this end, we will sim-

TABLE 8. GLITCHES FOR 8-BIT DACS AS PUBLISHED BY DIFFERENT AUTHORS.

f_c [MHz]	$V_g\tau$ [psV]	Authors	Logic family	V_{pp} [V]
100	80	Saul 1980 [23]		1.0
500	20	Maio 1985 [24]	ECL	1.0
80	100	Miki 1986 [14]	CMOS	1.0
400	3.5	Watanabe 1988 [25]		0.5
500	2	Kamato 1988 [26]		1.0
800	3	Nojima 1990 [4]		0.661
130	50	Fournier 1991 [27]	CMOS	0.8
150	100	Henriques 1994 [28]	CMOS	1.0

TABLE 9. GLITCHES FOR 8-BIT DACS AS PUBLISHED BY DIFFERENT MANUFACTURERS.

f_c [MHz]	$V_g\tau$ [psV]	Manufacturer	Logic family	V_{pp} [V]
25	200	Motorola MC10318	Silicon monolithic	1
40	30	Sony CXD1171M	CMOS	2
150	80	Plessey 976	ECL	1
300	25	Tri Quint TQ 6112	GaAs	1
450	20	Plessey 98608	ECL	1
500	25	Tri Quint TQ 6112	GaAs	1
500	20	Burn-Brown DAC650	Bipolar GaAs	1/12 bits

plify the situation to only one glitch during one output signal period T_x. In that case, we can use the Fourier series expansion of a train of narrow pulses

$$e_{sp}(t) = \frac{2V_g\tau}{T_x}\left[\frac{1}{2} + \sum_{n=1}^{n<r}\cos(n\omega_x t)\right]; \qquad r \approx \frac{T_x}{\tau} \qquad [23]$$

The output network exhibits nonlinear behavior, particularly at higher frequencies. First, we do not face only a simple RC load; an RLC circuit is much more probable. Second, there are deviations in characteristic impedance, dielectric and skin effect losses, and discontinuities at interfaces [29]. As a consequence, the glitch exhibits some ringing, as shown in Fig. 15. Thus, the Fourier expansion of the glitch is

$$e_{sp}(t) = \frac{2}{T_x}\Big[(V_g\tau)_1\cos(n\omega_x t)$$

$$- (V_g\tau)_2\sum\cos\big(n\omega_x(t-\delta_2)\big) + \cdots\Big]$$

$$\approx \frac{2}{T_x}V_g\tau\sum\cos(n\omega_x t) \qquad [24]$$

where $(V_g\tau)_1$ is the first positive part of the glitch in Fig. 15,

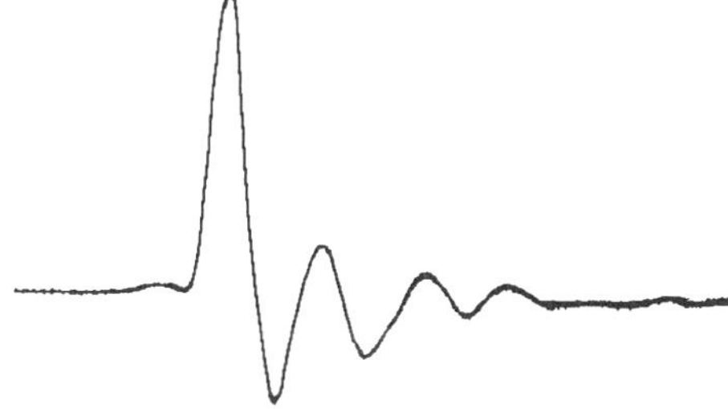

Fig. 15. Glitch occurring at major clock transition-reproduced from [30].

$(V_{g\tau})_2$ is the first negative part of the glitch, and so on. The result is that the effective glitch is the time integral of the whole transient. Note that in the literature the product, $V_g\tau$ is often called glitch energy. However, this term may be misleading since glitch is a pulse in the voltage–time space. Often we encounter another definition of the glitches, namely, LSB . τ in Vs.10^{-9}. The advantage is in the invariability in respect to the magnitude of the output voltage V_{pp}.

The final question, not yet answered, is about the acceptable magnitude of the glitch or glitches that would not appreciably degrade the output spectral purity of the respective DDFSs. By assuming that glitches behave as δ-functions, we can, without any loss of generality, represent them as follows

$$V_g\tau = V_{g,\mathrm{eff}}T_c \qquad [25]$$

where

$$T_c = \frac{1}{f_{\mathrm{clock}}} \qquad [26]$$

Since smaller glitches are encountered at nearly every DAC switching, we can estimate their variance (for a uniform distribution) as

$$\sigma_{v_g,\tau,\mathrm{eff}^2} \approx \frac{1}{12}\left(\frac{V_g\tau}{T_c}\right)^2 \qquad [27]$$

where $V_{g\tau}$ is defined in accordance with (23); usually, the first lobe of the glitch is the most important. In instances where the random error of DAC, with the resolution of D-bits, does not exceed $\pm$ one-half of the LSB, its variance is (by assuming the uniform distribution)

$$\frac{1}{12}(\mathrm{LSB})^2 = \frac{1}{12}\left(\frac{V_{pp}}{2^D}\right)^2 \qquad [28]$$

By equating the variance (27) to the variance due the DAC sampling process with S-bits, we get

$$\frac{1}{12}\left(\frac{V_g\tau}{T_c}\right)^2 = \frac{1}{12}\left(\frac{V_{pp}}{2^D}\right)^2 \qquad [29]$$

from which we have

$$V_g\tau \approx \frac{V_{pp}}{2^D} \cdot Tc \qquad [30]$$

This relation was first suggested as a rule of thumb by Nojima [4]. By assuming the validity of (27), the denominator in (10) is doubled, and we face a 3-dB reduction of the signal-to-noise ratio, S/N. In addition, relation (28) can serve for a qualitative comparison of different DACs; see the plot in Fig. 16.

3. Means for Glitch Reduction. Investigation of Fig. 14 and eq. (21) reveals that the glitch area $V_{g\tau}$ is proportional both to the voltage swing V_{pp} and to the duration of the erroneous code T_p. Cremonesi et al. [31] and Tien-yu Wu et al. [32] mention the major sources of glitches, such as

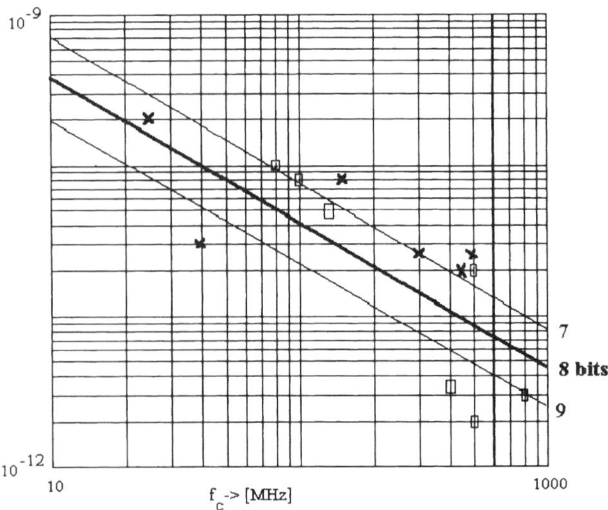

Fig. 16. Glitch $V_g\tau$ as function of the clock frequency, f_c, for an 8-bit DAC in accordance with the relation (28) for $V_{pp} = 1$ V; rectangles indicate published results from Table 8, and crosses have been taken from manufacturers' technical data from Table 9.

1. Imperfect synchronization of inputs
2. Channel-length modulation of current sources due to voltage fluctuation at ground line and output resistor load
3. Charge and discharge of parasitic capacitances associated with current sources
4. Feedthrough of digital input data to the output of current source
5. The time that the switching transistors are simultaneously in the off state

One of the very effective means is segmentation of the DAC (cf. Fig. 4). Sometimes additional reduction of output glitches can be achieved by adjusting the skew of the higher order bits of driving circuitry and by regulating the logic threshold. Tien-yu et al. state that simultaneous turn-on state of the switching transistors for a short time degrades the speed only a little, but the glitch is reduced substantially—to 3.9 psV for a 10-bit 75-MHz CMOS DAC.

A very effective deglitching circuit is a simple sampler (sample and hold) connected to the DAC output. One arrangement has been suggested by Hsieh et al. [29], and we demonstrate the art in Fig. 17. When a new analog output is formed, the DAC output current is initially diverted into ground with the assistance of transistors QS2 and QS3. After the output transition effectively has decayed, the sampler switches state and directs the current from the rails into the doubly terminated 50 Ω output network.

Another type of sampler, namely, a conventional series modulator, was suggested by Higgins [33] and at nearly the same time by Koen [22] for applications in digital waveform synthesizers. The principle is shown in Fig. 18.

The effectiveness of sample-and-hold circuit applications in spurious signal reduction is shown in Figs. 5 and 6 in the paper by Giffard and Cutler [34].

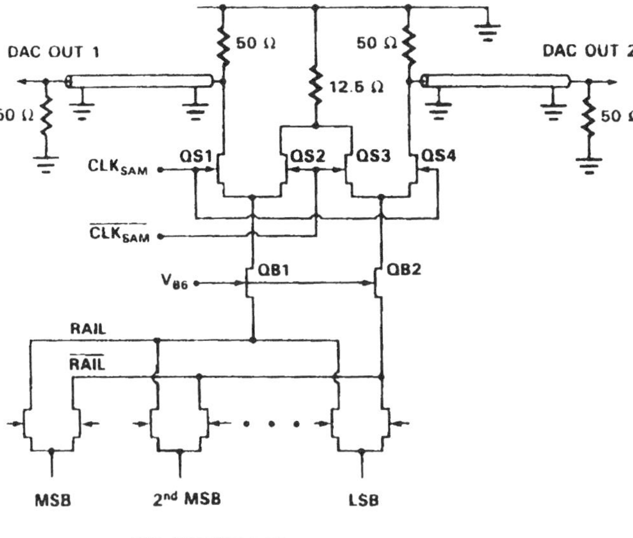

Fig. 17. Deglitching circuits suggested by Hsieh et al. [29].

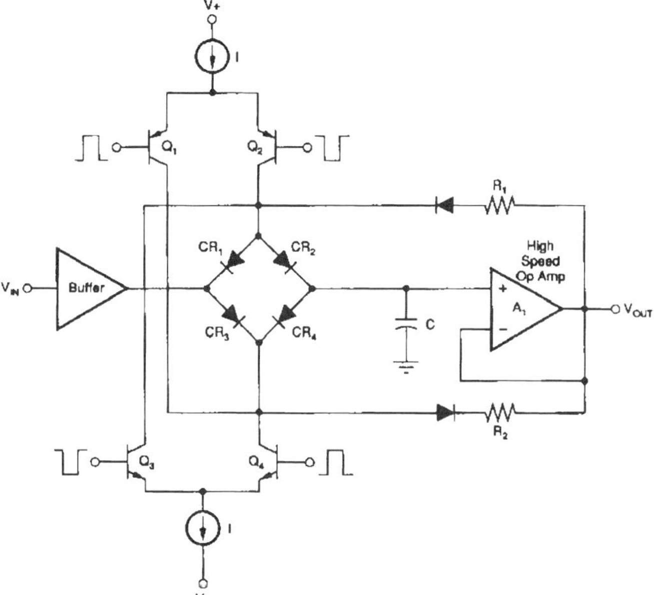

Fig. 18. Application of the conventional series modulator for deglitching, reproduced from [22]. © Burr-Brown. Used by permission.

VI. Conclusions

In this paper we have discussed properties of DACs, particularly those important for design and uses in DDFS.

First, we have investigated digital codes and pointed to the fact that in bipolar systems the analog output voltage is not symmetric around the zero, and we have discussed the problem with output waves. Furthermore, basic principles of DAC circuits have been dealt with. Major attention has been dedicated to the spurious signals generated in DACs. We have recalled the S/N due to the truncation of the sine words stored in lookup tables to S-bits only, and we have pointed to matching of the truncated sine words and to DAC resolution.

In short, we have also investigated DAC nonlinearities. The problems of the glitches have been discussed in some depth, together with guidelines for comparing the products from different manufacturers. In this connection, we recommend a minute investigation of the technical data before choosing a DAC for a particular DDFS.

References

[1] D. Sheingold, ed. *Analog-Digital Conversion Handbook.* Norwood, Mass.: Analog Devices, 1972.

[2] J. van de Plassche. "High-speed and high-resolution Analog-to Digital and Digital to Analog Converters," Ph.D. Thesis, University of Technology, Delft, 1989.

[3] D. J. Dooley. "A complete monolithic 10-b D/A converter." *IEEE Journal of Solid-State Circuits,* SC-8, pp. 404–8, December 1973.

[4] N. Nojima and Y. Gendai. "An 8-b 800-MHz DAC." *IEEE Journal of Solid-State Circuits,* SC-25, pp. 1353–59, December 1990.

[5] N. Caglio, J.-L. Degouy, D. Meignant, P. Rousseau, and B. Leroux. "An integrated GaAs 1.25 GHz clock frequency FM-CW direct digital synthesizer." *IEEE GaAs Symposium,* October 1993. (Reprinted in Part X.)

[6] G. A. Sunderland, R. A. Strauch, S. S. Wharfield, H. T. Peterson, and C. R. Cole. "CMOS/SOS frequency synthesizer LSI circuit for spread spectrum communications." *IEEE Journal of Solid-State Circuits,* vol. SC-19, pp. 479–506, August 1984. (Reprinted in Part III.)

[7] Stanford Research Systems. *Scientific & Engineering Instruments.* 1994–1995.

[8] *Data Converters & Voltage Reference IC Handbook.* Plessey Semiconductors, 1989.

[9] G. Kelson, H. H. Stellrecht, and D. S. Perloff. "A monolithic 10-b digital-to-analog converter using ion implantation." *IEEE Journal of Solid-State Circuits,* SC-8, pp. 396–403, December 1973.

[10] R. A. Schultz. "Monolithic current-switch DAC improvements." *IEEE Journal of Solid-State Circuits,* pp. 338–41, April 1976.

[11] M. Shoji and V. R. Saari. "A monolithic 8-bit D/A converter with a new scheme for error compensation." *IEEE Journal of Solid-State Circuits,* pp. 499–501, December 1975.

[12] *A/D and D/A Conversion Manual* (1988) and *Motorola Master Selection Guide,* Rev. 4, Motorola, Inc.

[13] *Sony Semiconductor IC Data Book,* 1993, A/D, D/A Converter.

[14] T. Miki, Y. Nakamura, M. Nakaya, S. Asai, Y. Akasaka, and Y. Horiba. "An 80-MHz 8-bit CMOS D/A converter." *IEEE Journal of Solid-State Circuits,* vol. 21, pp. 983–88, December 1986.

[15] J. A. Schoeff. "An inherently monotonic 12 bit DAC." *IEEE Journal of Solid-State Circuits,* vol. SC-14, pp. 904–11, December 1979. (Reprinted in this Part.)

[16] Burr-Brown. *Integrated Circuit Data Book,* supplement vol. 32c, 1992.

[17] P. Prazak. "What designers should know about data-convert drift." *Electronics,* pp. 111–14, November 10, 1977.

[18] Burr-Brown. *Applications Handbook,* 1994.

[19] *PMI Data Book Analog Integrated Circuits,* vol. 10, pp. 11–5.

[20] *Analog Devices,* CMOS Application Guide, 1984.

[21] J. R. Naylor. "Testing digital/analog and analog/digital converters." *IEEE Transactions on Circuit and Systems,* vol. CAS-25, pp. 526–38, July 1978.

[22] M. Koen. "High speed data conversion." *Burr-Brown Application Bulletin,* No. 27 (1993).

[23] P. H. Saul, P. J. Ward, and A. J. Fryers. "An 8-bit, 5 ns monolithic D/A converter subsystem." *IEEE Journal of Solid-State Circuits,* vol. SC-15, pp. 1033–39, December 1980. (Reprinted in this Section.)

[24] K. Maio, S.-I. Hayashi, M. Hotta, T. Watanabe, S. Ueda and N. Yokozawa. "A 500-MHz 8-bit D/A converter." *IEEE Journal of Solid-State Circuits,* vol. SC-20, pp. 1133–36, December 1985. (Reprinted in this Part.)

[25] T. Watanabe, K. Maio, K. Norisue, S. Hayashi, and S. Ueda. "A 400 MHz DA converter with a 4-bit color map for 2000-line display." *IEEE Journal of Solid-State Circuits,* vol. 23, pp. 147–51, February 1988.

[26] T. Kamoto, Y. Akazawa, and M. Shinagawa. "An 8-bit 2-ns monolithic DAC." *IEEE Journal of Solid-State Circuits,* vol. 23, pp. 142–46, February 1988. (Reprinted in this Section.)

[27] J. M. Fournier and P. Senn. "A 130 MHz 8-b CMOS video DAC for HDTV applications." *IEEE Journal of Solid-State Circuits,* vol. 26, pp. 1073–77, July 1991.

[28] B. G. Henriques and J. E. Franca. "A high-speed programmable CMOS interface system combining D/A conversion and FIR filtering." *IEEE Journal of Solid-State Circuits,* vol. 29, No. 8, pp. 972–77, August 1994.

[29] K. Hsieh, T. A. Knots, G. L. Baldwin, and T. Hornak. "A 12-bit 1-Gword/s GaAs digital-to-analog converter system." *IEEE Journal of Solid-State Circuits,* vol. 22, pp. 1048–54, December 1987.

[30] C. A. A. Bastiaansen, D. W. J. Groenveld, H. J. Schouwenaars, and H. A. H. Termeer. "A 10-b 40-MHz 0.8-mm CMOS current-output D/A converter." *IEEE Journal of Solid-State Circuits,* vol. 26, pp. 917–21, July 1991.

[31] A. Cremonesi, F. Maloberti, and G. Polito. "A 100-MHz CMOS DAC for video-graphic systems." *IEEE Journal of Solid-State Circuits,* vol. 4, pp. 635–39, June 1989.

[32] Tien-yu Wu, Ching-Tsing Jih, Jueh-Chi Chen, and Ching-Yu Wu. "A low glitch 10-bit 75 MHz CMOS video D/A converter." *IEEE Journal of Solid-State Circuits,* vol. 30, no. 1, pp. 68–72, January 1995.

[33] T. M. Higgins, Jr. "Analog output system design for a multifunction synthesizer." *Hewlett-Packard Journal,* pp. 66–69, February 1989.

[34] R. P. Giffard and L. S. Cutler. "A Low Frequency High Resolution Digital Synthesizer," *Proceedings of the 1992 IEEE International Frequency Control Symposium,* pp. 188–92. (Reprinted in Part V.)

A Monolithic 10-b Digital-to-Analog Converter Using Ion Implantation

GARY KELSON, MEMBER, IEEE, HANS H. STELLRECHT, MEMBER, IEEE, AND DAVID S. PERLOFF

Abstract—The design and fabrication of a self-contained 10-b monolithic digital-to-analog converter is described. To overcome the limitations of standard bipolar processing and to achieve a circuit with reasonably low process sensitivities, a new process incorporating ion implantation is used. The circuit has been designed in a manner to fully utilize the characteristics of this process with the objective of high performance along with simple wafer processing. System considerations as well as the design of each component block are discussed.

I. Introduction

A COMMON problem has plagued monolithic analog integrated circuits since their inception: poor component tolerance. However, early workers in the field quickly recognized the advantages of being able to fabricate large numbers of components, in close proximity to one another, on a single silicon substrate [1].

Designers began using the excellent matching characteristics of integrated components to their advantage as well as the relative economy of active components over passive components. During the ensuing years great strides forward in circuit design techniques and process sophistication have been made to the point where many monolithic circuits outperform their discrete counterparts, often at a lower cost [2]. In spite of these achievements, many circuit applications still exist to challenge both circuit and process engineer with virtually the same tolerance problems. The high-accuracy monolithic digital-to-analog converter (DAC) is such a circuit.

The monolithic circuit to be described in this paper is a completely self-contained 10-b DAC which requires no external precision components. Overall system accuracy (±0.05 percent) is ensured through the use of well-matched internal components. The unique characteristics of ion implantation have been combined with standard bipolar process technology to achieve the high precision required by this circuit. Thus, the DAC can be manufactured on the same process line with ultrahigh volume circuits such as operational amplifiers with only a minimum of extra handling.

Manuscript received May 18, 1973.
The authors are with the Signetics Corporation, Sunnyvale, Calif. 94086.

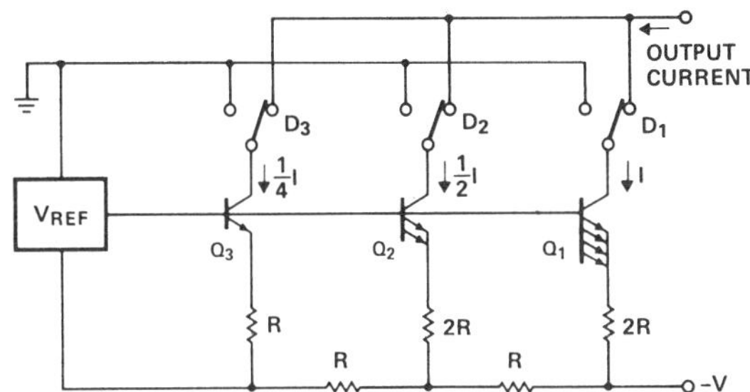

Fig. 1. Summing of binary weighted currents.

II. Circuit Approaches Suitable for Monolithic Realization

Many techniques are available for converting a parallel input digital signal into an analog signal. Most can be categorized into one of three groups:

1) pulsewidth modulation;
2) binary weighting;
3) binary attenuation.

The pulsewidth modulation method offers some advantages for high-resolution monolithic DAC's but is limited to low-speed applications [3].

The second and third categories are very similar but have some rather subtle differences especially when monolithic realization is considered. The binary weighting approach is shown in Fig. 1. Here a 3-b converter is formed by summing the collector currents of Q_1, Q_2, and Q_3 through three digitally controlled switches D_1, D_2, and D_3. These currents are weighted in a binary manner by the R-$2R$ resistor ladder network. Because each transistor operates with a different collector current, the geometry of each transistor must be scaled, as shown, to maintain equal emitter current density and therefore equal base–emitter voltage drops (V_{BE}). Differences in V_{BE} between current sources changes the termination voltage for the R-$2R$ ladder leading to errors in the binary weighting. The problem becomes especially acute for high-accuracy converters; a 10-b DAC requires a 512:1 current ratio. Due to the weighting of emitter currents, a binary weighted voltaged drop is generated across the series emitter resistors. This presents a subtle problem for monolithic integration since resistors are normally formed by reversed biased p and n-type semi-

Reprinted from *IEEE Journal of Solid-State Circuits*, Vol. SC-8, No. 6, pp. 396-403, December 1973.

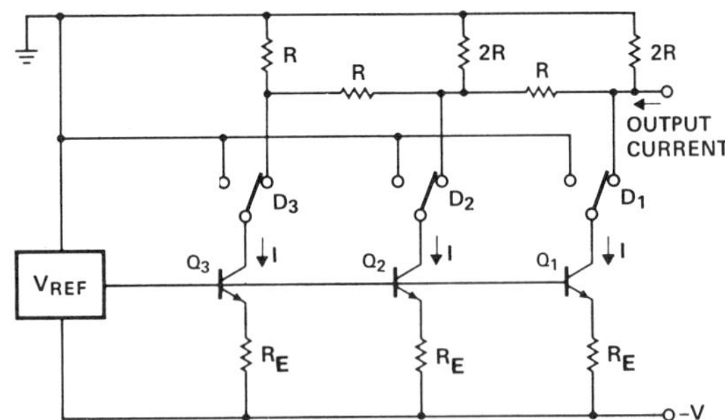

Fig. 2. Binary attenuation of equal-valued currents.

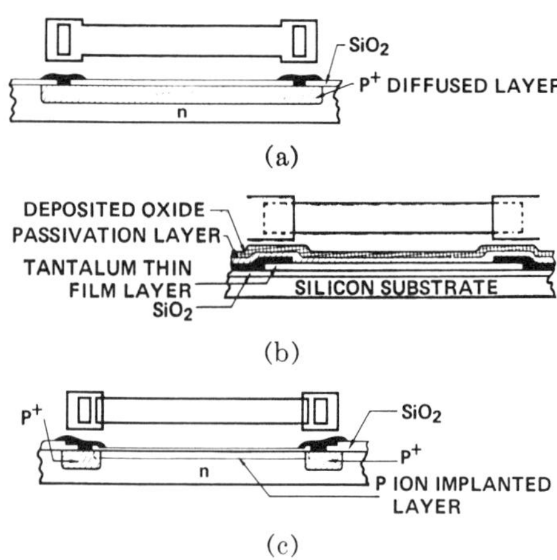

Fig. 3. Cross sections of monolithic resistor technologies. (a) Diffused resistor. (b) Thin-film resistor. (c) Ion-implanted resistor.

conductor regions which are voltage sensitive. The sheet resistance ρ_s for these resistors increases with the voltage drop across the resistor which produces nonlinearities in the DAC output. This phenomenon, characterized by a voltage coefficient of resistance (VCR), is similar to channel pinchoff in depletion mode FET's. Since VCR is proportional to sheet resistance, one can minimize this problem by using low sheet resistance base diffusions for the ladder network. (VCR $\simeq$ 0.025 percent/V for $\rho_s = 135$ $\Omega/\square$.) But the value of R in the ladder network cannot be made arbitrarily small since the voltage drop generated across the series emitter resistors $(I \cdot R)$ must be sufficiently large when compared to the V_{BE} mismatches of the current-source transistors. For a 10-b converter, this voltage drop should exceed approximately 1000 times the V_{BE} mismatch [4]. Thus, one must trade off tight transistor matching against increasing $I \cdot R$. Optimization of this tradeoff requires higher power dissipation and larger die areas or very tight masking and process control.

The binary attenuation method shown in Fig. 2 circumvents several of the problems of the binary weighted approach by separating the R-$2R$ ladder network and the series feedback current setting resistors. The current sources formed by Q_1, Q_2, and Q_3 along with the resistors R_E are now identical in value. These currents are switched into the binary attenuator to attain the weighted output current. With the voltage drop across R_E being the same for all bits, VCR will not produce any errors in current cell matching, thus allowing R_E to be made large using a high sheet resistance process. The resistors in the binary attenuator do have differing voltage drops but these can be made negligibly small since R can be made small without sacrificing accuracy. Thus, optimization of the $I \cdot R$ versus V_{BE} matching tradeoff can be made without requiring excess die area or high power dissipation. This scheme has two additional advantages stemming from the identical current cells. V_{BE} matching of identical transistors operating at the same current level is better than scaled transistors operating with scaled currents. Also, each cell has the same transient response when switched, giving faster settling for lower order bits and less output glitching [5].

III. Monolithic Resistor Technologies

In order to realize the advantages which the equal valued current approach offers, large resistance values are needed. This requires a resistor process with high sheet resistance. Other requirements of the resistor process are high matching accuracy, good stability, and ease of fabrication. Since resistors are such an important factor in DAC's, a statistical comparison was made between the most widely used monolithic resistor technologies with respect to sheet resistance, matching, and temperature sensitivity. The technologies that were examined are

1) diffusion;
2) thin film;
3) ion implantation.

Cross sections of these are shown in Fig. 3.

Diffused resistors, as shown in Fig. 3(a), represent the least complicated technology from a processing standpoint since no additional process steps are required in a standard bipolar process. However, sheet resistance is usually limited to under 200 $\Omega/\square$.

Thin-film resistors, shown in Fig. 3(b), are made by depositing a thin layer of resistive material, tantalum in this case, on an oxidized substrate. Being surface devices, thin-film resistors require careful passivation to maintain long-term stability. Typically, two passivation layers are required to obtain stability sufficient for DAC applications. From a processing standpoint, this is the most complex resistor technology.

An ion-implanted resistor cross section is shown in Fig. 3(c). Usually, the contact beds are formed by the base diffusion. Then a p-type layer is formed by ion implantation. Finally a post-implant heat treatment is used to electrically activate the ions, thereby obtaining the desired sheet resistance. The sheet resistance of the implanted layer is typically much higher than the diffused contact beds so that the resistance value is almost completely determined by the implantation.

Since the ion dose can be controlled very accurately, excellent definition of the absolute value of resistors, as well as good matching between resistor pairs, can be obtained. Also, since the device, unlike a thin-film resistor, is formed within the semiconductor substrate, it is passivated by silicon dioxide in the same manner as a diffused resistor. In other words, the ion-implanted resistor combines the advantages of bulk diffused devices with the high sheet resistance of thin-film devices.

The details of the resistor comparison are given in the Appendix along with a summary of the results. These results indicate that ion-implanted resistors match as well or better than diffused resistors while having nearly an order of magnitude higher sheet resistance. This can be achieved with a process which contains only one additional masking step in conjunction with an easily controlled ion-implantation step. Additionally, since the ion-implanted resistors do not require any high-temperature processing, all previous diffusion steps are unaffected. This is a key feature of the process since it allows all active components in the circuit to be fabricated by a standard diffusion process and the high-accuracy resistors to be formed at a subsequent time without interaction [6].

IV. Digital-to-Analog Circuit Design

A block diagram of the monolithic DAC is shown in Fig. 4. The ten equal valued current cells are biased in parallel from a reference current cell which is part of a negative feedback loop. The feedback loop sets the reference current at a value given by

$$I_{\text{ref}} = \frac{V_{\text{ref}} - V_{\text{mult}}}{R_{\text{ref}}}.$$

In this configuration, the current of each cell will be the same as I_{ref} so long as each cell matches the reference cell and therefore absolute parameter variations are automatically compensated. The multiplying input V_{mult} may be used to obtain 100-percent modulation of the reference voltage and thus the current of each cell. The binary R-$2R$ attenuator generates the weighted output current, I_{OA}. This current may be converted into a voltage V_{OA} which is given by

$$V_{OA} = \frac{R_F}{R_{\text{ref}}} (V_{\text{ref}} - V_{\text{mult}})$$

$$\cdot (D_1 + \tfrac{1}{2} D_2 + \tfrac{1}{4} D_3 + \cdots + \frac{1}{512} D_{10})$$

where D_n represents the binary inputs.

A simplified schematic drawing of one of the ten identical current cells and the emitter-coupled switch is shown in Fig. 5. The cell current is generated by transistor Q_5 and the series feedback resistor R_{E1}. The base of

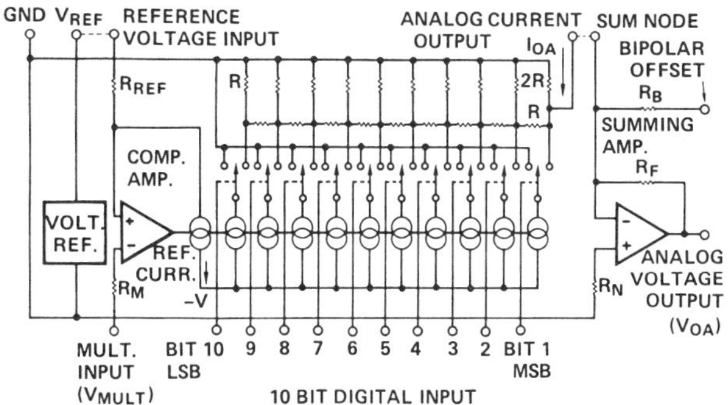

Fig. 4. Block diagram of monolithic 10-b DAC.

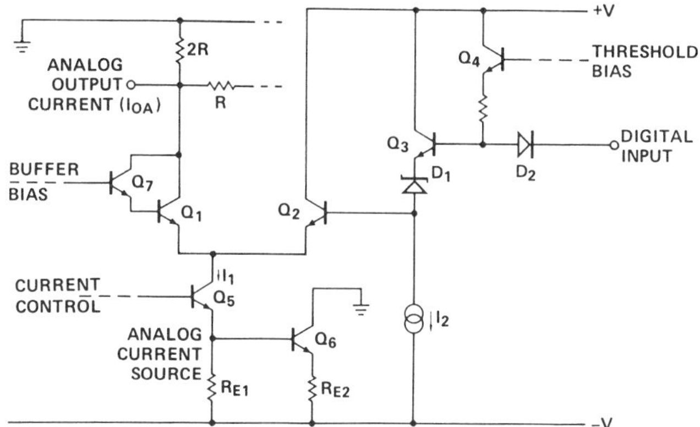

Fig. 5. Simplified schematic of current switching cell.

Q_5 is driven from a common current control bus by the compensation amplifier as shown in Fig. 4. Neglecting, for the moment, Q_6 and R_{E2},

$$I_1 = \frac{\beta}{\beta + 1} \left(\frac{V_{E5} + V}{R_{E1}} \right) = \alpha \left(\frac{V_{E5} + V}{R_{E1}} \right).$$

If the current cells in the DAC are to match with less than 0.1-percent error, the common-base current gain α must also match with less than 0.1-percent error. For a worst case β mismatch of 25 percent, a minimum absolute β of almost 400 would be required. While this is well within present technological limits, it does require some special processing and is not desirable on a circuit of this complexity. This minimum β constraint can be reduced by the addition of transistor Q_6 and resistor R_{E2} [7]. If I_{E6} is made equal to I_{E5} and $0.75 < \beta_{Q5}/\beta_{Q6} < 1.25$, then in the worst case, the error current flowing in R_{E1} due to the finite β of Q_5 is multiplied by at least 4 or the minimum absolute gain is reduced to $400/4 = 100$.

The emitter-coupled current switch, made up of transistors Q_1, Q_2, and Q_7, steers the cell current between the $+V$ bus and the binary attenuator. The Darlington connection of Q_1 and Q_7 minimizes current transmission errors to the attenuator and again reduces the minimum β for good cell matching. High switching speeds are achieved by limiting the voltage swing at the collector

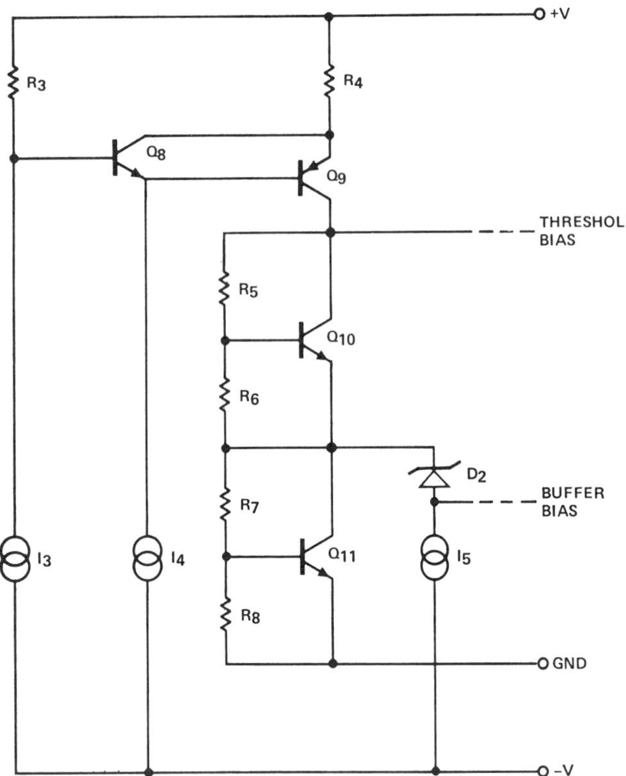

Fig. 6. Biasing circuit for current switching cells.

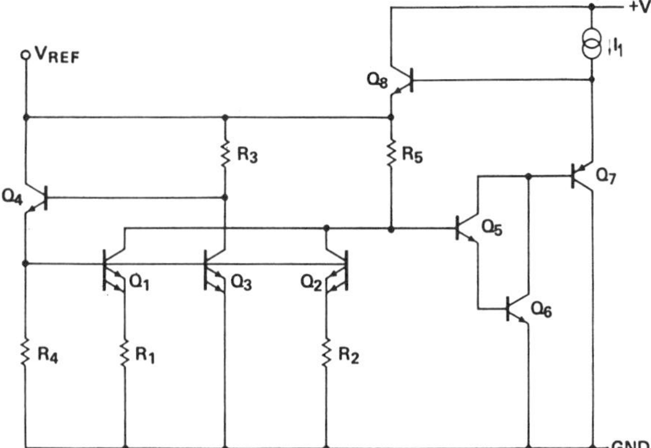

Fig. 7. Simplified schematic of voltage reference circuit.

of Q_5. In order to ensure complete switching under these conditions, the buffer bias V_{BB} and threshold bias V_{TB} must track both temperature and absolute variations of the Zener level-shift diode D_1. These bias voltages are generated in the bias circuit shown in Fig. 6 where

$$V_{BB} = V_{BE}(1 + R_7/R_8) - V_Z$$

$$V_{TB} = V_{BE}(2 + R_5/R_6 + R_7/R_8).$$

The emitter-coupled pair Q_1 and Q_2 (Fig. 5) will switch when $V_{B1} \simeq V_{B2}$. Thus

$$V_{in} \text{ (threshold)} - V_Z \simeq V_{BB} - V_{BE}$$

or

$$V_{in} \text{ (threshold)} \simeq V_{BE} (R_7/R_8)$$

which is independent of V_Z. In addition, the voltage swing ΔV at the collector of Q_5 is given by

$$\Delta V = V_{TB} - 3V_{BE} - V_Z - (V_{BB} - 2V_{BE})$$

$$= V_{BE}(R_5/R_6).$$

For TTL compatibility V_{in} (threshold) $= 1.5$ V. Thus $R_7/R_8 = 2.15$. If $\Delta V = 450$ mV then $R_5/R_6 = 0.69$. Under these conditions the propagation delay and settling time (to 0.1 percent) for the output current I_{OA} is approximately 10 ns and 100 ns, respectively.

The voltage reference for a 10-b DAC must have an extremely low temperature coefficient if the circuit is to have a wide operating temperature range. For example, if the DAC is to be $\frac{1}{2}$ LSB accurate over the commercial temperature range (0 to 75°C) the complete converter should not drift more than 10 ppm/°C. Since such low drift is difficult to achieve with present technology, it is customary to simply specify the full-scale drift of the DAC output.

Two kinds of reference circuits which are suitable for monolithic integration are Zener diode circuits and circuits which make use of the band-gap principle [8]. A band-gap-type circuit was chosen for the 10-b DAC because the temperature compensation depends, to a first order, only on V_{BE} matching and resistor ratios. Zener diode references normally are more surface sensitive and depend on the absolute value of the base sheet resistance, which is more difficult to control.

A simplified circuit diagram of the voltage reference circuit is shown in Fig. 7. This circuit uses the positive temperature drift of the V_{BE} difference (ΔV_{BE}) between transistors operating at different current densities to cancel the negative drift of the absolute base–emitter voltage. The combination of transistors Q_1, Q_2, and Q_3 provide the positive drift component by forcing ΔV_{BE} to appear across the parallel combination of resistors R_1 and R_2. This voltage is multiplied by $-R_5 (R_1 + R_2)/R_1R_2$ and compensates for the negative drift of the Darlington pair Q_5 and Q_6. The transistors Q_5, Q_6, Q_7 and Q_8 form a high gain feedback loop which provides regulation and low output impedance for the reference voltage. The approximate value of the reference voltage (V_{ref}) is given by

$$V_{ref} \simeq V_{BE_5} + V_{BE_6} + \frac{R_5(R_1 + R_2)}{R_1 R_2} \frac{kT}{q} \ln\left(\frac{J_3}{J_{1,2}}\right)$$

where $J_{1,2}$ and J_3 are the current densities in transistors Q_1, Q_2, and Q_3, respectively. These current densities are determined by the resistors R_1, R_2, and R_3 as well as by the ratio of emitter areas between Q_1, Q_2, and Q_3. To ensure tight thermal tracking of devices within the reference circuit, and to minimize the effects of chip thermal gradients, the transistors Q_1, Q_2, and Q_3 are made from groups of parallel connected transistors which are topologically symmetric about Q_5 and Q_6. Similar layout considerations were used for the resistors R_1, R_2, R_3, and R_5. With this circuit, temperature drifts of less than 10 ppm/°C have been realized over the 0 to 75°C temperature range.

To convert the DAC output current I_{OA} into a voltage, a transimpedance amplifier is required, as shown in Fig. 4. The transimpedance

$$\frac{V_{OA}}{I_{OA}} = R_F$$

is realized with an operational amplifier connected in the current summing mode. In order not to degrade the accuracy and speed of the basic DAC, the summing amplifier requires several important features:

1) high open-loop gain;
2) low input offset voltage;
3) high slew rate and low settling time.

From Fig. 4 the transfer function for the summing amplifier is given by

$$V_{OA} = \frac{R_F(I_{OA} + I_{os}) + V_{os}(1 + R_F/R_N)}{1 + (1/A_{VOL})(1 + R_F/R_N)}$$

where I_{os} and V_{os} are the input offset current and voltage, respectively; A_{VOL} is the open-loop gain; and R_N is the R-$2R$ ladder source resistance.

With a full-scale output voltage of 10 V the total error introduced by the summing amplifier must be much less than 5 mV ($\frac{1}{2}$ LSB). For typical operational amplifiers, I_{os} is negligible compared to I_{OA} for the least significant bit (1 μA). If the error introduced by the summing amplifier is split equally between the V_{os} term and the finite open-loop gain, then for I_{OA} (full scale) = 1 mA, $R_F = $ 10 kΩ and $R_N = 2$ kΩ,

$$V_{os} < 0.5 \text{ mV}$$

$$A_{VOL} > 90 \text{ dB}.$$

While this open-loop gain is not difficult to achieve in a two-stage amplifier, circuit yields would be low for a single operational amplifier requiring $V_{os} < 0.5$ mV. Fortunately, one has the freedom to null this offset initially, then the total drift in V_{os} with ambient changes must not exceed 0.5 mV. For the 0 to 75°C operating range

$$\frac{dV_{os}}{dT} < \frac{0.5 \text{ mV}}{75°C} = 6.7 \ \mu\text{V/°C}.$$

However, for a differential amplifier with balanced loads, there is a direct correlation between dV_{os}/dT and V_{os} since tightly matched transistors will track each other more closely [9]. It has been shown that for each millivolt of offset voltage, the offset drift will be approximately 3.3 μV/°C [9]. Thus the greatest input offset voltage which can be tolerated for the summing amplifier is

$$V_{os}' < 6.7 \ \mu\text{V/°C}/3.3 \text{ V/°C/mV} \simeq 2 \text{ mV}.$$

This limit should not appreciably diminish yields.

A simplified schematic drawing of the summing amplifier is shown in Fig. 8. A modified second-generation design is used with feed-forward compensation to enhance the transient response without degrading dc stability. The lateral p-n-p transistors Q_3 and Q_4 have split collectors to reduce the first-stage transconductance and simplify biasing. This technique provides high slew rates by allowing the first-stage bias current to be increased without increasing the compensation capacitor C_2 [10]. The second stage is a resistively loaded common-emitter amplifier comprised of Q_9 and R_4. A low parasitic ion-implanted resistor is used for R_4 to maintain wide bandwidth. High-frequency input signals are fed forward around the narrow band, first stage by C_1 and R_9. A frequency-compensation capacitor C_2 of approximately 5 pF is used to control the damping factor or output settling time. While dissipating less than 25 mW the summing amplifier has a slew rate of 20V/μs and settles to 0.05 percent in 2 μs.

V. Integrated-Circuit Layout and Performance

The inherent accuracy of the DAC described here is heavily dependent on component matching. This has been stated repeatedly in the descriptions of each of the blocks for the converter system. Integrating all of these blocks on a single silicon substrate presents numerous problems. High-speed switching signals must be isolated from high-accuracy reference signals. Power dissipation on the chip must be distributed so that variations of the output level do not produce thermal feedback which can seriously affect output linearity. To overcome these potential problems, the current switches and R-$2R$ ladder are arranged in a horseshoe pattern as shown in Fig. 9. With this arrangement the center line of power dissipation is independent of input code. Power buses are symmetrically arranged and fed so that ohmic metal drops do not affect accuracy. The summing amplifier is located symmetrically with respect to the reference and feedback resistors as well as the voltage reference and compensation amplifier. This configuration makes the output linearity virtually independent of load current. The chip size is 113 mil by 124 mil and contains 140 transistors, 25 diodes, and over 530 kΩ of precision resistance.

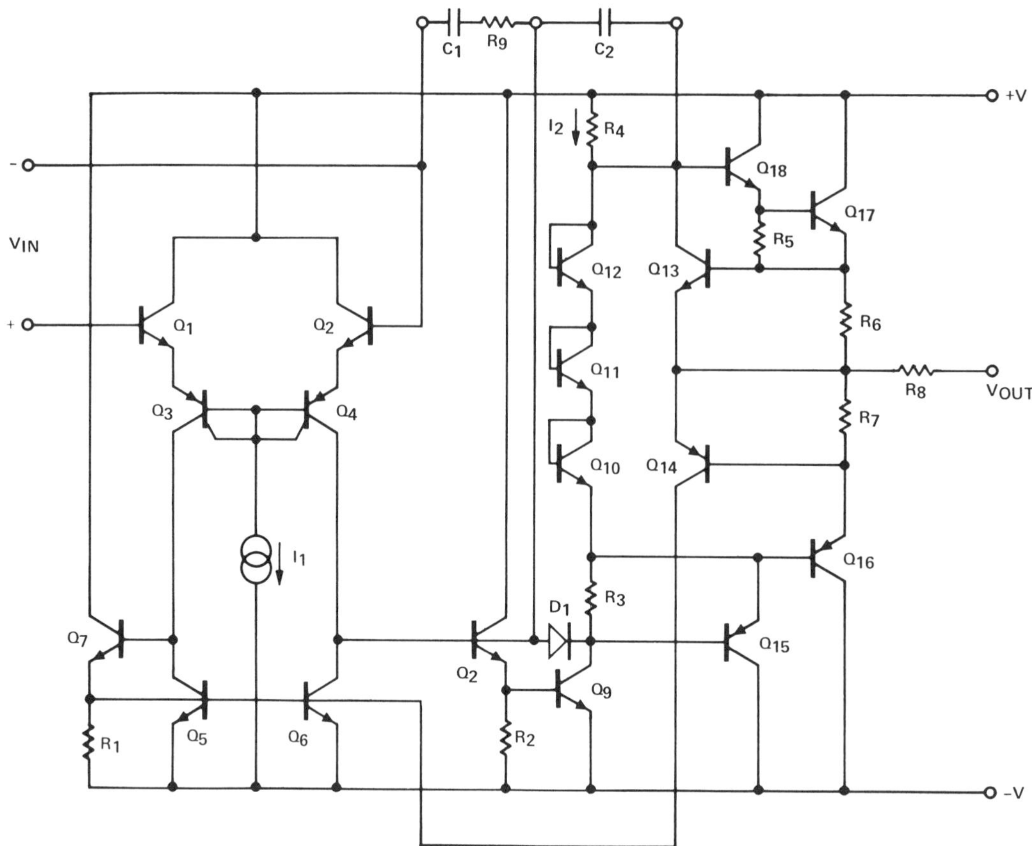

Fig. 8. Simplified schematic of summing amplifier.

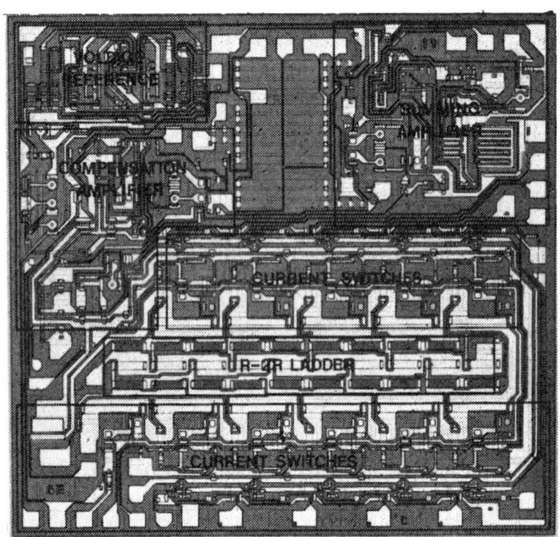

Fig. 9. Photomicrograph of 10-b DAC.

Fig. 10 shows the typical connection of external components for the DAC which requires only two external adjustments to provide 10-b accuracy. The zero adjustment nulls the system offsets, which are typically less than 5 mV, by varying the voltage on the noninverting input to the summing amplifier. Full scale is adjusted by varying the multipling input which changes the effective reference voltage. Due to the absence of thermal feedback, these two adjustments are sufficient to achieve $\frac{1}{2}$ LSB accuracy for the converter over a 0 to 50°C operating range. The performance of the system is summarized in Fig. 11. The temperature coefficient (TC) of the basic converter, with an external zero TC voltage reference, is less than 5 ppm/°C. The output slew rate when operating in the multiplying mode is limited to 5 V/µs by the compensation amplifier in the current cell bias system.

VI. Conclusion

The design and fabrication of a 10-b monolithic DAC has required the optimization of both circuit and process technologies. To meet the severe accuracy requirements (± 0.05 percent) in a complex system, a circuit has been designed around a process which complements the design without overly complicating processing.

This has been accomplished through the use of ion implantation to form the precision resistor ladder networks. Electrical and thermal interactions between components of the DAC have been analyzed to ensure that the system performance is not compromised by integration. The realization of a high-accuracy (10-b) monolithic DAC indicates that integrated-circuit technology has not only moved toward large-scale systems but also toward systems which can provide the precision only previously available with discrete or hybrid techniques.

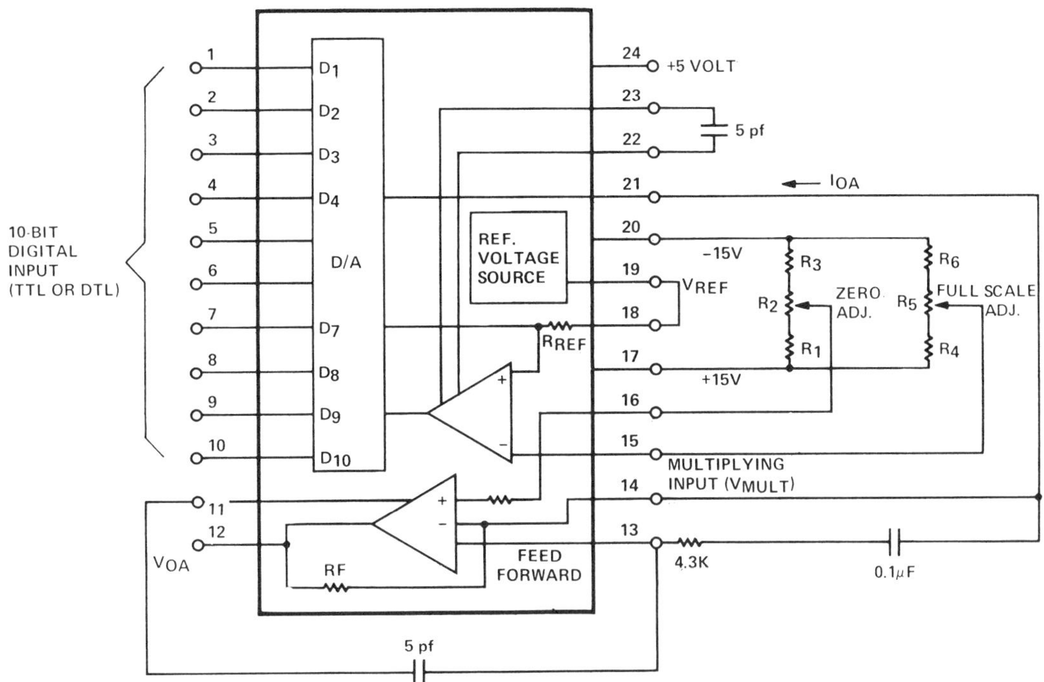

Fig. 10. Typical connection of external components for DAC.

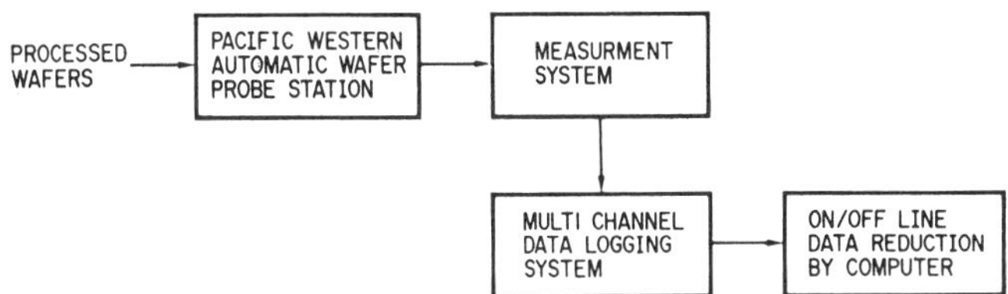

Fig. 12. Block diagram of measurement procedure.

```
OUTPUT VOLTAGE.........................0 TO ±5 V, 0 TO 10 V
LINEARITY..................................................0.05%
FULL SCALE TEMPERATURE DRIFT
    WITH INTERNAL REFERENCE..................±15 ppm/°C
VOLTAGE REFERENCE DRIFT....................±10 ppm/°C
POWER SUPPLY REJECTION........................005%/%
SWITCHING PERFORMANCE
SETTLING TIME
    CURRENT OUTPUT.......................200 ns TO .05%
    VOLTAGE OUTPUT.........................2 μs TO .05%
OUTPUT SLEW RATE
    DIGITAL INPUT...............................20 V/μs
    ANALOG INPUT (VMULT).........................5 V/μs
POWER CONSUMPTION...........................350 mW
```

Fig. 11. 10-b DAC performance.

Appendix

In order to determine a suitable resistor process for high-precision linear integrated circuits such as DAC's, a comparative evaluation of resistors made by the standard diffusion, by thin-film, and by ion-implantation processes, was carried out. The experiment was done with production processing, and a large number of devices was processed so that statistical analysis could be performed on the data. The statistical analysis included measurement of electrical parameters of resistor pairs with different geometries, calculation of distributions for absolute and matching tolerances, and calculation of percentile distributions. Fig. 12 shows a block diagram of the measurement procedure.

The results of this experiment are summarized in Table I. The nominal sheet resistance values of 1000 and 1250 $\Omega/\square$ for the thin-film and ion-implant process, respectively, were chosen for both process and layout compatibility. A wide range of sheet resistance values are available for both processes. The mean and standard deviation for matching are defined as follows:

TABLE I
Experimental Results of Comparative Resistor Evaluation

FABRICATION PROCESS	NOMINAL SHEET RESISTANCE OHMS/SQUARE	MATCHING TOLERANCE				TEMPERATURE COEFFICIENT ppm/°C
		σ (%)		MEAN (%)		
		10μ	40μ	10μ	40μ	
DIFFUSION	135	.44	.23	−0.1	0.07	1500
THIN FILM	1000	.24	.11	−0.1	−0.06	−200
ION IMPLANTATION	1250	.34	.12	−0.04	0.05	+400

$$\text{mean match} = \frac{\overline{\Delta R}}{R} = \frac{1}{N} \sum_{i=1}^{N} \frac{\Delta R_i}{R_i}$$

$$\sigma \text{ match} = \left[\frac{1}{N-1} \sum_{i=1}^{N} \left(\frac{\Delta R_i}{R_i} - \frac{\overline{\Delta R}}{R} \right)^2 \right]^{1/2}$$

where $\Delta R_i = R_{i1} - R_{i2}$ and $R_i = (R_{i1} + R_{i2})/2$, R_{i1} and R_{i2} being the ith matched pair. The four columns in Table I under matching tolerance give the sigma and the mean for resistors of 10 μ and 40 μ linewidth, respectively.

An independent calculation of mean and standard deviation allows a better comparison of the three processes. The mean reflects biased errors due to mask offsets and/or geometry and is relatively independent of the particular technology used. The standard deviation by contrast indicates statistical variations characteristic of the uniformity of the particular process. Also, the effect of resistor width on matching is more evident in this comparison in that wider resistors have significantly smaller standard deviations in all cases. The TC's of resistance which were obtained for the three different processes are given in the last column of Table I. It should be noted that the temperature sensitivity of ion-implanted resistors can be varied over a wide range, depending on the ion-implant process variables such as implant dose, energy, and anneal conditions [6].

ACKNOWLEDGMENT

The authors would like to thank Dr. J. Marley, Dr. J. T. Kerr, Dr. J. Conragen, J. Ennals, F. Coburn, and Dr. D. Kleitman for their technical contributions and support; H. Hart and R. Sahm for their characterization efforts; and M. Luft for her secretarial assistance.

REFERENCES

[1] R. J. Widlar, "Some circuit design techniques for linear integrated circuits," *IEEE Trans. Circuit Theory*, vol. CT-12, pp. 586–590, Dec. 1965.
[2] H. Johnson, "The anatomy of integrated-circuit technology," *IEEE Spectrum*, vol. 7, pp. 56–66, Feb. 1970.
[3] W. S. Schopfer, "A variable pulsewidth D/A," in *ISSCC Dig. Tech. Papers*, Feb. 1971.
[4] D. F. Hoeschele, *Analog to Digital and Digital to Analog Conversion Techniques*. New York: Wiley, 1968, pp. 108–120.
[5] D. H. Sheingold, *Analog-Digital Conversion Handbook*, Analog Devices, Inc., Norwood, Mass., p. I-50, 1972.
[6] H. H. Stellrecht, D. S. Perloff, and J. T. Kerr, "Precision ladder networks using ion implantation," in *1971 WESCON Tech. Papers*, Session 28.
[7] M. B. Rudin, R. O'Day, and R. Jenkins, "System/circuit device considerations in the design and development of a D/A and A/D integrated circuits family," in *ISSCC Dig. Tech. Papers*, Feb. 1967.
[8] R. J. Widlar, "New developments in IC voltage regulators," *IEEE J. Solid-State Circuits*, vol. SC-6, pp. 2–7, Feb. 1971.
[9] J. G. Graeme, G. E. Tobey, and C. P. Huelsman, *Operational Amplifiers*. New York: McGraw-Hill, 1971, pp. 54–55.
[10] R. W. Russell and T. M. Frederiksen, "Automotive and industrial electronic building blocks," *IEEE J. Solid-State Circuits*, vol. SC-7, pp. 446–454, Dec. 1972.

An Inherently Monotonic 12 Bit DAC

JOHN A. SCHOEFF, MEMBER, IEEE

Abstract—The design of 12 bit D/A converters has traditionally required precision thin-film resistors, a trimming method, and a binarily weighted ladder network. This paper describes a 12 bit DAC which uses diffused resistors and requires no trimming to guarantee monotonicity for all grades over the temperature range. The segmented ladder design, departing from the traditional R-$2R$ approach used in virtually all high-speed high resolution converters, provides inherent monotonicity and differential linearity as high as 13 bits. Also afforded is a more uniform step size over the temperature range than the trimmed 12 bit converters. The only critical resistor matching occurs at the major carries or midpoints of each of the eight segments, and the tolerances are equivalent to that of a 9 bit DAC, or eight times lower than the R-$2R$ approach. In essence, the problem has been divided into eight separate problems, each with tolerances eight times lower than that of a 12 bit DAC. The converter has been found to be immune to variations in temperature, time, process, and mechanical stress. The circuit also features differential high compliance current outputs, wide supply range, and a multiplying input.

INTRODUCTION

IN the 1950's and early 1960's, converters were built in large rack mounted units, first using vacuum tubes and later with discrete transistors. As components became smaller, it was possible to put an entire converter on a printed circuit board and later into a smaller module surrounded by potting compound. As integrated circuits became more powerful, converters were made with multiple chips and laser trimmed thin film resistors in a hybrid assembly [1]. More recently, monolithic converters using thin film resistors and laser trimming have been fabricated [2]. This paper will describe the first high speed 12 bit digital to analog converter to be built with a standard bipolar process and using diffused resistors. The converter requires no trimming to achieve performance exceeding 12 bit requirements.

DESIGN GOALS

The design goals for this converter are summarized in Table I. The prime design goal was to make the converter component tolerant and inherently monotonic, yielding 0.01 percent performance with 0.1 percent components over the military temperature range. It would require no trimming of component values to achieve 12 bit accuracy. In addition, no dynamic error correction techniques would be employed—the operation would be fully static. Also, no special test equipment, software, equipment operators, or the high level of maintenance normally required for trimming would be necessary. No burn-in of components would be needed to achieve stability. Since there would be no trimming, no extra chip area for a trim network would be consumed. No special technologies such as

Manuscript received May 10, 1979; revised July 30, 1979.
The author is with Advanced Micro Devices, Sunnyvale, CA 94086.

TABLE I
DESIGN GOALS FOR 12 BIT DAC

Component tolerant, inherently monotonic—
    0.01 percent performance with 0.1 percent components
    over temperature range
No trimming of any kind
No dynamic error correction—fully static operation
No special test equipment
    software
    operators
    maintenance
    burn-in
No special technology—
    thin film
    dielectric isolation
    double layer metal
    hybrid assembly
High speed
Low cost

thin film, ion implant, dielectric isolation, double layer metal, or hybrid assembly would be needed. Finally, the converter would have a fast settling time and be produced at a very low cost.

TECHNOLOGY

Since it was decided not to use any special processing, diffused resistors are the only remaining choice. Yet diffused resistors have a poor reputation for use in precision applications. Among the disadvantages of diffused resistors are a high temperature coefficient—approximately 2000 ppm/°C. They are pressure sensitive and have a high voltage coefficient which means they are nonlinear with applied voltage. They are not trimmable, and by reputation they match poorly. Are diffused resistors a practical choice?

Diffused resistors do have some advantages. They are built with a standard bipolar process so that they are simple and compatible with high volume manufacturing techniques. They are stable because they are diffused at 1000° and are thermally oxidized. Also, they are not altered by trimming which might disturb their long term stability. Diffused resistors are fabricated in single crystal material rather than with an amorphous substance which might change conductivity with time and temperature. Their contact resistance is stable since the contact is a simple alloying of aluminum and silicon rather than a sandwich involving polycrystalline films and barrier metals. They track very tightly with temperature since they are fabricated simultaneously in close proximity within the same material. No burn-in is required for stability. If laid out properly, for all practical purposes they match as well as thin film resistors. In short, they are economical, they match, and they stay put. Table II summarizes the advantages and disadvantages of diffused resistors.

Reprinted from *IEEE Journal of Solid-State Circuits*, Vol. SC-14, No. 6, pp. 904-911, December 1979.

Disadvantages:

  high temperature coefficient: 2000 ppm/°C
  pressure sensitive
  high voltage coefficient
  not trimmable
  "poor matching"

Advantages:

  standard process—simple, compatible
  stable—diffused at 1000°C
    thermally oxidized
  single crystal material
  contact resistance tracking
  tight temperature tracking
  no burn-in required
  match as well as thin film

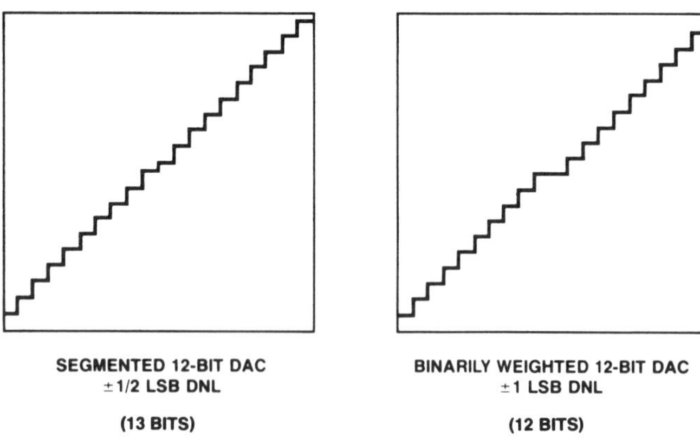

Fig. 1. Differential nonlinearity versus temperature.

PERFORMANCE GOALS

A converter is defined as monotonic if the output increases with an increasing digital input code. If the converter is monotonic, its output will yield the full number of levels indicated by the number of bits of resolution. Differential linearity is a measure of the uniformity of step size of the converter, and monotonicity is assured if the differential nonlinearity is less than ±1 LSB. The diagram on the right in Fig. 1 shows a converter which is just barely monotonic, having a differential nonlinearity of −1 LSB, indicated by the flat spot in the center. All existing available 12 bit converters are specified in this way over temperature. The dominant design approach used in these converters is the R-$2R$ ladder, which requires very precise resistor tracking to maintain monotonicity.

In many applications it is desirable to have all 4096 converter output levels separate and distinct, and, thus, a differential nonlinearity specification of $\frac{1}{2}$ LSB maximum is required as shown in the left portion of Fig. 1. The new design philosophy, called the segmented approach, was used to obtain 13 bit differential linearity over the temperature range. Note that it is not possible to determine linearity, or conformance to a straight line through zero and full scale, from either portion of the figure. Nonlinearity of ±0.01 percent maximum is required with an R-$2R$ structure to guarantee monotonicity, and this places very tight matching and tracking demands on

the resistors, increasing costs. As will be shown, ±0.01 percent nonlinearity and the tight resistor requirements are not necessary for the segmented converter to be monotonic.

Very few applications actually require ±0.01 percent nonlinearity, and those which do are usually measurements which can utilize the slow integrating methods of conversion. Even the best transducers have no better than 0.1 percent nonlinearity. In video display systems, the human eye has difficulty discerning less than 5 percent nonlinearity.

R-$2R$ LADDER

Fig. 2 shows the traditional binary weighted R-$2R$ ladder network design. Virtually every high-speed, high-accuracy converter available today uses this technique or a minor variation on it. The converter consists of 12 binary weighted current sources which are used in all possible binary combinations to produce 4096 analog output levels. The main advantage of the R-$2R$ technique is that it uses a minimum number of components, but these must match and track well over temperature. The most critical resistor in the circuit is the most significant bit (MSB) resistor. If the full scale current of this converter is 4 mA, then the MSB will be 2 mA and each succeeding bit will be divided by 2 all the way down to the least significant bit (LSB), which is 1 μA. At the major carry, the eleven LSB's will be turned on producing an output current of 1.999 mA for an input code of 011111111111. When the input code is incremented one count, becoming 100000000000, the lower order current sources turn off and the MSB source turns on and yields an output current of 2.000 mA. If this source has an error of more than −1 μA, the converter will be nonmonotonic. This corresponds to a resistor error of +0.05 percent, which must be maintained over the entire operating temperature range. Even when the resistors match well, laser trimming can alter their tracking characteristics and affect yield over temperature. Even though we have learned to match diffused resistors to within ±0.05 percent, they are not really practical with the R-$2R$ technique due to their high temperature coefficient, piezoresistance, and voltage coefficient.

2^nR LADDER

One design approach which provides monotonicity without requiring high linearity is the MOS switch-resistor string. This circuit is actually a full complement to a current switched R-$2R$ DAC since it is slower, has a voltage output, and, if implemented at the 12 bit level, would use 4096 low tolerance resistors rather than a minimum number of high tolerance resistors as in the R-$2R$ network. Its lack of speed and density for 12 bits are its drawbacks.

The technique described in this paper combines the advantages of both the R-$2R$ and 2^nR approaches. It is inherently monotonic, fast, and uses untrimmed resistors which are actually fewer in number than the classic R-$2R$ ladder.

SEGMENTED LADDER

Table III compares the resistor requirements for the standard R-$2R$ approach versus the new segmented design technique. The R-$2R$ requires 37 resistors, and the resistors must match to within ±0.05 percent to guarantee monotonicity, which is

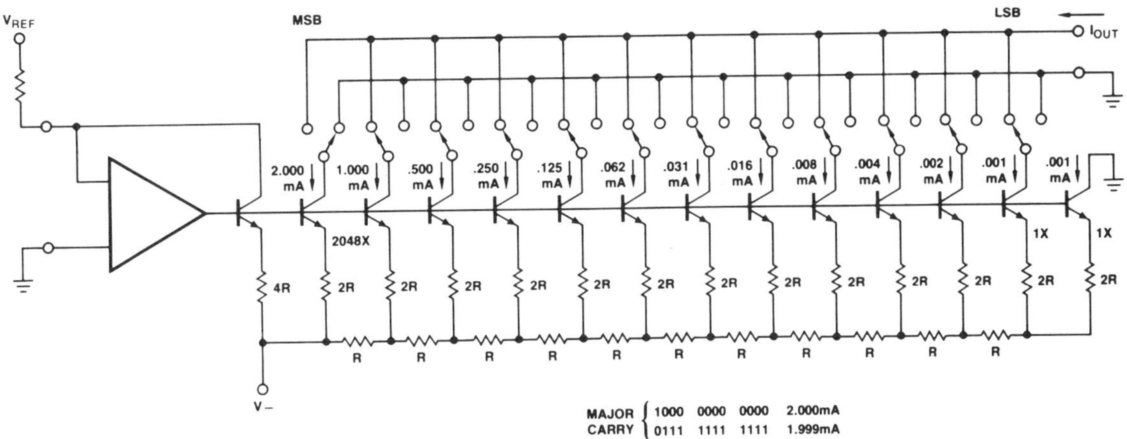

Fig. 2. Traditional R-$2R$ D/A converter.

TABLE III
LADDER RESISTOR COMPARISON

LADDER TYPE	NO. OF RESISTORS	INITIAL MATCHING REQUIRED FOR ±1 LSB DNL (%)	TRACKING REQUIRED FOR ±1 LSB DNL (ppm/°C)		TRACKING REQ'D FOR $\pm1/2$ LSB DNL (ppm°/C)
			0 INITIAL DNL	1/2 LSB INITIAL DNL	1/4 LSB INITIAL DNL
STRAIGHT R-2R	37	$\pm.05$	5	2.5	1.25
SEGMENTED 3 + 9 BITS	24	$\pm.4$	40	20	10

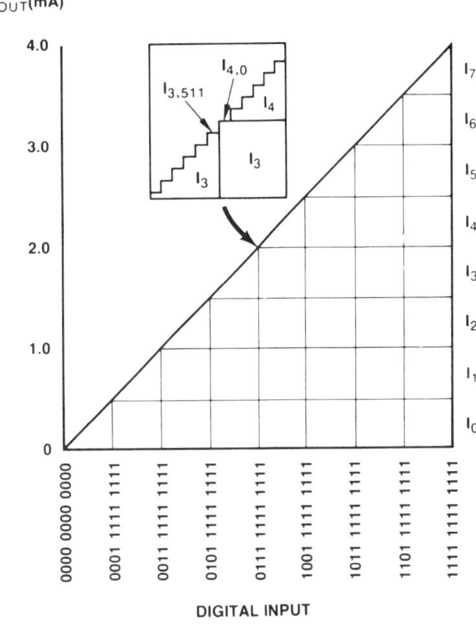

Fig. 3. Segmented DAC transfer characteristic.

defined by ±1 LSB differential nonlinearity. The remaining numbers indicate resistor temperature tracking requirements. If the converter is perfect at room temperature, resistor tracking within ±5 parts/million/°C would maintain monotonicity from 25 to 125°C. Some allowance must be made for error, however, so if the initial differential nonlinearity were $\frac{1}{2}$ LSB, a ±2.5 parts/million/°C tracking figure would be necessary. To obtain 13 bit performance over temperature, or $\pm\frac{1}{2}$ LSB DNL, allowing $\pm\frac{1}{4}$ LSB of initial error would produce a resistor tracking requirement of ±1.25 ppm/°C. This is difficult to maintain with almost any technology.

The new segmented approach requires only 24 resistors and the initial resistor matching is ±0.4 percent. Note that the numbers are relaxed by a factor of eight with respect to the R-$2R$ ladder. The tracking requirements for diffused resistors in the segmented ladder are ±40, ±20, and ±10 ppm/°C, respectively. Typical diffused resistor tracking is on the order of ±2 ppm/°C, so there is plenty of room to maintain 13 bit differential linearity over the military temperature range.

The DAC output current as a function of the digital input code is shown in Fig. 3. The output consists of 4096 analog levels divided into eight groups of 512 steps each. The key to the inherent monotonicity may be seen by examining the major carry, shown in the inset. Rather than switching in an entirely different current at the major carry as the R-$2R$ converter does, the current from the segment prior to the major carry I_3 is retained and the current to create additional steps is added to it in the form of increments of I_4. Thus, the converter is monotonic regardless of the relative slopes of the eight segments [4]. The only critical resistor matching occurs at the major carries or midpoints of each of the eight segments, and

the tolerances are equivalent to that of a 9 bit DAC, or eight times lower than the R-$2R$ approach. In essence, the problem has been divided into eight separate problems each with tolerances eight times lower than that of a 12 bit DAC.

DAC DESIGN

The functional diagram of the segmented 12 bit DAC is shown in Fig. 4. On the left is the segment generator and decoder, and on the right is the 9 bit DAC which generates the 512 levels within each segment. For a given 3 bit code at the input of the decoder, a segment, for example I_3, is selected and fed to the 9 bit DAC where it is divided into 512 levels. All the lower order segments are fed to I_{out} and summed with the 9 bit DAC output, and all the higher order segments are fed to $\overline{I_{out}}$ which in this description can be thought of as ground. At a segment carry such as with the input code 011111111111, 511 of the 512 levels of the 9 bit DAC appear at I_{out}. The 512th level is the remainder current on the right which is fed to ground. When the code is incremented one count, I_3 is switched away from the 9 bit DAC and fed to I_{out}, taking the

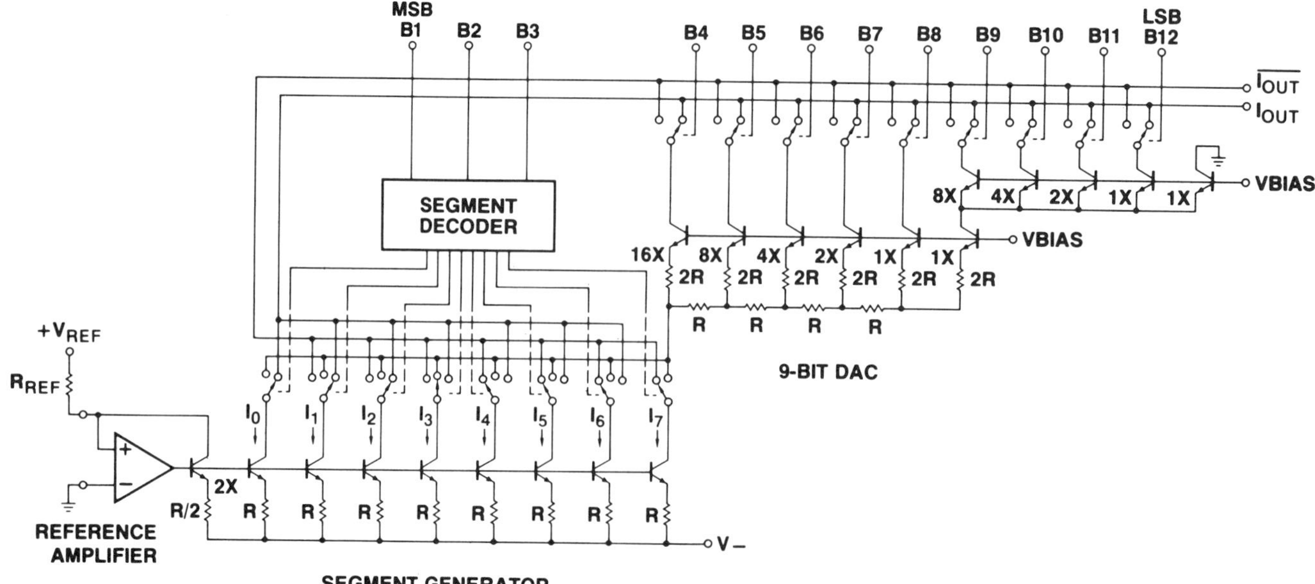

Fig. 4. Segmented DAC functional diagram.

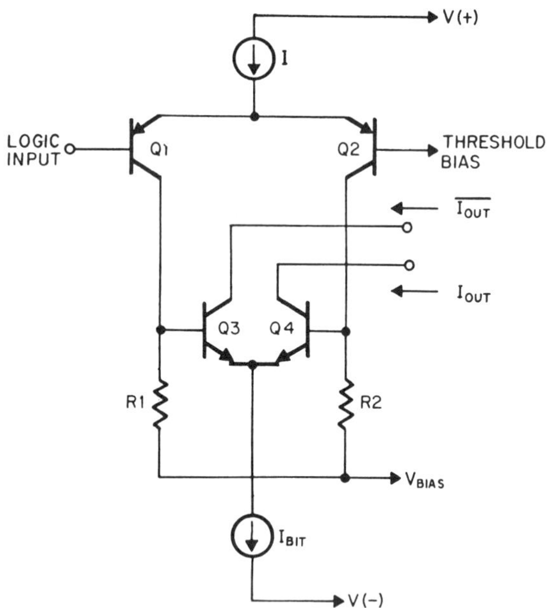

Fig. 5. High speed differential current switch.

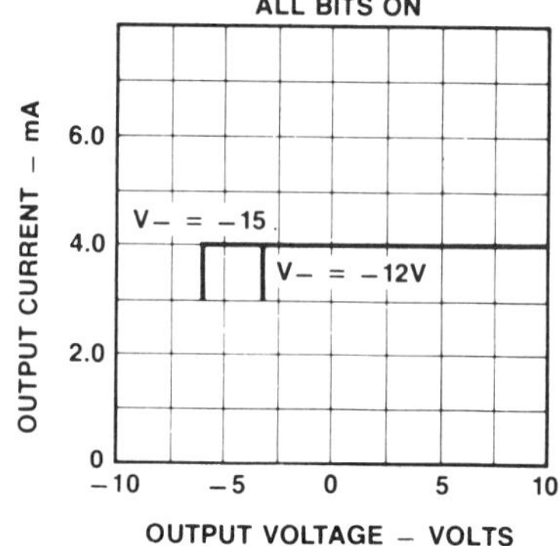

Fig. 6. Output voltage compliance.

remainder current with it. Thus, the output of the DAC increases by one LSB and is inherently monotonic. This of course assumes that the 9 bit DAC is monotonic, which can be achieved with high yield in a production environment. The segment carries do not depend on resistor matching at all. The most critical resistor in the circuit is now the MSB resistor of the 9 bit DAC, and its tolerance is eight times lower than that of a traditional R-$2R$ 12 bit DAC.

The currents in the 9 bit DAC are switched with fully differential current switches [1] which are capable of switching a 1 μA current in about 30 ns [4], [5]. The high speed current switch is shown in Fig. 5. The high speed is possible because the common emitter connection of the n-p-n differential pair remains at the same voltage regardless of logic state and the bit current need not charge or discharge the parasitic capacitance at this node. For a capacitance of 2 pF, a single ended switch which swings 0.7 V would require 1.4 μs to turn the LSB on.

The use of this switch makes possible the high voltage compliance and high impedance complementary differential outputs [4]-[6]. A curve of output compliance is shown in Fig. 6. The output impedance is typically 10 MΩ. The outputs I_{out} and $\overline{I_{out}}$ are analog complements of each other, which means that I_{out} increases and $\overline{I_{out}}$ decreases with an increasing digital input code. The sum of the two outputs is equal to the full scale current regardless of input code. Other advantages of the differential current switch include adjustable logic threshold by varying the base bias on the p-n-p opposite the input, a logic input range from below ground to above the positive power supply, fabrication with a standard bipolar process, and level shifting independent of the positive or negative power supply.

Because low currents may be switched at high speed, the ladder current may be binarily divided all the way down to the LSB with no need for artificial current boosting or switching of equal current sources with output attenuation. The 9 bit DAC in the segmented converter utilizes a "master-slave" ladder arrangement whereby the 5 MSB's are generated by a R-$2R$ ladder and the remainder current of this ladder is divided by active current splitting into the 4 LSB's [4], [5]. This minimizes the range of emitter scaling necessary to generate nine binarily weighted currents. It eliminates the ladder resistors in the lower order bits where the tolerance for error as a

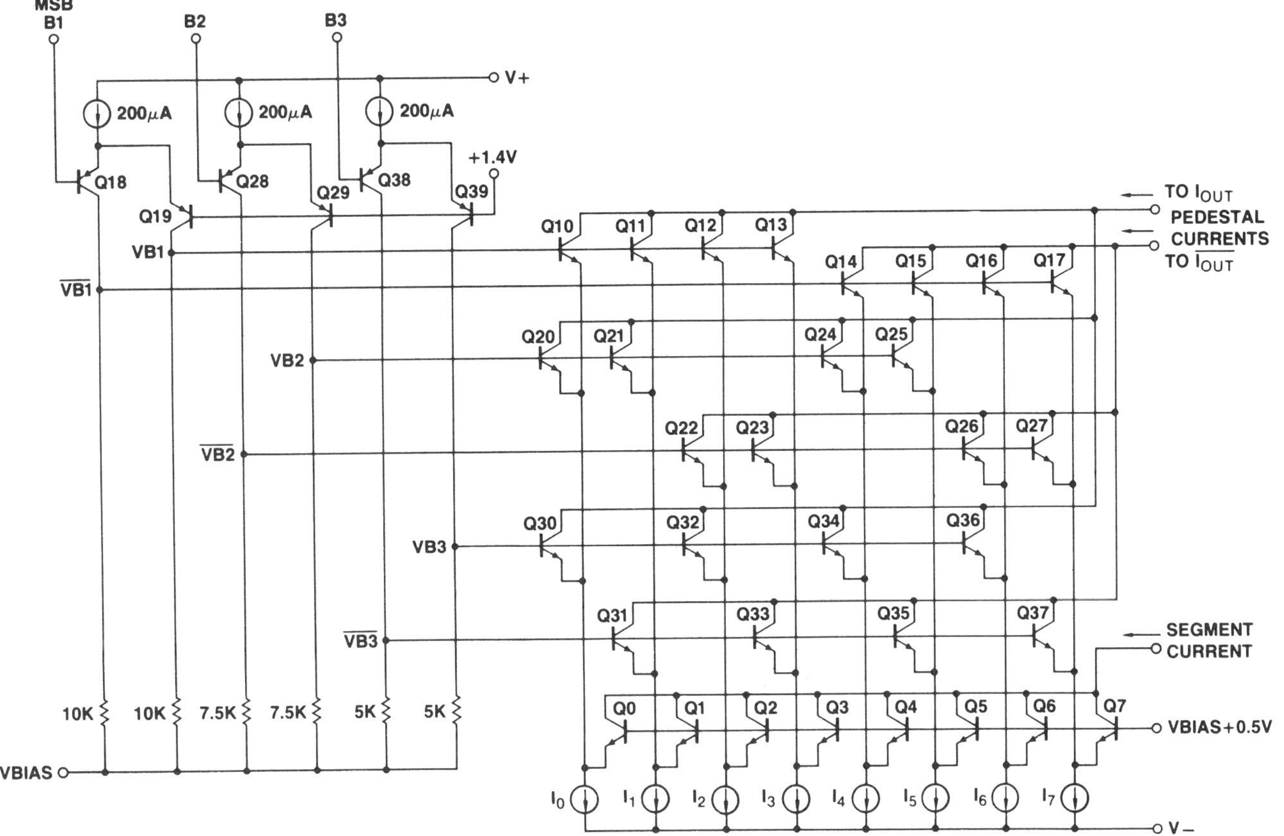

Fig. 7. Decoded segment switch.

percent of the bit current value is much higher. Ladder resistors in these bits would not be highly useful anyway, since the voltage drop across them would be very small.

The resistors in the segment generator have no bearing on the differential linearity of the converter, but determine the linearity or straight line conformance. Each resistor carries one-eighth of the full scale current, and thus its value is less critical than the MSB resistor of an R-$2R$ ladder. At the major carry, the left four resistors must match the right four resistors, so the linearity of this converter benefits from rms averaging and is actually higher for a given resistor tolerance than that of an R-$2R$ converter.

In review, the ladder network for the 12 bit DAC actually incorporates three philosophies: 1) active current scaling in the slave network for the LSB's where accuracy may be traded for area savings, 2) R-$2R$ current division in the master ladder where up to 9 bit accuracy is required, taking advantage of the precision diffused resistors, and 3) segment generation for the 3 MSB's where resistor matching and tracking of either diffused or thin film resistors would not be adequate.

DECODED SEGMENT SWITCH

The three segment switch drivers are in the left half of Fig. 7, and the segment current sources and decoded analog multiplexed switches are in the right portion. The overall function has three digital inputs which switch eight precision current sources in proper priority to three analog outputs. If standard binary building blocks were used to achieve this function, a one of eight decoder and two priority encoders would be necessary in addition to a large number of high speed current switches. The key to this circuit lies in the use of multiple

TABLE IV
SEGMENT DECODER LOGIC LEVELS IN VOLTS ABOVE V_{bias}

LOGIC INPUT	VB1	$\overline{\text{VB1}}$	VB2	$\overline{\text{VB2}}$	VB3	$\overline{\text{VB3}}$
111	2.0	0	1.5	0	1.0	0
110	2.0	0	1.5	0	0	1.0
101	2.0	0	0	1.5	1.0	0
100	2.0	0	0	1.5	0	1.0
011	0	2.0	1.5	0	1.0	0
010	0	2.0	1.5	0	0	1.0
001	0	2.0	0	1.5	1.0	0
000	0	2.0	0	1.5	0	1.0

logic levels [6]. The nodes $VB1$, $VB2$, and $VB3$ have "high" levels which are spaced 0.5 V apart, and including the low level makes a total of four logic levels. This adds an extra dimension to the switch matrix which simplifies it considerably. Table IV shows the node voltages in the decoder for all binary combinations of the 3 MSB's. These voltages are shown with reference to the V_{bias} line. Note that the underlying philosophy for this approach is to rank the bit switch driver output voltage levels according to the significance of each bit. In operation, the transistor group $Q10$ through $Q13$ is able to override all transistors beneath it, so that when the MSB is high I_0 through I_3 will be routed to I_{out}. For any given input code, seven of the eight segment currents will be switched to the output pair and the eighth will be presented to the 9 bit DAC at the segment current output. The complete truth table for the decoded segment switch is shown in Table V.

The segment decoder occupies only three isolation pockets since it makes liberal use of common collector connections. Many common base connections also enhance its density. The multilevel segment switch is very fast, since a switching

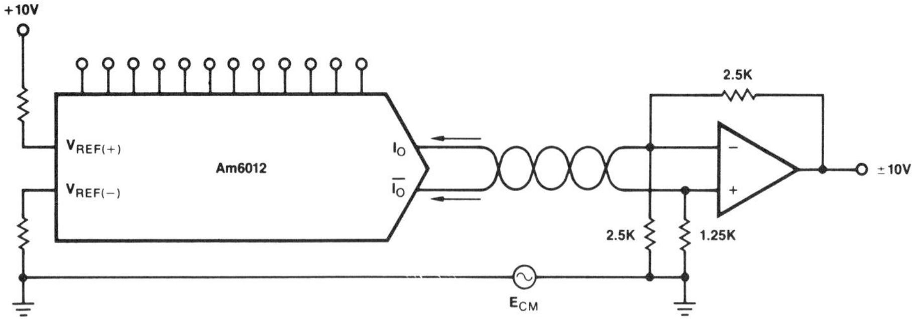

Fig. 8. Balanced load drive.

TABLE V
SEGMENT DECODER TRUTH TABLE

LOGIC INPUT	SEGMENT DECODER OUTPUT							
111	I_O	I_O	I_O	I_O	I_O	I_O	I_O	I_s
110	I_O	I_O	I_O	I_O	I_O	I_O	I_s	$\overline{I_O}$
101	I_O	I_O	I_O	I_O	I_O	I_s	$\overline{I_O}$	$\overline{I_O}$
100	I_O	I_O	I_O	I_O	I_s	$\overline{I_O}$	$\overline{I_O}$	$\overline{I_O}$
011	I_O	I_O	I_O	I_s	$\overline{I_O}$	$\overline{I_O}$	$\overline{I_O}$	$\overline{I_O}$
010	I_O	I_O	I_s	$\overline{I_O}$	$\overline{I_O}$	$\overline{I_O}$	$\overline{I_O}$	$\overline{I_O}$
001	I_O	I_s	$\overline{I_O}$	$\overline{I_O}$	$\overline{I_O}$	$\overline{I_O}$	$\overline{I_O}$	$\overline{I_O}$
000	I_s	$\overline{I_O}$	$\overline{I_O}$	$\overline{I_O}$	$\overline{I_O}$	$\overline{I_O}$	$\overline{I_O}$	$\overline{I_O}$
	I_0	I_1	I_2	I_3	I_4	I_5	I_6	I_7

TABLE VI
ELECTRICAL SPECIFICATIONS

Resolution	12 bits
Monotonicity	>12 bits
Differential nonlinearity	±0.01 percent (13 bits)
Nonlinearity	±0.01, 0.02, 0.05 percent
Settling time to ±0.01 percent	250 ns
Output impedance	10 MΩ, 20 pF
Output voltage compliance	–5 to +10 V
Full scale drift	±5 ppm/°C typ.
Reference input FS transition	500 ns
Logic inputs—threshold, input current	1.4 V
	4 μA
Power supply voltages	+5 to +18 V
	–12 to –18 V
Power supply sensitivity	0.001 %/% max
Power dissipation +5, –15 V	230 mW
Operating temperature range	–55 to +125°C

signal must propagate only through a p-n-p driver pair to reach the base of the current switching devices, making it possible to switch at approximately the same speed as a single fully differential bit current switch. A name for this circuit function might be a "high speed multilevel priority encoded analog multiplexer."

ELECTRICAL SPECIFICATIONS

Table VI lists the electrical specifications of the converter. It has 12 bit resolution and virtually all functional parts are monotonic. Differential nonlinearity is as low as ±0.01 percent or ±½ LSB, which defines 13 bit performance. The nonlinearity is distributed over a range of ±0.01 to ±0.05 percent. Again, all parts are monotonic, so that the user who needs only monotonicity or a high level of differential linearity does not need to pay for 0.01 percent integral linearity.

The settling time to within ±½ LSB typically is 250 ns and the outputs have a high impedance, typically 10 MΩ, with a voltage compliance of –5 to +10 V. The full scale drift is typically ±5 ppm/°C, and the multiplying reference input can undergo a full scale transition in 500 ns.

The logic inputs are transistor–transistor logic (TTL) compatible with 4 μA input current, and the device operates over a range of power supplies from +5 to +18 V and –12 to –18 V. The power supply rejection is 0.001 %/% maximum so that all the accuracy specifications are essentially independent of the supplies. The power dissipation is typically 230 mW.

In short, the device described here has speed close to that of dielectrically isolated laser trimmed thin film converters, and has higher differential linearity and lower power dissipation.

APPLICATIONS

The high impedance differential current outputs lend themselves to a number of applications which cannot be performed by either voltage or resistive output converters. Many converters which claim to have current outputs have a ladder network or resistor attenuator which interfaces with the load and lowers the output resistance. Thus, either the full scale current or the linearity is degraded when the output is operated away from ground. These outputs should be termed resistive and any voltage compliance specification given with them should be discounted. Offset binary operation with a balanced load is shown in Fig. 8. The balanced load converts the difference of the two output currents into a single ended voltage so that no offset resistor from the reference supply is required [5]. Thus, the analog ground at the load may be unrelated to ground at the converter, and a twisted pair may be used for current transmission if the converter and load are separated, as in a control system application. The extra 2.5 K load resistor to ground gives the amplifier an effective gain of 2 so that it may swing ±10 V even though the DAC output compliance is –5 V.

A 12 bit A/D converter with a conversion time of less than 10 μs is shown in Fig. 9. In the successive approximation process the load current for the sample and hold amplifier changes for every trial period. This disturbs the feedback loop which then must settle to within 0.01 percent of full scale in less than 1 μs. Another way to describe the problem is that the effective output impedance of the sample and hold is increasing with frequency. When using a DAC with differential high compliance outputs, $\overline{I_{out}}$ may be connected to present a constant load current to the sample and hold and thus increase A/D throughput.

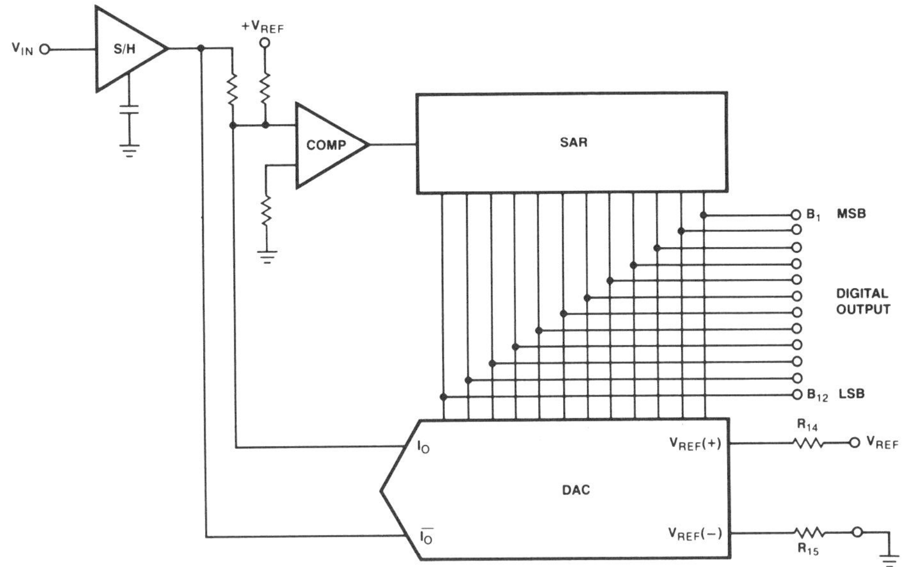

Fig. 9. High speed 12 bit A/D converter.

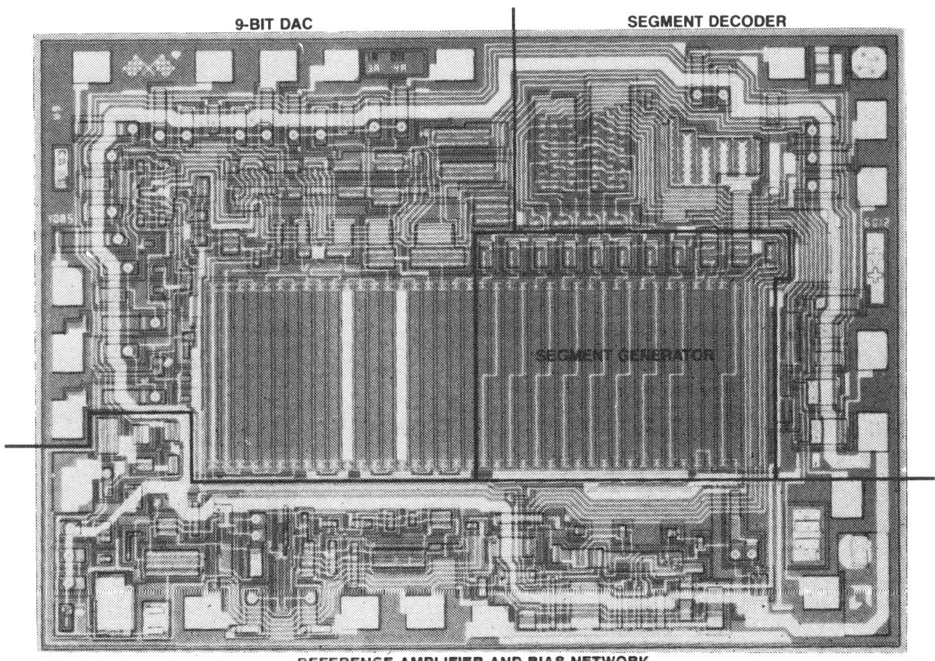

Fig. 10. Chip photograph.

Chip Layout

The converter is fabricated with a standard bipolar linear process and diffused resistors on a 12 500 square mil chip, and is packaged in a compact 20 pin dual inline package. The ladder network for the 9 bit DAC is in the left center of Fig. 10, and the segment generator resistors are in the right center. with the ROM-like analog multiplexer above them. The three decoder drivers are along the upper right of the chip, and the bias network and reference amplifier extend along the bottom. The 9 bit DAC switches and current sources occupy the upper left.

Device Performance

Fig. 11 is a plot of the nonlinearity of an actual device, with all 4096 digital input codes on the horizontal axis and the nonlinearity, or deviation from an ideal straight line, plotted in LSB's on the vertical axis. Each LSB corresponds to 0.025 percent of full scale, so it can be seen that this particular device lies within an error band of ±0.012 percent nonlinearity. Differential nonlinearity is a measure of the deviation from ideal step size, and is represented by individual jumps or steps in the curve. If the differential nonlinearity were zero, the curve would be smooth, but still could deviate from an

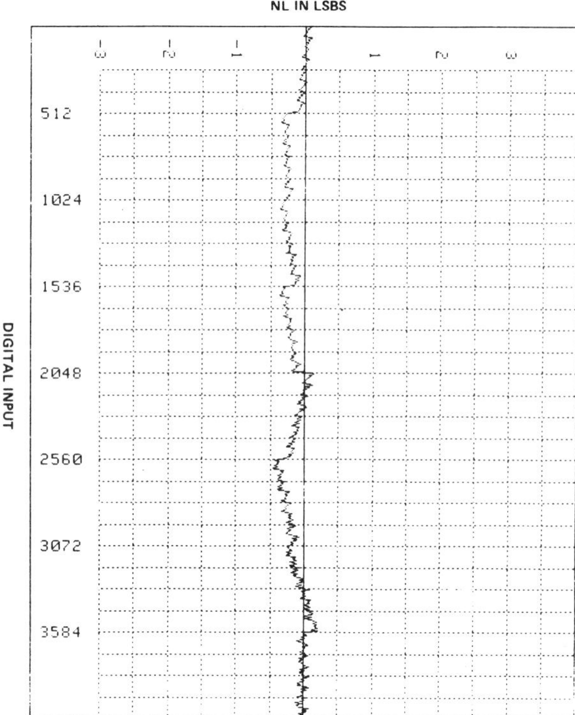

Fig. 11. Nonlinearity of actual device.

ideal straight line with the segmented converter. The largest differential error on this device is well under $\frac{1}{2}$ LSB for an overall differential nonlinearity of better than ±0.01 percent, which is in excess of 13 bit performance.

Summary

The design and fabrication of a 12 bit trimless DAC with a standard process will make high resolution converters widely available at low cost. In addition, a 12 bit DAC may be used as a building block in even larger integrated system applications.

In summary, a new algorithm has been shown for a high speed 12 bit converter and has been implemented with a high performance circuit design and a standard production process. The converter exhibits a low sensitivity to resistor absolute value and ratio and is resistant to temperature and mechanical stress. The device has a low component count, high speed, and superior differential linearity. It operates over the military temperature range and is now in high volume production.

Acknowledgment

The author wishes to express gratitude for the contributions of S. Zoehrer in mask design, G. Roflox in wafer fabrication, I. Sziebert in component evaluation, and M. Mullen in test hardware and programming.

References

[1] R. B. Craven, "An integrated circuit 12-bit D/A converter," in *ISSCC Dig. Tech. Papers*, Feb. 1975, pp. 40–41, 212.
[2] R. W. Webb, "A high-speed 12-bit monolithic D/A converter," in *ISSCC Dig. Tech. Papers*, Feb. 1978, pp. 142–143.
[3] A. R. Hamade and E. Campbell, "A single-chip 8-bit A/D converter," in *ISSCC Dig. Tech. Papers*, Feb., 1976, pp. 154–155.
[4] J. A. Schoeff, "A monolithic companding D/A converter," in *ISSCC Dig. Tech. Papers*, Feb. 1977, pp. 58–59.
[5] J. A. Schoeff and D. Soderquist, "DAC-08 applications collection," Precision Monolithics, Inc., Application Note AN-17, 1975.
[6] J. A. Schoeff, "A microprocessor compatible high-speed 8-bit DAC," in *ISSCC Dig. Tech. Papers*, Feb. 1978, pp. 132–133.

An 8-Bit, 5 ns Monolithic D/A Converter Subsystem

PETER H. SAUL, MEMBER, IEEE, PETER J. WARD, MEMBER, IEEE, AND A. J. FRYERS

Abstract—This paper describes a 5 ns settling time digital-to-analog converter device, which has been designed for use in video speed successive approximation analog to digital converters. The chip includes a precision reference source with a 25 ppm per degree C average temperature coefficient and a high-speed comparator.

The successive approximation approach, restricted to low-speed converters until now, has the advantages of low cost and straightforward drive requirements.

The achievement of the operating speeds described is dependent both on the circuit techniques used and the process employed. The DAC circuit, unlike most other devices, uses a multiple-matched current source array technique, which leads to a very linear, low glitch output. Without any form of trimming, most functional devices meet a $\pm\frac{1}{2}$ L.S.B. differential and integral linearity specification, and many are $\pm\frac{1}{4}$ L.S.B. or better.

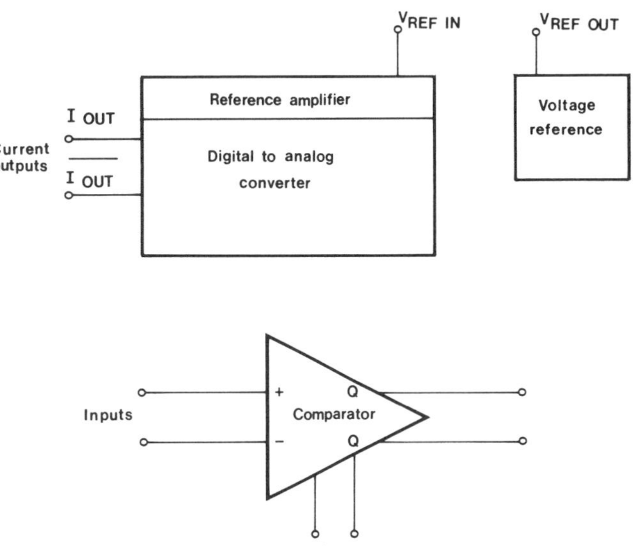

Fig. 1. The DAC subsystem. Chip contains the analog parts needed for a successive approximation ADC.

INTRODUCTION

THIS paper describes a user-orientated high-speed digital-to-analog converter subsystem intended for use in video analog-to-digital converters. The device was first described at ISSCC 1980 [1]. Until now, the market for video ADC's has been dominated by parallel or series parallel types. The chief problems associated with parallel conversion are the high-input capacitance, as much as 300 pF in some systems, and the high cost of an array of 2^n comparators, whether in monolithic or multichip form. Series-parallel converters only partially alleviate these problems. The well-known successive approximation format is less demanding on the external circuitry, and has the potential of being integrated in low-cost form.

The component parts of a successive approximation video ADC are a very fast digital-to-analog converter, a comparator, a successive approximation register, and a sample and hold. The first two of these have been integrated onto one chip and have a capability of operating in an 8-bit, 15 MHz sampling ADC. To achieve this performance, the DAC has a settling time, to $\pm\frac{1}{2}$ LSB of 5 ns, and the comparator has a propagation delay, at an equivalent $\frac{1}{2}$ LSB of overdrive, of 2.5 ns. Further expansion of the DAC design principles to 10 bits is practical, and 12-bit parts are under consideration. No trimming techniques are used in the DAC's; the designs rely on predictable yield tradeoffs to achieve the specified operational accuracy.

Manuscript received April 8, 1980; revised August 1, 1980. This work was supported by the Procurement Executive, Ministry of Defence, England and DCVD.

The authors are with Plessey Research (Caswell) Limited, Allen Clark Research Centre, Northhamptonshire, England.

COMPONENT PARTS OF THE DEVICE

The device is split into three essentially independent systems: the DAC, the voltage reference, and the comparator (Fig. 1). Although sharing common power supply rails, each of these components can be used independently and each is fully accessible via the pins. The digital-to-analog converter with the associated reference amplifier is the primary part of the device and occupies approximately 70 percent of the active chip area. Settling time of the current output DAC is 5 ns to $\pm\frac{1}{2}$ LSB into a 50 Ω load. Measurements of settling times of this order require specialized techniques, which will be described later. Occupying about 10 percent of the active chip area, the voltage reference shows an average 25 ppm per degree C temperature coefficient. Finally, the comparator, which takes 20 percent of the chip, has a propagation delay at 4 mV overdrive (the operational bit size) of 2.5 ns. Using the functional separation as a guideline, we can consider the parts as separate circuits.

THE COMPARATOR

The basic comparator circuit is shown in Fig. 2. Low propagation delay is achieved by operating in a latched, low gain configuration. The first stage is a transconductance amplifier with emitter follower inputs to give an input impedance of nominally 100 kΩ in parallel with 2 pF. Diodes in the first stage loads are used to avoid second-stage saturation; all stages

Reprinted from *IEEE Journal of Solid-State Circuits*, Vol. SC-15, No. 6, pp. 1033-1039, December 1980.

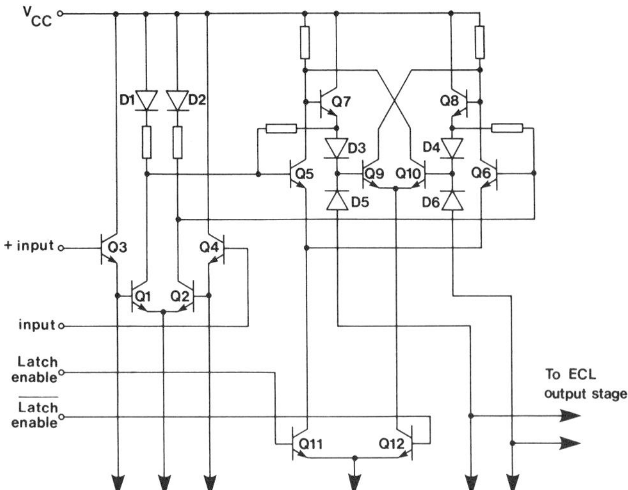

Fig. 2. Comparator circuit. Latched, low gain configuration gives low-propagation delay.

are nonsaturating, thus avoiding the need for Schottky clamp devices. The second-stage is latched transresistance amplifier with level shifting into a final, conventional ECL driver and output stage.

In the compare mode of operation, the input pair Q_3, Q_4 drive the differential pair Q_1, Q_2, which operate at a low-voltage gain set by the near-virtual-earth conditions of the inputs to Q_5 and Q_6. The feedback pairs Q_5, Q_7 and Q_6, Q_8 have a defined gain set by the resistors and the transistor parameters (primarily r_e). Tail current for Q_5 and Q_6 is supplied by Q_{11}, one of a long-tailed pair switched when necessary by the "latch enable" command. Signal output from the followers Q_7, Q_8 passes, via level shift diodes D_3, D_4 to the latch pair Q_9, Q_{10}, which are connected as a latch stage, enabled during the latch mode by Q_{12}. When the latch enable input goes high, current is switched from the pair Q_5, Q_6 to Q_9, Q_{10}, which form a positive feedback loop and, therefore, hold the state of the comparator at the time of switching of Q_{11}, Q_{12}. The differential voltage gain ahead of the latch stage is relatively low, about 16, and therefore the input offset voltage is affected to some extent by all the devices Q_1–Q_{10}, but careful layout and good process stability produce a standard deviation of 1 mV for the input offset about a mean of zero. Although the forward biased diodes D_3 and D_4 contribute to the offset term, the major level shift, 5.4 V, which comes from the zener diodes D_5 and D_6, occurs after the latched stage.

No observable thermal offset or hysteresis effects have been seen on the comparator at the high operating speeds used in measurement and in operation in a feedback ADC. At low-input slew rates in the compare mode oscillations of the output occur, indicating that the device has low hysteresis. With input offset trimming, an ultimate resolution of 100 μV is possible in the latched mode.

Outputs from the comparator are ECL compatible, with 1 ns rise and fall times into 50 Ω loads. In the unlatched mode, the input to output delay is 2.5 ns for a 100 mV signal giving 4 mV overdrive (an approximation to the operating conditions

in a successive approximation ADC). In the latched mode, the latch to output propagation delay is 1.3 ns, with an approximate setup time of 0.5 ns. Source impedance for these measurements was 50 Ω for both the analog input and the latch enable. Higher source impedances are usable provided the effects of input bias current (approximately 10 μA) on the offset voltage are taken into account, and provided that the source impedance is not so high as to limit the input slew rate to the point where oscillations occur (about 5 V/μs).

THE VOLTAGE REFERENCE

The primary requirements for the voltage reference in this circuit were a reasonably low-temperature coefficient and complete compatibility with the standard process. Typically, the reference for an 8-bit converter might be expected to have a temperature coefficient of 100 ppm per degree C, or, taking into account the usual parabolic temperature characteristics, about 1 percent change over the range $-55°C$–$+125°C$. Obviously, it is advantageous to aim for the best possible temperature coefficient, and then to estimate the production spread of the parameters affecting it. It will be shown that, using the conventional circuit of Fig. 3 and implantation processing techniques, adequate yields of low-temperature coefficient parts are achievable and, furthermore, selection of parts by temperature coefficient is possible by relation to the absolute value of the reference voltage.

In the configuration of Fig. 3, i.e., a negative voltage source referred to the ground rail, the output impedance is relatively high and temperature dependent, so a high-impedance input voltage to current converter is used in the DAC.

The circuit of Fig. 3 was laid out, with the associated temperature tracked current source, in such a way as to minimize the effects of process variation on the resistor ratios $R1:R2$ and $R2:R3$. With a nominal emitter ratio of 10:1 (well controlled due to the implant process used), R_2 and R_3 are realized using two series and five parallel equal valued resistors, respectively. The value of $R3$ was made mask adjustable, and was, in fact, stepped in two quads of circuits, one containing only the nominal value as calculated for $R3$, and a second mask set with four variants on the value of $R3$, set at ± 2 percent and ± 4 percent. The purpose of these variants was to explore a range of values and, in particular, to obtain an experimental value for the saturation current parameter.

Measurements on the batch showed that the -4 percent value of $R3$ was closest to the optimum, with an average temperature coefficient of 25 ppm per degree C. Other variants showed 40 ppm, 60 ppm, 85 ppm, and 110 ppm per degree C temperature coefficients. Repeatability within each group was extremely good; occasional devices were found with different coefficients, but these were all traced to individual die faults and were easily selected out by nonstandard absolute voltage values.

The temperature coefficient of the best variant of the above is illustrated in Fig. 4. The zero temperature coefficient point is still not optimum and some improvement could be made by a further modification to $R3$; this will be done when more batches are processed. Batch-to-batch repeatability is the key

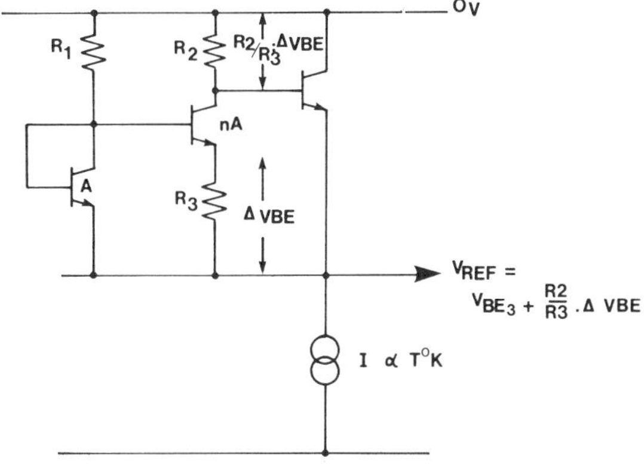

Fig. 3. Voltage reference. Full circuit includes temperature tracking current source and preregulation.

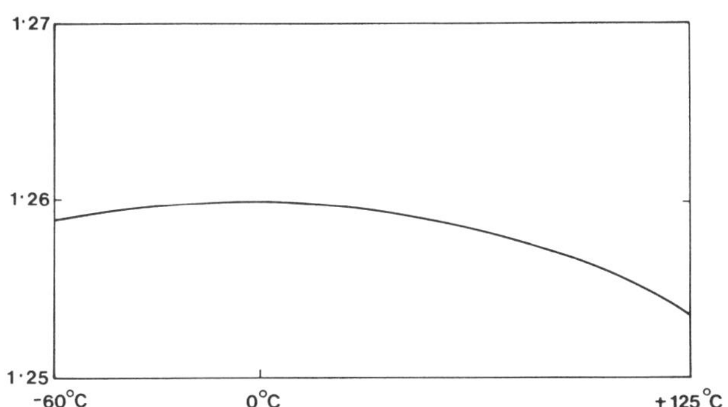

Fig. 4. Reference voltage variation with temperature.

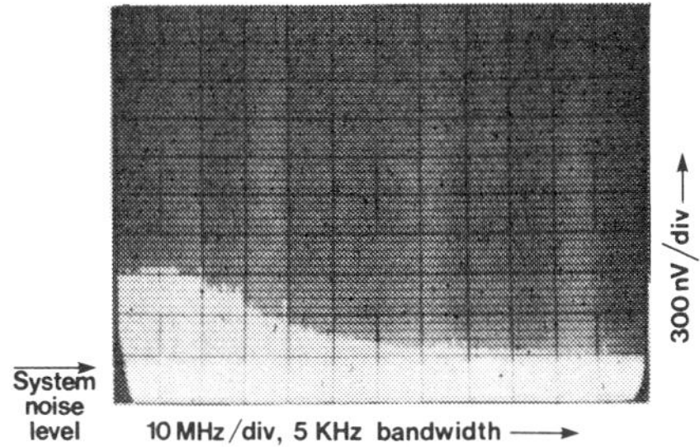

Fig. 5. Reference noise output voltage.

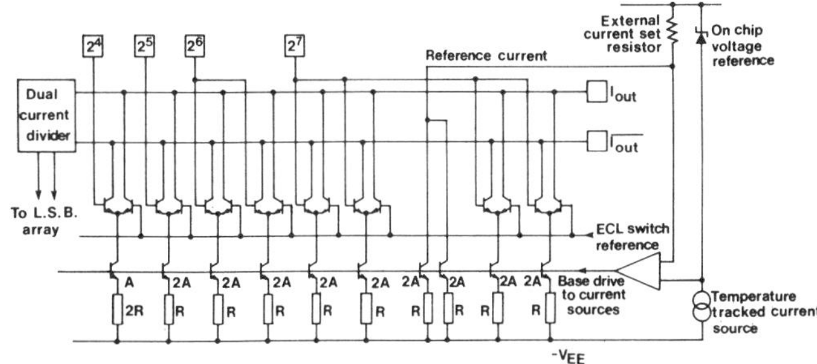

Fig. 6. The current source array technique. Feedback loop stabilizes output current against parameter variation and temperature.

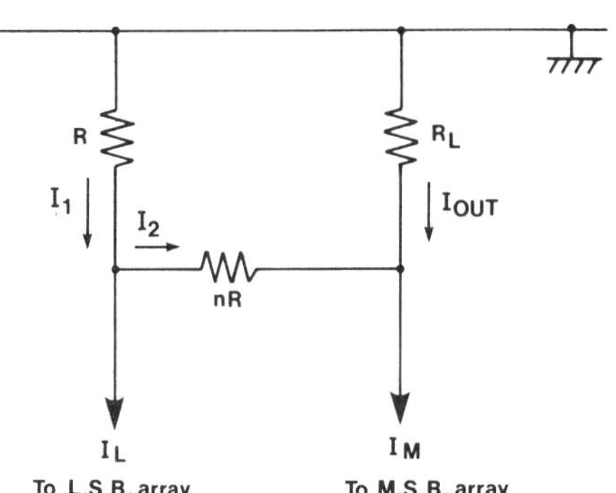

Fig. 7. The interstage current divider. Ratio used has $n = 15$.

to commercial success, so when a second version of the DAC device was designed, the same reference layout was used, including the modified $R3$. Initial results are excellent, with average temperature coefficients of less than 20 ppm per degree C. This represents an average batch in the middle of the process specification, and actually processed at a different location (Plessey Semiconductors Limited, Swindon). Process control parameters are such that all functional devices should achieve 100 ppm per degree C, and most should be less than 50 ppm per degree C.

The noise performance of bandgap voltage references has been critized in the past [3] but, in this design, the noise is measured to be 2.4 μV rms in a 100 kHz bandwidth. This is comparable with the best published data on buried zener references. The photograph (Fig. 5) shows the noise spectrum

of the reference without additional decoupling or filtering, measured in a 5 kHz bandwidth. Low-noise levels are a feature of the process, which is also used for low-noise amplifiers. The noise spectrum illustrates that the loop gain bandwidth of the reference source is relatively wide in order to maintain good supply voltage rejection in the presence of switching transients on the supplies.

THE DIGITAL-TO-ANALOG CONVERTER

The digital-to-analog converter, shown in Fig. 6, uses a multiple current source approach [4]. The three most significant bits contain four, two, and one equal value current sources, respectively. The fourth MSB has a half-value current source. A second current source array containing the four least significant bits connects to the group of four most significant bits via a dual-current divider. All devices in the array are matched and switch equal currents at equal current densities and, therefore, at very closely matched speeds on a given chip. This is not generally possible in fast R-$2R$ ladder networks, where each stage operates at half the current of its predecessor.

The interstage current divider, shown in Fig. 7, consists of an array of resistors which forms two 15 to 1 ratio current paths, returning $\frac{15}{16}$ of the least significant bit current to the ground rail and $\frac{1}{16}$ to the current outputs. This technique has the interesting property of being independent, in division ratio terms, of the output load. The load may be a resistor, a virtual earth input, or a comparator input in a feedback path.

The proof of the current division relationship is straightforward. With the notation of Fig. 7, we can write three simultaneous equations

$$I_{\text{OUT}} \cdot R_L = I_1 \cdot R + I_2 \cdot nR \tag{1}$$

$$I_1 = I_2 + I_L \tag{2}$$

and

$$I_2 = I_M - I_{\text{OUT}}. \tag{3}$$

From (2) and (3)

$$I_1 = I_M + I_L - I_{\text{OUT}}. \tag{4}$$

Substituting in (1)

$$I_{\text{OUT}} \cdot R_L = (I_M + I_L - I_{\text{OUT}}) \cdot R + (I_M - I_{\text{OUT}}) \cdot nR \tag{5}$$

So

$$I_{\text{OUT}} \cdot (R_L + R + nR) = I_M \cdot (n+1) \cdot R + I_L \cdot R \tag{6}$$

or

$$I_{\text{OUT}} = (I_M \cdot (n+1) + I_L) \cdot \frac{R}{(R_L + (n+1)R)}. \tag{7}$$

Equation (7) indicates that the contribution to I_{OUT} of I_M and I_L is in the relation $(n+1){:}1$ and is independent of R and R_L. Hence, if the interstage divider has a ratio of $15{:}1$, I_L contributes proportionately $\frac{1}{16}$ of I_{OUT}, while I_M contributes $\frac{15}{16}$. The relationship holds for any R and R_L with the limitation that $R_L \times I_{\text{OUT}}$ must not exceed the compliance limits of -1 V and $+3$ V.

The divider networks consist of eight equal resistors each, three in series and five in parallel, providing an accurate 15 to 1 ratio, independent of process or temperature variations. The division point is at $\frac{1}{16}$ of full scale, so the ratio need only be accurate to ± 3 percent, if this were the only source of LSB error. In practice, the layout is intended to produce a standard deviation of the ratio of less than 0.5 percent, so interstage division error is not normally a cause of devices failing testing.

The multiple current source technique has a number of significant advantages over alternative approaches. All the resistors are equal in value, have equal applied voltages, and are located in a single isolation area on the chip. This means that no differential voltage coefficients of resistance can exist. Equality of the currents ensures equal switching speeds in all devices. The number of devices used is such that no single current source contributes more than $\frac{1}{8}$ of full scale, so that statistically, yields are higher than would be the case in more conventional circuits. The interstage current divider makes true emitter area scaling easy to ensure that the temperature range performance of the device is good. Finally, distribution of the current source components through an array is helpful in reducing the effects of cross-chip temperature gradients.

Although the overall chip dissipation is not high, power usage inevitably concentrates in the comparator and, to some extent, in the reference areas. The current source transistor/ resistor combinations are distributed through the array to overcome the effects of photolithographic errors, resistivity gradients, and operational thermal gradients. The chip photograph in Fig. 8 shows how this is achieved. The matched

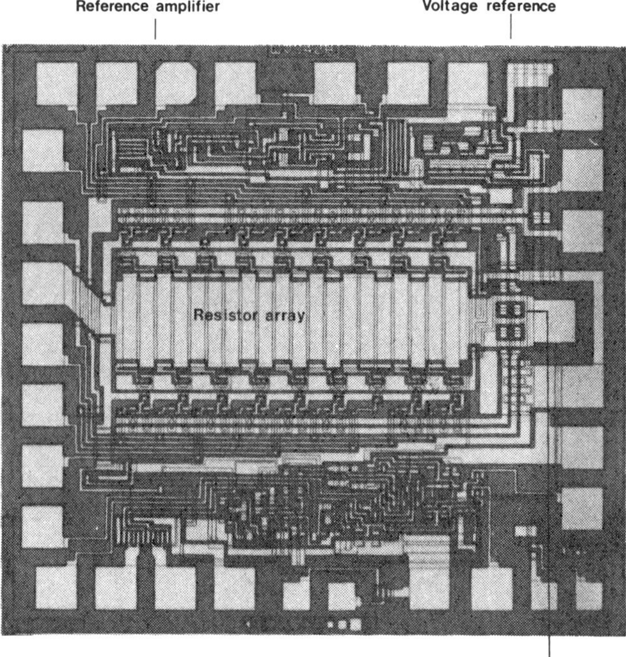

Fig. 8. The 8-bit DAC chip. Chip size is 1.7 mm $\times$ 1.8 mm.

resistor array shows clearly in the middle of the chip, covered by the negative supply rail. All resistors are equal in value and operate under the same voltage conditions to avoid differential temperature and field effects. The current source and switch transistors are distributed about the resistor array, with interdigitation of the resistors in the MSB and LSB arrays. All the MSB resistors connect to source transistors on the upper side of the array, while LSB resistors connect to sources on the lower part. Interdigitation in this way enables the multiple source bits to be more evenly spread through the array, and by putting LSB components at the ends of the array, avoids the need for dummy components.

The least significant bit transistors are located closest to the main heat source, the comparator, which is along the bottom of the chip. The voltage reference, with zener preregulator and temperature tracked current source, is at the top left, while the reference amplifier takes up the rest of the top of the chip. The MSB transistors are below the reference components and above the resistor array. The main output current lines pass across the chip at this point, with the interstage current divider on the right bringing in the LSB currents. The dividing resistors form the crossunders needed to traverse the other tracks. Only single-layer metallization was used, and the chip size is 1.7 mm $\times$ 1.8 mm.

Process Technology

Process technology plays a very significant part in a design of this kind. The cross section of a transistor is shown in Fig. 9.

The process uses a thin epitaxial layer with implants for isolation, p+, base and emitter. Base width is typically 0.15 μm and emitter depth 0.3 μm, with no push-on effect. Geometries used conform to 4 μm minimum feature sizes which,

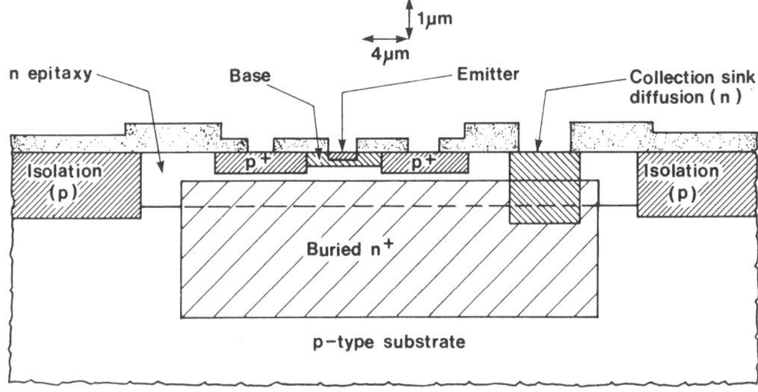

Fig. 9. Device structure. Geometries are based on 4 μm minimum feature size.

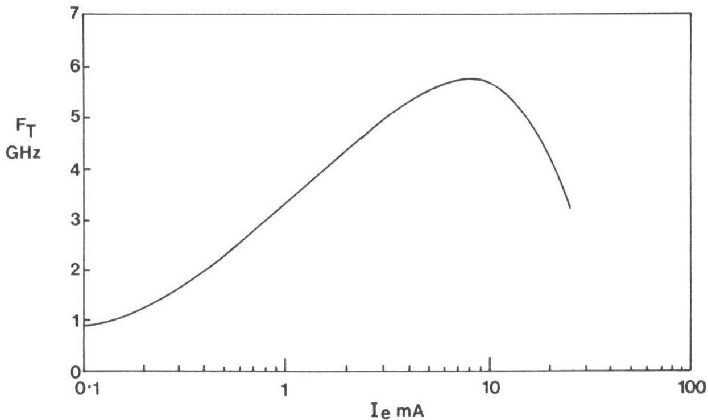

Fig. 10. F_T versus I_e for basic transistor.

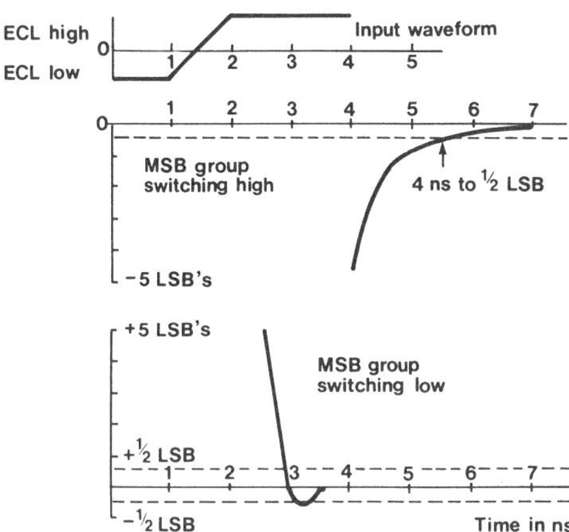

Fig. 11. CAD results for the DAC array.

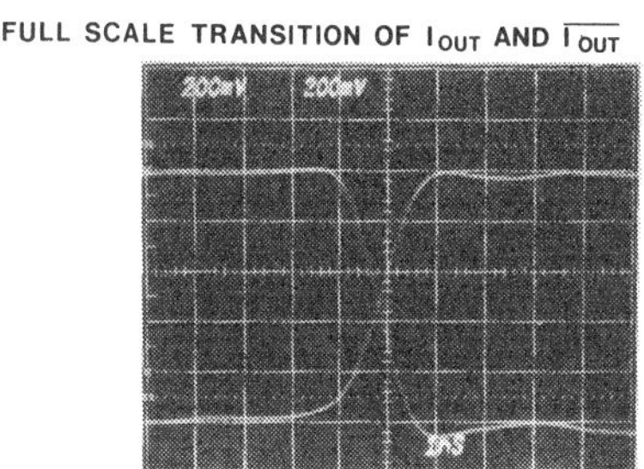

Fig. 12. Complementary full-scale output transitions. Oscilloscope is 500 MHz real-time.

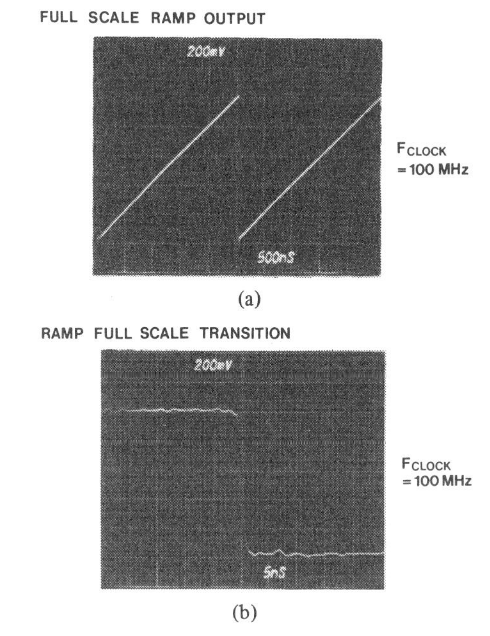

Fig. 13. (a) Full-scale ramp output, F_{CLOCK} = 100 MHz. (b) Full-scale ramp transition, F_{CLOCK} = 100 MHz.

with auto-registered contacts, produce a very compact layout with excellent matching characteristics. 40 μm wide resistors have a standard deviation (σ) of matching of 0.1 percent and dual four micron by ten micron emitters have a standard deviation of 0.25 mV. F_T peaks at 5.5 GHz and 8 mA on the devices used (Fig. 10); the operating point of the switch devices is close to the peak, and parasitic capacitances are small.

CAD

Computer modeling techniques were used extensively in the design phase, particularly in the optimization of the DAC settling time (Fig. 11). The diagram represents the CAD results obtained using ISPICE. The top plot is the input switching waveform, with a 50 percent point at 1.5 ns.

The lower plots show the settling predictions for the whole of the MSB group of the converter with all bits switched simultaneously. Two distinct conditions exist: the output settling "low" and the output settling "high." The latter case, the slower of the two, was predicted to take a total of 4 ns from the input switching 50 percent point to the output entering the $\pm\frac{1}{2}$ LSB bracket from the final settling point. Predictions were for settling into a 50 Ω and 2 pF load. Measurements show a total of 5 ns settling time due to the excess total load capacitance of the test equipment.

Experimental Results

The experimental results on a sample device are illustrated in Fig. 12, which shows the full-scale settling of the complementary outputs into the 50 Ω oscilloscope inputs. The settling time is evidently very fast, although this measurement is limited by the oscilloscope input VSWR. VSWR causes the reflections on the output traces at a time determined by the measurement cables, which is about 3 ns after the transition in this case.

The second oscilloscope photograph (Fig. 13) shows the output of the device driven by an 8-bit synchronous counter. The

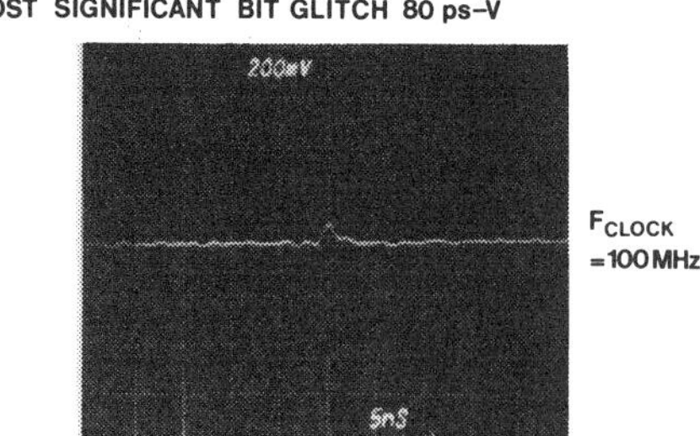

Fig. 14. Most significant bit "glitch." Glitch energy is 80 ps · V.

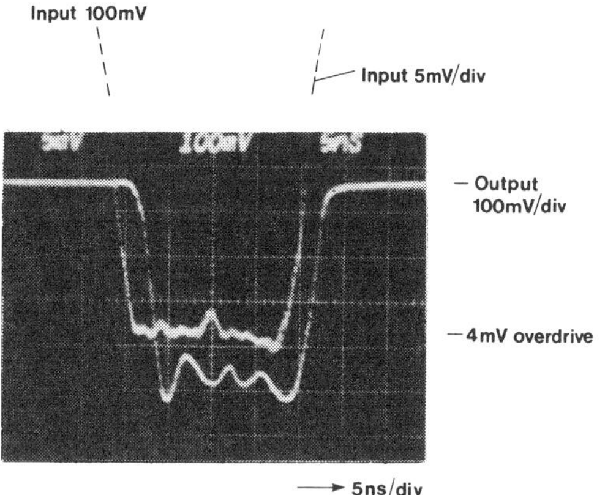

Fig. 15. Comparator delay in follow mode at 4 mV overdrive. Delay is 2.5 ns unlatched, latch to output delay 1.3 ns.

counter is clocked at 100 MHz, so the DAC input data are updated every 10 ns. The full-scale transition, Fig. 12(b), shows the overflow of the counter. The linearity is visually very good even at this speed, with no evidence of dynamic nonlinearity or false coding. No retiming was used for this measurement, but the glitch content is obviously very low, as shown by Fig. 14. The center point transition is the worst case glitch, and measures 80 ps · V, at 1 V full-scale. Part of the glitch is probably due to the drive counter, where precise synchronism of the edges to the accuracy required cannot be maintained using practical devices due to the counter carry delay between packages. This compares very favorably with the other products in this field, where 100 ps · V is regarded as "glitchless" and 200 ps · V is described as "low glitch." Conventional linearity checks showed that most functional devices meet a $\pm\frac{1}{2}$ LSB integral and differential linearity specification, and many are $\pm\frac{1}{4}$ LSB. Thermal matching of the current sources is also good. The temperature coefficient of the DAC output tracks very closely with the reference, and the devices show an overall variation of 1 LSB over the full military temperature range.

The comparator performance in the follow mode is illustrated in Fig. 15. The drive signal is 100 mV with a 4 mV overdrive arranged by the comparator input dc offset conditions. This drive level approximates to the operational characteristics of a successive approximation analog-to-digital converter. Under these conditions, where the comparator is not fully driven, the propagation delay is 2.5 ns for 4 mV overdrive. When the latch signal is applied, the output is fully driven, and the important parameter is the latch to output

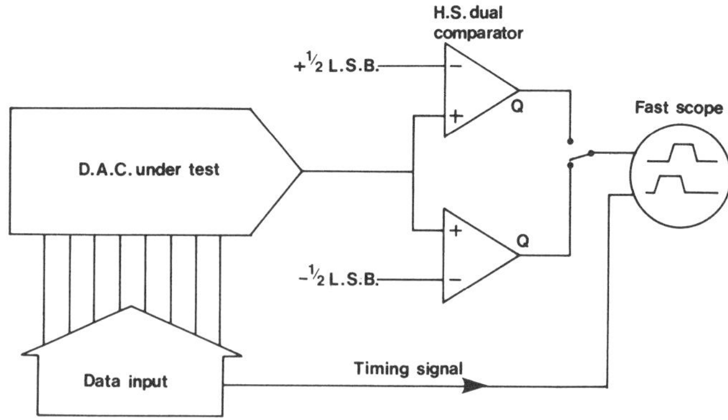

Fig. 16. DAC settling time measurement using high-speed dual comparator.

delay, which is typically 1.3 ns with 0.5 ns required for analog input setup time.

MEASUREMENT OF SETTLING TIME

Oscilloscopes, whether real-time or sampling, do not have sufficiently low-input VSWR or on-screen resolution for precise settling time measurements. Overload recovery time can also cause trouble when the oscilloscope amplifier is overdriven to increase resolution. Fig. 16 shows a measurement technique in which the DAC can settle into a nearly ideal 50 Ω load with minimal interconnection paths; this is also very closely related to the practical use of the device. Precision settling time measurements can be performed with a high-speed comparator. A dual device was chosen with a delay time of about 2.2 ns at 10 mV overdrive. Two references are set up to detect the DAC output settling within a defined window, which is referenced to ground. The circuit simulation predicts that this is the slower case. The lower comparator detects the DAC output coming within $-\frac{1}{2}$ LSB of the final settling point, while the upper device when switched in checks that there is no overshoot. It is found that the settling is very well behaved and, after correction for comparator delay, the results consistently show a DAC settling time of 5 ns $\pm\frac{1}{2}$ ns. This is defined from the 50 percent point of the DAC input switching waveform. No correction has been made for the excess loading capacitance at the output node, so these results are considered to be in very good agreement with the computer predictions. Clearly, the comparators used for the measurement need a particularly good transient response, with no hysteresis and good resolution. This requirement was satisfied by a commercially available dual comparator made on the same process as the DAC.

CONCLUSIONS

A new device has been described containing an 8-bit digital-to-analog converter with a 5 ns settling time, excellent integral and differential linearity, and a very low glitch output. The voltage reference included on the chip compares well with discrete reference devices using more complex technologies, and the comparator has a 2.5 ns propagation delay at 4 mV overdrive. These results have been achieved on a process which has been optimized for speed, rather than precision component matching, and illustrate that speed and precision

TABLE I

```
DAC

   Number of Bits                    8
   Full-scale Transition Time        <5 ns
   Multiply Mode Bandwidth (3 dB)    40 MHz
   Differential Non-linearity        <½ LSB (<¼ LSB by selection)
   Non-linearity Error               <½ LSB (<¼ LSB by selection)
   Nominal Full-scale Current
           (into 50 ohm)             20 mA


REFERENCE

   Stability                         <30 ppm/°C
   Supply Rejection (both supplies) >60 dB


COMPARATOR

   Propagation Delay (input-output) 2.5 ns    100 mV/4 mV overdrive
   Propagation Delay (latch-output) 1.3 ns    100 mV/4 mV overdrive
   Edge Rise/Fall Times             1 ns      100 mV/4 mV overdrive
```

are not mutually incompatible. Device parameters are summarized in Table I.

The chief area of use of the device described is seen in video speed analog-to-digital converters, in conjunction with the successive approximation register now being processed and an analog sample and hold. Target speed is a 15 MHz sample rate with a 135 MHz system clock. Digital-to-analog conversion is also possible with an update rate of 100 MHz, and a variant without the comparator has been produced for this function. Further extension of the design concepts to ten bits has been carried out. Ten-bit DAC's are now being evaluated.

Application areas of the 8-bit successive approximation system are in high-speed signal processing and television, where the low-power consumption and straightforward drive requirements should be a significant advantage.

ACKNOWLEDGMENT

This paper is presented with the kind permission of the Directors of Plessey Research (Caswell) Limited, Allen Clark Research Centre.

REFERENCES

[1] P. H. Saul, P. J. Ward, and A. J. Fryers, "A 5 ns monolithic D/A subsystem," in *ISSCC Dig. of Technical Papers*, 1980.
[2] J. G. Peterson, "A monolithic, fully-parallel, 8b A/D converter," in *ISSCC Dig. of Technical Papers*, 1979.
[3] P. Holloway and M. Timko, "Circuit techniques for achieving high speed resolution A/D conversion," in *ISSCC Dig. of Technical Papers*, 1979.
[4] P. H. Saul and A. J. Jenkins, "A 10-bit monolithic tracking A/D converter," in *ISSCC Dig. of Technical Papers*, 1978.

A 500-MHz 8-Bit D/A Converter

KENJI MAIO, SHIN-ICHI HAYASHI, MEMBER, IEEE, MASAO HOTTA, TOMOYUKI WATANABE, MEMBER, IEEE, SEIICHI UEDA, AND NORIO YOKOZAWA

Abstract —An ultra-fast monolithic 8-bit DAC is designed and fabricated. To realize this DAC, a new high-speed conversion technique, which we call the Data Multiplexing Method, and a variation of the segmented DAC [1] for low glitch are developed. The DAC is fabricated with shallow-groove-isolated 3-μm VLSI technology [2] with peak f_T's of 4.5 GHz. An experimental 8-bit DAC features a conversion rate of over 500-MHz, a full-scale settling time to 1 percent of 2 ns, rise/fall times of 0.6 ns, and a glitch energy of 20 ps·V without input latches or a deglitcher [3].

I. INTRODUCTION

RECENTLY, high-definition displays have become one of the key devices for CAD/CAM. The development of computer graphics requires higher definition displays. A 1000-line display is already commercially available. In the near future, it is obvious that this will be extended to higher definition displays of over 2000-line resolution. The high-definition display with 2000 lines will require very high performance on the part of the DAC.

A 60-Hz noninterlaced display with a resolution of 2448 $\times$ 2048 dots requires a DAC with a conversion rate of more than 400 MHz [4]. For a full-color display, 8-bit resolution is required. Moreover, to reduce switching noise, the glitch energy of the DAC must be less than 20 ps·V. Although conventional 8-bit DAC's with a conversion rate of 100–125 MHz, a settling time to $\pm 1/2$ LSB of 5–10 ns, and a glitch energy of 50–80 ps·V have already been developed [5], [6], their specifications cannot satisfy the requirements for 2000-line displays.

This paper describes a new DAC that can be applied to displays with over 2000 lines. An experimental 8-bit DAC has been fabricated through the use of a shallow-groove-isolated 3-μm bipolar VLSI technology. A conversion rate of over 500 MHz has been achieved.

II. DATA MULTIPLEXING DAC

A conventional DAC consists of a segmented DAC for upper bits and an R-$2R$ ladder network for lower bits. A schematic block diagram of a conventional DAC is shown in Fig. 1. The segmented DAC can achieve lower glitch and higher precision. The R-$2R$ DAC portion of the converter features high speed when low impedance of R and large enough bit currents are used. However, this conventional DAC has three problems. First, latches must be employed to prevent glitches caused by skews of input data. Second, the need for latches and segment decoders limits the conversion rate. It is difficult to achieve a conversion rate of higher than 350 MHz using the conventional high-speed bipolar technology. Third, the ECL-100K series, which is the fastest logic IC currently available on the market, cannot directly drive the DAC with a conversion rate of 400 MHz, because the maximum operation speed of the ECL is limited to 200–300 MHz. To solve those problems with existing technology, the authors propose a new method, the Data Multiplexing Method.

A schematic block diagram of the new DAC is shown in Fig. 2. Switching circuits instead of latches are employed at the segment decoders for the three MSB's and at the drivers for the five LSB's. Two pieces of input data (DATA 1, DATA 2) are exchanged alternately by the switching circuit.

Two other positions for the switching circuit are possible. One is the case where the switching circuit is located in front of the decoders. This is advantageous in that the number of decoders is decreased to one, but is disadvantageous in that bit skews in the decoders cause large glitches. The other is the case where the outputs of two DAC's are exchanged alternately by the switching circuit. With this method, a speed double that of a conventional DAC can be achieved. However, this method has some problems, namely, the requirement to precisely adjust the gain of the two DAC's and the high feedthrough noise of the output switching circuit. Thus, the position of the switching circuit shown in Fig. 2 is optimum.

The timing chart of the two data inputs and of the clock signal to exchange the data are shown in Fig. 3. The two pieces of input data have a phase difference of 180°. The switches driven by the clock signal select the two pieces of input data alternately. Therefore, the conversion frequency f_r is expressed as twice the data frequency.

The advantages of this method are as follows.

1) Since high-speed current switches to exchange the data mainly determine the conversion rate of the DAC, a high-speed DAC can be achieved.

2) Latches for data deskewing which limit the conversion rate are not necessary because of simultaneous switch-

Manuscript received April 12, 1985; revised July 22, 1985.

K. Maio, S.-I. Hayashi, M. Hotta, and T. Watanabe are with Central Research Laboratory, Hitachi Ltd., Kokubunji, Tokyo 185, Japan.

S. Ueda is with Takasaki Works, Hitachi Ltd., Takasaki, Gunma, Japan.

N. Yokozawa is with Hitachi Medical Co., Ltd., Kashiwa, Chiba, Japan.

Reprinted from *IEEE Journal of Solid-State Circuits*, Vol. SC-20, No. 6, pp. 1133-1136, December 1985.

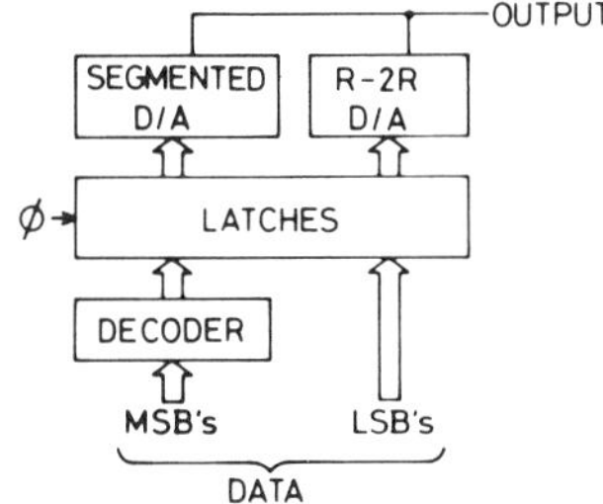

Fig. 1.   Schematic block diagram of a conventional DAC.

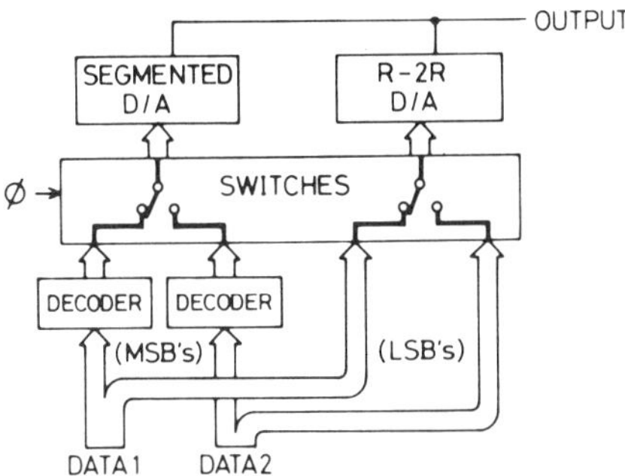

Fig. 2.   Principle block diagram of a DAC utilizing the Data Multiplexing Method.

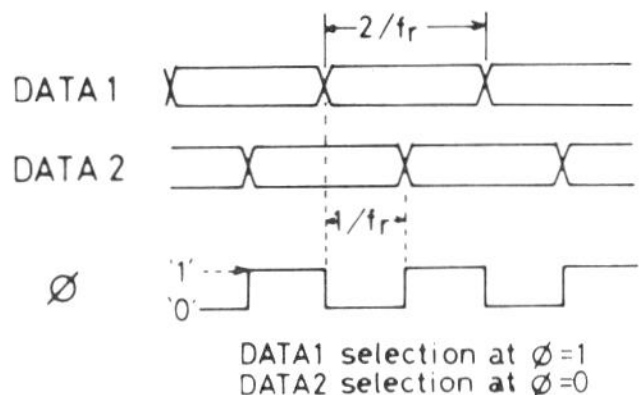

Fig. 3.   Timing chart of two input data (DATA 1, DATA 2) and switching signal (ϕ).

ing which occurs at the time the decoding operation has completely finished.

3) The update rate of each input datum need only be one half that of a conventional DAC, therefore the data generation composed of commercially available ECL's becomes useful for operation at 500 MHz.

III.   Circuit Design

A schematic of the Data Multiplexing DAC is shown in Fig. 4. This converter uses a segmented DAC for the three MSB's and an R-$2R$ ladder network for the five LSB's. The value of resistance R is 75 Ω and the value of each current source I is 2 mA. Therefore, the output impedance R_0 is 50 Ω and a full-scale output swing of about 0.8 V can be obtained. A common base transistor Q_0 shown in Fig. 4, which is located between the output of the segmented DAC and the R-$2R$ ladder network, decreases the parasitic capacitance at the summing output in the segmented DAC to ΔC (= 2 pF). This can improve the rise and fall times of the DAC. When it can be assumed that the response of the DAC is approximately first-order delay, the improved value of rise and fall times Δt_r, Δt_f is

$$\Delta t_r, \Delta t_f = 2.2\ R_0 \cdot \Delta C = 0.2\ \text{ns}.$$

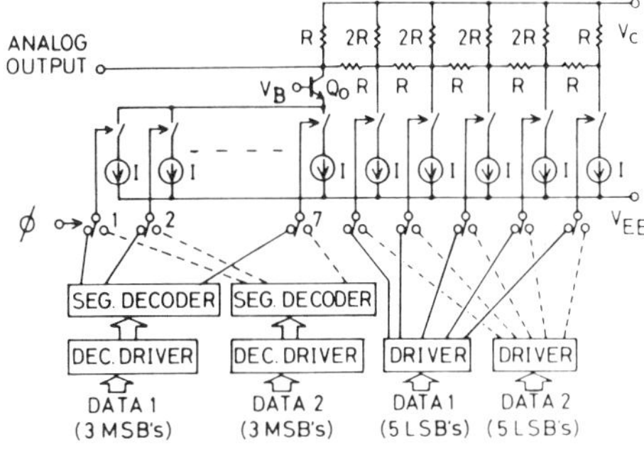

Fig. 4.   Circuitry of the Data Multiplexing DAC.

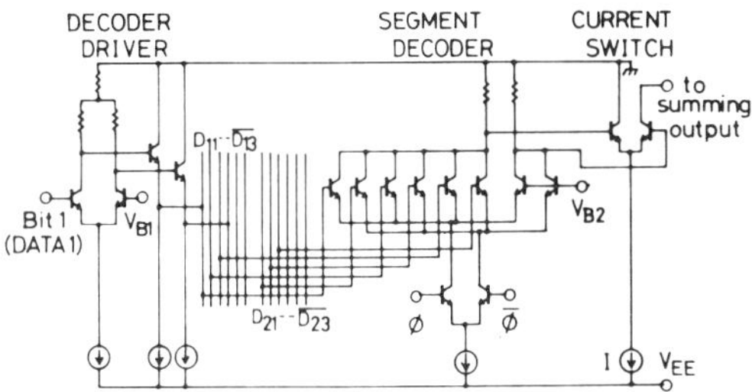

Fig. 5.   Circuitry of decoder driver, segment decoder, and output current switch.

A series comprising decoder driver, a segment decoder, and an output current switch is shown in Fig. 5. All stages are nonsaturating and operate without any clamping for maximum operation speed. The well-known analog decoding technique [1] has been used for segment decoding. To achieve higher speed operation, the decoder driver has been constructed using a differential circuit with only n-p-n transistors and a T-type low resistance load. Six decoder drivers are provided corresponding to three MSB's of two-channel input data. These drivers generate the output voltages with a four-level logic scheme. When the output levels for the MSB driver, the second bit driver, and the third bit driver are $V1$, $V2$, and $V3$, respectively, a four-level logic scheme can then be described by

$$(V1)_H > (V2)_H > (V3)_H > V_{B1}$$
$$(V1)_L = (V2)_L = (V3)_L < V_{B1}$$

where H and L are the high and low levels, respectively, and V_{B1} is the reference level.

Each level of $(V1)_H, (V2)_H, (V3)_H, V_{B1}, (Vi)_L$ has a voltage difference of 0.2 V, so that adequate switching and high-speed response can be achieved.

The outputs of the drivers are fed to the appropriate inputs of the seven segment decoders. Each segment decoder provides parallel input transistors for two-channel input. The current switch selects the two pieces of input data alternately and the output of each segment decoder switches a precise current I. This analog decoding approach, including the switch to select the two-channel input data, makes it possible to realize the Data Multiplexing Method with far fewer devices and lower propagation delay.

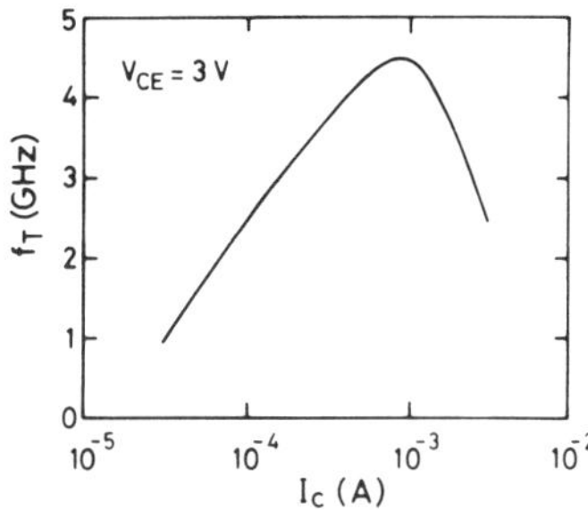

Fig. 6.   Relationship between cutoff frequency f_T and collector current I_c of an n-p-n transistor.

Fig. 7.   Chip photograph of 8-bit 500-MHz DAC.

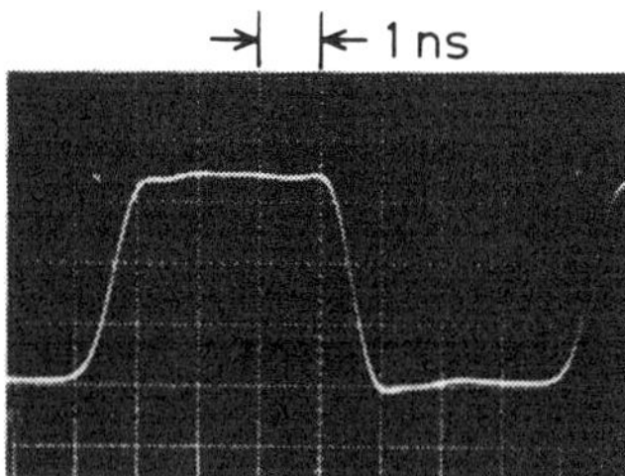

Fig. 8.   Full-scale output transition. Settling time to 1 percent is less than 2 ns.

IV.   Monolithic Process and Layout

An experimental DAC has been fabricated through the use of a shallow-groove-isolated 3-μm VLSI technology. The characteristics of the n-p-n transistors in the DAC are shown in Fig. 6. These transistors have low parasitic capacitance and f_T of 4.5 GHz at a low collector current of 0.7 mA. All of the switch devices operate at close to the peak of f_T. Fig. 7 is a chip photograph of this DAC. The chip size is 1.2 mm $\times$ 1.5 mm. All of the resistors are made by means of base diffusion. To achieve high precision, the width of the resistors used in the ladder network and the output current sources is more than 20 μm. The chip is packaged in a small (10 mm $\times$ 9 mm) 28-pin chip-carrier to decrease parasitic impedance.

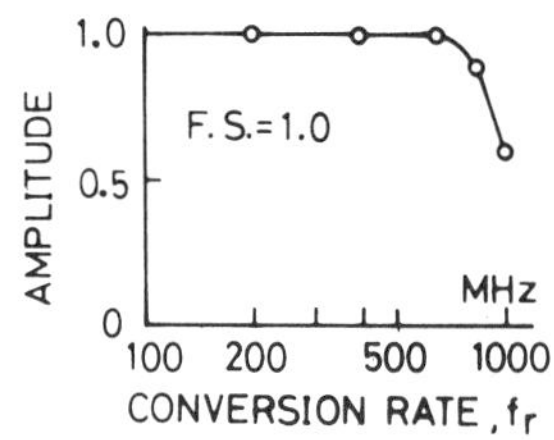

Fig. 9.   Relationship between conversion rate f_r and amplitude of full-scale output swing.

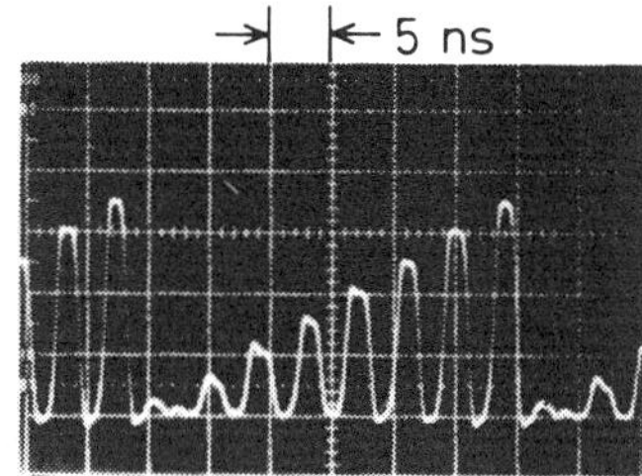

Fig. 10.   Output waveform of full-scale swing to low-level swing at 500-MHz operation.

V.   Performance of the IC

In the high-speed area, the high-precision measurement of waveforms is extremely difficult. One of the reasons is the reflection effect due to impedance mismatching among the DAC, coaxial cable, and measurement equipment. Another is the parasitic impedance in the LSI chip and the printed board. To avoid these problems, measurement was carried out utilizing a very carefully designed board and a 50-Ω measurement system.

A full-scale transition of the output is shown in Fig. 8. This waveform is generated by alternately exchanging DATA 1 of all 1's and DATA 2 of all 0's. A high-speed oscilloscope with a bandwidth of 1 GHz is used in the measurements. The rise and fall times are about 0.6 ns, and the settling time to 1-percent allowable in a DAC for video use is less than 2 ns. The other usual method of measuring settling time, the window comparison method [5], is also carried out. The measured settling time was about 1.8–2 ns. Therefore, the usable conversion rate of this DAC is higher than 500 MHz. The relationship between the conversion rate f_r and the amplitude of the full-scale output swing is shown in Fig. 9. It can be seen in the figure that the amplitude is flat up to a conversion rate of more than 500 MHz, and that this DAC can operate even at a rate of higher than 1 GHz, but with degraded full-scale amplitude. Fig. 10 shows the output waveform from full-scale swing to low-level swing at 500-MHz operation. Fig. 11 shows a full-scale ramp output of the DAC at a conversion rate of 500 MHz. The waveform is smooth throughout the full dynamic range. Fig. 12 shows the output of the converter being dithered up and down through the MSB, between $100,\cdots,0$ and $011,\cdots,1$. The glitch energy including the clock feedthrough noise is only 20 ps·V. The main characteristics of this DAC are summarized in Table I. The linearity error is less than $\pm 1/2$ LSB and the power consumption is typically only 350 mW. The performance

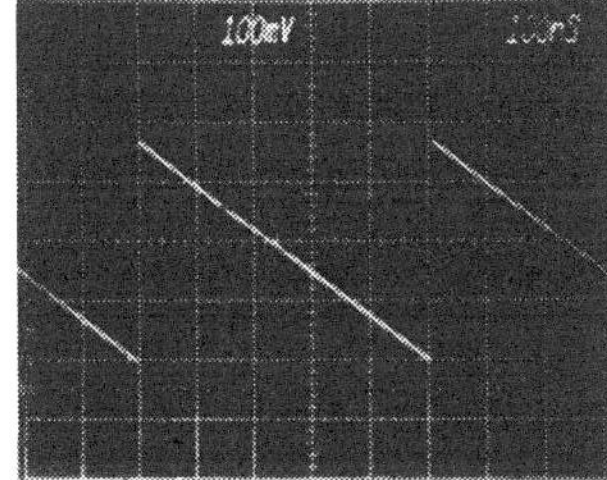

Fig. 11.   Full-scale ramp output at 500-MHz conversion rate.

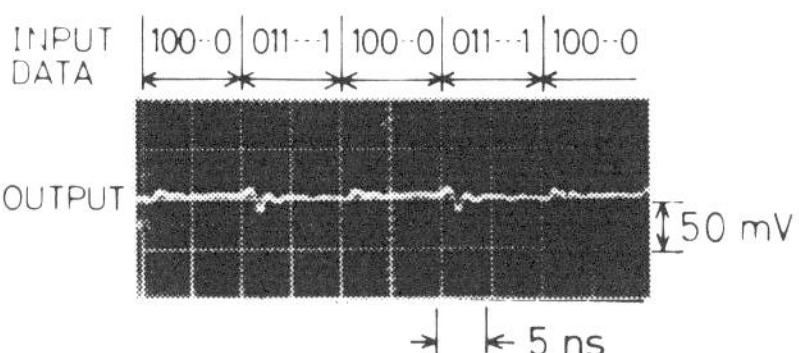

Fig. 12.   MSB glitch of the DAC.

TABLE I
MAIN CHARACTERISTICS OF THE DAC

Resolution	8 Bit
Conversion Rate	500 MHz
Settling Time (to 1%)	2 ns
Rise/Fall Time	0.6 ns
Glitch Energy	20 ps·V
Linearity Error	± 1/2 LSB
Power Consumption	350 mW

of this DAC is sufficient for use in a high-definition display with 2000 lines.

VI.   CONCLUSION

In order to achieve extremely high-speed conversion, a Data Multiplexing Method was developed. A monolithic 8-bit DAC has been built utilizing this method and the shallow-groove-isolated 3-μm VLSI technology. A conversion rate of over 500 MHz and extremely low glitch energy of 20 ps·V have been achieved with power consumption of only 350 mW. Moreover, this DAC has been applied to a color CRT display with 2000 lines [4]. Fine images corresponding to 2-ns pulsewidth and full-color images of 8-bit resolution without glitch noise have been achieved.

ACKNOWLEDGMENT

The authors wish to express their thanks to M. Nagata, S. Iguchi, S. Yamamoto, and M. Kubo for their constant support and encouragement.

REFERENCES

[1] J. A. Shoeff, "An inherently monotonic 12b DAC," in *ISSCC Dig. Tech. Pap.*, 1979, pp. 178–179.
[2] M. Hotta *et al.*, "An 120-mW, 8-bit, video frequency A/D converter with shallow grooved isolated bipolar VLSI technology," in *Symp. VLSI Technol.*, Sept. 1984.
[3] K. Maio *et al.*, "A 500 MHz, 8-bit DA converter," in *ISSCC Dig. Tech. Pap.*, 1985, pp. 78–79.
[4] K. Ando *et al.*, "A flicker-free 2448 × 2048 dots, color CRT display," in *SID Dig. Tech. Pap.*, 1985, 18.2, pp. 338–340.
[5] P. H. Saul *et al.*, "An 8-bit, 5-ns monolithic D/A converter subsystem," *IEEE J. Solid-State-Circuits*, vol. SC-15, no. 6, pp. 1033–1039, 1980.
[6] Analog Devices Data Catalog, AD9768.

An 8-bit 2-ns Monolithic DAC

TSUTOMU KAMOTO, YUKIO AKAZAWA, MEMBER, IEEE, AND MITSURU SHINAGAWA, MEMBER, IEEE

Abstract —An 8-bit resolution ultra-high-speed monolithic digital-to-analog converter (DAC) is fabricated using super self-aligned process technology (SST). In order to improve dynamic accuracy, which is determined by settling speed, clock feedthrough noise, and glitch, a number of new circuit technologies are developed including a rise- and fall-time control switch driver, a low-noise flip-flop, and a differential buffer configuration. In addition, a new chip assembly technology is developed employing a multilayer ceramic substrate. Performance measurements show a settling time to 8-bit accuracy of about 2 ns, a maximum conversion rate of 1 GHz, a glitch energy of 2 ps·V, and a 10-bit linearity error accuracy without trimming.

I. INTRODUCTION

THE RECENT progress in VLSI technology has accelerated the application of digital techniques to circuits which until now have utilized analog techniques. A key LSI in this technological trend is a high-speed and high-accuracy digital-to-analog converter (DAC). Practical applications for the DAC are high-speed analog LSI testers, high-resolution displays and electron beam pattern drawing systems, etc.

A state-of-the-art ultra-high-speed DAC, which utilizes Si bipolar or GaAs MESFET devices, has been developed. The fastest Si bipolar monolithic 8-bit DAC has achieved 500-MHz throughput [1], but this DAC does not satisfy 8-bit dynamic accuracy.

To obtain even higher accuracy and higher speed operation, two basic problems must be overcome: 1) unwanted ringing in the analog output waveform, and 2) clock noise in the flip-flop output. These problems are caused by stray inductance in the bonding wire and high-speed switching characteristics of transistors, respectively. In the design of high-speed analog circuits with high accuracy, not only highly stable assembly technology, but also circuit technology which can optimize the characteristics of the chip itself or between chip and assembly is important. The authors have, therefore, developed new circuit and chip assembly technologies which are mainly available for reducing the settling time and lowering the clock noise characteristics. Utilizing these technologies in combination with super self-aligned process technology (SST) [2], about 2-ns settling to 8-bit accuracy and 1-GHz throughput at 1-percent accuracy have been achieved.

This paper describes the circuit design concepts, the chip assembly technology, the measurement technology, and the characteristics of the DAC.

Manuscript received July 20, 1987; revised September 18, 1987.
The authors are with NTT LSI Laboratories, 3-1, Morinosato Wakamiya, Atsugi-shi, Kanagawa 243-01, Japan.
IEEE Log Number 8718249.

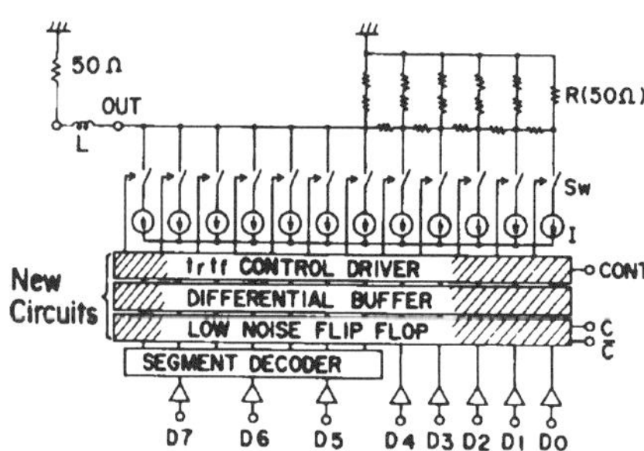

Fig. 1. Block diagram of the DAC.

II. CIRCUIT DESIGN CONCEPT

Fig. 1 shows a block diagram of the DAC. For the lower 5 bits, conventional current switches, binary weighted by an R-$2R$ ladder, are used. For the upper 3 bits, a current-summing configuration has been adopted. This configuration supplies the summing current of seven current switches, whose ON and OFF states are controlled according to the decoded signal in the segment decoder, to the output, and can achieve lower glitch noise. The value of the R is 50 Ω and the value of each current source I is 5 mA. Therefore, a full-scale output voltage of 1 V is obtained with external 50-Ω termination. To improve the settling time and lower the clock feedthrough, new circuit technologies have been adopted as shown in the shaded parts of Fig. 1. These include: 1) a rise- and fall-time control switch driver; 2) a low-noise flip-flop; and 3) a differential buffer.

For 1-GHz operation, unwanted ringing in the DAC output waveform and clock noise in the chip are fatal factors limiting the settling speed. Settling characteristics for the DAC output are strongly influenced by the chip assembly technique. The ringing is mainly caused by parasitic inductance and capacitance in the chip assembly, and also depends on the rise and fall times of the current switch output. These parasitic effects, which have a distributed constant condition, cannot be estimated precisely. As a practical matter, therefore, an effective method to minimize settling time is to adjust the current switch rise and fall times. To make a settling-time adjustment after chip assembly, a rise- and fall-time control switch driver, which can be adjusted electrically by the external terminal, has been developed.

When the clock signal changes a flip-flop to latching mode, spike noise occurs in the flip-flop output waveform.

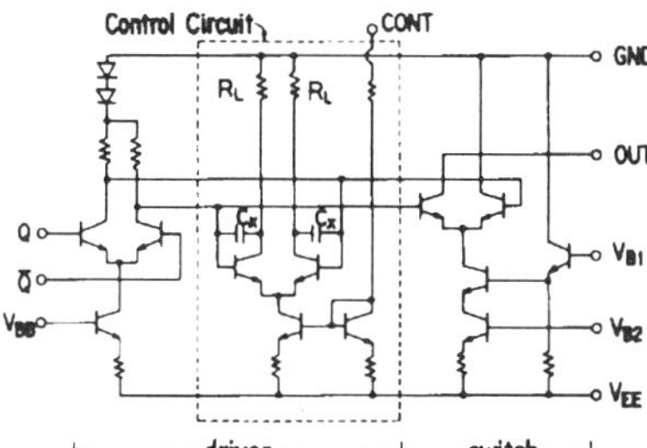

Fig. 2. Rise- and fall-time control switch driver.

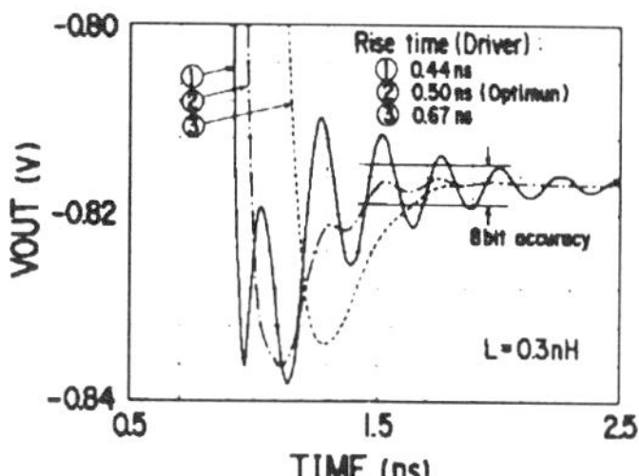

Fig. 3. Simulated DAC output waveforms. Transition starts at 0.5 ns.

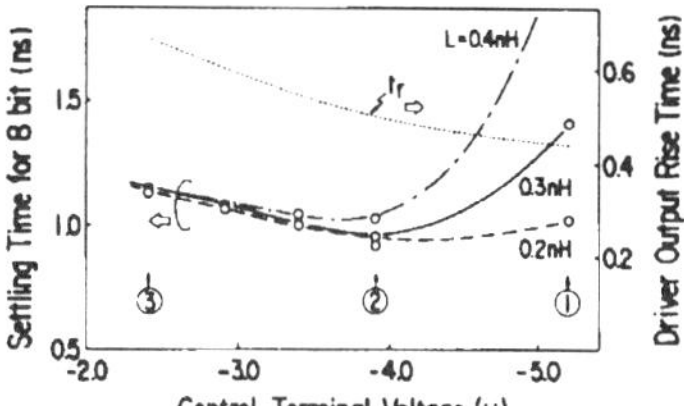

Fig. 4. Rise- and fall-time control effect.

This spike noise is transferred to the analog output through the switch driver and results in degradation of accuracy if it is too large. This spike noise is mainly due to crosstalk through the high-impedance emitter common node of the latching current switch in the flip-flop. This phenomenon is more noticeable at a high clock rate, such as about 1 GHz. For example, from the estimates based on circuit simulation, if an 80-mV_{PP} noise occurs in the flip-flop output, that noise is attenuated to 8 mV_{PP} in the analog output through the switch driver with 20-dB noise reduction characteristics. An 8-mV_{PP} noise limits the accuracy for 1-V full-scale analog output under 7 bits. To reduce this spike noise, a new flip-flop circuit configuration which suppresses the noise at its source has been developed. Moreover, to reduce the noise transfer to analog output, a differential buffer configuration between the flip-flop and switch driver has been adopted.

III. Circuit Technologies

A. Rise- and Fall-Time Control Switch Driver

The first circuit technology to be discussed is a switch driver that controls the rise and fall times of the current switch. Fig. 2 shows the switch driver and the current switch, and the driver includes a control circuit. This driver effectively controls the rise and fall times of the current switch by controlling the rise and fall times of its own output waveform.

In the driver, the rise and fall times can be adjusted by means of variable capacitance based on the Miller effect, which in turn can be controlled by an external voltage [3]. The variable capacitance value can be estimated with the equation

$$C_V = C_{je} + (1 + g_m R_L) \cdot (C_{jc} + C_x).$$

C_x is set to 0.1–0.2 pF to get a wide controllable range with minimum rise and fall times.

Fig. 3 shows blowups of one part of the simulated DAC output waveforms which change from high to low. In this simulation, the bonding-wire inductance was estimated to be 0.3 nH based on an estimated rate of 1 nH/1 mm for conventional bonding wire. The parasitic capacitance was estimated to be 0.4 pF based on a consideration of bonding pads and line capacitance. For waveform number 1, the driver output rise time is controlled to be 0.44 ns. For waveform 2, it is 0.5 ns, and for 3, it is 0.67 ns. From this figure it can be seen that unwanted ringing occurs at the fast rise time. Ringing is suppressed at slower rise times. If it is too slow, on the other hand, the slower response degrades the settling time.

A simulated effect of controlling the rise and fall times is shown in Fig. 4. Numbers 1–3 correspond to the waveform numbers in Fig. 3. The settling time changes according to the control voltage and has a minimum point. To the right of this minimum point, the settling time degrades because of ringing in the output. To the left, it also increases in accordance with the rise time of the driver. Moreover, the minimum point shifts to the left as the inductance increase.

Controlling the rise and fall times, therefore, proves to be a very effective and practical technique to improve settling characteristics.

B. Low-Noise Flip-Flop

A low-noise flip-flop was developed to reduce clock noise. Fig. 5 shows the low-noise flip-flop circuit and the simulated flip-flop output waveform. As mentioned above, the clock noise is caused by the high-impedance emitter common nodes which are indicated as points a and b in Fig. 5. The distinctive feature of this low-noise flip-flop is that an additional small bias current I_b flows through both sampling and latching current switches to lower the impedance of these emitter common nodes. As shown in Fig. 5, an I_b can suppress the clock noise to about half that of the case where no I_b is supplied. A larger I_b reduces the noise level even more, but has the adverse effect of causing the output logic level to deviate when the input data signal changes in the latching mode of the flip-flop. Therefore, the I_b is effectively optimized to be one-tenth of I_s.

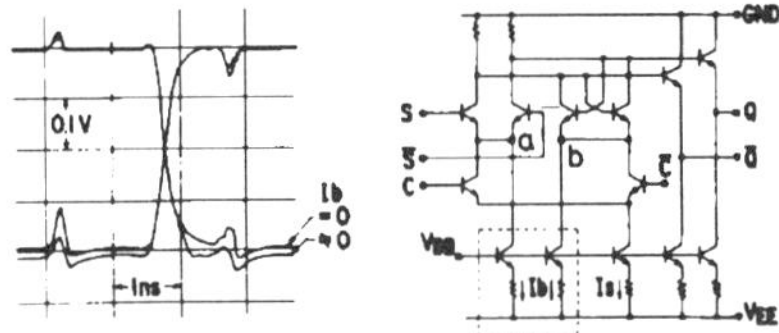

Fig. 5. Low-noise flip-flop circuit and simulated flip-flop output waveforms.

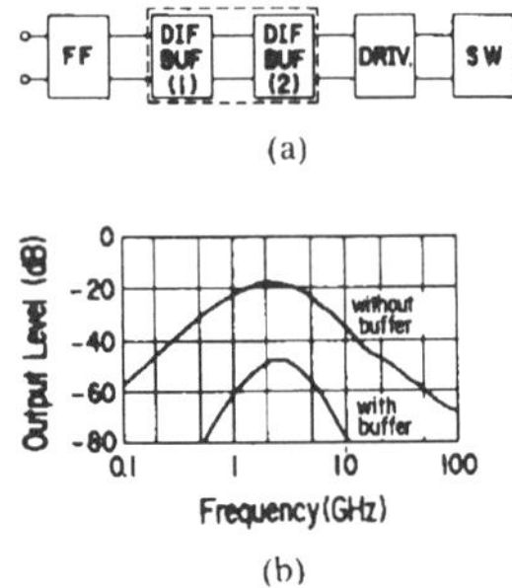

(a)

(b)

Fig. 6. (a) Differential buffer configuration and (b) clock-noise attenuation. *FF* = flip-flop, *DIF BUF* = differential buffer, *DRIV* = switch driver, and *SW* = current switch.

C. A Two-Stage Differential Buffer

The low-noise flip-flop is not sufficient by itself to eliminate the spike noise in the flip-flop output to satisfy 8-bit accuracy. Adding differential buffers between the flip-flop and the switch driver satisfy the requirement for a supplementary method of noise reduction as shown in Fig. 6(a). To eliminate spike noise on both high- and low-level flip-flop outputs, a two-stage configuration is needed. In the differential buffer, a low logic level input forces the output to cut off, and eliminates the spike noise on its own waveform. But, a high logic level input turns the output ON, and the clock noise is transmitted to the switch driver.

Fig. 6(b) shows simulated noise attenuation for the flip-flop output both with and without the buffer. Through the use of the two-stage buffer and switch driver, noise attenuation as high as 60 dB is obtained at 1 GHz. This value means that an 80-mV$_{PP}$ spike noise can be reduced to 0.08 mV$_{PP}$.

IV. Chip Assembly Technology

In the design of a high-speed high-precision analog LSI to effectively pull out chip performance without unwanted oscillation, ringing, and so on, it is important to consider ways of lowering the impedance of ground and power supply lines and reducing the coupling between digital and analog signals in a chip assembly [4], [5]. One way of obtaining these improvements is by utilizing multilayer ceramic substrate technology. A cross section of this substrate is shown in Fig. 7.

On the bottom layer, a ground metal plane is prepared. Ground patterns on the first and second layers are con-

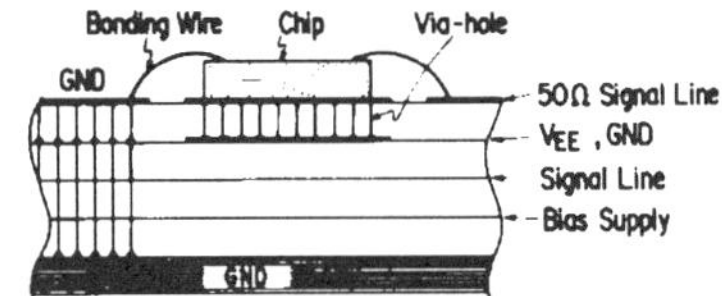

Fig. 7. Cross section of multilayer ceramic substrate.

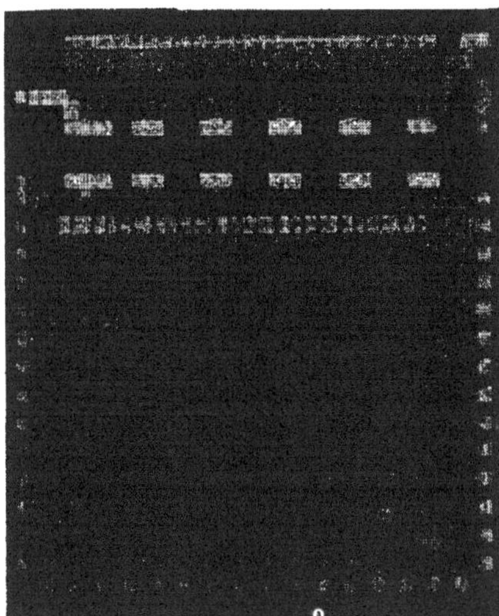

Fig. 8. DAC chip microphotograph.

nected to the ground plane through via holes with as short a path as possible. Fast signal patterns, which require a high-accuracy 50-Ω characteristic impedance, are assigned to the first layer on which a thin film pattern can be formed. Other signal patterns are assigned to the third layer. The dc bias patterns are formed on the fourth layer. By separating the patterns into different layers, lowering the impedance of the ground pattern and reducing the coupling between digital and analog signals can be achieved at the same time.

V. Implementation and Measurement Results

The chip was fabricated utilizing bipolar SST technology with a minimum emitter width of 0.5 μm and a maximum f_T of 18 GHz.

A microphotograph of the DAC chip, including about 1400 devices on a 3.4×3.0-mm chip area, is shown in Fig. 8. An important feature of the pattern layout is that the power supply lines for the digital and analog parts are separated at the top and bottom sides of the device to reduce coupling noise. The resistors in the ladder use wider patterns for higher accuracy. The DAC chip is mounted on the above-mentioned ceramic substrate.

Fig. 9 shows the linearity error for an 8-bit binary digital code. The errors are within ±1/8 LSB for all 256 codes without trimming. This is equivalent to 10-bit accuracy.

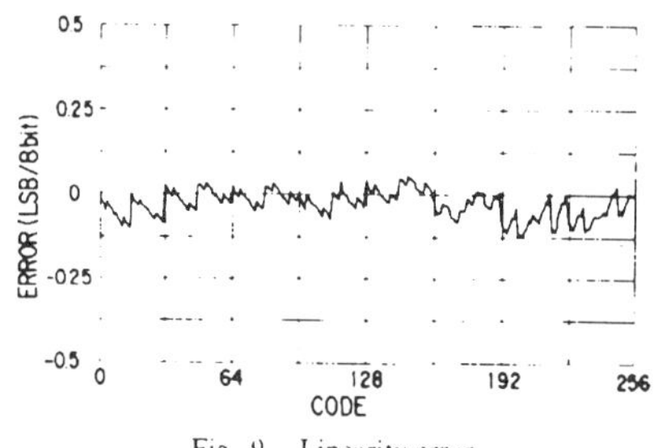

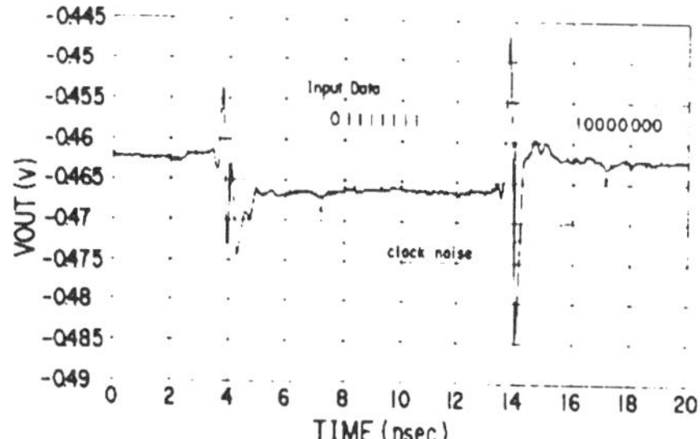

Fig. 9.   Linearity error.

Fig. 11.   DAC glitch measured by the digitizing system.

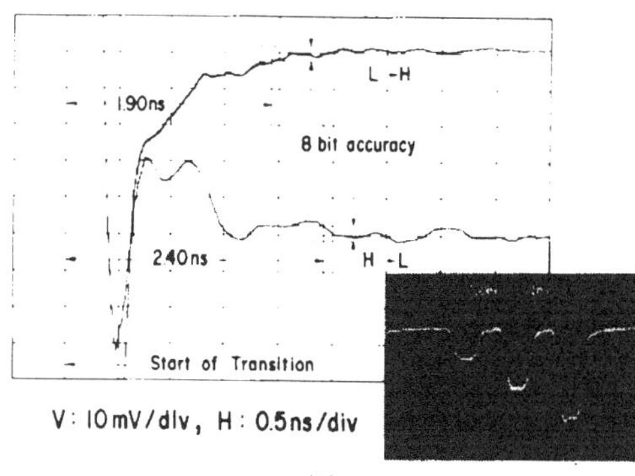

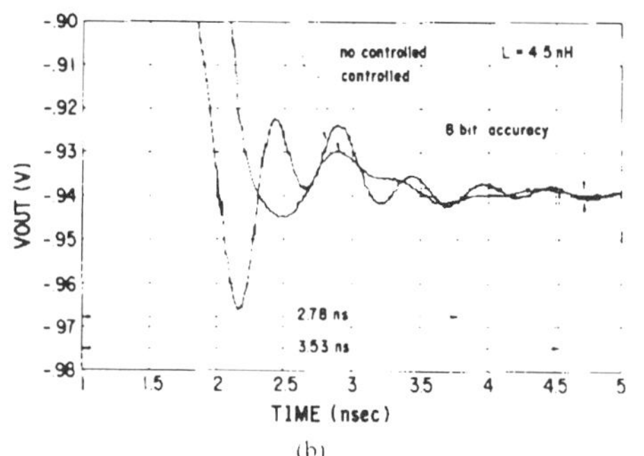

Fig. 10.   (a) DAC settling characteristics measured by the digitizing system, and output waveform of full-scale swing at a clock rate of 1 GHz. (b) Measured rise- and fall-time control effect.

TABLE I
CHIP PERFORMANCE

Resolution	8 Bit
Conversion Rate	1 GHz
Settling Time (to 8 Bit)	2 ns
Rise and Fall Time	0.3 ns
Glitch Energy	2 ps·V
Linearity Error	1/8 LSB
Power Dissipation	1.25 W

For the precise settling time measurement, a digitizing system was used. The system uses two frequencies, f_1 and f_2, which are related as follows

$$f_1 = n \cdot f_2 + \Delta f$$

where Δf is beat frequency. f_1 is applied to the DAC, and f_2 to the sampler. The output of the sampler is fed to a digitizer using a 16-bit A-to-D converter. The sampling head utilizes a GaAs Schottky-diode bridge configuration, which has a bandwidth of 3 GHz, a jitter of less than 2 ps, and a 12-bit accuracy.

Fig. 10(a) shows the fastest settling characteristics. Settling times for 8-bit accuracy of 1.9 ns for low-to-high and 2.4 ns for high-to-low transitions are obtained. For a 1-percent accuracy, they are 1.26 and 1.38 ns, respectively.

Fig. 10(a) also shows the DAC output waveform of a full-scale swing at a clock rate of 1 GHz, which is the fastest operation yet recorded for a bipolar DAC. In this operation, two upper digital codes alternate.

In this test fabrication, a rise- and fall-time variable range of 0.3 0.5 ns was attained, but optimum settling time was achieved at the fastest rise- and fall-time condition, which does not prove the effectiveness of the proposed rise- and fall-time control capability. On the other hand, this chip assembly performed ideally. According to the circuit simulation results in Fig. 4, it is estimated that parasitic inductance and capacitance were less than 0.2 nH and 0.4 pF, respectively.

The effectiveness of the rise- and fall-time control technique was clarified by another measurement. Fig. 10(b) shows measured output waveforms of the DAC, which was applied to a conventional assembly with parasitic inductance of 4.5 nH. When the rise time was not controlled, unwanted ringing degraded the settling time for 8-bit accuracy to 3.53 ns. Controlling the rise and fall times improved the settling to 2.78 ns, which is almost equal to that of the multilayer ceramic substrate. These results show that the rise- and fall-time controlling technique is effective to get higher dynamic accuracy over 8-bit using a lower cost assembly technique.

Fig. 11 shows a glitch in the analog output waveform that alternates between digital codes 10000000 and 01111111. These characteristics were measured with the same digitizing system that was used for the settling-time measurement. From this figure, a glitch energy of 2 ps·V was obtained. This output waveform is further evidence of the advantages of the new noise reduction techniques.

since the clock noise that occurs in the flip-flop output is only about 2 mV in this waveform.

Finally, the chip performance is summarized in Table I. The rise and fall times are 0.3 ns and the power dissipation is typically 1.25 W with a supply voltage of -5.2 V.

VI. Summary

An 8-bit monolithic DAC has been fabricated successfully with a settling time to 8-bit accuracy of about 2 ns, a conversion rate of 1 GHz, and a 10-bit accuracy without trimming.

In this design, two problems including a clock-noise and a settling-time degradation caused by fast output transition and chip assembly parasitic effect were pointed out. To solve these problems, new circuit technologies, namely, a rise- and fall-time control switch driver, a low-noise flip-flop, and a two-stage differential buffer, and a multilayer ceramic substrate assembly technology were proposed. The test fabrication results showed the effectiveness of these technologies.

It is expected that the chip discussed in this paper will result in significant improvements in high-speed analog LSI testers and high-resolution displays.

Acknowledgment

The authors are greatly indebted to T. Sudo, E. Arai, and A. Iwata for their useful suggestions and comments. The authors also wish to thank those who helped fabricate this LSI.

References

[1] K. Maio, S. Hayashi, M. Hotta, N. Yokozawa, T. Watanabe, and S. Ueda, "A 500 MHz 8 b DAC," in *ISSCC Dig. Tech. Paper*, 1985, pp. 78–79.

[2] T. Sakai, S. Konaka, Y. Kobayashi, M. Suzuki, and Y. Kawai, "Gigabit logic bipolar technology: Advanced super self-aligned process technology," *Electron Lett.*, vol. 19, pp. 283–284, 1983.

[3] M. Ohara, Y. Akazawa, N. Ishihara, and S. Konaka, "High gain equalizing amplifier integrated circuits for gigabit optical repeater," *IEEE J. Solid-State Circuits*, vol. SC-20, no. 3, pp. 703–707, June 1985.

[4] Y. Akazawa, N. Ishihara, T. Wakimoto, K. Kawarada, and S. Konaka, "A design and packaging technique for a high-gain, gigahertz-band single-chip amplifier," *IEEE J. Solid-State Circuits*, vol. SC-21, no. 3, pp. 417–423, June 1986.

[5] Y. Akazawa, A. Iwata, T. Wakimoto, T. Kamoto, H. Nakamura, and H. Ikawa, "A 400MSPS 8b flash AD conversion LSI," in *ISSCC Dig. Tech. Paper*, 1987, pp. 98–99.

Part IX

State of the Art and Some Applications

THE first attempts to realize practical DDFS were made in the early 1980s. According to Ian Hickman [IX-1] "A prototype DDS at the Plessey laboratories in 1981 occupied several complete boards of logic laid out on the bench and being clocked at 10 MHz."

Since then the technology has advanced rapidly. Nowadays DDFS chips using silicon technology provide output frequencies in excess of 300 MHz, and those based on GaAs already exceed the 1 GHz limit. The maturity of the technique is further demonstrated by the shift of the papers about DDFS from scientific to popular journals.

Thus, J. A. Gallant [IX-2], after a short introduction to the DDFS technique, discusses spurious signals and refers to properties of several commercial synthesizers, some of which are still marketed. In Fig. 2 he reproduces the block diagram of the type Q2334 from Qualcomm with "noise reduction circuits" following "sine lookup tables." The principle is the addition of several bit-long random numbers to the outputs of the sine words in accordance with a company patent [IX-3].

To the same category of articles belongs the series by I. Hickman and T. Stanley [IX-4], with a short survey of properties of high-frequency DDFS and their manufacturers. A similar paper by Saul and Coffey (1993) summarizes the state of the art, which is still valid, and refers to the most advanced DDFSs on the market. The authors offer some practical spectral measurements on output frequencies over 100 MHz. Unfortunately, the paper lacks a deeper discussion of spurious signals and was published with misplaced figure captions [IX-5].

The first reprinted paper, by Kent and Sheng (1995), provides information about the recent progress of DDFS design. The high-speed heterojunction bipolar transistors (AlGaAs/GaAs HBT technology), with $f\tau$ and f_{max} about 60 GHz, make possible use of 500-MHz clock frequency. After discussion of the architecture, a new measure for appreciation of the spurious signals is introduced. From their Fig. 8 one concludes that the theoretical spurious signal level due to the phase truncation ($-6W = -72$ dB) is exceeded nearly in the whole output range. Consequently, the "worst-case spurious" are dominated by intermodulation products. The dissipated power is 5 W.

At nearly the same time, Tan and Samueli (1995) discuss a 200-MHz "quadrature digital synthesizer/mixer" in CMOS technology consuming 2 W from 5 V only. The sine and cosine outputs are generated with application of the eighth wave symmetry. The modulation capabilities of the chip include frequency modulation (by directly modulating the frequency control word) and phase modulation (by adding a phase offset to the accumulator). Finally, quadrature amplitude modulation and single-sideband frequency translation are obtained by adding a complex multiplier block to sine and cosine outputs.

Another approach to the hundreds of MHz DDFSs was published again by Tan, Roth, Yee, and Samueli (1995). They used 0.8-μm CMOS technology, and with the assistance of the pipeline operation of sine ROMs they arrived at 800-MHz clock frequencies. The stated spectral purity is -84.3 dB due to the 14-bit truncated accumulator output (i.e., $-14 * 6.02$), the tuning latency 47 clock cycles, and power dissipation 3 W from a 5-V source.

T. McClelland et al. discuss an interesting application of DDFS. The quartz crystal oscillator is tuned to the direct submultiple (136th) of the Rb standard frequency as clock for the generation of the output frequencies in the range between 1 Hz and 20 MHz, with the assistance of a commercial DDFS [IX-6, IX-7].

Another field for DDFS applications provides Standard Time and Frequency laboratories [e.g. IX-8–IX-11], where the very high resolution is appreciated.

An interesting scheme was proposed by Rubiola, Del Casale, and De Marchi (1993) for field control of the caesium atomic resonator. The frequency synthesis chain starts with a high-resolution DDFS, followed by a three-stage direct synthesizer based on the Sylvester series model.

REFERENCES

[IX-1] Ian Hickman. "Direct digital synthesis." *Electronics World and Wireless World,* pp. 630–34, August; Part 2, pp. 746–48, September; Part 3: "Genuine solutions to spurious arguments," pp. 843–45, October; Part 4: "Putting DDS to work," pp. 937–41, November 1992.

[IX-2] J. A. Gallant. "Devices refine the art of frequency synthesis." *EDN,* pp. 95–104, November 1989.

[IX-3] R. J. Kerr and L. A. Weaver. Pseudorandom dither for frequency synthesis noise. *US patent 4,905,177,* Feb. 27, 1990, assignee to Qualcomm Inc.

[IX-4] T. Stanley. "Devices for frequency synthesis." *Electronics World and Wireless World,* pp. 999–1003, December 1992.

[IX-5] P. Saul and T. Coffey. "Current achievements in direct digital synthesis." *Microwave Engineering Europe,* pp. 31–36, December/January 1993.

[IX-6] M. Bloch, I. Pascaru, C. Stone, and T. McClelland. "Subminiature rubidium frequency standard for commercial applications." *1993 IEEE International Frequency Control Symposium, Proceedings,* pp. 164–77.

[IX-7] T. McClelland et al. "Subminiature rubidium frequency standard: Manufacturability and performance results from production units." *1995 IEEE International Frequency Control Symposium, Proceedings,* pp. 39–52.

[IX-8] E. M. Mattison and L. M. L. Coyle. "Phase noise in direct digital synthesizers." *Proceedings of the 42th Annual Frequency Control Symposium,* 1988. (Reprinted in Part VII.)

[IX-9] R. P. Giffard and L. S. Cutler. "A low-frequency, high resolution digital synthesizer." *1992 IEEE International Frequency Control Symposium, Proceedings,* pp. 188–92. (Reprinted in Part V.)

[IX-10] C. W. Nelson, F. L. Walls, F. G. Ascarrunz, and P. A. Pond. "Progress on prototype synthesizer electronics for $^{199}HG^+$ at 40.5 GHz." *Proceedings of the 1992 IEEE Frequency Control Symposium,* pp. 64–69, May 1992.

[IX-11] R. K. Karlquist. "A new RF architecture for cesium frequency standards." *Proceedings of the 1992 IEEE Frequency Control Symposium,* pp. 134–42, May 1992.

A High Purity, High Speed
Direct Digital Synthesizer

Gary W. Kent
Rockwell International Corporation
Collins Avionics & Communications Division
855 35th. St. NE, 137-153
Cedar Rapids, IA 52498
(319) 395-5346
FAX: 319-395-5248

Neng-Haung Sheng
Rockwell International Corporation
Science Center
1049 Camino Dos Rios, 083-A16
Thousand Oaks, CA 91360
(805) 373-4110
FAX: 805-373-4775

Abstract

A Direct Digital Synthesizer (DDS), that clocks at 500 MHz, has been constructed in a 70 pin hybrid circuit package that is 1.14 x 2.33 x 0.2 inches. Spectral purity is better than -55 dBc worst case spur, up to 245 MHz output frequency. TTL compatible, parallel lines are provided for 28 bits of frequency and 8 bits of phase control. The hybrid is based on two GaAs chips, a HBT digital to analog converter and a MESFET accumulator/ROM combination. Two silicon ECL chips are used for clock amplification and distribution. This allows an AC coupled sinusoidal input clock of 0 dBm nominal amplitude. Power dissipation is typically 5 watts using +5.0, -5.2, and -2.2 volt supplies.

This paper describes the DDS architecture, design of the GaAs chips, and special problems encountered during development of the hybrid.

Introduction

Specifications for new communications equipment increasingly require high speed Direct Digital Synthesizers (DDS). At the highest level, a DDS consists of three parts: a phase accumulator, a sine function read-only memory (ROM) look-up table, and a digital-to-analog converter (DAC), as shown in fig. 1. Since frequency is the rate of phase change, the command word which is added to the accumulator each clock period determines the output frequency. Because the fixed-length accumulator regularly overflows, the phase is periodic, with a frequency proportional to both the increment rate and the incremental amount. The largest value that can be stored in the accumulator represents 2π radians, so that overflow corresponds to a module 2π operation, and is not seen in the output. The accumulator addresses the ROM table to determine a digitized sine amplitude value. The DAC is then employed to convert the digital data to an analog sine-wave. For output frequencies close to one-half the clock frequency, a lowpass filter may be used to eliminate the alias frequency component created by the sampling process [1][5].

A DDS provides many significant advantages over phase-locked-loop (PLL) approaches. Fast settling time, sub-Hertz frequency resolution, continuous-phase frequency switching and low phase noise are features easily obtainable in DDS systems. However, this approach has

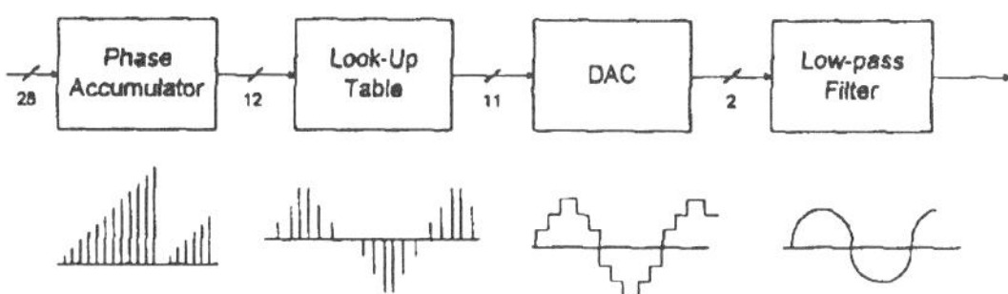

Figure 1. Direct Digital Synthesis

been limited to low frequency applications, since the component circuits, such as the ROM and DAC, have to support clock rates of about two times the synthesized frequency. Semiconductor technologies previously would not support the high clock rates required. Also, high spurious output made the concept unusable for many applications. High speed DDS technology has been dramatically improving. There are now several companies that offer products that clock at speeds in excess of 500 MHz and provide reasonable spurious performance. However, when available chips are reviewed for new radio applications, they have one or more of the following deficiencies:

1) Spurious is still too high.
2) Size is too large.
3) Power dissipation is too high.

HBT Technology

High bandwidth DDS systems can only be realized by the use of ultra-high speed transistors, such as Hetrojunction

Reprinted from *Proceedings of the IEEE International Frequency Control Symposium*, pp. 207-211, 1993.

Bipolar Transistors (HBTs). The AlGaAs/GaAs HBT technology has been developing at Rockwell for the past several years. A baseline process has recently been demonstrated for manufacturing LSI circuits of up to 8000 transistors at Rockwell's III-V facility. We use MOCVD-grown 4 inch GaAs substrates, with a Carbon-doped p+ base. Minimum transistor geometry is 1.4 x 2.0 μm^2. The base is very thin (700 angstroms) and doped with carbon to reduce base resistance. D.C. current gain for these transistors is measured to be in excess of 100. The f_T and f_{max} is about 60 GHz. Three years ago, Rockwell recognized that these HBTs would be advantageous in high speed DAC design and a research program was started to develop DAC prototypes that would clock at speeds in excess of 1 GHz.

DAC Architecture

The 12 bit digital-to-analog converter is a current steering type with complimentary outputs. Output levels are 500 mV into a 50 ohm load. The 4 most significant data bits are converted from binary codes into 15 thermometer codes. The remaining eight bits are sent to current sources that drive a R-2R ladder. Currents through all 23 sources are identical. This is called a segmented architecture (figure 2). It was chosen over an "all ladder" approach to minimize glitch energy at the code where the MSB switches from zero to one, and all the other bits switch from one to zero. This is traditionally the most difficult switching point to realize monotonic behavior. Data registers are used for time alignment at the DAC input and at the input to the current switches. This results in a latency of two clocks periods.

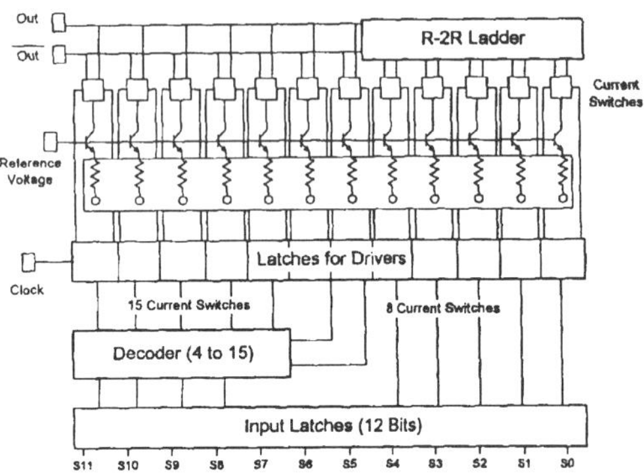

Figure 2. 12-bit, 4-bit segmented, 8-bit R-2R, current steering DAC.

The DAC circuit has been fabricated using the Rockwell HBT process described above. The resulting chip measures 88 x 88 mils. Power dissipation is 1.2 watts.

NCO Architecture

The phase accumulator and ROM sine wave mapping circuits are combined into a single MESFET chip called a numerical controlled oscillator (NCO). The 28-bit frequency select word is strobed into the input register and synchronized with the main clock, so that frequency changes are phase continuous. The accumulator uses seven four-bit adders in a pipeline configuration (figure 3). The output is truncated to the upper 12 bits. Phase modulation is provided by two four bit adders that sum the uppermost eight bits of the accumulator with a phase control word before they address the ROM.

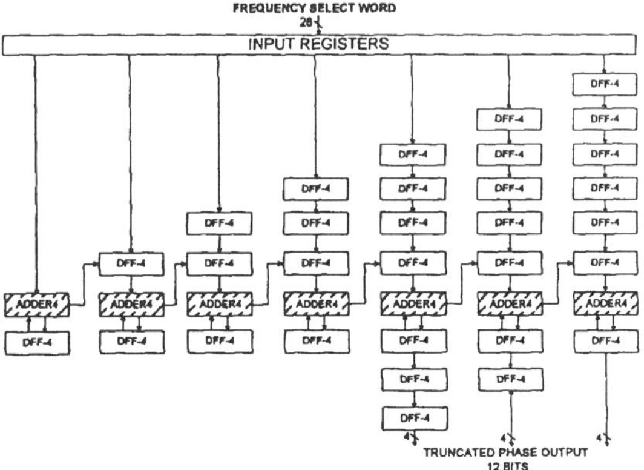

Figure 3. Fully pipelined Phase Accumulator.

It is impractical to map all 28 bits from the accumulator to a sine wave amplitude value. Reducing the number of ROM input bits introduces phase truncation error, and reducing the number of ROM output bits introduces amplitude truncation error. Both phase and amplitude truncation errors will generate unwanted spurs. Simulations show that 12 bits of phase and 11 bits of amplitude should result in -68 dBc worst case spurs.

Still, a look-up table with 12 bits of phase and 11 bits of amplitude resolution would correspond to a ROM size of 45K bits. To ensure adequate yield in fabrication and minimal power consumption, a compact architecture that minimizes storage requirements has been presented by Nicholas and Samueli [2]. The ROM used in the Rockwell DDS is based on a similar concept, although the architecture was designed in 1985 using a less sophisticated method. Great storage compression is achieved by partitioning the look-up table into three small ROMs. Phase arguments into the ROMs are segmented as a low phase resolution coarse ROM, with additional phase resolution provided by interpolation between coarse ROM samples using the slope of the sine wave at each coarse sample. The third ROM provides correction factors when needed. This approach reduces storage to about 272 bits, achieving a memory compression ratio of 1:165. The tradeoff is extra multipliers, adders and logic circuits required to decode the complete sine function [3].

The most elementary method of sine storage compression is to exploit the symmetry of the sine function about π and $\pi/2$. By properly inverting the phase and amplitude of the sine function, look-up table samples need only be stored for phase values between 0 and $\pi/2$ (figure 4).

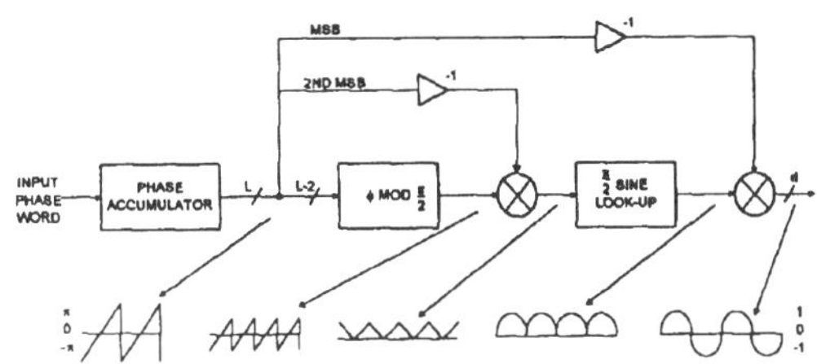

Figure 4. NCO architecture, with logic to exploit quarter-wave symmetry.

The complete sine wave can be constructed using hardware to truncate the phase MSB, and then using the second MSB to full wave rectify the magnitude of the phase. The remaining bits are used to address the $\pi/2$ look-up table that is partitioned into a coarse ROM, a slope ROM and a carry ROM. Several computer programs were used to optimize the partitioning for optimum ROM size and spur performance. Based on the simulation results, the quarter sine is divided into 16 segments, and a coarse ROM value of ten amplitude bits determined for the beginning of each segment. The remaining 64 values in each segment are calculated by multiplying the relative position in the segment by a segment slope, and adding it to the coarse value. The slope values are 5 bits. The amplitude error resulting from this approach can be as much as 2 LSB's, so in portions of some segments an extra bit is added in to correct the values to within 1 LSB. The information for this correction is stored in a 16x2 carry ROM that is combined with the slope ROM (figure 5).

Now that the ROMs are so small, (16x10 and 16x7) it is convenient to realize them using linear addressing and hard logic encoding instead of modern ROM architecture. The four address bits are decoded, and then multi-input

OR gates determine an output value for a particular address (figure 6). A simple mapping procedure is used to determine the logic necessary. If the circuitry is too slow, it is readily pipelined to increase the speed as necessary. The requirements of this design result in a pipeline of 22 stages for the entire NCO.

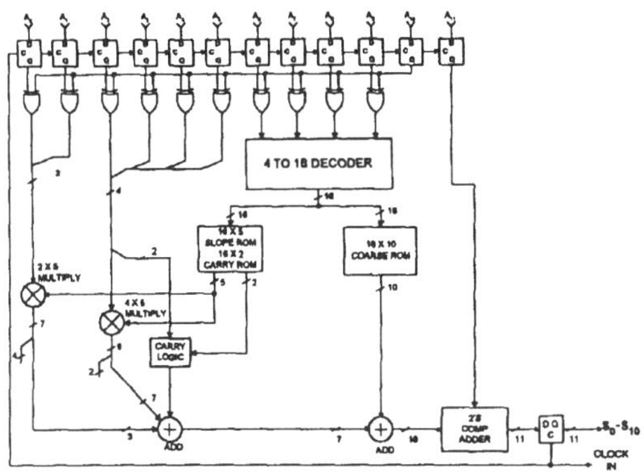

Figure 5. Sine-Converter logic.

The NCO circuit has been fabricated by Vitesse Semiconductor using their 0.6 micron H-GaAsIII MESFET process. The design uses the VGFX20K gate array that has a size of 218 x 135 mils. Power dissipation is 3 watts.

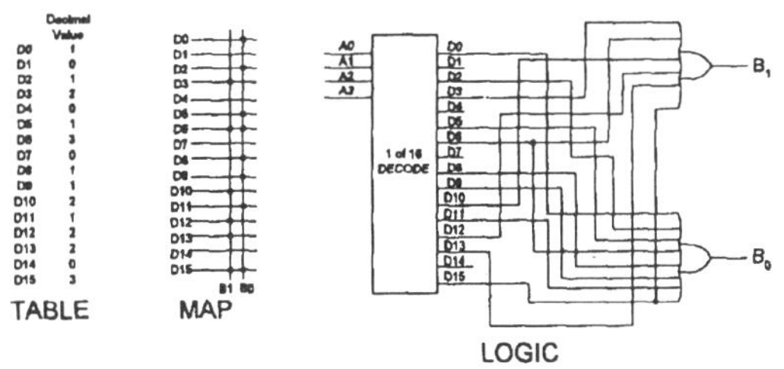

Figure 6. Carry ROM, with mapping procedure.

Hybrid Packaging

To combine the two chips into a complete DDS, a 70 pin hybrid, from National Hybrid Inc., has been selected. The package dimensions are 2.33 x 1.14 x 0.2 inches. It has a screw mount molybdenum flange for excellent heat sinking. The DAC chip is attached directly to the flange through a hole in the ceramic substrate. The NCO is attached to a beryllium oxide (BeO) insulator because the GaAs substrate must be connected to the Vee (-2.2V) supply and isolated from ground. Thermal analysis shows that the temperature rise from the hybrid flange to the

DAC junction is only 14 degrees Celsius. This is important because many DAC linearity effects are temperature sensitive.

Parallel lines provide the 28 frequency and 8 phase lines. A frequency or phase strobe is used to load the data after it has settled. Some DDS circuits load serially, or maybe several bits at a time, to minimize the number of I/O lines. The all-parallel approach allows high speed frequency-or-phase modulation.

Two power planes are used on the hybrid substrate to provide -5.2V and -2.2V to the chips. Even so, there is over 0.1V drop from the pins to the -2.0V Vee pad on the NCO chip, due to the 1.5A current needed. Care must be taken to prevent NCO switching currents from coupling onto the DAC clock and creating spurious output frequencies.

<u>High Speed Digital Interface</u>

The interface of a NCO and DAC is not trivial at clock speeds approaching 1 GHz. The set-up and hold times of the Rockwell DAC are approximately 180 picoseconds. Even though the hybrid circuit is small, line length differences can have a big impact on the DAC's ability to latch the correct NCO data. Ideally, the DDS clock should be fed to the NCO first and then passed through to the DAC with the proper clock to data timing relationships determined by NCO gate delays. Then, as process variations change the propagation delays of the NCO circuits, both the output data and output clock timing will be affected in the same manner. Their relationship to each other will remain somewhat constant. With proper design of the timing relationship, the DAC will clock correct data with the extremes of fast and slow NCO chips.

This approach was planned for the Rockwell DDS, but excessive jitter prevented the use of the NCO clock output as the DAC clock input. Since the DAC clock controls the register that drives the output current switches, it controls the digital-to-analog conversion process and must be considered an analog signal. It's purity has a direct effect on the output spurious. The NCO clock output circuit was automatically routed during the layout of the gate array. Little attention was paid to possible interaction with the other digital signals, especially the high current outputs. As a consequence, the clock edges are pulled due to parasitic coupling with the output data lines. This has no effect on proper digital operation of the NCO, but corruption of the clock output makes it unsuitable for driving the DAC.

To keep the DAC clock clean, the DDS system clock path is split by an ECL clock driver chip. One path goes directly to the NCO chip. The other must include a delay before reaching the DAC. The amount of delay is approximately equal to the clock-to-data delay of the NCO. Unfortunately, the delay of the NCO will vary from device to device, so the delay in the DAC clock path must be adjustable, to provide the proper timing relationships at the input of the DAC. The DDS uses an ECL programmable delay chip to provide the flexibility needed to prevent violations of the DAC set-up and hold times (figure 7).

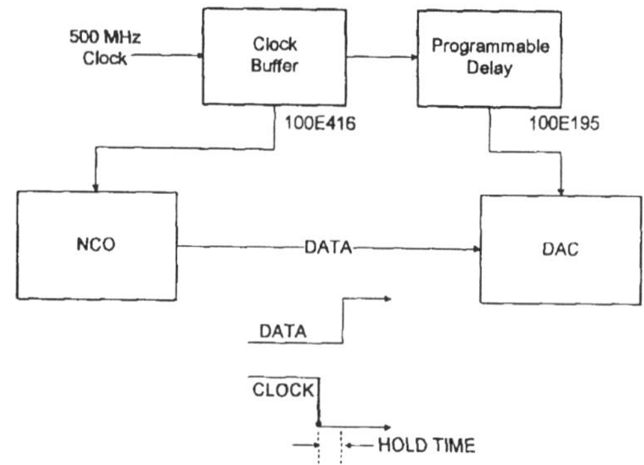

Figure 7. Hybrid DDS clock path. Programmable delay adjusts timing so that hold time is not violated.

<u>Performance Results</u>

Spurious performance and bandwidth are the most critical parameters of any DDS, since other features such as phase noise, frequency switching speed and modulation accuracy are almost ideal. The spurs are usually put in the best light possible by presenting spectrum results of only a few output frequencies, and letting the observer extrapolate to predict spurs at other frequencies/bit-patterns. A commercial manufacturer of synthesizers has presented a more definitive automated spur characterization technique [4]. The one spur with the highest amplitude and the spur with the second-highest amplitude are plotted on the y-axis and the output frequency on the x-axis. This permits evaluation of the unit's performance at multiple outputs up to the maximum bandwidth. A lowpass filter follows the DDS to attenuate out-of-band spurs. When the Rockwell Hybrid DDS is measured in this manner, the worst spur is -55 dBc (figure 8).

Looking at the two worst spurs is certainly a meaningful characterization technique. But some applications are interested in the spectral power of the spurs, where many small spurs can add up to be significant. Figure 9 shows the Total Integrated Spurious of the Rockwell Hybrid DDS. Here the output frequency was plotted from 100 to 245 MHz, and all of the spurs, within a bandwidth of 50 to 250 MHz, were added up. Each dot on the graph represents the total sum for one output frequency. The following formula calculates the value for each dot, in dBc:

$$\text{Total Integrated Spurious} = 10\log(10^{x_1/10} + 10^{x_2/10} + ...10^{x_n/10})$$
$$x^n = \text{spur in dBc}$$

When the Rockwell Hybrid DDS is measured in this manner, the worst total sum is -50 dBc.

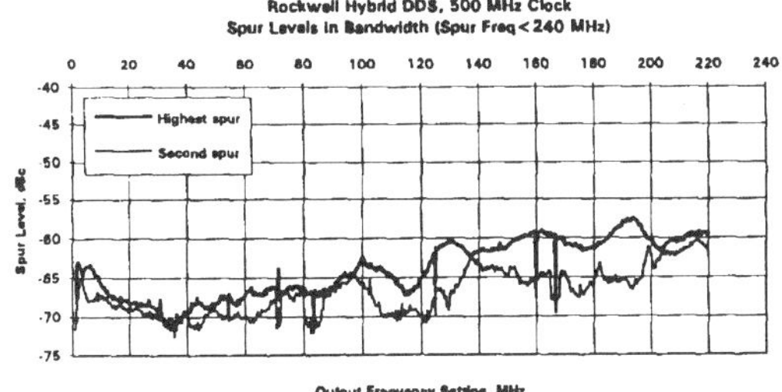

Figure 8. Worst Case Spurious

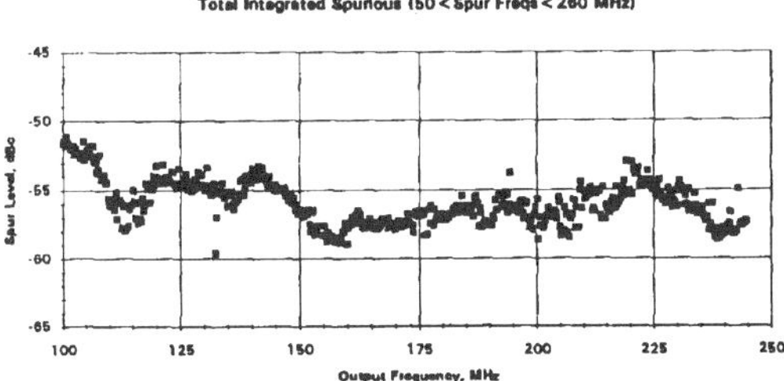

Figure 9 Total Integrated Spurious

Conclusion

A high speed DDS has been described that has significant improvements in spurious performance, size, and power dissipation. The NCO look-up table is compressed sufficiently to use logic gates instead of an actual ROM. The path of the DAC clock requires special attention to prevent timing jitter and the resulting degradation of spurious performance. Also presented is the concept of Total Integrated Spurious performance as a valuable measure of system performance.

AlGaAs/GaAs HBT technology has advanced to the point where a "next generation" DDS will be realized as a single large chip. Clock speeds in excess of 2 GHz should be possible.

Acknowledgments

The authors would like to thank Bob Marston, Glenn Doughty, and the rest of Rockwell Communications Systems Division management for their support of DDS research. We also acknowledge the contributions of the Rockwell Microelectronics Technology Center for fabricating and enhancing the performance of the HBT DAC. Special thanks goes to Scott Voyek for technical support of the test hardware.

References

[1] J. Tierney, C.M. Rader, B. Gold, "A Digital Frequency Synthesizer", IEEE Trans. Audio Electroacoustic., vol. AU-19, pp. 48-57, 1971.

[2] H.T. Nicholas, H. Samueli, "A 150-MHz Direct Digital Frequency Synthesizer in 1.25 μm CMOS with -90 dBc Spurious Performance", IEEE Journal of Solid-State Circuits, vol. 26, pp. 1959-1969, Dec. 1991.

[3] R.A. Freeman, "Digital Sine Conversion Circuit For Use In Direct Digital Synthesizers", U.S. Patent # 4,809,205, Feb. 28, 1989.

[4] "Frequency Synthesis & RF Subsystems Catalog", SCITEQ Electronics, Inc., San Diego, CA.

[5] D.A. Sunderland, R.A. Strauch, S.S. Wharfield, H.T. Peterson, and C.R. Cole, "CMOS/SOS Frequency Synthesizer LSI Circuit for Spread Spectrum Communications" IEEE Journal of Solid State Circuits, vol. SC-19, pp. 497-505, Aug. 1984.

A 200 MHz Quadrature Digital Synthesizer/Mixer in 0.8 μm CMOS

Loke Kun Tan and Henry Samueli, *Member, IEEE*

Abstract— A 200 MHz quadrature direct digital frequency synthesizer/complex mixer (QDDFSM) chip is presented. The chip synthesizes 12 b sine and cosine waveforms with a spectral purity of -84.3 dBc. The frequency resolution is 0.047 Hz with a corresponding switching speed of 5 ns and a tuning latency of 14 clock cycles. The chip is also capable of frequency, phase, and quadrature amplitude modulation. These modulation capabilities operate up to the maximum clocking frequency. The chip provides the capability of parallel operation of multiple chips with throughputs up to 800 MHz. The 0.8 μm triple level metal N-well CMOS chip has a complexity of 52 000 transistors with a core area of 2.6×6.1 mm^2. Power dissipation is 2 W at 200 MHz and 5 V.

I. INTRODUCTION

HIGH-PERFORMANCE direct digital frequency synthesizers (DDFS's) play an extremely important role in modern digital communications. They offer many advantages including fast continuous-phase switching response, fine frequency resolution, large bandwidth, and good spectral purity. This method of frequency synthesis also results in very low phase noise that is limited to be less than or equal to that of the reference clock source.

The architecture used in this design was originally introduced by Tierney, Rader, and Gold [1]. It utilizes an overflowing L-bit accumulator (or phase accumulator) to generate the phase argument of the sine function generator. Each overflow of the phase accumulator represents one period of a sine wave. The input word (Frequency Control Word) to the phase accumulator controls the frequency of the generated sine waveform. The sine function generator is a ROM look-up table which stores the sine samples. Inherent in this method of frequency synthesis is the high-frequency resolution attainable without the need to increase the size of the ROM look-up table. Frequency resolution is doubled by each addition of 1 b to the phase accumulator wordlength. For a given Frequency Control Word (FCW), clocking frequency (f_{clk}), and phase accumulator word length (L), the output frequency (f_{out}) of

Manuscript received July 15, 1994; revised November 6, 1994. This work was supported in part by research grants from the University of California MICRO Program, Hewlett–Packard, TRW Electronic Systems Group, Hughes Space and Communications Group, and the Advanced Research Projects Agency.

L. K. Tan is with Broadcom Corporation, Los Angeles, CA 90024 USA.

H. Samueli is with the Integrated Circuits and Systems Laboratory, University of California, Los Angeles, CA 90024 USA.

IEEE Log Number 9408738.

the synthesizer is given by

$$f_{out} = \frac{f_{clk} \cdot FCW}{2^L} \tag{1}$$

and the minimum frequency resolution is given by

$$\Delta f = \frac{f_{clk}}{2^L}. \tag{2}$$

In the design presented in this paper, the phase accumulator word length is 32 b and the maximum clocking frequency is 200 MHz. This gives a minimum frequency resolution of 0.047 Hz. Another advantage inherent in this architecture is fast switching speeds, with the capability to switch between two frequencies in 1 clock cycle. For 200 MHz operation, the switching speed is 5 ns.

This design is also based on the architectural optimizations of Nicholas and Samueli [2]. The same ROM look-up table compression techniques and phase accumulator design were used to simultaneously achieve good spectral purity and large bandwidth. In addition to the features inherent in the design in [2], the synthesizer presented in this paper provides quadrature outputs without increasing the size of the ROM look-up table. It also provides phase modulation, and quadrature amplitude modulation with full-complex multiplications for single-sideband frequency translation. Also, for very high throughput applications, 2 or 4 chips can be parallel to double or quadruple the maximum throughput rate to 400 MHz or 800 MHz.

II. APPLICATIONS AND DESIGN REQUIREMENTS

Applications for DDFS's range from instrumentation and measurement to modern digital communications. Since the area of instrumentation uses DDFS's as reference sources, the spurious performance requirement of the synthesizer is usually significantly higher than applications in other areas. The frequency, phase, and amplitude modulation capabilities together with the ability to operate at double or quadruple the throughput of a single device makes this design very suitable in the instrumentation field where performance is critical.

Two example applications of this chip in the area of digital communications are the implementation of tunable quadrature modulators and demodulators. A tunable quadrature amplitude modulation (QAM) modulator, as shown in Fig. 1, uses square-root Nyquist filters for pulse shaping and interpolation filters for removing in-band images and increasing the sampling rate. The shaded region is the QDDFSM chip which mixes the baseband signal up to IF. The dual of this

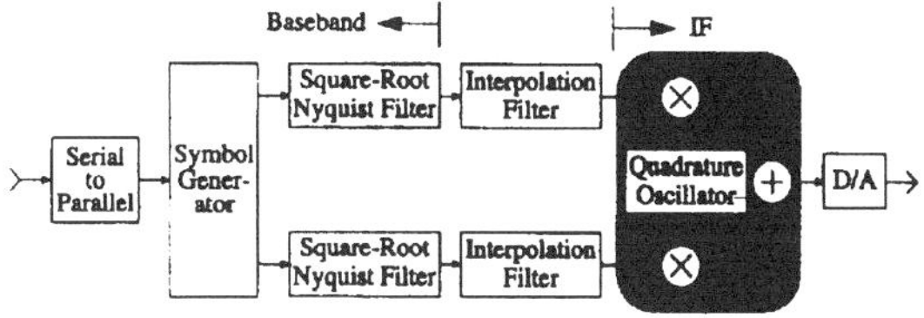

Fig. 1. Tunable QAM modulator.

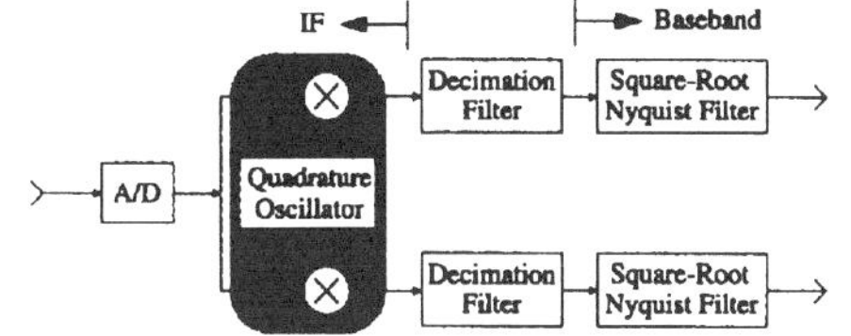

Fig. 2. Tunable QAM modulator.

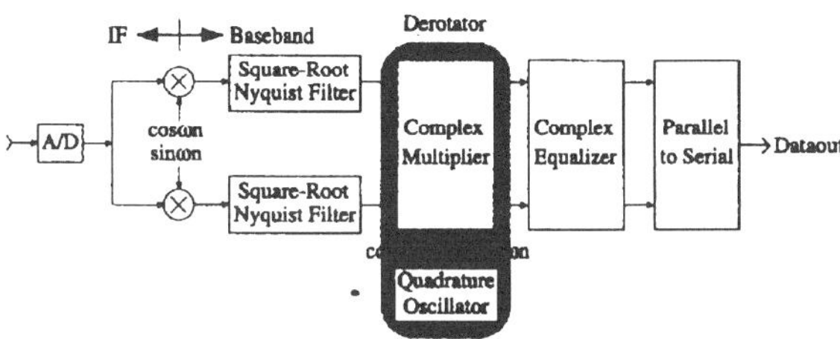

Fig. 3. Alternative QAM receiver architecture.

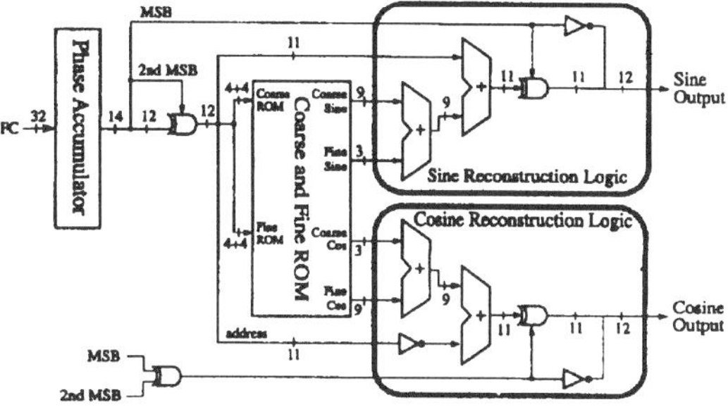

Fig. 4. Sine/cosine reconstruction logic.

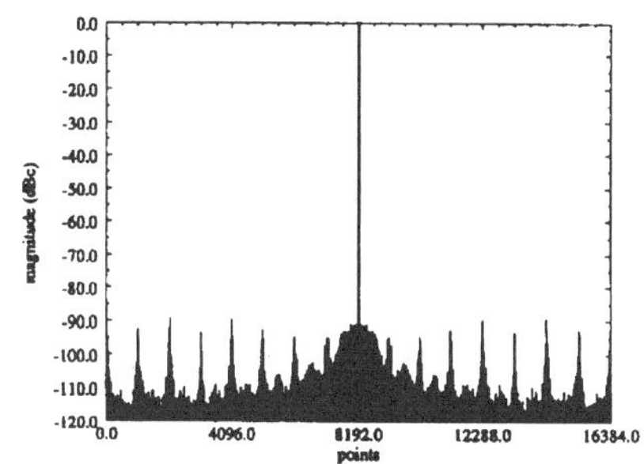

Fig. 5. Digital output spectrum.

architecture is the tunable QAM demodulator shown in Fig. 2. The QDDFSM chip here mixes the IF signal back to baseband where it is decimated and passed through matching square-root Nyquist receive filters. Another application of this chip is in a fixed IF QAM receiver [9] where it functions as a digital derotator (Fig. 3) to remove any residual frequency or phase errors in the system.

In order to satisfy a wide range of applications, it was required that the worst case digital spurious performance of this synthesizer be better than −80 dBc. Based on the results in [10], a phase resolution of 14 b results in a spurious performance due to phase accumulator truncation of −84.3 dBc. The total phase accumulator wordlength was chosen to be 32 b to achieve a frequency resolution of 0.047 Hz at a clock rate of 200 MHz and the upper 14 b of the phase accumulator are used by the table look-up circuitry.

III. CHIP ARCHITECTURE

A. ROM Compression

This design uses a generalized algorithm for the compression of arbitrary functions into coarse and fine ROM samples [5], [2]. Let $A + B + C$ be the total number of bits of the phase address, with A being the most significant bits, B the next most significant bits, and C the least significant bits. Then using this algorithm, the coarse ROM would have 2^{A+B} samples, and the fine ROM would have 2^{A+C} samples. Trial and error using the techniques described in [5] resulted in a ROM with $A = 4$, $B = 4$, and $C = 4$. The output wordlength of the coarse ROM is 9 b, and that of the fine ROM is 3 b. Thus, $2^{14} \times 12$ sine samples are compressed into $2^8 \times 9$ coarse samples and $2^8 \times 3$ fine samples resulting in a compression ratio of 64 : 1. Further compression of the ROM look-up table was obtained by not storing the sine samples, but rather by storing the difference between the sine samples and the phase $(\sin(\phi) - 2\phi/\pi)$. This results in the need for a final sine/cosine reconstruction stage that adds the coarse and fine ROM samples to the phase argument (Fig. 4). An FFT of the compressed ROM contents gives the worst case digital output spectral purity to be −89 dBc (Fig. 5). However, phase accumulator truncation to 14 b is still the dominant source of spurious noise, and thus the overall spectral purity of the synthesizer is −84.3 dBc.

B. Quadrature Outputs

A typical DDFS architecture takes advantage of the quarter-wave symmetry of a sine wave to reduce ROM storage requirements. Thus, only sine samples from 0 to $\pi/2$ are stored, and the second MSB of the phase accumulator is used to determine the quadrant, thereby synthesizing a sine wave from 0 to π. The MSB is then used as a sign bit to synthesize the complete sine wave from 0 to 2π. In the case of a cosine

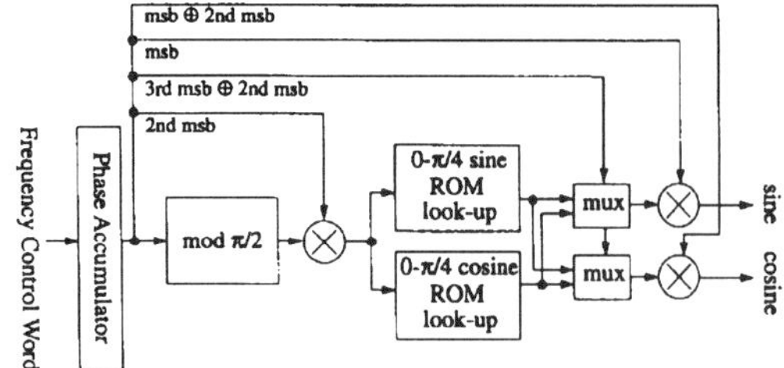

Fig. 6. Sine/cosine storage technique.

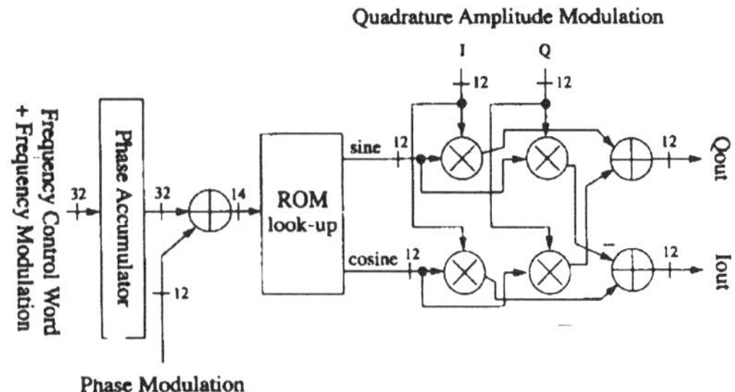

Fig. 7. Chip modulation capabilities.

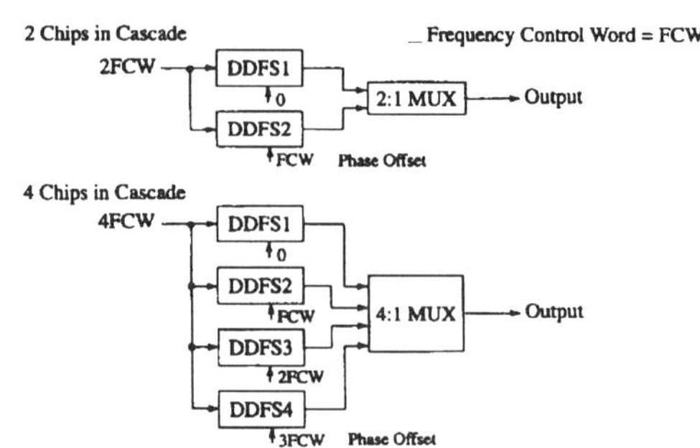

Fig. 8. Paralleling chips for high throughput.

waveform, its zero crossings are advanced by $\pi/2$ with respect to that of a sine waveform. To produce the sign bit for this case, the MSB is EXOR'ed with the second MSB.

For a design where quadrature outputs are desired, a brute force method would be to store both sine and cosine samples from 0 to $\pi/2$. This would double the size of the ROM look-up table. Instead, one could take advantage of eighth wave symmetry of a sine and cosine waveform, since sine samples from 0 to $\pi/4$ are the same as cosine samples from $\pi/4$ to $\pi/2$. Similarly, cosine samples from 0 to $\pi/4$ are the same as sine samples from $\pi/4$ to $\pi/2$. Hence, one need only store sine and cosine samples from 0 to $\pi/4$. The third MSB from the phase accumulator can be used to select between these samples. In the actual implementation, the third MSB is EXOR'ed with the second MSB to produce this signal. This is necessary so that the select signal is phase aligned with the eighth wave symmetry axis plane of the sine and cosine waveform. Hence, the only additional hardware cost to synthesize quadrature outputs is a 2-to-1 MUX. This architecture is illustrated in Fig. 6.

C. Modulation Formats

The modulation capabilities of this chip include frequency modulation, phase modulation, quadrature amplitude modulation, and single sideband frequency translation. Frequency modulation is performed by directly modulating the Frequency Control Word, thus no additional hardware is needed to implement this feature. Phase modulation is obtained by adding a phase offset to the phase accumulator output before addressing the ROM look-up table. In hardware, this amounts to incorporating an extra addition stage. This chip accepts a 12 b word for phase modulation. Finally, quadrature amplitude modulation and single sideband frequency translation are obtained by adding a complex multiplier block to the sine and cosine outputs of the quadrature DDFS as shown in Fig. 7. The wordlengths for the I and Q rails for amplitude modulation are 12 b each. The complex multiplier block is made up of four 12×12 real multipliers.

D. Utilizing Parallel Chips for High Throughput

The use of parallelism to attain high throughput is very common in VLSI design. Such techniques have been utilized in [3] and [4] for DDFS applications. Our chip has been designed with the capability to double or quadruple the single chip maximum frequency of operation by paralleling 2 or 4

chips. In the 2 chip case, this can be done by causing each chip to generate every other output sample. Hence, by using a 2-to-1 MUX and alternately selecting the outputs from the 2 chips, one can attain twice the maximum frequency. To generate every other sample of the output, each chip must use 2 times the Frequency Control Word. In addition, a phase offset must be added to one chip so that the outputs are not duplicated. Fig. 8 illustrates this idea for both the 2 chip case and the 4 chip case. In the actual chip, both the multiplication of the Frequency Control Word and the addition of the phase offset are implemented internally. The user merely sets specific control lines to configure the chips for a certain mode of operation.

IV. CIRCUIT DESIGN

A. Clock Distribution and Register Design

One problem that plagues high-speed CMOS designs is clock skew. Standard two-phase clocking schemes would be difficult to design since small amounts of clock skew would directly limit the speed of operation. To achieve the desired 200 MHz throughput, it was necessary to use a true single-phase clocking (TSPC) scheme. The register chosen for this design is a dynamic positive edge-triggered single-phase register [8] which is shown in Fig. 9.

Another problem inherent in high-speed CMOS chips is power supply clock switching noise. This problem exists due to the large number of simultaneously switching register elements, fast switching speeds, and power supply wiring inductance. This makes the design of the clock distribution

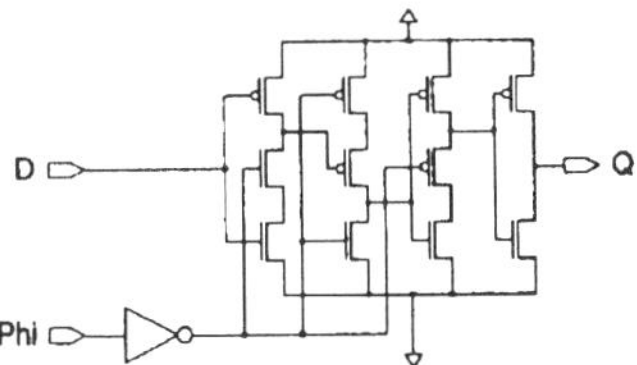

Fig. 9. Single-phase positive edge triggered register.

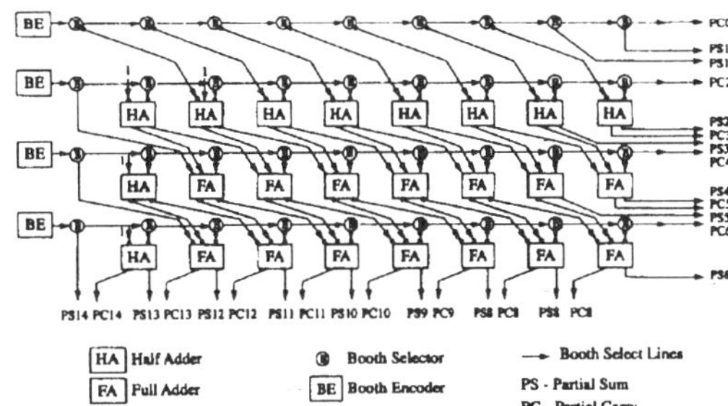

Fig. 10. Multiplier architecture.

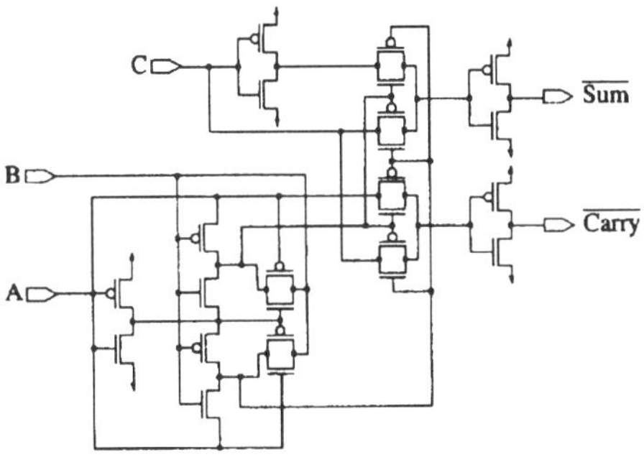

Fig. 11. Inverting transmission gate full adder.

network and its associated capacitive loading critical. Hence, it is important that registers that minimally load the clock distribution line be used. The equivalent clock loading of the register in Fig. 9 is one minimum sized inverter. This resulted in the lowest possible loading of the clock distribution lines. To further reduce power/ground switching noise, and to ensure that power supply levels never collapse during clock transitions, on-chip thin-oxide decoupling capacitors were placed directly across the power/ground lines connected to the clock buffer. Although on-chip decoupling capacitors reduce di/dt noise generated by on-chip circuitry, they do not reduce noise generated by simultaneously switching off-chip drivers. This is due to the fact that the off-chip ground plane is not common to the on-chip ground plane, but rather they are connected through inductive bonding wires and package pins. To reduce off-chip di/dt noise from coupling back to on-chip circuitry, on-chip power supplies were isolated from power supplies for off-chip drivers. This is accomplished by partitioning the chip power supplies into two unconnected ring halves—one-half for on-chip circuitry and the other half for off-chip drivers. To reduce the effective inductance of the off-chip drivers, one pair of power/ground pins were used for every 4 output pins. Double bonds were also used for all power/ground supply pins, further reducing effective inductance.

Another issue in high-speed designs is data race. To alleviate this problem, one large localized clock buffer was used to drive all the registers on the chip. This reduces clock skew and provides more reliable operation. In addition, exceptional care was taken to ensure that all clocks are distributed opposite to the flow of data, hence eliminating the possibility of data race.

One disadvantage of the single-phase register in Fig. 9 is that it requires clock edges with fast rise and fall times. To ensure that the RC time constant along the clock lines is minimal, all main clock lines were routed using third-level metal. The use of third-level metal also resolved the conflicting need for wide clock lines to reduce electromigration effects since its current carrying capability was greater than second- or first-level metal. Simulations were performed across model, temperature, and power supply variations to ensure that, under all conditions, the RC time constant on the clock distribution lines would not limit the speed of operation.

B. Multiplier Architecture

The 12 × 12 multipliers are parallel array multipliers that use radix-4 modified Booth encoding [6] and a carry-save partial product reduction scheme. To attain greater than 200 MHz throughput, two pipeline stages were inserted into the multiplier array. An 8 × 8 multiplier that is representative of this architecture is shown in Fig. 10. The full adder cell used in this multiplier is an inverting transmission gate full adder (Fig. 11). This circuit has the advantage of being fast and compact. It is compact since all 24 transistors can be laid out without any breaks in the diffusion regions. Another important aspect of this adder is that both the sum and carry outputs have partial propagation delays. This is important since the cumulative speed of the partial product reduction array depends equally on both the propagation delay of the sum and carry outputs. SPICE simulations with nominal models, 5 V power supplies, and at 25°C predict the full adder propagation delay to be 1.2 ns.

The circuit diagrams of the Booth encoder and Booth selector are shown in Fig. 12. The Booth encoder implements 2 b scanning (Y_{i-1}, Y_i) with 1 overlap bit (Y_{i+1}). It maps these 3 b to a signed-digit set represented by the signals Comp, $\overline{\text{Comp}}$, Shift, $\overline{\text{Shift}}$, and Zero. Comp and $\overline{\text{Comp}}$ determine if the partial product should be complemented, Shift and $\overline{\text{Shift}}$ determine if it should be shifted left by 1 b, and Zero determines if it should be nulled. The Booth selector circuit implements a 5-to-1 MUX using only 14 transistors. This circuit has been designed to operate with the Booth encoder by processing data with as much parallelism as possible. SPICE simulations estimate that the propagation delay of the outputs of the Booth encoder are approximately 0.7, 0.8, and 1.0 ns, respectively, for the signals Comp/$\overline{\text{Comp}}$, Shift/$\overline{\text{Shift}}$, and

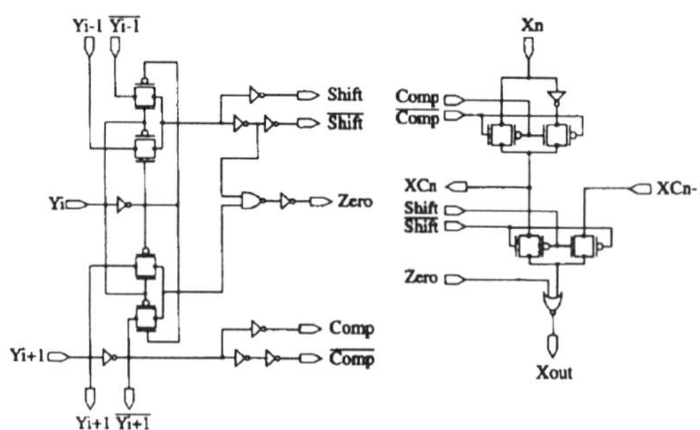

Fig. 12. Booth encoder and Booth selector.

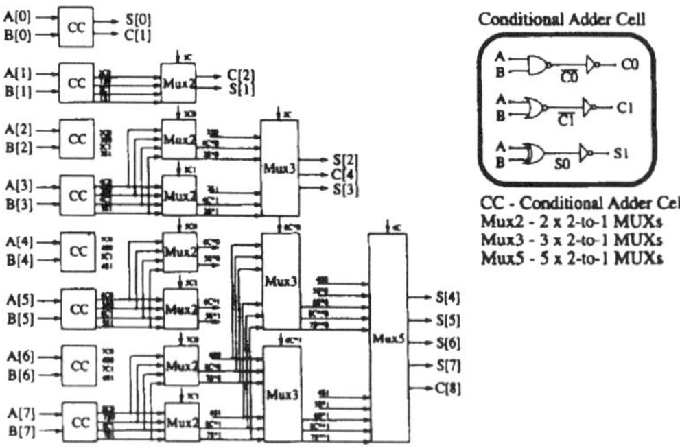

Fig. 13. The 8 b conditional sum adder.

Zero. The important point here is that these signals complete their transitions in that sequence, hence the Booth selector has been designed to operate on data in that same sequence by placing the Comp/$\overline{\text{Comp}}$ MUX, first, the Shift/$\overline{\text{Shift}}$ MUX second, and the Zero logic last.

C. Carry Propagate Adder

One of the speed critical paths in this design exists in the phase accumulator. A carry propagate adder with a wordlength of 32 b is necessary to produce the final address for the ROM look-up table. To achieve the 200 MHz throughput, a sophisticated carry propagate adder had to be used. One possible candidate for this design is a pipelined carry ripple adder. Due to the large wordlength needed, this adder would have to be extensively pipelined. This would result in the use of many registers and would impact the loading of the clock distribution network. Instead, a conditional sum adder [7] was chosen. An example 8 b version of such a carry propagate adder is shown in Fig. 13. The 32 b version of this carry propagate adder was simulated in SPICE using nominal models, 5 V power supplies, and at 25°C to have a speed of 3.1 ns.

D. ROM Block Design

The decoders for the word and bit lines use pseudo-NMOS logic. A representative 3-to-8 decoder is shown in Fig. 14. This design has the advantage of being small and fast at

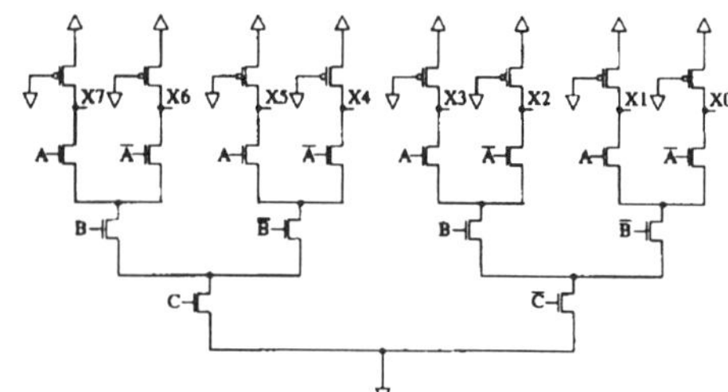

Fig. 14. The 3-to-8 decoder.

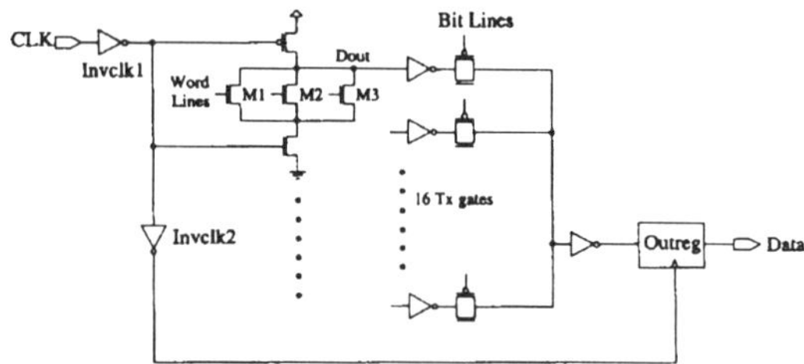

Fig. 15. Domino ROM circuit.

the expense of some dc power dissipation. The ROM basic cell uses dynamic Domino logic. Fig. 15 shows a simplified diagram of the ROM cell together with its associated word and bit lines. On the rising edge of CLK, the output of the ROM cell Dout is precharged high. The evaluate phase occurs when CLK goes low hence conditionally discharging Dout. Dout is pulled low during this phase if any of the transistors M1, M2, or M3 exist. The high-to-low transition of this node has to be completed within the remaining half a clock cycle. Due to the large loading along this path, it is with difficulty that this transition meets the specification of 200 MHz operation. This problem was solved by routing the clock input of the output register Outreg through two inverter delays Invclk1 and Invclk2. This essentially delays the clock to the output register, therefore buying more time for this transition to complete. This modification was simulated using SPICE over temperature, power supply, and model variations to ensure that a data race condition does not exist. The price paid for delaying the clock signal to Outreg is that the next pipeline stage would have less than 1 clock period to complete all transitions. The critical path through the entire ROM block has been simulated to be 4 ns inclusive of setup and hold times of pipeline registers using nominal models, 5 V power supplies, and at 25°C.

V. Fabrication and Test Results

The Quadrature Digital Synthesizer/Mixer chip was fabricated in a 0.8 μm triple-level metal N-well CMOS process. The core of the chip contains 52 000 transistors occupying an area of 2.6 × 6.1 mm^2. Power dissipation is 2 W at a clock rate of 200 MHz. The maximum operating frequency of the chip was measured to be 210 MHz.

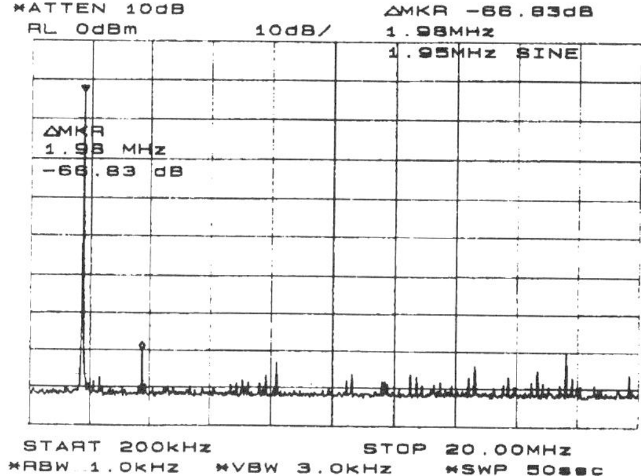

Fig. 16. Spectrum of 1.95 MHz sine wave.

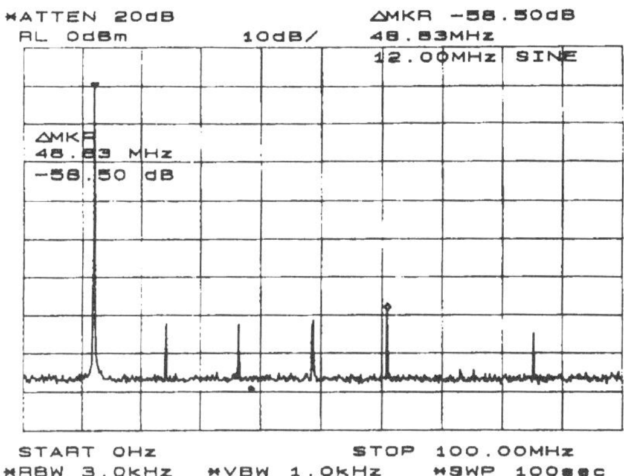

Fig. 17. Spectrum of 12.0 MHz sine wave.

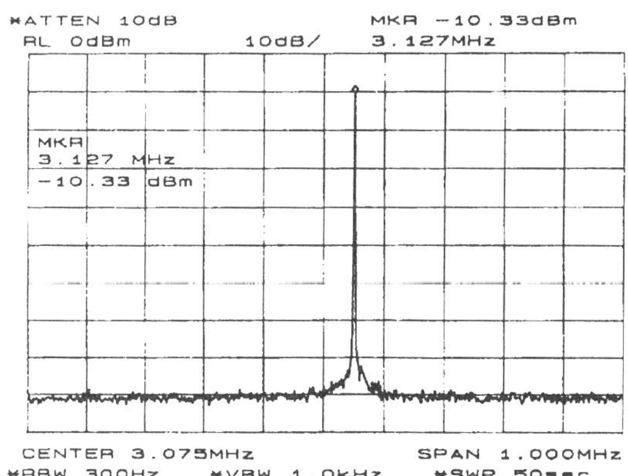

Fig. 18. Spectrum of 3.127 MHz sine wave.

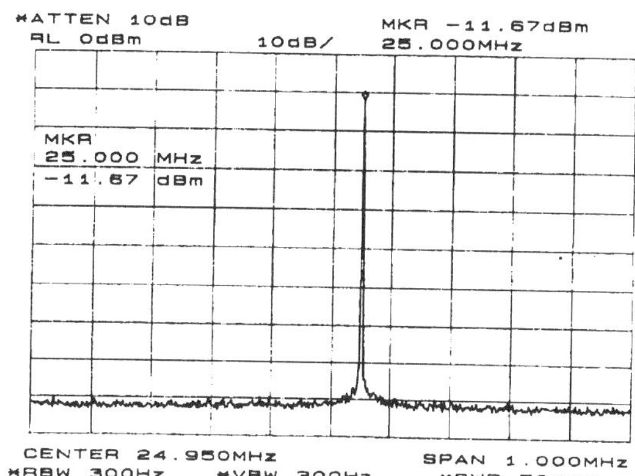

Fig. 19. Spectrum of 25.0 MHz sine wave.

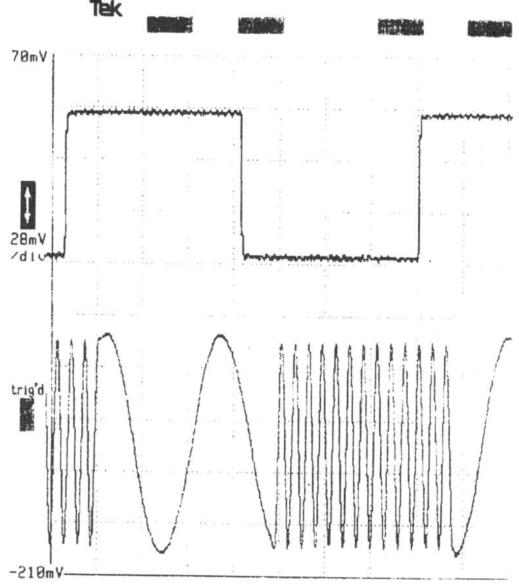

Fig. 20. Frequency modulation.

All functional vector tests were performed with the Tektronix LV500 ASIC tester. High-speed testing was accomplished by building a breadboard containing two 10 b D/A converters. The converter used was the Analog Devices AD9720 which has a maximum operating frequency of 400 MHz.

Figs. 16 and 17 show spectrum plots of a 1.95 MHz and a 12MHz output sinewave, respectively. These spectrum plots span a large frequency range to show the spurious and harmonic components resulting from using a 10 b DAC. In Fig. 16, the worst case spurious component is at −66.83 dB, and in Fig. 17 it is at −58.50 dB. Figs. 18 and 19 show spectrum plots of a 3.127 MHz and a 25 MHz output sinewave, respectively. These plots have very narrow frequency spans and fine resolution bandwidths for viewing spurs near the fundamental. The DDFS produces a −84.3 dBc spectrally pure signal over its entire tuning range, thus the spurious response in the analog outputs is primarily due to the D/A conversion

TABLE I
QDDFSM Chip Specifications

Technology	0.8μm HPCMOS26B, Triple Level Metal, N-well Process
Maximum Clock Frequency	200 MHz
Switching Time	5ns (at 200-MHz f_{clk})
Tuning Latency	14 clock cycles
Tuning Bandwidth	100 MHz (at 200-MHz f_{clk})
Frequency Resolution	0.047 Hz (at 200-MHz f_{clk})
Worst Case Spurious	-84.3 dBc
Phase Modulation Wordlength	12 bits
Quadrature Amplitude Modulation Wordlength	12 bits for I, and 12 bits for Q
Output Word Length	12 bits
Power Dissipation	2 Watts with I/Os switching (at 200-MHz f_{clk})
Transistor Count	52,000 (including I/Os)
Die Size	5.8mm x 7.8mm (Pad Limited: 7 mil Spacing)
Core Size	2.6mm x 6.1mm

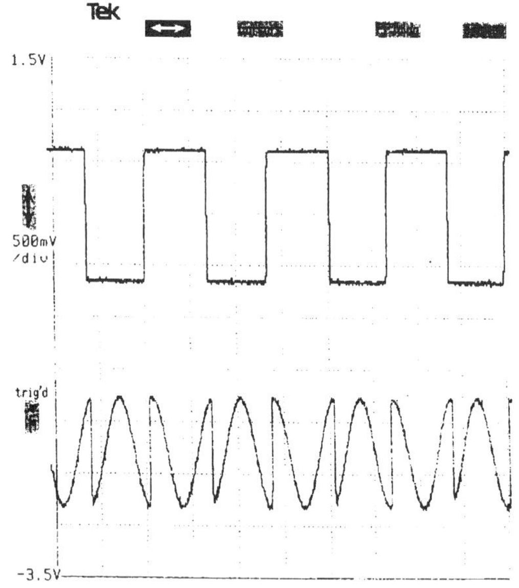

Fig. 21. Phase modulation.

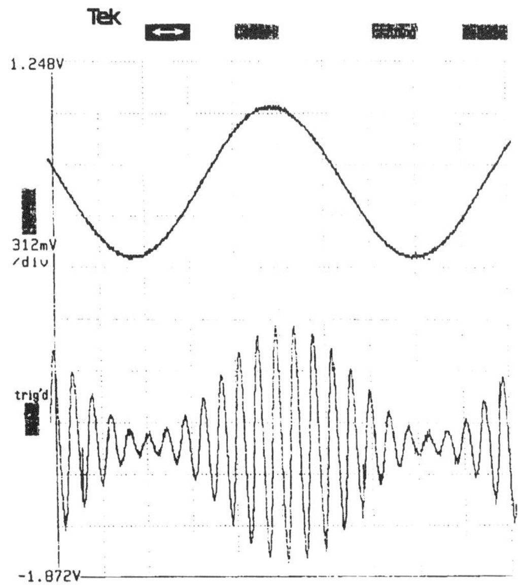

Fig. 22. Amplitude modulation.

process; however, it is quite good considering that only a 10 b D/A converter was used. The reference clock frequency for all four cases is 200 MHz.

The modulation capabilities of the QDDFSM chip were also tested. Fig. 20 shows an example of frequency modulation. The modulating signal is displayed above and the modulated signal below it. The modulating signal is a square wave at 323 kHz. Fig. 21 illustrates the phase modulation capability of the chip. The modulating signal is a square wave at a frequency of 94 kHz, and therefore this is an example of Binary Phase Shift Keying (BPSK). Fig. 22 illustrates the amplitude modulation capability of the chip where the modulating signal is a sine wave at 200 kHz. For all three cases, the clock frequency of

the chip was 200 MHz. Table I summarizes the performance of the QDDFSM chip.

VI. Conclusions

A Quadrature Digital Synthesizer/Mixer chip has been designed that operates at 200 MHz and synthesizes −84.3 dBc spectrally pure sine and cosine digitized waveforms. This chip exhibits large bandwidth (dc to 100 MHz), high spectral purity, fast switching speed, and fine frequency resolution (0.047 Hz). Fig. 23 shows a photomicrograph of this chip. By taking advantage of sine and cosine symmetries, the size of the ROM look-up table is no larger than that of a DDFS which only generates sine outputs. The chip also incorporates modulation

302

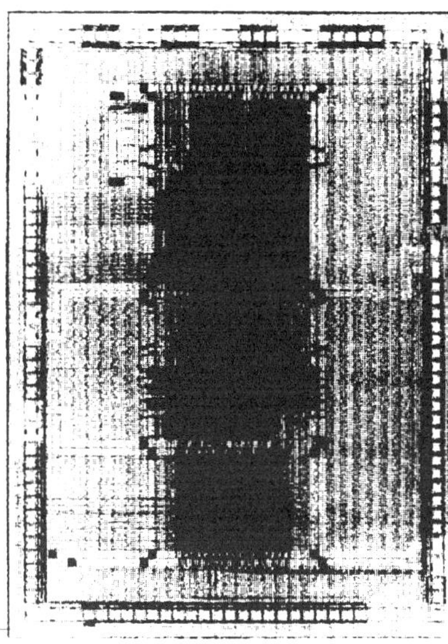

Fig. 23. Photomicrograph of QDDFSM.

capabilities. The modulation formats include frequency modulation, phase modulation, quadrature amplitude modulation, and single sideband frequency translation. Hence, this chip encompasses and provides many basic features required in digital communication systems. For very high throughput applications, the quadrature DDFS can be paralleled to provide twice or four times the maximum throughput. Hence, 400 MHz or 800 MHz clocking speeds are possible by paralleling two or four chips.

ACKNOWLEDGMENT

The authors wish to thank F. Lu for the design of the Booth encoder, G. Yee and S. Olafson for work on the ROM, and also S. Liu and E. Roth for assistance in testing the chip.

REFERENCES

[1] C. Tierney, M. Rader, and B. Gold, "A digital frequency synthesizer," *IEEE Trans. Audio Electroacoust.*, vol. AU-19, pp. 48–57, 1971.
[2] H. T. Nicholas, III and H. Samueli, "A 150 MHz direct digital frequency synthesizer in 1.25-μm CMOS with -90 dBc spurious performance," *IEEE J. Solid-State Circuits*, vol. 26, pp. 1959–1969, Dec. 1991.
[3] M. Thompson, "Low latency, high-speed numerically controlled oscillator using progression-of-states technique," *IEEE J. Solid-State Circuits*, vol. 27, pp. 113–117, Jan. 1992.
[4] R. Hassun and A. W. Kovalick, "Waveform synthesis using multiplexed parallel synthesizers," U.S. Patent 4 454 486, June 12, 1984.
[5] H. T. Nicholas, III, H. Samueli, and B. Kim, "The optimization of direct digital frequency synthesizer performance in the presence of finite word length effects," in *Proc. 42nd Annu. Frequency Cont. Symp. USER-ACOM*, May 1988, pp. 357–363.
[6] A. D. Booth, "A signed binary multiplication technique," *Quart. J. Mech. Appl. Math.*, vol. IV, pt. 2, pp. 236–240, Aug. 1950.
[7] J. Sklansky, "Conditional sum additional logic," *IRE Trans. Electron. Comput.*, vol. EC-9, no. 2, pp. 226–231, June 1960.
[8] J. Yuan and C. Svensson, "High speed CMOS circuit technique," *IEEE J. Solid-State Circuits*, vol. 24, pp. 62–69, Feb. 1989.
[9] B. C. Wong and H. Samueli, "A 200 MHz all-digital QAM modulator and demodulator in 1.2μm CMOS for digital radio applications," *IEEE J. Solid-State Circuits*, vol. 26, pp. 1970–1979, Dec. 1991.
[10] H. T. Nicholas, III, and H. Samueli, "An analysis of the output spectrum of direct digital frquency synthesizers in the presence of phase-accumulator truncation," in *Proc. 41st Annu. Frequency Cont. Symp. USERACOM*, May 1987, pp. 495–502.

An 800-MHz Quadrature Digital Synthesizer with ECL-Compatible Output Drivers in 0.8 μm CMOS

Loke Kun Tan, Edward W. Roth, Gordon E. Yee and Henry Samueli, *Member, IEEE*

Abstract— An 800 MHz quadrature direct digital frequency synthesizer (QDDFS4) chip is presented. The chip synthesizes 12 b sine and cosine waveforms with a spectral purity of -84.3 dBc. The frequency resolution is 0.188 Hz with a corresponding switching speed of 5 ns and a tuning latency of 47 clock cycles. The chip is also capable of frequency and phase modulation. ECL-compatible output drivers are provided to facilitate I/O compatibility with other high speed devices. A high gain amplifier at the clock input enables the QDDFS4 chip to be clocked with ac-coupled RF signal sources with peak-to-peak voltage swings as small as 0.5 V. The 0.8 μm triple level metal N well CMOS chip has a complexity of 94 000 transistors with a core area of 5.9 $\times$ 6.7 mm^2. Power dissipation is 3 W at 800 MHz and 5 V.

I. INTRODUCTION

TRADITIONAL designs of high bandwidth frequency synthesizers employ the use of a phase-locked-loop (PLL). However, the presence of a feedback loop in a PLL implementation significantly increases the acquisition time hence impacting its ability to simultaneously provide fast frequency switching and high frequency resolution [1]. Direct digital frequency synthesizers (DDFS's) do not exhibit these disadvantages since their design is purely feedforward. Thus, the switching speed of a DDFS is essentially instantaneous. Other advantages inherent in the DDFS approach include its ability to provide fine frequency resolution and phase coherent frequency switching. Furthermore, the amount of phase noise in a DDFS is primarily limited by that of the reference clock source. The primary drawback to the DDFS approach is the requirement for high-speed, high resolution D/A converters. In times past, DDFS's have been considered a low speed method of frequency synthesis since their performance is directly limited by the availability of fast digital integrated circuit (IC) processes. Fortunately, recent advances in IC technology have brought about considerable progress in this area. Advanced GaAs and Si-bipolar IC processes have resulted in DDFS designs with very high operating frequencies [2], [3]. However, the esoteric nature as well as the lower yields of these processes result in relatively costly solutions. This paper presents a high performance 800-MHz monolithic quadrature DDFS design that achieves a spectral purity of -84.3 dBc

Manuscript received May 22, 1995; revised July 25, 1995. This work was supported in part by research grants from the University of California MICRO Program, Hewlett-Packard, TRW Electronic Systems Group, Hughes Space and Communications Group, and the Advanced Research Projects Agency.

L. K. Tan is with Broadcom Corporation, Irvine, CA 92718 USA.

E. W. Roth, G. E. Yee, and H. Samueli are with Integrated Circuits and Systems Laboratory, University of California, Los Angeles, CA 90024 USA.

IEEE Log Number 9416003.

using a generic 0.8-μm CMOS process. The high performance of the QDDFS4 was achieved through the use of a parallel architecture.

II. CHIP ARCHITECTURE

A. Overall Architecture

The basic architecture of the QDDFS4 chip consists of an overflowing L-bit accumulator (or phase accumulator) to generate the phase argument of the sine function generator. Each overflow of the phase accumulator represents one period of a sine wave. The input word (Frequency Control Word) to the phase accumulator controls the frequency of the generated sine waveform. The sine function generator is a ROM lookup table that stores the sine samples. The use of parallelism to attain high throughput is common in VLSI design and has been utilized in [4] and [5] for DDFS applications. Our chip has been designed with four parallel ROM lookup tables to achieve four times the throughput of a single DDFS. The size of the ROM table is $2^8 \times 12$ b. Each ROM table generates every fourth sample of a sine waveform. This is performed by applying four times the Frequency Control Word (FCW) to each ROM table and also adding a phase offset equal to 0, 1, 2, and 3 times the Frequency Control Word to each corresponding ROM table. Last, a 4-to-1 MUX is used to alternately select the outputs of the four ROM's to reconstruct the final sine waveform. Thus, the only block that operates at the maximum clock rate of the chip is the 4-to-1 MUX. All other blocks operate at a quarter of the reference clock frequency. The detailed architecture of this design is illustrated in Fig. 1. A chip that uses only one ROM table has been fabricated and tested to perform at 200 MHz [6]. Using the parallel architecture with four ROM tables, the QDDFS4 chip attains the targeted speed of 800 MHz.

The QDDFS4 has a phase accumulator wordlength of 32 b. When clocked at 800 MHz, a minimum frequency resolution of 0.188 Hz is attained. The 32 b phase accumulator output is truncated to 14 b prior to addressing the ROM lookup table, resulting in a -84.3 dBc worst case spurious rejection based on the fact that phase accumulator truncation is the dominant source of spurious noise [7].

B. Quadrature Outputs

The QDDFS4 chip takes advantage of the quarter wave symmetry of a sine wave to reduce ROM storage requirements. Thus, only sine samples from 0 to $\pi/2$ are stored. To produce

Reprinted from *IEEE Journal of Solid-State Circuits*, Vol. 30, No. 12, pp. 1463-1473, December 1995.

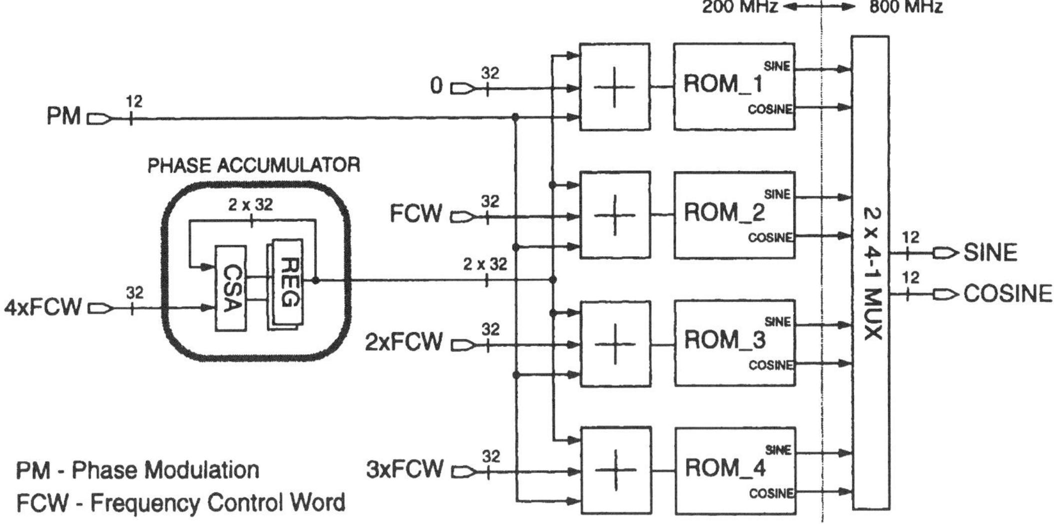

Fig. 1. Detailed top level architecture of QDDFS4.

quadrature outputs, rather than having a separate ROM lookup table for the cosine waveform, this design takes advantage of the eighth wave symmetry of the sine and cosine functions. Since sine samples from 0 to $\pi/4$ are identical to cosine samples from $\pi/4$ to $\pi/2$, and cosine samples from 0 to $\pi/4$ are identical to sine samples from $\pi/4$ to $\pi/2$ thus, one ROM lookup table containing only sine samples has sufficient information to produce both sine and cosine waveforms. Thus, the only additional hardware cost of generating quadrature outputs is the use of a 2-to-1 MUX to select the appropriate ROM table contents to produce the sine and cosine samples [6].

C. Modulation Formats

This chip has modulation capabilities that include frequency modulation and phase modulation. Frequency modulation is performed by directly modulating the Frequency Control Word, thus, no additional hardware is needed to implement this feature. Phase modulation is accomplished by adding a phase offset to the phase accumulator output before addressing the ROM lookup table. In hardware, this amounts to incorporating an extra addition stage. The QDDFS4 accepts a 12 b word for phase modulation. Fig. 1 illustrates these modulation capabilities. To reduce input pad count, the 12 b phase modulation word is multiplexed with the 12 MSB's of the Frequency Control Word.

III. CIRCUIT DESIGN

The size of the ROM lookup table is $2^8 \times 12$ b. The decode logic of the ROM was designed using pseudo-NMOS logic while the ROM basic cell uses dynamic Domino logic. The carry propagate adder that produces the final phase word to address the ROM lookup table uses a conditional sum architecture. Details pertaining to the design of the conditional sum adder and the ROM lookup table are described in [6].

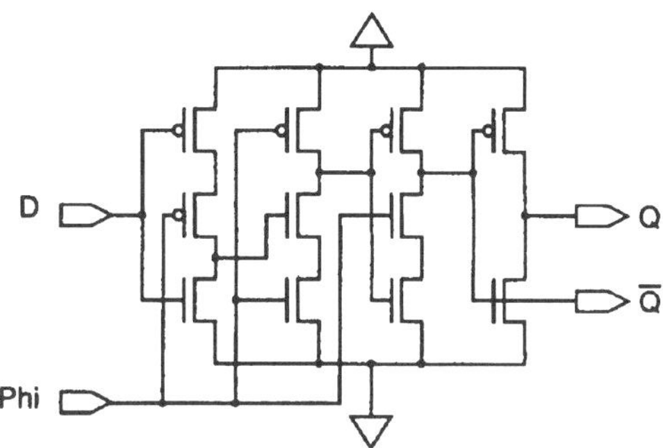

Fig. 2. Single-phase positive-edge triggered register.

A. Register Design

The QDDFS4 chip uses only dynamic positive-edge triggered single-phase registers. A schematic of this register is illustrated in Fig. 2. Standard two-phase clocking schemes for high speed operation are difficult to design since small amounts of clock skew directly limit speed. Single-phase registers do not exhibit a sensitivity to clock skews and are thus able to achieve higher clock frequencies [8]. SPICE simulations of this register were performed over variations in transistor models, temperature, and power supply levels to compute T_s and $T_{\mathrm{clk-Q}}$, where T_s is the register setup-time and $T_{\mathrm{clk-Q}}$ is the register clock-to-Q delay. A plot of $T_s + T_{\mathrm{clk-Q}}$ versus T_s is illustrated in Fig. 3. Each line in the plot describes a particular simulation condition with a fixed transistor model, fixed temperature, and fixed power supply voltage. The entire collection of simulation cases characterizes the spread of timing information associated with this register. The data obtained from this plot is summarized in Table I. From this table, it can be seen that the worst case setup time is $T_s = 0.13$ ns. If the setup-time T_s were reduced below this value, at least one simulation case fails to latch data correctly. Using $T_s = 0.13$ ns as a reference point, $T_{\mathrm{clk-Q}} + T_s$ varies

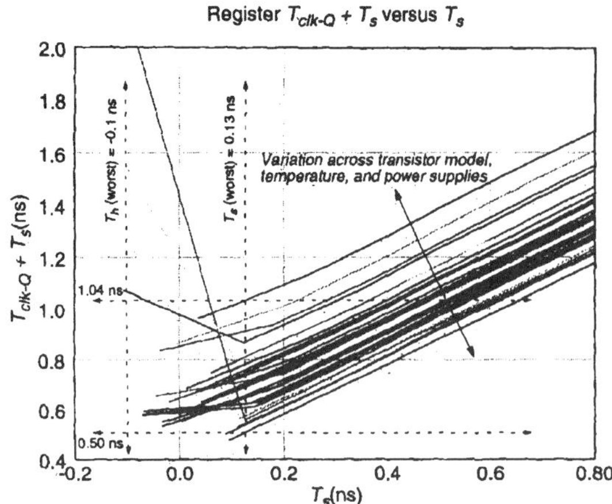

Fig. 3. SPICE simulation of single-phase positive-edge triggered register.

TABLE I
SUMMARY OF REGISTER TIMING PARAMETERS BASED ON SPICE SIMULATIONS

Timing Parameter	Fast Case (ns)	Nominal Case (ns)	Slow Case (ns)
$T_{clk-Q} + T_s$ (@T_s = 0.13ns)	0.50	0.77	1.04
T_{clk-Q} (linear region)	0.37	0.64	0.91

Timing Parameter	Worst Case (ns)		
T_h	0.10		
T_s	0.13		

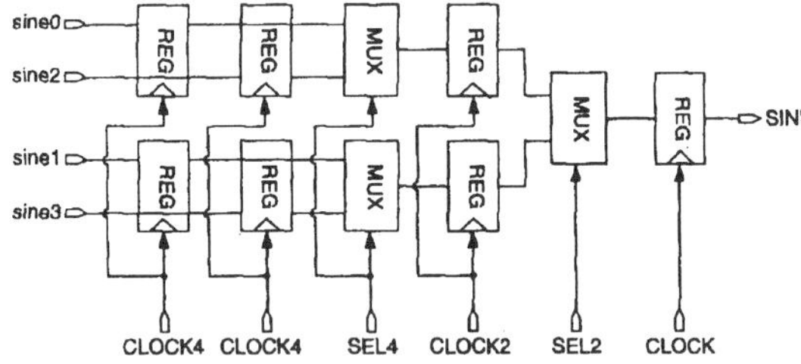

Fig. 4. Block diagram of 4-to-1 MUX.

from 0.50–1.04 ns across all simulation conditions. As long as the register setup time is not violated, the relationship between $T_{clk-Q} + T_s$ and T_s is linear. Hence, T_{clk-Q} can be calculated for this region of operation. Subtracting $T_s = 0.13$ ns from the first row of Table I results in T_{clk-Q}, which ranges from 0.37–0.91 ns. The other important timing parameter in this simulation is the worst case hold time $T_h = 0.10$ ns. If this hold time is violated, at least one simulation case will exhibit data flash-through.

B. 4-to-1 MUX Design

The design of the 4-to-1 MUX is challenging since it is the only block operating at the 800 MHz clock rate. The circuit schematic of this 4-to-1 MUX is shown in Fig. 4. Two such MUXes are used; one for generating sine waveforms and one for cosine waveforms. The signals Sine0, Sine1, Sine2, and Sine3 are the four staggered sine samples used

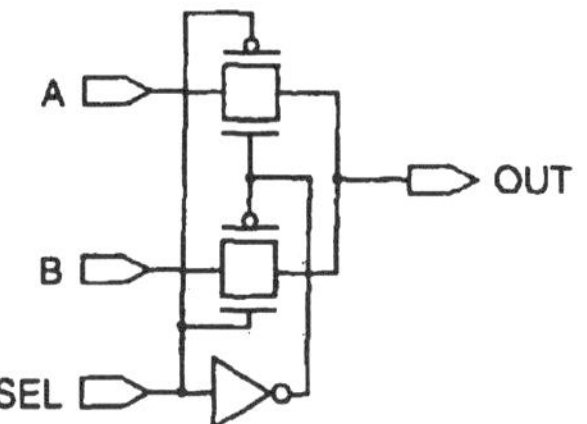

Fig. 5. Transistor schematic of 2-to-1 MUX.

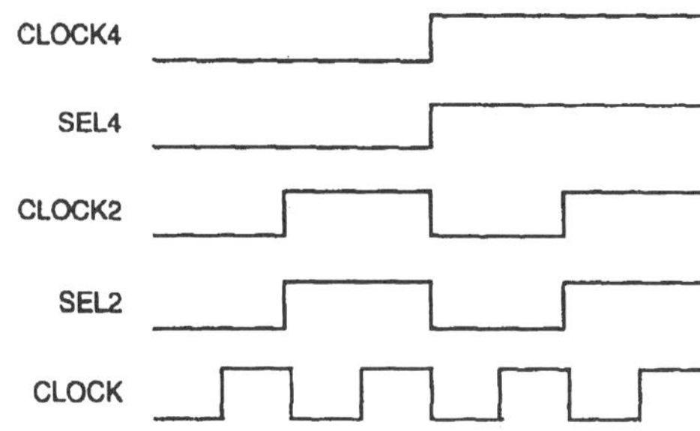

Fig. 6. Timing of clock and select signals.

to reconstruct the final sine waveform. The 4-to-1 MUX consists of a cascade of two 2-to-1 MUXes. Pipeline stages are placed in between these two stages for maximum throughput. All pipeline registers are single-phase positive-edge triggered registers. A transistor schematic of the 2-to-1 MUX is shown in Fig. 5. The clock signals CLOCK, CLOCK2, and CLOCK4 are the input clock, input clock divided by 2, and input clock divided by 4, respectively. The timing relationship of these clocks and select signals are shown in Fig. 6.

C. Clock Generation

The timing of the clock and select lines is extremely critical. The circuit used for the generation of these signals is illustrated in Fig. 7. A divide-by-2 flip-flop is used twice to generate the CLOCK2 and CLOCK4 signals. These two circuits are cascaded to create an asynchronous divide-by-4 circuit for the generation of the CLOCK4 signal. The use of an on-chip high gain amplifier enables this chip to be clocked with RF signal sources that are capable of generating small voltage swing ac-coupled signals in the GHz frequency range. This circuit accepts voltage swings as small as 0.5 V_{p-p} and also exhibits a low sensitivity to noise [9]. The inset in Fig. 7 illustrates the transistor schematic of this design. The capability of comparing the threshold of the clock signal to a dc input reference voltage V_{ref} gives this design an insensitivity to common-mode threshold shifts. The output of the clock amplifier labeled N2 drives a CMOS inverter that further increases the gain of the combination. The manner in which this clock amplifier is connected to the clock generation circuit of the chip is illustrated in Fig. 7. Back-annotated SPICE simulations predicted that the clock generation circuit is capable of producing on chip CMOS clocks in excess of 800 MHz. This result is illustrated in Fig. 8 where the top trace is the 0.5 V_{p-p} input reference clock. The final

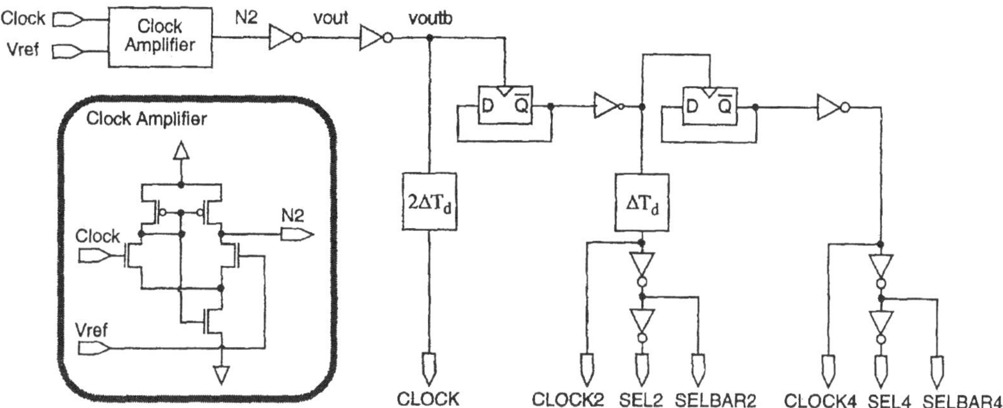

Fig. 7. Block diagram of clock generation circuit.

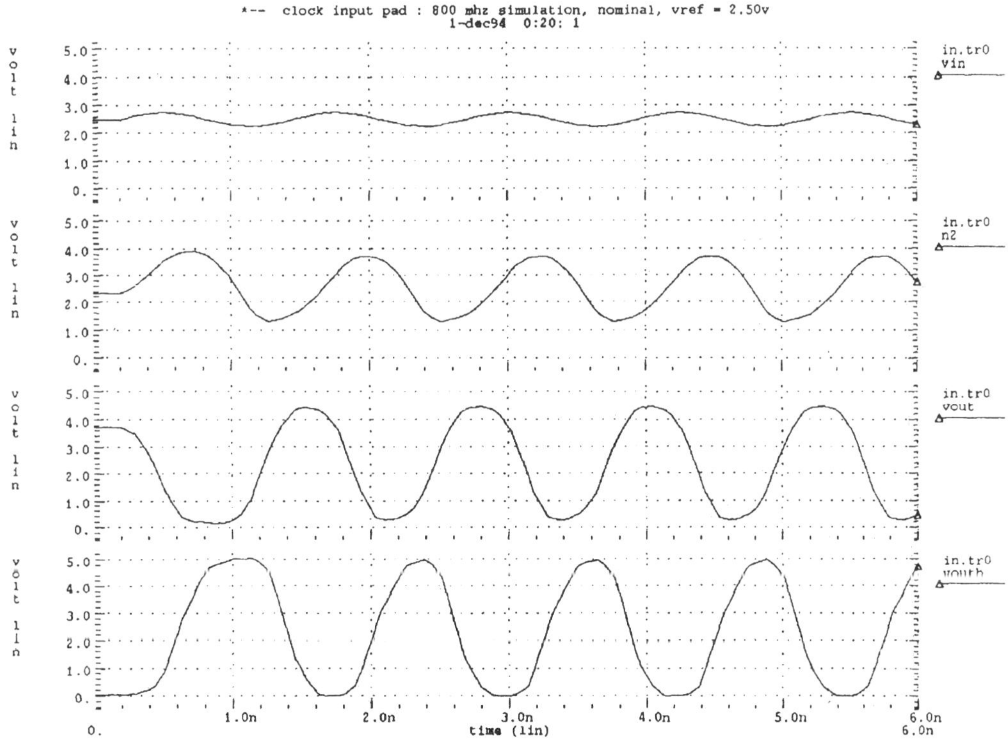

Fig. 8. SPICE simulation of clock amplifier at 800 MHz.

waveform Voutb has full 0–5 V signal swing and is thus fully CMOS compatible. The simulation was performed using nominal transistor models, at 65°C, 5 V power supplies, and worst case extracted parasitics.

The use of an asynchronous divide-by-4 circuit results in register clock-to-Q delays between the three clock signals CLOCK, CLOCK2, and CLOCK4. A synchronous design would have been a safer choice where minimizing clock skews is concerned; however, such a design could never achieve the necessary 800 MHz throughput. The divide-by-2 circuit is essentially a dynamic positive-edge triggered register of Fig. 2 with the $\bar{Q}$ output feeding back to the D input. This circuit was simulated across models, temperature, and power supply voltages to ensure robust operation. The blocks labeled ΔT_d and $2\Delta T_d$ are used to compensate for the asynchronous divider delays. A simple timing analysis indicates that it is possible to increase the throughput of the 4-to-1 MUX by over-compensating these delays. This timing analysis can be described using a block diagram showing the final stage of the 4-to-1 MUX and its associated timing diagram. This is illustrated in Fig. 9. The timing delays in this figure assume that the delay through a 2-to-1 MUX is negligible. This is a valid assumption since the critical path through this block is an inverter driving the drain-source region of a transmission gate that is connected to a purely capacitive load on its other side. Hence, the delay through the 2-to-1 MUX is very small. From Fig. 9, two critical timing parameters T_{sb} and T_{hb} must be met in order for the circuit to function properly. T_{sb} is the remaining amount of setup time, and T_{hb} is the hold time available when data is latched from node B. These two parameters must be smaller than the actual required setup (T_s) and hold (T_h) times of a register. From Fig. 9, the timing relations governing T_{hb} and T_{sb} are

$$T_{\text{clk}-Q} - T_d + 2T_{\text{inv}} = T_{hb} \geq T_h \qquad (1)$$

$$T_p - (T_{\text{clk}-Q} - T_d + T_{\text{clk}-Q}) = T_{sb} \geq T_s \qquad (2)$$

where $T_{\text{clk}-Q}$ is the clock-to-Q delay of a register, and T_{inv} is the delay of an inverter gate, and T_d is the delay inserted between CLOCK and CLOCK2 to compensate for the clock-to-Q delay that exists between these two signals. T_{inv} has been simulated to be 0.058 ns, 0.088 ns, and 0.165 ns under fast, nominal, and slow conditions, respectively. Table II summarizes T_{inv} and $T_{\text{clk}-Q}$ for these cases. The task here is to determine an appropriate T_d measured in number of inverter delays such that T_p is minimized and furthermore, both (1) and (2) are satisfied. By substituting T_h, T_s, T_{inv} and $T_{\text{clk}-Q}$ from Tables I and II into both (1) and (2), one can determine a range of T_d such that setup and hold times are satisfied. Plots of frequency ($1/T_p$) versus T_d (normalized to number of inverter delays) for slow, nominal, and fast conditions are illustrated in Fig. 10. The shaded sections in these three plots indicate acceptable regions of operation where both (1) and (2) are not violated. Choosing an operating point further to the right or equivalently using a larger T_d results in increasing maximum throughput of the 4-to-1 MUX. The limit of this choice is the right boundary of operation, which indicates that a hold time requirement is violated. To accommodate process variations as well as varying operating conditions, it is important that the value of T_d be chosen so as to satisfy all three plots. Furthermore, T_d has to be an even number of inverter delays; otherwise, the clock signal will be inverted. Thus, based on these results, T_d is chosen to be six inverter delays. Also based on the results of Fig. 10, the maximum frequency of operation f_{max} ranges from 880 MHz–1.9 GHz depending on process variations and operating conditions. It is doubtful that a throughput of 1.9 GHz can be achieved since other factors not accounted for in the timing model might prove to be dominant under this condition.

D. Output Driver Design

Due to the chip's very high operating frequency, the design of the output drivers is crucial. Since the large switching currents of CMOS drivers generate substantial di/dt noise, ECL compatible drivers are not only desirable, but also

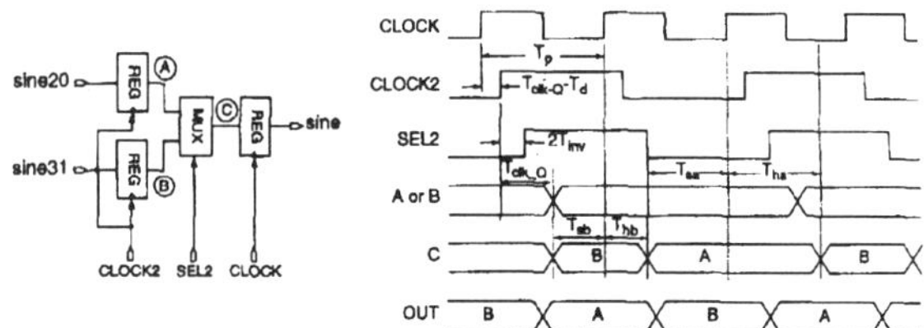

Fig. 9.   4-to-1 MUX final stage block diagram and associated timing diagram.

TABLE II
SUMMARY OF SIMULATION RESULTS OF T_{inv} AND $T_{\text{clk}-Q}$

Simulation Condition	T_{inv} (ns)	T_{clk-Q} (ns)
Fast, 0°C, 5.5V	0.058	0.37
Nominal, 25°C, 5.0V	0.088	0.64
Slow, 125°C, 4.5V	0.165	0.91

necessary. ECL drivers also facilitate direct I/O compatibility with commercial high-speed D/A converters.

The ECL compatible output driver (Fig. 11) consists of output buffers and a control unit similar to the design described in [10]. A single pull-up transistor at the output stage of the buffer cannot be used because it may not satisfy the small tolerances of an ECL level output (typically within 0.08 V) due to process variations. Fig. 12 shows the detailed circuit of the ECL output driver where a current source is used as the output stage. When V_{in} is *low*, the current mirror ($M6$ and $M7$) is off, and V_{out} is simply $V_{L(\text{ECL})}$ (~ -1.7). When V_{in} is *high*, the current source is switched on, thus causing a constant voltage across a matching resistor R_L. A second current mirror ($M1$ and $M2$) is used to decouple the switched current mirror from I_{ref}, the input reference current generated by the control unit.

One of the improvements made to this circuit was the addition of M3. This PMOS transistor acts as a current steering device to dampen the ground bounce on the power supply. $M3$ and $M7$ are complementary relative to each other, that is, if one is on, the other is off, and vice versa. As a result, the output buffer draws finite current in the *low* state. This effectively reduces the difference in current drawn between the *high* and *low* states of the output buffer and thus reduces inductive ground bounce. Simulations have indicated that the addition of $M3$ can reduce inductive ground bounce by as much as 1 V.

Unacceptable variations in the *high* output level may be caused by the degeneration of the current mirror due to channel-length modulation. Therefore, a feedback control unit is used to compensate for technology parameter variations. The control unit consists of the transistor schematic of Fig. 13 as well as a dummy output driver (Fig. 11) whose input is fixed at Vdd to generate a permanent *high* output level (Out[0]). That *high* output level is compared with the desired level $V_{H(\text{ECL})}$, and the result is used to control the reference current source (I_{ref0}) at its output regardless of current mirror imperfections. Since the other output drivers receive the same reference current ($I_{\text{ref2}} = I_{\text{ref1}} = I_{\text{ref0}}$), their *high* output level is also controlled by $V_{H(\text{ECL})}$.

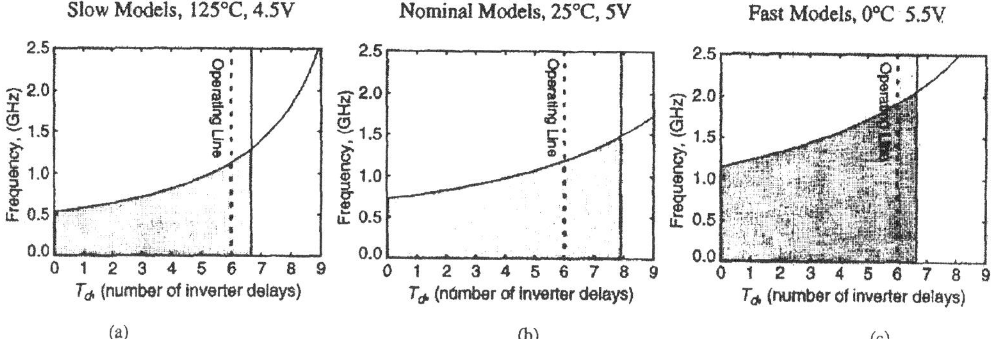

Fig. 10. SPICE simulation summary of 4-to-1 MUX region of operation.

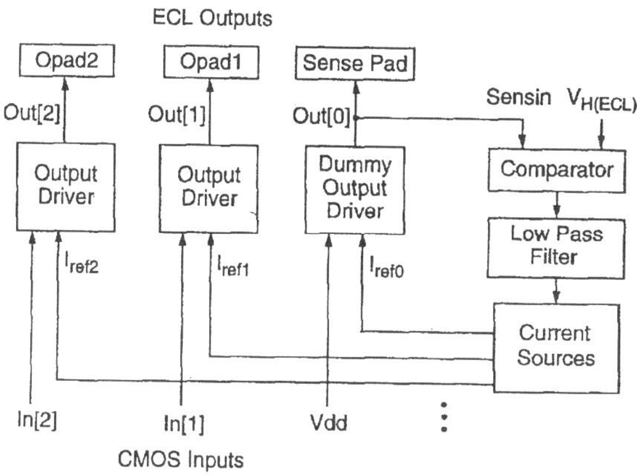

Fig. 11. Block diagram of CMOS-to-ECL level output drivers.

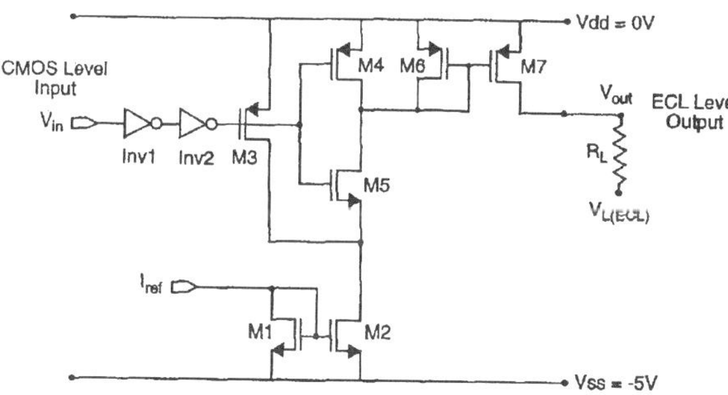

Fig. 12. Transistor schematic of output buffer.

The configuration of the control circuit was modified to ensure that the comparator of Fig. 13 possesses only one dominant pole located at node Z. Additional current sources will only move that pole down in frequency and further away from any secondary pole. Thus, system stability is maintained while keeping the number of desired current sources flexible. Moreover, the connection of a capacitor from node Z to V_{dd} helps maintain a constant gate-source voltage V_{gs} of the PMOS current sources and thus a constant I_{ref} for the output drivers. This capacitor has no effect on the speed of the driver since the control unit itself is not operating at switching frequencies.

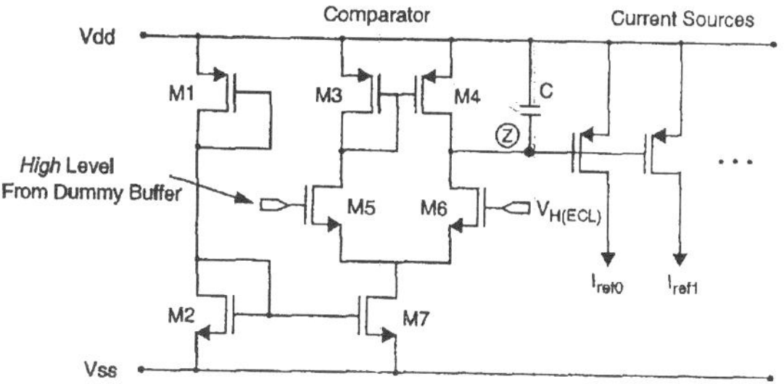

Fig. 13. Transistor schematic of control unit for ECL output drivers.

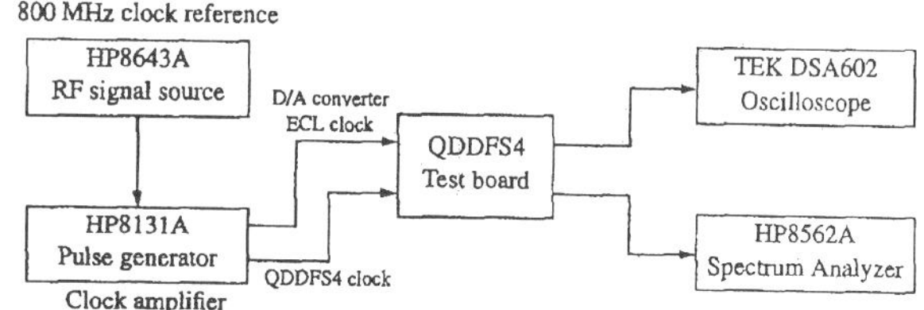

Fig. 14. QDDFS4 test setup.

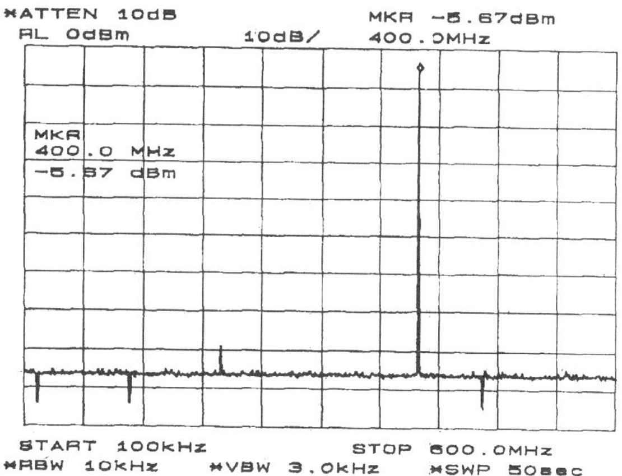

Fig. 15. 400 MHz spectrum.

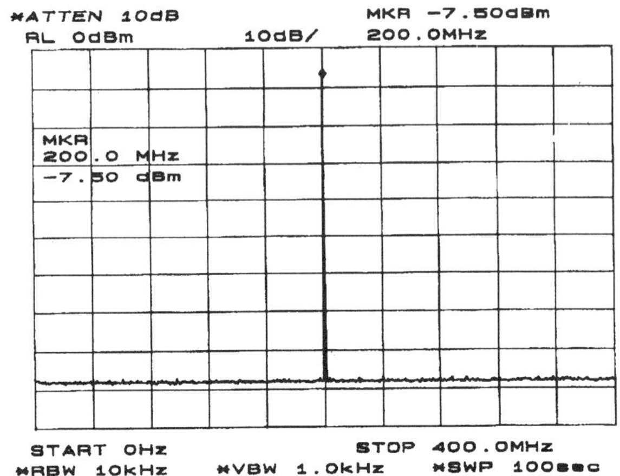

Fig. 16. 200 MHz spectrum.

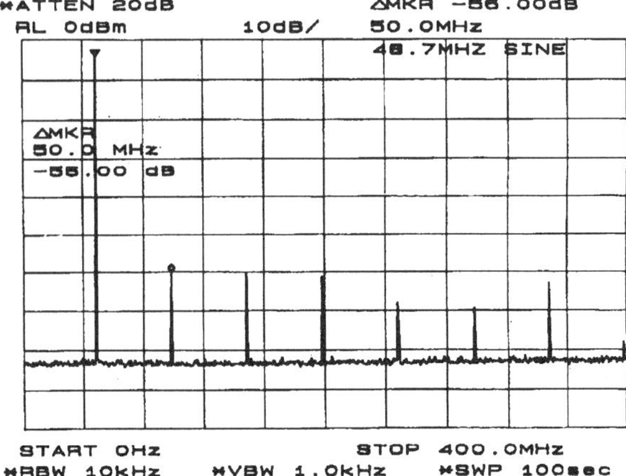

Fig. 17. 48.7 MHz spectrum.

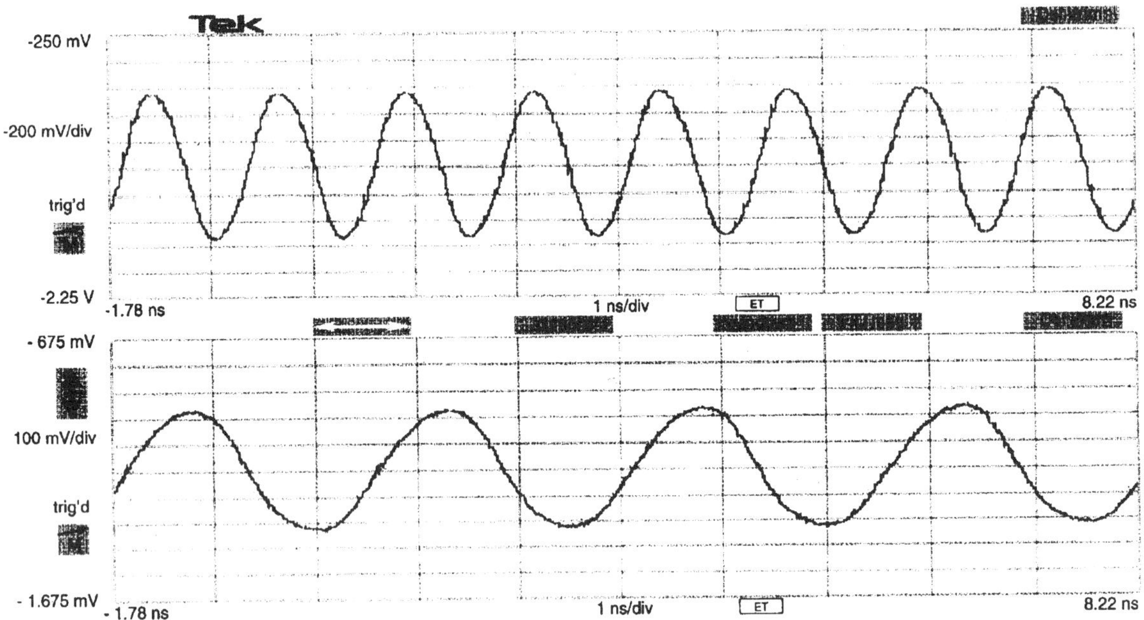

Fig. 18. 800 MHz clock and MSB of cosine from ECL output driver switching at 400 MHz.

IV. FABRICATION AND TEST RESULTS

Testing of the DDFS chip was accomplished by fabricating a testboard containing two GaAs 10-b D/A converters. The device used was the Rockwell RI61008, which has a maximum operating frequency of 1.2 GHz. The CMOS DDFS die is mounted in a 132 position, quad flat pack leaded ceramic package featuring controlled impedance 50 Ω lines for all signal paths. This package is installed in a low profile test socket utilizing wire button with plunger contact technology for a very low inductance and short electrical path interconnect to signal, power, and ground paths. The wire button contacts then mate with small gold plated pads located on microstrip lines that exist on the top layer of a six layer epoxy fiberglass (FR-4) circuit board that was designed and fabricated to evaluate the QDDFS4 chip. The other five layers of the circuit board were used to provide ac and dc ground planes, power and bias distribution paths, and control signal paths.

The block diagram of Fig. 14 illustrates the test setup. An HP8643A RF signal source was used to produce the reference clock. An HP8131A dual channel pulse generator reference was used in slave mode to amplify the reference clock to appropriate ECL levels. Analog outputs of the QDDFS4 and D/A converter combination were viewed using a TEK DSA602 digitizing signal analyzer. Spectrum measurements were per-

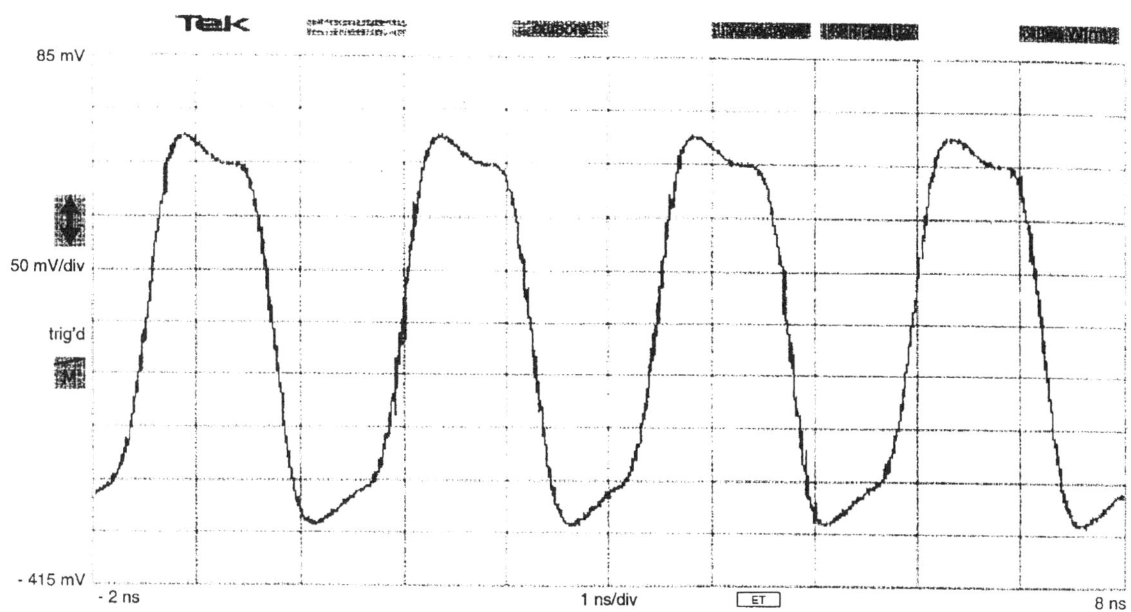

Fig. 19. 400 MHz output signal from D/A converter.

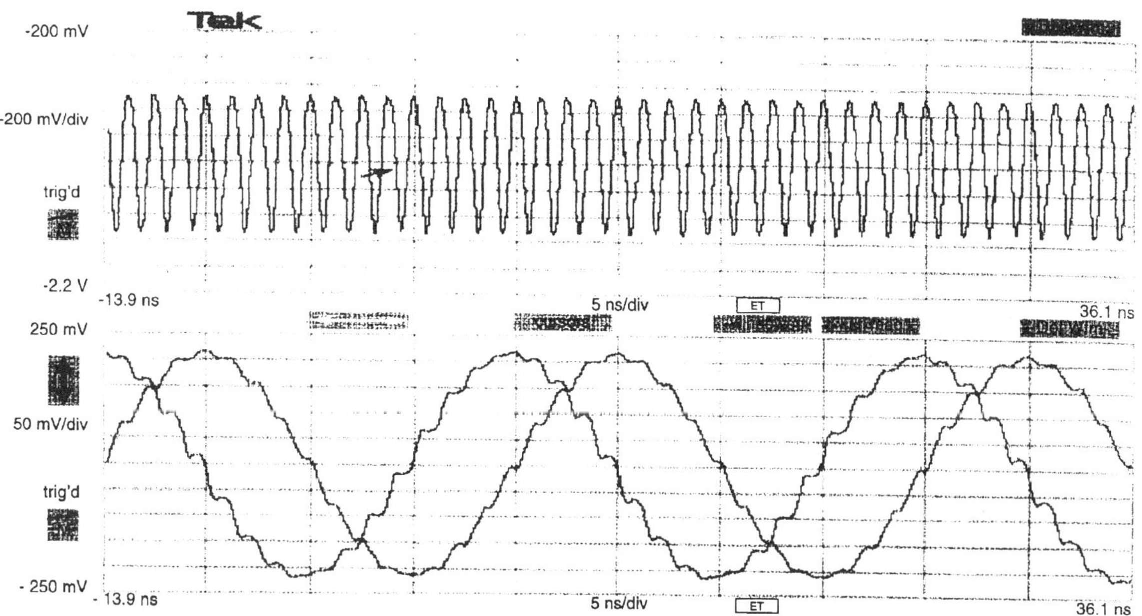

Fig. 20. Unfiltered 50 MHz output sine and cosine waveform from D/A converter.

formed with an HP8562A spectrum analyzer. All tests results presented here used a reference clock frequency of 800 MHz. Figs. 15, 16, and 17 illustrate spectrum plots of a 400 MHz, a 200 MHz, and a 48.7 MHz output sinewave, respectively. These plots span a wide frequency range to facilitate the viewing of harmonics produced by the nonlinearity of the D/A converter. The 400 MHz spectrum represents the case

where $f_{\text{out}} = f_s/2$, which is the highest possible synthesizable frequency. The bottom trace of Fig. 18 illustrates the MSB of a cosine ECL output-driver switching at 400 MHz. The top trace is the corresponding reference clock at 800 MHz. The waveform is produced by setting the 32 b FCW of the QDDFS4 chip to equal 80 000 000(H), thus generating an $f_{\text{out}} = 400$ MHz. The loading of the output-driver is 50 Ω

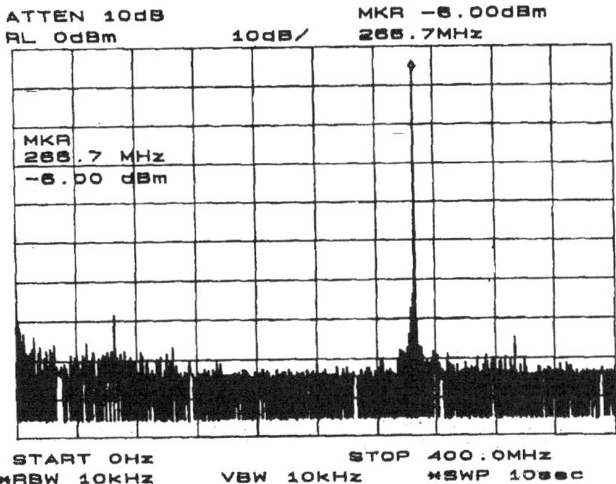

Fig. 21. Spectrum of $F_s/3 + \Delta$, where $\Delta = 65.1$ kHz, 400 MHz span.

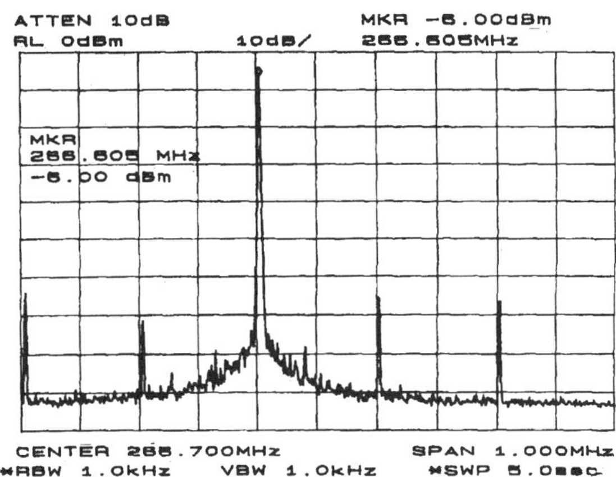

Fig. 22. Spectrum of $F_s/3 + \Delta$, where $\Delta = 65.1$ kHz, 1 MHz span.

in parallel with the loading of a 450 Ω TEK 6231 active probe used to sample the signal. Fig. 19 is the corresponding oscilloscope trace of the 400 MHz sinewave at the output of the D/A converter. The plug-in amplifier used has a 1 GHz bandwidth and thus partially filters the 400 MHz D/A converter output. Fig. 20 displays oscilloscope traces of a pair of unfiltered quadrature sine and cosine D/A converter output waveforms at 50 MHz. The signal trace above the quadrature signals is the 800 MHz reference clock. As shown in this figure, the sine and cosine waveforms are comprised of 16 discrete levels. Figs. 21 and 22 illustrate the QDDFS4 chip producing a $f_s/3 + \Delta$ fundamental where $\Delta = 65.1$ kHz. A frequency offset equal to Δ from $f_s/3$ produces a spreading of the aliased harmonic components spaced 3Δ apart. The worst case harmonic component is approximately 58 dB below the fundamental. The QDDFS4 produces a -84.3 dBc spectrally pure signal over its entire tuning range. Hence, the harmonic dibstortion in the analog outputs is primarily due to the D/A conversion process.

V. CONCLUSION

A quadrature digital synthesizer chip utilizing a parallel architecture has been designed that operates at 800 MHz and synthesizes -84.3 dBc spectrally pure sine and cosine digitized waveforms. This chip exhibits large bandwidth (dc to 400 MHz), high spectral purity, fast switching speed, and fine frequency resolution (0.188 Hz). The use of parallelism is an effective technique for increasing throughput in CMOS technology. Parallelism can be used successfully in CMOS processes simply because the yields are very high. Furthermore, effective use of parallelism provides a designer with the freedom to perform area versus speed trade-offs thus enabling production costs to be optimized for a specific targeted performance criterion. This technique is less effective in other high performance GaAs or silicon bipolar technologies since these fabrication processes are not able to match the yields of CMOS.

TABLE III
QDDFS4 CHIP SPECIFICATIONS

Technology	0.8 μm BiCMOS26B, Triple Level Metal, N well Process
Maximum Clock Frequency	800 MHz
Switching Time	4 clock cycles or 5 ns (at 800 MHz f_{clk})
Tuning Latency	47 clock cycles
Tuning Bandwidth	400 MHz (at 800 MHz f_{clk})
Frequency Resolution	0.188 Hz (at 800 MHz f_{clk})
Worst Case Spurious	-84.3 dBc
Phase Modulation Wordlength	12 b
Output Word Length	12 b
Clock Input	Small Signal Swing Clock ≤ 0.5 V
Output Drivers	ECL-Compatible levels
Power Dissipation	3 Watts @5 V with I/Os switching (at 800 MHz f_{clk})
Power Efficiency	3.75 mW/MHz @5 V
Transistor Count	94,000 (including I/Os)
Die Size	7.2 mm x 7.9 mm
Core Size	5.9 mm x 6.7 mm

The QDDFS4 chip dissipates 3 W at 800 MHz operation thus providing a power efficiency of 3.75 mW/MHz. The power dissipation achieved is the lowest of any reported DDFS of the same speed performance class in any fabrication process. By taking advantage of sine and cosine symmetries, the size of the ROM lookup table is no larger than that of a DDFS, which only generates sine outputs. The QDDFS4 chip attains a fourfold speed increase over the previous fastest CMOS DDFS [6] and is the fastest 12 b output single-chip DDFS ever reported in any technology. Table III summarizes the chip specifications. A chip photomicrograph is shown in Fig. 23.

ACKNOWLEDGMENT

The authors wish to thank F. Lu for valuable discussions, E. Berg for work on the ROM, D. Kruse for register characterization, and R. Patel for laying out the printed circuit board. High speed testing was facilitated by equipment donations from Hewlett Packard and Tektronix. N. H. Sheng and D. Mehrotra of Rockwell International are gratefully acknowledged for technical assistance and for supplying samples of the RI61008 D/A converter.

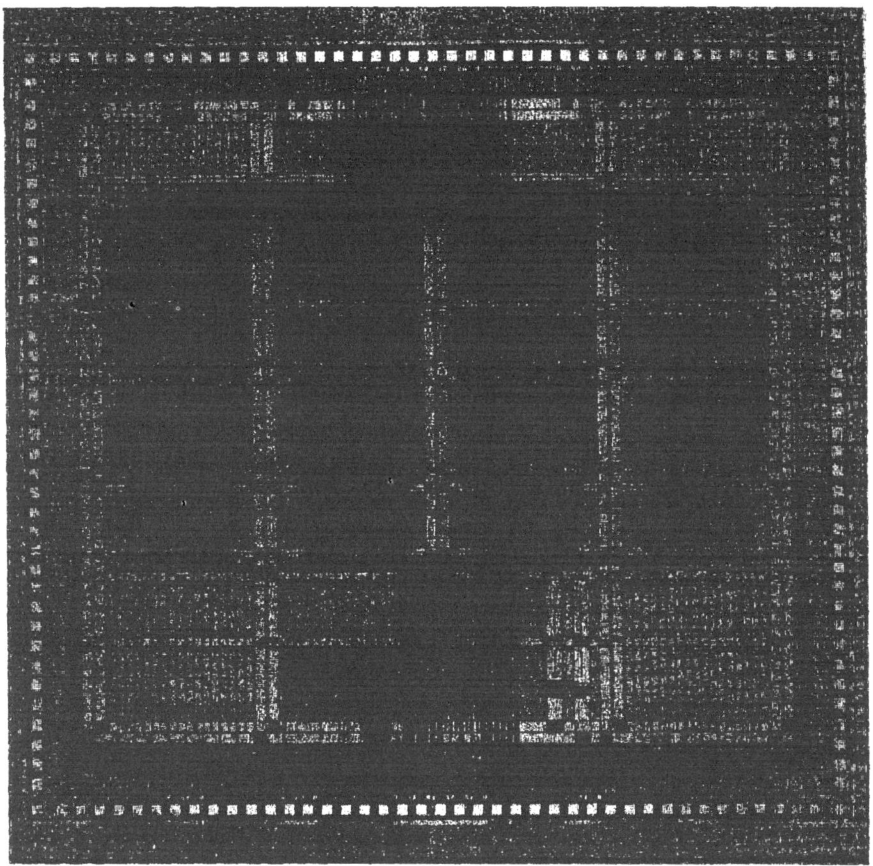

Fig. 23. QDDFS4 chip photomicrograph.

References

[1] V. Manassewitch, *Frequency Synthesizers, Theory and Design*, 2nd ed. New York: Wiley, 1989.

[2] P. H. Saul and D. G. Taylor, "A high speed direct frequency synthesizer," *IEEE J. Solid-State Circuits*, vol. 25, no. 1, pp. 215–219, Feb. 1990.

[3] V. Andrews, C. T. M. Chang, J. D. Cayo, S. Sabin, W. A. White, and M. P. Harris, "A monolithic digital chirp synthesizer chip with I and Q channels," *IEEE J. Solid-State Circuits*, vol. 27, pp. 1321–1326, Oct. 1992.

[4] M. Thompson, "Low latency, high-speed numerically controlled oscillator using progression-of-states technique," *IEEE J. Solid-State Circuits*, vol. 27, pp. 113–117, Jan. 1992.

[5] R. Hassun and A. W. Kovalick, "Waveform synthesis using multiplexed parallel synthesizers," U.S. Patent no. 4, 454, 486, Hewlett Packard, June 12, 1984.

[6] L. K. Tan and H. Samueli, "A 200-MHz quadrature frequency synthesizer/mixer in 0.8 μm CMOS," *IEEE J. Solid-State Circuits*, vol. 30, pp. 193–200, Mar. 1995.

[7] H. T. Nicholas (III) and H. Samueli, "An analysis of the output spectrum of direct digital frequency synthesizers in the presence of phase-accumulator truncation," in *Proc. 41st Annual Frequency Control Symp. USERACOM*, Ft. Monmouth, NJ, May 1987, pp. 495–502.

[8] J. Yuan and C. Svensson, "High-speed CMOS circuit technique," *IEEE J. Solid-State Circuits*, vol. 24, no. 1, pp. 62–70, Feb. 1989.

[9] B. A. Chappell, T. I. Chappell, S. E. Schuster, H. M. Segmuller, J. W. Allan, R. L. Franch, and P. J. Restle, "Fast CMOS ECL receivers with 100-mV worst case sensitivity," *IEEE J. Solid-State Circuits*, vol. 23, no. 1, pp. 59–67, Feb. 1988.

[10] S. R. Meier, E. De Man, T. G. Noll, U. Loibl, and H. Klar, "A 2-μm CMOS digital adaptive equalizer chip for QAM digital radio modems," *IEEE J. Solid-State Circuits*, vol. 23, no. 5, pp. 1212–1217, Oct. 1988.

A Dual Frequency Synthesis Scheme for a High C-Field Cesium Resonator

E. RUBIOLA, A. DEL CASALE, A. DE MARCHI

Politecnico di Torino, Dipartimento di Elettronica
c.so Duca degli Abruzzi n. 24, I-10129 Torino, Italy

Abstract

The high C-field concept for a Cesium resonator is based on the idea that a true two level system can be obtained, without having to observe excessive field stabilities, if the C-field is set at the turning point of the frequency of a Zeeman component. This is due to the fact that the reference transition has no linear Zeeman effect at that field.

Nevertheless, because of the second order Zeeman coefficient, a $5 \cdot 10^{-7}$ field accuracy must be realized for a Zeeman bias effect smaller than 10^{-14}. Therefore the C-field must be under closed loop control based on frequency measurements on a field dependent transition.

In this paper a synthesis scheme is proposed, which can provide two frequencies with the necessary resolution: two different frequency lock loops are used to measure the ν_{-1} and ν_0 transitions. The former controls a high stability quarts oscillator with a relatively long time constant. The latter is used to measure C-field and can be faster, if needed to enact a tight control of the magnet's current power supply.

1 High C-field physics and its implications on electronics

In the high C-field standard, as proposed in [1], the field level ($B_0 \simeq 82$ mT) is chosen in such way as to minimize the frequency ν_{-1} of the $m_F = -1$ transition, which is used as the reference. In this conditions the 1st order Zeeman effect vanishes and, from the Breit Rabi Formula, the clock frequency turns out to be

$$\nu_{-1\,\text{min}} = \nu_{00}\sqrt{15/16}$$
$$\nu_{-1}(B) \approx 8.900,727,438,257 \cdot 10^9 \, \text{Hz} +$$
$$+ (B - B_0)^2 \cdot 4.41 \cdot 10^{-10} \, \text{Hz/T}^2 \quad (1)$$

where ν_{00} is the resonance frequency of the unperturbed atoms.

In the error budget, a bias of $3 \cdot 10^{-14}$ due to C-field can be tolerated for a residual uncertainty below 10^{-14}. For this target, the field must be set to within $5 \cdot 10^{-7}$. Therefore, the C-field must be under closed loop control based on frequency measurement of a field dependent transition. The selected one is $m_F = 0$, whose frequency at B_0 is given by

$$\nu_0 = \nu_{00}\sqrt{15/16}$$
$$\nu_0(B) \approx 9.475,548 \cdot 10^9 \text{Hz} +$$
$$+ (B - B_0) \cdot 2.1 \cdot 10^{10} \text{Hz/T} \quad (2)$$

A cavity which resonates at both frequencies ν_{-1} and ν_0 was designed and realized [2]. Because of the wide frequency separation of the two modes, clock operation and C-field measurement can take place at the same time without appreciable accuracy loss due to AC Zeeman effect. In order to accomplish this, two separate sinthesizers, each one with its own modulation frequency, are needed.

The synthesizers' main specifications can be drawn from the results of a theoretical analysis

Reprinted from *Proceedings of the IEEE International Frequency Control Symposium*, pp. 105-108, 1993.

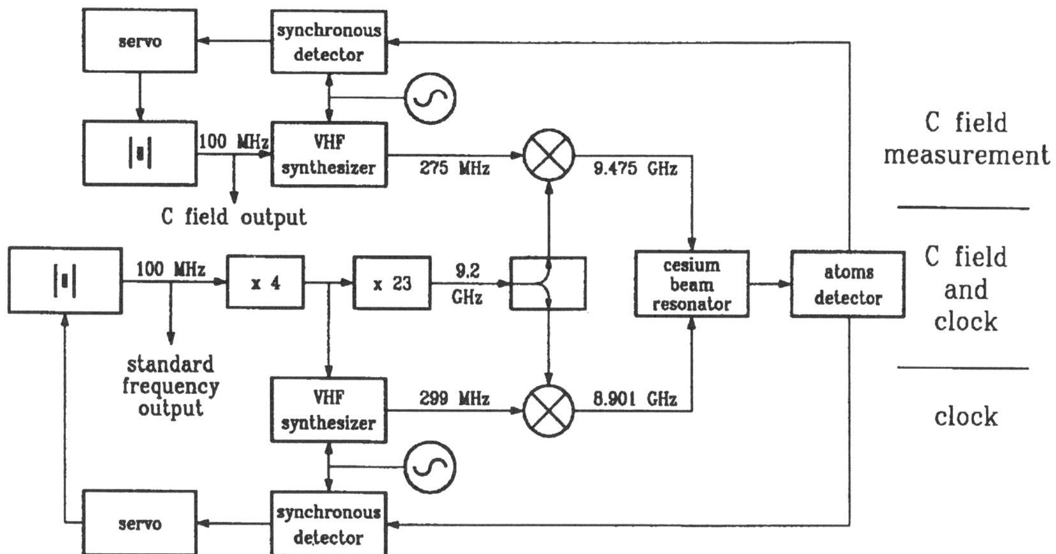

Figure 1: Block diagram of the dual frequency synthesis system under development.

of the beam tube which leads to the following expected parameters: accuracy in the 10^{-14} range, white frequency noise limitation for σ_y at several $10^{-13}/\sqrt{\tau}$, atomic linewidth of 160 Hz and cavity bandwidth of 300 kHz.

Resolution is not a critical parameter for ν_0. From equation (2), it can be stated that 100 Hz it all what's needed if a C-field accuracy of 10^{-7} must be achieved. The clock frequency ν_{-1} should be synthesized at least with the same resolution as for the accuracy limit of the standard, i.e., 10^{-4} Hz, although in principle this is not mandatory.

2 Synthesis scheme

The synthesis scheme is based on the idea that the cesium beam tube can be seen as two separate resonators. Then, two quartz oscillators are frequency locked for clock operation and C-field measurement.

The practical scheme, shown in fig. 1, uses a single multiplication chain for rising the quartz output frequency to 9.2 GHz, and two separate synthesizers for obtaining the final frequencies.

The value was chosen to be 9.2 GHz because it is close to $(\nu_0 + \nu_{-1})/2$ and a suitable multiple of the main quartz (100 MHz). Moreover, most of this scheme can be reused for low field operation at 9.193 GHz, if this option is needed.

The multiplication from 100 MHz to 9.2 GHz needs two steps. The first stage ($\times 4$), which can be made with low noise double balanced mixers or with a classic transistor multiplier, is filtered for the minimum 100 MHz spurious output, and the second one ($\times 23$) is based on a SRD for the lowest noise. A single stage 100 MHz $\rightarrow$ 9.2 GHz must be avoided because the SRD multiplier would produce an appreciable power at any frequency multiple of 100 MHz, including 8.9 GHz; this last, close to ν_{-1}, is almost in the cavity bandwidth.

A selection on high performance commercially available oscillators shows that one can choose a low frequency device (5–10 MHz) for the best frequency flicker, or a high frequency one (≥ 100 MHz) for the lowest phase noise. The phase noise S_ϕ expected at 8.9 GHz from two representative oscillators is shown in Fig. 2, together with the expected white frequency noise of the tube. Since at all frequencies below the atomic linewidth the

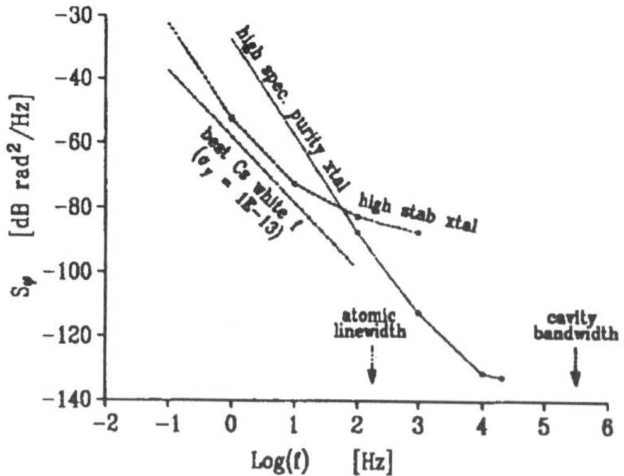

Figure 2: Phase noise of the oscillators at 8.9 GHz, compared to the expected tube noise.

considered oscillators' noise is higher than the white noise expected in closed loop from the beam intensity limitation, the overall noise will be limited by the oscillator. Then it was decided to take full benefits of the technology by using a 100 MHz low noise oscillator phase locked to a 5 MHz low flicker one. This PLL is seen as the main 100 MHz oscillator. A buffer, with a forward/backward gain ratio of 140 dB [3] provides the clock output.

The 299 MHz VHF synthesizer provides a modulated signal and it is responsible of the standard frequency tuning. In a first stage of the experimental work it can be replaced by commercial equipment, loosing in resolution.

The second oscillator is locked to the ν_0 line, thus providing continuous monitoring of the C-field. When the C-field loop is open, this oscillator allows numerically controlled precision scanning of the ν_{-1} Zeeman curve.

3 VHF synthesizer

The main concerns in design the 299 MHz VHF synthesizer are:

1. Spectral purity. The estimated microwave noise floor is −130 dB rad^2/Hz, due to the quartz (see Fig. 2), plus the noise of the multiplication chain. This synthesizer should not degrade the spectral purity of the microwave signal. A limit of −135 dB rad^2/Hz seems reasonable.

2. Spurious signals. They should be kept at a negligible level in the cavity bandwidth. Moreover, spurious signal around 100 MHz are to be carefully avoided because a multiplication by three can take place when mixing this signal with the 9.2 GHz microwave, thus yelding to spurious near ν_{-1}.

3. Resolution should be about 10^{-4} Hz.

The proposed scheme is shown in Fig. 3, where an high resolution 207 kHz synthesizer is up converted to 299 MHz in several steps.

The number of conversion stages and the intermediate frequencies have been chosen for the lowest spurious signals, expecially for those which are close to the carrier, avoiding high Q filters or resonators because of their phase instability. These unwanted signals have been evaluated using double balanced mixers intermodulation tables reported in [9, 10] and experimental data, taking into account all harmonics up to the 11th order.

In order to keep the output clean from 100 MHz spurious signals, the last stage is driven by a 266.$\bar{6}$ MHz signal made by a regenerative converter. This converter is in principle a regenerative divider in which the output is taken from the multiplier output instead of the input. The regenerative scheme is well known for its low noise performances [4, 5].

The lower frequency conversion stages are less critical and can be based on digital dividers. With these dividers, a white noise floor less than −140 to −150 dB can be achieved [5, 6, 7, 8]. Flicker and higher slope phase noise types are not a problem, as compared to the quartz multiplied to 9.2 GHz.

The low frequency synthesizer design suffers from the difficulty of obtaining sub-millihertz resolution and a sufficient spectral purity. NCO based schemes, that have excellent resolution, show a pseudo random distribution of spectral lines spaced from the carrier by multiples of the

316

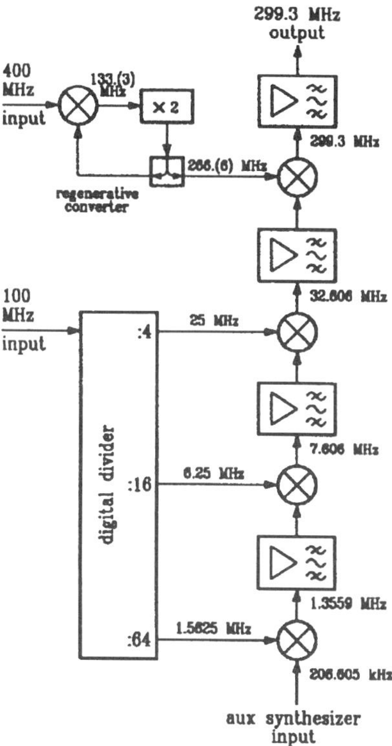

Figure 3: Main VHF synthesizer.

resolution; these signals are within the main FLL bandwidth. In order to overcome these problems, two schemes are under study.

Following the first approach, the 207 kHz signal is obtained dividing the output of a commercial synthesizer by a suitable factor, say 1000, and shaping the output with a digital to analog converter. In the second one, the 207 kHz is implemented as a infinite Q IIR filter in a DSP driven by the standard frequency; a slow software gain control avoids saturation problems.

References

[1] A. De Marchi, "The high C-field concept for an accurate cesium beam resonator," *Proc. 7th European Frequency and Time Forum*, Neucâtel, Switzerland, March 1993 (to be printed).

[2] A. De Marchi, E. Bava, P. Tavella, "Accuracy analysis of the atom-microwave interaction in a TE_{017} cylindrical Rabi cavity for a high C-field cesium beam resonator," *Proc. 7th European Frequency and Time Forum*, Neucâtel, Switzerland, March 1993 (to be printed).

[3] A. De Marchi, F. Mussino, M. Siccardi, "A high isolation low noise amplifier with near unity gain up to 100 MHz," *Proc. 47th Frequency Control Symposium*, 1993.

[4] E. Rubiola, M. Olivier, J. Groslambert, "Phase noise in the regenerative frequency dividers," *IEEE Trans. Instr. & Meas.*, vol. 41 no. 3, pp. 353–360, Jun 1992.

[5] M. M. Driscoll, T. D. Merrell, "Spectral performance of frequency multipliers and dividers" *Proc. 46th Frequency Control Symposium*, 1992, pp. 193–200.

[6] W. F. Egan, "Modeling phase noise in frequency dividers" *IEEE Trans. on Ultrasonics, Ferroelectrics and Frequency Control*, vol. UFFC-37 no. 4, July 1990, pp. 307–315.

[7] W. F. Egan, "Phase noise modeling in frequency dividers" *Proc. 45th Annual Frequency Control Symposium*, May 1991, pp. 629–635.

[8] F. W. Walls, C. M. Felton "Low noise frequency synthesis," *Proc. 41st Annual Frequency Control Symposium*, May 1987, pp. 512–518.

[9] B. C. Henderson, "Mixers: part 1 — characteristic and performances," "Mixers: part 2 — theory and technology," *RF microwave components designers' handbook*, Watkins-Johnson Company, 1990-91.

[10] D. Cheandle "Selecting mixers for best intermod performance" (part 1 and 2), *RF microwave components designers' handbook*, Watkins-Johnson Company, 1990-91.

Part X

New Ideas for the DDFS Design

THE largest number of DDFS devices are still based on the classic idea by Tierney et al. [X-1]. However, attempts to get to higher frequency ranges and to reduce power or spurious signals result in larger or smaller modifications of the original approach.

A numerically controlled oscillation (NCO) using a circuit design technique called true single-phase clock pipelined CMOS has been developed by Fang Lu et al. (1993) in the standard 1.2-μm CMOS technology. The maximum clock rate is 700 MHz, which results in an output frequency range from DC to 350 MHz. The output waveform is a rectangular one with a peak jitter of 1.4 ns, that is, one clock period (cf. the paper by the editor in Part V).

Caglio, Degouy, Meignant, Rousseau, and Leroux (1993) report on a GaAs DDFS with maximum clock frequency of 1.25 GHz, which can supply chirp signals up to a 100-MHz range. Instead of lookup tables, they generate the output sine wave with the assistance of the Johnson counter and weighted currents.

To save DC power, Kushner (1993) suggested combining a low-speed, high-resolution DDFS with a high-speed, low-resolution accumulator, both combined via a serrodyne modulation technique. In the proposed device, the first part provides the fine tuning, while the high-speed circuitry provides the coarse tuning of the output frequency. The design of an 800-MHz serrodyne IC modulator is discussed in detail later [X-2].

An interesting approach to DDFS with sine-wave output and high spectral purity was proposed by Karlquist in two papers (1995, 1996). The first paper describes a synthesizer that generates a narrow range of frequencies around 10 MHz, with extremely small step size and high spectral purity. The idea is to exploit inexpensive ceramic filters that are normally used in the IF stages of radio receivers. In the second paper, the basic 10.7-MHz band is applied to an M/N synthesizer to produce an output band from 3 to 30 MHz.

Lo Presti, Cardamone, De Marchi, and Rubiola (1994) propose a completely new approach to DDFS with sine-wave output and good spectral purity by using a second-order IIR resonator, the impulse response of which is a sine wave. Since this impulse response can be expressed in a second-order recursive form, sample generation only requires a few multiplications and additions. Few instructions with a fast floating-point DSP microprocessor implement the recursive equation in real time. Some actual spectral measurements are enclosed. This reprinted paper is a revised version of [X-3].

Finally we mention a paper by Chren who suggested application of the "Residue Number System" for reduction both of the frequency switching latency and the chip area [X-4].

REFERENCES

[X-1] J. Tierney, Ch. M. Rader, and B. Gold. "A digital frequency synthesizer." *IEEE Transactions on Audio and Electroacoustics,* pp. 48–57 March 1971. (Reprinted in Part III.)

[X-2] L. J. Kushner, G. Van Andrews, W. A. White, J. B. Delaney, M. A. Vernon, M. P. Harris, and D. A. Whitmire. "An 800-MHz monolithic GaAs HBT serrodyne modulator." *IEEE Journal of Solid-State Circuits,* vol. 30, pp. 1041–50, October 1995.

[X-3] L. Lo Presti, G. Cardamone, A. de Marchi, and E. Rubiola. "A new architecture for a sinewave output DDS with a high spectral purity." *Proceedings of the 8th European Frequency and Time Forum,* pp. 646–55, March 1994.

[X-4] W. A. Chren, Jr. "RNS-based enhancements for direct digital frequency synthesis." *IEEE Transactions on Circuit and Systems—II,* pp. 516–24, August 1995.

A 700-MHz 24-b Pipelined Accumulator in 1.2-μm CMOS for Application as a Numerically Controlled Oscillator

Fang Lu, *Member, IEEE,* Henry Samueli, *Member, IEEE,* Jiren Yuan, and Christer Svensson

Abstract— To accomplish timing recovery/synthesis in high-speed communication systems, a 24-b numerically controlled oscillator (NCO) IC using a circuit design technique called true single-phase clock (TSPC) pipelined CMOS has been fabricated in a standard 1.2-μm CMOS process. The device achieves a maximum tested input clock rate of 700 MHz, which results in an output frequency tuning range from dc up to 350 MHz with a 41.7-Hz tuning resolution and a peak-to-peak phase jitter of 1.4 ns. The 1.7 $\times$ 1.7-mm^2 IC dissipates 850 mW with a single 5-V supply, which is substantially lower than similar ECL and GaAs devices.

I. INTRODUCTION

IN modern communication circuit designs, all-digital implementations of the signal processing functions are becoming more popular as advances in VLSI technology facilitate system-level integration into a small number of chips. Direct digital frequency synthesis schemes were originally proposed in 1971 [1], and the primary focus of recent work in this field has been to extend the bandwidth of these devices. Even indirect phase-locked-loop (PLL) based frequency synthesizers often incorporate digital synthesis techniques to achieve fine frequency resolution [2], [3]. In these schemes, the frequency control signal is represented as a binary data word, and a fairly pure sine wave is synthesized at the outputs.

In digital timing recovery/synthesis schemes, pure sine waves are not always required. For example, in a digital modem receiver, only a rectangular waveform is required to represent the bit-rate or symbol-rate clock. Two approaches are often used to translate the digital control word into a timing waveform. Fig. 1(a) shows an approach using a digital-to-analog converter (DAC) followed by a voltage-controlled oscillator (VCO). The drawbacks of this approach are: 1) it is difficult for the DAC to achieve a resolution greater than 16 b, and a large chip area is required; 2) it is even more difficult for an analog VCO to have an exactly linear V-to-f characteristic; 3) the performance of both the DAC and VCO

Manuscript received July 6, 1992; revised April 5, 1993. This work was supported in part by grants from the University of California MICRO Program, TRW, Inc., and DARPA.

F. Lu is with Broadband Telecom, Inc., Los Angeles, CA 90024.

H. Samueli is with the Integrated Circuits and Systems Laboratory, Electrical Engineering Department, University of California, Los Angeles, CA 90024.

J. Yuan and C. Svensson are with the Department of Physics and Measurement Technology, Linkoping University, S-581 83 Linkoping, Sweden.

IEEE Log Number 9209808.

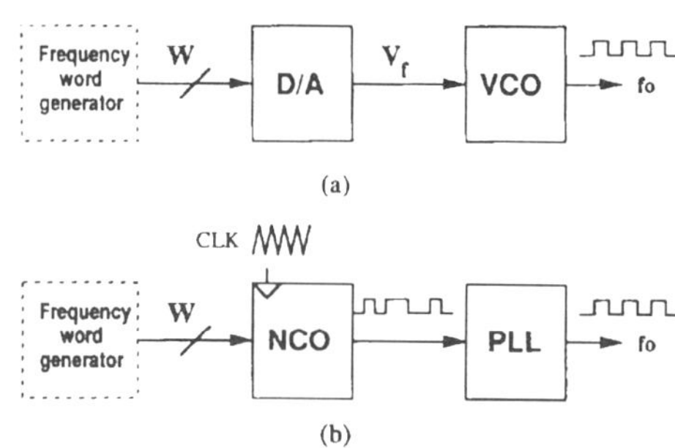

Fig. 1. Approaches for synthesizing digital timing waveform.

will drift due to aging and environmental changes; and 4) the VCO has a limited frequency tuning range. An alternative solution is shown in Fig. 1(b), which contains a numerically controlled oscillator (NCO) followed by an optional smoothing PLL. Its merits include: 1) the NCO circuit consumes little chip area and its tuning resolution is limited only by the high accuracy of the crystal oscillator; 2) as long as the oscillator in the PLL is monotonic, it does not need to be linear; 3) if the NCO clock frequency is much higher than the generated waveform, the NCO output phase jitter is negligible and the smoothing PLL is no longer needed; and 4) without the PLL, the NCO output frequency can range from dc up to *half* the reference clock rate.

An NCO can be implemented with a digital accumulator, which is basically a two-operand adder followed by a register with its output fed back to the adder, as shown in Fig. 2. The input to the accumulator is the frequency control word, and the rectangular waveform of the MSB output is used as the oscillator output. If the frequency control word takes on power-of-2 values, the MSB output of a free-running accumulator will have a 50% duty cycle. If the frequency control word is not a power of 2, the durations of the "high" and "low" states of the MSB output will have a difference of at most one clock cycle, which is exactly the maximum phase jitter of the NCO output. Both of the above two cases are exemplified in Fig. 3 assuming the register is initialized to zero. In order to obtain a low phase jitter, the computation rate (clock rate) of the accumulator has to be much higher than the frequency of the synthesized output waveform. In other words, the clock rate of this type of NCO will inevitably be much faster than the clock rate of most of the other digital circuits in a communication/signal-processing system.

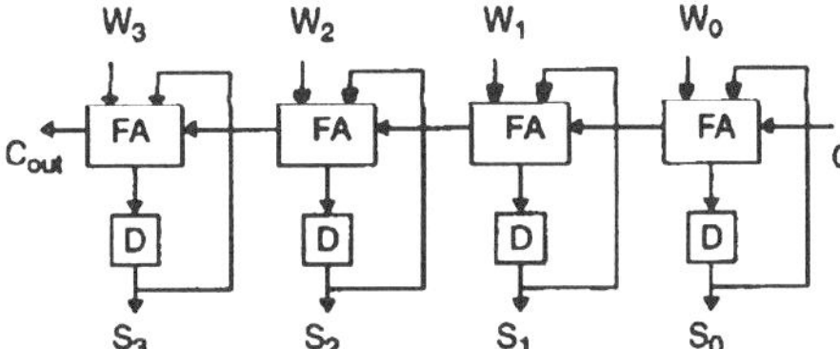

Fig. 2. Primitive accumulator structure.

In this paper we describe the implementation of a very high-speed CMOS NCO chip by applying a previously proposed circuit design technique called true single-phase clock (TSPC) pipelined CMOS [4], [5] to the accumulator design. Under nominal operating conditions (5-V supply and room temperature), the maximum input clock rate was measured to be 700 MHz, which is the fastest CMOS NCO ever reported. The performance is approaching, and in some aspects even superior to, its GaAs counterparts.

In Section II, the architecture of the NCO chip is described. Section III discusses the crucial circuit design issues such as the TSPC full-adder cell, the frequency updating mechanism, high-speed clock buffering, and on-chip power-rail decoupling capacitors. The fabrication and test results are presented in Section IV.

II. CHIP OVERVIEW

The primitive structure of the phase accumulator shown in Fig. 2 is not well suited for implementing a high-speed NCO. The flip-flops cannot be triggered until the internal carry signal propagates through all the full-adder (FA) cells. By partitioning the carry propagation path with flip-flops, a pipelined accumulator can be obtained as shown in Fig. 4(a), in which the black circles represent D flip-flops. Preskewing and deskewing registers are used to provide a bit-parallel I/O interface to the user. The *accumulation loop* at each bit position still performs a 1-b full addition every clock cycle.

In a nonpipelined accumulator, a fast adder scheme such as a carry-lookahead adder (CLA) can reduce the propagation delay. In a bit-level pipelined system, however, throughput is determined only by the single accumulation loop delay. Thus, using any fast-carry strategy would in fact *degrade* the clock speed since its accumulation loop delay would be much longer than one FA cell. To provide the reset (initialization) mechanism, AND gates can be added on the feedback paths of the accumulation loops, as shown in Fig. 4(b). The signal $\overline{RES}$ also has to be pipelined to reset and release the accumulator, bit by bit, thereby compensating for the carry-pipeline delay.

Fig. 5 shows the block diagram of the NCO chip. It contains a 24-b fine-grain pipelined accumulator, a 24-b frequency-word updating register pair (double buffer), a true single-phase clock buffer, and a microprocessor handshaking interface control circuit. The first and second MSB outputs of the accumulator are fed to output buffers. The contents of the accumulator is incremented at each clock cycle with a step size equal to the 24-b frequency control word W. The accumulator's overflow characteristic makes its MSB toggle between ZERO and ONE with a frequency $f_{\text{out}} = [f_{clk}/(2^{24})] \times W$.

	Binary	Decimal		Binary	Decimal	
Freq. Ctrl. Word	0 0 1 0	2		0 0 1 1	3	
Accumulator	0 0 0 0	0		0 0 0 0	0	
Output vs. Time	0 0 1 0	2	<--\	0 0 1 1	3	<---\
	0 1 0 0	4	|	0 1 1 0	6	|
	0 1 1 0	6	|	1 0 0 1	9	|
	1 0 0 0	8	|	1 1 0 0	12	|
	1 0 1 0	10	|	1 1 1 1	15	|
	1 1 0 0	12	|	0 0 1 0	2	|
	1 1 1 0	14	|	0 1 0 1	5	|
	0 0 0 0	0	>---/	1 0 0 0	8	|
	v			1 0 1 1	11	|
	MSBa			1 1 1 0	14	|
				0 0 0 1	1	|
				0 1 0 0	4	|
CLK				0 1 1 1	7	|
				1 0 1 0	10	|
MSBa				1 1 0 1	13	|
				0 0 0 0	0	>---/
MSBb				v		
				MSBb		

Fig. 3. Effects of the frequency control word on the accumulator MSB output.

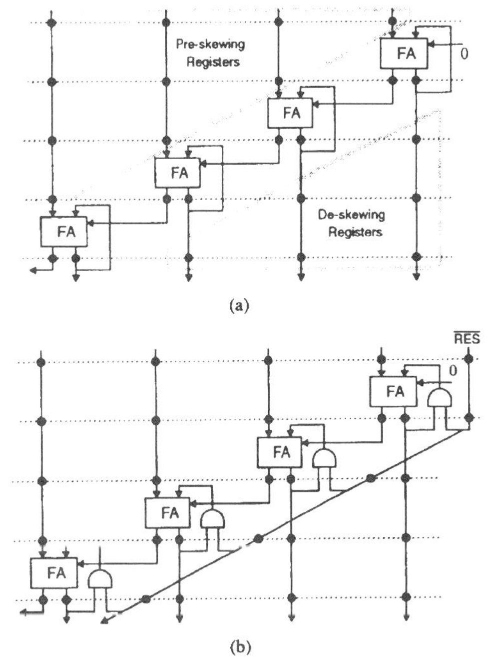

(a)

(b)

Fig. 4. Pipelined accumulator with (a) free-running structure, and (b) reset mechanism.

The frequency tuning resolution is $f_{clk}/(2^{24})$, and the highest synthesized frequency is $f_{clk}/2$.

In Fig. 5, a double-buffer structure is used to store the frequency control word, which eliminates the need for the large number of flip-flops normally required by the preskewing registers (see Fig. 4(a)). The deskewing registers are not needed since the NCO chip uses only the MSB to synthesize its output. In the double buffer, the first-stage register latches

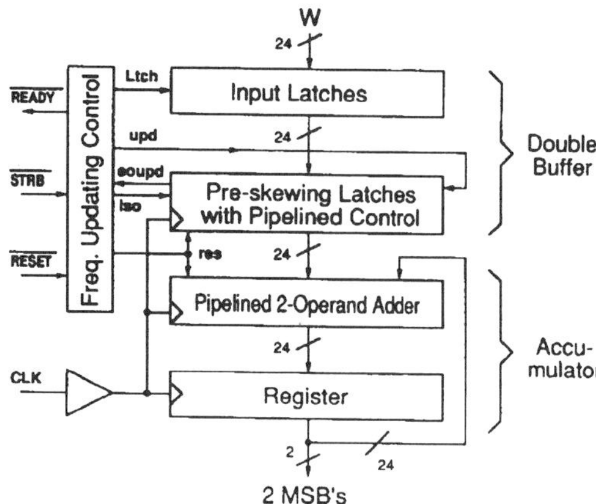

Fig. 5. NCO chip block diagram.

in the 24-b word from the external pins in parallel, then the frequency control word is reclocked into the second-stage register called the preskewing latches. The on-chip pipelined frequency updating scheme is crucial for applications requiring real-time frequency tracking or frequency hopping since it can change the output frequency on the fly, while maintaining phase continuity of the output waveform.

To provide for application flexibility of the NCO chip, a control circuit was designed so that the chip can either interface to a microprocessor/microcontroller through handshaking, or it can operate by itself and accept a new frequency control word every 55 clock cycles if the $\overline{STRB}$ input is held "low." The 55 clock cycles of total latency includes the pipelined preskewing register and the frequency updating control circuit. If the NCO is used in frequency hopping applications, the maximum hopping rate will be 700 MHz/55, or 13 million hops per second.

III. CIRCUIT DESIGN ISSUES

A. Pipelined Accumulator

The high-speed performance of the NCO chip is achieved by using the true single-phase clock (TSPC) pipelined CMOS circuit design technique. The pipelined accumulator contains 24 special TSPC full-adder cells as shown in Fig. 6(a), in which B_n represents the nth bit of the frequency control word W stored on chip. The operation is best explained by the equivalent static version of the full-adder cell depicted in Fig. 6(b). The lower right group of logic gates in Fig. 6(b) performs the first-stage half-adder (HA) function, the lower left group of logic gates performs the second-stage HA function, and the other circuitry generates the carry output. Working with clocked switches, a high *Reset* signal can force S_n and $\overline{C}_n$ to be "low" and "high," respectively. The topological relations among the logic gates in Fig. 6(b) are the same as those in Fig. 6(a), and the truth table of the static adder is shown in Fig. 6(c).

To reach a truly high speed, the full-adder cell itself is finely pipelined down to the logic-gate level. Each gate has a dynamic (domino) logic structure that contains precharging/pre-

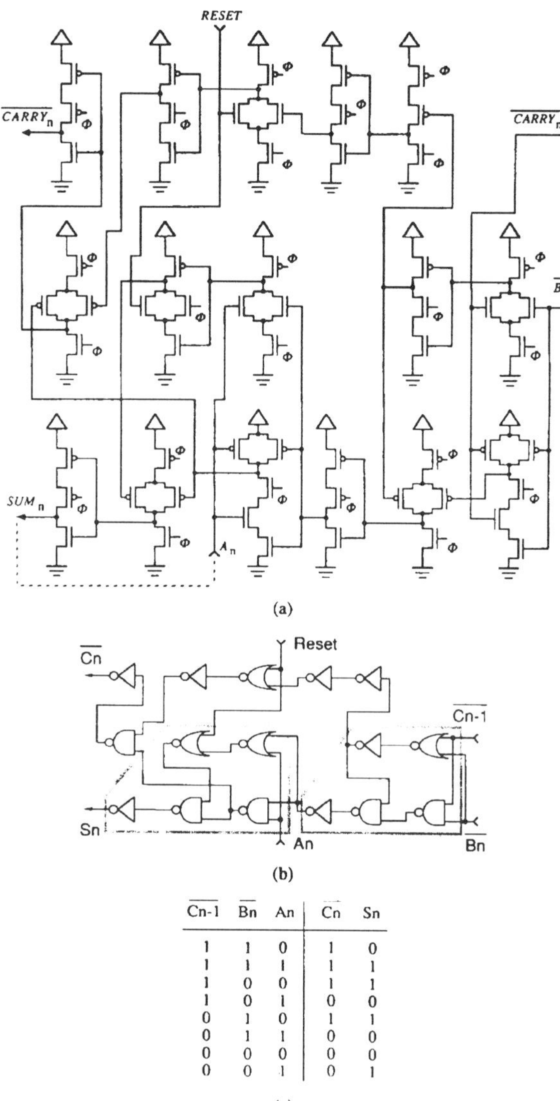

(a)

(b)

C_{n-1}	$\overline{B}_n$	An	C_n	S_n
1	1	0	1	0
1	1	1	1	1
1	0	0	1	1
1	0	1	0	0
0	1	0	1	1
0	1	1	0	0
0	0	0	0	0
0	0	1	0	1

(c)

Fig. 6. (a) TSPC full-adder cell with sum-output feedback. (b) Equivalent static version of adder cell. (c) Truth table.

discharging switches and evaluating switches. All pipeline stages contribute half a clock cycle to the system latency, and they are separated by N-CMOS gates (those with N-switches only) and P-CMOS gates (with P-switches) alternately to achieve a true single-phase clocking scheme. At every bit location, accumulation has to be performed within each clock cycle; thus, the output of the full-adder cell is fed back to one of the three input nodes (A_n) such that the loop contains a latency of only one clock cycle. On the other hand, the carry propagation path of the full-adder cell is partitioned into four pipelined stages to maximize the throughput rate, which contributes two clock cycles per bit to the accumulator latency. Consequently, the TSPC accumulator has twice the clock cycles in its carry propagation latency as that in the

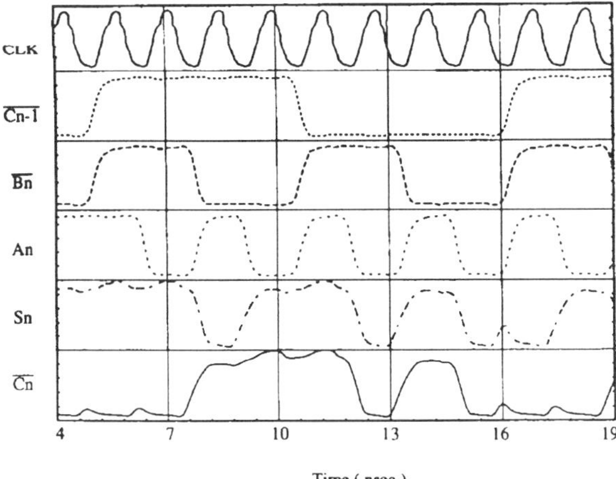

Fig. 7.   I/O waveforms of the TSPC full-adder cell.

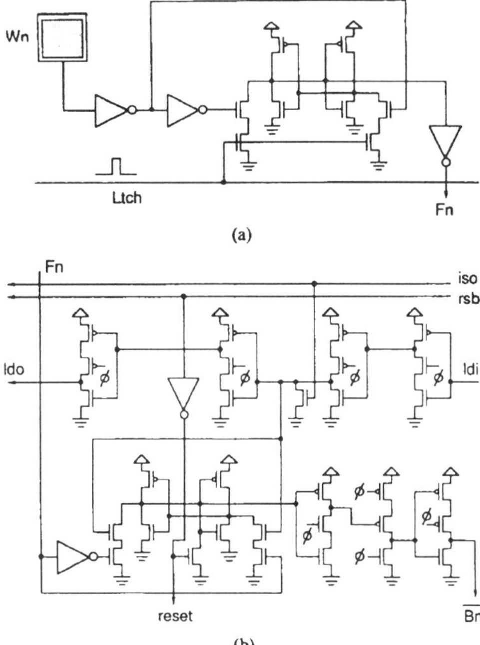

Fig. 8.   Double-buffer updating circuitry: (a) first-stage latch, and (b) second-stage latch with pipelined control.

structure of Fig. 4(b) with the same word length, but the phase accumulation can still be performed *every* clock cycle.

To verify the function and performance of the TSPC full-adder cell, the simulated I/O waveforms using SPICE are shown in Fig. 7 when the TSPC full-adder cell is in normal operating mode (i.e., when *Reset* is set "low"). The topmost curve is the clock waveform, the three waveforms beneath are the adder inputs, and the lowest two waveforms are the adder outputs which are loaded with capacitors to mimic the actual loading. By delaying the $\overline{C}_{n-1}$ and $\overline{B}_n$ waveforms by two clock cycles, and delaying the A_n waveform by one clock cycle, the truth table in Fig. 6(c) can be easily verified. The timing relations also demonstrate the one clock-cycle latency from A_n to the adder outputs, and the two clock-cycle latency from $\overline{B}_n$ and $\overline{C}_{n-1}$ to the adder outputs. The clock period in the simulation is 1.4 ns, which is equivalent to 714 MHz.

The reasons that the TSPC circuit can achieve extremely high clock rates are threefold. First, all logic gates in the TSPC full-adder cell use a dynamic structure, which is well known to be much faster than static CMOS logic. Second, the gate-level pipeline stage has at most two gate delays (including the N- or P-CMOS gate) with their fan-ins no more than two, which minimizes the required clock period. Third, the true single-phase clocking scheme is used throughout the chip, which eliminates the clock skew problems of complementary-phase or multiphase clocking schemes, and not any portion of the clock cycle is wasted. Furthermore, each transistor size has been carefully optimized to minimize the propagation delay.

B. Double Buffer

As mentioned in Section II, a double-buffer structure is used to update the frequency control word in the NCO chip (see Fig. 5). To facilitate off-chip interfacing, the first-stage register latches in the 24-b word in parallel. The frequency control word is then reclocked into the second-stage register (preskewing latches) bit by bit, sequentially starting from the LSB. Assuming the clock period is T and the LSB is updated at $t = 0$, then the second LSB will be updated at $t = 2T$,

the third LSB will be updated at $t = 4T$, and so on. In this way the frequency updating function is properly aligned with the carry-propagation latency ($2T$ per bit) of the 24-b pipelined accumulator, and phase continuity at the accumulator output can be maintained. This feature is desirable in real-time frequency tracking applications.

Fig. 8(a) shows the first-stage latch of the double buffer, which is laid out in the spare area of the input pad cell. All 24 first-stage latches will fetch the entire frequency control word when they are triggered by a short *Ltch* pulse. Fig. 8(b) shows the second-stage latch with a pipelined control. Signal *rsb* resets all 24 second-stage latches and 24 TSPC adder cells simultaneously. Signal *ldi* loads the first-stage latch data into the second-stage latch. Its rising edge is then delayed by two clock cycles and transferred to the next higher bit position through *ldo*. This delay is required to compensate for the carry latency of one TSPC adder cell. $\overline{B}_n$ has been resynchronized through a TSPC flip-flop so that the slower static latch will not affect the speed of a TSPC adder cell. Once all 24 second-stage latches have been updated, signal *iso* is activated to isolate the second-stage latch from the first by disabling the load function of all second-stage latches simultaneously.

C. Interface Control

The details of the frequency updating control block of the NCO chip (see Fig. 5) are depicted in Fig. 9. It has two operating modes: 1) handshaking mode and 2) self-timed mode.

In the handshaking mode, the very high-speed NCO chip can interface with a lower speed processor (or any circuit that

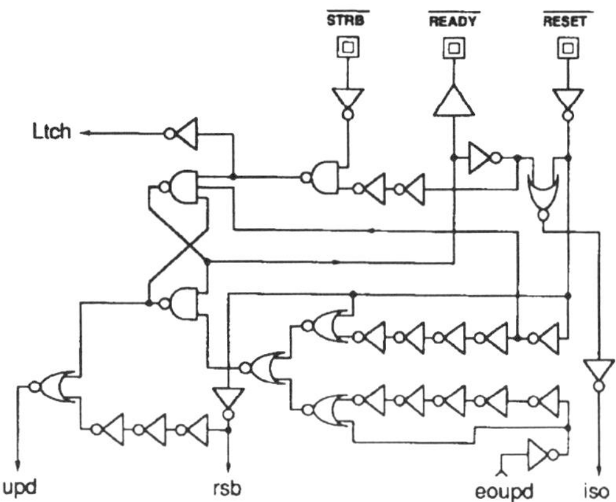

Fig. 9. Handshaking/self-timed control block.

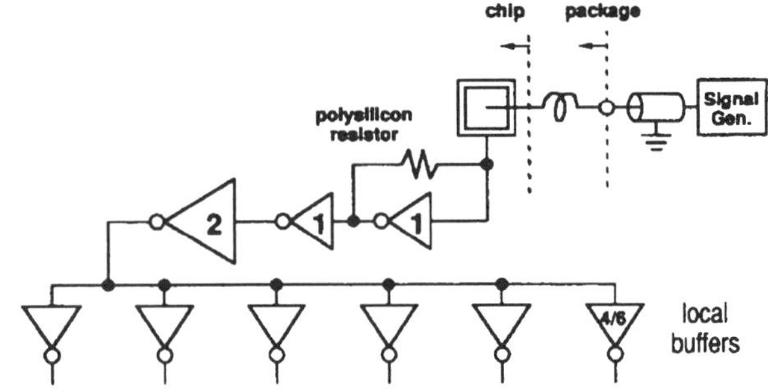

Fig. 10. Clock buffer with on-chip termination.

generates the frequency control word). The processor pulls down the $\overline{STRB}$ input of the NCO chip when a frequency control word is available. Then, the control block loads the frequency word into the first-stage latches of the double buffer. Meanwhile, the control block pulls up the $\overline{READY}$ output to notify the processor that the data have been fetched and the NCO chip is busy in updating the second-stage latches of the double buffer. During this interval, the processor can pull up the signal $\overline{STRB}$. After all second-stage latches have been updated sequentially, the control block pulls down $\overline{READY}$ to inform the processor that the NCO chip is ready to accept a new frequency control word.

In the self-timed mode, the $\overline{READY}$ output can be directly fed back to the $\overline{STRB}$ input, or $\overline{STRB}$ can be simply wired to Gnd. In this mode, the NCO chip will fetch its first frequency control word after the rising edge of the $\overline{RESET}$ input. Then, the chip will automatically fetch new frequency control words every 55 clock cycles. Data will be loaded correctly as long as they are stable around the falling edge of the $\overline{READY}$ output.

In Fig. 9, the signals $Ltch, rsb$, and iso are the same as those in the double-buffer that were described earlier. Signal upd is sent to the ldi input of the second-stage latch at the LSB position, and $eoupd$ is received from the ldo output of the second-stage latch at the MSB position (see Fig. 5).

D. On-Chip Termination for Clock Input

It has been shown that on-chip terminations can reduce the input signal waveform distortion [6]. To assure the high-speed performance of the NCO chip, its clock buffer was designed so that a polysilicon resistor ($\sim 250\,\Omega$) is added between the I/O nodes of the first inverter in the buffer. This strategy makes the effective input impedance close to 50 Ω in order to match the clock input transmission line. The ratios among the buffer stages (see Fig. 10) are chosen to minimize the buffer delay while maintaining a nearly 5-V on-chip swing at the maximum clock rate of the NCO chip. This on-chip termination technique greatly reduces clock signal distortion and the power/ground line disturbances. Furthermore, if a dc-decoupling capacitor is inserted in front of the clock input pin, then the feedback resistor automatically

biases the center level of the clock input at the buffer's logic threshold (~ 2.3 V). Hence, low-swing signals can be accepted and an off-chip biasing network is not needed.

Fig. 11(a) shows the SPICE simulation on the NCO clock buffer with "off-chip" termination, which is accomplished by removing the on-chip polysilicon resistor (see Fig. 10) and adding a 50-Ω resistor between the cable termination and Gnd. Since the cable termination is not always physically close to the IC clock input on the circuit board, noise often exists between the on-chip ground plane and the ground side of the off-chip resistor. In the SPICE simulation, noise is provided to the on-chip V_{dd} and Gnd nodes by using two sine-wave sources with a 150-MHz frequency and a 0.2-V amplitude, as shown by the topmost and the third curves in Fig. 11(a). The 700-MHz clock source has an amplitude of 2 V and is in series with a 50-Ω resistor to mimic the cable impedance. The clock input waveform seen by the on-chip buffer is shown by the second curve in Fig. 11(a), and the buffer output with respect to "on-chip" Gnd is shown by the bottom curve. It can be seen that the duty cycle of the buffer is very sensitive to on-chip V_{dd}/Gnd noise, which in turn degrades the NCO speed performance significantly.

On the other hand, the duty cycle of the clock buffer output with "on-chip" termination is much more balanced and stable, as is shown by the bottom curve in Fig. 11(b). The reason for this improvement is that the on-chip V_{dd}/Gnd noise shifts the input logic threshold of the clock buffer away from the dc center of the clock input when an off-chip termination is used, while the on-chip termination helps the clock input partially track the fluctuating logic threshold of the clock buffer since the equivalent 50-Ω resistor is "terminated" at that threshold, thus the buffer output duty cycle is kept close to 50%.

E. On-Chip Power-Rail Decoupling Capacitors

Due to the compact layout of the NCO circuitry, the chip size is determined by the number of bonding pads and the minimum required pad pitch, which results in unused silicon area between the core circuit and the pad frame. In these regions, shunt capacitors containing thin oxide, poly-to-metal1 thick oxide, and metal1-to-metal2 insulator were placed between the power and ground lines. As shown in Fig. 12, the three types of capacitors are stacked and connected such that the effective capacitance is the parallel combination of the three. The total on-chip shunt capacitance is approximately 500

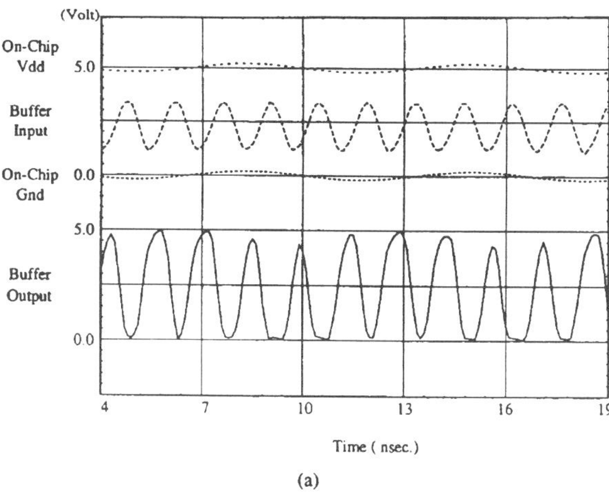

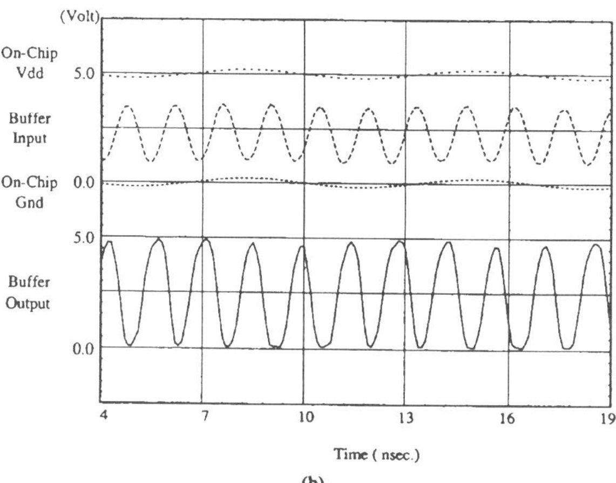

Fig. 11.   (a) Clock buffer waveforms with off-chip termination. (b) Clock buffer waveforms with on-chip termination.

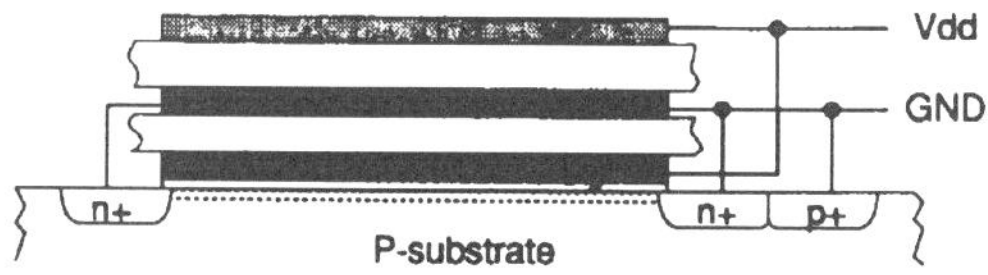

Fig. 12.   On-chip shunt capacitor.

pF. This technique helps to avoid the speed degradation caused by the power and ground line bounce due to the package-lead inductance. SPICE simulations projected a reduction in the internal V_{dd} and Gnd bounce from 1 V_{p-p} down to about 500 mV$_{p-p}$ as a result of the on-chip decoupling capacitors.

IV. Fabrication and Test Results

The NCO chip has been fabricated through MOSIS in a 1.2-μm single-polysilicon double-metal n-well CMOS technology. Fig. 13 shows the chip photomicrograph. The total transistor count is 3900. The 40-pad chip size is $1.7 \times 1.7 \, \text{mm}^2$, and the core circuit occupies a silicon area of only $0.9 \times 0.9 \, \text{mm}^2$. A large amount of diffusion contact holes were distributed evenly throughout the shunt–capacitor area, thus the power-stabilizing

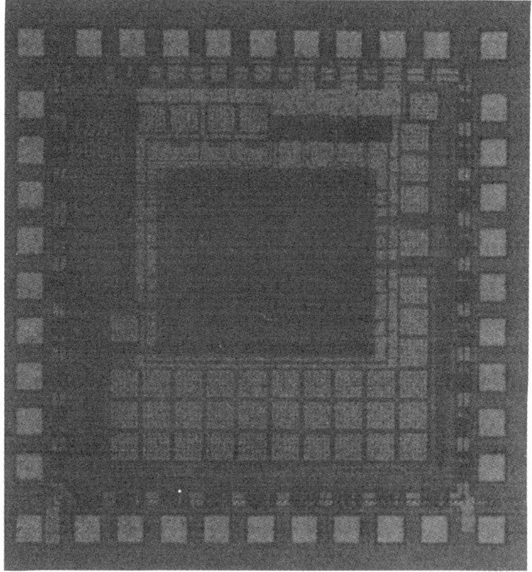

Fig. 13.   NCO chip photomicrograph.

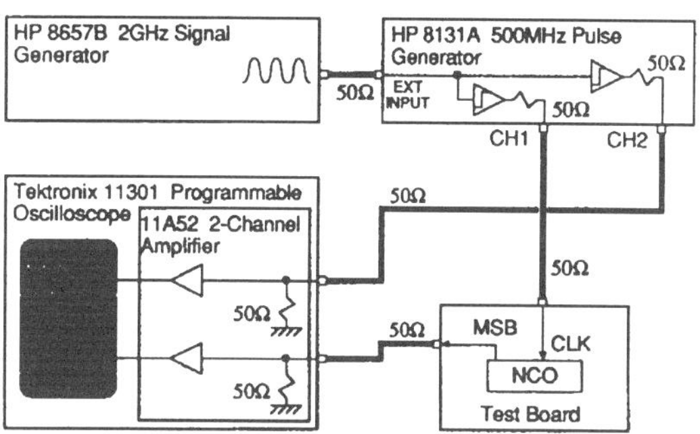

Fig. 14.   Complete setup for testing NCO.

effect of the capacitors is not degraded by the large n-channel resistance. Fifty chips bonded in standard 40-pin ceramic dual-in-line packages were received from MOSIS and tested.

The test setup of the NCO chip is depicted in Fig. 14. An HP8657B 2-GHz signal generator provides a sine-wave output with a maximum peak-to-peak swing of 1.5 V and a fixed center level of 0 V. The NCO chip requires a clock input with a larger voltage swing and a positive dc offset. Thus, an HP8131A 500-MHz pulse generator was adopted in the setup. Although its internal oscillator has a maximum frequency of 500 MHz, its two-channel amplifier has a bandwidth of 1 GHz, and its output amplitude and dc offset are variable between 0 and 5 V. Hence, the HP8131A was set to the external trigger mode to amplify and level shift the sine-wave output from the HP8657B. The NCO chip reached its maximum speed when the peak-to-peak swing and center level of the HP8131A outputs were set at about 2 and 2.3 V, respectively. The MSB output of the NCO chip and the clock waveform were displayed on a Tektronix 11301 oscilloscope to verify device performance.

The average measured maximum clock rate is 700 MHz when the chip is powered with a single 5-V supply at room

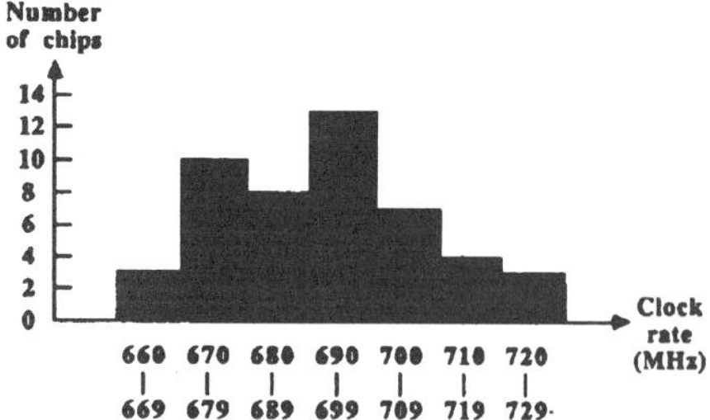

Fig. 15. Histogram of NCO speed performance.

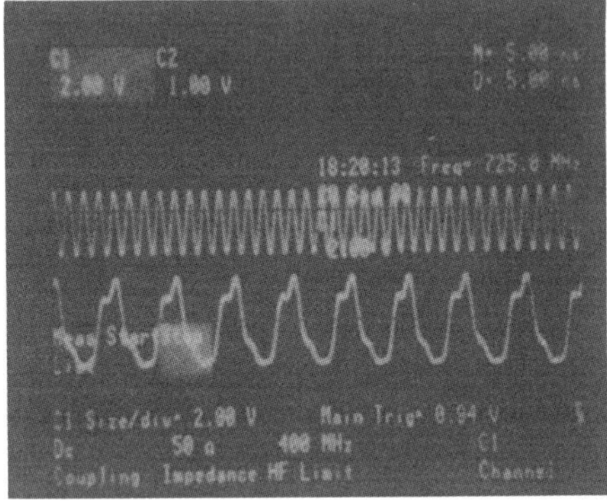

(a)

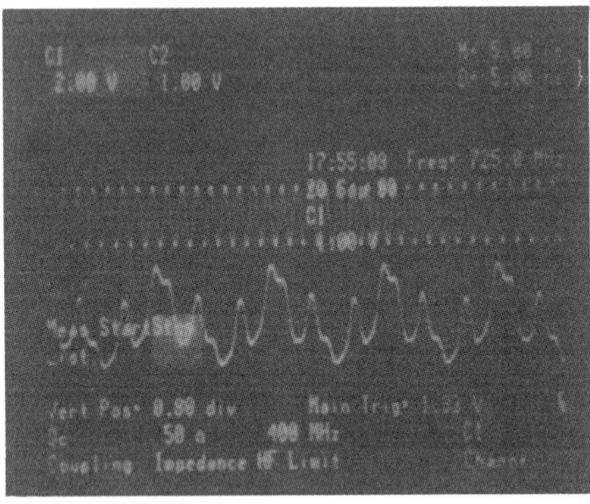

(b)

Fig. 16. Examples of NCO MSB outputs (a) with frequency control word $\langle 0100\cdots 0\rangle(f_o = (1/4)f_{clk})$, and (b) with frequency control word $\langle 0110\cdots 0\rangle(f_o = (3/8)f_{clk})$.

temperature. This results in a frequency tuning resolution of $f_{clk}/(2^{24}) = 41.7$ Hz, and a tuning range from dc up to 350 MHz with perfect linearity. The peak output phase jitter is $1/f_{clk} = 1.4$ ns. The yield of the 50 samples of NCO chips was 96%. A histogram of the speed performance of the 48 working chips is depicted in Fig. 15. The speed ranges from 660 to 725 MHz, and the average power consumption at 700 MHz is 850 mW with the MSB output loaded by the 50-Ω input resistance of the oscilloscope. Examples of the clock input and MSB output waveforms are shown in Fig. 16.

Fig. 17 shows a typical NCO chip maximum speed performance and the corresponding power consumption versus V_{dd} variations. When the supply voltage is reduced to 3 V, the chip can still operate at a clock rate above 400 MHz.

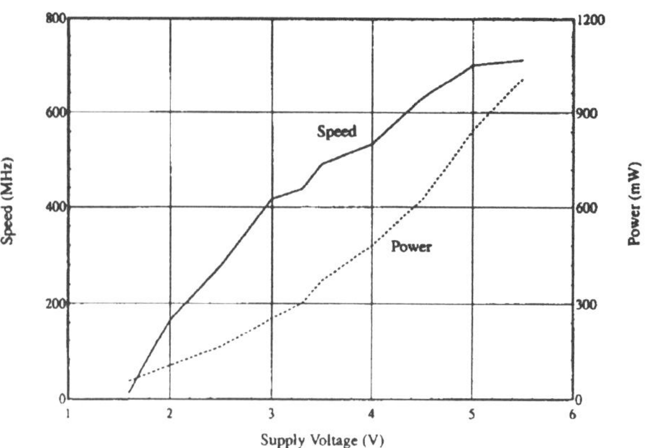

Fig. 17. Maximum speed and power consumption versus supply voltage.

TABLE I
MAJOR FEATURES OF NCO CHIP

IC process	1.2-um N-well CMOS
Max. clock rate	700 MHz
Input freq. wordlength	24 bits
Freq. tuning range	DC to 350 MHz
Peak output phase jitter	1.4 ns
Freq. switching latency	55 clock cycles
Chip area	1.7 x 1.7 mm^2
Core area	0.9 x 0.9 mm^2
Transistor count	3,900
Power supply	+ 5V
Power consumption	850 mW @ 700 MHz
Package	40-pin ceramic DIP

The power consumption includes the portion consumed by the clock buffer with negative feedback and by the output buffer with a 50-Ω load. The chip keeps operating until the supply voltage is reduced to the sum of the threshold voltages of the N- and P-devices ($\sim$1.6 V). This demonstrates the robustness of the TSPC circuit design technique. The major features of the NCO chip are listed in Table I.

Based on the above measurement results, this NCO chip is over five times faster than any CMOS phase accumulator reported to date. In [7] a lookahead technique, rather than pipelining, was incorporated into a phase accumulator to reduce the frequency-tuning latency, which resulted in a 130-MHz phase accumulator in 1.0-μm CMOS. Furthermore, the NCO chip reported in our paper is more than twice as fast as commercial ECL bipolar phase accumulators (300 MHz) [8] and nearly as fast as comparable GaAs devices (1 GHz) [9]. The power/speed ratio of this CMOS NCO chip is 1.2 mW/MHz, which is much more efficient than the ECL and GaAs devices if they were scaled to the same word length [8], [9]. The comparisons among NCO's with different technologies are summarized in Table II.

If a high-speed pipelined/parallel look-up table is available, the upper part of the phase accumulator output can be used as an address to read out sine-wave samples from the memory, and a direct digital synthesizer (DDS) can be constructed. The fastest commercial monolithic CMOS DDS to date achieves a speed in excess of 100 MHz [10], and research prototypes have demonstrated speeds up to 150 MHz [11].

[11] H. T. Nicholas, III, and H. Samueli, "A 150-MHz direct digital frequency synthesizer in 1.25-μm CMOS with -90-dBc spurious performance," *IEEE J. Solid-State Circuits*, vol. 26, pp. 1959–1969, Dec. 1991.

TABLE II

COMPARISONS AMONG DIFFERENT PHASE ACCUMULATORS

Chip / Feature	UCLA CMOS NCO (24-bit)	Analog Devices AD9950 ECL NCO (32-bit)	GigaBit Logic 10G102 GaAs NCO (32-bit)
Maximum Clock Rate	700 MHz	300 MHz	1 GHz
Power Supplies	+5 V	+5V, -5.2V	+5V, -3.4V, -5.2V
Power Efficiency at Max F_{CLK}	1.2 mW / MHz	~ 3.5 mW / MHz (scaled to 24 bits)	~ 3 mW / MHz (scaled to 24 bits)

V. CONCLUSIONS

To accomplish timing recovery/synthesis in high-speed communication systems, a numerically controlled oscillator (NCO) chip, using a circuit design technique called true single-phase clock (TSPC) pipelined CMOS, has been developed in standard 1.2-μm CMOS technology. The typical maximum input clock rate is up to 700 MHz, which is the fastest CMOS NCO chip ever reported. The digital NCO technique results in a frequency-tuning resolution and an open-loop control linearity that cannot be achieved by analog approaches. Owing to the fine-grain pipelining scheme, subhertz tuning resolution can be achieved by simply extending the accumulator word length without any speed degradation. Through the use of advanced circuit design techniques in conjunction with submicrometer CMOS technology, an NCO device with clock rates in excess of 1 GHz is indeed possible.

ACKNOWLEDGMENT

The authors wish to thank the Hewlett-Packard and Tektronix corporations for their generous donations of high-speed test equipment.

REFERENCES

[1] J. Tierney, C. M. Rader, and B. Gold, "A digital frequency synthesizer," *IEEE Trans. Audio Electroacoust.*, vol. 19, pp. 48–57, 1971.

[2] U. L. Rohde, *Digital PLL Frequency Synthesizers Theory and Design*, 2nd ed. New York: Wiley, 1979.

[3] H. Hikawa and S. Mori, "A digital frequency synthesizer with a phase accumulator," in *Proc. IEEE Int. Symp. Circuits Syst.*, 1988, pp. 373–376.

[4] J. Yuan and C. Svensson, "High-speed CMOS circuit technique," *IEEE J. Solid-State Circuits*, vol. 24, pp. 62–70, Feb. 1989.

[5] F. Lu, H. Samueli, J. Yuan, and C. Svensson, "A 700-MHz 24-bit pipelined accumulator in 1.2-μm CMOS for application as a numerically controlled oscillator," in *Proc. IEEE CICC*, May 1991, pp. 25.3.1–25.3.4.

[6] J. Hauenschild *et al.*, "Influence of transmission-line interconnections between gigabit-per-second IC's on time jitter and instabilities," *IEEE J. Solid-State Circuits*, vol. 25, pp. 763–766, June 1990.

[7] M. Thompson, "Low-latency, high-speed numerically controlled oscillator using progression-of-states technique," *IEEE J. Solid-State Circuits*, vol. 27, pp. 113–117, Jan. 1992.

[8] "300 MHz phase accumulator for direct digital synthesis," AD9950 Data Sheet, Analog Devices, Inc., Greensboro, NC, 1991.

[9] "32-bit DDS phase accumulator, 1.0 GHz clock rate," 10G102 Data Sheet, Gigabit Logic, Inc., Newbury Park, CA, May 1988.

[10] "100 MHz direct digital synthesizer," AD9955 Data Sheet, Analog Devices, Inc., Mar. 1992.

327

An Integrated GaAs 1.25 GHz Clock Frequency FM-CW Direct Digital Synthesizer

Nathalie CAGLIO*, Jean-Luc DEGOUY**, Didier MEIGNANT*
Patrick ROUSSEAU*, Bruno LEROUX*

*LEP / PHILIPS MICROWAVE LIMEIL,
BP 15 - 94453 Limeil-Brévannes Cedex, France

**THOMSON CNI
BP 402 - 92103 Boulogne-Billancourt Cedex, France

Abstract

An integrated GaAs FM-CW Direct Digital Synthesizer (DDS) has been developed using Philips Microwave Limeil (PML) standard ER07AD technology.
The DDS is composed of two successive functional blocks:

- A double phase accumulator which produces the right synchronizing sequences
- A Digital to Analog Sine Converter (DASC) which generates the linear chirp waveform

The double phase accumulator has been implemented with 5 chips while the DASC is monolithic.
The maximum measured clock frequency on the phase accumulator is 1.25 GHz and the power consumption is 320 mW. For the DASC, the maximum measured clock frequency is 1.5 GHz and the associated consumption is 600 mW.

Introduction

Linear FM (chirps) waveform generators are often used in high compression ratio radars or radio-altimeters. In the last case, the receiver matched filter can be very easily implemented with an homodyne mixer.
The mixer generates a CW signal, the duration of which is equal to that of the ramp (ΔT).
Consequently, its spectrum width is $1/\Delta T$. The compression ratio (spectrum width of the input ΔF over spectrum width of the output) then writes : $\Delta F * \Delta T = R$.
In radio-altimeters, ultra high compression ratio ($R \simeq 10^8$) allows on the one hand to keep the peak output power to reasonable value ($\simeq 1$ W), and on the other hand to achieve excellent distance accuracy ($\simeq 1$ m for $\Delta F = 150$ MHz).

General architecture

The linear frequency ramp is generally implemented with a VCO, but this approach requires complex corrections to reach modest linearity factors ($\simeq 1$ %) resulting in poor compression ratio (typically 10^3 to 10^5) due to increased spectrum width of the beat signal.
The second weakness of VCO's is the phase noise which increases also the width of the beat signal spectrum and therefore introduces degradation of the signal to noise ratio.

The FM-CW Direct Digital Synthesizer (DDS) described in this work improves by several orders of magnitude these two limitations, compared with a conventional approach using VCO's.
The use of digital synthesis results per definition in a perfect linearity [1].
The phase noise of a DDS is caused by the reference clock phase noise and by the quantization noise of the DA converter.
The first one can be easily reduced by classical PLL methods and the second one is a white noise spread over a very broadband (DC to clock frequency).
A perfectly quantized signal has quantization noise close to the clock frequency and consequently can be easily filtered.
Conversely, inaccuracy in phase and amplitude generates white noise in the band.
In our architecture, the ratio "N" between the clock frequency (typically 1 GHz depending on the process) and the maximum frequency of the ramp ($\simeq 100$ MHz for radio-altimeters) determines the maximum phase error around $2\,\Pi/10$. Thus, a good trade-off is to use a DAC having a number of states close to N ($\simeq 10$).
The traditional DDS architecture [1, 2] uses 10 to 14 MSBs coming from a 30- bit phase accumulator to drive a ROM (2 to 8 Kbits). The output of the ROM is an 8 to 12 bit word representing an amplitude for the following DAC.

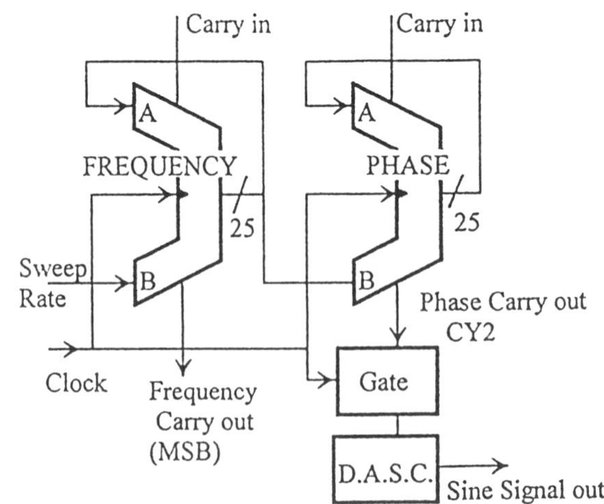

Figure 1 : Block diagram of DCS

Reprinted from *IEEE GaAs Symposium*, pp. 167-170, October 1993.

In our case, since we are using a linear FM-CW ramp as a correlation system, the phase noise requirements are less stringent than in a CW DDS system. Consequently, in order to optimize the trade-off between computation speed and power consumption, a full pipe-line architecture is used for the phase accumulator and an incremental DAC for the sine generation (**Figure** 1).

The incremental DAC needs only one clock coming from the carry of the phase accumulator, thus avoiding complex time alignment of the 10 to 14 MSBs going to ROM. Furthermore, the discrete levels of this DAC are sine coded, thus avoiding the use of a large ROM.

Phase accumulator architecture

The digital synthesis of the chirp waveform is based on the fact that the quadratic time base :

$$\phi(t) = \frac{1}{2} \cdot \frac{\Delta F}{\Delta T} . t^2$$

can be generated numerically at high speed using adders only.

The phase generator needs 25 pipelined stages (**Figure 2**), each stage consists of 2 full adders realizing a double integration of the input sweep rate $\Delta F/\Delta T$. The carry signals of each stage are stored in registers, thus enabling the clock period to be equal to the computation time of a full adder (plus the acquisition time of a register).

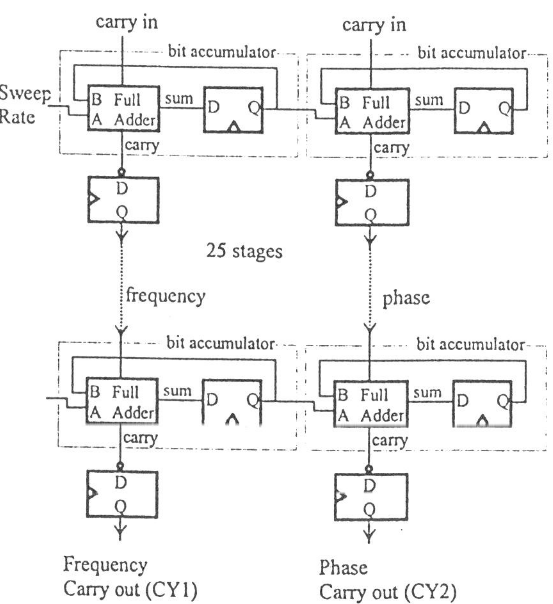

Figure 2 : Accumulator structure

Circuit fabrication

The phase accumulator and digital to analog sine converter are manufactrured using Philips Microwave Limeil (PML) production proven 0.7 μm gate ER07AD MESFET process [3].
The process uses 3″ LEC wafer, gate recessing, Si$^+$ implantation, boron isolation, standard photolithography, dry etching and enhanced lift-off techniques. The process makes available 3 levels of gold interconnect and MIM capacitors. Two pinch-off voltages are available in the process, + 0.2 V for the enhancement FET and - 2 V for the depletion FET. This second threshold voltage is mostly used for power applications and in our case only enhancement FET have been used with GaAs resistors as loads for BDCFL (Buffered DCFL) logic gates. The typical transconductance and F_T are 280 mA/V/mm and 20 GHz (measured at Vgs = 0.7 V).

Circuit design

The DCS module is implemented as five 5-bit accumulator cells and one DASC.
Thanks to the pipeline structure and the choice of the incremental DASC, the number of connections between chips is minimized.
The logic gates used in the design of the phase accumulator are high speed, medium power, high performance BDCFL standard cells from PML library [4].
The advantage of this logic is to have a low fan-out sensitivity (typically less than 20 ps per fan-out).

Phase accumulator design

Furthermore, the BDCFL logic allows the full adder to have only one logic layer between the inputs and the carry/sum signals. The maximum speed of a 1-bit accumulator is 1.2 GHz with an associated power consumption of 14 mW.
The pipeline registers use standard 6-NOR gate resetable D flip-flop implemented also with BDCFL. The maximum speed of this flip-flop is 1.2 GHz and the power consumption 13 mW.

Phase accumulator results

Fully fonctional phase accumulators were fabricated and characterized. The **Figure 3** shows the chip photograph of a single 5-bit phase accumulator cell.

Figure 3 : Photograph of the 5 bit
phase accumulator chip

It contains 800 FETs and the chip size is 1.9 mm x 1 mm. In order to test the 5-bit accumulator chip, we have added low output impedance buffers at the carry outputs (CY1 and CY2).

The CY1 carry has been used to feed the clear input, thus ensuring a synchronous clear for easy oscilloscope observations.

Correct operation of the chip has been obtained for the different tested input patterns.

The maximum operating frequency of the phase accumulator is 1.25 GHz and the associated measured power consumption is 320 mW.

Digital to analog sine converter design

The **Figure 4** shows the DASC structure : it is composed of a gate input circuit used to strobe the carry out signal (NRZ to RZ transform) followed by a Johnson counter driving directly the sine coded currents into a resistor.

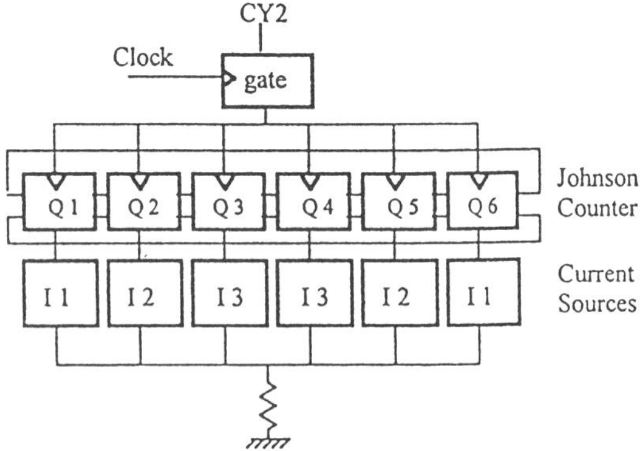

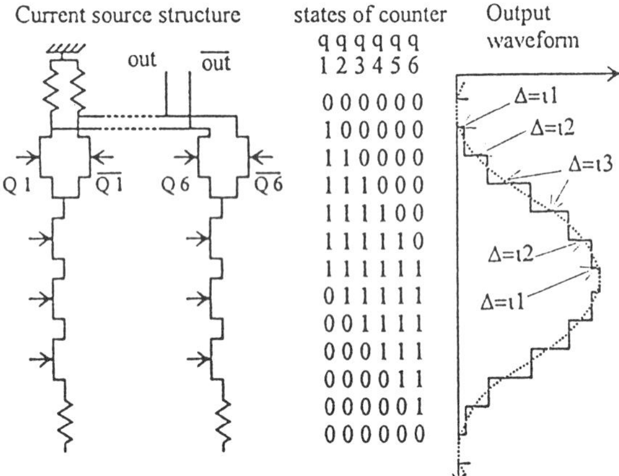

Figure 4 : Digital to analog sine converter structure

For the targetted application, the maximum frequency of the synthesizer is 100 MHz. A clock frequency of 1.2 GHz therefore results in an output frequency having 12 phase states during one period.

These 12 phases are generated by a 6-stage Johnson counter. The symmetry of the sine function and the Johnson counter properties allow to use directly each output of the counter to build a complete sinewave having 12 phase states with only 6 calibrated current sources.

The current sources are built using a 3 stages cascode structure in order to get a better stabilization. The gate width of this current source is sine coded. Each of the 6 current sources is followed by a differential pair used as switches driven by the outputs of the Johnson counter, then the currents are summed into a common 50 Ω resistor connected to the ground.

DASC results

The **Figure 5** shows the DASC chip.

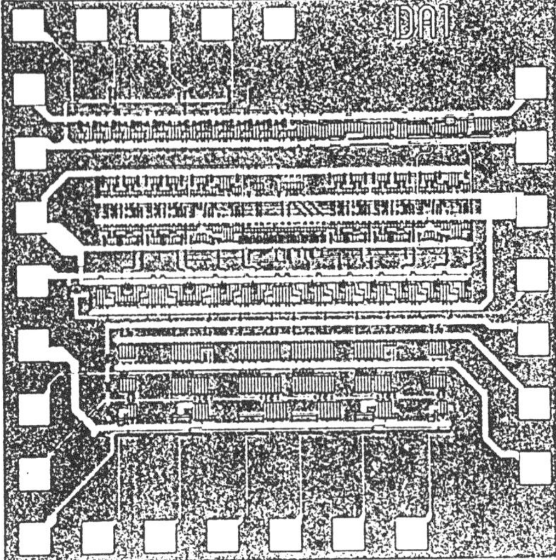

Figure 5 : Photograph of the DASC chip

The maximum operating frequency is 1.5 GHz and the associated power consumption is 600 mW. An example of synthesized sinewave with a clock at 1.2 GHz is shown on **Figure 6**.

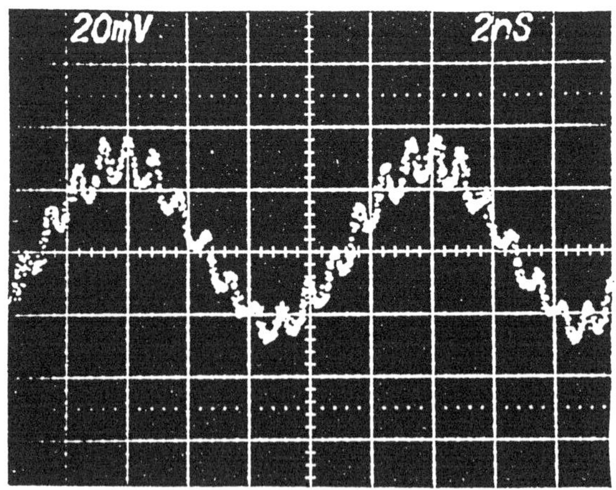

Figure 6 : Output of the DASC at 1.2 GHz clock frequency

The corresponding spectrum is shown on **Figure 7**. The fundamental frequency is then 100 MHz and the harmonics are typically in the range of -40 dBc indicating an effective number of bits close to 7.

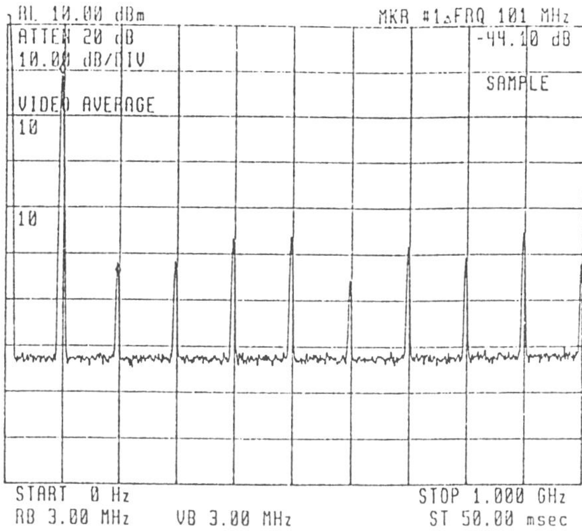

Figure 7 : Spectrum of the synthesized sinewave

Chirp synthesizer results

We have associated 5 5-bit accumulator chips with a DASC on a specially designed alumina substrate (**Figure 8**) and this DCS system has been loaded with a positive sweep rate. **Figure 9** shows the chirp output waveform from the DCS system, with a clock of 810 MHz.

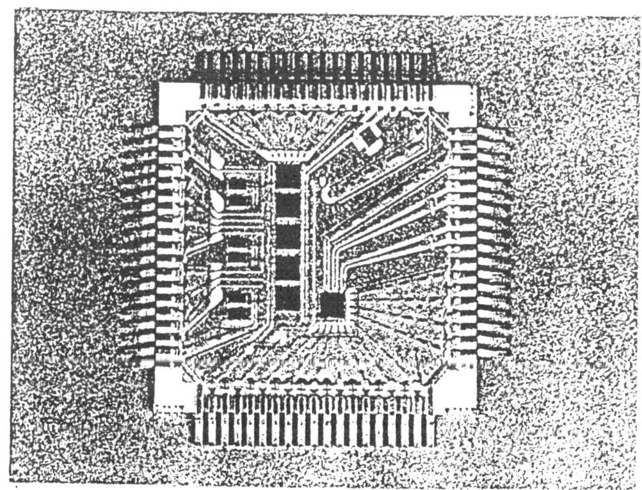

Figure 8 : Photograph of the DCS on alumina substrate

In this case, the sweep rate was quite high (162 MHz/μs) in order to be easily observed on the oscilloscope. But it can be varied between 4 kHz/μs to 1.2 to 10^8 KHz/μs.

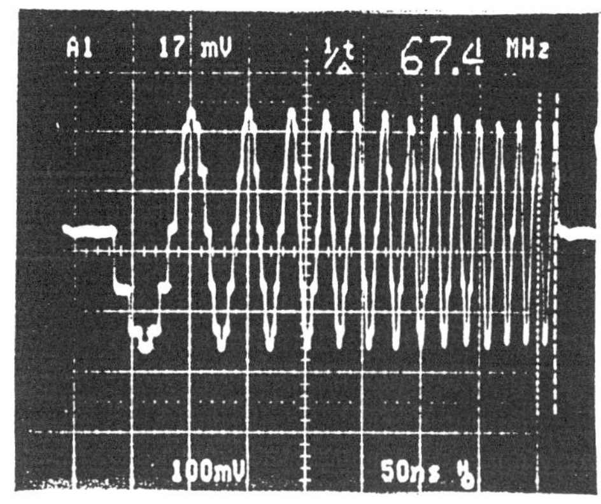

Figure 9 : Chirp synthesizer waveform

Conclusion

The chip synthesizer results shown in this paper demonstrate the feasibility of such a circuit for radio-altimeters or radars. Proper operation of the basis circuits has been demonstrated up to 1.25 GHz. The total power consumption of the 25-bit DDS system is 2.2 W and is able to generate chirp signals having a maximum frequency in the 100 MHz range.

Acknowledgements

The authors wish to thank Marc Grillaud and Jean-Pierre Damour for their tireless effort in providing alumina substrates.

References

[1]   J. Brown, "All GaAs fuels high-speed digital synthesizer"
Microwave and RF, July 1988, pp. 129-131

[2]   G. van Andrews et al., "A monolithic digital chirp synthesizer chip with I and Q channels"
GaAs IC Symposium, 1991, pp. 19-22

[3]   T. Ducourant et al., "A 5 bit, 2.2 Gs/s monolithic A/D converter with gigahertz bandwidth, and 6 bit A/D converter system"
GaAs IC Symposium 1989, pp. 337-340

[4]   D. Meignant et al., "A 0.1 - 4.5 GHz, 20 mW GaAs prescaler operating at 125°C"
GaAs IC Symposium 1986, pp. 129-132

The Composite DDS–A New Direct Digital Synthesizer Architecture

LAWRENCE J. KUSHNER

LINCOLN LABORATORY

MASSACHUSETTS INSTITUTE OF TECHNOLOGY

LEXINGTON, MA 02173-9108

Abstract–A new, low-power, high-speed, Direct Digital Synthesizer (DDS) architecture is presented, called the Composite DDS (CDDS). A low-speed, high-resolution DDS is combined with a high-speed, low-resolution phase accumulator and phase shifter via the serrodyne modulation technique. The low-speed circuitry provides the fine tuning while the high-speed circuitry provides coarse tuning. By minimizing the amount of circuitry required to clock at high-speeds, dc power is conserved.

The results from numeric simulations and a low-frequency proof-of-concept breadboard are presented. Progress on our 800 MHz CDDS development effort is described, followed by proposed enhancements to the CDDS architecture which promise improved performance.

Introduction

A LTHOUGH Direct Digital Synthesizers (DDSs) were invented decades ago [1], they did not come to play a dominant role in wideband frequency generation until recent years. Initially, DDSs were limited to producing narrow bands of closely spaced frequencies, due to the limitations of digital logic and Digital-to-Analog Converter (DAC) technologies. Larger bandwidths were generated by other means, such as phaselock loops, direct analog synthesis, or frequency multiplication, using the DDS to generate the fine resolution. For applications requiring fast tuning, these approaches resulted in expensive, complex, power-hungry, frequency generators.

Due to the rapid advance of digital technology, there is now an alternative approach. Direct Digital Synthesizers clocking at 1 GHz have recently become available [2]. Most, if not all, of the needed bandwidth can be generated by the DDS, resulting in reduced system complexity, size, and weight. While it is likely that DDS technology will continue to improve as digital technology advances, Figure 1 illustrates a key tradeoff that will remain: the faster the clock rate, the higher the dc power consumption.

While dc power may be unimportant in some systems, it is of utmost concern in many applications, such as portable, battery-powered communications equipment, and satellites. Additionally, even systems with power to burn may still have difficulty removing the heat generated by a high-power DDS.

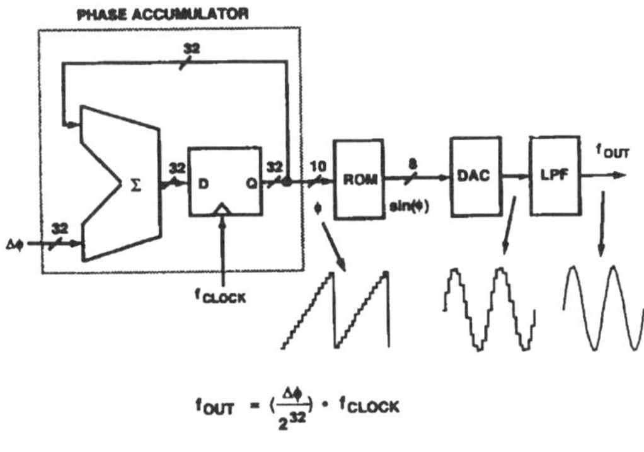

$$f_{OUT} = \left(\frac{\Delta\phi}{2^{32}}\right) \cdot f_{CLOCK}$$

PRACTICALLY, $\quad f_{OUT} \leq 0.4 \cdot f_{CLOCK}$

Figure 2 - Direct Digital Synthesizer Architecture.
(Typical data-path widths shown)

The high-level architecture of a conventional DDS is shown in Figure 2. Since the frequency resolution of a DDS is equal to $f_{clock}/(2^N)$, where N is the number of bits in the phase accumulator, the greater the clock frequency, the more bits are needed to maintain a given resolution. 32-bit wide phase accumulators are commonly used in today's commercial DDSs. For a 1.6 GHz clock, 32 bits yields 0.37 Hz resolution.

The problem with this conventional approach is that on every clock cycle, the phase accumulator has to accumulate, the ROM has to produce a new value, and the DAC must convert another sample from digital to analog. All of this high-speed activity results in high dc power consumption.

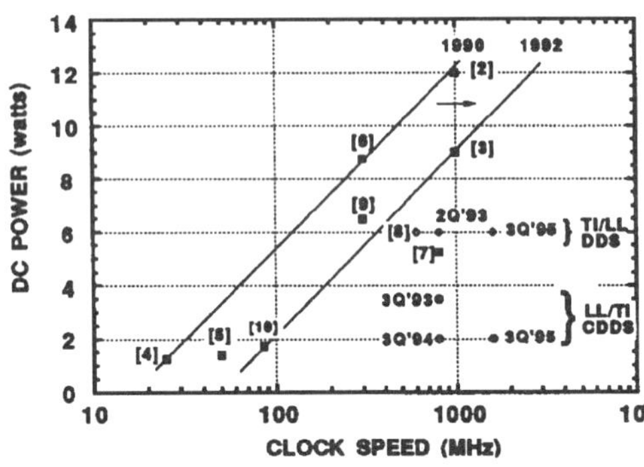

Figure 1 - DDS Power Consumption vs. Clock Speed.
Data points based on references [4] - [10], and
Lincoln / Texas Instruments projections (dates).

This work was sponsored by the US Army under Air Force Contract F19628-90-0002

<u>Composite Direct Digital Synthesizer (CDDS)</u>

In contrast, the Composite DDS [11] uses a new architecture (Figure 3), allowing it to achieve high speed and low power consumption simultaneously. Instead of clocking the entire circuit at the maximum clock rate, the CDDS partitions the circuitry into a small, high-speed circuit that generates coarse frequency steps, and a larger, low-speed circuit to generate the fine frequency resolution. The fine-resolution frequency (f_1) is upconverted by the coarse-resolution frequency (f_2) via the serrodyne modulation technique (Figure 4).

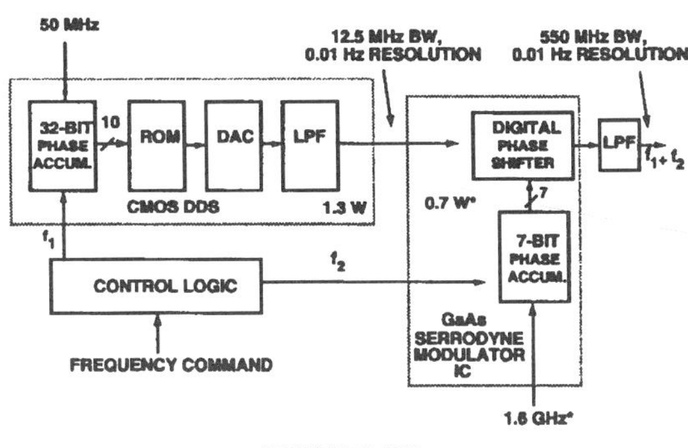

Figure 3 - Composite DDS Architecture (nominal clock frequencies and data-path widths shown for clarity).

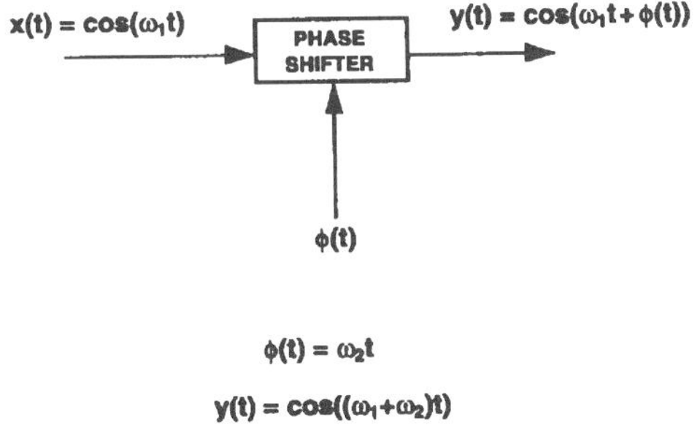

Figure 4 - Serrodyne Modulation Principle [12]-[14].

Since its inception in the late 1950's, the serrodyne principle has been used in radar applications to chirp fixed, microwave carriers at a relatively low rate [12]-[18]. In modern implementations of these serrodyne systems, a fixed microwave carrier is passed through microwave phase-shifters driven from a digital phase-accumulator, resulting in small frequency shifts and a narrow-band (< 1 octave) output.

In contrast, the CDDS uses a coarsely tuned, high-speed phase accumulator driving a fast-switching, low-frequency phase-shifter to translate a finely-tuned, low-frequency "carrier" by large amounts. The CDDS uses the serrodyne technique for broadband bandwidth expansion, resulting in a decade or more of output bandwidth. For the specific clock frequencies and bandwidths shown in Figure 3, the low-frequency DDS produces 12.5 MHz of bandwidth with 0.01 Hz of resolution, which is then upconverted to one of 44 different "bins" by the Serrodyne Modulator, resulting in 550 MHz of output bandwidth.

For GigaHertz operation with today's technology, the CDDS's high-speed circuitry could be implemented with GaAs MESFETs, GaAs HBTs, or high-speed Si bipolar transistors, while the low-speed circuit might be built in CMOS. Alternatively, a lower-frequency CDDS might be created entirely in CMOS, with a high frequency clock of ≈80 MHz and low-frequency clock of ≈2.5 MHz, again resulting in considerable power savings over a conventional DDS in which all of the logic clocks at the maximum clock frequency (80 MHz).

As digital logic technologies advance, so will these clock rates, but the underlying advantage of the CDDS will remain. By minimizing the amount of circuitry clocking at high speeds, dc power is conserved. With the addition of control circuitry (Figure 3), the CDDS will appear to the outside world as a standard DDS . By choosing the two CDDS clock frequencies to be related by a factor of 2, the amount of control logic needed can be minimized.

While phase-shifters could be built at these low frequencies using broadband all-pass networks [19], [20], the low-frequency poles and zeros associated with these networks would limit the phase-shifters' switching speeds. Instead, the Composite DDS (Figure 5) employs a vector modulator phase-shifter [21], made up of simple switches and attenuators (or amplifiers). These components are inherently broadband, and switching speed is limited only by energy storage in the switches' parasitic capacitances.

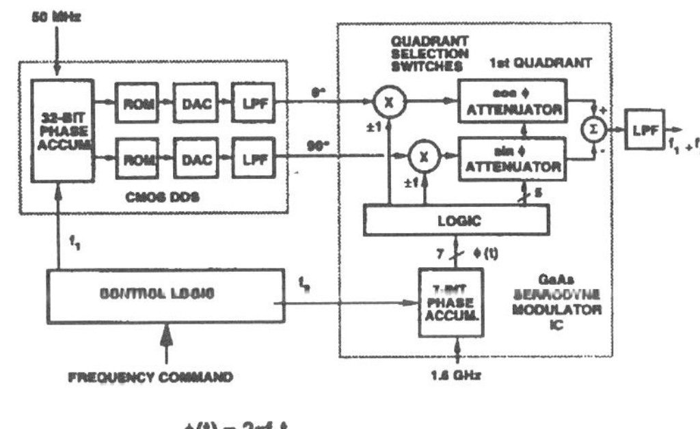

$$\phi(t) = 2\pi f_2 t$$

$$\cos(2\pi f_1 t + \phi(t)) = \cos(2\pi f_1 t) \cdot \cos(\phi(t)) - \sin(2\pi f_1 t) \cdot \sin(\phi(t))$$

Figure 5 - Composite DDS Architecture with 4-Quadrant Vector-Modulator Digital Phase-Shifter.

Additionally, this approach is much more amenable to integrated-circuit implementation than an allpass network-based design. The primary drawback to the vector modulator approach is that there are now two channels that must closely track each other for optimum performance. As discussed below, channel mismatch is the source of one of the close-in spurious signals in the Composite DDS's output spectrum.

Simulations and Frequency-scaled Proof-of-Concept Hardware

In order to study the spurious performance of a "real world" Composite DDS, a frequency-scaled, proof-of-concept CDDS was constructed (Figure 6). In this

Figure 6 - Proof-Of-Concept Breadboard.

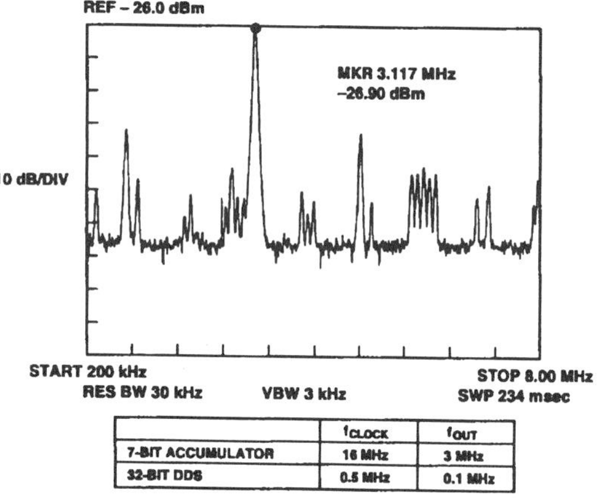

	f_{CLOCK}	f_{OUT}
7-BIT ACCUMULATOR	16 MHz	3 MHz
32-BIT DDS	0.5 MHz	0.1 MHz

Figure 7 - Proof-Of-Concept Breadboard Output Spectrum.

demonstration, the high-speed accumulator was clocked at 16 MHz while the low-speed circuitry was clocked at 500 kHz. Figure 7 shows the typical measured output spectrum from this proof-of-concept CDDS. Besides the desired 3.1 MHz output signal, a number of spurious components are also present. None of these would be present in a perfect CDDS implementation. Before attempting to build a higher frequency CDDS, it was essential to understand how all of these spurious frequency components were generated.

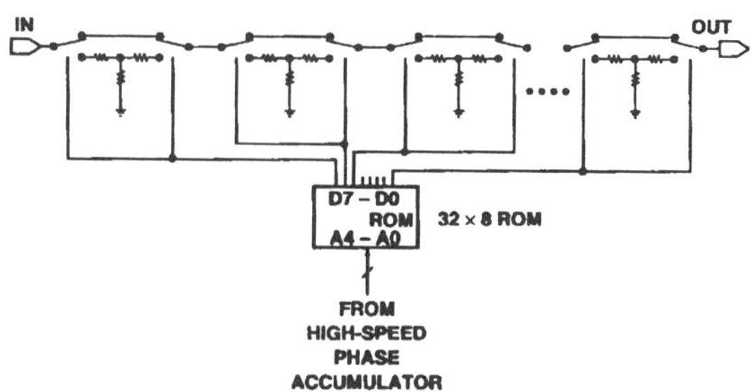

Figure 8 - Switched-Attenuator – ROM / Cascade Implementation.

. In this breadboard CDDS, the vector modulator's switched-attenuator was implemented as a cascade of eight switched-attenuator sections controlled by a ROM lookup table (Figure 8). This approach has two fundamental problems: amplitude quantization (analogous to a DDS's DAC) and timing skew. Amplitude quantization leads to amplitude errors, even with perfect attenuators. For our breadboard CDDS these quantization effects were relatively small. We used a 7-bit high-speed phase accumulator, with the bottom 5 bits driving the switched-attenuator. We chose our eight attenuator bit weights to be approximately 16 dB, 8 dB, 4 dB, ... 0.125 dB, so with ideal attenuators, our worst case amplitude error would be 0.0625 dB.

The timing problem associated with this switched-attenuator implementation is even more insidious. Due to the finite time delay required for the signal to pass through each attenuator section, the switch control-signals must be skewed in time to line up with the signal as it flows through the attenuator chain. Even if the control-signal skews could be optimized at one temperature, maintaining the desired skew over temperature would present a difficult challenge. Additionally, there is typically some dispersion through the switched-attenuator sections due to their finite bandwidth, further complicating matters.

Figure 9 identifies most of the CDDS spurious sources. Besides the two sources described above, other sources include attenuator-section VSWR interaction, switch distortion products, switch glitches, quadrant-switch LO leakage, and channel imbalance. A numeric simulation of the CDDS was written (using MATLAB [22]) to further study the spurious sources. The simulation modeled phase and amplitude errors, harmonic distortion, intermodulation distortion, leakage, and imbalances. The simulation did not include the effects of switching glitches or timing errors.

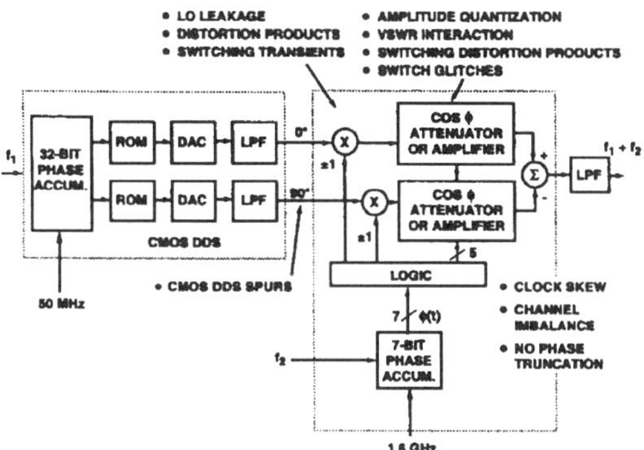

Figure 9- CDDS Spurious Sources.

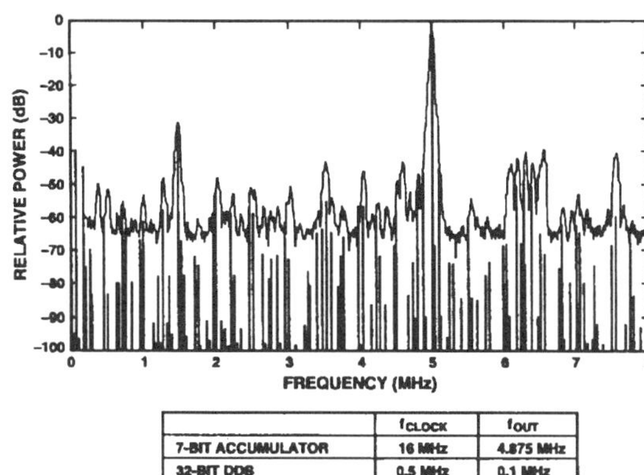

Figure 10 - Measured and Simulated Spectrum,
f_{OUT} = 4.875 MHz.

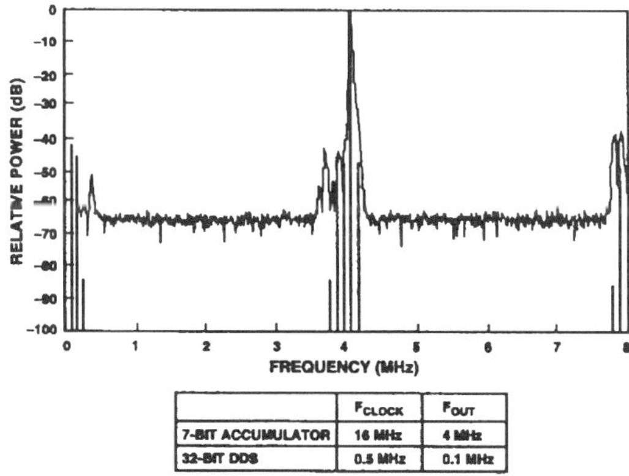

Figure 11 - Measured and Simulated Spectrum,
f_{OUT} = 4.1 MHz.

Figures 10 and 11 show simulated spectrum superimposed on measured output spectrum, for a "dirty" and "clean" frequency, respectively. Since the simulation neglected glitch transients and timing skews, it was overly optimistic in its prediction of the amplitudes of some of the spurious signals. However, it did a good job of predicting their locations, allowing most of the sources of the spurious signals to be identified. The simulation also allowed parameters to be varied in order to study the output spectrum's sensitivity to them.

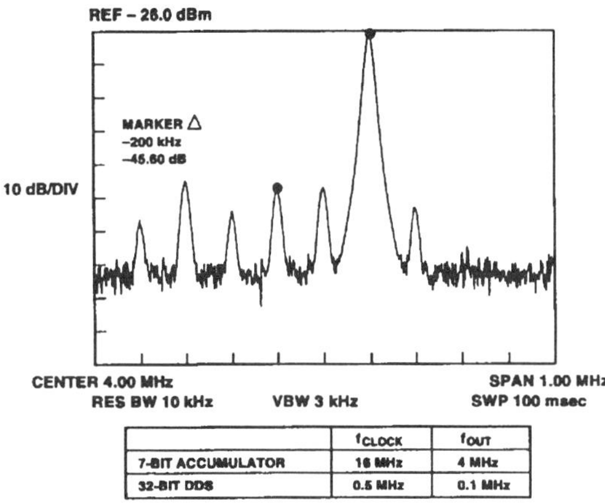

Figure 12 - Measured Close-in Spectrum, f_{OUT} = 4.1 MHz.

Most of the spurious sources associated with the quadrant switches, along with channel imbalance errors, result in close-in spurious components. Figure 12 shows a zoomed-in plot of the spectrum of Figure 11. This 4.1 MHz output is generated from a 0.1 MHz fine-tune signal from the low-speed DDS, upconverted by a 4 MHz signal from the 7-bit high-speed accumulator. Notice the 4 MHz LO leakage, the 3.9 MHz lower sideband, and the 2-1 and 3-1 mixer products. Improving the mixer symmetry reduces the absolute level of the LO leakage, while increasing the rf signal level reduces the relative amplitude of this LO leakage. This rf signal level cannot be increased arbitrarily, however, since the larger the signal level, the larger the 2-1 and 3-1 spurs.

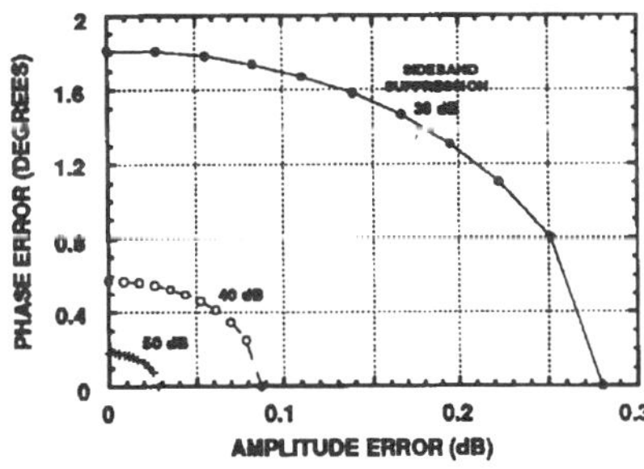

Figure 13 - Single-Sideband Mixer Phase/Amplitude Tracking Requirements.

The 3.9 MHz lower sideband level is determined totally by channel matching, just as in a single-sideband mixer. Figure 13 shows the channel tracking requirements that have to be maintained for a given level of sideband rejection. In order to achieve -40 dBc lower sideband levels, amplitude and phase tracking must be within $\pm$ 0.05 dB and $\pm$ 0.4°, respectively. While this level of matching would be difficult to achieve in discrete form, we hope to meet these goals by building the two channels on the same integrated circuit chip.

<u>800 MHz Development Plans</u>

Having learned from our low-frequency breadboard and simulations, we decided to replace the cascade switched-attenuator implementation of Figure 8 with the decoded-amplifier architecture shown in Figure 14. In this scheme, every phase state has its own fixed-gain feedback amplifier, eliminating any amplitude quantization effect. Since the signal now only passes through a single switch, the timing issue associated with the attenuators is eliminated. The VSWR interaction problem has been virtually eliminated, since there are now much fewer stages that the signal must pass through, and since the amplifiers provide input/output isolation. One last advantage of this switched feedback amplifier approach is that each switch sees the same drive impedance, which should improve transient response uniformity and push much of the switching noise up to the clock frequency.

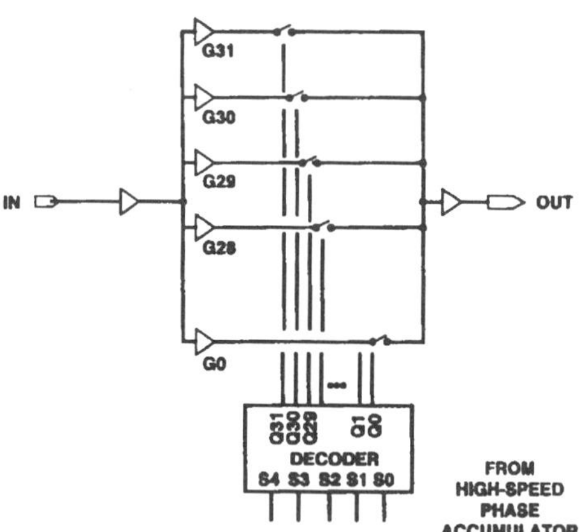

Figure 14 - Switched-Attenuator – Decoded Amplifier Implementation.

An 800 MHz CDDS "breadboard" has been designed and is currently being fabricated. This first high-speed CDDS consists of a commercially-available CMOS low-speed DDS and a custom GaAs single-chip Serrodyne Modulator IC (Figure 15). This GaAs chip is a collaboration between Lincoln Laboratory and the GaAs Signal Processing Technology Group of Texas Instruments, using TI's HI^{2}L HBT logic [23]. The architecture and analog design of this chip are Lincoln's responsibility, while TI is responsible for the digital design and I.C. fabrication. The first GaAs chips are expected to be delivered by July, 1993, and will be described in a later paper. The goal of this first breadboard

is an 800 MHz CDDS, consuming <4W of dc power, and delivering -35 dBc worst-case spurious performance.

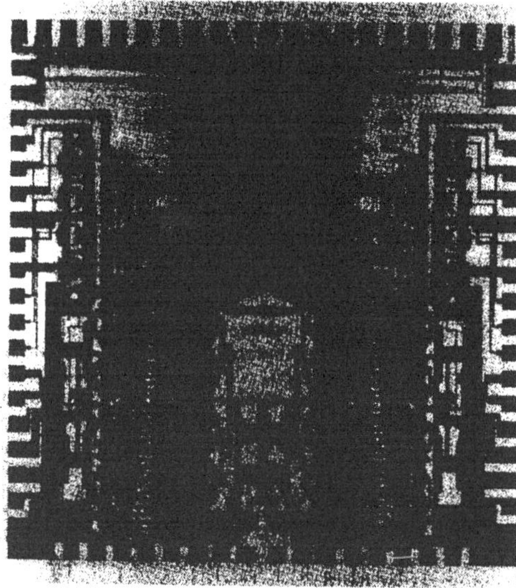

Figure 15 - 800 MHz GaAs HBT Serrodyne Modulator Integrated Circuit – CDDS1.

<u>CDDS Architecture Enhancements:</u>

Figure 16 shows a proposed improvement to the CDDS architecture. The order of the quadrant switches and the

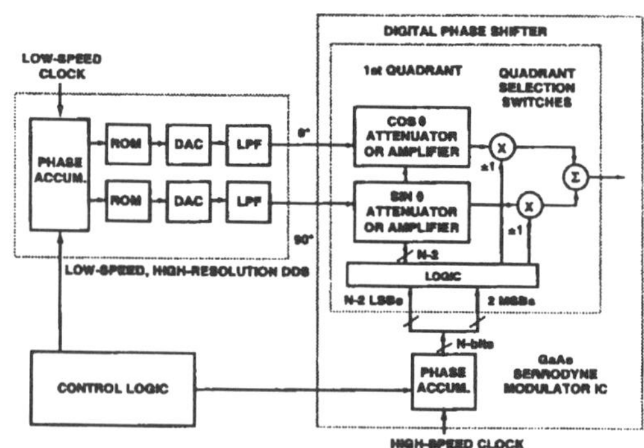

Figure 16 - CDDS – Improved Architecture.

first-quadrant switched-amplifiers has been reversed. While at first glance this seems like a trivial change, this architecture provides at least two improvements over our earlier approach: reduced clock-skew error, and reduced switched-amplifier bandwidth. While the clock skews inside the switched-attenuators themselves have been eliminated by going to the decoded-amplifier version of

Figure 14, there is still the timing of the quadrant-switches relative to the switched-amplifiers to contend with. Since the switched-amplifier sections have the N-way switch on their outputs, reversing the order as in Figure 16 puts these amplifier switches immediately preceding the quadrant switches, simplifying the timing requirements.

Since in this new implementation the switched-amplifiers only must pass the low frequencies of the low-speed DDS, the hardware implementation is made much easier. For example, in our 800 MHz CDDS, the amplifier bandwidth drops from 350 MHz to 20 MHz. This is quite significant for two reasons: it makes tight amplitude / phase tracking through the two channels easier to achieve, and it allows higher impedances to be used, resulting in reduced dc power consumption.

A second proposed improvement is shown in Figure 17. In addition to the order-reversal of Figure 16, the design has been made fully balanced. While the switched-amplifiers and quadrant-switches could be individually balanced, instead, the two circuits were merged, resulting in a balanced amplifier/switch combination. This merging reduces dc power consumption and complexity, eliminating the need for two broadband 180° signal splitters. Balancing the design has a number of advantages, including common-mode rejection (switching glitches, digital noise, and dc level shifts) and even-order spur. cancellation. The complexity of the balanced design may not be any higher than that of the unbalanced design, since the dc-level compensation circuitry that was present in the unbalanced design may be eliminated.

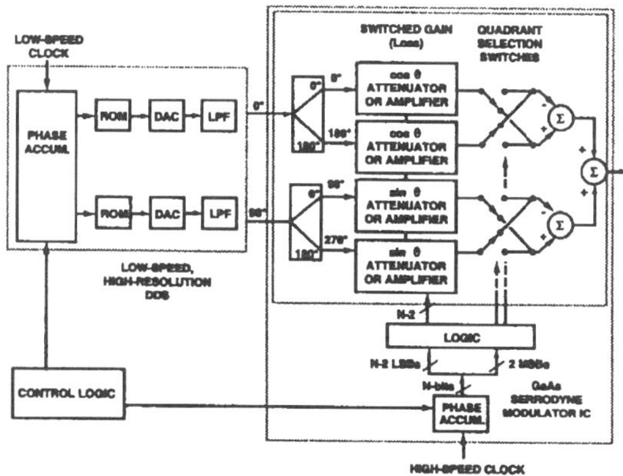

Figure 17 - Balanced CDDS.

After simulating these enhancements, a second pass GaAs Serrodyne Modulator I.C. design will be fabricated that will incorporate them along with additional improvements. By eliminating unneeded circuitry and by optimizing the design, the dc power of this second pass design is expected to drop to the 2 W goal, with the CMOS DDS consuming 1.3 watts, and the GaAs Serrodyne modulator IC consuming 0.7 watts. Advances in technology will continue to push the power down, and the speed up. We hope to achieve our 1.6 GHz, 2 W, CDDS goal by 1995.

Acknowledgments:

The author would like to thank Dave Snider, Ron Bauer, and the rest of Lincoln's management for their support and encouragement, and Marcus Ainsworth for building all of the CDDS breadboard hardware. The author would also like to acknowledge the contributions of Texas Instruments in making the GaAs Serrodyne Modulator chip: Van Andrews for the digital design along with the overall TI program management responsibilities, Bill White for consultation with the analog design, and Paul Garner and Dallas White for the layout of the chip.

References:

[1] J. Tierney, C. M. Rader, and B. Gold, "A Digital Frequency Synthesizer,", **IEEE Transactions on Audio Electroacoust.**, AU-19, 43-57, March 1971.

[2] Stanford Telecom, STEL-2373 data sheet, May 1991.

[3] Stanford Telecom STEL-2373, personal communication with P. Chizinski of STEL, Dec. 1992.

[4] Sciteq DDS-1 data sheet, 30 April 1991.

[5] Stanford Telecom STEL-1177 NCO data sheet, February, 1990, and Fujitsu 40978 DAC data sheet, Edition 3, September 1990.

[6] Stanford Telecom STEL-2172 NCO and Honeywell HDAC51400, as measured by Harry Wolfson & John Murphy of Lincoln Laboratory, October 1989.

[7] Plessey SP2002 data sheet (effective clock) May 1991.

[8] V. Andrews, M. Brown, W. White, P. Wang, R. Meyer, J. Vaal, "A Monolithic GaAs DDS for a Digital Radio Application," **1992 GOMAC Conference Digest**, pp. 431-434.

[9] Motorola 120DDSEVK data sheet, 1992.

[10] Analog Devices AD9955 data sheet, Rev.0, and AD9712B data sheet, Rev.0.

[11] patent pending.

[12] R. C. Cumming, "The Serrodyne Frequency Translator," **Proc. of the IRE**, Feb. 1957, pp. 175-186.

[13] —, "Method and Apparatus for Translating the Frequency of a Signal," U.S. Patent # 2,927,280, March 1960.

[14] —,, "Serrodyne Performance and Design," **The Microwave Journal**, Sept. 1965, pp. 84-87.

[15] G. Klein and L. Dubrowsky, "The DIGILATOR, a New Broadband Microwave Frequency Translator," **IEEE Transactions on Microwave Theory and Techniques**, vol. MTT-15, March 1967, pp. 172-179.

[16] S. Mitchell, J. Wachsman, G. Lizama, F. Ali, A. Adar, "Wideband Serrodyne Frequency Translator," **Applied Microwave**, Summer 1990, pp. 58-67.

[17] M. Topi, "An Eight-Phase Broadband Serrodyne Modulator," **1983 IEEE MTT-S Symposium Digest**, pp. 432-434.

[18] I. W. Smith and M. J. Schindler, "Serrodyne Modulator," U.S. Patent # 5,063,361, Nov. 5, 1991.

[19] R. B. Dome , "Wideband Phase Shift Networks," **Electronics**, December 1946, pp. 112-115.

[20] F. R. Shirley, "Shift phase independent of frequency," **Electronic Design** 18, 1 September, 1970, pp. 62-66.

[21] M. Tuckman, "I-Q Vector Modulator–The Ideal Control Component?," **Microwave System News and Communications Tech.**, May 1988, pp. 105-115.

[22] MATLAB, The MathWorks, Inc., Natick MA.

[23] H-T. Yuan, H-D Shih, J. Delaney, C. Fuller , "The Development of Heterojunction Integrated Injection Logic," **IEEE Transactions on Electron Devices**, vol. 36, no. 10, October 1989, pp. 2083-2092.

A Narrow Band High-Resolution Synthesizer using a Direct Digital Synthesizer Followed by Repeated Dividing and Mixing

RICHARD KARLQUIST

HEWLETT-PACKARD COMPANY, SANTA CLARA DIVISION

5301 STEVENS CREEK BLVD., MS 52U/7

SANTA CLARA, CA 95052

Abstract

A synthesizer is described that generates a narrow band of frequencies around 10 MHz with extremely small step size and high spectral purity. A DDS running at 667 kHz is upconverted to 10.667 MHz and divided back to 667 kHz several times to improve the resolution and spurious suppression of the DDS. The final 10.667 MHz offset frequency is then mixed with 667 kHz resulting in a 10 MHz output. The post mixer filtering is done with inexpensive 10.7 MHz ceramic filters of the type used in FM broadcast receivers.

Applications

Metrology

It is frequently necessary to add a small, precisely controllable offset to a standard frequency signal. In some cases, the purpose is to "steer" a free running clock to agree with a known time reference such as UTC without losing the information about the error between the clock's own time and the reference. In other cases, the purpose is to get a beat note between the clock being tested and a reference clock, in order to apply resolution enhancement techniques to the frequency difference measurement [1]. A third related use would be as the "vernier" section of a synthesizer used in an atomic frequency standard.

In these applications, it is usually unsatisfactory to directly steer the clock in question. It may not be steerable at all, or may use a suboptimal method such as variations in the C field of an atomic standard. Even if none of these problems is present, it is desirable to have both a corrected and uncorrected version of the clock available simultaineously so that the uncorrected version can be compared to UTC. Hence what is called for is an external two-port frequency offset device.

Communications

A non-metrology use would be to facilitate a technique known in the paging industry as "simulcasting" where a paging receiver may receive pages from multiple base station transmitters at the same time. If two stations are in "zero beat," destructive interference will result. To get a controlled beat note, a small offset is added to the base station's frequency reference. Since this reference is frequency multiplied to 800 MHz or more, the offset synthesizer needs to have very good spectral purity. Another communications use might occur when building modulators. With some modifications to the architecture, it is possible to build frequency stable analog modulators. In many cases, analog modulation is still simpler and more accurate than digital modulation, except for the frequency drift problem.

Prior art

A popular method of generating frequency offsets with moderate performance is the phase microstepper. It makes use of the fact that a small frequency offset is equivalent to a slow phase ramp. The microstepper controls a digitally programmable phase shifter, initiating phase increments at appropriate intervals. When 360° of phase is reached, it resets to 0°, which is equivalent. It is subject to phase jumps, phase drift and spectral impurities. To improve on this performance, sophisticated multiloop phase locked VCXO synthesizers have been built [2]. While the performance of these systems is excellent, it depends on the use of high quality VCXO's. In some cases, general purpose synthesized signal generators have been pressed into service for frequency offset work, but even the best of these is only marginally useful in metrology work and in any case they are not very cost effective for this type of task. The most appropriate general purpose synthesizers for this application are ones fashioned after the repeated mix and divide architecture popularized in the Hewlett-Packard 5100A produced in the 1960's. Even these architectures tend to have a problem with temperature coefficient of phase. They are also very complex and expensive to implement.

A synthesizer will be described below that is essentially a simplified 5100 type architecture with just enough functionality to implement small frequency offsets of a 10 MHz reference. The important issue of im-

Reprinted from *Proceedings of the IEEE International Frequency Control Symposium*, pp. 217-226, 1995.

proving the temperature induced phase drift will be covered in some detail.

Requirements

Frequency tuning range

A tuning range of ±1 Hz is quite adequate for many metrology applications and will work for simulcasting. For generating beat notes to measure frequency differences, up to ±1 kHz may be needed. In modulator work, a range up to ±5 kHz is sufficient for most FM voice and data channels. As shown below, there is a tradeoff: more resolution and spectral purity come at the price of reducing tuning range.

Spectral purity

In metrology applications, it is customary to require non-harmonically related spurious sidebands to be better than -80 dBc as a minimum, with -100 to -120 dBc being desirable. When observed with phase noise test instrumentation, spurs may be visible down to -140 dBc or even less. In communications applications, a good rule of thumb is to keep all spurs below -40 dBc *after multiplication.* In many cases, regulations require -60 to -80 dBc performance. With multiplication factors in the range of 100 to 1000, the original spurs are enhanced 40 to 60 dB. Hence the synthesizer needs to hold spurious sidebands to -80 dBc to -140 dBc as measured at the 10 MHz reference.

Resolution (step size)

For metrology, resolutions of at least 1 part in 10^{15} are desirable, which means steps measured in nanoHz at 10 MHz. Less critical applications may only need to have steps in the milliHz range. The synthesizer to be described utilizes a direct digital synthesizer (DDS) at the front end of the synthesizer to generate steps, so step size is to some extent a non-issue. By extending the DDS accumulator size, a relatively easy thing to do, it is possible to achieve any desired step size. While the divide and mix sections to be described later do add 4 bits of resolution apiece, this is not their major function, a least in the case where a DDS is used. In modulator applications where the DDS is replaced by an analog frequency/phase modulator, this 4 bits of resolution is more properly stated as 24 dB of additional dynamic range in the analog domain., which is a non-trivial improvement.

Phase stability vs temperature

By definition, frequency is the time rate of change of phase. If phase is temperature sensitive, and the ambient temperature changes with time, there will effectively be a frequency error [4]. The required phase stability depends on the assumed rate of change of temperature and the frequency error tolerance, but a good rule of thumb is to keep the overall phase drift expressed as a change in delay time to under 1 nanosecond over the full expected temperature range, such as 0 to 70°C.

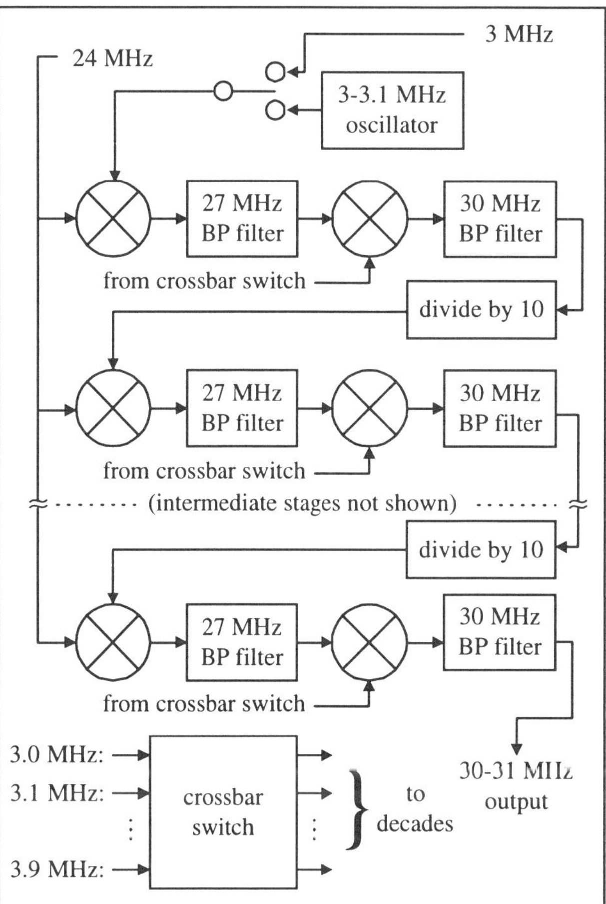

Figure 1. HP 5100 type architecture.

The classical mix and divide architecture

As an introduction, a brief review of the HP 5100 type of synthesizer will be given, which represents the mix and divide family of architectures. Figure 1 shows the 5100 block diagram. A synthesizer driver system on a separate chassis (not shown) generates reference frequencies at 24 MHz and 3.0, 3.1, 3.2, ... 3.9 MHz from the 5 MHz reference frequency input. The 24 MHz signal goes to all modules. A crossbar switch routes the 3 to 3.9 MHz signals to the appropriate divide and mix modules according to the frequency selected on the front panel. Each module is connected to a digit switch on the front panel which has a possible setting of 0 through 9. If this setting is denoted "N," then the switch will feed 3.N MHz to the corresponding module. In a two-stage upconversion process, each module converts an incoming tone in the 3 to 3.1 MHz band to a tone in the 30 to 31 MHz band, then divides the resulting frequency by 10 to get an output that again falls in the 3 to 3.1 MHz band. Although the input and output frequencies fall in the same band, they are different because a decade of resolution has been added effectively. For example, if the input frequency to a module is 3.89 MHz, the output frequency will be 3.N89 MHz, where N is the setting of the digit for that module. E.g., if N=5, then the frequency would be 3.589 MHz. The modules are cascaded so that any amount of resolution can be realized by using a sufficient number of modules. The last module omits the ÷10, resulting in a 30 to 31 MHz output band. The first module's input that would normally come from the previous module is either hardwired to 3 MHz or can be connected to a "search oscillator" that tunes continuously from 3 to 3.1 MHz. This allows "infinite" resolution with good stability because any frequency drift in this oscillator is reduced by a factor of 10^M, where M is the number of mix and divide modules. Of course the total tuning range (originally 100 kHz) is also divided by 10^M. Each module improves the spectral purity of the signal it receives from the previous module by 20 dB due to its ÷10 frequency divider.

<u>The 5100 architecture as an offset synthesizer</u>

Suppose a tuning range of only 1 Hz were needed for a typical metrology application. Then a search oscillator tunable over 1 MHz could be followed by six mix and divide modules, and one mix-only module, resulting in the required coverage. If it could be specified that the output frequency would be 30.000000 to 30.000001 MHz, then all modules could be hardwired to 3 MHz thereby eliminating the crossbar switch and 3.1 through 3.9 MHz generators. Having done this, the result would be that each module's first mixer would combine 3 MHz with 24 MHz to get 27 MHz. Hence these mixers could be dispensed with in all modules by instead feeding 27 MHz directly to the second mixer. The result is the reduction of the 5100 architecture to little more than the decade modules, and they are simplified to have only one stage of mixing and filtering instead of two. Finally, the search oscillator can be replaced by its modern equivalent, the direct digital synthesizer, to make the system a true synthesizer. The advantage of this arrangement compared to using the DDS output directly is that each module reduces DDS spurs by 20 dB when the signal is frequency divided and adds 4 bits of resolution to the DDS. The resulting configuration is shown in Figure 2.

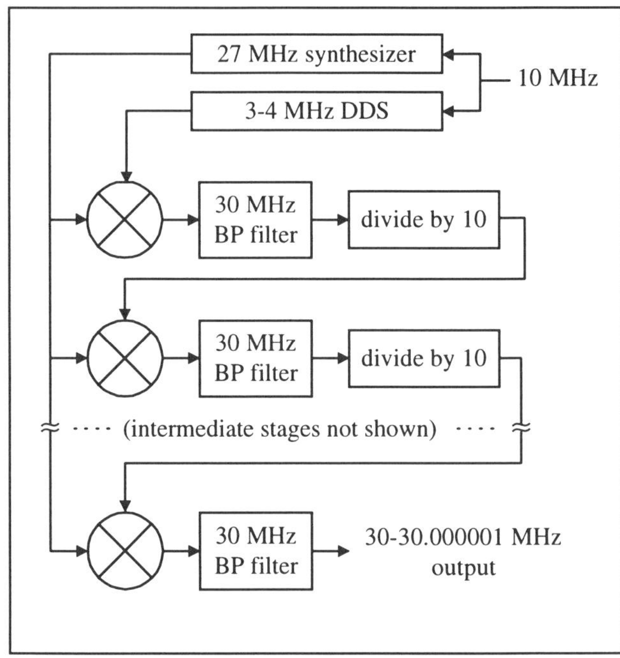

Figure 2. Narrowband architecture with DDS.

<u>Filtering</u>

The bandpass filter after each mixer is essential to getting high spectral purity. In the original 5100 and in the initial stages of the simplified architecture of Figure 2, the bandpass filter must pass 30 to 31 MHz, a fractional bandwidth of 3.3%. In the later stages of the simplified architecture, the passband can be much narrower, if non-identical modules are acceptable. In all cases, the filter must be capable of rejecting the 24 MHz image and the 3 and 27 MHz input frequencies. The 9:1 ratio of the mixer input frequencies results in a situation where the only mixer products that fall within the 3.3% passband are high order ones that decrease rapidly as the mixer drive level is backed off. It is very important that the out-of-band mixer products not be allowed to reach the divider. If they do, the non-linear nature of the divider will alias them back into the passband and compromise spectral purity. Hence the main factor in maintaining spectral purity is proper filtering.

The tradeoffs in filtering that apply here are bandwidth limitations, temperature drift, microphonics, phase noise, and cost. With the 3.3% bandwidth required for the 5100 architecture, crystal filters are ruled out because they are incapable of achieving such a wide bandwidth. On the other hand, 3.3% is so narrow as to be barely feasible for LC filters, and there isn't much prospect for improvements in passive filter technology. Besides the manufacturing problems of producing such narrow filters, LC filters have a substantial problem with temperature stability. This could cause a temperature dependent phase shift that would be most damaging in the last stage. One possible approach to this problem would be to use a crystal filter for the last filter only, since the required passband at that stage would be narrow enough to implement with a crystal filter. This would not necessarily alleviate temperature stability problems because of the long delay time through the filter caused by the narrow bandwidth. A long delay time represents many radians of phase shift and hence a small change in bandwidth or center frequency could cause a large change in delay time. Other problems with crystal filters are phase noise and possible microphonics.

<u>Ceramic filters</u>

What is needed is a moderate bandwidth filter with good stability. A so-called "ceramic" piezoelectric filter is a good match for this requirement. It is optimal for applications such as this one requiring bandwidths too narrow for LC filters and too wide for crystal filters. Ceramic filters are more stable than LC filters but are wide enough to avoid the close in phase noise and microphonic problems of crystal filters. Additionally, ceramic filters can be much cheaper than either of the alternatives. They are used worldwide in FM broadcast receivers for IF filtering and cost a fraction of a dollar in production quantities. These filters are generally available only with a center frequency of 10.7 MHz and with various bandwidths from less than 100 kHz to 500 kHz. Although these filters have been in use for decades, their performance has improved in recent years. When receivers were changed from analog tuning to synthesized digital tuning, it became necessary to improve the tolerance and stability of the center frequency. This is because it was no longer possible to compensate for an error in the IF filter center frequency by merely offsetting the analog tuning slightly. It is now easy to buy filters that are centered very precisely at 10.7 MHz and have low temperature drift. Recently, larger bandwidths such as 400 or 500 kHz have been implemented for use in DBS receivers, where the deviation is greater than with FM broadcast.

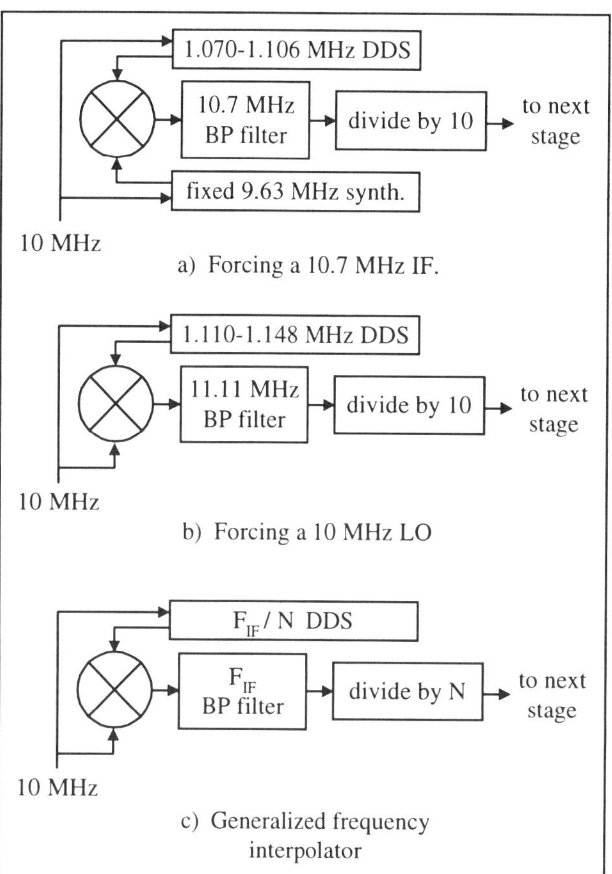

a) Forcing a 10.7 MHz IF.

b) Forcing a 10 MHz LO

c) Generalized frequency interpolator

Figure 3. Transformations of 5100 architecture.

<u>10.7 MHz architecture</u>

The key to using 10.7 MHz filters is to determine a technique for scaling the simplified 5100 type architecture such that the filter frequency becomes 10.7 MHz. If all the frequencies in the architecture in figure 2 were merely scaled by a factor of $^{10.7}\!/_{30}$, the filters would of course change to 10.7 MHz, but now the 27 MHz reference frequency scales to 9.63 MHz, which is quite inconvenient to synthesize from 10 MHz. (figure 3a). If instead the frequencies are scaled by a factor of $^{10}\!/_{27}$, then the 10 MHz could be used directly as a reference, but the filters would need to be changed to 11.111 MHz (figure 3b). The problem is that the reference and filter frequencies are too far apart. The mix and divide process can be generalized as shown in figure 3c. The frequencies obey the equation: $F_{ref} = (F_{IF})(N-1)/N$, where N is an integer equal to the frequency divider ratio. Table 1 shows the results of this equation for values of N of interest, assuming F - IF = 10.7 MHz. The values of N that give the closest reference frequency to 10 MHz are 15 and 16, which result in 9.987 and 10.031 MHz respectively. It is now possible to take advantage of the fact that it is acceptable to operate the filter slightly off center, if a small decrease in tuning range can be tolerated. If the N=15 and N=16 configurations are scaled in frequency to get the refererence frequency to exactly 10 MHz, then the resulting IF frequencies are $10\tfrac{5}{7}$ =10.714286... MHz and $10\tfrac{2}{3}$ =10.66666... MHz respectively (figures 4a and 4b). With several hundred kHz of bandwidth available, sacrificing 14 or 33 kHz is a reasonable compromise. Although the IF is closer to 10.7 MHz with N=15, it is usually preferrable to use N=16 for three reasons:

N:	Ref. freq.	N:	Ref. freq.	N:	Ref. freq.
4	8.025	14	8.560	24	10.254
5	8.560	15	9.987	25	10.272
6	8.917	16	10.031	26	10.288
7	9.171	17	10.071	27	10.304
8	9.363	18	10.106	28	10.318
9	9.511	19	10.137	29	10.331
10	9.630	20	10.165	30	10.343
11	9.727	21	10.191	31	10.355
12	9.808	22	10.214	32	10.366
13	9.877	23	10.235	33	10.376

Table 1. Possible ref. frequencies for 10.7 MHz IF.

1. The divider is easier to implement in digital hardware and gives an output with very low even harmonics because the divide ratio is a power of 2.

2. It is usually easier to deal with fractional frequencies involving factors of $\tfrac{1}{3}$ than $\tfrac{1}{7}$, assuming it is necessary to convert the repeating decimal frequency to a more useful number. For example, $10\tfrac{2}{3}$ can be used to drive a frequency tripler to produce 32 MHz. With $10\tfrac{5}{7}$, a difficult-to-build frequency septupler would be required to produce 75 MHz.

3. A lower DDS frequency is possible as will be explained below.

<u>Front and back ends for the mix and divide chain</u>

A chain constructed of the modules shown in figure 4c requires an input near $\tfrac{2}{3}$ MHz ≈ 667 kHz for the first module. In order to have the capability of acheiving zero offset, the input must be at exactly $\tfrac{2}{3}$ MHz, not merely a good approximation. Most DDS's are binary or decimal based. Neither of these can generate exactly $\tfrac{2}{3}$ MHz starting from 10 MHz., regardless of the number of bits of resolution. A ternary (base 3) DDS would be required, which is possible, but highly unlikely to be available. Instead, the DDS is operated in a band centered at 2 MHz and is followed by a ÷3 frequency divider (figure 5). In most cases, a DDS output of exactly 2 MHz is easily realized. If the N=15 architecture is chosen, then the DDS would need to generate 5 MHz instead of 2 MHz and a divide by 7 circuit would need to be used to obtain $\tfrac{5}{7}$ MHz ≈ 714.286 kHz.

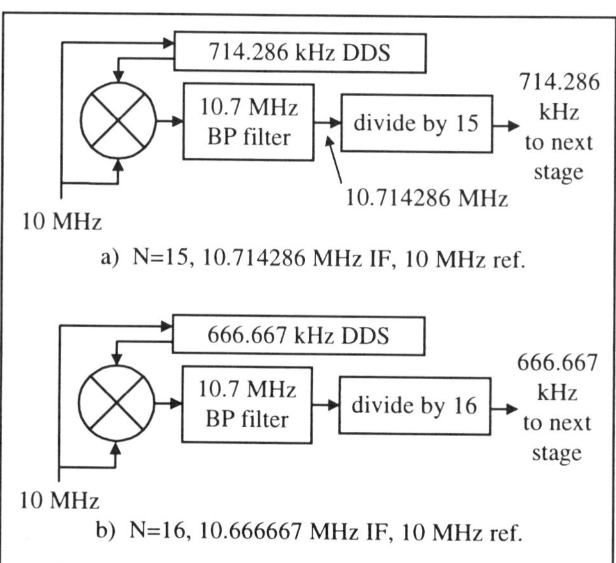

Figure 4. Usable frequency schemes for 10 MHz ref.

At the end of the chain, 10.667 MHz is available with programmable offset, but of course it needs to be converted back to 10 MHz to be useful. This is accomplished with one additional mixer and filter stage. However in this case the LO for the mixer is the 10.667 MHz from the previous stage. and the filter is tuned to 10 MHz.

The output stage raises the issue of filtering, since the standard 10.7 MHz filters are no longer applicable. At the end of the chain, the filter bandwidth required is negilgible so it is possible to use a crystal filter. This filter should be as wide as possible to minimize phase noise and microphonics effects. Another possibility is to use 10 MHz ceramic resonators in combination with discrete capacitors to build a 10 MHz filter. These resonators are similar to the ones in the 10.7 MHz filters,and, fortunately, 10 MHz is one of the standard frequencies for them. They are normally used as low priced replacements for microprocessor clock crystals. Their moderate Q is a good match to this requirement since the nearest frequency to be rejected is 7% away from the passband center. It is also possible to use a phase locked VCXO as a cleanup loop to get pure 10 MHz out.

Figure 5 shows the complete design with the blocks regrouped into iterated divide/mix/filter modules, which will be henceforth be referred to as **"interpolators."** This organization is more logical for this design than the 5100 partitioning into mix/filter/divide modules.

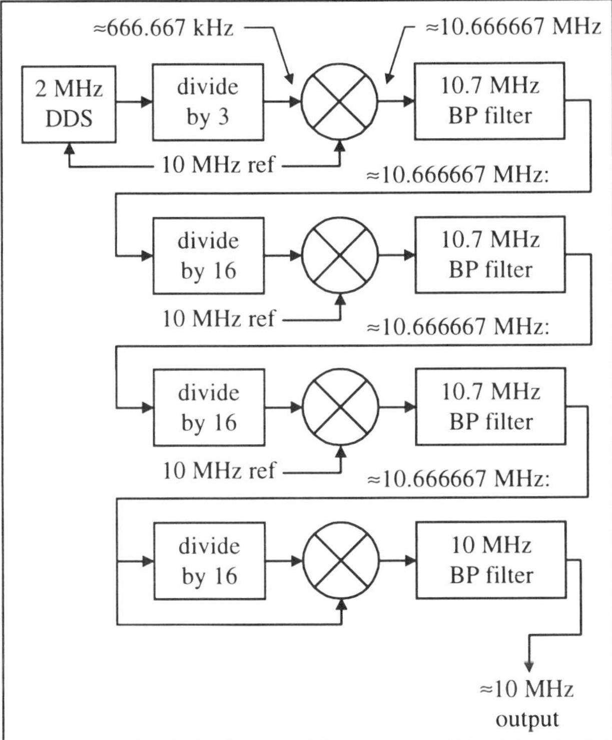

Figure 5. 10 MHz output, 10 MHz ref. synthesizer.

Design constraints

The architecture of figure 5 raises the question of how many interpolators are appropriate. Each iteration decreases step size by a factor of 16 (which may or may not be very significant) and it also improves spectral purity (i.e. phase noise, spurious sidebands, jitter, etc.) by 24 dB = 20 log 16, which usually is of great importance. After a sufficient number of stages, the total tuning range may become too small to be usable or the 24 dB per stage improvement in spectral purity may cease to occur because the noise floor of the interpola-

tors or the reference frequency source becomes the limiting factor. The limit on tuning range depends on the number of interpolators and the bandwidth of the IF filters used. With the DBS type filters, about 400 kHz of range is available. This results in final tuning ranges of 25 kHz, 1.5 kHz, 100 Hz, 6 Hz, and 0.4 Hz for 1, 2, 3, 4, and 5 interpolations respectively. Since 5 interpolations gives a theoretical spectral improvement of 120 dB (in addition to the 10 dB from the divide by 3 after the DDS), it is unlikely that more than 5 interpolators would ever be used, even if less tuning range were acceptable.

Critical system design details

To get maximum performance for the overall design, it is important that the 10 MHz distribution amplifier (not shown in figure 5) have high isolation between outputs to keep the slightly different IF frequencies of each interpolator from intermixing. Also, it is helpful for it to have low distortion, especially even order, in order to minimize mixer distortion. As usual, it should also have low phase noise and good short term stability so as not to degrade the reference source. The various interpolators should be well isolated from each other by proper ground plane management and PC board layout, or installed in individual shielded boxes if necessary. Finally, adequate power supply filtering should be used to preclude coupling through the power distribution system.

Critical module design details

At the front end of the interpolator, a high quality sine wave to logic level converter should be used. A excellent treatment of this subject is given in reference [5]. After converting to logic levels, a low jitter logic family such as 74ACXXX should be used for the divider, which is preferably of the synchronous type, not the ripple type. The divider must be followed by a low pass filter to suppress harmonics; an elliptic function filter is probably the most appropriate type for this purpose, with 5, 7, or 9 poles depending on the required performance. A pad should be used between the filter and the mixer to properly terminate both the filter and the IF port of the mixer, which is driven with the filtered 667 kHz. The amount of attenuation in the pad can be adjusted to give the desired mixer drive level. A drive level of -30 to -10 dBm is reasonable for a +7dBm LO mixer depending on the required level of performance. For extremely high performance systems, it may be appropriate to use high drive level mixers.

The output of the mixer goes through two to four stages of filtering. It is convenient to insert amplifiers between the filters to prevent interaction between the ceramic filters since they are not designed to be cascaded directly. The amplifiers not only must make up for the filter loss of 3 to 6 dB but also provide a net gain to help raise the low level signal coming out of the mixer to a higher amplitude to drive the clock of the next divider. The filters, when used as recommended, have a rounded ("stoop shouldered") amplitude response in order to get constant group delay, which is critical in FM demodulation. When several of them are cascaded, the resulting response has an even more pronounced rounding. Fortunately, it is possible to fix this problem by deliberately mismatching the filters. It is also possible to change the center frequency somewhat by adding capacitors (or inductors) in series with the input and output terminals. The effect of temperature on phase for a typical filter is 3 milliradians/°C. The phase does not accumulate through the interpolators, but rather is only critical in the last interpolator and the final frequency translation stage filtering.

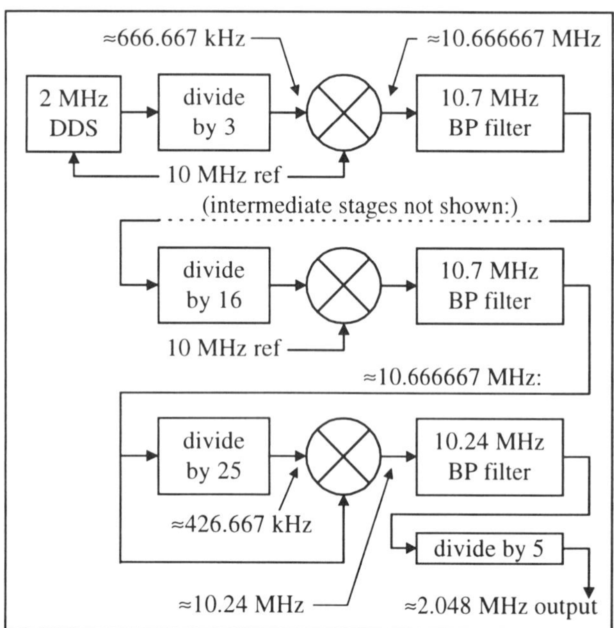

Figure 6. 2.048 MHz output, 10 MHz ref. synthesizer.

<u>Input and output frequencies other than 10 MHz</u>

The final translation stage can be changed to produce frequencies of interest other than 10 MHz. For example, 2.048 MHz can be synthesized from a 10 MHz reference. Figure 6 shows the last divider changed to ÷25 to convert 10.667 MHz to 10.24 MHz. This can then be divided by 5 to get an offsettable 2.048 MHz

output. In figure 7, the last divider has been returned to ÷16 but the previous ones have been changed to ÷25. This makes it possible to multiply a 2.048 MHz input by 5 to get 10.24 MHz, which is used as the LO for the mixers. The IF frequency is still 10.667 MHz and the output stage of figure 5 produces offset 10 MHz.

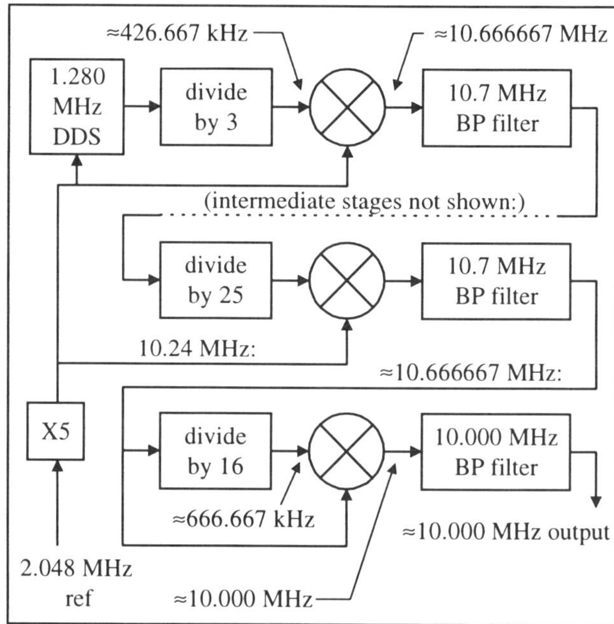

Figure 7. 10 MHz output, 2.048 MHz ref synthesizer.

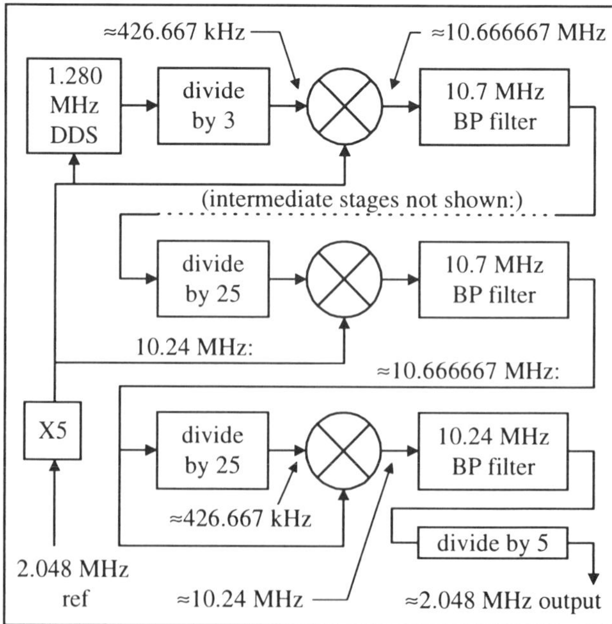

Figure 8. 2.048 MHz out, 2.048 MHz ref. synthesizer.

Figure 8 combines the techniques of the two previous figures to get a synthesizer with 2.048 MHz reference frequency and an offset 2.048 MHz output. In figure 9, the North American telecom frequency of 1.544 Mhz is multiplied by ⅔ to get 2.058667 MHz. This is close enough to 2.048 MHz to use the scheme of figure 8, but it is advantageous to increase the divide ratio from 25 to 26 to get the signal closer to the center of the passband. The result in figure 9 is the North American equivalent of figure 8, with 1.544 MHz in and offset 1.544 MHz out.

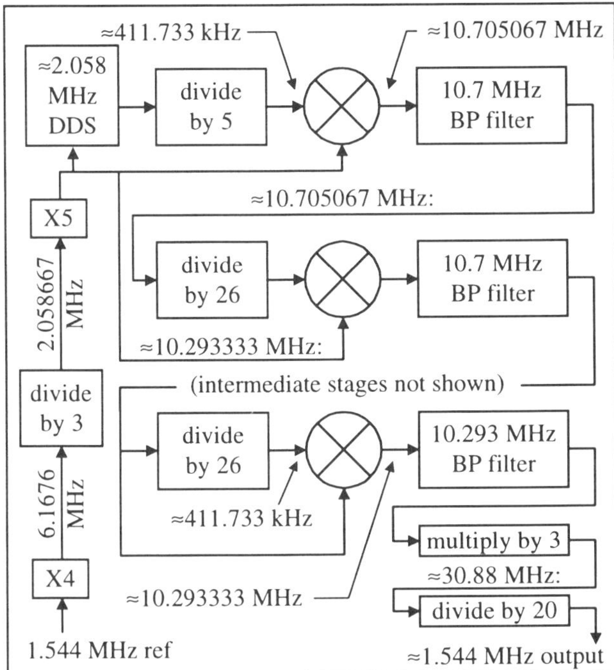

Figure 9. 1.544 MHz out, 1.544 MHz ref. synthesizer.

In figure 10, output sections are shown for generating the Sonet/SDH frequency reference of 51.84 MHz or 1.544 MHz starting with either the 10 MHz input synthesizer shown in figure 5 or the 2.048 MHz input synthesizer shown in figure 7. Figure 11 shows a synthesizer for exciting cesium beam tubes (or absorption cells) with a direct multiply frequency scheme. The synthesizer produces 10.714 MHz which can be multiplied by 78, then multiplied by 11 resulting in the cesium transition frequency of 9.192 GHz. There is sufficient tuning range to move 50 kHz off the main line to the Zeeman line for measuring the effective C field magnitude [6]. The output of the ×78 multiplier in figure 11 is conveniently filtered by an off the shelf 836.5 MHz dielectric resonator filter, of the type used for cellular telephone transceivers.

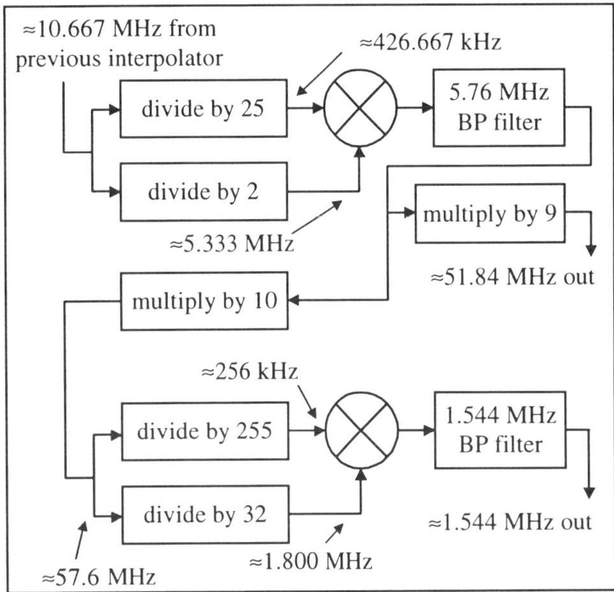

Figure 10. Output sections for 51.84 and 1.544 MHz.

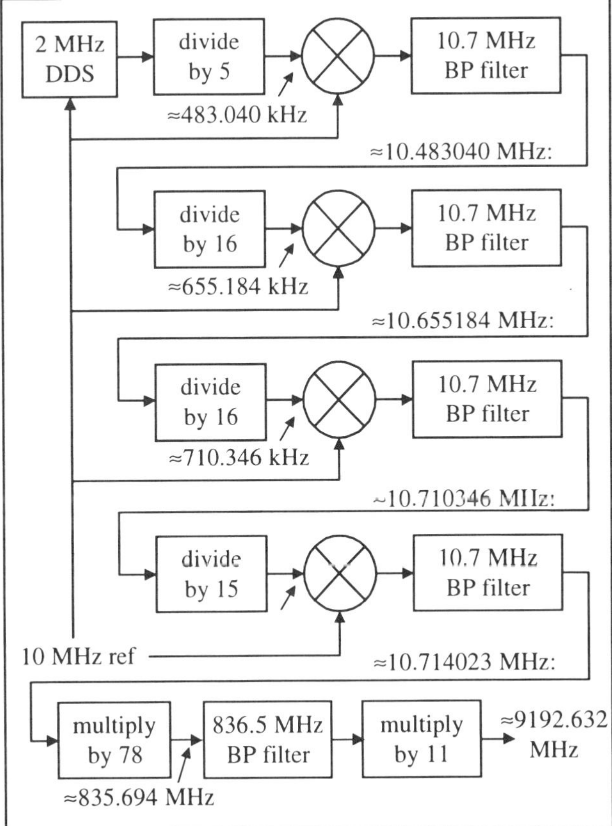

Figure 11. RF chain for cesium standard.

Figure 12 shows an alternative output architecture where the ÷N in the path from the previous stage is changed to ÷(N - 1) and relocated to the reference path. This modification can also be adapted to the other frequency schemes previously described. Figure 13 shows a 10.23 MHz output synthesizer for use as a GPS timebase locked to a 10 MHz reference. It is adapted from figure 6.

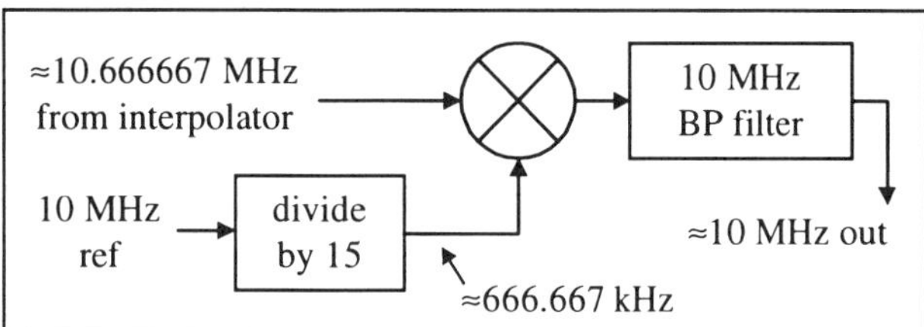

Figure 12. Alternative 10 MHz output architecture.

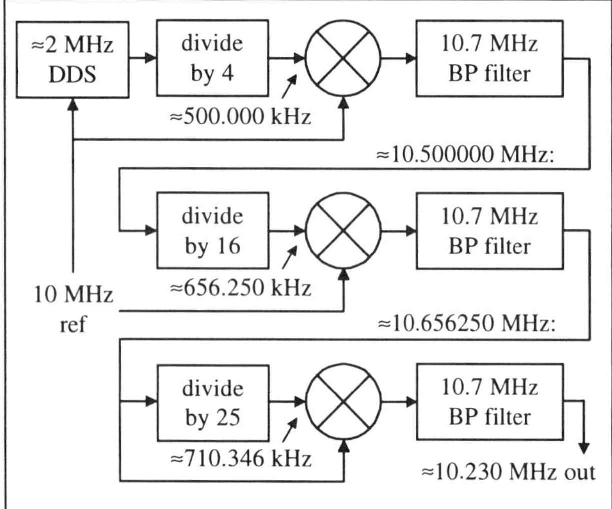

Figure 13. 10.23 MHz out, 10 MHz ref synthesizer.

<u>Frequency modulator applications</u>

If the DDS is replaced by a frequency or phase modulator, then the synthesizer can be used to reduce the modulation index. This allows generating the signal with much wider deviation than the actual signal. The advantage is that any center frequency drift of the carrier is reduced correspondingly. This drift reduction permits the use of a true frequency modulator with DC coupling utilizing a VCO, rather than an AC coupled technique such as an Armstrong modulator. DC coupled modulation is important in asynchronous data transmission. Current workarounds for this problem involve questionable schemes such as putting a very narrow phase locked loop on the VCO so that the data is "almost" DC coupled or using a DDS to generate mark and space frequencies. The problem with using a DDS is that the data pulse shaping filters have to be digital filters. Some systems implement a workaround consisting of a look up table that describes the transition from mark to space or

vice versa. This method is suboptimal with respect to intersymbol interference and spectrum utilization.

<u>Experimental verification</u>

A experimental model was built using two interpolators with 3 filters each and a final 10 MHz filter with 5 discrete ceramic resonators. A synthesized signal generator was used in place of the DDS. With a moderate amount of attention to the details enumerated above, it was easy to keep spurious sidebands down at least 120 dB. The ceramic filters did not significantly increase phase noise.

<u>References</u>

[1] D. W. Allan and H. Daams, "Picosecond time difference measurement system," in <u>Proceedings of the 29th Annual Symposium on Frequency Control</u>, 1975, pp. 404-411.

[2] R. L. Hamell, P. F. Kuhnle and R. L. Sydnor, "An Improved Offset Generator Developed for Allan Deviation Measurement of Ultra Stable Frequency Standards," in <u>Proceedings of the 23rd Annual Precise Time and Time Interval (PTTI) Applications and Planning Meeting (NASA confereence publication 3159)</u>, 3-5 Dec. 1991, pp. 209-218.

[3] V. E. Van Duzer, "A 0-50 Mc Frequency Synthesizer with Excellent Stability, Fast Switching, and Fine Resolution," <u>Hewlett-Packard Journal</u>, vol. 15, no. 9, May 1964, pp. 1-6.

[4] R. K. Karlquist, "A New RF Architecture for Cesium Frequency Standards," in <u>Proceedings of the 1992 IEEE Frequency Control Symposium</u>, 27-29 May 1992, pp. 134-142.

[5] G. J. Dick, P. F. Kuhnle and R. L. Sydnor, "Zero-Crossing Detector with Sub-Microsecond Jitter and Crosstalk," in <u>Proceedings of the 22nd Annual Precise Time and Time Interval (PTTI) Applications and Planning Meeting (NASA Conference Publication 3116)</u>, 4-6 Dec. 1990, pp. 269-282.

[6] L. S. Cutler and R. B. Giffard, "Architecture and Algorithms for New Cesium Beam Frequency Standard Electronics," in <u>Proceedings of the 1992 IEEE Frequency Control Symposium</u>, 27-29 May 1992, pp. 127-133.

A 3 to 30 MHz High-Resolution Synthesizer Consisting of a DDS, Divide-and-Mix Modules, and a M/N Synthesizer

RICHARD KARLQUIST

HEWLETT-PACKARD COMPANY, SANTA CLARA DIVISION
5301 STEVENS CREEK BLVD., MS 52U/7
SANTA CLARA, CA 95052

Abstract

A 3 to 30 MHz synthesizer is described that uses divide and mix modules to clean up the output of a DDS resulting in a building block signal source that generates an output at 10.7 MHz with a 4.2% tuning range. The divide and mix modules are based on a previously described narrowband design that has been modified for greater tuning range. Both the old and new designs exploit inexpensive ceramic filters that are normally used in the IF stages of radio receivers. The basic 10.7 MHz band is applied to an M/N synthesizer to produce an output band from 3 to 30 MHz. Various techniques for filtering harmonics from the frequency divider output are discussed, as well as circuit design details of the 10.7 MHz mixers and filters. A PLL version of the M/N synthesizer is described that covers the band from 240 to 700 MHz, again starting from the 10.7 MHz input.

Introduction

A narrow band synthesizer using divide and mix modules based on 10.7 MHz ceramic filters previously described had the capability of extremely good spectral purity and high resolution [1]. This is in contrast to fractional N techniques, which have limited resolution; and DDS techniques, which have limited spectral purity. However, the architecture was limited to applications requiring a very small tuning range.

The divide and mix modules had input and output frequencies in the vicinity of 10.7 MHz to allow them to be cascaded, with each having typically a 10 MHz reference frequency input. The narrow bands of frequencies available at the output of the final divide and mix module needed to be centered at frequencies that were offset from the reference by a submultiple of the reference. For example, with a 10 MHz. reference, and an offset of $\frac{1}{15}$ of the reference, the center frequency is $10\,\frac{10}{15}$, i.e.: $10\,\frac{2}{3}$ MHz. Similarly, $10\,\frac{10}{14}$, i.e.: $10\,\frac{5}{7}$ MHz is also possible.

In figure 11 of [1], a scheme is shown for generating the particular frequency used for exciting cesium atoms. Of interest here is the fact that it depended on generating the frequency 10.714023 MHz, which is slightly different from $10.7142857... = 10\,\frac{5}{7}$ MHz. The 263 Hz difference far exceeds the tuning range of the synthesizer. The solution, shown here in simplified form in fig. 1, was to cascade two modules having a "natural" frequency of $10\,\frac{5}{7}$ MHz (i.e.: ÷15) following a module having a "natural" frequency of $10\,\frac{2}{3}$ MHz (i.e.: ÷16.)

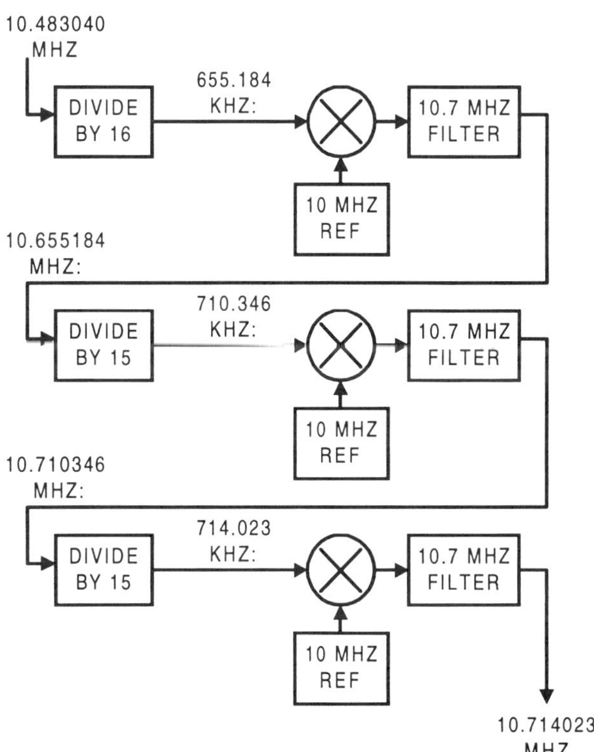

Fig. 1. Cesium architecture of [1].

Reprinted from *Proceedings of the IEEE International Frequency Control Symposium*, pp. 928-933, 1996.

Enhanced divide and mix modules

This raises the question as to what other frequencies might be generated this way by fortuitious combinations of divide ratios. The ceramic filters (described below) have a usable bandwidth of 10.47 to10 10.93 MHz, hence both the input and output frequencies of the modules are constrained to lie in this band. For division by values of N from 12 to 23, a non contiguous set of frequency bands can be generated that yield over 50% coverage of the 10.47-10.93 MHz band For example, if N=16, then the input band of 10.47-10.93 MHz is compressed into an output band of 10.654 to 10.683 MHz The ≈50% coverage could be increased to 100% if N were allowed to take on the in-between values of 11.5, 12.5, 13.5 ... 20.5, 21.5, 22.5. This fractional N division can be accomplished best by first multiplying the frequency by 2, then dividing by 2N. This scheme is shown in fig. 2.

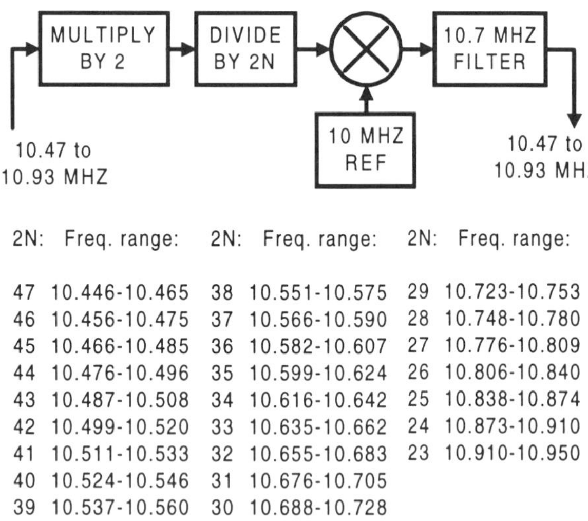

2N:	Freq. range:	2N:	Freq. range:	2N:	Freq. range:
47	10.446-10.465	38	10.551-10.575	29	10.723-10.753
46	10.456-10.475	37	10.566-10.590	28	10.748-10.780
45	10.466-10.485	36	10.582-10.607	27	10.776-10.809
44	10.476-10.496	35	10.599-10.624	26	10.806-10.840
43	10.487-10.508	34	10.616-10.642	25	10.838-10.874
42	10.499-10.520	33	10.635-10.662	24	10.873-10.910
41	10.511-10.533	32	10.655-10.683	23	10.910-10.950
40	10.524-10.546	31	10.676-10.705		
39	10.537-10.560	30	10.688-10.728		

Fig. 2. Wideband frequency interpolator.

Any number of these frequency interpolator modules can be cascaded in order to get sufficient spectral purity and/or resolution. The scheme for choosing the value of N for each module is as follows. The final module output frequency will lie in one or two of the bands shown in the table in fig. 2. One or two values of N for the final module can then be determined from the table. Knowing N, it is then possible to determine the input frequency corresponding to the desired output frequency. This input frequency is then the output frequency for the second to last module. The process can then be repeated to find the value of N and input frequency for this module. Working backward

through the modules, eventually the input frequency of the first module is determined, and the DDS is set to this first input frequency as described. For example, to generate 10.808 MHz, the DDS is set to 10.868 MHz and followed by three interpolator modules having values of N of 13, 21.5, and 13, and generating frequencies of 10.836, 10.504, and 10.808 MHz respectively. As before, each module improves spectral purity by at least 22 dB and increases resolution by over an order of magnitude. This permits continuous, high-resolution coverage of the 10.47 to 10.93 MHz band.

M/N Synthesizer

The 10.47 to 10.93 MHz front end can be used as a drive source for an M/N synthesizer to generate 3 to 30 MHz as shown in fig. 3. The input frequency is multiplied by 64 using six cascaded frequency doublers resulting in an output frequency band of 670.08 to 699.52 MHz. The frequency in this band is then divided by an integer between 23 and 255 using a commercially available IC with a guaranteed clock rate of 700 MHz.

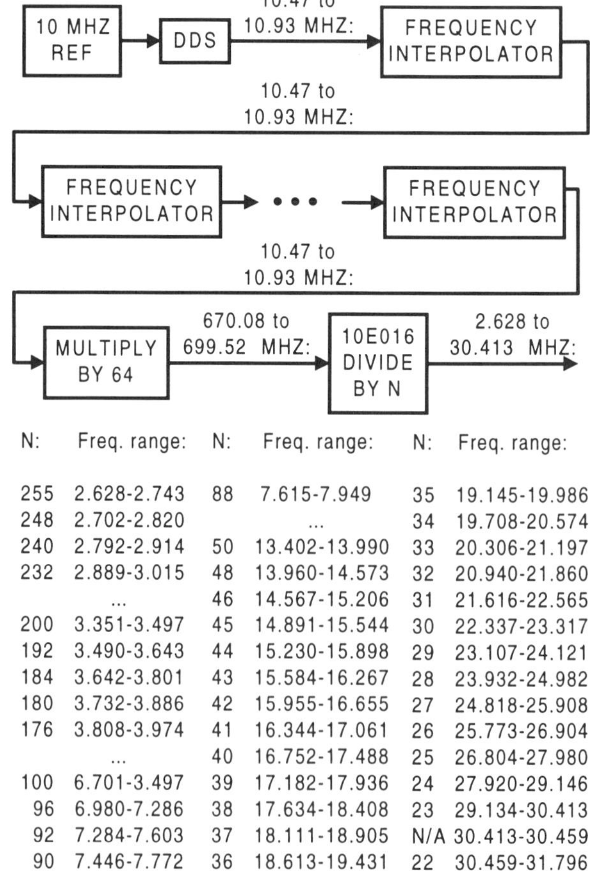

N:	Freq. range:	N:	Freq. range:	N:	Freq. range:
255	2.628-2.743	88	7.615-7.949	35	19.145-19.986
248	2.702-2.820		...	34	19.708-20.574
240	2.792-2.914	50	13.402-13.990	33	20.306-21.197
232	2.889-3.015	48	13.960-14.573	32	20.940-21.860
	...	46	14.567-15.206	31	21.616-22.565
200	3.351-3.497	45	14.891-15.544	30	22.337-23.317
192	3.490-3.643	44	15.230-15.898	29	23.107-24.121
184	3.642-3.801	43	15.584-16.267	28	23.932-24.982
180	3.732-3.886	42	15.955-16.655	27	24.818-25.908
176	3.808-3.974	41	16.344-17.061	26	25.773-26.904
	...	40	16.752-17.488	25	26.804-27.980
100	6.701-3.497	39	17.182-17.936	24	27.920-29.146
96	6.980-7.286	38	17.634-18.408	23	29.134-30.413
92	7.284-7.603	37	18.111-18.905	N/A	30.413-30.459
90	7.446-7.772	36	18.613-19.431	22	30.459-31.796

Fig. 3. M/N synthesizer and drive source.

This results in a series of overlapping output frequency bands giving continuous coverage from 2.628 to 30.413 MHz. Divide ratios below 23 generate non-contiguous bands above 30 MHz that may be useful in certain applications. Additional frequency dividers can be cascaded to generate frequencies below 2.7 MHz.

<u>Output harmonic filtering techniques</u>

The architecture of fig. 3 is sufficient to generate digital clock signals, but requires additional circuitry to add sine wave output capability. The first step is to suppress even order harmonics by reconfiguring the 700 MHz divider as a two modulus (P, P+1) prescaler followed by a divide by 2 flip flop (fig. 4), where P is $^N\!/_2$ (integer division.) For example, if N=23, P=11, and P+1=12. For odd values of N, this generates an output with a duty cycle within 2% of 50% resulting in even harmonic levels of around -25 dBc in these cases. For even values of N, the harmonic suppression is limited only by circuit balance, with typical values of -40 to -60 dBc. Note that odd values of N are only needed above 15 MHz.

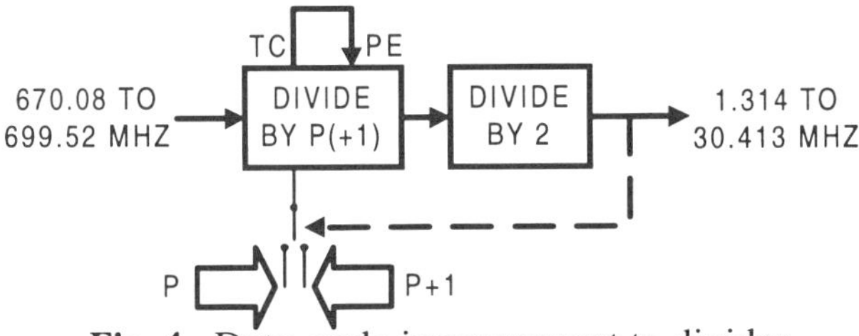

Fig. 4. Duty cycle improvement to divider.

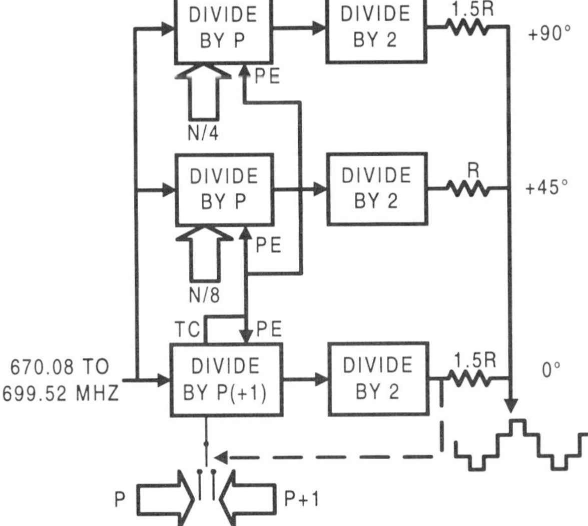

Fig. 5. Stepped sine wave generator.

It is also possible to gang together three dividers and program them to have phases spaced at approximately 45 degrees (fig. 5). They are then summed together to generate a stepped sine wave as in [2]. This suppresses the 3rd, and 5th harmonics, making filtering easier. The suppression improves as N increases. For values of N that are multiples of 8, the suppression is optimal.

<u>Mixer circuit design</u>

The operation of the mixer is very important to the performance of the divider and mix module. The mixer is the only source of spurious signals that cannot be filtered down to an arbitrarily low level. Hence the level of in band spurs at the output of the final mixer establishes a lower bound on the overall spur spec. Fortunately, the in-band spurs are all of a fairly high order, the lowest being a 12th order spur that zero beats with the desired signal at 10.909 MHz. Other in band spurs may be as high as the 22nd order spur at 10.47619 MHz. According to basic mixer theory, an nth order spur should decrease in amplitude n dB for every dB decrease in the desired signal, once the signal levels are low enough for the mixer to be operating quasi-linearly.

Three different mixers were tested: the ASK-1 (+7 dBm. LO), the SRA-3H (+17 dBm LO), and the M9E (+27 dBm. LO) It was found that the rate of decrease of high order spurs as the I.F. drive level was backed off was nowhere near the value based on the order number, at least for any level likely to be used in this synthesizer. Fig. 6 shows the results for the SRA-3H for 15th order spurs. This results show that a tradeoff must be made between spur levels and phase noise floor, since phase noise will increase as the mixer drive level decreases. Another surprising discovery was that the ASK-1 attained about the same spur to carrier ratios at merely 3 dB less drive than the SRA-3H, despite the 10 dB. decrease in LO level. Also, the M9E wasn't significantly better than the SRA-3H, despite the 10 dB. increase in LO level. It is believed that this was due to insufficient harmonic filtering of the M9E drive signals. Further study is probably warranted.

Input (dBm):	Spurious level (dBc):	Input (dBm):	Spurious level (dBc):
-2	-99	-8	-112
-3	-100	-9	-115
-4	-103	-10	-117
-5	-104	-11	-120
-6	-107	-12	-124
-7	-109	-13	-126

Fig. 6. Mixer spurious levels

Putting resistive attenuators on the mixer ports to assure that they were matched with a 50 ohm system impedance didn't help the mixer performance. Another interesting phenomenon is that if the mixer is connected directly to the ceramic filter, its apparent conversion loss decreases by 3 dB compared to a 50 ohm load. This is believed to be due to the fact that the 9.03 to 9.53 MHz image is reflected back to the mixer by the high input impedance of the filter at that frequency. This decrease in conversion loss compensates in part for the filter insertion loss, so the optimum architecture appears to be to have the filter immediately following the mixer and before any amplifier. This also has the advantage that the amplifier is only dealing with a pure sine wave and there is no chance of it generating its own intermods.

Filter circuit design

The 10.7 MHz filters are intended for use in the IF stages of direct broadcast satellite receivers and have a minimum bandwidth of 400 kHz and a typical bandwidth of 500 kHz Fig. 7 shows the manufacturer's recommended circuit for a single filter. It was determined experimentally that, for a single filter, this was indeed optimum for the purposes of this synthesizer. For example, it was not possible to increase the bandwidth or change the center frequency by creative use of "matching" circuitry as it is with the narrow FM broadcast type filters. Optimum coupling circuits for multiple filters with intervening amplifiers were determined experimentally as shown in fig. 7.

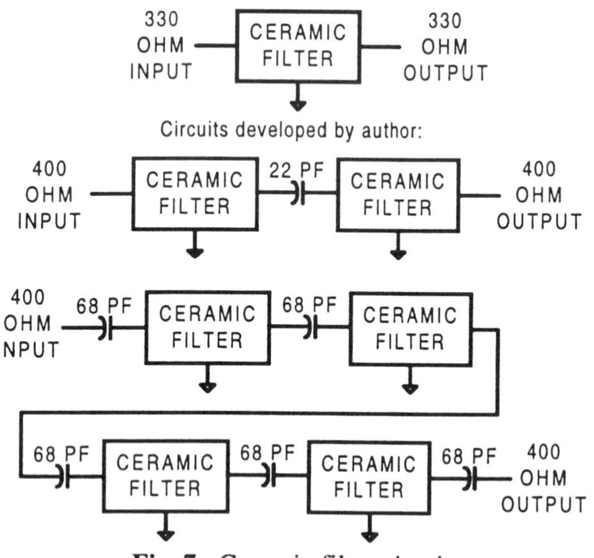

Fig. 7. Ceramic filter circuits

Up to at least four filters can be cascaded. It was found empirically that when multiple filters are cascaded, the frequency response is flatter if small (less than 100 pF) capacitors are used to couple them together, rather than a direct connection. It also is helpful to raise the source and load impedances to 400 to 500 ohms. Fig. 8 and fig. 9 show the frequency response of the single filter and double filter of fig. 7.

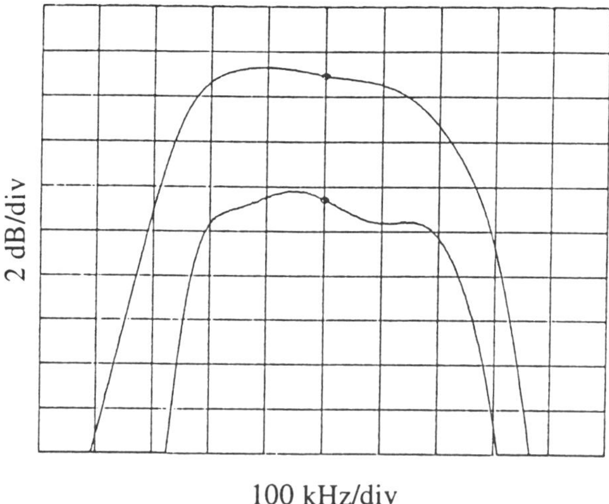

Fig. 8. Close-in selectivity of ceramic filters.

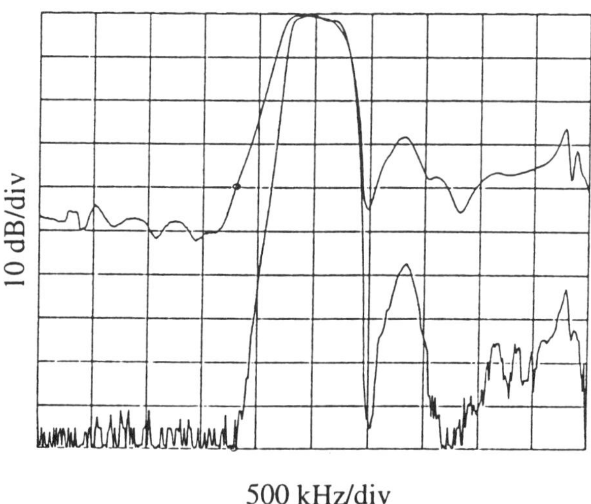

Fig. 9. Ultimate rejection of ceramic filters.

In most cases it doesn't make sense to use more than two filters without an intervening amplifier because with two filters the response has already rolled off to the ultimate attenuation floor (100 dB) of the pair of filters before reaching 10 MHz, where the stop band begins. It is unlikely that the theoretical 150 dB. for three cascaded filters could be reached in practice in a single stage because of the extraordinary shielding that would be necessary.

Figure 10 shows the overall configuration of the mixer and filter sections. The 1.8 µH. inductor and 100 pF. capacitor combination forms a 50 ohm to 400 ohm matching network. A third ceramic filter at the output was added to remove harmonics generated in the amplifiers and also to give additional spur reduction. The signal out of this filter drives the frequency doubler in the next module or the beginning of the X64 multiplier. The doublers are of a conventional "full-wave rectifier" design.

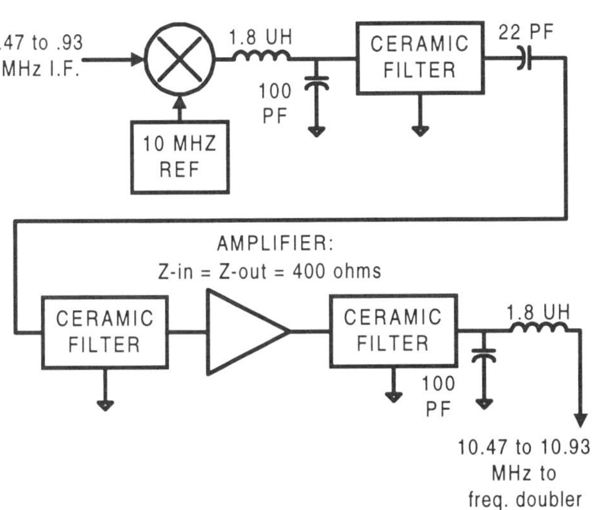

Fig. 10. Mixer and filter configuration.

It was found that the presence of second harmonic energy at the input to the doubler degraded the suppression of fundamental and third harmonic signals at the output of the full wave rectifier.

The programmable divider was connected by selector switches to one of a set of three, ⅓ octave bandpass filters used to suppress harmonics of the IF signal. As explained in [1], high order harmonics of this signal, if present, would cause spurious signals in the mixer output. An experimental synthesizer was constructed that was able to achieve a phase noise floor of less than -140 dBc/Hz with less than -110 dBc spurs. The spurs probably could have been somewhat lower if better shielding had been used.

A PLL type M/N synthesizer

Another way the 10.47 to 10.93 MHz output can be used is to shown in fig. 11. The X64 multiplier can be replaced by a PLL type multiplier using the same 700 MHz programmable counter that was used before in fig. 3. If the VCO has sufficient tuning range, the multiplication factor can be programmed over the range of 23 to 65 giving continuous tuning from 240.81 to 700 MHz. (In practice, several VCO's would probably be used to cover the range). This octave plus band can be divided down or multiplied up by powers of 2 to get continuous coverage over any desired frequency range. The phase detector reference frequency in excess of 10 MHz allows the PLL multiplier to have excellent spectral purity.

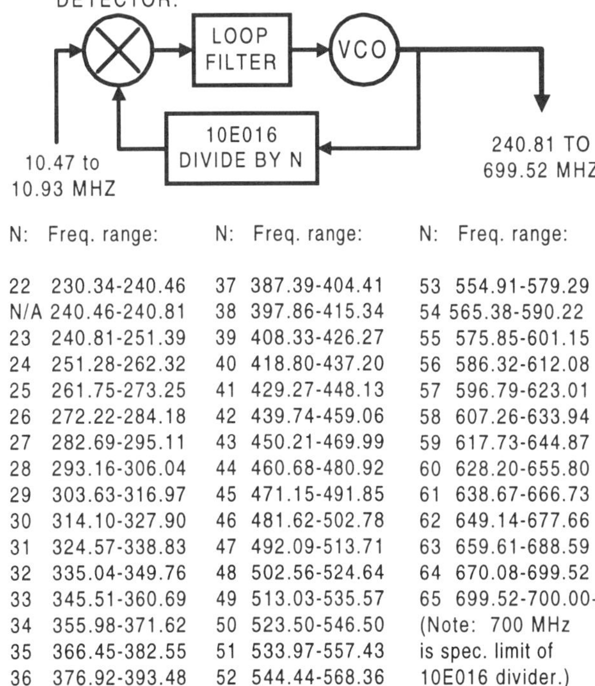

N:	Freq. range:	N:	Freq. range:	N:	Freq. range:
22	230.34-240.46	37	387.39-404.41	53	554.91-579.29
N/A	240.46-240.81	38	397.86-415.34	54	565.38-590.22
23	240.81-251.39	39	408.33-426.27	55	575.85-601.15
24	251.28-262.32	40	418.80-437.20	56	586.32-612.08
25	261.75-273.25	41	429.27-448.13	57	596.79-623.01
26	272.22-284.18	42	439.74-459.06	58	607.26-633.94
27	282.69-295.11	43	450.21-469.99	59	617.73-644.87
28	293.16-306.04	44	460.68-480.92	60	628.20-655.80
29	303.63-316.97	45	471.15-491.85	61	638.67-666.73
30	314.10-327.90	46	481.62-502.78	62	649.14-677.66
31	324.57-338.83	47	492.09-513.71	63	659.61-688.59
32	335.04-349.76	48	502.56-524.64	64	670.08-699.52
33	345.51-360.69	49	513.03-535.57	65	699.52-700.00+
34	355.98-371.62	50	523.50-546.50		(Note: 700 MHz
35	366.45-382.55	51	533.97-557.43		is spec. limit of
36	376.92-393.48	52	544.44-568.36		10E016 divider.)

Fig. 11. PLL M/N synthesizer scheme.

References

[1] R. K. Karlquist, "A Narrow Band High-Resolution Synthesizer Using a Direct Digital Synthesizer Followed by Repeated Dividing and Mixing," in Proceedings of the 1995 IEEE Frequency Control Symposium, 31 May-2 June, 1995, pp.217-235.

[2] R. K. Karlquist, "A New RF Architecture for Cesium Frequency Standards," in Proceedings of the 1992 IEEE Frequency Control Symposium, 27-29 May 1992, pp. 134-142.

A New Architecture
for a Sine-Wave Output DDS
with High Spectral Purity

LETIZIA LO PRESTI, GIUSEPPE CARDAMONE,
ANDREA DE MARCHI, ENRICO RUBIOLA

Abstract: **Direct digital frequency synthesizers (DDFSs) are generally implemented by successively scanning through a lookup table stored in a read-only memory (ROM) and converting the recalled sine samples to an analog waveform via a digital-to-analog converter (DAC).**

This paper describes a new architecture of a DDFS using a second-order IIR resonator, whose impulse response is a sine wave. Since this impulse response can be expressed in a second-order recursive form, sample generation only requires a few multiplications and additions. Therefore, the idea is to use a fast floating-point DSP microprocessor that can implement the recursive equation in "real time."

This technique guarantees a larger set of output frequencies, a high spectral purity, and a very simple, compact, and reliable hardware design.

I. INTRODUCTION

THE applications of frequency synthesizers are numerous. They are used in areas such as communication, control, music synthesis, and instrumentation. Designers must choose between using analog or digital synthesizers. Although stable synthesizers with low harmonic distortion can be built with analog circuits, automatic control of the parameters of the sine wave, namely amplitude, frequency, and phase, is difficult with this approach. Direct digital frequency synthesizers (DDFSs) have several advantages, including ease of control, flexibility, excellent temperature and aging stability, and low cost. They have the disadvantage of increased harmonic distortion caused by different numerical errors. Using a digital signal processor (DSP) provides complete control and the speed required to generate a wide range of frequencies with the ability to perform distortion reduction calculations in real time.

The most popular technique for direct digital frequency synthesis is the sine lookup table method first introduced by Tierney, Rader, and Gold [1]. This method synthesizes a sine wave by successively scanning through a lookup table stored in a read-only memory (ROM) and converting the recalled sine samples to an analog waveform via a digital-to-analog converter (DAC).

Using this architecture, we see that both the frequency and phase resolution of the synthesizer are determined by the wordlength of the phase accumulator. This technique has some limitations:

- The set of output frequencies is limited.
- The spectral purity is strongly related to the wordlength of the phase accumulator.
- A large memory is required to achieve an acceptable frequency and phase resolution.

A theoretical analysis of the DDFS output spectrum is available [2], which also suggests a method for improving the worst-case performance. In more recent work [3], the phase noise is related to resolution of the digital-to-analog converter (DAC) and to the intermodulation between output and clock frequencies.

The DDFS technique seems to be attractive for atomic clocks frequency synthesis. It has been studied for hydrogen masers [4], for rubidium cell standards [5], and for both laboratory [6] and commercial [7] cesium beam standards. This interest is due to the high resolution and to the fair spectral purity of the DDFS devices.

This paper describes a new architecture of a sine-wave DDFS. The basic idea is to use a second-order IIR resonator, whose impulse response is a sine wave; since this impulse response can be expressed in a second-order recursive form, sample generation only requires a few multiplications and additions. Therefore, the idea is to use a fast floating-point DSP microprocessor that can implement the recursive equation in "real time" [8]. This technique guarantees a large set of output frequencies, a high spectral purity, and a very simple, compact, and reliable hardware design.

The crucial point of the approach presented here is that the finite precision of calculations performed with the microprocessor does not guarantee that, in the numerical implementation of the IIR filter, the two complex conjugate poles in the z-plane are exactly on the unit circle. As a result, the amplitude of the output sine wave may not be stable in time and may increase or decrease according to the random direction toward which truncation errors move the poles from the unit circle, that is, outside or inside respectively. A number of techniques are described in this paper, which have been experimentally

Dipartimento di Elettronica, Politecnico di Torino, Corso Duca degli Abruzzi, 24 10129 Torino–Italy "Work funded by ASI and INFM"

tested and shown to be adequate to keep the output sine wave stable. These techniques are referred to as *reset methods.*

Numerical and experimental results are presented in the following, which show that the spectral purity of the analog output is limited by the resolution of the digital-to-analog converter. A comparison with the results obtainable with the lookup table method is also given.

II. THE SINE LOOKUP TABLE METHOD

In all DDFSs that use some variation of the lookup table technique, one sample of the sine wave is recalled from a ROM at each clock pulse of a stable frequency reference $f_s \cdot 2^L$ sinusoidal samples are stored in the ROM, and the sine frequency is assigned by means of a *frequency control word* F_r, which is an integer number, expressed as a binary word, in the range $1 \leq F_r \leq 2^{L-1}$. The phase increment $\theta(n) = nF_r$ addressing the ROM, is normally implemented by accumulating the binary frequency control word F_r, using a standard two's complement adder and a register (the *phase accumulator*). Therefore, in the ideal case of infinite precision memory and no phase quantization, the output sequence of the DDFS is given by

$$S(n) = \sin\left(2\pi \frac{F_r n}{2^L}\right) = \sin\left(2\pi \frac{\theta(n)}{2^L}\right) \qquad [1]$$

The phase accumulator operates on the principle of overflow arithmetic using the Modulo 2^L property of an L bit, periodically overflowing the accumulator register to simulate the Modulo 2π property of the sine function by exploiting the relationship

$$\sin\left(2\pi \frac{\langle\theta(n)\rangle_{2^L}}{2^L}\right) = \sin\left(2\pi \frac{\theta(n)}{2^L}\right) \qquad [2]$$

where $\langle\cdot\rangle_{2^L}$ stands for the integer residue of a number modulo 2^L.

A. Frequencies of the Output Sine Wave

The frequency of the output sine wave depends on the DAC clock frequency $f_s = 1/T_s$ and on the ROM dimension 2^L. The highest possible output frequency is

$$f_h = \frac{1}{2T_s} = \frac{f_s}{2}$$

while the lowest one is related to the ROM dimension L, that is

$$f_l = \frac{1}{2^L T_s} = \frac{f_s}{2^L}$$

The other output frequencies, f_o, can be obtained by decimating the sine wave at frequency f_{sl}; the decimation step is related to the frequency control word F_r, that is,

$$f_l = F_r \cdot f_{sl} = F_r \cdot \frac{f_s}{2^L}$$

B. Bit Truncation in the Phase Accumulator

Using this architecture, we determine both the frequency and phase resolution of the synthesizer by the wordlength of the phase accumulator. For most applications, a large-phase accumulator and, hence, a prohibitively large ROM are required to achieve an acceptable resolution. Therefore, in actual practice, all sine DDFSs, which require fine phase resolution, must truncate part of the phase accumulator output when addressing the ROM. In this way, smaller ROMs are needed.

Therefore, if the output of the phase accumulator is truncated to W bits, only 2^W (instead of 2^L) ROM positions can be addressed. In this case, if $B = L - W$ is the number of truncated bits, the phase value addressing the ROM will be the integer part of $(F_r n)/2^B$. Hence, the output sequence of the DDFS becomes

$$S_t(n) = \sin\left(\frac{2\pi}{2^W}\left[\frac{F_r n}{2^B}\right]\right) = \sin\left(\frac{2\pi}{2^L}2^B\left[\frac{F_r n}{2^B}\right]\right) \qquad [3]$$

where $[\cdot]$ stands for integer part; it follows that the sequence addressing the ROM can be written as

$$\left\langle\left[\frac{F_r n}{2^B}\right]\right\rangle 2^W$$

where $\langle\cdot\rangle_m$ means Modulo m.

The phase accumulator truncation has the effect of producing undesired spurs in the line spectrum of $S_t(n)$, as described in [2].

In the next sections, we present an altogether new approach for the generation of a sine wave, which does not require any ROM. A greater frequency resolution is achieved, and the errors, due to the implementation in a finite precision microprocessor, can be efficiently controlled.

III. THE IIR FILTER METHOD

A fast and efficient method to generate a digital sine wave is to implement a digital resonator in a recursive form. The basic idea is to design the digital resonator as an IIR filter with two complex conjugate poles on the unit circle on the z-plane, and to implement it by means of its input-output relationship, which is a simple recursive form. In fact, it is well known from the z-transform theory that the impulse response of a filter with two poles on the unit circle is a sine wave, whose frequency depends on the pole locations. Such sinusoidal impulse response can be written in a recursive form and generated by initializing the recursive equation with a digital impulse at time $n = 0$.

The theoretical impulse response cannot be perfectly achieved because of round-off errors. Therefore, the recursive algorithm that implements the IIR resonator must be modified in order to control the propagation of numerical errors during the sine-wave generation and to guarantee the filter stability and a low spur level.

Both the idea of using an IIR resonator and the methods implemented to reduce the effect of using a finite precision arithmetic are totally innovative with respect to the most popular technique for direct digital frequency synthesis described in Part II.

A. IIR Resonator

A digital resonator, implemented as an IIR filter with two complex conjugate poles $(e^{\pm j\varphi})$ on the unit circle, has a transfer function $H(z)$ of the form

$$H(z) = \frac{(\sin\varphi)z^{-1}}{1 - 2(\cos\varphi)z^{-1} + z^{-2}} \qquad [4]$$

while its impulse response, $h(n)$, is an undamped sinusoid with unitary amplitude, which can be written in the recursive form

$$h(n) = c_0\delta(n-1) + c_1 h(n-1) - h(n-2) \qquad [5]$$

where $\delta(n)$ is the Dirac delta sequence, defined as $\delta(n) = 1$ for $n = 0$ and $\delta(n) = 0$ for $n \neq 0$, $c_0 = \sin\varphi$ and $c_1 = 2\cos\varphi$. The implementation of eq. (5) only requires a few operations that can be executed using a fast DSP microprocessor.

The output frequency, f_o, of the sine wave is related to the pole locations by the equation

$$\varphi = 2\pi f_o T_s$$

where T_s is the DAC clock period. Therefore, f_o is related to the filter coefficients by the equations

$$c_0 = \sin 2\pi f_o T_s$$
$$c_1 = 2\cos 2\pi f_o T_s$$

These two coefficients must be evaluated by the DSP microprocessor before starting the sine-wave generation.

In our implementation of eq. (5), we impose some constraints on the value of f_o, writing it in the form

$$f_o = \frac{M}{N} f_s \qquad [6]$$

where $f_s = 1/T_s$ is the DAC clock frequency and M, N are integers. These two integer parameters are introduced to control the error propagation, as will be explained in section IV. By using eq. (6), the resonator coefficients can be written as

$$c_0 = \sin\left(2\pi\frac{M}{N}\right) \qquad c_1 = 2\cos\left(2\pi\frac{M}{N}\right) \qquad [7]$$

Hence, different frequencies may be synthesized by changing the values of M and N.

The maximum sampling frequency f_s turns out to be limited by DAC speed in high spectral applications, where about 16 bits are required. Were DAC speed limits removed, the maximum sampling frequency, in our implementation based on the TMS320C30 DSP, would be between 600 kHz and 3 MHz, depending on the choice of the reset method, which is discussed below.

B. Sine-wave Generation with a TMS320C30

The DSP microprocessor, used to implement eq. (5), is a third-generation Texas Instruments floating-point microprocessor (TMS320C30). The DDFS algorithm does not require a large memory; therefore, both instructions and data can be stored in the on-chip memory of the TMS320C30.

The procedure for the sine-wave generation can be divided into three phases:

1. Microprocessor initialization. Prior to the execution of a digital signal processing algorithm, it is necessary to initialize the processor. Generally, initialization takes place any time the processor is reset. After reset, the processor should be initialized to meet the requirements of the system. Instructions should be executed that set up operational modes, memory pointers, interrupts, and the remaining functions needed to meet system requirements.
2. Algorithm initialization. In this phase, the IIR coefficients c_0 and c_1 are calculated. This calculation is performed as follows:

   - Calculation of $\varphi = 2\pi(M/N)$
   - Calculation of $\cos\varphi$ and $\sin\varphi$

   The calculation of φ is complicated by the restriction that division should be avoided since the TMS320C30 has no divide instructions; therefore, good approximations of φ can be obtained using the Newton-Raphson algorithm [9], while $\sin\varphi$ and $\cos\varphi$ are calculated using the CORDIC algorithm [10]. These algorithms are optimized for DSP applications.
3. The third phase is the dynamic one, where the sinusoid samples are evaluated and sent to the DAC every T_s seconds. If τ_m is the TMS320C30 clock period, the microprocessor will spend $\tau_D = k_D \tau_m$ seconds, where $\tau_D < T_s$, to produce one sine sample. During this time interval, the arithmetic operations of the recursive equation and the instructions necessary to control the propagation error (see section IV) must be executed.

   In our implementation, we will try to reduce τ_D as much as possible. In fact, a low value of τ_D will give the microprocessor the possibility of spending some time $(\tau_R = T_s - \tau_D)$ to implement other functions.

IV. Reset Methods

In theory, eq. (5) allows the digital generation of an infinite precision sine wave; but in practice, its implementation, based on the TMS320C30, is affected by different numerical errors (round-off errors, propagation errors, drift errors, . . .). These errors can produce damping or divergence of the signal amplitude. Therefore, different techniques to control the amplitude have been conceived, which are described in the following sections.

A. IIR Filter Without Stability Control

This technique can be used if a short segment (few periods) of the sine wave has to be generated. In this case, no control of the numerical errors is generally necessary. In fact, the drifts toward an unstable or damped output are generally slow in a floating-point implementation of eq. (5), and they produce negligible errors for many periods of the sine wave. Therefore, in this case, a very simple software routine can be written in order to achieve a fast execution time and a small value of τ_D.

B. IIR Filter with Stability Control

A simple method to control the error propagation, when generating the sinusoidal impulse response, is to reset the recursive eq. (5) at each period of the generated signal. Notice that such period is generally different for the inverse of the sine-wave frequency f_o. It is very easy to show that the period of the sequence is given by

$$R = \frac{N}{(M, N)} \qquad [8]$$

where (M, N) is the greatest common divider of M and N. Therefore, the sinusoidal sequence repeats itself every R samples. Hence, it is possible to make the DDFS stable by resetting the recursive equation every R output samples. The evaluation of R exploits the Euclidean algorithm.

This technique has the following characteristics:

- High stability and ease of control,
- τ_D greater than the one of the previous technique; in fact, as it is necessary to implement a reset routine, a higher number of software instructions are needed,
- A longer time necessary to initialize the DDFS. But this is not a problem if a high switching frequency is not required.

C. IIR Filter with Stability Control and a Low Value of τ_D

With a technique similar to the previous one, it is possible to reduce τ_D by storing the first R sine-wave samples into the on-chip microprocessor memory.

After calculation of the reset period R, the signal generation can be divided into the following steps:

1. R sine samples are calculated and stored in the on-chip microprocessor memory.
2. The R samples, previously stored in memory, are outputted.

This technique is limited by the on-chip memory dimension of the microprocessor.

D. DCSW Method

The Dynamic Correction of Sine Waves method allows the user to get higher performances in the generation of a sine wave with an IIR resonator.

As each sine-wave sample is calculated using a recursive equation [see eq. (5)], the drift error is responsible for the performance degradation. In particular, the reset methods described previously do not guarantee a spur level of -135 dBc in their whole frequency range.

The basic idea of this method is to correct numerical errors by calculating some sine-wave samples with a higher accuracy than the one obtained using eq. (5). The number of higher accuracy samples calculated in a period of the output digital sequence of the DDFS can be chosen by the user.

The output samples of the DDFS can be divided into two sets:

1. Sine-wave samples calculated by means of the recursive equation.
2. Sine-wave samples calculated with an algorithm that guarantees high accuracy and fast generation.

Both sets are computed in "real time"; since the last group of samples requires a more complex process for their generation, and therefore a longer τ_D, at first sight it seems to be the main cause that limits the maximum output frequency, f_o. However, in an optimized implementation, the high accuracy sample calculation can be spanned over the dead time between the recursive equation samples.

This method guarantees an output signal with a high spectral purity, and it becomes possible to maintain the spur level below -135 dBc, independently of the output frequency.

E. The IIR-ALT Method

The IIR-Adaptive Lookup Table method is based on the same idea of the DCSW method but realizes it in a different way.

In the DCSW method, correction is made in "real time," while in the IIR-ALT method higher accuracy samples are calculated during the DDFS initialization. Then these samples are stored in a memory that can be accessed by the microprocessor. In this way, the maximum output frequency is higher than that of the DCSW method, for only a few microprocessor instructions are required to generate an output sample.

In the IIR-ALT method, the choice of the samples to be corrected is fundamental; it has been found that the best results are obtained if the correction is made before and after the waveform reaches its maximum values. Therefore, if the sine-wave argument is $2\pi(M/N)n$, then the selected samples must satisfy the following inequalities

$$2\pi \frac{M}{N} n \le \frac{\pi}{2} + 2k\pi \qquad 2\pi \frac{M}{N}(n+1) > \frac{\pi}{2} + 2k\pi$$

and

$$2\pi \frac{M}{N} n \le \frac{3\pi}{2} + 2k\pi \qquad 2\pi \frac{M}{N}(n+1) > \frac{3\pi}{2} + 2k\pi$$

Hence, if the number of periods of the sine wave at frequency f_o contained in the period of the output digital sequence is

$$P = \frac{M}{(M, N)}$$

then the number of samples, calculated and stored in memory during the microprocessor initialization, is $4P$.

The spectral purity of the sine wave generated with this technique is very high (the spur level is below -135 dBc).

V. Experimental Results

A block diagram of a DDFS implemented with the IIR filter method is shown in Fig. 1. The output frequency can be selected through a software program loaded in the on-chip memory of the TMS320C30. The digital samples generated by the microprocessor pass through a digital-to-analog interface, and finally the DAC constructs the analog output waveform.

In this section we will examine the spectral characteristics of the DDFS output signal referring to both the digital and analog domain. In the first case, the spur level is due to the numerical errors of the algorithm, while in the second one, the spur level is determined by the DAC's number of bits.

A. Numerical Output

In Figs. 2 and 3 the spectrum of a digital sine wave at frequency $f_o = (3/32) \cdot f_s$ is shown as obtained from a simulation with the ROM method using two different ROM sizes.

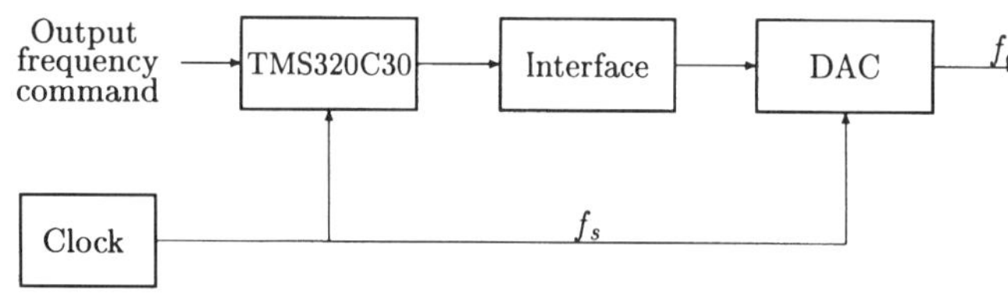

Fig. 1. DDFS block diagram.

ROM method

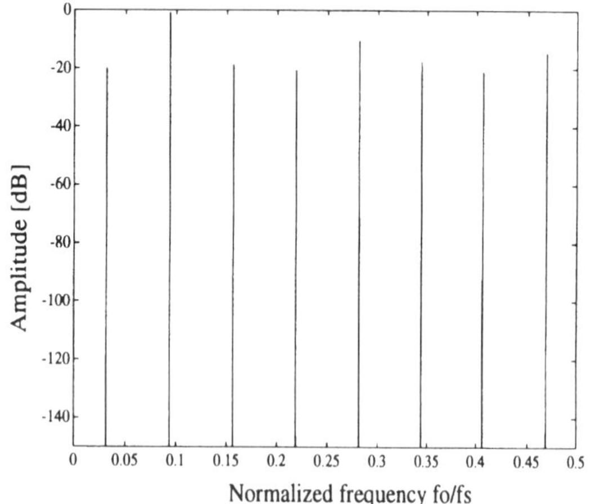

Fig. 2. Simulated amplitude spectrum of a sine wave at frequency $f_o = (3/32)f_s$, where f_s is the sampling frequency; the relevant synthesis parameters are $L = 5$ (phase accumulator wordlength) and $B = 3$ (phase accumulator discarded bits). The DAC's truncation error is not taken into account.

ROM method

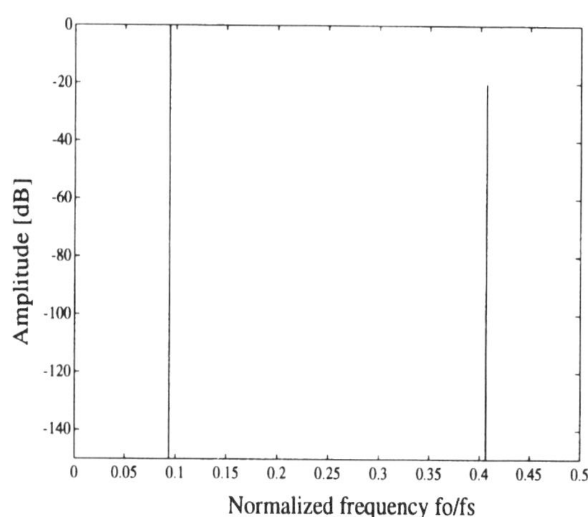

Fig. 3. Simulated amplitude spectrum of a sine wave at frequency $f_o = (3/32)f_s$, where f_s is the sampling frequency; the relevant synthesis parameters are $L = 5$ (phase accumulator wordlength) and $B = 1$ (phase accumulator discarded bits). The DAC's truncation error is not taken into account.

Since the simulation involves only the digital circuitry of the synthesizer, thus neglecting the DAC's truncation error, in principle we expect the same results from an experiment, if we consider the spectrum of the digital signal before the DAC.

The frequency $f_o = (3/32) \cdot f_s$ belongs to the set of frequencies that can be generated with both the IIR and the sine lookup table method. This choice was determined by the need to compare the two methods.

The spectrum of this same frequency, as obtained from a simulation with the IIR method, is shown in Fig. 4. As in the ROM method simulations, the DAC's truncation error has been neglected. Consequently, this last result can be compared directly to the results of Figs. 2 and 3. Spectra calculated for the IIR method have been found to be very similar in spectral purity for any frequency.

IIR method

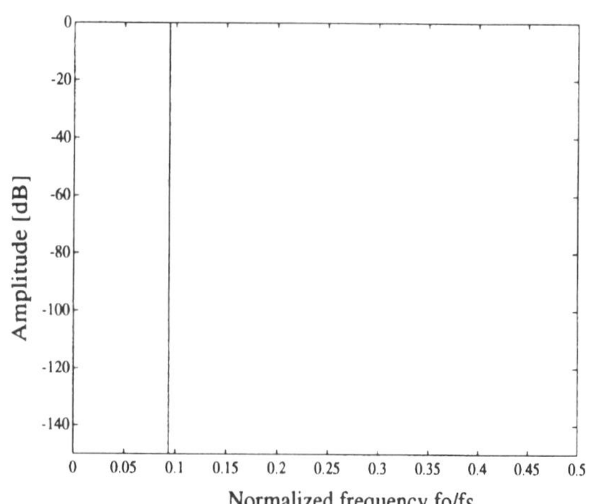

Fig. 4. Simulated amplitude spectrum of a sine wave at frequency $f_o = (3/32)f_s$. The DAC's truncation error is not taken into account.

B. Analog Output

Experimental results are given as obtained by synthesizing a 1.21-kHz sine wave with a 16-bit DAC. This frequency was chosen because it is in the range needed by our final application, in a primary cesium frequency standard. It is shown in Figs. 5 and 6 that the spur level and the spectral purity of the output signal are limited by the resolution of the DAC. If a smaller number of bits were used in the output DAC, for example, for speed reasons, the spectral purity would accordingly turn out to be worse. Reported spectra were obtained with an FFT analyzer. It must be pointed out that the latter is also a 16-bit machine.

VI. Conclusions

In this paper a new architecture for a sine-wave output DFS was described. Its implementation is based on a very simple recursive equation representing the impulse response of an IIR resonator. Few instructions of the TMS320C30 allow the generation of a larger set of output frequencies than the one obtained with the sine lookup table method. Besides, other frequencies have been tested; the results validate the IIR resonator as an interesting candidate for a new hardware architecture for DDFS implementation.

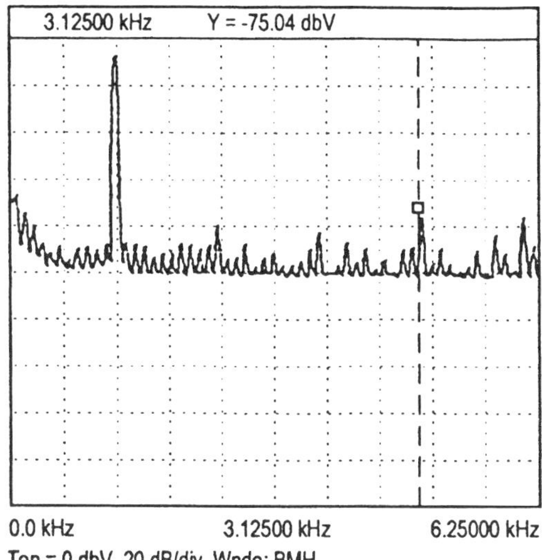

Fig. 5. Amplitude spectrum of a 1.21 kHz sine wave. The relevant spectrum analyzer settings are: top of screen 0 dBV, vert. span 20 dB/div, hor. span 625 Hz/div.

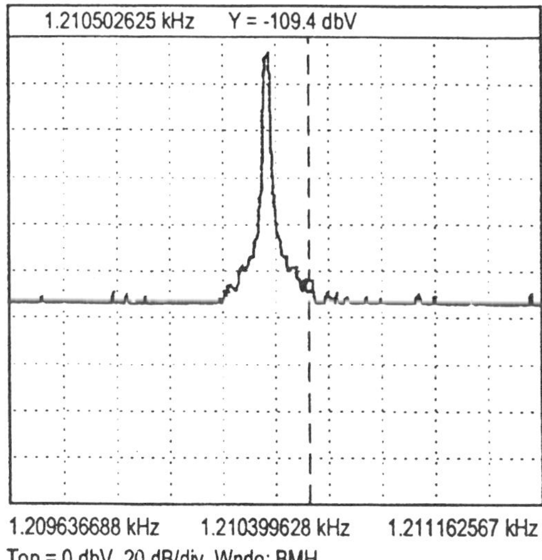

Fig. 6. Amplitude spectrum of a 1.21 kHz sine wave. The relevant spectrum analyzer settings are: top of screen 0 dBV, vert. span 20 dB/div, hor. span 153 mHz/div.

References

[1] J. Tierney, C. M. Rader, and B. Gold. "A digital frequency synthesizer." *IEEE Transactions on Audio Electroacoustics,* vol. AU-19, pp. 48–57, 1971.

[2] H. T. Nicholas III and H. Samueli. "An analysis of the output spectrum of direct digital frequency synthesizers in the presence of phase-accumulator truncation." *Proceedings of the 41st Annual Frequency Control Symposium,* pp. 495–502, IEEE, 1987.

[3] V. F. Kroupa. "Intermodulation signals in direct digital frequency synthesizers." *Proceedings of the 8th European Frequency and Time Forum,* Weihnstephan, Germany, March 9–11, 1994.

[4] E. M. Mattison and L. M. Coyle. "Phase noise in direct digital synthesis." *Proceedings of the 42nd Annual Frequency Control Symposium,* pp. 352–56, IEEE, 1988.

[5] P. Schumacher. "A narrow band frequency synthesizer with a NCO circuit." *Proceedings of the 5th European Frequency and Time Forum,* pp. 123–28, Besançon, France, March 12–14, 1991.

[6] J. Viennet. "Synthetiseur digital de fréquence pour standards atomique de fréquence." *Proceedings of the 3rd European Frequency and Time Forum,* pp. 333–36, Besançon, France, March 21–23, 1989.

[7] R. P. Giffard and L. S. Cutler. "A low frequency, high resolution digital synthesizer." *Proceedings of the 46th Annual Frequency Control Symposium,* pp. 188–92, Philadelphia, USA, May 27–29, 1987.

[8] G. Cardamone and L. Lo Presti. "A direct digital frequency synthesizer using an IIR filter implemented with a DSP microprocessor." *Proceedings of the 1994 International Conference on Acoustics, Speech & Signal Processing,* Adelaide, Australia, April 19–22, 1994.

[9] Texas Instruments. *Theory, Algorithms, and Implementations.* © Texas Instruments Inc., 1990.

[10] M. D. Ercegovac and T. Lang. "Fast cosine/sine implementation using on-line CORDIC." *Proceedings of the 21st Annual Asilomar Conference on Signals, Systems and Computers,* 1987.

[11] Texas Instruments. *Third-Generation TMS320 User's Guide.* © Texas Instruments Inc., 1989.

Part XI

Mathematical Background

THIS last Part of the book presents two specially written papers by the editor. The first one deals with the theory of quasiperiodic omission of pulses supported with several examples. In the second paper, the editor summarizes several computer programs useful for investigating the theoretical properties of DDFS.

The aliasing problem is closely related to the sampling process, the mathematics of which is based on the z-transform. For readers willing to delve into the necessary mathematics, we recommend the paper by Angelo [XI-1] which can serve as a simple introduction to the properties of the z-transform.

REFERENCE

[XI-1] E. J. Angelo, Jr. "A tutorial introduction to digital filtering." *The Bell System Technical Journal,* vol. 60, No. 7, pp. 1455–1546, September 1981.

Quasiperiodic Omission of Pulses

VĚNCESLAV F. KROUPA

I. Introduction

AS a result of the progress in frequency synthesis in the last years [I-1 through I-5] and the extensive applications of digital IC technology, the frequency synthesizer is no longer an expensive piece of laboratory equipment but a commonly used electronic device and electronic component. From this point of view, systems based only on IC-logic circuits would have many advantages.

The difficulty with practical realization is poor spectral purity. This need not be a problem in some instances [1], but otherwise we must look for the means to reduce the undesirable output signals either with the assistance of an output divider [2] or even filter [3]. In all cases, knowledge of the power level of spurious components, together with their frequencies, is advantageous. The problem will be discussed in detail later in this paper.

II. Principles

A. Pulse Train with Periodically Subtracted Pulses

We will consider a pulse train in which pulses are placed in regular distances T_i on the time axis. If we remove each q_1 pulse (see Fig. 1), then one period is missing and the phase undergoes a step change equal to 2π, that is,

$$\phi_{s1}(t) = \omega_i t - 2\pi U(t_k) \qquad [1]$$

where $U(t_k)$ is the step function [6] occurring at instances

$$t_k = kq_1 T_i$$
$$(k = \cdots - 2, -1, 0, 1, 2, \ldots) \qquad [2]$$

After differentiation of eq. (1), we get the instantaneous frequency

$$\omega_o(t) = \omega_i - 2\pi \delta(t - kq_1 T_i) \qquad [3]$$

which with the assistance of the Poisson Sum Formula [7] may be expressed as

$$\omega_o(t) = \omega_i - \frac{\omega_i}{q_1} \sum_{m=-\infty}^{\infty} e^{jm\frac{\omega_i}{q_1}t}$$

$$= \omega_i - \frac{\omega_i}{q_1} - \frac{\omega_i}{q_1} \sum_{m=-\infty/m\neq 0}^{\infty} e^{jm\frac{\omega_i}{q_1}t} \qquad [4]$$

$$= \omega_{s1}\left[1 - \dot{s}_1(t)\right]$$

The instantaneous frequency (4) has a steady-state component ω_{s1}

$$\omega_{s1} = \omega_i \frac{q_1 - 1}{q_1} \qquad [5]$$

and frequency-modulated part

$$\omega_{s1}\dot{s}_l(t) = \frac{\omega_{sl}}{q_1 - 1} \sum_{-\infty/m\neq 0}^{\infty} e^{jm\frac{\omega_i}{q_1}t} \qquad [6]$$

The waveform of the output phase modulation is found by integration of eq. (6)

$$\phi_{s1} = \omega_{s1}s_1(t) = \frac{1}{j} \sum_{-\infty/m\neq 0}^{\infty} \frac{1}{m} e^{jm\frac{\omega_i}{q_1}t} \qquad [7]$$

In the trigonometric form we get

$$\omega_{s1}s_1(t) = 2 \sum_{m=1}^{\infty} \frac{1}{m} \sin\left(m\frac{\omega_i}{q_1}t\right) \qquad [8]$$

By comparing the above relation with the published Fourier series, we find that it is identical with the expansion of the sawtooth periodic wave with the amplitude $A = 2\pi$ [cf. 6]. See Fig. 2.

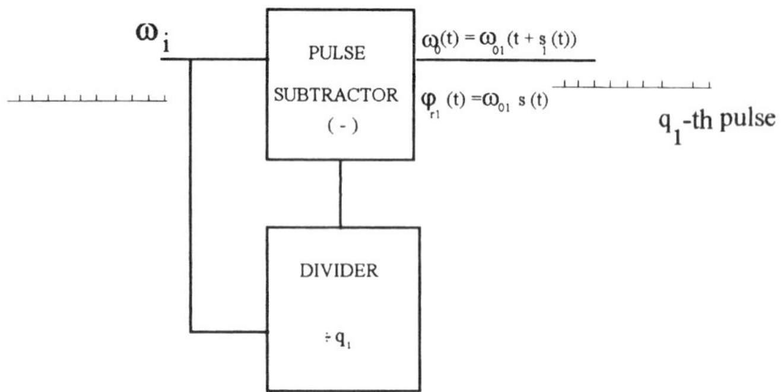

Fig. 1. Idealized arrangement for "swallowing" of each q_1-th pulse from a periodic pulse train.

(This paper was specially written for this volume.)

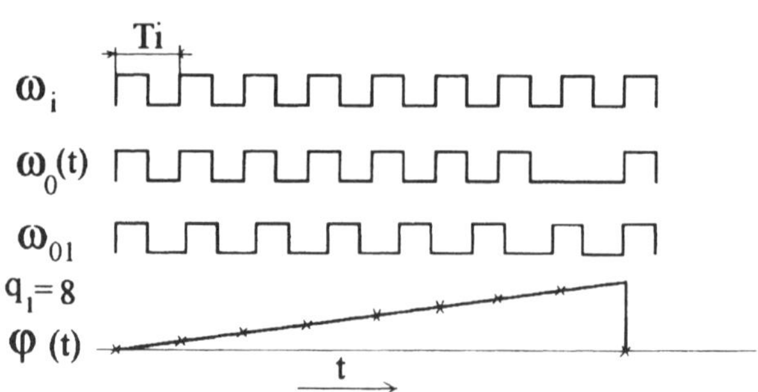

Fig. 2. Spurious phase modulation by quasiperiodic omission of each 8th pulse $(q_1 = 8)$.

B. Position-Modulated Pulse Train Passed Through a Digital Divider

Here, we will simplify the operation of the digital divider, with the division ratio q_2, in such a way that its output appears only at each q_2-th pulse (see Fig. 3). In this case, the idealized output time function $f_\delta(t)$ is

$$f_\delta(t) = \delta(t - q_2 t_r) \qquad [9]$$

As the pulses at the output of the divider q_2 must coincide with pulses of the original pulse train, they exhibit a time modulation $s_2(t_r)$. Thus, with the assistance of the relation (5), we get

$$q_2 t_r = q_2 r \frac{q_1}{q_1 - 1} T_i - s_2(t_r) \qquad [10]$$

where the amplitude of the modulation function $s_2(t_r)$ cannot exceed one T_i. From this condition, we easily deduce the analytical form of $s_2(t_r)$, that is,

$$s_2(t_r) = T_i \left[r \frac{q_1 q_2}{q_1 - 1} - \mathrm{int}\left(r \frac{q_1 q_2}{q_1 - 1} \right) \right] \qquad [11]$$

It is clear that in instances where q_2 and $q_1 - 1$ are relatively prime the fundamental period of the modulation function $s_2(t_r)$ is

$$T_{\mathrm{mod}} = q_1 q_2 T_i \qquad [12]$$

Otherwise

$$T_{\mathrm{mod}} = \frac{q_1 q_2}{p} T_i \qquad [13]$$

where p is the largest common divisor of q_2 and $q_1 - 1$.

C. Divided Position-Modulated Pulse Train Subtracted from the Original Pulse Train

We will repeat the same procedure as above, namely, remove from the original pulse train all the pulses that coincide with the output pulses of the divider q_2 (see Fig. 3). Evidently, the phase at the output of the second pulse subtracter undergoes step changes at times $q_2 t_r$, specified by eq. (10)

$$\phi_{s2}(t) = \omega_i t - 2\pi U(q_2 t_r) \qquad [14]$$

The instantaneous frequency is again found by differentiating the above equation (see Appendix):

$$\omega_{s2}(t) = \omega_i - 2\pi \delta \left[t - r \frac{q_1 q_2}{q_1 - 1} T_i + s_2(t_r) \right]$$

$$= \omega_i - \omega_i \frac{q_1 - 1}{q_1 q_2} [1 + \dot{s}_2(t)] \sum_{-\infty}^{\infty} e^{jn\omega_i \frac{q_1-1}{q_1 q_2}[t+s_2(t)]} \qquad [15]$$

Its steady-state component is

$$\omega_{s2} = \omega_i \frac{q_1 q_2 - q_1 + 1}{q_1 q_2} \qquad [16]$$

the fundamental period of the modulation function is that of $s_2(t_r)$—see eq. (11)—and the phase modulation index again never exceeds 2π. Eventually, this is one way to subtract from each group of $q_1 q_2$ pulses the required $(q_1 - 1)$ pulses that would result in a smallest spurious phase modulation. This conclusion follows immediately from the analytical form of the modulation function $s_2(t_r)$ in eq. (11).

After an inductive reasoning, we will find the mean output frequency ω_{s3}, and the modulation function $s_3(t_r)$ at the next pulse subtracter

$$\omega_{s3} = \omega_i - \frac{\omega_{s2}}{q_3}$$

$$= \omega_i \left[1 - \frac{1}{q_3} \left(1 - \frac{q_1 - 1}{q_1 q_2} \right) \right] \qquad [17]$$

$$= \omega_i \left\{ 1 - \frac{1}{q_3} \left[1 - \frac{1}{q_2} \left(1 - \frac{1}{q_1} \right) \right] \right\}$$

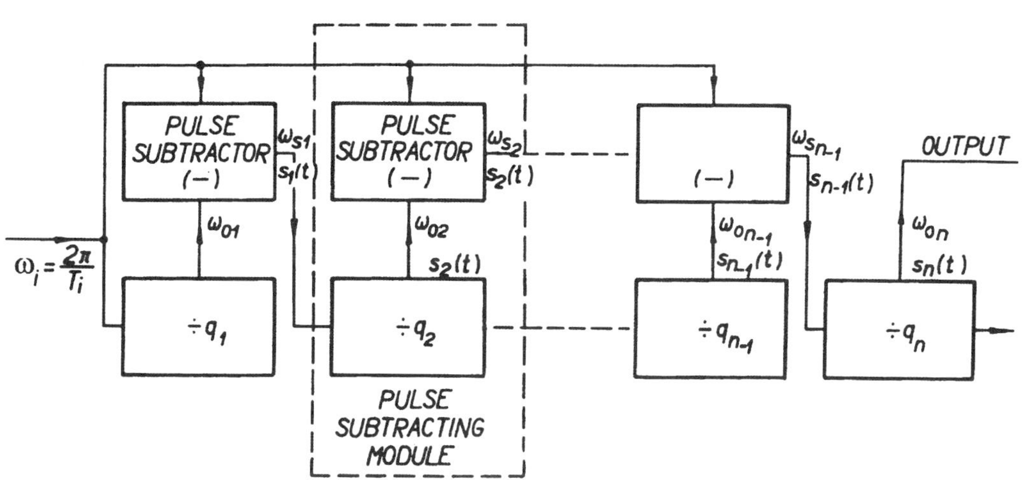

Fig. 3. Position-modulated pulse train passed trough a digital divider. See output of ω_{02} and $s_2(t)$; further divided position-modulated pulse train subtracted from the original pulse train.

$$s_3(t_r) = T_i \left[r \frac{q_1 q_2 q_3}{q_1 q_2 - q_1 + 1} - \mathrm{int}\left(r \frac{q_{q_2} q_3}{q_1 q_2 - q_1 + 1} \right) \right]$$

$$= T_i \left[r \frac{\omega_i}{\omega_{o3}} - \mathrm{int}\left(r \frac{\omega_i}{\omega_{03}} \right) \right] \qquad [18]$$

Both eqs. (17) and (18) can be generalized to any number of pulse-subtracting (swallowing) modules of which the last one is, usually, reduced to a mere divider (see Figs. 3 and 4). Thus, the normalized output frequency is

$$\frac{\omega_{on}}{\omega_i} = \frac{X}{Y}$$

$$= \frac{1}{q_n}\left(1 - \frac{1}{q_n} - 1\left(1 - \cdots - \frac{1}{q_2}\left(1 - \frac{1}{q_1} \right) \cdots \right) \right) \qquad [19]$$

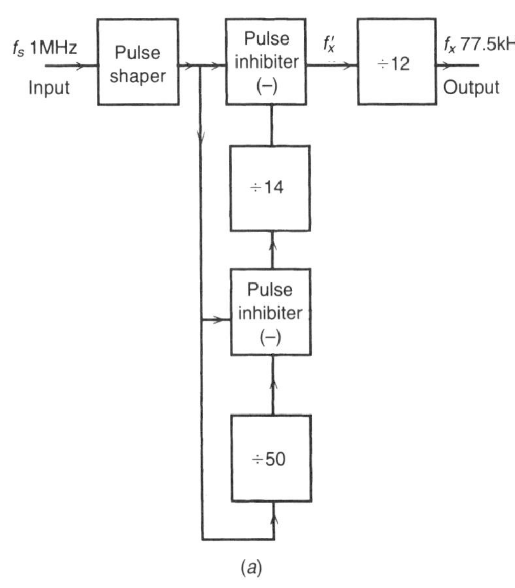

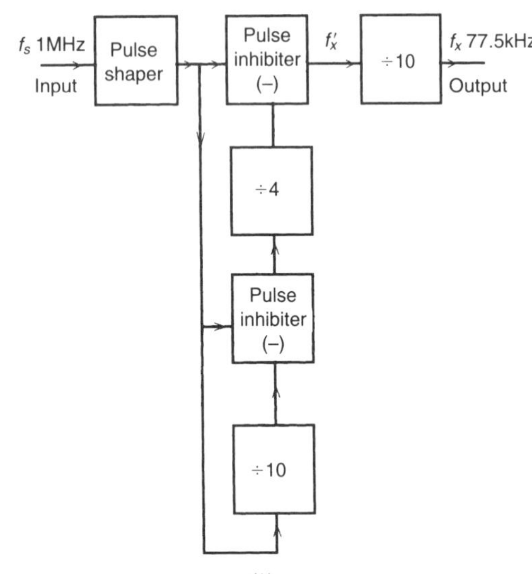

Fig. 4. Block diagram of DDFS for generation of the carrier frequency, 77.5 kHz from the standard frequency 1 MHz.
(a) Application of the relation (24).
(b) Modification of ξ_x into $(\gamma_o/10) \cdot \gamma'_o$ in accordance with the relation (25).

III. Systems Based on the Modified Engel Series Expansions

Up to now, no constrains have been imposed on the division factors q_1, q_2, q_3 besides the condition that the sum of the series (19) must be equal to the normalized output frequency X/Y. However, this expansion resembles the modified Engel series introduced by the editor [4], and there are no reasons why this algorithm could not be used as the mathematical model for the pulse rate frequency synthesizers. On the contrary, there are significant advantages: any rational number may be expanded into a finite modified Engel series. The expansion is unique and converges rapidly.

Unfortunately, the last property would result in unnecessarily large division factors $q_n = p_1, q_{n-1} = p_2, \ldots, q_1 = p_n$ in all instances where

$$\frac{X}{Y} \ll 1 \qquad [20]$$

which is, however, a situation often encountered. By application of the Engel series algorithm, we get for the first division factor p_1

$$p_1 = \mathrm{int}\left(\frac{Y}{X} \right) \qquad [21]$$

whereas all others p_k $(k > 1)$ are larger than p_1.

The difficulty can be overcome by finding two factors of Y

$$Y = Y_1 Y_2 \qquad [22]$$

in such a way that the ratio X/Y_2 is smaller but as close as possible to 1 and by subjecting only this term to the modified Engel series expansion. Eventually, the mathematical model for a pulse rate frequency synthesizer is as follows [4]:

$$\frac{f_x}{f_s} = \frac{X}{Y} = \frac{1}{Y_1} * \frac{X}{Y_2} = \frac{1}{Y_1} \sum_{k=1}^{n} \frac{(-1)^k}{p_1 p_2 \cdots p_n} \qquad [23]$$

Example 1. Let us synthesize the carrier frequency of the German standard frequency and time signal transmitter (in accordance with the [2]):

$$f_x = 77.5 \text{ kHz}$$

from the standard frequency

$$f_s = 1 \text{ MHz}$$

Evidently, the normalized output frequency γ_o is equal to

$$\gamma_o = 77.5/1000 = 31/400$$

From the paper by Kroupa [4] we get

$$c_o = 1$$
$$\gamma_1 = 1/(c_0 - \gamma_o) = 400/369$$

from which

$$p_1 = \mathrm{int}(\gamma_o) = 1$$

Further

$$\gamma_2 = \gamma_1/(\gamma_1 - p_1) = 400/31$$

and

$$p_2 = \text{int}(400/31) = 12$$

Similarly, we find that $p_3 = 14$ and $\gamma_4 = 50 = p_4$. Since γ_4 is an integer, the expansion into the modified Engel series is terminated and the mathematical model of the synthesizer might be as follows:

$$\gamma_o = 1 - 1 + \frac{1}{12}\left(1 - \frac{1}{14}\left(1 - \frac{1}{50}\right)\right)$$
$$= \frac{1}{12}\left(1 - \frac{1}{14}\left(1 - \frac{1}{50}\right)\right) \qquad [24]$$

The respective block diagram is shown in Fig. 4a.

Example 2. As stated above, we can arrive at a smaller division factors in instances where the ratio of the nominator and denominator of the effective γ_o is as close as possible but smaller than 1 (cf. eq. 22). After examination of the above example, the evident choice for Y_1 is 10 and the expansion of the considered normalized frequency simplifies to

$$\gamma_o = \frac{1}{10}\left[1 - \frac{1}{4}\left(1 - \frac{1}{10}\right)\right] \qquad [25]$$

This pulse rate frequency synthesizer was suggested by Becker, and its block diagram is shown in Fig. 4b; note the simplification in respect to Fig. 4a.

Example 3. Let us generate the European PAL color subcarrier frequency

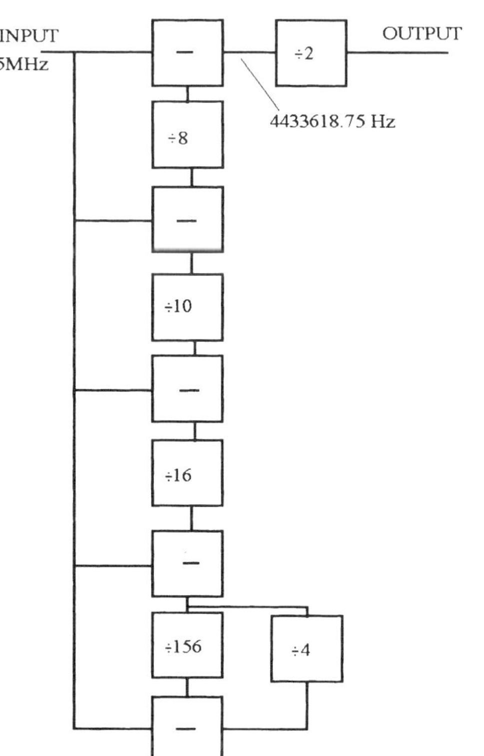

Fig. 5. Block diagram of DDFS for generation of one-half of the color subcarrier frequency in the European PAL TV system.

$$f_x = 4.433\ 618\ 75 \text{ MHz}$$

from the standard frequency

$$f_s = 5 \text{ MHz}$$

For filtering of spurious signals, we use a PLL system with a crystal VCO. Since the spurious phase modulation may be as high as 2π (in which case the possibility of cycle-slipping in the filtering PLL cannot be excluded), we will reduce the reference frequency with a divider by two. Consequently, the normalized frequency is

$$f_o/2f_x = (1/2) * (709379/800000)$$

and the expansion into a modified Engel series leads to

$$\frac{f_o}{2f_s} = \frac{1}{2}\left(1 - \frac{1}{8}\left(1 - \frac{1}{10}\left(1 - \frac{1}{16}\left(1 - \frac{1}{156}\left(1 - \frac{1}{625}\right)\cdots\right)\right)\right)\right) \qquad [26]$$

The resulting pulse rate frequency synthesizer is shown in Fig. 5 where the last two terms have been combined with the assistance of the continued fraction algorithm to spare the large divider by 625. The procedure will be explained in detail in the next section.

IV. Systems Based on Continued Fraction Expansions

In the previous sections, when investigating the quasiperiodic omission of pulses, we have arrived at the modified Engel series expansions. However, since continued fraction expansion provides the best possible step-by-step approximation of real numbers, it must also be a useful model for frequency synthesizers, particularly of the pulse rate and direct digital versions.

A. Simple Fractional Dividers

By assuming that the output frequency f_x is smaller than the input standard frequency f_s, the partial quotient b_o (in accordance with the APPENDIX I in the previous paper by Kroupa [4]) is equal to zero. Consequently, the first convergent is

$$\xi_x = \frac{X}{Y} \approx \frac{1}{b_1} \qquad [27]$$

and the second convergent is

$$\xi_x \approx \frac{b_2}{b_1 b_2 + 1} = \frac{A_2}{B_2} \qquad [28]$$

In instances where the normalized frequency is just equal to the second convergent, the rearrangement of the above relation reveals

$$f_s b_2 = f_x(b_1 b_2 + 1) \qquad [29]$$

from which

$$f_s - \frac{f_x}{b_2} = f_x b_1 \qquad [30]$$

and the block diagram of the respective DDFS is shown in

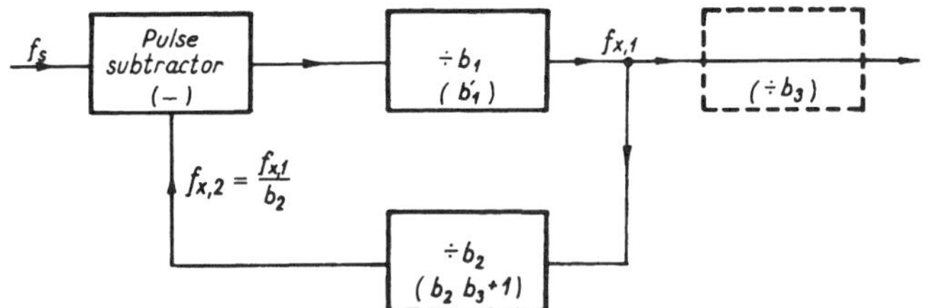

Fig. 6. Block diagram of DDFS based on the second convergent of the continued fraction expansion only.

Fig. 6. Note that with such an arrangement we can realize a rectangular output as symmetrical as possible, even in instances where the output division factor is odd, by putting $b_2 = 1$ and by decreasing the actual b_1 by one.

Furthermore, note that the exceptionally third-order convergent can be applied in this simple arrangement, for example, in instances where b_3 is a factor of b_1, then the mathematical model is

$$\xi_x = \frac{b_2 b_3 + 1}{b_3 \left[\dot{b}_1 (b_2 b_3 + 1) + 1 \right]} \equiv \frac{A_3}{B_3}; \quad \text{where} \quad \dot{b}_1 b_3 = b_1$$

[31]

For an actual arrangement, again see Fig. 6 and a simple numerical example:

Example 4. Let the normalized frequency be

$$\xi_x = 5/22 = [0, 4, 2, 2];$$
$$\text{i.e., } b_0 = 0, b_1 = 4, b_2 = 2, b_3 = 2$$

(for the continued fraction expansion we have used notation introduced by Perron; see Appendix I in [4]). With the assistance of eq. (31) we arrive at

$$\xi_x = (1/2) * \left(5/(2 * 5 + 1) \right)$$

B. Product Approximations

Another solution, resulting in a chain of fractional dividers, is provided by a suitable factorizaton of X and Y in relation (27) to meet conditions (31). An actual example provides synthesis of $10/2\pi$ kHz from 5 MHz discussed by Small [3]. We review it here in the following example.

Example 5. The fourth-order approximation of the normalized frequency, with the assistance of the continued fractions, reveals

$$\xi_x \doteq 27/84823$$

with the error

$$R_4 \approx -6.2 * 10^{-12},$$

i.e., $31 * 10^{-6}$ Hz approximately or 195 parts in 10^{10}

By computing the prime factors of the denominator [e.g. with the assistance 8] we get

$$84823 = 271 * 313$$

Consequently, the normalized frequency can be expanded into the product of two partial fractions:

$$\xi_x = (27/271) * (1/313)$$

Expansion of the first partial fraction reveals $b_1 = 10$ and $b_2 = 27$. To get a nearly rectangular output, we apply on the 1/313 the above-mentioned algorithm and arrive at the block diagram in Fig. 3 in the paper by Small [3].

C. Application of the Complete Continued Fraction Expansion

The procedure discussed in Example 5 is again a trial-and-error method (gear box method). To get a more rigorous solution, we will replace the partial quotient b_2 in eq. (30) by its better approximation

$$\xi_2 \approx b_2 + \frac{1}{b_3}$$

[32]

in accordance with Fig. 6 and generate the needed auxiliary frequency $f_{x,2}$ with the assistance of another fractional frequency divider from $f_{x,1}$. Proceeding further and further in the same way, we will eventually arrive at the last partial quotient b_n [7].

The block diagram of such a frequency synthesizer is shown in Fig. 7.

The governing recurrence formulas are

$$f_s - f_{x,2} = b_1 f_{x,1}$$
$$f_{x,1} - f_{x,2} = b_2 f_{x,2}$$
$$\vdots$$

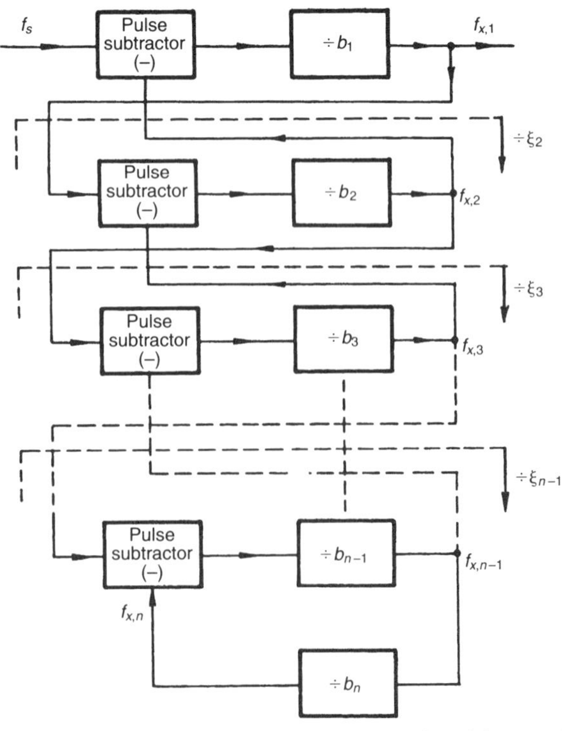

Fig. 7. Block diagram of DDFS employing the expansion of the normalized frequency f_{x1}/f_s into a continued fraction.

$$f_{x,k} - f_{x,k+2} = b_{k+1} f_{x,k+1} \qquad [33]$$

$$\vdots$$

$$f_{x,n-2} - f_{x,n} = b_{n-1} f_{x,n}$$

$$f_{x,n-1} = b_n f x, n$$

and

$$\frac{f_{x,k}}{f_{x,k+1}} = \xi_{k+1} \qquad [34]$$

The advantage of this approach is a rather simple hardware even in instances where X and Y in eq. (19) are products of large prime numbers or are prime numbers themselves. The difficulty is in instances where we encounter several b_n equal to 1. Solution of this problem is discussed in the next section.

V. Systems Based on Combinations of Expansions

We have already seen, in Example 1, that a rigorous application of the modified Engel series need not necessarily lead to the simplest hardware arrangement. The same may also happen when the continued fraction algorithm is used. We will illustrate the problem with the following example.

Example 6. Design a generator of the sidereal time from the mean solar time. Their relation is given approximately by

$$\xi_x \approx \frac{366.24219430}{365.24219430} \qquad [35]$$

In the past, only mechanical gear trains were available for this purpose (e.g. [9]). However, the technological progress made it possible to solve the problem with the assistance of the electronic means [1,10,11]. Our task in this Example is to design a "sidereal time" generator with the output frequency in the kHz range for driving astronomical instruments.

Let the input frequency be 100 kHz in the UTC (Coordinated Universal Time) time scale and the output frequency approximately 10 kHz in the sidereal time scale. In that case, the normalized frequency is

$$\xi_x = 10.027\,379\,093\,507\,95/100$$

(The nominator was computed in the Astronomical Institute of the Czechoslovak Academy of Sciences.) The expansion into a continued fraction is reprinted in Table 1. Note the slow convergence caused by the consecutive b_n's equal to 1. By following the rigorous solution as explained in Section IV, the frequency synthesizer hardware would require a large number of pulse inhibitors (cf. Fig. 7). However, we can simplify the system by applying the modified Engel series expansion from $1/\xi_4$ downward. Note that

$$\xi_o = \cfrac{1}{9 + \cfrac{1}{1 + \cfrac{1}{35 + \frac{1}{\xi_4}}}} \qquad [36]$$

and

$$\frac{1}{\xi_4} = \frac{201}{322} = 1 - \frac{1}{2}\left[\left(1 - \frac{1}{4}\left(1 - \frac{1}{161}\right)\right)\right] \qquad [37]$$

TABLE 1. CONTINUED FRACTION EXPANSION OF THE NORMALIZED FREQUENCY IN ξ_X IN RELATION (35).

n	b_n	A_n	B_n	R_n
0	0	0	1	
1	9	1	9	-10^{-3}
2	1	1	10	$2.7*10^{-4}$
3	35	36	359	$-4.8*10^{-6}$
4	1	37	369	$2.8*10^{-6}$
5	1	73	728	$-1.0*10^{-6}$
6	1	110	1097	$3.2*10^{-7}$
7	1	183	1825	$-1.8*10^{-7}$
8	1	293	2922	$5.8*10^{-9}$
9	19	5750	57343	$-1.1*10^{-10}$
10	2	11793	117608	$3.6*10^{-11}$

In addition, after rearranging the last two terms in the above modified Engel series with the assistance of the continued fraction algorithm, we get

$$\frac{1}{4}\left(1 - \frac{1}{161}\right) = \frac{40}{161} = \frac{40}{40*4+1} = \frac{1}{4 + \frac{1}{40}} \qquad [38]$$

This approach will save one pulse inhibitor, since we introduce a simple fraction divider in accordance with Fig. 6. The final block diagram of the realized sidereal time generator is shown in Fig. 8 [11].

VI. Conclusions

In the present paper we have discussed a theoretical approach to the problem of quasiperiodic omission of pulses. The advantages are a simple DDFS with good spectral purity, actually with the best one in a pulse train or rectangular waves output.

First, we have shown that a pulse train with periodically subtracted pulses exhibits phase modulation by a sawtooth wave.

Then we have investigated the modulation function of such a pulse train passed through a digital divider.

In the third step, we have removed pulses from the original train coinciding with the output pulses of the divider.

This step-by-step approximation eventually resulted in the chain of pulse "swallers" (inhibitors and subtracters—[12]) and dividers in the hardware and modified Engel series in the software approach.

The modified Engel series converges rapidly; that is, it provides very good step-by-step approximations of real numbers. By having this in mind, we have started with investigation of continued fractions, which present the best step-by-step approximations of real numbers, as mathematical models for DDFS and we have arrived at a feasible hardware model. However, in some instances combination with modified Engel series expansion may result in some simplifications.

For better understanding, we have supported our theoretical investigations with practical examples.

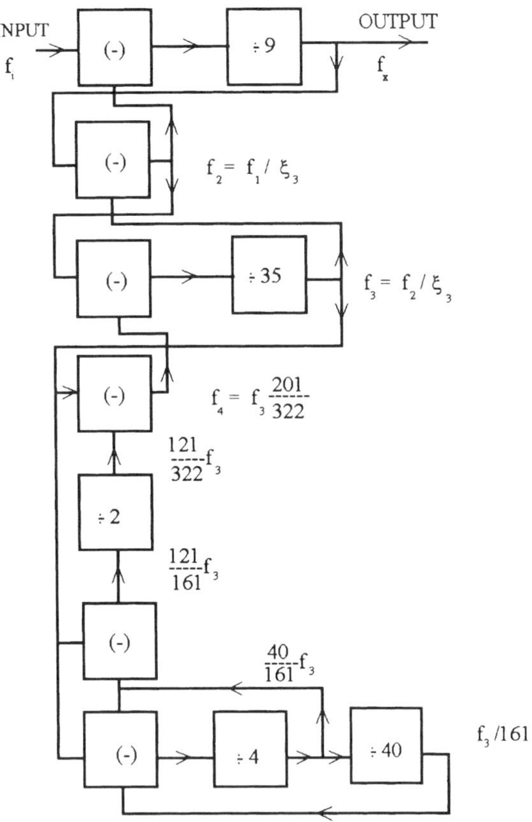

Fig. 8. Block diagram of DDFS generating the sidereal time from mean solar time.

VII. Appendix

1. Pulse Trains

By considering a train of very narrow pulses, we get for the Fourier expansion

$$f(t) = \frac{S}{T} \sum_{-N}^{N} e^{j2\pi n \omega t} + R_{\text{emainder}} \qquad [39]$$

where S is the area of one pulse, T the repetition period, and $\omega = 2\pi/T$. In the limiting case the effective current-flow angle $\gamma \to 0$ and the pulse train change effectively into a train of δ-functions. For $S = 2\pi$ we get

$$f_\delta(t) \doteq \frac{2\pi}{T} \sum_{n=-\infty}^{\infty} e^{jn\omega t}$$

$$= 2\pi \sum_{k=-\infty}^{\infty} \delta(t - kT) \qquad [40]$$

In instances where the pulse train $f_\delta(t)$ is position modulated (natural sampling), we must replace $\delta(t - kT)$ with $\delta[g(t)]$ where

$$g(t) \equiv t - kT - s(t) = 0 \qquad [41]$$

with t_k being zeros of a function $g(t)$. Consequently, we can transform eq. (40) by using the following properties of δ functions [13]:

$$\delta[g(t)] = \frac{\delta(t - t_k)}{|\dot{g}(t)|} \qquad [42]$$

By substituting eqs. (41) and (42) into (40), we obtain the desired result:

$$f_\delta(t) = \sum_{k=-\infty}^{\infty} \delta(t - t_k) =$$

$$= \frac{2\pi}{T} [1 - \dot{s}(t)] \sum_{n=-\infty}^{\infty} e^{jn\omega_i[t - s(t)]} \qquad [43]$$

References

[1] J. Engdahl. "Conversion électronique de temps moyen en temps sidéral." *Colloque international de chronométrie,* Paris, 1969.

[2] G. Becker. "Quasiperiodic frequency synthesis." *Proceedings of the 26th Annual Frequency Control Symposium,* pp. 279–91.

[3] G. W. Small. "A frequency synthesizer for $10/2\pi$ kHz." (Reprinted in Part II.)

[4] V. F. Kroupa. "Approximating frequency synthesizer." (Reprinted in Part II.)

[5] R. Shroeder. "Contribution to quasiperiodic frequency division." *Frequenz,* pp. 50–55, March 1976 (in German).

[6] G. A. Korn and T. M. Korn. *Mathematical Handbook.* New York: McGraw-Hill, 1961.

[7] A. Papoulis. *The Fourier Integral and Its Application.* New York: McGraw-Hill, 1962.

[8] V. F. Kroupa. "Synthesis techniques." Section III/C. (Reprinted in Part I.)

[9] G. Jeansaume. "Elaboration du temps sidéral avec un diviseur de fréquence programme." *Annales francaisés de chronométrie et de micromécanique,* pp. 191–204, 1972.

[10] G. R. Hovey. "A phase-locked solar to sidereal frequency converter." *Proceedings of the IEEE Australia,* pp. 59–60, March 1973.

[11] V. F. Kroupa. "Converter of the solar mean time into the sidereal time." Report Z-1642/A, Institute of Radio Engineering and Electronics, March 1987 (in Czech).

[12] V. F. Kroupa. "Principles of fractional-N frequency synthesizers." (A specially written paper for Part VI, Appendix A.)

[13] B. Friedman. *Principles and Techniques of Applied Mathematics.* New York: John Wiley, 1956.

[14] H. S. Wall. *Analytic Theory of Continued Fractions.* Princeton, N.J.: D. Van Nostrand, 1948.

Useful Computer Programs
for Investigation of Spurious Signals
in DDFS

VĚNCESLAV F. KROUPA

I. Introduction

IN this paper, we summarize several computer programs we have used when simulating spectral properties of idealized DDFS. Actually, measured spectra can differ substantially from the idealized ones due mainly to DAC imperfections (nonlinearities, glitches, etc.). Nevertheless, we believe that a theoretical investigation of the designed DDFS, or those intended for applications, should be performed as one of the first steps for better understanding of their properties.

II. Useful Programs

A. Spurious Signals in DDFS with the Rectangular Output Wave

In this first program, we consider cases where the denominator in the normalized frequency is not a power of 2. Therefore, we had to use for computation of the output spectrum ordinary Fourier series expansion with the repetition period equal to Yf_c. After some rearrangements of the fundamental

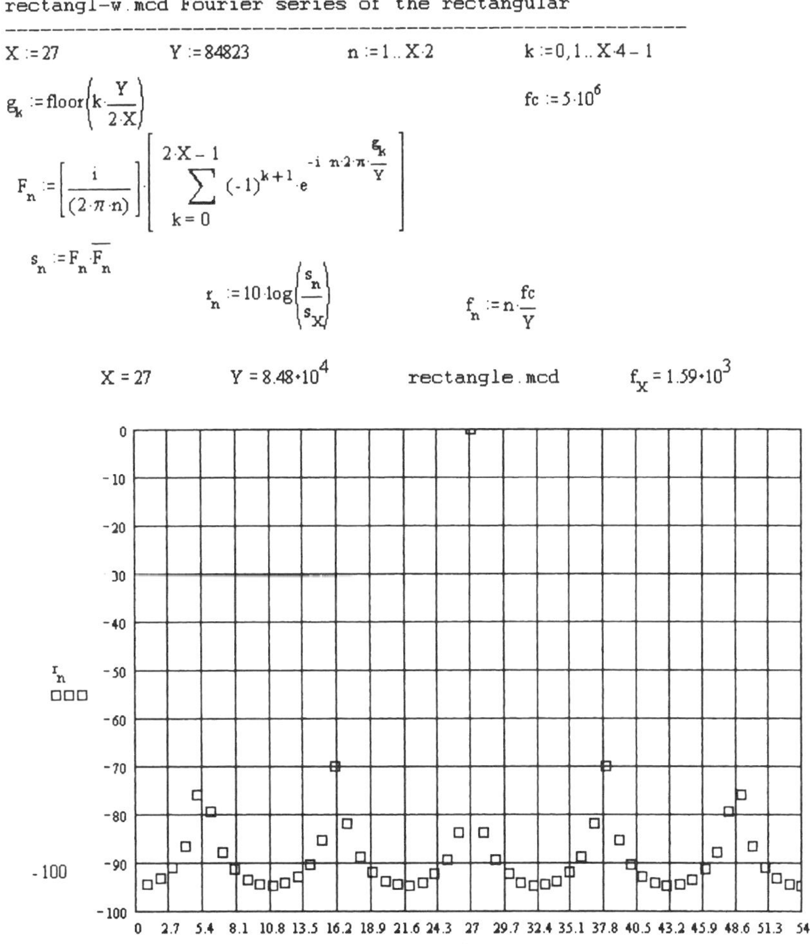

Fig. 1. MathCAD program for computation of output signals of the quasi-periodic rectangular wave with the normalized frequency $\xi_x = 27/84823$.

(This paper was specially written for this volume.)

367

equations, we have found for individual harmonics simple closed-form relations that are easily recognized in the Math-CAD program shown in Fig. 1.

Today the only advantage of the specially designed approximating frequency synthesizers is that the approximation error can be made much smaller in short observation time in contradistinction to commercial DDFS with the same accuracy. For the numerical example, we have chosen the normalized frequency solved in Part II by Small [1].

$$\xi_x = \frac{1}{(\pi \cdot 10^3)} \qquad [1]$$

First, we have started with the continued fraction expansion of the normalized frequency (1) recalled in Table 1.

TABLE 1. CONTINUED FRACTION EXPANSION OF NORMALIZED FREQUENCY $\xi_x = 1/(\pi*10^3)$, WHERE A_K/B_K IS THE KTH ORDER APPROXIMATION AND R_K THE RESPECTIVE APPROXIMATION ERROR.

k	A_k	B_k	R_k
0	0	1	
1	1	3142	$4.13*10^{-8}$
2	2	6283	$9.39*10^{-10}$
3	5	15708	$7.44*10^{-12}$
4	27	84823	$-6.18*10^{-12}$

TABLE 2. ESTIMATED, COMPUTED, AND MEASURED SPECTRAL LINES OF THE ABOVE EXAMPLE.

Estimated Level [2]	Computed Levels	Measured Levels [dB]
$20\log(X/Y) = -69.94$	$r_{16} = -69.94$	$20\log(318*10^{-6}) = -69.95$
$20\log(X/5Y) = -83.92$	$r_{26} = -83.8$	$20\log(64*10^{-6}) = -83.88$
	$r_1 = -94.28$	$20\log(5018*10^{-6}) = -45.99$

The computer program, together with the output spectrum, is shown in Fig. 1. Note a very good agreement between the computed levels of major spectral lines with the estimation in accordance with the editor's paper in Part V [2] and with the author's measurements in Table 2. The only difficulty is with the lowest spurious signal which was probably spoiled with the mains frequency.

B. Expansion of the Decimal Fraction into the Binary Number

Since nearly all commercial DDFSs are based on binary systematic fractions, it is often helpful to be able to make a quick change of the decimal fraction into the respective binary one. To this end, we publish a MathCAD software in Fig. 2. [3]. The numerical example presented is taken from the paper by Giffard and Cutler [4].

C. Spurious Signals in DDFS with the Sine-Wave Output

Practically all commercial DDFSs are built in such a way that all normalized output frequencies have the denominator

dekabin.mcd-Expansion of the decimal fraction into binary number

$$X := 131770 \qquad Y := 1000000$$

$$ksi := \frac{X}{Y} \qquad R := 33 \qquad ksi = 0.13177$$

$$c := floor\left[ksi \cdot 2^R\right]$$

$$z := \log(c) - floor\left[\frac{\log(c)}{\log(2)}\right] \cdot \log(2)$$

$$K := if\left[z \approx 0, floor\left[\frac{\log(c)}{\log(2)}\right] + 1, ceil\left[\frac{\log(c)}{\log(2)}\right]\right] \qquad K = 31$$

$$k := 0 \,..\, K \qquad a_0 := c$$

$$\begin{bmatrix} a_{k+1} \\ r_{k+1} \\ b_{k+1} \end{bmatrix} := \left[floor\left[\frac{a_k}{2}\right] \quad a_k - 2 \cdot floor\left[\frac{a_k}{2}\right] \quad until\left[a_k - 2 \cdot floor\left[\frac{a_k}{2}\right], floor\left[a_k - 2 \cdot floor\left[\frac{a_k}{2}\right]\right]\right] \right]$$

$$j := 0 \,..\, K - 1 \qquad i\alpha_j := b_{j+1} \qquad in_j := j$$

$$\alpha := reverse(i\alpha) \qquad n := reverse(in)$$

$$K = 31$$

$$c(bin) = \alpha(K-1)*2\hat{} (K-1) + ...+ \alpha(1)*2\hat{}1 + \alpha(0)*2\hat{}0$$

$$\beta := n - R \qquad rx := 11$$

$$n = \begin{bmatrix} 30 \\ 29 \\ 28 \\ 27 \\ 26 \\ 25 \\ 24 \\ 23 \\ 22 \\ 21 \\ 20 \\ 19 \\ 18 \\ 17 \\ 16 \\ 15 \\ 14 \\ 13 \\ 12 \\ 11 \\ 10 \\ 9 \\ 8 \\ 7 \\ 6 \\ 5 \\ 4 \\ 3 \\ 2 \\ 1 \\ 0 \end{bmatrix} \quad \alpha = \begin{bmatrix} 1 \\ 0 \\ 0 \\ 0 \\ 0 \\ 1 \\ 1 \\ 0 \\ 1 \\ 1 \\ 1 \\ 0 \\ 1 \\ 1 \\ 1 \\ 0 \\ 1 \\ 0 \\ 1 \\ 1 \\ 0 \\ 1 \\ 1 \\ 1 \\ 0 \\ 0 \\ 0 \\ 0 \\ 0 \\ 0 \\ 1 \end{bmatrix} \quad \beta = \begin{bmatrix} -3 \\ -4 \\ -5 \\ -6 \\ -7 \\ -8 \\ -9 \\ -10 \\ -11 \\ -12 \\ -13 \\ -14 \\ -15 \\ -16 \\ -17 \\ -18 \\ -19 \\ -20 \\ -21 \\ -22 \\ -23 \\ -24 \\ -25 \\ -26 \\ -27 \\ -28 \\ -29 \\ -30 \\ -31 \\ -32 \\ -33 \end{bmatrix}$$

$$\left[2^{-3} + 2^{-8} + 2^{-9} + 2^{-11}\right] = 0.131348$$

Fig. 2. MathCAD program for changing a decimal number into a binary one.

equal to the power of 2. Therefore, the computation of spurious signals is much simpler than with the first reproduced program since we can apply the FFT procedure.

The respective MathCAD program is shown in Fig. 3. Note that we have taken into account quantization due to the finite length, S, of sine words in a lookup table as well as quantization due to the DAC resolution. The numerical example is taken from Fig. 2. However, ξ_x was approximated with the first nine terms, resulting in $\xi_x = 67/512$. Investigation of Fig. 3b reveals large harmonic signals, third, fifth and seventh ($r_{201}, r_{177}, r_{43}, \ldots$) as one would expect for the truncation-quantization process [2] in contradistinction to the mean spurious frequency level.

$$S(r_j) \approx -102 \text{ dB} \qquad [2]$$

On the other hand, in Fig. 3c we have computed the output spectrum for the rounded values in the sine lookup table (note the reduced harmonic signals). A set of simulations revealed that for matching of the sine lookup tables and DAC, the wordlength (S bits) of the stored rounded sine values should be equal to $(D - 1)$ bits where D is the DAC resolution.

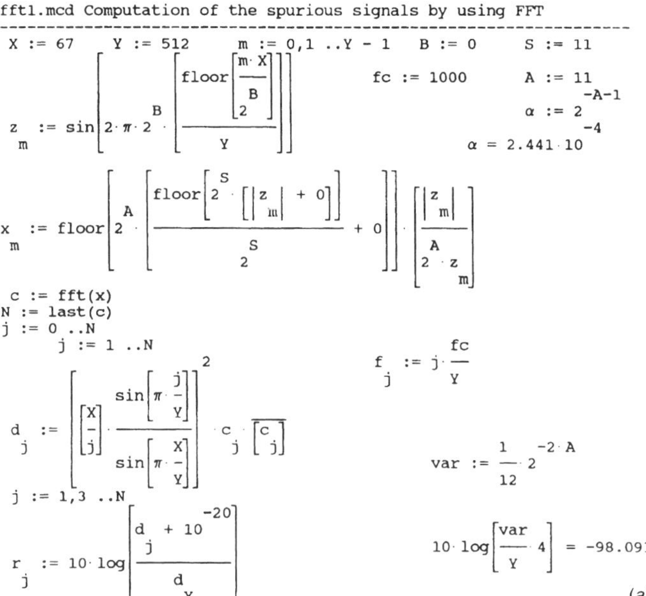

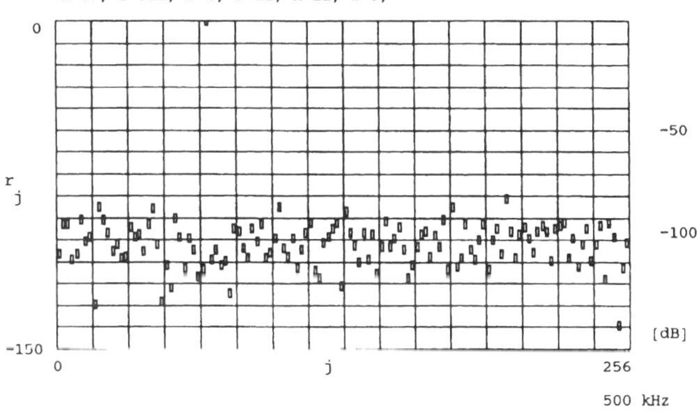

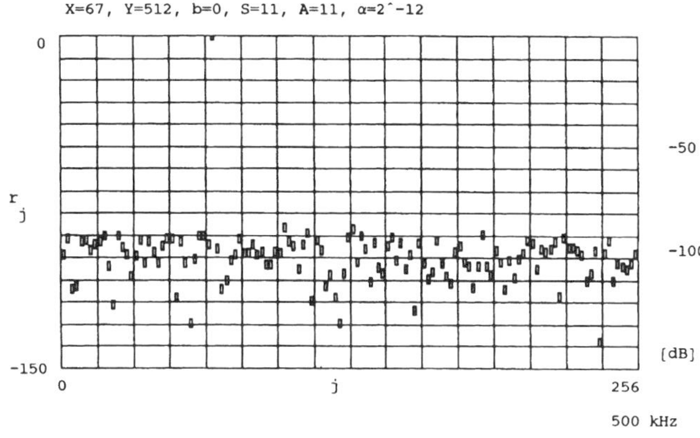

$$\frac{\sum_j r_j}{\frac{Y}{4} - 1} = -100.677 \qquad 10 \cdot \log\left[\frac{\left[\sum_j d_j\right] - d_X}{\frac{Y}{2} - 1} \cdot \frac{1}{d_X}\right] = -98.418$$

$$\text{mean}(r) \cdot 2 = -99.891$$

$$r_{201} = -108.796 \quad r_{177} = -94.331 \quad r_{43} = -102.235 \quad r_{91} = -103.034$$

$$r_{225} = -95.168 \qquad\qquad (c)$$

Fig. 3. *Continued.*

    (c.) Output spectrum for sine values rounded to 11 bits ($\alpha = 2^{-12}$)

D. Computation of Intermodulation Signals in DDFS

DDFSs are plagued with intermodulation signals that are well known, particularly from direct analog synthesizers and other rf devices. They are generated in nonlinear circuits (mixers). The major culprit in DDFS is DAC. Intermodulation signals in DDFS are defined as follows:

$$|r \cdot f_x + s \cdot f_c| \le \frac{1}{2} f_c \qquad [3]$$

where r and s are integers, either positive or negative. Since solution of (2) with earlier "mixer charts," for example [5], is time consuming, we have written the software shown in Fig. 4. The numerical example in Fig. 4b is again taken from [4]. The second example in Fig. 4c verifies the spectral lines close to the carrier in the case where the normalized frequency ξ_x is a ratio of "nearly" small integers—see Fig. 9 in the paper by Bloch et al. [6]. Note that for simplicity we use only positive values of r and s. The correct signs are easily estimated from the output frequency (see examples in Fig. 4).

E. Spurious Signals in DDFS Due to the Truncation of the Accumulator Bits

In the neighborhood of the desired carrier, we encounter spurious signals the origin of which is not the direct mixing of the reference, f_c, and the output frequency f_x. However, these spurious signals are generated by the phase modulation caused by discarding some information stored in the DDFS accumula-

```
intermod.mcd - computation of intermodulation signals
-----------------------------------------------------------

f1 := 131.77        f2 := 1000

        f2          f1
esp := --          -- = 0.13177
        2           f2
r := 0 ..20         s := 0 ..6

f      := f1· (r) - (s)· f2
 r,s

g      := if⌈|f    | ≤ esp,f    ,0⌉                                  (a)
 r,s        ⌊| r,s|        r,s    ⌋

        f1/f2 = 131.77/1000

      ⌈   0         0         0      0  0  0  0⌉
r=1   | 131.77      0         0      0  0  0  0|
      | 263.54      0         0      0  0  0  0|
      | 395.31      0         0      0  0  0  0|
      |   0      -472.92      0      0  0  0  0|
g =   |   0      -341.15      0      0  0  0  0|
      |   0      -209.38      0      0  0  0  0|
      |   0       -77.61      0      0  0  0  0|
      |   0        54.16      0      0  0  0  0|
      |   0       185.93      0      0  0  0  0|
r=10  |   0       317.7       0      0  0  0  0|
      |   0       449.47      0      0  0  0  0|
      |   0         0      -418.76   0  0  0  0|
      |   0         0      -286.99   0  0  0  0|
      |   0         0      -155.22   0  0  0  0|
      |   0         0       -23.45   0  0  0  0|
      |   0         0       108.32   0  0  0  0|
      |   0         0       240.09   0  0  0  0|
      |   0         0       371.86   0  0  0  0|
      |   0         0         0   -496.37 0 0 0|
      ⌊   0         0         0   -364.6  0 0 0⌋

    s=0       example     12· f1 - 2· f2 = -418.76
                                                                    (b)

        f1/f2 = 10/50.255

      ⌈  0        0         0        0       0  0  0⌉
r=1   | 10        0         0        0       0  0  0|
      | 20        0         0        0       0  0  0|
      |  0     -20.255      0        0       0  0  0|
      |  0     -10.255      0        0       0  0  0|
g =   |  0      -0.255      0        0       0  0  0|
      |  0       9.745      0        0       0  0  0|
      |  0      19.745      0        0       0  0  0|
      |  0        0      -20.51      0       0  0  0|
      |  0        0      -10.51      0       0  0  0|
r=10  |  0        0       -0.51      0       0  0  0|
      |  0        0        9.49      0       0  0  0|
      |  0        0       19.49      0       0  0  0|
      |  0        0         0     -20.765    0  0  0|
      |  0        0         0     -10.765    0  0  0|
      |  0        0         0      -0.765    0  0  0|
      |  0        0         0       9.235    0  0  0|
      |  0        0         0      19.235    0  0  0|
      |  0        0         0        0    -21.02 0 0|
      |  0        0         0        0    -11.02 0 0|
      ⌊  0        0         0        0     -1.02 0 0⌋

    s=0       example    4· f1 - 1· f2 = -10.255
                                                                    (c)
```

Fig. 4. MathCAD program for computation of intermodulation signals in DDFS.

   (a.) The program

   (b.) Numerical example for $\xi_x = 131.77/1000$ [4]

   (c.) Numerical example for $\xi_x = 10/50.255$ [6]

tor. The problem was discussed by the editor in [7]. The computer program (in the MathCAD version) is reproduced in Fig. 5. The numerical example was again taken from [4]. Note the good agreement for the first-order sidebands both with the measured and computed levels.

F. Estimating the Output Phase Noise of DDFS

The problem has been discussed in some detail in Part VII. For quick qualitative information, we have written a simple MathCAD program shown in Fig. 6. For its application we shall make a short introduction.

f_c is the frequency of the reference oscillator.

f_{out} is the actual output frequency.

a_r is an experimental constant of the quartz resonator [8].

a_e is the $1/f$ phase noise at 1 Hz for the used amplifier in the oscillator feedback loop.

a_{0d} is its white noise constant.

Qf_o is the quality times frequency product of the quartz resonator (it can be even one order lower than indicated in the program in Fig. 6).

a_{1d} is the expected or interpolated $1/f$ phase noise of the investigated DDFS at 1 Hz [see Table 3].

D_m is the expected additive phase noise of DDFS including an estimated white noise level.

H_m is the oscillator stability measure invariable to frequency multiplications and divisions.

S_m is the output phase noise measure.

F_m is the output phase noise measure including DDFS degradation.

The scarce published data about the additive flicker phase-noise constant, a_{1d} and white phase constant a_{0d} in DDFS, are summarized in Table 3.

TABLE 3. PUBLISHED FLICKER PHASE NOISE CONSTANT, A_{1D}, AND WHITE PHASE NOISE CONSTANT, A_{0D}, ENCOUNTERED IN DDFS WITH THE RESPECTIVE CLOCK AND OUTPUT FREQUENCIES.

Author	f_c [MHz]	f_{out} [MHz]	a_{1d}	a_{0d}
Mattison & Coyle [9]	5	0.4	$10^{-10.5}$	
Saul & Mudd [10]	328	82	10^{-8}	
Goldberg [11]	500	230 (max)	10^{-8}	10^{-14}
Sciteq ADS-63 [12]		250	$10^{-8.5}$	$10^{-13.5}$
Zavrel & Edwards [13]	20	8	10^{-12}	10^{-14}
Walls [14]	40	2.5	10^{-11}	

III. Conclusions

The programs reproduced were written with the assistance of the MathCAD software. Their use here does not signify endorsement of this type of mathematical solution in DDFS applications.

IV. Acknowledgment

I appreciate the cooperation of Dr. J. Štursa and J. Netolický in the preparation of this article.

$$X := .13177 \qquad R := 32 \qquad Y := 1 \qquad W := 14$$

$$\mathrm{ksi} := \frac{X}{Y} \qquad B := R - W \qquad B = 18 \qquad fc := 10^6$$

$$\mathrm{ksi} = 0.13177 \qquad \mathrm{ksi} \cdot fc = 1.318 \cdot 10^5$$

$$a := \mathrm{ksi} \cdot 2^W$$

$$ra := a - \mathrm{floor}(a) \qquad a = 2.159 \cdot 10^3$$

$$r := 0, 1 .. 8 \qquad\qquad rb := 1 - ra$$

$$s := 0, 1 .. 4 \qquad ra = 0.92$$

$$\lim := \frac{Y}{2} \qquad\qquad rb = 0.08$$

$$rc := \mathrm{if}(ra \leq 0.5, ra, rb) \qquad rc = 0.08$$

$$n_{r,s} := X + (rc \cdot (r - 4) + s - 2) \cdot Y \qquad fx := X \cdot \frac{fc}{Y} \qquad fx = 131770$$

$$g_{r,s} := \mathrm{if}\!\left(|n_{r,s}| \leq \lim, n_{r,s}, 0\right)$$

$$h_{r,s} := g_{r,s} \cdot \frac{fc}{Y} \qquad r := 1, 2 .. 8 \qquad u_r := 20 \cdot \log\!\left(\frac{2^{-W}}{r}\right)$$

$$g = \begin{bmatrix} 0 & 0 & -0.19 & 0 & 0 \\ 0 & 0 & -0.109 & 0 & 0 \\ 0 & 0 & -0.029 & 0 & 0 \\ 0 & 0 & 0.051 & 0 & 0 \\ 0 & 0 & 0.132 & 0 & 0 \\ 0 & 0 & 0.212 & 0 & 0 \\ 0 & 0 & 0.292 & 0 & 0 \\ 0 & 0 & 0.373 & 0 & 0 \\ 0 & 0 & 0.453 & 0 & 0 \end{bmatrix} \qquad h = \begin{bmatrix} 0 & 0 & -189510 & 0 & 0 \\ 0 & 0 & -109190 & 0 & 0 \\ 0 & 0 & -28870 & 0 & 0 \\ 0 & 0 & 51450 & 0 & 0 \\ 0 & 0 & 131770 & 0 & 0 \\ 0 & 0 & 212090 & 0 & 0 \\ 0 & 0 & 292410 & 0 & 0 \\ 0 & 0 & 372730 & 0 & 0 \\ 0 & 0 & 453050 & 0 & 0 \end{bmatrix} \qquad u = \begin{bmatrix} 0 \\ -84.288 \\ -90.309 \\ -93.831 \\ -96.33 \\ -98.268 \\ -99.851 \\ -101.19 \\ -102.35 \end{bmatrix}$$

Fig. 5. MathCAD program for computation of spurious signal due to the phase modulation caused by neglecting a lot of LSBs accumulator bits (in columns (g) are non modulated frequencies (h) actual frequencies, and (u) the expected spurious levels).

$$ar := 1 \cdot 10^{-12.75} \qquad ae := 1 \cdot 10^{-11.8} \qquad ao := 1 \cdot 10^{-14} \qquad Qfo := 1.3 \cdot 10^{13}$$

$$fo := 327.68 \cdot 10^6 \qquad m := 1 .. 20 \qquad ald := 1 \cdot 10^{-8}$$

$$fout := 81.925 \cdot 10^6 \qquad f_m := 2^{m-1} \qquad \frac{Qfo}{fo} = 3.967 \cdot 10^4$$

$$H_m := \frac{1}{3} \cdot \frac{ar + ae}{f_m^2} \cdot \frac{fo^2}{Qfo} + \frac{1}{2} \cdot \frac{1}{f_m^2} \cdot \left[\frac{ar}{Qfo} + \frac{ao}{Qfo^2} \cdot fo^2\right] + \left[\frac{ae}{f_m} + ao\right] \cdot \frac{1}{fo^2}$$

$$d_m := 1 \cdot \frac{ald}{f_m} + 1 \cdot 10^{-13}$$

$$D_m := 10 \cdot \log\!\left[d_m\right]$$

$$S_m := 10 \cdot \log\!\left[H_m \cdot fout^2\right] \qquad\qquad F_m := 10 \cdot \log\!\left[H_m \cdot fout^2 + d_m\right]$$

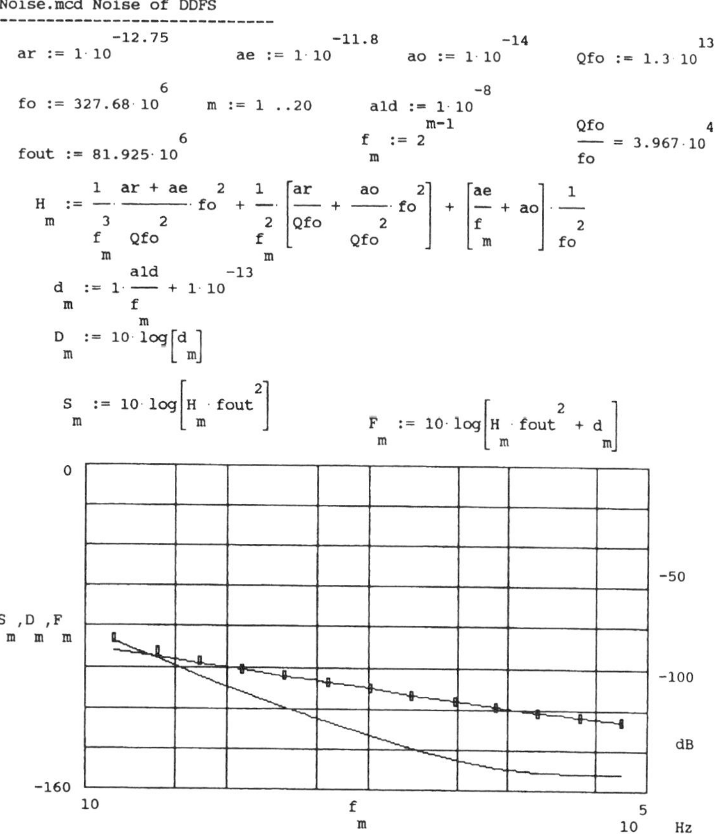

Fig. 6. MathCAD program for quick qualitative information about the DDFS output phase noise (the numerical example in taken from [10]). (S_m is plotted by unmarked full line, F_m is marked by ($\square$).

REFERENCES

[1] Greig W. Small. "A frequency synthesizer for 10/2 kHz." *IEEE Transactions on Instrumentation and Measurement,* pp. 34–37, March 1973. (Reprinted in Part II.)

[2] V. F. Kroupa. "Spectral properties of DDFS: Computer simulations and experimental verifications." *Proceedings IEEE International Frequency Control Symposium,* pp. 613–23, June 1994.

[3] J. Stursa, private communication. 1994.

[4] R. P. Giffard and L. S. Cutler. "A low-frequency, high resolution digital synthesizer." *1992 IEEE International Frequency Control Symposium Proceedings,* pp. 188–92. (Reprinted in Part V.)

[5] V. F. Kroupa. *Frequency Synthesis: Theory, Design and Applications.* London: Ch. Griffin, 1973; New York: John Wiley, 1973.

[6] M. Bloch, I. Pascaru, C. Stone, and T. McClelland. "Subminiature rubidium frequency standard for commercial applications." *IEEE Proceedings of the International Frequency Control Symposium, 1993,* pp. 164–77.

[7] V. F. Kroupa. "Discrete spurious signals and background noise in direct digital frequency synthesizers." *IEEE Proceedings of the International Frequency Control Symposium,* pp. 242–50, 1993.

[8] V. F. Kroupa. "Flicker frequency noise in BAW and SAW quartz resonators." *IEEE Transactions on Ultrasonics, Ferroelectrics, and Frequency Control,* UFFC-35, pp. 406–20, 1988.

[9] E. M. Mattison and L. M. L. Coyle. "Phase noise in direct digital frequency synthesizers." *Proceedings of the 42nd Annual Frequency Control Symposium,* pp. 352–56, 1988.

[10] P. H. Saul and M. S. J. Mudd. "A direct digital synthesizer with 100 MHz output capability." *IEEE Journal of Solid-State Circuits,* pp. 819–21, June 1988.

[11] Bar-Giora Goldberg. "Linear frequency modulation-theory and practice." *RF Design,* pp. 39–46, September 1993.

[12] Editorial: "Flexibility dominates synthesizer architectures," *Microwave Engineering Europe,* pp. 25–29, December/January 1995.

[13] R. J. Zavrel and Gwyn Edwards, eds. *The DDFS Handbook.* Santa Clara: Stanford Telecommunications, 1990.

[14] F. L. Walls, private communication, 1994.

Author Index

Subject Index

About the Editor

Věnceslav F. Kroupa has been affiliated with the Institute of Radio Engineering and Electronics of the Academy of Sciences of the Czech Republic since 1955. He received his degree (M. Sc. diploma) in electrical engineering from Prague Technical University. After an additional year of postgraduate study in radio engineering, he studied theoretical physics for three years at Charles University, Prague, Czechoslovakia. In 1967 he received the Ph.D. degree from the Czechoslovak Academy of Sciences, Prague with his thesis "Numerical Theory of Frequency Synthesis." In 1991, he received the DrSc. degree completing 24 papers and two books on frequency synthesis and frequency stability.

Dr. Kroupa began his career as development and research engineer in the Transmitter Department of the TESLA Research Institute. Later, he worked in the Research Institute for Electronics and Physics, and in 1955 he joined the Institute of Radio Engineering and Electronics of the Czechoslovak Academy of Sciences (now the Academy of Sciences of the Czech Republic).

His major scientific and research interests have been standard time and frequency; frequency stability and noise; frequency synthesis; and precise frequency measurements. More recently, his interest has focused on flicker noise problems in electronic devices. He has written four books and published over 100 technical papers and reports on these subjects. He holds 15 patents and has received two awards from the Czechoslovak Academy of Sciences. In 1969 he received an award for his CSc. dissertation "General Theory of Frequency Synthesis," and in 1974 he received an award for a collection of works on circuit theory and computer applications. He is a member of the IEEE Ultrasonics, Ferroelectrics, and Frequency Control Society.